ANNUAL PLANT REVIEWS, VOLUME 49

ANNUAL PLANT REVIEWS, VOLUME 49

The Gibberellins

Edited by

Peter Hedden
Plant Biology and Crop Science Department, Rothamsted Research, UK

Stephen G. Thomas
Plant Biology and Crop Science Department, Rothamsted Research, UK

Annual Plant Reviews

WILEY Blackwell

This edition first published 2016 © 2016 by John Wiley & Sons Ltd

Registered office: John Wiley & Sons, Ltd, The Atrium, Southern Gate, Chichester, West Sussex, PO19 8SQ, UK

Editorial offices: 9600 Garsington Road, Oxford, OX4 2DQ, UK
The Atrium, Southern Gate, Chichester, West Sussex, PO19 8SQ, UK
111 River Street, Hoboken, NJ 07030-5774, USA

For details of our global editorial offices, for customer services and for information about how to apply for permission to reuse the copyright material in this book please see our website at www.wiley.com/wiley-blackwell.

Library of Congress Cataloguing-in-Publication Data

The Gibberellins / edited by Peter Hedden & Stephen G. Thomas.
p. cm. – (Annual plant reviews; v. 49)
Includes bibliographical references and index.
ISBN 978-1-119-21042-9 (hard cover : alk. paper)
1. Gibberellins. I. Hedden, Peter. II. Series: Annual plant reviews; v. 49.
QK898.E8P573 2012
571.7_42–dc23
2011035805

A catalogue record for this book is available from the British Library.

Wiley also publishes its books in a variety of electronic formats. Some content that appears in print may not be available in electronic books.

Set in 9.5/13pt Meridien LT Std by SPi Global, Chennai, India

1 2016

Annual Plant Reviews

A series for researchers and postgraduates in the plant sciences. Each volume in this series focuses on a theme of topical importance and emphasis is placed on rapid publication.

Titles in the series:

1. **Arabidopsis**
   Edited by M. Anderson and J.A. Roberts
2. **Biochemistry of Plant Secondary Metabolism**
   Edited by M. Wink
3. **Functions of Plant Secondary Metabolites and their Exploitation in Biotechnology**
   Edited by M. Wink
4. **Molecular Plant Pathology**
   Edited by M. Dickinson and J. Beynon
5. **Vacuolar Compartments**
   Edited by D.G. Robinson and J.C. Rogers
6. **Plant Reproduction**
   Edited by S.D. O'Neill and J.A. Roberts
7. **Protein–Protein Interactions in Plant Biology**
   Edited by M.T. McManus, W.A. Laing and A.C. Allan
8. **The Plant Cell Wall**
   Edited by J.K.C. Rose
9. **The Golgi Apparatus and the Plant Secretory Pathway**
   Edited by D.G. Robinson
10. **The Plant Cytoskeleton in Cell Differentiation and Development**
    Edited by P.J. Hussey
11. **Plant–Pathogen Interactions**
    Edited by N.J. Talbot
12. **Polarity in Plants**
    Edited by K. Lindsey
13. **Plastids**
    Edited by S.G. Moller
14. **Plant Pigments and their Manipulation**
    Edited by K.M. Davies

15. **Membrane Transport in Plants**
    Edited by M.R. Blatt
16. **Intercellular Communication in Plants**
    Edited by A.J. Fleming
17. **Plant Architecture and Its Manipulation**
    Edited by C.G.N. Turnbull
18. **Plasmodeomata**
    Edited by K.J. Oparka
19. **Plant Epigenetics**
    Edited by P. Meyer
20. **Flowering and Its Manipulation**
    Edited by C. Ainsworth
21. **Endogenous Plant Rhythms**
    Edited by A. Hall and H. McWatters
22. **Control of Primary Metabolism in Plants**
    Edited by W.C. Plaxton and M.T. McManus
23. **Biology of the Plant Cuticle**
    Edited by M. Riederer
24. **Plant Hormone Signaling**
    Edited by P. Hedden and S.G. Thomas
25. **Plant Cell Separation and Adhesion**
    Edited by J.R. Roberts and Z. Gonzalez-Carranza
26. **Senescence Processes in Plants**
    Edited by S. Gan
27. **Seed Development, Dormancy and Germination**
    Edited by K.J. Bradford and H. Nonogaki
28. **Plant Proteomics**
    Edited by C. Finnie
29. **Regulation of Transcription in Plants**
    Edited by K. Grasser
30. **Light and Plant Development**
    Edited by G. Whitelam
31. **Plant Mitochondria**
    Edited by D.C. Logan
32. **Cell Cycle Control and Plant Development**
    Edited by D. Inzé
33. **Intracellular Signaling in Plants**
    Edited by Z. Yang
34. **Molecular Aspects of Plant Disease Resistance**
    Edited by J. Parker
35. **Plant Systems Biology**
    Edited by G.M. Coruzzi and R.A. Guti'errez
36. **The Moss *Physcomitrella patens***
    Edited by C.D. Knight, P.-F. Perroud and D.J. Cove
37. **Root Development**
    Edited by T. Beeckman
38. **Fruit Development and Seed Dispersal**
    Edited by L. Østergaard

39. **Function and Biotechnology of Plant Secondary Metabolites**
Edited by M. Wink
40. **Biochemistry of Plant Secondary Metabolism**
Edited by M. Wink
41. **Plant Polysaccharides**
Edited by P. Ulvskov
42. **Nitrogen Metabolism in Plants in the Post-genomic Era**
Edited by C. Foyer and H. Zhang
43. **Biology of Plant Metabolomics**
Edited by R.D. Hall
44. **The Plant Hormone Ethylene**
Edited by M.T. McManus
45. **The Evolution of Plant Form**
Edited by B.A. Ambrose and M.D. Purugganan
46. **Plant Nuclear Structure, Genome Architecture and Gene Regulation**
Edited by D.E. Evans, K. Graumann and J.A. Bryant
47. **Insect-Plant Interactions**
Edited by C. Voelckel and G. Jander
48. **Phosphorus Metabolism in Plants**
Edited by W.C. Plaxton and H. Lambers

CONTENTS

List of Contributors xv
Preface xvii

1 Signal Achievements in Gibberellin Research: The Second Half-Century 1
Valerie M. Sponsel

1.1 Introduction 1
1.2 Gibberellin biosynthesis 6
1.3 Gibberellin signalling 17
1.4 Physiological responses to gibberellins 25
References 29

2 Gibberellin Biosynthesis in Higher Plants 37
Peter Hedden

2.1 Introduction 37
2.2 Synthesis of *ent*-kaurene 39
2.2.1 Formation of *trans*-geranylgeranyl diphosphate 39
2.2.2 Formation of *ent*-kaurene from *trans*-geranylgeranyl diphosphate 40
2.3 Reactions catalysed by cytochrome P450 mono-oxygenases 42
2.4 Reactions catalysed by 2-oxoglutarate-dependent dioxygenases 45
2.5 Sites of gibberellin biosynthesis 49
2.6 Regulation of gibberellin biosynthesis 50
2.6.1 Developmental control 50
2.6.2 Gibberellin homoeostasis 51
2.6.3 Regulation by other hormones 54
2.6.4 Regulation by environmental factors 55
2.7 Concluding remarks 59
Acknowledgements 60
References 60

3 Inactivation Processes 73
Hiroshi Magome and Yuji Kamiya

3.1 Introduction 73
3.2 Gibberellin inactivation 75
3.2.1 Gibberellin 2-oxidase 75
3.2.2 Gibberellin methyltransferase 77
3.2.3 Gibberellin 16,17-oxidase 78

3.2.4 Gibberellin 13-oxidase and 12α-oxidase 78
3.2.5 Conjugation with sugar 80
3.3 Regulation of gibberellin inactivation 80
3.3.1 Developmental regulation 81
3.3.2 Gibberellin homoeostasis 82
3.3.3 Regulation by other hormones 83
3.3.4 Environmental regulation 84
3.4 Concluding remarks 87
References 88

4 Gibberellin Transport 95
Jonathan Dayan
4.1 Introduction 95
4.2 Gibberellins can be translocated along plant bodies 96
4.3 Gibberellin transport in seeds 100
4.4 Pattern of gibberellin biosynthesis in transport analysis 101
4.5 Grafting experiments 103
4.6 Significance for secondary growth 104
4.7 Orientation of gibberellin signal flow: source and sink tissues 107
4.8 Monitoring intra- and intercellular gibberellin concentration 110
4.9 Conclusion: new aspects for gibberellin transport 111
4.9.1 Potential transporters 111
4.9.2 Analysis through perception 112
4.9.3 Links to sugar transport 112
Acknowledgements 113
References 114

5 Gibberellins in Fungi, Bacteria and Lower Plants: Biosynthesis, Function and Evolution 121
Bettina Tudzynski, Lena Studt and María Cecilia Rojas
5.1 Introduction 122
5.2 Gibberellin biosynthesis in fungi 122
5.2.1 The biosynthetic pathway in *F. fujikuroi*: genes and enzymes 122
5.2.2 Gibberellin production in distantly related fungi 126
5.2.3 Evolution of the gibberellin biosynthetic gene cluster in fungi 128
5.2.4 The role of gibberellins in plant infection 131
5.2.5 Strain improvement 132
5.3 Gibberellin biosynthesis in bacteria 133
5.3.1 Free-living rhizobacteria 133
5.3.2 Symbiotic rhizobacteria: genes and reactions of the gibberellin biosynthetic pathway 134
5.3.3 Function and evolution 137
5.4 Gibberellin biosynthesis and signalling components in lower plants 139

5.5 Concluding remarks 143
References 144

6 Gibberellin Hormone Signal Perception: Down-Regulating DELLA Repressors of Plant Growth and Development 153
Sven K. Nelson and Camille M. Steber

6.1 Introduction 154
6.2 DELLA proteins are repressors of gibberellin responses 154
6.3 Gibberellin signalling lifts DELLA repression of gibberellin responses 157
6.4 The gibberellin receptor GID1 (GA-INSENSITIVE DWARF1) 159
6.5 The structural requirements for gibberellin binding by GID1 161
6.6 The structural requirements for the GID1-DELLA protein–protein interaction 162
6.7 The DELLA destruction model: negative regulation of DELLA repressors by SLY1/GID2 and the ubiquitin-proteasome pathway 166
6.8 Regulation of DELLA by phosphorylation and *O*-GlcNAc modification 169
6.9 Evidence for gibberellin-independent DELLA regulation 173
6.10 Evidence for gibberellin signalling without DELLA destruction 175
6.11 Concluding remarks 177
Acknowledgements 179
References 179

7 DELLA Proteins: Master Regulators of Gibberellin-Responsive Growth and Development 189
Stephen G. Thomas, Miguel A. Blázquez and David Alabadí

7.1 Introduction 190
7.2 DELLAs regulate downstream gibberellin signalling 191
7.3 Gibberellins relieve DELLA-growth repression by targeting their degradation 193
7.4 Functional diversification of *DELLA* genes 194
7.5 DELLA activity invokes rapid changes in the transcriptome 197
7.6 DELLA proteins activate transcription 198
7.7 DELLAs regulate transcription by physical interaction with transcriptional regulators 199
7.7.1 DELLAs sequester *bona fide* TFs by physical interaction 200
7.7.2 DELLAs interact with TFs in the context of promoters 204
7.7.3 DELLAs interact with other transcriptional regulators 206
7.7.4 DELLAs regulate chromatin dynamics 208
7.8 A non-genomic response regulated by DELLAs 209
7.9 Analysis of DELLA protein structure-function 210

7.10 GAMYB: A transcriptional regulator of gibberellin responses during cereal grain germination and pollen development 213
7.10.1 GAMYB positively regulates gene expression in cereal aleurone cells 214
7.10.2 GAMYB regulates gibberellin-dependent anther development 216
7.11 Concluding remarks 217
Acknowledgements 218
References 218

8 Interactions Between Gibberellins and other Hormones 229
John J. Ross, Asemeh Miraghazadeh, Amelia H. Beckett, Laura J. Quittenden and Erin L. McAdam
8.1 Introduction 229
8.2 Interactions involving effects of other hormones on gibberellin levels 230
8.2.1 Auxin promotes gibberellin biosynthesis 230
8.2.2 Ethylene inhibits gibberellin biosynthesis 231
8.2.3 Do gibberellin and abscisic acid inhibit each other's synthesis? 232
8.2.4 Do brassinosteroids act by affecting gibberellin levels? 234
8.2.5 Possible effects of other hormones on gibberellin synthesis 234
8.3 Interactions between hormone signal transduction pathways 234
8.3.1 Do other hormones affect DELLA stability? 235
8.3.2 DELLAs interact with proteins from the signalling pathways of other hormones 237
8.4 Gibberellins and auxin transport 245
8.5 Conclusion 246
Acknowledgements 247
References 247

9 Gibberellins and Seed Germination 253
Terezie Urbanova and Gerhard Leubner-Metzger
9.1 Introduction 254
9.2 Spatiotemporal expression of gibberellin metabolism during Brassicaceae seed germination 254
9.3 Gibberellin signalling and seed germination 264
9.3.1 The GID1ac and GID1b pathways in seeds 264
9.3.2 DELLA proteins and seed germination 268
9.4 Gibberellin and abiotic stress factors: thermoinhibition of seed germination 270
9.5 Gibberellin and biotic stress factors: allelochemical interference of gibberellin biosynthesis during seed germination 273

9.6 Conclusions and perspectives 276
Acknowledgements 277
References 277

10 Gibberellins and Plant Vegetative Growth 285
Cristina Martínez, Ana Espinosa-Ruiz and Salomé Prat
10.1 Introduction 285
10.2 Gibberellins and shoot development 288
10.2.1 Control of SAM function and leaf size 289
10.2.2 Elongation of the hypocotyl 290
10.2.3 Apical hook formation 295
10.3 Gibberellin function in root development 298
10.3.1 Hormonal control of root growth 298
10.3.2 Gibberellin signalling from the endodermis 302
10.3.3 DELLAs downstream signalling in the root 304
10.3.4 DELLAs promote mycorrhizal symbiosis 306
10.4 Growth under unfavourable conditions 308
10.4.1 DELLAs promote resistance to abiotic stress 308
10.4.2 DELLAs and biotic stress 310
10.5 Concluding remarks 311
References 312

11 Gibberellins and Plant Reproduction 323
Andrew R.G. Plackett and Zoe A. Wilson
11.1 Introduction 323
11.2 The floral transition 324
11.2.1 Gibberellin promotes flowering through multiple interacting pathways 324
11.2.2 Sites of gibberellin biosynthesis and action during the floral transition 329
11.2.3 Gibberellin and flowering in perennial species 331
11.3 Floral development 331
11.3.1 Floral patterning and early development 332
11.3.2 Gibberellin and fertility 334
11.4 Seed and fruit development 340
11.4.1 Fruit development 341
11.4.2 Embryo and seed development 345
Acknowledgements 348
References 348

12 Chemical Regulators of Gibberellin Status and their Application in Plant Production 359
Wilhelm Rademacher
12.1 Introduction 359
12.2 Gibberellins 361

12.3 Inhibitors of gibberellin biosynthesis 363
12.3.1 Quaternary ammonium compounds 365
12.3.2 Compounds with a nitrogen-containing heterocycle 366
12.3.3 Structural mimics of 2-oxoglutaric acid 369
12.3.4 16,17-Dihydro-gibberellins 371
12.4 Uses for gibberellins and inhibitors of gibberellin biosynthesis in crop production 372
12.4.1 Wheat, barley, rye, oats and other small-grain cereals 373
12.4.2 Rice 376
12.4.3 Sugarcane 377
12.4.4 Pasture and turf grasses 377
12.4.5 Oilseed rape 379
12.4.6 Cotton 379
12.4.7 Peanuts 381
12.4.8 Opium poppy 382
12.4.9 Fruit trees growing in temperate climate 382
12.4.10 Fruit and nut trees growing in subtropical and tropical climates 385
12.4.11 Grapevines 387
12.4.12 Ornamentals 389
12.4.13 Hybrid seed production 391
12.5 Outlook 391
References 391

13 Genetic Control of Gibberellin Metabolism and Signalling in Crop Improvement 405
Andrew L. Phillips

13.1 Introduction 405
13.2 The *REDUCED HEIGHT-1* (*Rht-1*) alleles of wheat 406
13.2.1 Pleiotropic effects of *Rht-1* alleles 410
13.2.2 *Rht-1* orthologues in other crop species 412
13.3 The *SEMI-DWARF-1*(*SD-1*) alleles of rice 413
13.4 The *ELONGATED UPPERMOST INTERNODE (EUI)* gene of rice 415
13.5 Commercially useful alleles of other genes from the gibberellin pathway 416
13.6 Transgenic approaches to manipulation of gibberellin-dependent processes in crops 419
13.6.1 Cereals 419
13.6.2 Other crop species 420
13.7 Conclusions 423
Acknowledgements 424
References 424

Appendix The structures of the gibberellins 431

Index 437

LIST OF CONTRIBUTORS

David Alabadí
Instituto de Biología Molecular y Celular de Plantas (CSIC-UPV)
Spain

Amelia H. Beckett
School of Biological Sciences
University of Tasmania
Australia

Miguel A. Blázquez
Instituto de Biología Molecular y Celular de Plantas (CSIC-UPV)
Spain

Jonathan Dayan
Department of Biology
Duke University
USA

Ana Espinosa-Ruiz
Centro Nacional de Biotecnología (CNB-CSIC)
Spain

Peter Hedden
Plant Biology and Crop Science Department
Rothamsted Research
UK

Yuji Kamiya
RIKEN Center for Sustainable Resources
Japan

Gerhard Leubner-Metzger
Royal Holloway University of London
School of Biological Sciences
Plant Molecular Science and Centre for Systems and Synthetic Biology
UK

Hiroshi Magome
RIKEN Center for Sustainable Resources
Japan
Current address: Japan Tobacco Inc.
Leaf Tobacco Research Center
Japan

Cristina Martínez
Centro Nacional de Biotecnología (CNB-CSIC)
Spain

Erin L. McAdam
School of Biological Sciences
University of Tasmania
Australia

Asemeh Miraghazadeh
College of Medicine, Biology and Environment
Australian National University
Australia

Sven K. Nelson
Molecular Plant Sciences Program
Washington State University
USA
Current address: USDA-ARS Plant Genetic Research Unit
University of Missouri
USA

Andrew L. Phillips
Plant Biology and Crop Science Department,
Rothamsted Research
UK

Andrew R.G. Plackett
Department of Plant Sciences
University of Oxford
UK

Salomé Prat
Centro Nacional de Biotecnología (CNB-CSIC)
Spain

Laura J. Quittenden
School of Biological Sciences
University of Tasmania
Australia

Wilhelm Rademacher
BASF SE
Global Research Crop Protection
Germany

María Cecilia Rojas
Laboratorio de Bioorgánica
Departamento de Química
Facultad de Ciencias
Universidad de Chile
Chile

John J. Ross
School of Biological Sciences
University of Tasmania
Australia

Valerie M. Sponsel
Department of Biology
The University of Texas at San Antonio
USA

Camille M. Steber
USDA-ARS
Wheat Health, Genetics, and Quality Unit and the Department of Crop and Soil Science
Washington State University
USA

Lena Studt
Westfälische Wilhelms-Universität Münster
Institut für Biologie und Biotechnologie der Pflanzen
Germany

Stephen G. Thomas
Plant Biology and Crop Science Department
Rothamsted Research
UK

Bettina Tudzynski
Westfälische Wilhelms-Universität Münster
Institut für Biologie und Biotechnologie der Pflanzen
Germany

Terezie Urbanova
Laboratory of Growth Regulators
Faculty of Science
Palacký University and Institute of Experimental Botany AS CR
UP&IEB AVCR and Centre of the Region Haná for Agricultural and Biotechnological Research
Czech Republic

Zoe A. Wilson
Department of Plant and Crop Science
University of Nottingham
UK

PREFACE

It is now nine years since the publication in 2006 of the Annual Plant Reviews volume on plant hormone signalling, which included a chapter on gibberellin (GA) metabolism and signal transduction. At the time of this publication the GA receptor GID1 had just been discovered, opening up a rich vein of research on GA perception. Since 2006 there have been substantial advances in our understanding of GA signalling and, although there have been several reviews covering aspects of this topic in the intervening years, a volume covering all facets of GA research is now timely. The last volume dedicated to the GAs, which contained the proceedings of a conference in Tokyo to commemorate the retirement of Professor Nobutaka Takahashi, was published as along ago as 1991.

We have included an appendix providing the structures of the 136 chemically characterized GAs. It is noteworthy that it is over 10 years since that last novel GA was identified, although further uncharacterised GAs are present in plants and some may have physiological importance. Due to the very low abundance of GAs in plant tissues, identification of novel compounds has necessitated the synthesis of proposed structures for comparison with the natural metabolites. Regrettably there are now very few laboratories engaged in GA chemistry, making this task increasingly less feasible. The GA research community owes considerable debt to the pioneering chemists, such as Jake MacMillan, who sadly died in 2014, Nobutaka Takahashi and Lewis Mander. In particular, the isotopically labelled GA standards produced by Professor Mander have provided an enormous boost to GA research. It is crucial to the GA field that it continues to receive adequate chemical support.

As described in the following chapters, there have been numerous highlights in GA research in the last nine years. In terms of GA biosynthesis, the cloning of 13-hydroxylases from rice, provided an important piece missing from our understanding of the metabolic pathway. The determination of the X-ray crystal structure of the GID1 receptor and the identification of many of the transcription factors and other proteins that interact with the DELLA GA signalling components are key advances. The establishment of DELLAs as hubs that integrate GA signalling with that of other hormones is of particular note, although the physiological relevance of these observations still needs to be fully explored. These topics will continue to occupy scientists interested in GA research in the coming years, as will the emerging interest in GA transport, which, with the identification of GA transporters and the observed structural specificity of GA movement, is providing evidence to suggest that transport is not dependent solely on membrane diffusion as previously assumed. Although there have been advances in localising the sites of GA synthesis, catabolism and action, further refinement in analytical methods is required to define these at the cellular level. The development of *in situ* methods for visualising GA, as has been reported for auxin and jasmonate, is a high priority. Such approaches will ensure that GA research remains an active and exciting field in the next nine years and beyond.

Peter Hedden and **Stephen G. Thomas**

Annual Plant Reviews (2016) **49**, 1–36
doi: 10.1002/9781119210436.ch1 http://onlinelibrary.wiley.com

Chapter 1

SIGNAL ACHIEVEMENTS IN GIBBERELLIN RESEARCH: THE SECOND HALF-CENTURY

Valerie M. Sponsel
Department of Biology, The University of Texas at San Antonio, USA

Abstract: Chapter 1 briefly recounts the discovery of gibberellins (GAs) as natural products of the fungus *Gibberella fujikuroi* in the early part of the twentieth century, and provides a historical overview of GA research from the late 1950s to the present day. It describes how biosynthetic pathways to GAs in *Gibberella* and higher plants were defined, and how stem length mutants of cereals and legumes were instrumental in establishing which GAs are biologically active and have hormonal function. The chapter presents an overview of the cereal aleurone system in which GA signalling was first studied, and describes how more recent use of *Arabidopsis* and rice led to the characterisation of a GA receptor (GID1) and downstream regulatory proteins (DELLAs). A number of DELLA-interacting proteins are described, illustrating how it is that GA-induced degradation of DELLAs facilitates downstream responses including cell elongation. Other 'classical' GA responses include germination and flowering in some species.

Keywords: Cereal aleurone, DELLA proteins, *Gibberella fujikuroi*, gibberellin biosynthesis, gibberellin receptor, gibberellin signalling, stem length mutants

1.1 Introduction

Gibberellins (GAs), once known only as fungal products, comprise a group of over 136 structurally related compounds that are natural constituents of plants. Just a small number of GAs have intrinsic biological activity, and they regulate many aspects of growth and development throughout the plant life cycle. Other GAs are biosynthetic precursors or inactivation products of the

Annual Plant Reviews, Volume 49: The Gibberellins, First Edition.
Edited by Peter Hedden and Stephen G. Thomas.

bioactive GAs, or may be metabolic by-products with no known function. Commercial-scale microbiological production of gibberellic acid (GA_3) facilitates its use in agriculture, particularly in fruit production, and there are also important uses for synthetic inhibitors of GA biosynthesis that act as dwarfing agents (discussed in Chapter 12).

Gibberellins were first identified in *Gibberella fujikuroi*, which is a fungal pathogen of rice.[1] The 'bakanae' or 'foolish seedling' disease, which has been known to rice farmers in the Orient for at least 200 years, causes supra-optimal elongation of seedlings and reduced yield of grain. At the end of the nineteenth century, Shotaro Hori, a mycologist working at the Imperial Agricultural Experiment Station in Nishigahara, Tokyo, induced these symptoms in healthy rice seedlings by infecting them with the 'bakanae' fungus. More than two decades later, Eiichi Kurosawa, a Japanese scientist working in Taipei, Taiwan, succeeded in producing sterile filtrate from *G. fujikuroi* cultures which, when applied to uninfected rice seedlings, could duplicate the pathological symptoms. The race was then on to identify the chemical substances that were secreted by *Gibberella*, and which caused overgrowth and reduced grain yield of infected seedlings. Phinney, who has documented the early history of GAs, reported the publication of more than 50 articles on the subject between 1927 and 1940 (Phinney, 1983). Teijiro Yabuta, an organic chemist working with Kurosawa, who had moved from Taipei to Nishigahara in 1933, obtained a semi-purified non-crystalline material from culture filtrates, which he termed 'gibberellin'. It could stimulate stem elongation not only in rice, but in several other important crops, including barley, buckwheat and soybean. The material was crystallised two years later (Yabuta and Sumiki, 1938), yielding two biologically active components, which they named gibberellin A and B.

After World War II interest in these growth-promoting factors from *Gibberella* reached the West, and two research groups, one at the Imperial Chemical Industries (ICI) Akers Research Laboratory in Welwyn, UK and the other at the United States Department of Agriculture (USDA) Laboratory in Peoria, Illinois, took on the task of chemical characterisation of the compounds secreted by *Gibberella fujikuroi*. It culminated in the isolation of gibberellic acid by the UK group (Cross, 1954) and gibberellin X by the US group (Stodola *et al.*, 1955). It was soon discovered that gibberellic acid and gibberellin X were the same, and the latter name was dropped. Gibberellic acid (see GA_3, Figure 1.1) was defined as a tetracyclic-dihydroxy-lactonic acid with the molecular formula $C_{19}H_{22}O_6$ (Cross, 1954). A reinvestigation

[1]The fungus has had a succession of names, being identified initially as the Deuteromycete, *Fusarium moniliforme*, prior to the discovery of its perfect (sexual) stage whereupon it was reclassified as the Ascomycete *Gibberella fujikuroi*. Recently the name *Fusarium fujikuroi* has been adopted. Because the fungus has been known predominantly as *Gibberella* for the period of time this history covers, that is the name used throughout this chapter.

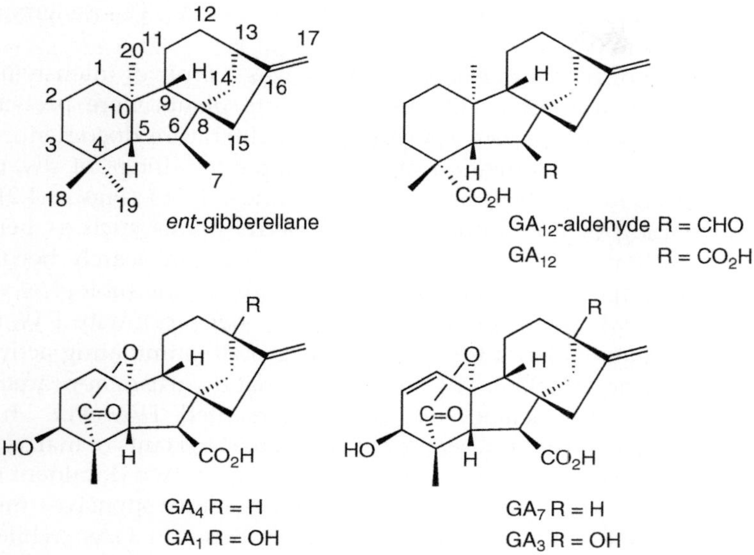

Figure 1.1 The *ent*-gibberellane skeleton shows the carbon atom numbering scheme used for gibberellins. GA_{12}-aldehyde is the first-formed GA in fungal and plant pathways. It is oxidised to the C-7 acid, GA_{12}. C_{20}-GAs, such as GA_{12}, contain the full complement of carbon atoms. They are precursors of C_{19}-GAs in which carbon-20 has been lost by metabolism. GA_1, GA_3, GA_4, and GA_7 are biologically active C_{19}-GAs, each possessing a 3β-hydroxyl group and a γ-lactone.

by Japanese chemists of the gibberellin 'A' sample that had been isolated more than a decade earlier yielded three components, which were termed gibberellins A_1, A_2 and A_3 (Takahashi *et al.*, 1955). An additional GA, GA_4, was isolated from *Gibberella* culture filtrate in 1959. Thus began the nomenclature of this large class of structurally related compounds that has now reached gibberellin A_{136}. The trivial name gibberellin A_x is now commonly abbreviated to GA_x, with GA used as a general abbreviation for gibberellin. GA is often used erroneously to represent gibberellic acid, which is identical to gibberellin A_3 (GA_3). Both names are still in use for this compound. It is the major product of GA biosynthesis in *Gibberella* (discussed in Chapter 5) and is produced commercially for horticultural and agronomic use.

A review of the extensive series of publications from the UK group in the late 1950s and early 1960s summarises the evidence for the structure of GA_3, particularly that of ring A, including the location of the hydroxyl group, the olefinic double bond and the lactone (Cross *et al.*, 1961). The C numbering scheme used at that time has been superseded by that shown on the *ent*-gibberellane skeleton in Figure 1.1. The structural determination of the other fungal GAs that were known at the time, namely GA_1, GA_2, GA_4,

GA_7 and GA_9, was also reviewed, with reference to GA_3. The assignment of stereochemistry to GA_3 was discussed too.

The response of plants to exogenous GA_3 was a topic of intense interest beginning in the mid-1950s. Dwarf and rosette plants were particularly responsive, and many papers appeared in the literature documenting the spectacular internode elongation in, for example, seedlings of dwarf pea (Brian and Hemming, 1955) and maize (Phinney, 1956) (Figure 1.2), and the rapid bolting of non-induced photoperiodic plants such as henbane (*Hyoscyamus*) (Lang, 1956). Almost immediately the search began for endogenous compounds in plants that could mimic the biological effects of applied GA_3. Margaret Radley, following up her work with P.W. Brian, provided bioassay evidence for endogenous growth-stimulating activity in pea seedlings (Radley, 1956). The observation that dwarfism appeared to be associated with GA-deficiency was also documented. However, Phinney, who had produced many different non-allelic dwarf mutants of maize, noted that while most recessive mutants responded to GA_3, two dominant dwarf mutants did not. Three decades later the recessive (responsive) mutants were used to determine metabolic sequences between GAs, while four decades later dominant (non-responsive) mutants were used to investigate GA signalling.

The first definitive characterisation of GA from plants came after the extraction of kilograms of developing bean seeds. It was a wise choice of plant material as immature seeds of both *Phaseolus coccineus* (formerly *multiflorus*, runner bean) and *Ph. vulgaris* (French bean) are rich sources of many GAs compared to vegetative tissue, though the task was still mammoth. Working at the ICI Akers Laboratory, Jake MacMillan and P.J. Suter identified GA_1 (which had been isolated previously from *Gibberella*) from *Ph. coccineus* seeds, obtaining 2 mg of the crystalline GA_1 from 87.3 kg of immature seeds that had been harvested from approx. 2 metric tons of locally grown pods (MacMillan and Suter, 1958). In a prescient comment in the final paragraph to their paper, MacMillan and Suter wrote, 'The occurrence of gibberellin A_1 in higher plants adds new significance to the gibberellins and their growth promoting properties. It leaves little doubt that at least gibberellin A_1 participates directly in the growth regulating system of higher plants.' Indeed, GA_1 is now known to have intrinsic bioactivity and to be the major bioactive GA in most plants studied to date. In subsequent papers, the same research group characterised several additional GAs, namely GA_5, GA_6 and GA_8 from the same extract.

Concurrent work taking place at the University of California at Los Angeles (UCLA) led to the isolation of bean factors I and II from *Ph. vulgaris* seeds. Factor I was shown to be GA_1 and it was equally active on *dwarf-1* and *dwarf-5* mutants of maize, whereas factor II was a new GA with less bioactivity than GA_1 when assayed on *dwarf-1* (West and Phinney, 1959). Their inference that 'the genetically controlled enzymatic block in *dwarf-1* would be between the production of factor II and the active gibberellin' predated by more than 20 years the characterisation of *DWARF-1* as encoding

Figure 1.2 The ability of exogenous GA_1 applied to *dwarf-1* maize seedlings to restore a normal (wild-type) phenotype was one of the earliest demonstrations of the growth-promoting activity of bioactive GAs. Note that GA_1 has no effect on normal (wild-type) seedlings. (After B.O. Phinney. © Sinauer Associates, Inc. 2015.)

a 3β-hydroxylase that is necessary for GA bioactivity (Spray *et al.*, 1984). Bean factor II was shown to be GA_5 (MacMillan *et al.*, 1959).

The continued report of new GAs from *Gibberella* and *Phaseolus* by scientists, many of whom had worked in the Akers Laboratory at ICI or at the University of Tokyo, brought the number to 17 (GA_1–GA_{17}) by 1967. However, the proposal that additional GAs be assigned trivial names based on the plant source (for example *Canavalia* GAs I and II) was anticipated to 'result in complete confusion' by Jake MacMillan and Nobutaka Takahashi, since the same GA was often present in more than one species. For this reason they proposed assigning 'A numbers' in approximate chronological order of discovery to 'naturally occurring, fully characterised compounds which possess the gibbane skeleton and the appropriate biological properties' (MacMillan and Takahashi, 1968). The provision of infrared and mass spectra was required to ensure that each compound assigned an A number had a unique structure. The allocation of gibberellin A numbers by these organic chemists worked exceedingly well, and was a service to the plant biology community, though the need to prove 'appropriate biological properties' was not strictly enforced. Many of the 136 known GAs[2] do not have biologically activity *per se*, and the class of hormones is defined by chemical structure rather than bioactivity. However, because of the early reliance on bioassay for GA discovery and isolation, the GAs with the highest biological activities (e.g. GA_1, GA_3, GA_4 and GA_7) were among the first to be characterised (see Figure 1.1). Gibberellins contain either 19 or 20 carbon atoms. The C_{20}-GAs contain the full diterpenoid complement of 20 carbon atoms, whereas the C_{19}-GAs have lost one carbon through metabolism.

The remainder of this chapter focuses on the history of gibberellin research from the late 1950s to the present day. Due to the amount and scope of research during this period the review must be selective. Broadly, the chapter describes our acquisition of knowledge of GA biosynthetic pathways, both in *Gibberella* and in flowering plants. The specific pathways are described in detail in Chapters 5 and 2, respectively. The chapter documents our knowledge of the biosynthetic enzymes and the genes that encode them, and our current understanding of their regulation. It describes the discovery of the GA receptor, and the body of information on DELLA proteins that repress GA response (see Chapter 6). The current identification of DELLA-interacting proteins is moving the field forward in exciting ways as we discover the downstream events that mediate GA responses that lead, for example, to seed germination, stem growth, and reproductive growth, which are discussed later in the book. The chapter closes with a brief review of the research that established these physiological responses to GA.

1.2 Gibberellin biosynthesis

The biosynthesis of GAs, which are tetracyclic diterpenes, was studied initially in *Gibberella*. There were many reasons for using the fungus as a model system: it is easy to grow in defined liquid media, substrates can be administered in and products can be extracted from the medium with ease, and

[2]Continuing the UK–Japanese partnership in assigning A numbers, Peter Hedden and Yuji Kamiya took over this responsibility for the plant biology community in the late 1990s.

the levels of GAs in *Gibberella* are several orders of magnitude higher than those in plants. Moreover, the major end product, GA_3, accumulates, facilitating its isolation for determination of site-specific labelling, which can be diagnostic for assigning structure and biosynthetic origin. Although the end products of the fungal and plant pathways are not the same (GA_3 does not occur universally in higher plants, and even in those plants in which it occurs it is a usually minor metabolite), the assumption was made that GA biosynthetic pathways in the fungus and in higher plants would be similar. What we *now* know about the similarities and differences between the fungal and plant pathways is discussed at the end of this section.

Feeds of radiolabelled substrates to *Gibberella* cultures, followed by degradation and analysis of the resulting [^{14}C]GA_3, showed that it is formed from 12 molecules of [^{14}C]acetate or from four molecules of [^{14}C]mevalonic lactone (MVL) (Birch *et al.*, 1958). The pathway proceeds from MVL to isopentenyl diphosphate, the five-carbon building block of all terpenoids, and thence to the linear diterpene geranylgeranyl diphosphate (GGPP). The pathways from GGPP in *Gibberella* are shown in Figure 1.3. The conversion of GGPP to the bicyclic intermediate *ent*-copalyl diphosphate, and the subsequent conversion of this intermediate to tetracyclic *ent*-kaurene was demonstrated in a cell-free system from *Gibberella* (Shechter and West, 1969). The two-stage reaction was shown to be catalysed by *ent*-kaurene synthase A and B (Fall and West, 1971). This terpene cyclase appeared to be a single protein with two separate catalytic activities, since the two cyclisation reactions, from GGPP to CPP, and from CPP to *ent*-kaurene, had different pH optima, metal ion requirements and sensitivities to plant growth retardants (Fall and West, 1971).

Earlier studies (Cross *et al.*, 1964) had shown *ent*-kaurene to be on the pathway to GA_3. The oxidative steps beyond *ent*-kaurene were identified in *Gibberella* concurrently with research on GA biosynthesis in cell-free systems from plants (discussed below). Much of this early work, on both *Gibberella* and plants, was conducted at UCLA in the laboratory of Charles West. *ent*-Kaurenol, *ent*-kaurenal and *ent*-kaurenoic acid were all individually shown to be precursors of GA_3, inferring the sequential oxidation of the CH_3 group at C-19 in *ent*-kaurene to CH_2OH (*ent*-kaurenol), to CHO (*ent*-kaurenal), and to COOH (*ent*-kaurenoic acid) (Figure 1.3). The enzymes catalysing these steps were shown to be microsomal cytochrome-P450-dependent mono-oxygenases.

The steps in the pathway after *ent*-kaurenoic acid constitute a branch-point, with one branch being the committed pathway to GAs, and the other (not shown) being a route to poly-oxygenated kaurenoids that accumulate in the fungus and some plants, and for which there is no known function. The dedicated pathway to GAs requires the contraction of the six-membered B-ring, with extrusion of C-7, giving GA_{12}-aldehyde (see Figure 1.1), which is the first-formed GA in all systems studied. Considerable work on the mechanism of the ring contraction in the fungus and higher plants was conducted.

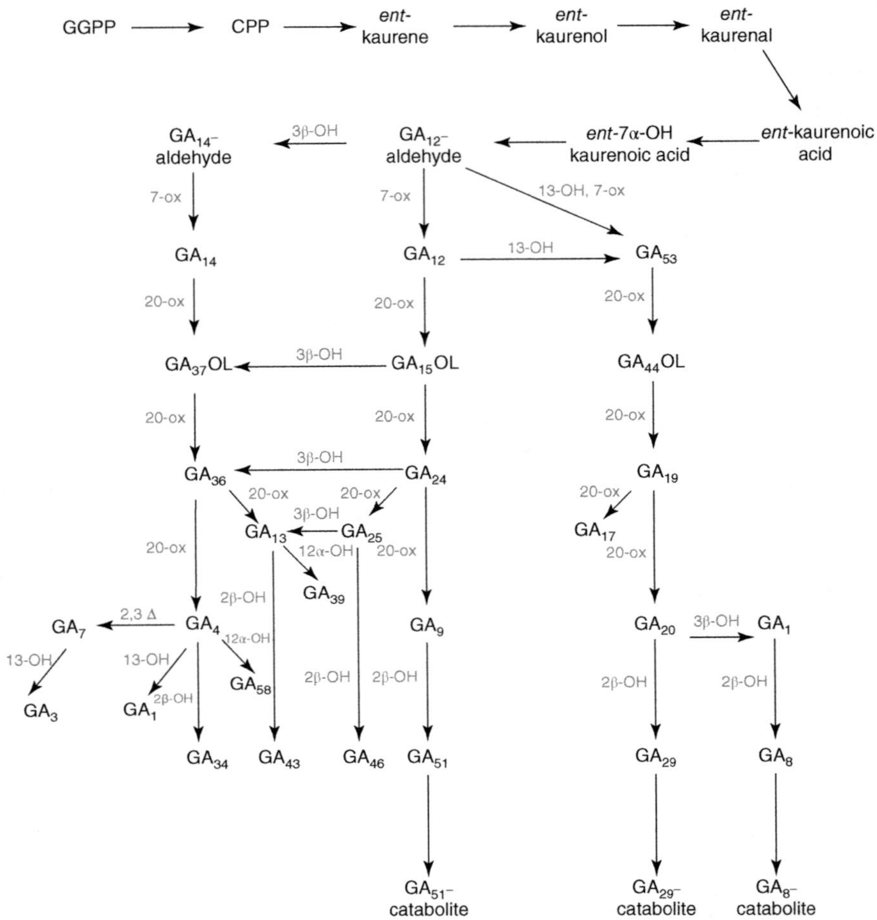

Figure 1.3 Composite scheme showing the conversion of GGPP to the first-formed GA, GA_{12}-aldehyde, and the predominant GA metabolic pathways from GA_{12}-aldehyde in three model systems: *Gibberella* (early 3-hydroxylation pathway, left), pumpkin (late 3-hydroxylation pathway, centre and left), and pea (early 13-hydroxylation pathway, right, and non-hydroxylation pathway, center). Note that GA_1 occurs on both left and right sides of the scheme. GGPP geranylgeranyl diphosphate, CPP copalyl diphosphate, OL open lactone (CH_2OH at C-20). (See insert for colour representation of this figure.)

Potential intermediates between *ent*-kaurenoic acid and GA_{12}-aldehyde, with stereospecific ^{14}C or ^{3}H labeling of atoms in the B ring, were tested. Feeds of labelled *ent*-7α-hydroxykaurenoic acid produced labelled GA_3 in sufficiently high yield (4%) after 2 days to anticipate that it was an intermediate on the GA pathway (Lew and West, 1971). The intermediacy of *ent*-7α-hydroxykaurenoic acid was subsequently confirmed (Hanson *et al.*, 1972).

Gibberellin A_{12}-aldehyde is on the main pathway to GAs in *Gibberella*, (see Figure 1.3), whereas the C-7 acid, GA_{12}, is not (Bearder *et al.*, 1973). Hydroxylation of GA_{12}-aldehyde at C-3 gives GA_{14}-aldehyde, feeds of which produce 3-hydroxylated C_{19}-GAs (Figure 1.3). Gibberellin A_3, a 3,13-dihydroxylated C_{19}-GA, is the major end product of GA biosynthesis in *Gibberella*, and it accumulates. Geissman had previously obtained evidence from feeds of *ent*-kaurenoic acid that GA_4, the first C_{19}-GA on the pathway, was a precursor of GA_7 (1,2-dehydro-GA_4), and GA_3 (13-OH GA_7) (Geissman *et al.*, 1966). This and all other evidence suggested that 13-hydroxylation occurs late in the pathway in *Gibberella*. Intermediates between GA_{14}-aldehyde and GA_4 did not accumulate in the fungus. In a separate, though minor, pathway in *Gibberella*, the C-7 acid, GA_{12}, is the precursor of non-hydroxylated GAs, including GA_9 (Bearder *et al.*, 1973; Bearder *et al.*, 1975).

The highly vigorous wild-type strain of *Gibberella*, GF-1a, was shown by combined gas chromatography-mass spectrometry (GC-MS) to contain at least 25 diterpenes, including 15 known or putative GAs (MacMillan and Wels, 1974). Gibberellins are not required for the growth of *Gibberella* in culture, although they may facilitate pathogenesis by affecting the host plant. The B1-41a strain of *Gibberella*, which was isolated by Bernard Phinney after UV irradiation of GF-1a, was shown to be essentially GA-deficient and yet its growth and morphology was indistinguishable from that of GF-1a. *ent*-Kaurene oxidation is blocked in B1-41a (Bearder *et al.*, 1974), and the absence of downstream metabolites meant that GA metabolic studies could be conducted without the need for isotopic labelling of substrates.

The ability to identify products in complex mixtures using GC-MS, which was pioneered by the laboratory of Jake MacMillan (Binks *et al.*, 1969), was revolutionary to the field. Individual products, if they were known compounds, could be identified unequivocally, even in complex mixtures, without the need for isolation. In addition, detection of ^{14}C or stable isotopes in the mass spectra of products could prove the biogenic origin of metabolites. Furthermore, mass spectral information of unknown compounds was often very informative for structural determination. For all these reasons, exceedingly rapid progress was made in the mid-1970s defining naturally occurring pathways beyond GA_{12}-aldehyde in *Gibberella* (Bearder *et al.*, 1975).

Concurrent with these early studies using *Gibberella* were concerted efforts to study GA biosynthesis in plants. The plants most frequently used for metabolic work were cucurbits, legumes, and cereals. The major pathways were defined well before *Arabidopsis thaliana* became the model system of choice. Much of the earliest work focused on *in vitro* systems from plants, and was conducted by Charles West's group, which included Jan Graebe. The tissue selected for use was liquid endosperm from seeds of members of the Cucurbitaceae, notably *Marah macrocarpus* (Californian wild cucumber, previously called *Echinocystis macrocarpa*) and later *Cucurbita maxima* (pumpkin).

For plants, *in vitro* systems, such as those using liquid endosperm, have advantages over *in vivo* studies – substrates can be administered to cell-free systems without concerns about differential uptake, the products can be extracted with ease, incubation conditions can be defined, and individual enzymatic reactions can be studied by including or excluding a particular cofactor, or adding an inhibitor.

Up to 1 mL of gelatinous acellular endosperm can be squeezed from each developing seed of wild cucumber, and after filtration, with perhaps additional purification by dialysis, the preparation is ready for use. Initial studies with *Marah* confirmed the conversion of MVA to *ent*-kaurene, and its sequential oxidation (Graebe *et al.*, 1965). On a historical note, the ease of isolating intermediates from feeds to cell-free systems from *Marah* facilitated the preparation of labelled compounds for subsequent feeds to *Gibberella* (Graebe *et al.*, 1965). Some years later, the ease of feeding derivatives and analogs to GA-deficient cultures of the B1-41a fungal mutant allowed for the preparation of labelled GAs, such as 12- and 13-hydroxylated GAs, for feeding to plant systems (Gaskin *et al.*, 1984).

A cell-free extract from pumpkin endosperm, with which much pioneering work was done by the research group established by Jan Graebe in Göttingen, was the first plant system in which the conversion of MVA to GA_{12}-aldehyde was achieved (Graebe *et al.*, 1972). Feeding of intermediates confirmed the sequence of *ent*-kaurene oxidation described for the fungus, with each conversion shown to be enzymatic. When *ent*-7α-hydroxykaurenoic acid was fed, it was completely converted, giving GA_{12}-aldehyde, GA_{12}, and two unidentified compounds that were later identified as *ent*-kaurenoids. Graebe and Hedden further examined the ring-contraction mechanism, by which the gibbane skeleton in GA_{12}-aldehyde and all other GAs is formed.

Subsequent conversion of GA_{12}-aldehyde to GA_{12}, GA_{15}, GA_{24}, GA_{36} and GA_{37} in the pumpkin system demonstrated oxidation at C-7, C-20, and C-3 was occurring *in vitro* (Graebe *et al.*, 1974a) (Figure 1.3). Feeds of GA_{12} gave GA_{15}, GA_{24}, GA_{36} and GA_{37} too, unlike the fungal system in which GA_{12}-aldehyde and GA_{12} give different products. The 3- and 20-oxidation of both GA_{12}-aldehyde and GA_{12} required different incubation conditions from earlier enzymatic reactions that are catalysed by mono-oxygenases, notably the omission of Mn^{2+}. In a breakthrough the same year the first conversion in a plant system of MVA to a C_{19}-GA, namely GA_4, was achieved (Graebe *et al.*, 1974b). C_{20}-GA products also identified in these incubations were the tricarboxylic acids GA_{13} and its metabolite, GA_{43} (Figure 1.3). This 2β-hydroxylated C_{20}-derivative was diluted by endogenous GA_{43}, underscoring that the metabolic conversions observed *in vitro* reflected those occurring in pumpkin seeds. Subsequently a comprehensive examination by GC-MS of both endosperm and embryo extracts of pumpkin seeds of several different developmental stages showed over 30 compounds, including eleven GAs, and many poly-hydroxylated *ent*-kaurenoid derivatives (Blechschmidt *et al.*, 1984). In addition to the GAs that had been identified as products in metabolic studies, four new GAs were identified, namely 12α-hydroxylated derivatives of GA_{12}, GA_{14}, GA_{37} and GA_4, the last of which was named GA_{58} (Blechschmidt *et al.*, 1984). Gibberellin A_{58} accumulates more than GA_4. To aid in structural determination of new GAs, *ent*-12α-and 12β-hydroxylated kaurenoic acids were fed to *Gibberella* to obtain reference samples for comparison (Gaskin *et al.*, 1984). These compounds were later obtained as metabolites of GA_{12}-aldehyde in the pumpkin cell-free system, but only when the pH during incubation was between 6 and 7. Above pH 7, hydroxylation at C-12 was not observed, with GA_{12}-aldehyde being converted predominantly to GA_{43} instead (Hedden *et al.*, 1984).

One of the novel features of GA metabolism in pumpkin seeds is that C_{20}-tricarboxylic acids, such as GA_{13} and GA_{43}, accumulate to a much greater extent than in other plants that were also being used for GA metabolic studies, such as pea and corn. In addition, the 13-hydroxylation pathway, which would turn out to be the predominant pathway in many plants is of minor importance in pumpkin (Hedden *et al.*, 1984).

Work with *in vitro* systems from other plants besides pumpkin provided additional useful information. Studies using cell-free systems from pea shoot tips were novel in that they sought to relate *ent*-kaurene biosynthesising activity with seedling phenotype (Coolbaugh *et al.*, 1973), but correlation of enzymatic activity with altered seedling growth in wild-type and dwarf cultivars grown in dark and light gave equivocal results (Ecklund and Moore, 1974). In contrast, cell-free systems from shoots of the *dwarf-5* maize produced less *ent*-kaurene and more *ent*-isokaurene (which would not be a precursor of bioactive GAs) than preparations from wild-type seedlings (Hedden and Phinney, 1979).

The properties and cofactor requirements for GA-metabolizing enzymes in plants were studied most comprehensively by Jan Graebe's research group. Similar to the situation in *Gibberella*, the enzymes catalysing the oxidation of *ent*-kaurene and derivatives are endoplasmic-reticulum-localised cytochrome-P450-dependent mono-oxygenases. So too are the enzymes that catalyse the oxidation of GA_{12}-aldehyde at C-7 and C-13. In contrast, enzymes that catalyse oxidation at C-20, C-3, and C-2 were shown to be soluble 2-oxoglutarate-dependent dioxygenases (2ODDs) (Hedden and Graebe, 1982; Smith and MacMillan, 1984). This is in contrast to the enzymes that oxidise GAs in *Gibberella*, which, like earlier enzymes in the pathway, are also mono-oxygenases.

In vivo metabolic studies with plants began in the early 1970s, and the main focus was on developing seeds, predominantly from legumes. From a historical perspective, the discovery process was different from that with Cucurbits in which, as described previously, work with cell-free systems in the 1970s was predictive of GAs that would later be found as endogenous components.

With pea, for example, analyses of native GAs and *in vivo* metabolic studies in the 1970s were predictive of the pathways that would later be confirmed with cell-free systems.

Work in the MacMillan group on pea seeds began by identifying the major C_{20}- and C_{19}-GAs in immature seeds at different developmental stages (Frydman *et al.*, 1974). *In vivo* metabolic studies were conducted using intact plants, by injecting labelled substrates through the pod wall into the cotyledons of developing seeds. The results of these feeds predicted the presence of two parallel pathways, one with 13-hydroxylation occurring early (at the C_{20}-GA stage), giving GA_{20} as the first C_{19}-GA, and one pathway in which 13-hydroxylation does not occur, giving GA_9 as the first C_{19}-GA (Sponsel and MacMillan, 1977) (see Figure 1.3). The presence of 13-hydroxylated C_{20}-GAs as endogenous components of developing pea seeds supported this contention.

The presence of the early 13-hydroxylation pathway as the major pathway in pea was later confirmed in cell-free systems from developing seeds (Kamiya and Graebe, 1983). Both GA_{12}-aldehyde and GA_{12} could be 13-hydroxylated by a microsomal preparation, yielding GA_{53}. Feeds of GA_{53} to soluble enzymes gave GA_{44}, GA_{19} and GA_{20} (see Figure 1.3). Refeeding all intermediates (GA_{44} was refed in the open lactone form) gave the sequence GA_{53} to GA_{44} to GA_{19} to GA_{20}. GA_{20} was 2β-hydroxylated in preparations from older seeds, giving GA_{29}. Gibberellin A_{12} fed to a soluble enzyme preparation gave non-13-hydroxylated C_{20}-intermediates and GA_9 and GA_{51} (the later step was demonstrated predominantly in preparations from older seeds). Thus the two parallel pathways inferred from *in vivo* studies were demonstrated in entirety *in vitro* (Kamiya and Graebe, 1983) (Figure 1.3).

No evidence of 3-OH was observed in either feeds to maturing pea seeds (10 days from anthesis and older) or in these cell-free systems. Later studies utilising younger fruits of pea showed that 3-hydroxylated C_{19}-GAs (GA_1 and GA_3) do occur transiently in both developing seeds and pericarps shortly after pollination and may well be necessary for the earliest stages of seed development, and for pod elongation (Garcia-Martinez *et al.*, 1991).

In vivo studies of pea seeds also showed the importance of 2β-hydroxylation during the later stages of seed maturation, and the production of novel α, β-unsaturated ketone derivatives called GA-catabolites that accumulated predominantly in the testa (Sponsel, 1983) (Figure 1.3). The accumulation of biologically inactive GA catabolites in pea and in the closely related species *Vicia faba* was seen as an alternative to GA-conjugation, which is observed in other legumes. For example, the multiplicity of free GAs in developing *Ph. vulgaris* seeds, and the accumulation of GA conjugates in mature seeds has been documented (Hiraga *et al.*, 1974). In feeds to older seeds, GAs were conjugated to glucose, either through ether or ester linkages. Evidence for hydrolysis of GA_{20}-glucosyl ether to liberate GA_{20}, which was itself further metabolised to GA_1 when it was fed to maize plants, suggested the conjugate could represent a form for temporary sequestration of GA for later use.

However, conjugates of already inactive GAs would be permanently inactive (Schneider and Schliemann, 1994). These enzymatic reactions in plants that inactivate GAs have not evolved in *Gibberella*, in agreement with the proposition that GAs have no biological activity in the fungus. Mechanisms for GA-inactivation are described in Chapter 3.

Not only did metabolic studies in plants demonstrate that GAs can be inactivated by metabolism, they indicated that many GAs may show bioactivity only because they are converted to an active GA in the plant material used for bioassay. Structure/activity relationships, coupled with metabolic studies, revealed the requirement for certain functional groups for intrinsic activity (Reeve and Crozier, 1974). Bioactive GAs possess 19 rather than 20 carbon atoms, and have a γ-lactone between C-19 and C-10 (Figure 1.1). They possess an exocyclic methylene at C-16, and carboxylic acid at C-6. 3β-Hydroxylation or other functionality at C-3 is required for bioactivity. 13-Hydroxylation neither enhances nor inhibits activity except in certain plants such as members of the Cucurbitaceae and, as shown later, in *Arabidopsis* in which 13-hydroxylated GAs have less activity than their 13-deoxy-counterparts. On the other hand, 2β-hydroxylation (as in GA_8, GA_{29}, GA_{34}, GA_{51}) always reduced bioactivity or the potential to be metabolised to an bioactive GA. Gibberellins with a 1,2 double bond (GA_7 and GA_3) are not inactivated by 2β-hydroxylation. Gibberellin derivatives such as 2,2-dimethyl GA_4 and 2β-methyl GA_4 were synthesised and tested to see whether they would have higher bioactivity than GA_4, since 2β-hydroxylation should not occur for these GAs (Hoad *et al.*, 1981). The results varied by test material, but with bioassays using monocotyledonous plants (e.g. oat first leaf, dwarf rice, and *dwarf-5* maize assays) and with extended duration of testing, the GA derivatives in which 2β-hydroxylation is blocked displayed longer-lasting activity than GA_4.

Extraction of seeds of many different species increased the number of known GAs very substantially during the 1970s and 80s. All GAs had to have confirmed chemical structures before A numbers could be assigned. For some species there was a characteristic pattern of hydroxylation. For example, immature seeds of moonflower, *Caloniction aculeatum* (now *Ipomoea alba*) were shown to contain three GAs that possess 12α-hydroxyl groups, and after structural determination they were assigned the numbers GA_{30}, GA_{31}, and GA_{33} (Murofushi *et al.*, 1988). Developing grain of wheat (*Triticum aestivum*) was shown to contain GAs that are hydroxylated at C-1, two of which were named GA_{60}, and GA_{61} after preparation of authentic reference compounds (Gaskin *et al.*, 1980). Sunflower, *Helianthus annuus*, contains many GAs that are hydroxylated at C-15. After structural determination they were assigned the numbers, GA_{64}, GA_{65}, GA_{66}, GA_{67} and GA_{72} (Hutchison *et al.*, 1988).

The numerous poly-hydroxylated (and thus very polar) GAs that accumulate in developing seeds have little bioactivity is seedling assays, and are not known to have physiological function in seed development. Why such a diversity of GA structures occurs in maturing seeds, and why they accumulate to very high levels during development and decline during the later stages of maturation, is still something of a mystery.

Continuing improvements in the sensitivity of GC-MS instrumentation were being made over time. The MacMillan group was one of the leaders in this area, with Paul Gaskin assembling a large array of reference spectra of naturally occurring GAs, kaurenoids and synthetic analogs, as the methyl esters and trimethylsilyl ether derivatives. Eventually GC-MS instrumentation had the requisite sensitivity to make comprehensive analysis of GAs in vegetative material feasible. Simultaneously work proceeded on pea and corn seedlings.

Two groups led by geneticists Ian Murfet in Hobart, Tasmania and Bernard Phinney (UCLA) had, over time, been isolating single gene dwarf mutants of pea and corn, respectively (Phinney, 1956; Reid *et al.*, 1983). The early 13-hydroxylation pathway was known to be the major pathway in pea seeds (Kamiya and Graebe, 1983) and all GAs that were identified in maize tassels were 13-hydroxylated (Hedden *et al.*, 1982), thus, feeding studies focused on the metabolism of GA_{20}, which is the first-formed C_{19}-GA in that pathway.

GC-MS analyses of seedlings of GA-responsive dwarf mutants of pea and maize helped to define the enzymatic steps that were blocked by each genetic lesion. Researchers fed labelled GA_{20} to *LE* and *le* pea seedlings (Ingram *et al.*, 1984), and to *DWARF-1* and *dwarf-1* maize seedlings (Spray *et al.*, 1984). Results showed that the *le* mutation of pea and the *dwarf-1* mutation of maize both prevent 3β-hydroxylation, thus blocking the conversion of GA_{20} to GA_1 (Figure 1.3). This is a crucial step – the responses of *le* and *dwarf-1* mutants to exogenous GA application indicated that GA_{20} has no activity *per se*, and the metabolite of GA_1, GA_8, is inactive. Thus, GA_1 must have hormonal function for internode elongation in both species. That the LE/le gene difference defines Mendel's tall and dwarf lines of pea made the discovery particularly exciting (Ingram *et al.*, 1984). Additional work on both pea and maize GA biosynthesis mutants have subsequently revealed the locations in the GA biosynthetic pathway at which other mutations block (Fujioka *et al.*, 1988; Davidson *et al.*, 2003; 2004).

Reviewing these three decades of GA metabolic studies in plants, it became evident that there was a multiplicity of pathways beyond the first-formed GA, GA_{12}-aldehyde, especially in developing seeds, which produced a plethora of GAs with many interesting functional features, but of unknown function. In time, the near universality of the early-13-hydroxylation pathway in vegetative tissue, and the importance of GA_1 as a 'hormone' was substantiated. The comment made by MacMillan and Suter in 1956 that 'the occurrence of gibberellin A_1 in higher plants… leaves little doubt that at least gibberellin A_1 participates directly in the growth regulating system of higher plants' was indeed prescient. The observation that in some plants 13-hydroxylation may reduce biological activity (Magome *et al.*, 2013) identifies GA_4 as another GA with intrinsic hormonal activity in members of the Cucurbitaceae, *Arabidopsis*, and rice.

The advent of *Arabidopsis thaliana* as a model system from the 1980s moved our knowledge of GA biosynthesis further as it facilitated the study of genes encoding biosynthetic enzymes. The endogenous GAs in *Arabidopsis* were first identified by Jan Zeevaart's research group (Talon *et al.*, 1990). Twenty GAs were identified by GC-MS in shoots of the Landsberg ecotype. The GAs were representative of three pathways, non-hydroxylated, early-3-hydroxylation, and early-13-hydroxylation. In contrast to most plants previously studied, and in fact in contrast to most crop plants studied to date, the early 13-hydroxylation pathway in *Arabidopsis* is a minor pathway. The non-hydroxylation pathway predominates.

A series of GA-responsive dwarf mutants of *Arabidopsis* had been generated by Maarten Koornneef at Wageningen, the Netherlands, in the 1980s (Koornneef and van der Veen, 1980). He named the mutant loci *ga1*, *ga2*, *ga3*, *ga4*, and *ga5* based on epistasis tests. *GA1* was cloned by Sun *et al.* using the *ga1-3* mutant that Koornneef had generated by fast neutron bombardment. Because *ga1-3* has a large deletion they were able to use a novel technique of genomic subtraction to identify the sequence present in the wild-type that was missing from the mutant (Sun *et al.*, 1992). GA1 is a terpene cyclase that catalyses the conversion of GGPP to the bicyclic intermediate *ent*-copalyl-diphosphate (Sun and Kamiya, 1994). To clone GA2, Yamaguchi and co-workers used pumpkin *ent*-kaurene synthase cDNA to isolate a homologous cDNA from *Arabidopsis* that when expressed as a fusion protein in *E. coli* had *ent*-kaurene synthase activity (Yamaguchi *et al.*, 1998). The *ga2-1* mutant contains a truncated protein and could be complemented with the wild-type cDNA, confirming that *GA2* encodes *ent*-kaurene synthase. *ent*-Kaurene oxidase, encoded by *GA3*, was cloned by conventional map-based cloning and random sequencing (Helliwell *et al.*, 1998). Expressing the cDNA in yeast confirmed that the enzyme can catalyse the three sequential steps in the oxidation of *ent*-kaurene to *ent*-kaurenoic acid (Helliwell *et al.*, 1999). Intriguingly, although GA1 and GA2 are expressed in chloroplasts, GA3 is localised on the outer face of the chloroplast membrane (Helliwell *et al.*, 2001b), and may direct the catalytic product, *ent*-kaurenoic acid, to the next enzyme in the pathway, *ent*-kaurenoic acid oxidase. This enzyme, originally defined by the *grd5* mutant of barley and the *dwarf-3* mutant of maize, was cloned from barley (Helliwell *et al.*, 2001a). *Arabidopsis* contains two genes encoding *ent*-kaurenoic acid oxidase with overlapping function (Regnault *et al.*, 2014), and this redundancy precluded a mutant phenotype in *Arabidopsis*. Like GA3 *ent*-kaurenoic acid oxidase is a multi-functional cytochrome-P450-dependent mono-oxygenase. It catalyses the three-step oxidation from *ent*-kaurenoic acid to GA_{12}. In *Arabidopsis* this enzyme is localised to the endoplasmic reticulum (Helliwell *et al.*, 2001a).

The mutants *ga1*, *ga2* and *ga3* are extreme dwarfs. As *GA1* and *GA2* are the only genes encoding CPS and KS, respectively, it has been assumed that these dwarf seedlings are completely GA-deficient, though traces of GAs of unknown origin are apparent. However, *ga1*, *ga2* and *ga3* are all sterile dwarfs that can be rescued by treating with an *ent*-kaurenoid or GA beyond the metabolic block.

Talon *et al.* analysed the GA content of *ga4* and *ga5* mutants, and proposed that *GA4* encodes a 3β-hydroxylase, and that *GA5* encodes a multi-functional GA 20-oxidase responsible for catalysing the formation of C_{19}-GAs (Talon *et al.*, 1990). They also recognised the importance of 3β-hydroxylation, reporting that GA_9 had no biological activity on *ga4* mutant seedlings, and that GA_1 and GA_4 were probably the active hormones. The *GA4* gene was cloned by Chiang *et al.* and *GA5* was cloned by Phillips *et al.* and Xu *et al.*, providing important information on the enzymatic reactions catalysed by the enzymes, their specificity, and their regulation by feedback repression (Chiang *et al.*, 1995; Phillips *et al.*, 1995; Xu *et al.*, 1995). The notable cloning of the first plant GA 20-oxidase had been reported the previous year by Lange *et al.* from pumpkin (Lange *et al.*, 1994). It was shown to be a dioxygenase that could indeed catalyse the multi-step conversion of GA_{12} to a C_{19}-GA.

Several years later the completion of the *Arabidopsis* genome (2000) revealed that the GA 20-, 3- and 2-oxidases are all encoded by small gene families, as described in detail in Chapter 2. Because of redundancy, albeit it partial in some cases, a severely dwarf phenotype only results when mutations exist in multiple members of the GA 20-oxidase or GA 3-oxidase gene families.

In concluding this section on GA metabolism it is worthwhile reflecting on the use of *Gibberella* for initial studies. In many ways it was a wise choice for the practical reasons mentioned earlier. Moreover, it provided a useful model on which to base the *in vitro* and *in vivo* plant studies. But 50 years on, it is now known that GA biosynthesis in *Gibberella* and in plants is not the same. In fact there are many differences. For example, an alternative to the mevalonic acid pathway for producing IPP, namely the methyl erythritol phosphate (MEP) pathway was identified in plant plastids, and although it occurs in some bacteria and algae it does not occur in fungi (Rohmer, 1999). The MEP pathway appears to be the predominant route for the production of IPP to serve as a precursor for GAs in plants, at least in vegetative tissues, though a minor contribution of the MVA pathway cannot be ruled out (Kasahara *et al.*, 2002). Furthermore, over the past two decades, information has been obtained by the group of Bettina Tudzynski on the enzymes that catalyse *ent*-kaurenoid and GA metabolism in *Gibberella* (see Chapter 5). Many of the fungal enzymes have different properties from those encoding similar steps in the pathway in plants, including some fungal enzymes that demonstrate remarkable multi-functionality. Even the mechanism to produce GA_3 from its immediate precursor differs between *Gibberella* and plants (Albone *et al.*, 1990). It is evident from this work that the pathways in *Gibberella* and in plants evolved separately (Bömke and Tudzynski, 2009). The identification of GAs in a small number of other fungi provides evidence there may have been horizontal gene transfer from one fungus to another, but horizontal gene transfer from *Gibberella* to plants is ruled out by the fundamental differences in the nature of the pathways. Continued study of the fungal pathway, and its regulation, is timely because of the continued commercial production of GA_3 using *Gibberella.*

1.3 Gibberellin signalling

Classically, there have been two major foci for research on GA signal transduction: the cereal aleurone and the stem apex (Paleg, 1965). Germinating cereal grain has been the subject of scientific study for nearly two centuries with a view to enhancing the malting of grain for the brewing industry. It had been known for some time that the presence of the embryo enhanced amylolytic activity in the endosperm, and that barley and malt (germinated grain) contained GA-like biological activity. In 1960 Yomo and Paleg independently showed that pre-incubation of embryo-less half seeds of barley with GA_3 increases the amounts of amylase and reducing sugars released from the endosperm. Historical aspects of this groundbreaking work have been reviewed in detail (Paleg, 1965). In intact grain the embryo supplies the GA for induction of starch breakdown in the endosperm.

The origin of the α-amylase in cereal grains was shown to be the aleurone, the outermost layer of living cells that surrounds the dead, starch-filled cells of the mature endosperm. The synthesis and release of α-amylase by isolated aleurone layers matched that of intact endosperm as long as the incubation buffer contained calcium (Chrispeels and Varner, 1967). Experiments utilising $H_2{}^{18}O$ elegantly demonstrated that essentially all of the α-amylase required for breakdown of stored starch arises by *de novo* synthesis (Filner and Varner, 1967), and Varner and Chandra noted 'it is a delightful nicety that the key to these reserves is kept by the embryo, the only tissue capable of growth' (Varner and Chandra, 1964).

Thus began several decades of productive research on the biochemical mechanism whereby the 'key' (GA) from the embryo induces *de novo* synthesis of several isoforms of α-amylase in the aleurone to 'unlock' (hydrolyse) starch in non-living cells of the endosperm. The advantages of this system for studying GA action are manifold – aleurone layers, which can be readily separated from the rest of the endosperm, provide a population of uniform differentiating cells from which protoplasts can be prepared. The cytology of these cells/protoplasts has been studied in detail, including the effects of GA on the number and appearance of protein storage vacuoles, oleosomes and endomembranes, and eventual programmed cell death (Bethke *et al.*, 1999). Moreover, unlike other GA responses like internode elongation, the GA response in aleurone cells has a well-defined and measurable biochemical end point – the production of α-amylase.

The nature of the GA receptor in aleurone cells is somewhat controversial. Several lines of evidence suggested that it was in the plasma membrane. For example, GA_4 that had been covalently linked to agarose beads to prevent its uptake into oat aleurone protoplasts was still able to induce the synthesis of amylase, though it was inactive on aleurone cells (Hooley *et al.*, 1991). Furthermore, if GA is injected directly into the cytosol of barley aleurone protoplasts it is inactive (Gilroy and Jones, 1994). Despite this convincing evidence, a GA receptor from aleurone plasma membranes has not been identified. The identification of GID1, which is a soluble GA receptor (see later), raised the possibility that there may be two types of GA receptor, one that is plasma-membrane-localised and one that is soluble. Recent convincing evidence that GID1 is the *only* GA receptor in rice (Yano *et al.* 2015) does not preclude the existence of an additional plasma-membrane-localised receptor in barley and oat.

The involvement of second messengers in GA response in aleurone cells has been extensively studied. Applied GA induces both Ca^{2+}-independent and Ca^{2+}-dependent events. The induction of amylase synthesis by GA does not require Ca^{2+}, whereas secretion of the enzyme does (Jones and Carbonell, 1984). In addition, evidence for the involvement of G-proteins, cyclic GMP, and protein phosphorylation is reviewed in detail (Bethke *et al.*, 1997) (Figure 1.4). In the pathway leading to amylase production GA acts primarily by increasing the transcription of amylase genes. The purification of *α-amylase* mRNA, which is produced in relatively large amounts in aleurone cells, enabled the isolation of genomic clones containing both the structural gene for α-amylase and its upstream promoter sequences. The partial deletion of known sequences of bases from *α-amylase* promoters indicates that sequences conferring GA responsiveness, termed GA response elements (GREs), are 200–300 base pairs upstream of the transcription start site. Identical GREs were found to occur in all cereal *α-amylase* promoters so far examined, and their presence was shown to be essential for the induction of *α-amylase* gene transcription by GA.

The sequence of the GRE (TAACAAA) in the *α-amylase* gene promoter resembles a motif in the binding site for MYB transcription factors. *GAMYB* mRNA increases in aleurone cells as early as 1 hour after GA treatment, preceding the increase in *α-amylase* mRNA by several hours (see Figure 1.4). These and other data discussed in Chapter 6 are consistent with GAMYB regulating *α-amylase* gene expression (Gubler *et al.*, 1995). Cycloheximide has no effect on the production of *GAMYB* mRNA, indicating that protein synthesis is not required for *GAMYB* expression, and that *GAMYB* can therefore be defined as a primary or early response gene. In contrast, the *α-amylase* gene is a secondary or late response gene.

Turning to the second focus of research on GA signalling, namely that on stem apices and internode elongation, single gene dwarf mutants whose internode growth was *not* correlated with endogenous GA levels were crucial to gaining insight into GA signal transduction. It had been known for some

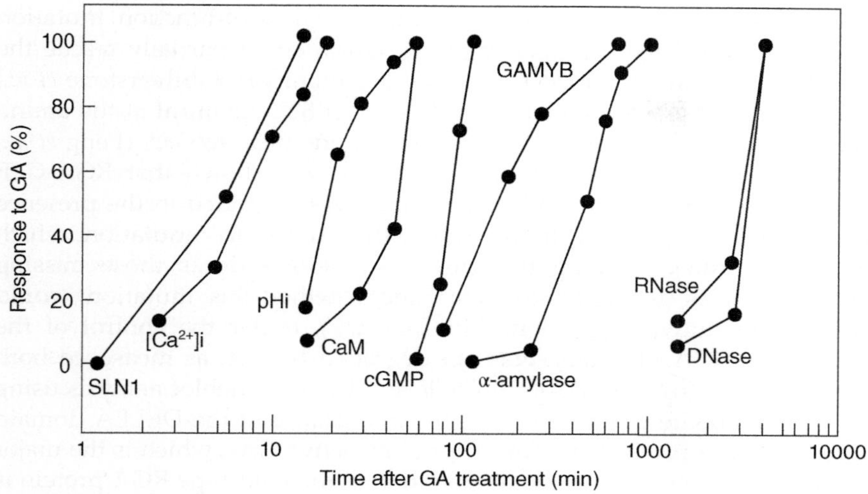

Figure 1.4 Following the addition of bioactive GA to barley aleurone protoplasts, a multiple-component signalling pathway is initiated. CaM calmodulin. (Sun and Gubler, 2004. Reproduced with permission from Annual Reviews.)

time that some GA non-responsive, semi-dominant, dwarf mutants of maize (*Dwarf-8*), wheat (*Reduced height, Rht*), and *Arabidopsis* (*gai-1*) accumulated high levels of endogenous GAs and yet were still dwarf. In addition, other stem length mutants e.g. barley (*sln*) and pea (*la cry*) were characteristically taller than their respective wild-type seedlings. These so-called 'slender' mutants were resistant to inhibitors of GA biosynthesis, and continued to have a slender phenotype even if they were lacking endogenous GAs. The characterisation of these two types of mutants, in which (a) the GA response was irretrievably repressed, or (b) it was constitutively expressed, defined genes that were involved in the GA signal transduction in stem growth, and spurred an exciting phase of GA research.

Work on the GA-insensitive dwarf mutants of *Arabidopsis* utilised, at first, the semi-dominant dwarf *gai-1* mutant (Koornneef *et al.*, 1985). Cloning of *GAI*, together with a gene referred to as *GRS* (*GAI Related Sequence*) determined that these genes encode putative transcription factors each with a nuclear localisation sequence. A deletion of 17 amino acids in the N-terminal region of GAI (gai-1) gave a semi-dominant GA-resistant dwarf phenotype (Peng *et al.*, 1997). The deletion included a five-amino-acid motif, DELLA, though the significance of this motif was not immediately recognised. Peng *et al.* concluded that GAI is a repressor of GA responses, and that GA can release the repression by the wild-type protein, but not that imposed by the gain-of-function mutation *gai-1*. Intriguingly, other mutant alleles of *GAI*, rather than giving gain-of-function phenotypes like *gai-1*, gave loss-of-function phenotypes such that mutant plants appeared wild-type.

Silverstone *et al.* independently identified a loss-of-function mutation which they called *rga* (*repressor of ga1-3*), which could partially rescue the semi-dwarf phenotype of the GA-deficient mutant *ga1-3* (Silverstone *et al.*, 1997). Cloning the gene showed that RGA was 82% identical at the amino acid level to GAI, and that *RGA* was, in fact, identical to *GRS* (Peng *et al.*, 1997; Silverstone *et al.*, 1998). Significantly, it was shown that RGA-GFP fusion protein localised to the nucleus, and was degraded in the presence of GA_3 (Dill *et al.*, 2001; Dill and Sun, 2001). An *rga-Δ17* mutation, which encoded a protein missing the same 17 amino acids as those missing from gai-1, was generated to determine whether this mutation would lead to constitutive repression. GFP-(rga-Δ17), under the control of the endogenous RGA promoter, was not degraded by GA, as measured both by confocal microscopy of root tip cells and by immunoblot analysis using anti-GFP antibodies, demonstrating the importance of the DELLA domain for GA-induced proteolysis. The effect of bioactive GA_4, which is the major bioactive GA in *Arabidopsis*, on degradation of the wild-type RGA protein is comparatively rapid, being visible in plants that have a GA-deficient (*ga1-3*) background in less than 30 minutes (Dill *et al.*, 2001). Generation of a triple mutant line containing null alleles at both *RGA* and *GAI* loci (*rga-24* and *gai-t6*) along with *ga1-3* could completely rescue the GA-deficient phenotype of *ga1-3* (Dill and Sun, 2001). Thus it was confirmed that the ground state in wild-type individuals is one of growth repression caused by GAI and RGA. Repression can be relieved by bioactive GA, but GA is not required for some aspects of growth in the absence of GAI and RGA.

Shortly after the cloning of *GAI* and *RGA*, it established that *DWARF-8* in maize and *Rht-1* in wheat are their functional orthologues (Peng *et al.*, 1999). Semi-dominant, gain-of-function mutations *dwarf-8*, *Rht-B1b*, and *Rht-D1b*, gave GA-insensitive dwarf phenotypes, and were all shown to have deletions in the N-terminus regions of the respective proteins. The observation that orthologous proteins regulate GA response in monocots and dicots underscored the importance and potential universality of these regulatory proteins. Moreover, the *Rht-B1b* and *Rht-D1b* mutations were those selected in wheat to give the short-stemmed, lodging-resistant, high-yielding strains introduced as part of the Green Revolution almost 50 years earlier demonstrating the immensely valuable agronomic benefits that can be attained by modulating the activity of these regulatory proteins (Peng *et al.*, 1999). Semi-dwarf varieties of wheat containing the *Rht-B1b* and *Rht-D1b* mutations, when grown with fertiliser and irrigation, have been credited with saving billions of lives. Norman Borlaug, one of the scientists most closely involved in the breeding program, received a Nobel Peace Prize in 1970, and on March 25 2014, the 100th anniversary of his birth, a statue of Borlaug was unveiled in the US Capitol.

GAI, RGA, DWARF-8 and RHT1 belong to the plant-specific GRAS family of putative transcriptional regulators, named after the first three to be discovered GAI, RGA and SCARECROW (SCR). They all contain a GRAS domain at the carboxy-terminus to which is ascribed the transcriptional regulatory function. Three additional homologues to GAI and RGA in *Arabidopsis*, RGA-LIKE 1 (RGL1), RGL2, and RGL3 have been identified. The five homologues have some overlapping and some unique functions (Lee *et al.*, 2002; Cheng *et al.*, 2004; Tyler *et al.*, 2004). The N-terminal regions of all five have highly conserved DELLA and VHYNP motifs, which are required for GA-induced proteolysis (see later). The importance of the DELLA motif has led to the N-terminal region being referred to as the DELLA domain, and also gives the name to this entire sub-family of GRAS proteins. In the GRAS domain there are leucine heptad repeats (LHR), which were anticipated to be a site of protein/protein interaction, together with the nuclear localisation signal.

Unlike *Arabidopsis*, rice and barley were shown to each have a single DELLA protein, SLENDER1 (SLR1) in rice and SLENDER1 (SLN1) in barley (Ikeda *et al.*, 2001; Chandler *et al.*, 2002). The genes encoding these proteins were initially defined by loss-of-function mutations that caused the plants to have taller than wild-type phenotypes even in the absence of GA, underscoring the contention that GA signalling was constitutive in these slender mutants because they were lacking a repressor of GA response (see Figure 1.5). In contrast, a dominant gain-of-function mutation, *Sln1d*, in barley and a deletion of 17 amino acid in the DELLA domain of SLR1 in rice gave dwarf phenotypes. The contrast between the slender and dwarf phenotypes of loss- and gain-of-function mutations, respectively, was particularly striking in these cereals (Figure 1.5). In contrast, in *Arabidopsis*, in which there is considerable redundancy between the five DELLA proteins, a slender phenotype is not evident if only one DELLA is deleted.

Work on signalling in cereal aleurone and on stem growth converged after the observation that embryo-less half seeds of the loss-of-function *slr1* and *sln1* mutants produced amylase in the absence of exogenous GA, and that aleurone cells from the dominant dwarf mutants were far less responsive to GA than wild-type. Furthermore, GA treatment of aleurone cells from wild-type barley caused the reduction in SLN levels within 5 minutes of treatment, almost 2 hours prior to the buildup of GAMYB (Gubler *et al.*, 2002). Figure 1.4 provides an integrated view of aleurone response (Sun and Gubler, 2004). Thus not only do DELLA proteins appear to be universal, but more and more information was appearing to link them to GA signalling in multiple response pathways.

The next piece in the puzzle of GA signal transduction fell into place when two genes defined by gain-of-function mutations in *Arabidopsis* (*sly1*) and rice (*gid2*) were cloned (McGinnis *et al.*, 2003; Sasaki *et al.*, 2003). These mutations gave GA-insensitive semi-dwarf phenotypes, and defined genes that encode F-box sub-units that are each part of an SCF E3-ubiquitin ligase complex. Poly-ubiquitination of DELLA proteins and their proteolysis by the SCF $^{SLY1/GID2}$-proteasome pathway in *Arabidopsis*/rice relieves growth repression caused by DELLA proteins. These proteins cannot be degraded in *sly1* and

Figure 1.5 Two-week old wild-type (WT) and two mutants of barley demonstrating the opposite effects on growth of two different mutations in the same gene, *SLN1*. Left, the *slnc* loss-of-function mutation confers a slender, GA-constitutive phenotype. Right, the gain-of-function *sln1d* mutation gives a dominant dwarf phenotype. The mutations are in different domains in the single DELLA protein in barley. (From Chandler *et al.*, 2002, courtesy of P. M. Chandler.)

gid2 mutants, keeping the repression of growth in place. However, why the DELLA domain is so important had to await the characterisation of the GA receptor. This breakthrough came shortly after when a soluble protein for rice that had all the hallmarks of a GA-receptor was cloned (Ueguchi-Tanaka *et al.*, 2005).

Rice contains just a single DELLA protein, and it also contains just a single gene encoding a GA receptor (Ueguchi-Tanaka *et al.*, 2005). Cloning of GID1

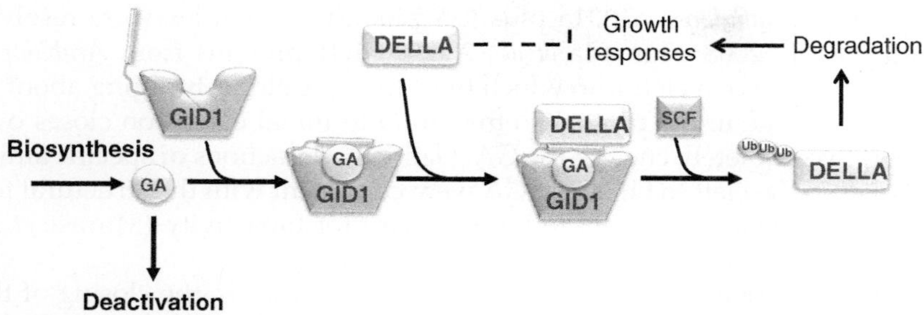

Figure 1.6 Overview of GA signalling. Bioactive GA binds the GID1 receptor bringing about an allosteric change that allows a DELLA protein that represses GA responses to bind to the GID1-GA complex. The DELLA protein then undergoes an allosteric change that allows for its polyubiquitination by the SCF complex. It can then be degraded by the 26S proteasome, relieving repression, and facilitating GA responses.

showed it to have close homology to a family of hormone-sensitive lipases (HSLs). The binding of radiolabelled GA_4 (the 16,17 diol was used) to the wild-type GID1 protein was rapid and saturable, and could be competed out by non-labelled GA equally rapidly. Moreover there was low to no binding of inactive GAs to the wild-type protein, nor was there binding of bioactive GAs to the mutant protein.

Genetic tests placed SLR1 and GID1 on the same pathway, with GID1 being upstream of SLR1 and essential for the GA_3-induced proteolysis of SLR1. SLR1 is resistant to GA-induced proteolysis in the *gid1* mutant. Moreover, in yeast two-hybrid assays GID1 and SLR1 were shown to interact, but *only* in the presence of GA_3 (Ueguchi-Tanaka *et al.*, 2005). GA-dependent binding of GID1 and SLR was later confirmed *in planta* (Ueguchi-Tanaka *et al.*, 2007). The GA signalling pathway is outlined in Figure 1.6, and described in detail in later chapters.

Analysis of the *Arabidopsis* genome revealed that it contains three orthologues of *GID1* (Ueguchi-Tanaka *et al.*, 2005), and these have been termed *GID1a, GID1b,* and *GID1c* (Nakajima *et al.*, 2006). Mutation in any one of these homologues does not give a discernible phenotype (Griffiths *et al.*, 2006), explaining perhaps why identification of the GA receptor eluded *Arabidopsis* researchers. Double mutants show reduced growth and fertility, but phenotypes differ somewhat depending on which two of the three homologues are knocked out, whereas triple mutants are extremely severe, sterile, non-GA-responding dwarfs, indicating that the GID1 homologues are the only receptors in *Arabidopsis* (Griffiths *et al.*, 2006). This work is discussed in further detail in Chapter 6.

Binding of the *Arabidopsis* receptors to DELLA proteins was demonstrated by several workers (Griffiths *et al.*, 2006; Nakajima *et al.*, 2006; Willige *et al.*, 2007), but the actual mechanism of GA-GID1-DELLA interaction was best defined when the crystal structures of the rice GID1 receptor protein plus

GA, and an *Arabidopsis* GID1a plus GA plus RGA complex were resolved (Murase *et al.*, 2008; Shimada *et al.*, 2008). GID1 proteins from *Arabidopsis* and rice both have a cleft into which bioactive GA binds, bringing about an allosteric change in the protein so that an N-terminal extension closes over the cleft to completely enclose the GA. Defined interactions of specific amino acids within the cleft to bioactive GA were consistent with the structural features of GAs that are known to be necessary for bioactivity (Murase *et al.*, 2008; Shimada *et al.*, 2008).

For GID1 in rice, and GID1a and GID1c in *Arabidopsis*, the closing of this N-terminal extension over a bioactive GA is absolutely required before any of the DELLA proteins can bind to the receptor. The DELLA/VHYNP motifs within the N-terminal domain of the DELLA proteins interact with hydrophobic amino acids on the upper surface of the GID1 'lid'. In contrast, the GID1b lid can partially close in the absence of GA within the cleft, allowing some binding of DELLA even in the absence of GA.

The binding of DELLA proteins to the receptor brings about an allosteric change in the GRAS domain of those proteins, that facilitates them binding to the F-box components of, for example, SCF^{SLY1} in *Arabidopsis* and SCF^{GID2} in rice, resulting in ubiquitination and proteolysis. Thus the degradation of DELLA proteins is a crucial step in transducing a GA signal. Anything that prevents this process occurring is likely to block GA signalling. The absence of the DELLA motif in the *gai-1*, *rga-Δ17*, and *sln1d* mutants prevents the binding of these proteins to the receptor, precluding the conformational change in their GRAS domains that is required for them to be substrates for SLY1 and GID2. Thus these mutants are dwarf, and the repression by mutated proteins cannot be relieved by application of GA.

In certain circumstances it is possible to discern an unusual scenario: that of GA signalling in the absence of DELLA protein degradation. This was encountered first by scientists working with the *sly1* mutant, in which DELLA proteins are not targeted for proteolysis (Ariizumi and Steber, 2007; Ariizumi *et al.*, 2008) and is discussed in detail in Chapter 6.

Since DELLA proteins have no recognisable DNA-binding domain it had been postulated for some time that they would bring about transcriptional regulation through protein–protein interaction. Several microarray investigations conducted within the past decade have been important for recognising DELLA-interacting proteins. For example, Zentella, working with 8-day-old seedlings of the *ga1-3* mutant, looked at alterations in gene expression as a consequence of applied GA_4 or of inducing *rga-Δ17* expression (Zentella *et al.*, 2007). Expression of 14 genes was down-regulated by GA and up-regulated by the DELLA protein. Several targets of GA and DELLA were genes involved in GA homeostasis, including those encoding the GA 3- and 20-oxidases and GID1a and c. Other genes encoded bHLH, MYB and WRKY transcription factors. Supporting the notion that DELLA proteins are involved in regulating hormone crosstalk, XERICO, an inducer of abscisic acid biosynthesis, was down-regulated by GA and up-regulated by DELLA (Zentella *et al.*, 2007). More detailed discussion of GA perception and the early events in signal transduction are described in Chapter 6.

Gibberellin research has entered an exciting phase over the past few years with the characterisation of several DELLA-interacting proteins, clarifying how DELLAs can mediate so many fundamental changes in growth and development. This is an extremely active area of research at the present time, which is discussed in Chapter 7. For example, DELLA proteins interact with phytochrome interacting factors (PIFs). These are bHLH transcription factors that can mediate the transition from skotomorphogenesis to photomorphogenesis (de Lucas *et al.*, 2008; Feng *et al.*, 2008). In dark-grown seedlings, PIFs bind to promoter sequences to activate transcription of genes encoding expansins and other factors that promote cell elongation and hypocotyl growth. If DELLA proteins are abundant (for example in light) they can bind to PIFs to prevent the transcription of their target genes. In the presence of bioactive GA DELLA proteins are degraded and cannot bind PIFs, so that PIFs activate transcription and hypocotyls elongate. Thus the classic GA response of hypocotyl elongation in light-grown *Arabidopsis* seedlings occurs because GAs disrupt the DELLA–PIF interaction, thereby releasing PIFs to activate transcription of genes whose products induce cell elongation.

In another example, in *Arabidopsis* hypocotyls the DELLA protein, GAI, interacts with prefoldin5 (PFD5) which is one of the sub-units of a chaperone protein involved in α/β-tubulin dimerisation (Locascio *et al.*, 2013). Tubulin dimerisation facilitates the assembly of microtubules whose orientation in the cortical cytoplasm will, in turn, direct the orientation of cellulose microfibrils in cell walls. When PFD5 is bound to GAI, the chaperone–GAI complex is localised in the nucleus and is inactive. In the presence of bioactive GA_4 the DELLA protein is degraded, allowing for cytoplasmic localisation of the chaperone. The now active cytoplasmic chaperone facilitates the assembly of tubulin dimers into microtubules and their orientation in a transverse direction. This is turn directs the laying down of cellulose microfibrils in a transverse orientation, which is conducive to cell elongation (Locascio *et al.*, 2013).

Many other examples of DELLA-interacting proteins have been described recently. This research provides an avenue for further investigating downstream events that will define the biochemical mechanisms for GA responses.

1.4 Physiological responses to gibberellins

Gibberellins are active in regulating growth and development throughout the entire life cycle. Sometimes they act alone, more often they act in concert with other hormones, either synergistically or in an antagonistic manner. The action of GA in particular target cells is a result of an intricate series of events such as hormone biosynthesis, transport, presence and accessibility of the receptor, occurrence of DELLA proteins and interacting protein partners, and the machinery to degrade DELLA proteins. Often environmental factors such as presence or absence of light, light quality and duration, or temperature can impinge on any part of parts of this sequence of events.

Gibberellins break dormancy in seeds, especially those that have a light or cold requirement for germination. They can speed up germination of grain by promoting the hydrolysis of reserves, and can aid seedling establishment by promoting hypocotyl and internode growth. Gibberellins are important for shortening the juvenile phase of some species, and they induce flowering in certain species by transducing the effect of the appropriate temperature or photoperiod. They are necessary for pollen formation, pollen tube growth, and fruit and seed development. These processes and phenomena are considered in detail in several subsequent chapters.

As mentioned earlier in this chapter, the clearly defined biochemical events in the cereal aleurone system make it an ideal system for studying mechanism of GA action. In most other instances the responses were described in morphological terms, and it is only comparatively recently that they have been 'dissected' to reveal the biochemical events that are responsible for the macroscopic changes in plant size or form.

Historically, the physiological effects of GAs were recognised even before the first GAs had been identified as natural components of plants. Stowe and Yamaki reviewed the effects of GA (obtained from fungal cultures) when applied to 80 different species, leading them to attest that the action of GAs corresponds to that of naturally occurring compounds in higher plants (Stowe and Yamaki, 1957). As a way of distinguishing GAs from the already well-characterised auxin, they defined GAs as a class of compounds that causes internode elongation when applied to certain intact genetically dwarfed plants, with the elongation of monocot leaves as a supporting definition. Although, as described earlier in this chapter, it was the chemical structure not the biological activity that became the defining feature of GAs, internode elongation remains one of the most notable effects.

Stowe and Yamaki inferred that the action of GAs in shoot elongation, leaf expansion, growth of dwarfs, parthenocarpy, bolting of long-day plants, or reversal of light-inhibition involved the 'removal of certain limitations on cell elongation'. Citing the possibility of suppression of an inhibitor as a way to remove the 'normal limitation' Stowe and Yamaki predated by several decades the recognition of the GA-induced proteolysis of the DELLA class of transcriptional regulators as the way to remove the limitation to cell elongation. The suggestion that GAs exerted their growth-promoting effects by enhancing the levels of auxin received considerable attention, but the multiplicity of responses that were unique to GAs ensured that these compounds were duly recognised as a second distinct class of hormones.

Internode elongation is the basis for many GA bioassays by which 'GA-like substances' (i.e. biologically-active compound(s) that had not yet been chemically characterised) were first recognised and later 'quantified' in plant extracts in order to discern their potential roles in growth and development (Phinney and West, 1960). Correlation between the level of GA-like substances in extracts and plant growth were, in some instances, comparatively easy to obtain. For example, the amount of GA-like material in extracts of single gene dwarf mutants of corn was half (or less) of that in their wild-type seedling counterparts. *Hyocyamus niger* (black henbane) plants that had received inductive long days (LDs), and exhibited the earliest stages of bolting and had microscopic flower primordia, contained more GA-like activity than extracts of non-induced plants (Lang, 1960). The judicious use of GA biosynthesis inhibitors and the demonstration that further application of GA can normalise inhibitor-treated plants (Zeevaart, 1964) supported the contention that GAs are naturally occurring regulators of processes such as stem growth and flowering.

When the steady-state levels of GA-like substances in plant extracts did not match the amount of growth, scientists postulated that the rate of GA turnover may be more important than the static size of the GA pool. At the time, a lack of information about GA pathways of biosynthesis and degradation, and the paucity of labelled substrates precluded many metabolic studies. However, an estimate of the dynamic GA status within plants could be obtained by allowing GA-like material to diffuse from plant parts into agar gel over several hours or days (Jones and Phillips, 1964). Using a combination of diffusate and tissue extraction it was shown that not all GA-like substances were mobile, in some cases the non-diffusible GA-like substances appeared to be precursors of the mobile substances and external conditions could influence GA metabolism.

In due course, once many more GAs had been characterised, the pathways for GA biosynthesis and deactivation had been defined, and the methods for isotopically labelling GAs had been developed, it became feasible to study GA occurrence, biosynthesis, and degradation more directly. Evidence quickly accumulated that plants contain many different GAs, not just two or three observed as zones of bioactivity after separation by thin layer chromatography. The scientific inquiry that established that it is the 3β-hydroxylated C_{19}-GAs that have intrinsic biological activity was described in detail earlier in this chapter. In addition, specific metabolic steps could be altered by inductive photoperiods. For example, in the long-day plant (LDP) spinach, GA_1 is the bioactive GA that causes stem growth in LDs. The C_{20}-GA, GA_{53}, is a substrate for two competing enzymes, a GA 20-oxidase that is upregulated in LDs and can lead to elevated levels of bioactive GA_1 (Gilmour *et al.*, 1986; Lee and Zeevaart, 2007) or a GA2-oxidase that converts it to GA_{97} in reaction that is predominant in SDs (Lee and Zeevaart, 2005). In a similar way, low temperatures are also inductive for flowering in some species. For example, vernalisation of *Thalaspi arvense* (field pennycress) leads to elevated levels of bioactive GA and flowering, though in this instance the inductive treatment enhances an earlier step in GA biosynthesis, namely the oxidation of *ent*-kaurene (Hazebroek *et al.*, 1993).

The effect of GA on flowering is particularly complex. Not only has there been intense academic interest in the subject for decades, but the ability to manipulate flowering by GAs, inhibitors of GA biosynthesis, or environmental factors has profound practical applications in agronomy and horticulture (see Chapters 12 and 13). The flowering responses of many types of plants have been tested over the years (Pharis and King, 1985; King and Evans, 2003 and see Chapter 11). Although for a brief period of time the notion was considered that bioactive GA might be the flower-inducing substance 'florigen' that travelled from leaf to apex, this idea quickly lost favour when it became evident that the positive effects of GA on flowering were not universal. Moreover, there was no consensus on whether GA induced bolting *and* flowering in LDPs, or whether flowering was a consequence of bolting.

After decades of intense research we now have a profound understanding of the regulation of flowering. The transcriptional regulator CONSTANS (CO) accumulates in the light in companion cells of leaves in inductive conditions, leading to the production of a phloem-mobile signal FLOWERING LOCUS T (FT) (or its orthologue, Hd3a in rice) (Corbesier *et al.*, 2007; Tamaki *et al.*, 2007). The sequence of events transduced by the arrival of FT or Hd3a at the stem apex is considered in Chapter 11. FT/Hd3 are phosphatidylethanolamine-binding proteins that are transcriptional regulators themselves, and their production in leaves, transport in the phloem and their action in the apex fulfill the criteria laid down long ago for 'florigen'. GA acts downstream of FT, and activates *LFY*, one of the floral meristem identity genes (Blazquez *et al.*, 1998).

In the grass, *Lolium temulentum*, detailed studies have provided evidence for a florigen-like role for GA_5. King *et al.* traced the movement of GA_5 from induced leaves to apices after a single long day, and demonstrated its arrival at the stem apex prior to the appearance of floral primordia (King *et al.*, 2006). This interesting scenario not only indicates a novel situation with regard to a florigenic role for a GA in *Lolium*, it reveals that different bioactive GAs can have unique roles – thus in *Lolium* GA_5 (but not GA_4) is florigenic, whereas GA_4 (but not GA_5) induces stem elongation. This is thought not to be due to inherent differences in the biochemical functions of GA_4 and GA_5, but to differences in their susceptibility to deactivation. In *Arabidopsis* GA_4 is the primary bioactive GA both for *LFY* transcription and stem elongation (Eriksson *et al.*, 2006).

Application of GA to woody gymnosperms alters reproductive behavior differently from that in woody angiosperms. In conifers, GA_3 can promote strobilus formation in members of the Cupressaceae and Taxodiaceae, whereas less polar GAs such as GA_4 are more effective in members of the Pinaceae (Pharis and King, 1985). In contrast, GAs tend to inhibit flowering in woody angiosperms such as apple. Research has revealed that these effects of exogenous application reflect the GA status of these plants, and GAs and GA biosynthesis inhibitors are used commercially in forestry and fruit-growing industries.

The mechanism by which GAs could stimulate stem elongation has been a topic of sustained inquiry, since it is one of the most notable manifestations of GA response. From the earliest studies it was evident that GAs promoted *cell* elongation, though whether cell elongation alone was sufficient to account for observed increases in plant height was not clear. Experiments with gamma-irradiated wheat seedlings, in which cell division cannot occur, showed a normal response to GA by cell elongation alone. In contrast GA-induced bolting of rosette plants was shown to involve mitotic activity in the sub-apical meristem to provide a source of cells for subsequent elongation (Sachs, 1965).

Defining the biochemical and physical factors that facilitate irreversible cell elongation has helped to distinguish the action of GA from that of auxin, which also causes cell elongation. GA was found to have little effect on turgor pressure, but instead affected the wall-yielding properties of the cell wall (Cosgrove and Sovonick-Dunford, 1989). The anisotropic growth that occurs in cells of pea epicotyls in response to both GAs and auxin was compared. Both hormones enhance the rate of relaxation, but only GAs affect the value of the yield threshold. The requirement of GA for pea pollen tube growth (Singh *et al.*, 2002) suggests that GAs are also important in regulating tip growth in certain types of cells.

Although 50 years ago GAs were initially considered to exert their biological effects through increasing auxin levels, half a century of research has revealed the unique and vital roles of GAs in plant growth and development. However, as we understand more fully the molecular events that underlie the changes in plant size and form, and especially the impact of the external environment, we are aware that integration of signalling pathways plays a vital role. This aspect is considered in more detail in Chapter 8.

References

Albone, K.S., Gaskin, P., MacMillan, J. *et al.* (1990). Biosynthetic origin of gibberellin A_3 and gibberellin A_7 in cell-free preparations from seeds of *Marah macrocarpus* and *Malus domestica*. *Plant Physiology* **94**, 132–142.

Ariizumi, T. and Steber, C.M. (2007). Seed germination of GA-insensitive *sleepy1* mutants does not require RGL2 protein disappearance in *Arabidopsis*. *The Plant Cell* **19**, 791–804.

Ariizumi, T., Murase, K., Sun, T.P. and Steber, C.M. (2008). Proteolysis-independent downregulation of DELLA repression in Arabidopsis by the gibberellin receptor GIBBERELLIN INSENSITIVE DWARF1. *The Plant Cell* **20**, 2447–2459.

Bearder, J.R., MacMillan, J. and Phinney, B.O. (1973). 3-Hydroxylation of gibberellin A_{12}-aldehyde in *Gibberella fujikuroi* Strain REC-193A. *Phytochemistry* **12**, 2173–2179.

Bearder, J.R., MacMillan, J., Wels, C.M. *et al.* (1974). Position of the metabolic block for gibberellin biosynthesis in mutant B1-41a of *Gibberella fujikuroi*. *Phytochemistry* **13**, 911–917.

Bearder, J.R., MacMillan, J. and Phinney, B.O. (1975). Fungal Products. Part XIV. Metabolic pathways from *ent*-kaurenoic acid to fungal gibberellins in mutant B1-41a of *Gibberella fujikuroi*. *Journal of the Chemical Society, Perkin Transactions* **1**, 721–726.

Bethke, P.C., Schuurink, R. and Jones, R.L. (1997). Hormonal signalling in cereal aleurone. *Journal of Experimental Botany* **48**, 1337–1356.

Bethke, P.C., Lonsdale, J.E., Fath, A. and Jones, R.L. (1999). Hormonally regulated programmed cell death in barley aleurone cells. *The Plant Cell* **11**, 1033–1045.

Binks, R., MacMillan, J. and Pryce, R.J. (1969). Plant hormones. VIII: Combined gas chromatography-mass spectrometry of methyl esters of gibberellins A_1 to A_{24} and their trimethylsilyl ethers. *Phytochemistry* **8**, 271–284.

Birch, A.J., Rickards, R.W. and Smith, H. (1958). The biosynthesis of gibberellic acid. *Proceedings of the Chemical Society of London*, 192–193.

Blazquez, M.A., Green, R., Nilsson, O. *et al.* (1998). Gibberellins promote flowering of *Arabidopsis* by activating the *LEAFY* promoter. *The Plant Cell* **10**, 791–800.

Blechschmidt, S., Castel, U., Gaskin, P. *et al.* (1984). GC-MS analysis of the plant hormones in seeds of *Cucurbita maxima*. *Phytochemistry* **23**, 553–558.

Bömke, C. and Tudzynski, B. (2009). Diversity, regulation, and evolution of the gibberellin biosynthetic pathway in fungi compared to plants and bacteria. *Phytochemistry* **70**, 1876–1893.

Brian, P.W. and Hemming, H.G. (1955). The effect of gibberellic acid on shoot growth of pea seedlings. *Physiologia Plantarum* **8**, 669–681.

Chandler, P.M., Marion-Poll, A., Ellis, M. and Gubler, F. (2002). Mutants at the *Slender1* locus of barley cv Himalaya. Molecular and physiological characterization. *Plant Physiology* **129**, 181–190.

Cheng, H., Qin, L.J., Lee, S.C. *et al.* (2004). Gibberellin regulates *Arabidopsis* floral development via suppression of DELLA protein function. *Development* **131**, 1055–1064.

Chiang, H.H., Hwang, I. and Goodman, H.M. (1995). Isolation of the *Arabidopsis GA4* locus. *The Plant Cell* **7**, 195–201.

Chrispeels, M.J. and Varner, J.E. (1967). Gibberellic acid-enhanced synthesis and release of α-amylase and ribonuclease by isolated barley aleurone layers. *Plant Physiology* **42**, 398–406.

Coolbaugh, R.C., Moore, T.C., Barlow, S.A. and Ecklund, P.R. (1973). Biosynthesis of *ent*-kaurene in cell-free extracts of *Pisum sativum* shoot tips. *Phytochemistry* **12**, 1613–1618.

Corbesier, L., Vincent, C., Jang, S.H. *et al.* (2007). FT protein movement contributes to long-distance signaling in floral induction of *Arabidopsis*. *Science* **316**, 1030–1033.

Cosgrove, D.J. and Sovonick-Dunford, S.A. (1989). Mechanism of gibberellin-dependent stem elongation in peas. *Plant Physiology* **89**, 184–191.

Cross, B.E. (1954). Gibberellic acid. Part I. *Journal of the Chemical Society*, 4670–4676.

Cross, B.E., Grove, J.F. and Morrison, A. (1961). Gibberellic acid. Part XVIII. Some rearrangements of ring A. *Journal of the Chemical Society*, 2498–2515.

Cross, B.E., Hanson, J.R. and Galt, R.H.B. (1964). The biosynthesis of the gibberellins. Part I. (–)-Kaurene as a precursor of gibberellic acid. *Journal of the Chemical Society*, 295–300.

Davidson, S.E., Elliott, R.C., Helliwell, C.A. *et al.* (2003). The pea gene *NA* encodes *ent*-kaurenoic acid oxidase. *Plant Physiology* **131**, 335–344.

Davidson, S.E., Smith, J.J., Helliwell, C.A. *et al.* (2004). The pea gene *LH* encodes *ent*-kaurene oxidase. *Plant Physiology* **134**, 1123–1134.

de Lucas, M., Daviere, J.M., Rodriguez-Falcon, M. *et al.* (2008). A molecular framework for light and gibberellin control of cell elongation. *Nature* **451**, 480–484.

Dill, A., Jung, H.S. and Sun, T.P. (2001). The DELLA motif is essential for gibberellin-induced degradation of RGA. *Proceedings of the National Academy of Sciences USA* **98**, 14162–14167.

Dill, A. and Sun, T.P. (2001). Synergistic derepression of gibberellin signaling by removing RGA and GAI function in *Arabidopsis thaliana*. *Genetics* **159**, 777–785.

Ecklund, P.R. and Moore, T.C. (1974). Correlations of growth rate and de-etiolation with rate of *ent*-kaurene biosynthesis in pea (*Pisum sativum* L.). *Plant Physiology* **53**, 5–10.

Eriksson, S., Bohlenius, H., Moritz, T. and Nilsson, O. (2006). GA_4 is the active gibberellin in the regulation of *LEAFY* transcription and *Arabidopsis* floral initiation. *The Plant Cell* **18**, 2172–181.

Fall, R.R. and West, C.A. (1971). Purification and properties of kaurene synthetase from *Fusarium moniliforme*. *Journal of Biological Chemistry* **246**, 6913–6928.

Feng, S.H., Martinez, C., Gusmaroli, G. *et al.* (2008). Coordinated regulation of *Arabidopsis thaliana* development by light and gibberellins. *Nature* **451**, 475-479.

Filner, P. and Varner, J.E. (1967). A test for de novo synthesis of enzymes: Density labeling with H_2O^{18} of barley α-amylase induced by gibberellic acid. *Proceedings of the National Academy of Sciences USA* **58**, 1520–1526.

Frydman, V.M., Gaskin, P. and MacMillan, J. (1974). Qualitative and quantitative analyses of gibberellins throughout seed maturation in *Pisum sativum* cv. Progress No. 9. *Planta* **118**, 123–132.

Fujioka, S., Yamane, H., Spray, C.R. *et al.* (1988). Qualitative and quantitative analyses of gibberellins in vegetative shoots of normal, *dwarf-1*, *dwarf-2*, *dwarf-3*, and *dwarf-5* seedlings of *Zea mays* L. *Plant Physiology* **88**, 1367–1372.

Garcia-Martinez, J.L., Santes, C., Croker, S.J. and Hedden, P. (1991). Identification, quantitation and distribution of gibberellins in fruits of *Pisum sativum* L. cv. Alaska during pod development. *Planta* **184**, 53–60.

Gaskin, P., Kirkwood, P.S., Lenton, J.R. *et al.* (1980). Identification of gibberellins in developing wheat grain. *Agricultural and Biological Chemistry* **44**, 1589–1593.

Gaskin, P., Hutchison, M., Lewis, N. *et al.* (1984). Microbiological conversion of 12-oxygenated and other derivatives of *ent*-kaur-16-en-19-oic acid by *Gibberella fujikuroi*, mutant B1-41a. *Phytochemistry* **23**, 559–564.

Geissman, T.A., Verbiscar, A.J., Phinney, B.O. and Cragg, G. (1966). Studies on biosynthesis of gibberellins from (–)-kaurenoic acid in cultures of *Gibberella fujikuroi*. *Phytochemistry* **5**, 933–947.

Gilmour, S.J., Zeevaart, J.A., Schwenen, L. and Graebe, J.E. (1986). Gibberellin metabolism in cell-free extracts from spinach leaves in relation to photoperiod. *Plant Physiology* **82**, 190–195.

Gilroy, S. and Jones, R.L. (1994). Perception of gibberellin and abscisic acid at the external face of the plasma membrane of barley (*Hordeum vulgare* L.) aleurone protoplasts. *Plant Physiology* **104**, 1185–1192.

Graebe, J.E., Dennis, D.T., Upper, C.D. and West, C.A. (1965). Biosynthesis of gibberellins. I. Biosynthesis of (–)-kaurene, (–)-kauren-19-ol and trans-geranylgeraniol in endosperm nucellus of *Echinocystis macrocarpa* Greene. *Journal of Biological Chemistry* **240**, 1847–1854.

Graebe, J.E., Bowen, D.H. and MacMillan, J. (1972). Conversion of mevalonic acid into gibberellin A_{12}-aldehyde in a cell-free system from *Cucurbita pepo*. *Planta* **102**, 261–271.

Graebe, J.E., Hedden, P., Gaskin, P. and MacMillan, J. (1974a). Biosynthesis of gibberellins A_{12}, A_{15}, A_{24}, A_{36} and A_{37} by a cell-free system from *Cucurbita maxima*. *Phytochemistry* **13**, 1433–1440.

Graebe, J.E., Hedden, P., Gaskin, P. and MacMillan, J. (1974b). Biosynthesis of a C_{19}-gibberellin from mevalonic acid in a cell-free system from a higher plant. *Planta* **120**, 307–309.

Griffiths, J., Murase, K., Rieu, I. *et al.* (2006). Genetic characterization and functional analysis of the GID1 gibberellin receptors in *Arabidopsis*. *The Plant Cell* **18**, 3399–3414.

Gubler, F., Kalla, R., Roberts, J.K. and Jacobsen, J.V. (1995). Gibberellin-regulated expression of a *MYB* gene in barley aleurone cells: evidence for MYB transactivation of a high-pI α-amylase gene promoter. *The Plant Cell* **7**, 1879–1891.

Gubler, F., Chandler, P.M., White, R.G. *et al.* (2002). Gibberellin signaling in barley aleurone cells. Control of SLN1 and GAMYB expression. *Plant Physiology* **129**, 191–200.

Hanson, J.R., Hawker, J. and White, A.F. (1972). Studies in terpenoid biosynthesis. Part IX. The sequence of oxidation on ring B in kaurene-gibberellin biosynthesis. *Journal of the Chemical Society, Perkin Transactions* **1**, 1892–1895.

Hazebroek, J.P., Metzger, J.D. and Mansager, E.R. (1993). Thermoinductive regulation of gibberellin metabolism in *Thlaspi arvense* L. 2. Cold induction of enzymes in gibberellin biosynthesis. *Plant Physiology* **102**, 547–552.

Hedden, P. and Graebe, J.E. (1982). Cofactor requirements for the soluble oxidases in the metabolism of C_{20}-gibberellins. *Journal of Plant Growth Regulation* **1**, 105–116.

Hedden, P. and Phinney, B.O. (1979). Comparison of *ent*-kaurene and *ent*-isokaurene synthesis in cell-free systems from etiolated shoots of normal and *dwarf-5* maize seedlings. *Phytochemistry* **18**, 1475–1479.

Hedden, P., Phinney, B.O., Heupel, R. *et al.* (1982). Hormones of young tassels of *Zea mays*. *Phytochemistry* **21**, 391–393.

Hedden, P., Graebe, J.E., Beale, M.H. *et al.* (1984). The biosynthesis of 12α-hydroxylated gibberellins in a cell-free system from *Cucurbita maxima* endosperm. *Phytochemistry* **23**, 569–574.

Helliwell, C.A., Sheldon, C.C., Olive, M.R. *et al.* (1998). Cloning of the *Arabidopsis ent*-kaurene oxidase gene *GA3*. **95**, 9019–9024.

Helliwell, C.A., Poole, A., Peacock, W.J. and Dennis, E.S. (1999). Arabidopsis *ent*-kaurene oxidase catalyses three steps of gibberellin biosynthesis. *Plant Physiology* **119**, 507–510.

Helliwell, C.A., Chandler, P.M., Poole, A. *et al.* (2001a). The CYP88A cytochrome P450, *ent*-kaurenoic acid oxidase, catalyses three steps of the gibberellin biosynthesis pathway. *Proceedings of the National Academy of Sciences USA* **98**, 2065–2070.

Helliwell, C.A., Sullivan, J.A., Mould, R.M. *et al.* (2001b). A plastid envelope location of *Arabidopsis ent*-kaurene oxidase links the plastid and endoplasmic reticulum steps of the gibberellin biosynthesis pathway. *The Plant Journal* **28**, 201–208.

Hiraga, K., Kawabe, S., Yokota, T. *et al.* (1974). Plant growth substances in seeds of *Phaseolus vulgaris*. 2. Isolation and characterization of plant growth substances in immature seeds and etiolated seedlings of *Phaseolus vulgaris*. *Agricultural and Biological Chemistry* **38**, 2521–2527.

Hoad, G.V., Phinney, B.O., Sponsel, V.M. and MacMillan, J. (1981). The biological activity of 16 gibberellin A_4 and gibberellin A_9 derivatives using 7 bioassays. *Phytochemistry* **20**, 703–713.

Hooley, R., Beale, M.H. and Smith, S.J. (1991). Gibberellin perception at the plasma-membrane of *Avena fatua* aleurone protoplasts. *Planta* **183**, 274–280.

Hutchison, M., Gaskin, P., MacMillan, J. and Phinney, B.O. (1988). Gibberellins in seeds of *Helianthus annuus*. *Phytochemistry* **27**, 2695–2701.

Ikeda, A., Ueguchi-Tanaka, M., Sonoda, Y. *et al.* (2001). Slender rice, a constitutive gibberellin response mutant, is caused by a null mutation of the *SLR1* gene, an ortholog of the height-regulating gene *GAI/RGA/RHT/D8*. *The Plant Cell* **13**, 999–1010.

Ingram, T.J., Reid, J.B., Murfet, I.C. *et al.* (1984). Internode length in *Pisum* : The *Le* gene controls the 3β-hydroxylation of gibberellin A_{20} to gibberellin A_1. *Planta* **160**, 455–463.

Jones, R.L. and Carbonell, J. (1984). Regulation of the synthesis of barley aleurone α-amylase by gibberellic acid and calcium ions. *Plant Physiology* **76**, 213–218.

Jones, R.L. and Phillips, I.D. (1964). Agar-diffusion technique for estimating gibberellin production by plant organs. *Nature* **204**, 497–499.

Kamiya, Y. and Graebe, J.E. (1983). The biosynthesis of all major pea gibberellins in a cell-free system from *Pisum sativum*. *Phytochemistry* **22**, 681–689.

Kasahara, H., Hanada, A., Kuzuyama, T. *et al.* (2002). Contribution of the mevalonate and methylerythritol phosphate pathways to the biosynthesis of gibberellins in *Arabidopsis*. *Journal of Biological Chemistry* **277**, 45188–45194.

King, R.W. and Evans, L.T. (2003). Gibberellins and flowering of grasses and cereals: Prizing open the lid of the 'florigen' black box. *Annu Rev Plant Biol* **54**, 307–328.

King, R.W., Moritz, T., Evans, L.T. *et al.* (2006). Regulation of flowering in the long-day grass *Lolium temulentum* by gibberellins and the *FLOWERING LOCUS T* gene. *Plant Physiology* **141**, 498–507.

Koornneef, M. and van der Veen, J.H. (1980). Induction and analysis of gibberellin sensitive mutants in *Arabidopsis thaliana* (L.) heynh. *Theoretical and Applied Genetics* **58**, 257–263.

Koornneef, M., Elgersma, A., Hanhart, C.J. *et al.* (1985). A gibberellin insensitive mutant of *Arabidopsis thaliana*. *Physiologia Plantarum* **65**, 33–39.

Lang, A. (1956). Bolting and flowering in biennial *Hyoscyamus niger*, induced by gibberellic acid. *Plant Physiology* **31 (suppl)**, 35.

Lang, A. (1960). Gibberellin-like substances in photoinduced and vegetative *Hyoscyamus* plants. *Planta* **54**, 498–504.

Lange, T., Hedden, P. and Graebe, J.E. (1994). Expression cloning of a gibberellin 20-oxidase, a multifunctional enzyme involved in gibberellin biosynthesis. *Proceedings of the National Academy of Sciences USA* **91**, 8552-8556.

Lee, D.J. and Zeevaart, J.A. (2005). Molecular cloning of *GA 2-oxidase3* from spinach and its ectopic expression in *Nicotiana sylvestris*. *Plant Physiology* **138**, 243–254.

Lee, D.J. and Zeevaart, J.A. (2007). Regulation of *gibberellin 20-oxidase1* expression in spinach by photoperiod. *Planta* **226**, 35–44.

Lee, S.C., Cheng, H., King, K.E. *et al.* (2002). Gibberellin regulates *Arabidopsis* seed germination via *RGL2*, a *GAI/RGA*-like gene whose expression is up-regulated following imbibition. *Genes & Development* **16**, 646–658.

Lew, F.T. and West, C.A. (1971). (-)-Kaur-16-en-7β-ol-19-oic acid, an intermediate in gibberellin biosynthesis. *Phytochemistry* **10**, 2065–2076.

Locascio, A., Blazquez, M.A. and Alabadi, D. (2013). Dynamic regulation of cortical microtubule organization through prefoldin-DELLA interaction. *Current Biology* **23**, 804–809.

MacMillan, J. and Suter, P.J. (1958). The occurrence of gibberellin A_1 in higher plants – isolation from the seed of runner bean (*Phaseolus multiflorus*). *Naturwissenschaften* **45**, 46.

MacMillan, J. and Takahashi, N. (1968). Proposed procedure for the allocation of trivial names to the gibberellins. *Nature* **217**, 170–171.

MacMillan, J. and Wels, C.M. (1974). Detailed analysis of metabolites from mevalonic lactone in *Gibberella fujikuroi*. *Phytochemistry* **13**, 1413–1417.

MacMillan, J., Seaton, J.C. and Suter, P.J. (1959). A new plant growth promoting acid – gibberellin A_5 from the seed of *Phaseolus multiflorus*. *Proceedings of the Chemical Society of London*, 325–326.

Magome, H., Nomura, T., Hanada, A. *et al.* (2013). CYP714B1 and CYP714B2 encode gibberellin 13-oxidases that reduce gibberellin activity in rice. *Proceedings of the National Academy of Sciences USA* **110**, 1947–1952.

McGinnis, K.M., Thomas, S.G., Soule, J.D. *et al.* (2003). The Arabidopsis *SLEEPY1* gene encodes a putative F-box subunit of an SCF E3 ubiquitin ligase. *The Plant Cell* **15**, 1120–1130.

Murase, K., Hirano, Y., Sun, T.P. and Hakoshima, T. (2008). Gibberellin-induced DELLA recognition by the gibberellin receptor GID1. *Nature* **456**, 459–463.

Murofushi, N., Nakayama, M., Takahashi, N. *et al.* (1988). 12-Hydroxylation of gibberellin A_{12} and gibberellin A_{14} by prothallia of *Lygodium japonicum* and identification of a new gibberellin, GA_{74}. *Agricultural and Biological Chemistry* **52**, 1825–1828.

Nakajima, M., Shimada, A., Takashi, Y. *et al.* (2006). Identification and characterization of *Arabidopsis* gibberellin receptors. *The Plant Journal* **46**, 880–889.

Paleg, L.G. (1965). Physiological effects of gibberellins. *Annual Review of Plant Physiology* **16**, 291–322.

Peng, J.R., Carol, P., Richards, D.E. *et al.* (1997). The Arabidopsis *GAI* gene defines a signaling pathway that negatively regulates gibberellin responses. *Genes & Development* **11**, 3194–3205.

Peng, J., Richards, D.E., Hartley, N.M. *et al.* (1999). 'Green revolution' genes encode mutant gibberellin response modulators. *Nature* **400**, 256–261.

Pharis, R.P. and King, R.W. (1985). Gibberellins and reproductive development in seed plants. *Annual Review of Plant Physiology* **36**, 517–568.

Phillips, A.L., Ward, D.A., Uknes, S. *et al.* (1995). Isolation and expression of three gibberellin 20-oxidase cDNA clones from Arabidopsis. *Plant Physiology* **108**, 1049–1057.

Phinney, B.O. (1956). Growth response of single-gene dwarf mutants in maize to gibberellic acid. *Proceedings of the National Academy of Sciences USA* **42**, 185–189.

Phinney, B.O. (1983). The history of gibberellins. In *The Biochemistry and Physiology of Gibberellins*, (ed. A. Crozier). Praeger, New York, pp. 19–52.

Phinney, B.O. and West, C.A. (1960). Gibberellins as native plant growth regulators. *Annual Review of Plant Physiology* **11**, 411–436.

Radley, M. (1956). Occurrences of substances similar to gibberellic acid in higher plants. *Nature* **178**, 1070–1071.

Reeve, D.R. and Crozier, A. (1974). Assessment of gibberellin structure-activity relationships. *Journal of Experimental Botany* **25**, 431–445.

Regnault, T., Daviere, J.M., Heintz, D. *et al.* (2014). The gibberellin biosynthetic genes AtKAO1 and AtKAO2 have overlapping roles throughout *Arabidopsis* development. *The Plant Journal* **80**, 462–474.

Reid, J.B., Murfet, I.C. and Potts, W.C. (1983). Internode length in *Pisum*. 2. Additional information on the relationship and action of loci *Le, La, Cry, Na* and *Lm*. *Journal of Experimental Botany* **34**, 349–364.

Rohmer, M. (1999). The discovery of a mevalonate-independent pathway for isoprenoid biosynthesis in bacteria, algae and higher plants. *Natural Product Reports* **16**, 565–574.

Sachs, R.M. (1965). Stem elongation. *Annual Review of Plant Physiology* **16**, 73–96.

Sasaki, A., Itoh, H., Gomi, K. *et al.* (2003). Characterization of rice dwarf mutant, *GA insensitive dwarf 2* (*gid2*). *Plant and Cell Physiology* **44**, S186.

Schneider, G. and Schliemann, W. (1994). Gibberellin conjugates – an overview. *Plant Growth Regulation* **15**, 247–260.

Shechter, I. and West, C.A. (1969). Biosynthesis of gibberellins. IV. Biosynthesis of cyclic diterpenes from trans-geranylgeranyl pyrophosphate. *Journal of Biological Chemistry* **244**, 3200–3209.

Shimada, A., Ueguchi-Tanaka, M., Nakatsu, T. *et al.* (2008). Structural basis for gibberellin recognition by its receptor GID1. *Nature* **456**, 520–523.

Silverstone, A.L., Mak, P.Y., Martinez, E.C. and Sun, T.P. (1997). The new *RGA* locus encodes a negative regulator of gibberellin response in *Arabidopsis thaliana*. *Genetics* **146**, 1087–1099.

Silverstone, A.L., Ciampaglio, C.N. and Sun, T.P. (1998). The *Arabidopsis RGA* gene encodes a transcriptional regulator repressing the gibberellin signal transduction pathway. *The Plant Cell* **10**, 155–169.

Singh, D.P., Jermakow, A.M. and Swain, S.M. (2002). Gibberellins are required for seed development and pollen tube growth in *Arabidopsis*. *The Plant Cell* **14**, 3133–3147.

Smith, V.A. and MacMillan, J. (1984). Purification and partial characterization of a gibberellin 2β-hydroxylase from *Phaseolus vulgaris*. *Journal of Plant Growth Regulation* **2**, 251–264.

Sponsel, V.M. (1983). The localization, metabolism and biological activity of gibberellins in maturing and germinating seeds of *Pisum sativum* cv. Progress No. 9. *Planta* **159**, 454–468.

Sponsel, V.M. and MacMillan, J. (1977). Further studies on the metabolism of gibberellins (GAs) A_9, A_{20} and A_{29} in immature seeds of *Pisum sativum* cv. Progress No. 9. *Planta* **135**, 129–136.

Spray, C., Phinney, B.O., Gaskin, P. *et al.* (1984). Internode length in *Zea mays* L. – the *dwarf-1* mutation controls the 3β-hydroxylation of gibberellin A_{20} to gibberellin A_1. *Planta* **160**, 464–468.

Stodola, F.H., Raper, K.B., Fennell, D.I. *et al.* (1955). The microbial production of gibberellins A and X. *Archives of Biochemistry and Biophysics* **54**, 240–245.

Stowe, B.B. and Yamaki, T. (1957). The history and physiological action of the gibberellins. *Annual Review of Plant Physiology* **8**, 181–216.

Sun, T.P. and Gubler, F. (2004). Molecular mechanism of gibberellin signaling in plants. *Annual Review of Plant Biology* **55**, 197–223.

Sun, T.P. and Kamiya, Y. (1994). The Arabidopsis *GA1* locus encodes the cyclase *ent*-kaurene synthetase A of gibberellin biosynthesis. *The Plant Cell* **6**, 1509–1518.

Sun, T.P., Goodman, H.M. and Ausubel, F.M. (1992). Cloning the Arabidopsis *GA1* locus by genomic subtraction. *The Plant Cell* **4**, 119–128.

Takahashi, N., Kitamura, H., Kawarada, A. *et al.* (1955). Biochemical studies on 'Bakanae' fungus. Part XXXIV. Isolation of gibberellins and their properties. *Bulletin of the Agricultural Chemical Society of Japan* **19**, 267–277.

Talon, M., Koornneef, M. and Zeevaart, J.A. (1990). Endogenous gibberellins in *Arabidopsis thaliana* and possible steps blocked in the biosynthetic pathways of the semidwarf *ga4* and *ga5* mutants. *Proceedings of the National Academy of Sciences USA* **87**, 7983–7987.

Tamaki, S., Matsuo, S., Wong, H.L. *et al.* (2007). Hd3a protein is a mobile flowering signal in rice. *Science* **316**, 1033–1036.

Tyler, L., Thomas, S.G., Hu, J. *et al.* (2004). DELLA proteins and gibberellin-regulated seed germination and floral development in Arabidopsis. *Plant Physiology* **135**, 1008–1019.

Ueguchi-Tanaka, M., Ashikari, M., Nakajima, M. *et al.* (2005). *GIBBERELLIN INSENSITIVE DWARF1* encodes a soluble receptor for gibberellin. *Nature* **437**, 693–698.

Ueguchi-Tanaka, M., Nakajima, M., Katoh, E. *et al.* (2007). Molecular interactions of a soluble gibberellin receptor, GID1, with a rice DELLA protein, SLR1, and gibberellin. *The Plant Cell* **19**, 2140–2155.

Varner, J.E. and Chandra, G.R. (1964). Hormonal control of enzyme synthesis in barley endosperm. *Proceedings of the National Academy of Sciences USA* **52**, 100–106.

West, C.A. and Phinney, B.O. (1959). Gibberellins from flowering plants. I. Isolation and properties of a gibberellin from *Phaseolus vulgaris* L. *Journal of the American Chemical Society* **81**, 2424–2427.

Willige, B.C., Ghosh, S., Nill, C. *et al.* (2007). The DELLA domain of GA INSENSITIVE mediates the interaction with the GA INSENSITIVE DWARF1A gibberellin receptor of *Arabidopsis*. *The Plant Cell* **19**, 1209–1220.

Xu, Y.L., Li, L., Wu, K. *et al.* (1995). The *GA5* locus of *Arabidopsis thaliana* encodes a multifunctional gibberellin 20-oxidase: molecular cloning and functional expression. *Proceedings of the National Academy of Sciences USA* **92**, 6640–6644.

Yabuta, T. and Sumiki, T. (1938). Communication to the editor. *Journal of the Agricultural Chemistry Society of Japan* **14**, 1526.

Yamaguchi, S., Sun, T., Kawaide, H. and Kamiya, Y. (1998). The *GA2* locus of *Arabidopsis thaliana* encodes *ent*-kaurene synthase of gibberellin biosynthesis. *Plant Physiology* **116**, 1271–1278.

Yano, K., Aya, K., Hirano, K. *et al.* (2015). Comprehensive gene expression analysis of rice aleurone cells: probing the existence of an alternative gibberellin receptor. *Plant Physiology* **167**, 531–44.

Zeevaart, J.A. (1964). Effects of the growth retardant CCC on floral initiation and growth in *Pharbitis nil*. *Plant Physiology* **39**, 402–408.

Zentella, R., Zhang, Z.L., Park, M. *et al.* (2007). Global analysis of DELLA direct targets in early gibberellin signaling in *Arabidopsis*. *The Plant Cell* **19**, 3037–3057.

Annual Plant Reviews (2016) **49**, 37–72
doi: 10.1002/9781119210436.ch2

Chapter 2

GIBBERELLIN BIOSYNTHESIS IN HIGHER PLANTS

Peter Hedden
Plant Biology and Crop Science Department, Rothamsted Research, UK

Abstract: Biosynthesis of the major biologically active gibberellins (GAs) in higher plants proceeds first in plastids via the tetracyclic diterpene hydrocarbon *ent*-kaurene, which is oxidised by membrane-associated cytochrome P450 mono-oxygenases to produce GA_{12} and GA_{53} on the endoplasmic reticulum. These intermediates are oxidised further by the soluble 2-oxoglutarate-dependent dioxygenases GA 20-oxidase and GA 3-oxidase to the active C_{19}-GA end products, GA_4 and GA_1, respectively. The genes encoding most of the enzymes of the pathway have been identified in several species and the mechanisms by which their expression is regulated in response to developmental and environmental signals are being elucidated. The chapter will review the pathway, discuss the function of the biosynthetic enzymes and highlight recent understanding of their regulation.

Keywords: 2-oxoglutarate-dependent_dioxygenase, biosynthesis, cytochrome_P450_monooxygenase, diterpene_cyclase, gibberellin_homoeostasis, multifunctional_enzymes, pathway_compartmentation

2.1 Introduction

The outline pathway to the main biologically active gibberellins (GAs) in higher plants, broadly acknowledged to be GA_1 and GA_4, has been known for about 30 years (Graebe, 1987). By that time the enzymes had also been identified and some aspects of their regulation, particularly by light, were already being documented. However, with the cloning of the genes encoding these enzymes from the early 1990s it has become possible to probe the molecular mechanisms underlying this regulation, as highlighted in recent reviews of this topic (Yamaguchi, 2008; Hedden and Thomas, 2012).The cloning

Annual Plant Reviews, Volume 49: The Gibberellins, First Edition.
Edited by Peter Hedden and Stephen G. Thomas.

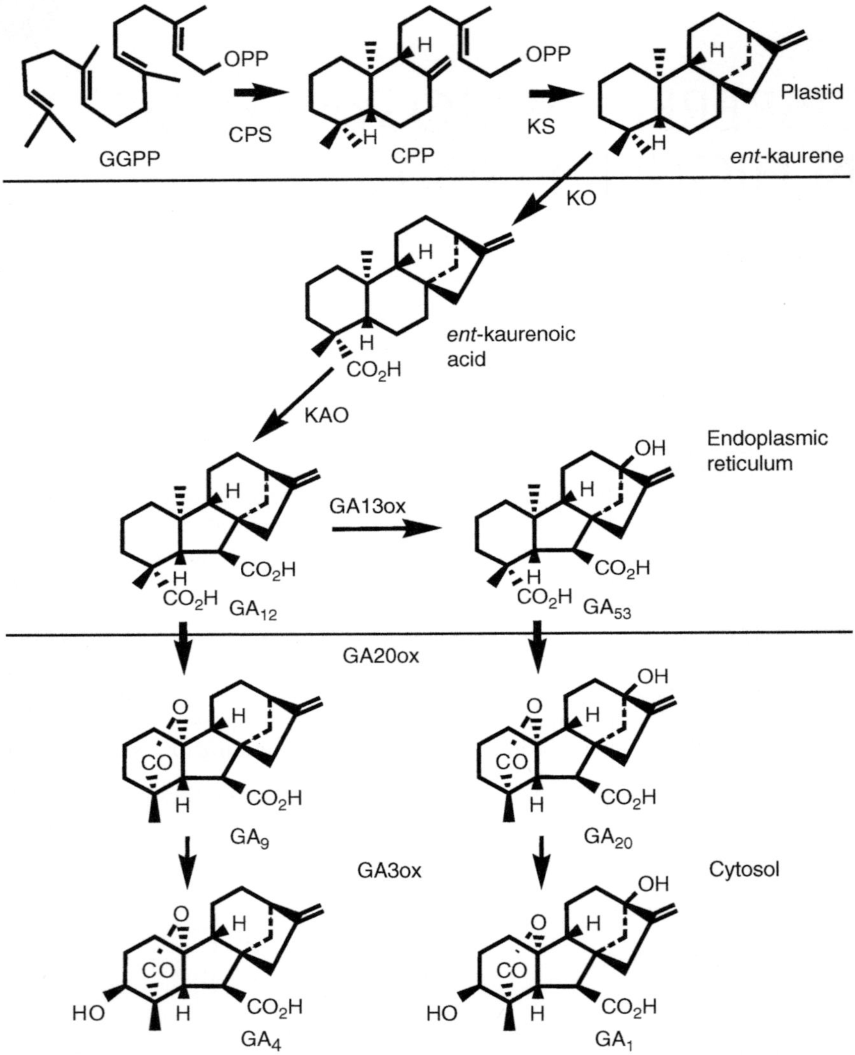

Figure 2.1 Summary of the biosynthetic pathway from *trans*-geranylgeranyl diphosphate (GGPP) to the biologically active end-products GA_4 and GA_1, indicating the subcellular compartmentation.

of GA 13-hydroxylase genes from rice, described in 2012, allowed one of the last missing pieces in the GA-biosynthetic jigsaw to be put in place. The major features of the biosynthetic pathway are presented in Figure 2.1, which illustrates its sub-cellular division into the plastidic, endomembrane and cytosolic phases. The diterpene *ent*-kaurene is produced in plastids, mainly via the 2-C-methyl-D-erythritol-4-phosphate (MEP) pathway, at least in green tissues. Conversion of *ent*-kaurene to bioactive GAs involves progressive oxidation through initially membrane-localised cytochrome P450 mono-oxygenases (P450s) and then soluble 2-oxoglutarate-dependent dioxygenases (2ODDs). The following sections describe each of these three phases in detail.

2.2 Synthesis of *ent*-kaurene

The intermediacy of the tetracyclic hydrocarbon *ent*-kaurene, also known as (−)-kaurene, in GA biosynthesis was originally demonstrated by conversion of *ent*-[^{14}C]kaurene into GA_3 by cultures of the fungus *Fusarium moniliforme* (*Gibberella fujikuroi*) (Cross *et al.*, 1964; Graebe *et al.*, 1965). Furthermore, *ent*-kaurenol and *ent*-kaurenoic acid, oxidised metabolites of *ent*-kaurene, could restore growth in the GA-deficient dwarf maize mutants *d5* and *an1* (Katsumi *et al.*, 1964), which have been subsequently shown to be defective in *ent*-kaurene synthesis (Hedden and Phinney, 1979; Fujioka *et al.*, 1988a; Bensen *et al.*, 1995). Numerous studies have shown that *ent*-kaurene biosynthesis is associated with plastids (Simcox *et al.*, 1975; Moore and Coolbaugh, 1976; Railton *et al.*, 1984; Aach *et al.*, 1995), with work in wheat demonstrating that pro-plastids, rather than mature chloroplasts, possess this activity (Aach *et al.*, 1995; Aach *et al.*, 1997). Uniquely, plants utilise both the MEP and mevalonic acid (MVA) pathways for isoprenoid biosynthesis, with the MEP pathway predominating in plastids (Vranova *et al.*, 2013). Consistent with this, Kasahara *et al.* (2002) found that *ent*-kaurene was produced in seedlings of *Arabidopsis thaliana* (*Arabidopsis*) mainly via the MEP pathway, with a small contribution from the MVA pathway, particularly when the MEP pathway was limited, indicating some cross-flow between them. Ironically the first demonstration of *ent*-kaurene biosynthesis was from radiolabelled MVA in a cell-free homogenate of endosperm of the Californian wild cucumber, *Marah macrocarpus* (formerly *Echinocystis macrocarpa*) (Graebe *et al.*, 1965). This and similar preparations from pumpkin (*Cucurbita maxima*) endosperm (Lange and Graebe, 1993) convert MVA to *ent*-kaurene and later intermediates of the GA-biosynthetic pathway with very high efficiency. Although it is not possible to draw firm conclusions on sub-cellular localisation from such preparations, it is possible that the non-photosynthetic leucoplasts, in which *ent*-kaurene is synthesised in the endosperm (Aach *et al.*, 1995), obtain a significant proportion of their isoprenoid precursors from the MVA pathway. Nevertheless, it is clear that in green tissues the MEP pathway is the major source of isoprenoid precursors for *ent*-kaurene biosynthesis.

2.2.1 Formation of *trans*-geranylgeranyl diphosphate

The MEP pathway from pyruvic acid and D-glyceraldehyde 3-phosphate and the MVA pathway from acetyl CoA are shown in Figure 2.2. These pathways have been discussed in a recent review (Vranova *et al.*, 2013) and will not be described in detail here. The MEP pathway results in the formation of isopentenyl diphosphate (IPP) and dimethylallyl diphosphate DMAPP), which are produced as a mixture from 4-hydroxy-3-methylbutenyl diphosphate (HMBPP) by the action of HMBPP reductase, with IPP predominating (Tritsch *et al.*, 2010). The MVA pathway forms IPP as sole product; IPP and DMAPP are interconverted by IPP isomerase (IPPI), which is encoded by two, partially redundant genes in *Arabidopsis*, each gene producing alternative transcripts, which encode proteins with or without transit sequences (Phillips *et al.*, 2008). Thus the enzymes are present in plastids, mitochondria and the cytosol. Loss of both *IPPI* genes in *Arabidopsis* caused dwarfism and a substantial loss of sterol production, but there was no effect on production of tocopherol, which is biosynthesised via the MEP pathway (Okada *et al.*, 2008).

trans-Geranylgeranyl diphosphate (GGPP), the precursor of all diterpenoids, including GAs, carotenoids and chlorophyll, is formed by sequential condensations of IPP, first with DMAPP, to form geranyl diphosphate (GPP; C_{10}), and then with GPP to form farnesyl diphosphate (FPP; C_{15}) and finally with FPP to form GGPP (C_{20}) (Figure 2.2). These reactions are catalysed by GGPP synthase (GGPPS), of which there are ten functional paralogues in *Arabidopsis* (Beck *et al.*, 2013). Expression of the *GGPPS* genes as fusions with green fluorescent protein determined that seven are located in plastids, two in the endoplasmic reticulum and one in mitochondria (Zhu *et al.*, 1997; Okada *et al.*, 2000; Beck *et al.*, 2013). GGPPS enzymes are homodimers and, in dicots, are distinct from GPP synthases (GPPS), which are also present in plastids. van Schie *et al.*, (2007) found that when they silenced the single *GPPS* gene in tomato and *Arabidopsis* they reduced GA formation without affecting chlorophyll or carotenoid production, implying that biogenesis of GAs and the major chloroplast pigments require different GGPPS paralogues, with that responsible for GA biosynthesis accepting GPP, but not DMAPP as substrate. This needs to be tested by determining the substrate specificities of the different GGPPS paralogues, but if correct would suggest that GAs and the pigments are formed from different GGPP pools. Indeed, this would be consistent with the finding that *ent*-kaurene is produced in pro-plastids, but not in mature chloroplasts (Aach *et al.*, 1997). On the basis of the tissue distribution of its transcript (Dugardeyn *et al.*, 2008), Beck *et al.* (2013) suggested that GGPPS2 may be responsible for GA biosynthesis in *Arabidopsis* roots. Interestingly, while this gene is expressed in a wide range of organs, its expression level is relatively low compared with the highly expressed GGPPS11, which is a strong candidate for being responsible for the biogenesis of carotenoids and chlorophyll.

2.2.2 Formation of *ent*-kaurene from *trans*-geranylgeranyl diphosphate

The cyclisation of GGPP to *ent*-kaurene occurs in two steps, with the bicyclic compound *ent*-copalyl diphosphate (CPP) being produced as an

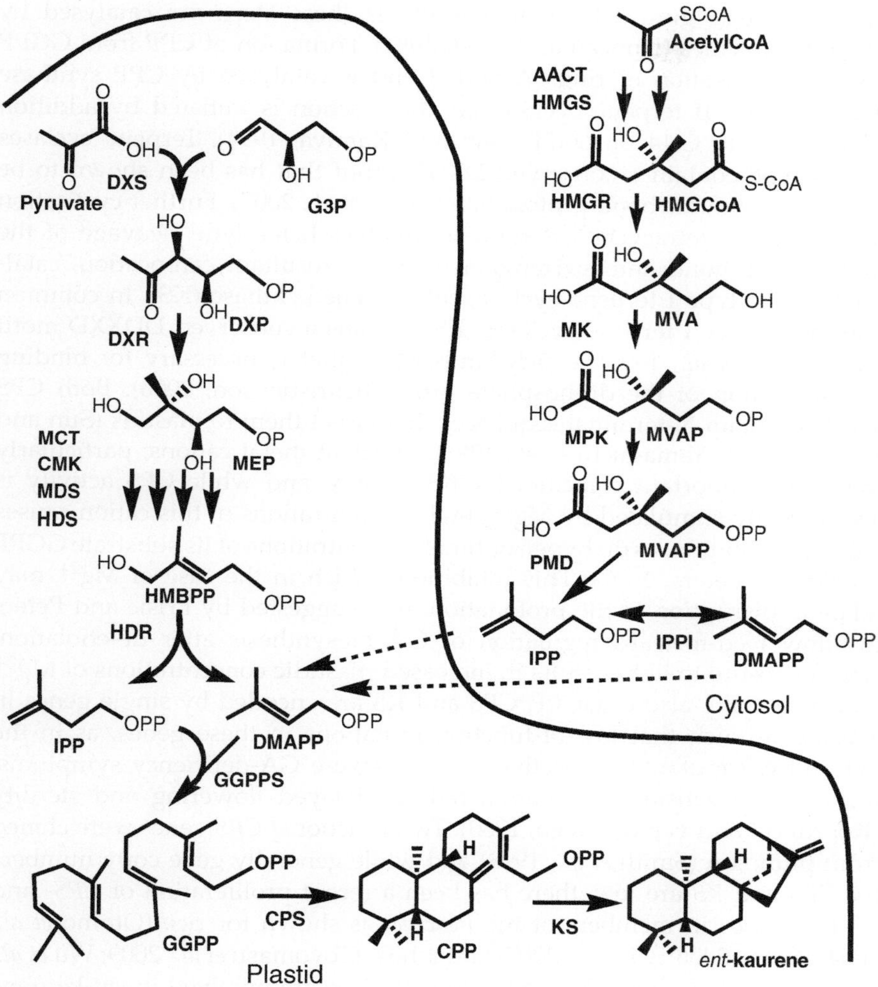

Figure 2.2 The biosynthesis of *ent*-kaurene. The major contribution is from the plastid-localised methylerythritol 4-phosphate (MEP) pathway from pyruvate and glyceraldehyde 3-phosphate (G3P), while in most tissues there is a minor contribution from the cytosolic mevalonate (MVA) pathway via plastid import of intermediates. Intermediates: acetylCoA, acetyl coenzyme A; CPP, *ent*-copalyl diphosphate; DXP, 1-deoxy-D-xylulose 5-phosphate; DMAPP, dimethylallyl diphosphate; HMBPP, 4-hydroxy-3-methylbut-2-enyl diphosphate; HMGCoA, 3-hydroxy-3-methylglutaryl coenzyme A; IPP, isopentenyl diphosphate; MVAP, mevalonate 5-phosphate; MVAPP, mevalonate 5-diphosphate. Enzymes: AACT, acetylCoA C-acetyltransferase; CMK, 4-(cytidine 5′-diphospho)-2-methylerythritol kinase; CPS, CPP synthase; DXR, DXP reductoisomerase; DXS, DXP synthase; GGPPS, GGPP synthase; HDR, HMBPP reductase; HDS, HMBPP synthase; HMGR, HMGCoA reductase; HMGS, HMGCo synthase; IPPI, IPP isomerase; KS, *ent*-kaurene synthase; MCT, MEP cytidyltransferase; MDS, 2-*C*-methyl-*D*-erythritol 2,4-cyclodiphosphate synthase; MK, MVA kinase; MPK, MVAP kinase; PMD, MVAPP decarboxylase.

intermediate (Figure 2.2). In higher plants these steps are catalysed by separate enzymes (Duncan and West, 1981). Formation of CPP from GGPP involves cyclisation of rings A and B and is catalysed by CPP synthase (CPS), a class II terpene cyclase, i.e. the reaction is initiated by addition of a proton to C-14 of GGPP (Sun and Kamiya, 1994). Terpene cyclases of this type contain a conserved DXDD motif that has been shown to be involved in double bond protonation (Prisic *et al.*, 2007). Further cyclisation of CPP to the tetracyclic *ent*-kaurene involves heterolytic cleavage of the diphosphate bond and rearrangement of the resultant carbocation, catalysed by the type I terpene cyclase, *ent*-kaurene synthase (KS). In common with other type I terpene cyclases, KS contains a conserved DDXXD motif (Yamaguchi *et al.*, 1996), which binds Mg^{2+} that is necessary for binding and ionisation of the diphosphate group (Christianson, 2006). Both CPS and KS contain N-terminal sequences that target them to plastids (Sun and Kamiya, 1994; Yamaguchi *et al.*, 1996). Divalent metal cations, particularly Mg^{2+}, are important cofactors for KS activity, and while CPS activity is also strongly stimulated by Mg^{2+}, high concentrations of this cation causes enzyme inhibition, as do hyper-optimal concentrations of its substrate GGPP (Prisic and Peters, 2007). This inhibition, which in the case of Mg^{2+} may relate to disruption of the protonation, was suggested by Prisic and Peters to allow feed-forward regulation of GA biosynthesis after de-etiolation, since exposure to light results in increased plastidic concentrations of Mg^{2+} and potentially also of GGPP. CPS and KS are encoded by single genes in *Arabidopsis* such that loss-of-function mutations in these genes, as in the *ga1* and *ga2* mutants, respectively, cause severe GA-deficiency symptoms, including dwarfism, germination failure, delayed flowering and sterility (Koornneef and van der Veen, 1980). Two functional *CPS* genes were cloned from pumpkin (Smith *et al.*, 1998) and while generally gene copy numbers for *CPS* and *KS* are low, there has been a recent proliferation of *CPS*- and *KS*-like genes in members of the *Poaceae*, as shown for rice (Otomo *et al.*, 2004; Peters, 2006; Xu *et al.*, 2007) and wheat (Toyomasu *et al.*, 2009; Wu *et al.*, 2012; Zhou *et al.*, 2012). However, few of these are involved in *ent*-kaurene synthesis, whereas the rest have undergone functional diversification for production of allelochemicals and plant defence (Xu *et al.*, 2012).

2.3 Reactions catalysed by cytochrome P450 mono-oxygenases

All steps beyond *ent*-kaurene are oxidative, the initial oxidations being carried out by membrane-associated cytochrome P450 mono-oxygenases. Two such enzymes convert *ent*-kaurene to GA_{12} (Figure 2.3): *ent*-kaurene oxidase (KO), which belongs to the CYP701A family, oxidises the C-4α methyl group (C-19) to the carboxylic acid in three steps via the alcohol

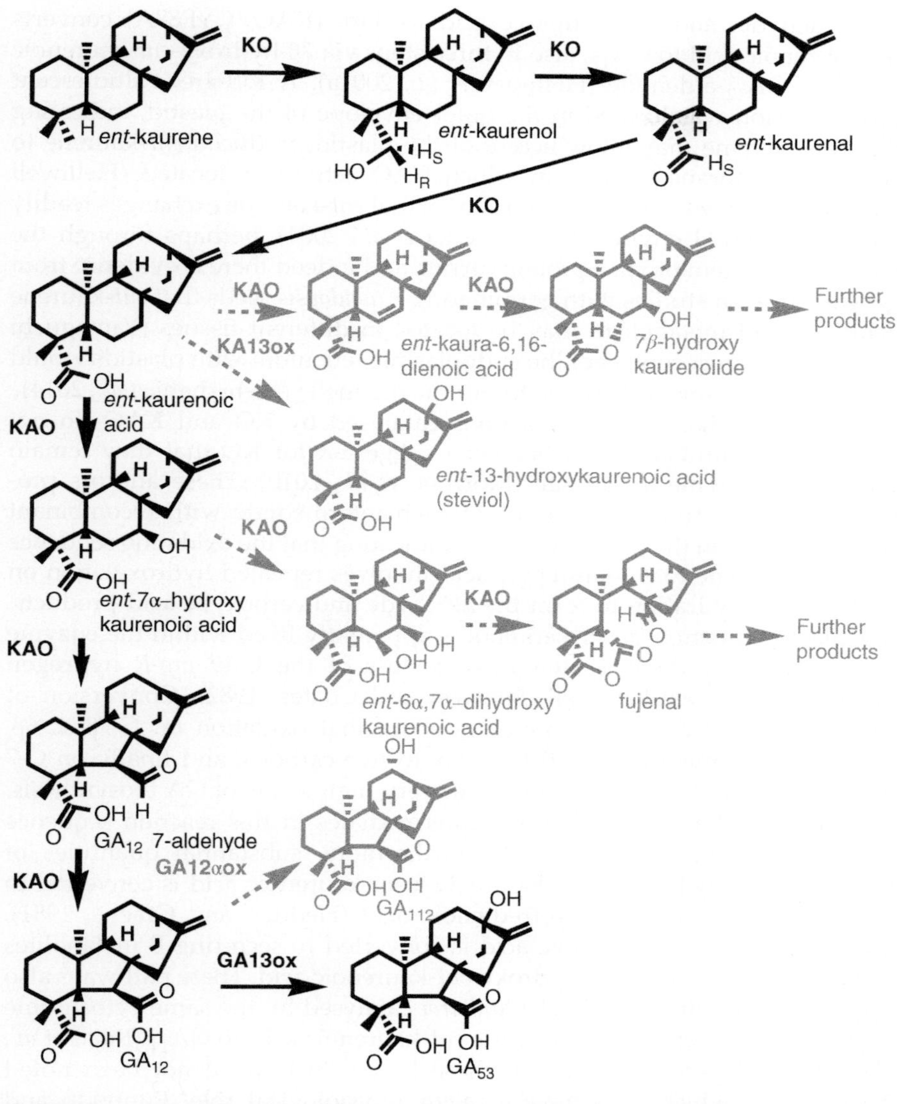

Figure 2.3 Reactions catalysed by membrane-associated cytochrome P450 monooxygenases. Reactions in bold indicate the main pathway from *ent*-kaurene to GA_{53}; reactions in grey with dotted arrows denote minor side reactions present in some tissues/species. GA12αOH, GA12α-hydroxylase (CYP714A2 in *Arabidopsis*); GA13ox, GA 13-hydroxylase (CYP714B); KAO, *ent*-kaurenoic acid oxidase (CYP88A); KA13ox, *ent*-kaurenoic acid 13-hydroxylase (CYP714A2 in *Arabidopsis*, CYP716D in *Stevia rebaudiana*) KO, *ent*-kaurene oxidase (CYP701A).

and aldehyde, and *ent*-kaurenoic acid oxidase (KAO; CYP88A) converts *ent*-kaurenoic acid to GA_{12}, also in three steps via 7β-hydroxy-*ent*-kaurenoic acid and GA_{12}-aldehyde (Helliwell *et al.*, 2001a). A KO-green fluorescent protein fusion was located in the outer envelope of the plastid, indicating that this enzyme may allow access of the plastid-produced *ent*-kaurene to the endomembrane system, in which KAO activity is located (Helliwell *et al.*, 2001b). However, it has been shown that *ent*-kaurene exchanges readily with the external environment (Otsuka *et al.*, 2004), perhaps through the membrane system or via a protein carrier, and indeed there is evidence from gene expression studies with germinating *Arabidopsis* seeds that *ent*-kaurene synthesis and metabolism may be located in different tissues (Yamaguchi *et al.*, 2001). Direct contact of the endoplasmic reticulum with plastids would provide a route for *ent*-kaurene to exit the organelle (Mehrshahi *et al.*, 2014).

The intermediates in the reactions catalysed by KO and KAO do not normally accumulate and it has been suggested for KO that they remain within the enzyme active site (Morrone *et al.*, 2010). These authors provided evidence from incubations of each intermediate with recombinant *Arabidopsis* KO in the presence of $^{18}O_2$ indicating that the oxidation sequence from *ent*-kaurene to *ent*-kaurenoic acid involves repeated hydroxylation on C-19 with dehydration to form the aldehyde and carboxylic acid products. The alcohol function in *ent*-kaurenol is apparently fixed within the enzyme active site since there is stereospecific loss of the C-19 *pro-R* hydrogen on oxidation to *ent*-kaurenal (Sherwin and Coates, 1982). Conversion of *ent*-kaurenoic acid to GA_{12} involves sequential oxidation on C-7β, C-6β, with rearrangement of ring B from six to five carbons, and finally on C-7 (Castellaro *et al.*, 1990). In tissues with very high levels of GA biosynthesis, such as pumpkin endosperm, the intermediates in this reaction sequence accumulate (Graebe *et al.*, 1972). Furthermore, substantial quantities of by-products may be formed (Figure 2.3): *ent*-kaurenoic acid is converted to kaurenolides via *ent*-kaura-6,16-dienoic acid (Hedden and Graebe, 1981), and 7β-hydroxy-*ent*-kaurenoic acid is converted to seco-ring B metabolites such as fujenal via 6β, 7β-dihydroxy-*ent*-kaurenoic acid. These pathways also operate in the fungus *Gibberella fujikuroi* catalysed by the same cytochrome P450 mono-oxygenase that converts *ent*-kaurenoic acid to GA_{14} (Rojas *et al.*, 2004; see Chapter 5). However, these by-products have not been noted in tissues in which GAs have a major physiological role. Pumpkin and cucumber (*Cucumis sativa*) contain 2ODDs that convert GA_{12}-aldehyde to GA_{12} (Lange, 1997; Lange *et al.*, 2013), although, so far the report for potato (*Solanum tuberosum*) (Fixen *et al.*, 2012) provides the only example of these GA 7-oxidases (GA7ox) occurring outside of the Cucurbitaceae. The physiological relevance of GA7ox activity is unclear, although its over-expression in *Arabidopsis* resulted in growth stimulation, indicating that KAO activity may be limiting for GA biosynthesis in this case (Radi *et al.*, 2006).

Arabidopsis contains a single *KO* gene, such that the *ko* (*ga3*) mutant is severely GA-deficient (Koornneef and van der Veen, 1980). A cluster of five *KO-like* (CYP701A) genes is present in rice on chromosome 6, but only one of the genes, *OsKO2* (CYP701A6), is required for GA biosynthesis (Sakamoto *et al.*, 2004). One of the others has been shown to act on different diterpenes in the biosynthesis of phytoalexins (Wang *et al.*, 2012). Mutations in *OsKO2* are responsible for the agriculturally important Tan-ginbozu/d35 semi-dwarf phenotype (Itoh *et al.*, 2004). There are two fully redundant *KAO* genes in *Arabidopsis* (Helliwell *et al.*, 2001a), requiring loss of both genes to produce a dwarf phenotype (Regnault *et al.*, 2014), whereas, while pea and sunflower also contain two *KAO* genes, these differ in expression domains with lesions in the dominant gene causing severe dwarfism (Davidson *et al.*, 2003; Fambrini *et al.*, 2011). Cereals contain a single *KAO* gene, in which dwarfing mutations have been described for rice and maize (Helliwell *et al.*, 2001a; Sakamoto *et al.*, 2004).

Gibberellin A_{12} lies at a branch-point in the pathway, undergoing oxidation on C-20, described in the following section, or on C-13 to form GA_{53}. While a soluble 2ODD-type GA_{12} 13-hydroxylase has been described for spinach leaves (Gilmour *et al.*, 1987), most 13-hydroxylases that have been characterised are cytochrome P450s, for example those from pumpkin endosperm (Lange *et al.*, 1993) and pea cotyledons (Kamiya and Graebe, 1983). Two P450-type 13-hydroxylases (CYP714B1 and CYP714B2) have recently been cloned from rice (Magome *et al.*, 2013) and one from *Arabidopsis* (CYP714A2) (Nomura *et al.*, 2013), although the substrate for 13-hydroxylation by CYP714A2 is *ent*-kaurenoic acid rather than GA_{12}. CYP714A2 converts GA_{12} to 12α-hydroxyGA_{12}, while a closely related paralogue, CYP714A1, converts GA_{12} to the C-16 carboxylic acid by oxidation of C-17 (Nomura *et al.*, 2013). Thus the CYP714 family of enzymes, which also includes the rice enzyme EUI (CYP714D1) that epoxidises the C-16,17 double bond (Zhu *et al.*, 2006), oxidise GAs or *ent*-kaurenoids on rings C and D, and, since their over-expression reduces growth, it has been proposed that their activity constitutes inactivation (Zhang *et al.*, 2011a; Magome *et al.*, 2013). This topic is covered in detail in Chapter 3. The suggestion that 13-hydroxylation results in a partial inactivation is somewhat controversial, because in most plant species the 13-hydroxylation pathway and its end-product GA_1 predominate, but is consistent with the properties of the GA receptor, which, at least in rice, has a higher affinity for the non-13-hydroxylated GA_4 than for GA_1 (Ueguchi-Tanaka *et al.*, 2005). The presence of the 13-hydroxyl group would increase water solubility so it is also possible that it serves to enhance long-distance transport.

2.4 Reactions catalysed by 2-oxoglutarate-dependent dioxygenases

Parallel pathways convert GA_{12} and GA_{53} to the bioactive end-products, GA_4 and GA_1, respectively (Figure 2.4a). This requires the activities of the

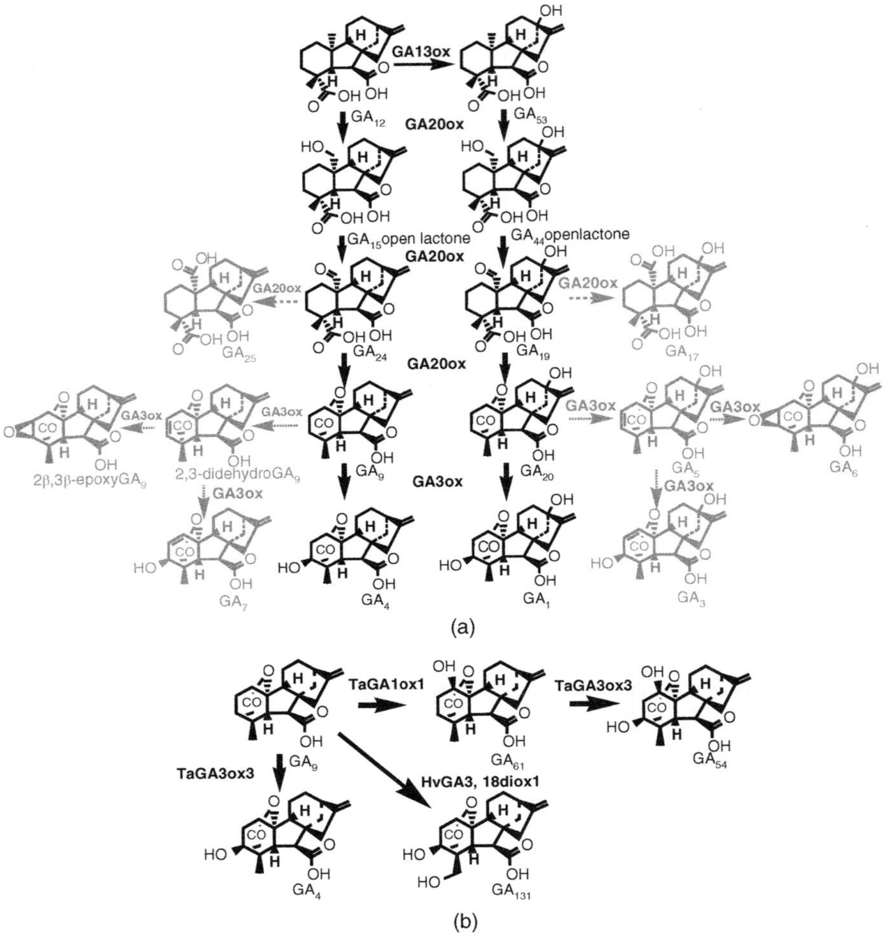

Figure 2.4 Reactions catalysed by 2-oxoglutarate-dependent dioxygenases. A. Most widespread reactions. Those in bold indicate the main pathway to the biologically active end-products, GA_1 and GA_4; reactions in grey with dotted arrows denote side reactions catalysed by some enzymes. B. Reactions catalysed by GA 3-oxidase-like enzymes in developing seed of bread wheat (*Triticum aestivum*) and barley (*Hordeum vulgare*). GA20ox, gibberellin 20-oxidase; GA3ox, gibberellin 3-oxidase; HvGA3,18diox1; *Hordeum vulgare* gibberellin 3β,18-dihydroxylase 1; TaGA1ox1, *T. aestivum* gibberellin 1β-hydroxylase 1; TaGA3ox3, *T. aestivum* gibberellin 3β-hydroxylase 3.

2ODDs GA 20-oxidase (GA20ox), which removes C-20 to form a C-4–C-10 lactone, and GA 3-oxidase (GA3ox), which inserts a 3β-hydroxyl group, thus providing the structural elements required for binding to the GID1 receptor (see Chapter 6). These are soluble enzymes that are present in the cytosol and nucleus (Helliwell *et al.*, 2001b; Chen *et al.*, 2014). GA20ox catalyses the sequential oxidation of C-20 from a methyl to the alcohol and then to the aldehyde, from which it is removed with formation of the C_{19} γ-lactone GAs. Thus GA_{12} is converted to GA_9 via GA_{15} and GA_{24}, and GA_{53} is converted to GA_{20} via GA_{44} and GA_{19}. The alcohol intermediates GA_{15} and GA_{44} readily form δ-lactones with the C-4α carboxylic acid group and are always isolated as lactones under the acidic conditions used. The extent to which lactonisation occurs *in planta* is not known, but it is noteworthy that GA20ox requires that the lactone is open for further oxidation to the aldehyde (Ward *et al.*, 2002). The natural occurrence of the lactones is indicated by the presence in spinach leaves (Gilmour *et al.*, 1986) and barley seedlings (Grosselindemann *et al.*, 1992) of a soluble enzyme that catalyses only the alcohol to aldehyde conversion and is able to accept the lactone as substrate. This activity has some characteristics of an ODD, but it has not been isolated and its properties are not known in detail. While the 'regular' GA20ox removes the 20-*pro*-R H atom in oxidation of the alcohol to the aldehyde, the 20-*pro*-S H atom is removed from the lactone by the spinach enzyme (Ward *et al.*, 1997).

The mechanism for C-20 cleavage from the aldehyde is not known precisely; it does not involve oxidation to the carboxylic acid, but C-20 is lost as CO_2, requiring two rounds of oxidation from the aldehyde (Kamiya *et al.*, 1986). This requires the existence of an intermediate, but none has been detected and it may remain bound to the enzyme. There is evidence for the formation of a free radical intermediate on C-10 after cleavage of the C-10–C-20 bond, capture of the C-4 carboxylic acid group by the radical then forming the lactone (Ward *et al.*, 2002). This is supported by older work with the fungus *G. fujikuroi*, in which the GA20ox is a cytochrome P450 (see Chapter 5), showing that both oxygen atoms in the C-4–C-10 lactone originate from the C-4 carboxylic acid group (Bearder *et al.*, 1976). Oxidation of the C-10 aldehyde to the carboxylic acid can also occur, usually as a minor side reaction of GA20ox activity, but in some cases it is the major product. Indeed, the first GA20ox to be characterised and cloned, from immature seed of pumpkin, produces the biologically inactive carboxylic acid as the main product (Lange, 1994; Lange *et al.*, 1994a).

In *Arabidopsis* and many other species, GA20ox activity limits the production of bioactive GAs, such that over-expression of GA20ox results in increased GA content and growth (Huang *et al.*, 1998; Coles *et al.*, 1999; Carrera *et al.*, 2000; Vidal *et al.*, 2001; Radi *et al.*, 2006; Bhattacharya *et al.*, 2012). Consistent with this, *GA20ox* genes are important sites of regulation in the GA-biosynthetic pathway. In contrast to the enzymes catalysing the earlier reactions of the GA-biosynthetic pathway, the GA20ox enzymes are encoded by gene families, with family members exhibiting differential expression patterns and regulation. The *GA20ox* gene family has four members in rice and five in *Arabidopsis* (Sakamoto *et al.*, 2004; Rieu *et al.*, 2008), although one of the *Arabidopsis* enzymes, AtGA20ox5, is functionally abnormal in carrying out only the first two steps to the aldehyde intermediate (Plackett *et al.*, 2012). Due to partial gene redundancy, loss of a single *GA20ox* gene, for example through mutation, typically has only a small effect on GA content and phenotype. For example, mutation of the *Arabidopsis AtGA20ox1* gene, as in *ga5*, results in reduced stem height, but has only a minor effect on fertility, whereas loss of a gene encoding an early enzyme results in extreme dwarfism and sterility (Koornneef and van der Veen, 1980). Genetically diverse, natural mutations in *AtGA20ox1* giving rise to semi-dwarfism have been found in different *Arabidopsis* populations, indicating that such mutations may have selective advantages in certain environments (Barboza *et al.*, 2013). Indeed, semi-dwarf varieties of crop species, such as rice and barley, which have been selected in breeding for lodging resistance and high yield, were found to have mutations in *GA20ox* genes (see Chapter 13).

In the final step in the formation of the biologically active hormones, the C_{19}-GAs GA_9 and GA_{20} are converted by 3β-hydroxylation to GA_4 and GA_1, respectively, by GA3ox ODDs. Enzymes present in vegetative tissues of dicotyledonous species such as *Arabidopsis* and pea have high regiospecificity, producing only the 3β-hydroxylated products from GA_9 and GA_{20} (Lester *et al.*, 1997; Martin *et al.*, 1997; Williams *et al.*, 1998), whereas the rice and wheat enzymes OsGA3ox2 and TaGA3ox2, respectively, produce also small amounts of 2β-hydroxy and 2,3-didehydro side products (Itoh *et al.*, 2001; Appleford *et al.*, 2006). These monocot enzymes also convert the 2,3-didehydroGAs to 3β-hydroxy-1,2-didehydro products, such that GA_3 is produced from GA_{20} via GA_5 (Figure 2.4a). Thus GA_3 is a minor by-product of GA3ox activity in cereal shoots (Spray *et al.*, 1996; Itoh *et al.*, 2001; Appleford *et al.*, 2006), while it is not produced in shoots of *Arabidopsis* and pea. In contrast to the cereal enzymes, the *Arabidopsis* enzyme AtGA3ox1 converts 2,3-didehydroGAs to epoxides (Williams *et al.*, 1998) (Figure 2.4a). The wheat enzyme TaGA3ox2 was shown also to 13-hydroxylate GA_9 to a small extent, requiring the substrate to be rotated through 180° within the active site (Appleford *et al.*, 2006).

Several *GA3ox* paralogues are expressed, although fewer than for the *GA20ox* genes, with four genes in *Arabidopsis* and two in rice (Hedden *et al.*, 2002; Sakamoto *et al.*, 2004; Kawai *et al.*, 2014). The rice gene *OsGA3ox2* is expressed in GA-producing tissues throughout the plant, while the second gene *OsGA3ox1* is restricted to the tapetum and scutellum epithelium, in which both genes are expressed, allowing high levels of GA biosynthesis for GA export to other tissues (Kaneko *et al.*, 2003). In *Arabidopsis, AtGA3ox1* and *AtGA3ox2* are involved in vegetative development, while all four *GA3ox* genes are expressed in reproductive tissues (Mitchum *et al.*, 2006; Hu *et al.*, 2008). It has been shown for several species that expression of certain *GA20ox* and *GA3ox* paralogues is specific to developing seeds. These genes may be expressed at very high levels and in some cases encode enzymes producing atypical product profiles. The GA3ox paralogues in particular are responsible for many of the unusual GA structures found in immature seeds. For example, developing grain of *Triticum aestivum* (bread wheat) contain two closely related GA3ox-like enzymes, one of which (TaGA1ox1) acts as a 1β-hydroxylase, converting GA_9 to GA_{61}, while the second (TaGA3ox3) has conventional 3β-hydroxylase activity and converts GA_{61} to the highly abundant GA_{54} (Figure 2.4b; Pearce *et al.*, 2015). A closely related orthologue in *Hordeum vulgare* (barley) grain (HvGA3,18diox), has both 3β- and 18-hydroxylase activities, forming GA_{131} (3β, 18-dihydroxy GA_9) from GA_9 (Figure 2.4b). Immature seeds of *Malus pumila* (apple) and *Marah macrocarpus* (Californian wild cucumber) contain, in addition to GA_4, high levels of GA_7, which in parallel reactions to those for GA_3 biosynthesis, is produced from GA_9 via 2,3-didehydroGA_9 (Albone *et al.*, 1990; Figure 2.4a). *M. macrocarpus* seeds express at least three *GA3ox* genes; two of the encoded enzymes are required to produce GA_7 from GA_9, MmGA3ox2 catalysing the conversion of GA_9 to 2,3-didehydroGA_9 and MmGA3ox1 converting this intermediate to GA_7 (Ward *et al.*, 2010). In common with the wheat enzyme TaGA3ox2, MmGA3ox2 possesses some 13-hydroxylase activity against C_{19}-GA substrates, which may account for the presence of 13-hydroxylated C_{19}-GAs in pumpkin and *M. macrocarpus* seeds, whereas 13-hydroxylated C_{20}-GAs are absent (Blechschmidt *et al.*, 1984; MacMillan and Gaskin, 1996). Another unusual feature of some GA3ox enzymes present in Cucurbitaceae seeds is the ability to 3β-hydroxylate C_{20}-GAs (Lange *et al.*, 1994b; Lange *et al.*, 2013).

2.5 Sites of gibberellin biosynthesis

On the basis of expression levels of GA biosynthetic genes, expanding organs have been identified as major sites of GA biosynthesis (Silverstone *et al.*, 1997; Kaneko *et al.*, 2003). Fully expanded leaves are also active for GA-biosynthesis, although it was shown in pea that they contain low levels of bioactive GAs due to high rates of inactivation by 2β-hydroxylation (Ross *et al.*, 2003). Expression of *CPS* in mature leaves was found in the vasculature and it was suggested that the leaves produce GAs for export to other tissues via the phloem (Silverstone *et al.*, 1997; Sun and Kamiya, 1997). In this case, inactivation would protect the mature leaf blade from diffusion of GAs from the veins. Movement of GAs from leaves to the shoot apex has been demonstrated in *Arabidopsis* (Eriksson *et al.*, 2006) and the grass *Lolium temulentum* (King *et al.*, 2001) and it is proposed that they act as mobile signals for flower induction in these species (see Chapter 4). The presence of leaves is required for GA-induced stem extension (Garcia-Martinez and Rappaport, 1982; Dayan *et al.*, 2012), although it is not certain that GA is the mobile signal in this case. In the *Arabidopsis* root, distribution of transcripts for the most highly expressed GA-biosynthesis genes indicates that the meristem and elongation zones are primary sites of GA biosynthesis, while GA-inactivation occurs in the mature root (Dugardeyn *et al.*, 2008).

Kaneko *et al.* (2003) compared the sites of expression of *GA20ox* and *GA3ox* genes with those of genes encoding GA signal transduction components in rice and concluded that in most tissues the sites of GA synthesis and action are coincident. However, two tissues, the scutellum epithelium and the tapetum, which are particularly rich sources of GAs, are proposed to provide GAs for the non-GA-autonomous aleurone and other floral organs, respectively. Petals are dependent on an external source of GAs, probably the anthers, while the stamen filament requires provision of GA precursors from the anther (Weiss and Halevy, 1989; Silverstone *et al.*, 1997; Hu *et al.*, 2008). In some species, the suspensor is also a rich source of GAs, which may be required for early embryo development (reviewed in Yeung and Meinke, 1993).

2.6 Regulation of gibberellin biosynthesis

The formation of CPP by CPS has been referred to as the gateway to GA biosynthesis, and since *CPS* is generally expressed at very low levels and CPS activity limits the flux of metabolites into the GA pathway, it is an obvious site of regulation (Silverstone *et al.*, 1997). However, over-expression of *CPS* in *Arabidopsis* resulted in elevation of C_{20}-GAs levels, particularly of GA_{12} and GA_{24}, but not of the C_{19}-GAs, consistent with bioactive GA production being limited by GA20ox activity in this species (Fleet *et al.*, 2003). Indeed the genes encoded ODDs in the GA pathway are major sites of regulation and their expression is precisely controlled. In particular, they are subject to regulation by environmental stimuli and form part of a homoeostatic mechanism in GA signalling.

2.6.1 Developmental control

With the exception of certain organs that act as particularly rich sources of GA (discussed above), levels of bioactive GAs are highest in actively growing organs (Smith *et al.*, 1992), which contain the highest levels of *CPS* expression (Silverstone *et al.*, 1997). The synthesis of *ent*-kaurene occurs in pro-plastids, rather than mature chloroplasts (Aach *et al.*, 1997), the former being present in dividing cells, but also in the leaf vasculature. CPS would compete more effectively in pro-plastids for its substrate GGPP in the presence of the much more prolific pigment pathways. As discussed earlier, there is an indication that GA and chlorophyll/carotenoid production utilise different GGPP synthases (van Schie *et al.*, 2007), which invites the possibility of a GA-specific *GGPPS* that could be independently regulated.

Despite the evidence for developmental regulation of *CPS*, later enzymes in the pathway, particularly the ODDs may have a greater influence on GA content (Fleet *et al.*, 2003). There are numerous reports of growth and/or GA concentrations being limited by the levels of GA20ox or GA3ox activity, with recent reports of ODD genes underlying major growth QTLs (Abe *et al.*, 2012; Teng *et al.*, 2013). Furthermore, tissue- and development-specific expression of the different ODD paralogues highlight the importance of these genes in regulating GA-mediated development. For example, stem height in *Arabidopsis* is controlled predominantly by a single *GA20ox* paralogue,

AtGA20ox1, while *AtGA20ox2* has a greater influence on flowering time and silique length (Rieu *et al.*, 2008). Spatial and temporal coordination of expression of the biosynthetic genes, particularly with genes involved in GA inactivation, is an important element of GA biosynthesis, as was highlighted by a report on work with the pea *lele* mutant (Reinecke *et al.*, 2013). Rescue of this dwarf mutant, which is deficient in PsGA3ox1 activity, by expression of *35S::PsGA3ox1* was far less successful at restoring GA_1 content and internode growth than the native *PsGA3ox1* gene when introgressed into the same background. This was explained by higher levels of GA2ox-catalysed inactivation in the former case, when the lack of tissue specificity resulted in the *PsGA2ox1* gene being strongly up-regulated. It was concluded that cellular or temporal segregation of GA3ox1 and GA2ox1 is necessary for effective GA production in growing tissues.

Several transcription factors that regulate the patterning of GA-biosynthetic gene expression have been identified. The homoeodomain Class-I KNOX transcription factors suppress expression of *GA20ox* genes in the shoot apical meristem, from which GAs are excluded to maintain a pool of undifferentiated stem cells (Sakamoto *et al.*, 2001; Hay *et al.*, 2002). In tobacco, the homoeobox transcription factor NTH15 was shown to interact within the first intron of *NtGA20ox1* (Sakamoto *et al.*, 2001), while a heterodimer of the KNOX factor POTH1 and the BEL-type homoeodomain protein StBEL5 binds to the promoter of the potato *StGA20ox1* gene (Chen *et al.*, 2004). During embryogenesis, GA biosynthesis is regulated to allow correct tissue specification. Two transcription factors LEC2 and FUS3 control this process in *Arabidopsis* by down-regulating *AtGA3ox2* expression, FUS3 being shown to bind the RY elements in the promoter (Curaba *et al.*, 2004). Expression of *AtGA3ox1* is up-regulated by the MADS domain protein AGAMOUS in developing flowers to promote floral organ development (Gomez-Mena *et al.*, 2005) and in the silique valve margins and septa by the bHLH protein INDETERMINATE to enable pod opening (Arnaud *et al.*, 2010), the transcription factors acting directly in both cases.

2.6.2 Gibberellin homoeostasis

The first clues that the action of GAs regulated their biosynthesis were present already in 1970 when Radley (1970) showed that Norin-10 dwarf lines of wheat are insensitive to applied GA and have elevated GA levels. Norin-10 contains gain-of-function mutations in the *Rht-1* gene, and it was shown subsequently that near isogenic dwarf lines containing the *Rht1* (*Rht-B1b*) or *Rht3* (*Rht-B1c*) mutant alleles possess elevated levels of C_{19}-GAs, but reduced C_{20}-GAs levels compared with the tall (*Rht-B1a*) line (Appleford and Lenton, 1991). Similar results were shown for the GA-insensitive *dwarf-8* mutant of maize (Fujioka *et al.*, 1988b) and *Arabidopsis gai* mutant (Talon *et al.*, 1990), which like the *Rht-1* semi-dwarf wheat lines contain mutations in DELLA proteins rendering them resistant to GA-induced degradation (see Chapter 6). Furthermore, the accumulation of GA_{20} in the GA-sensitive *dwarf-1* maize mutant, in which the 3β-hydroxylation of GA_{20} is blocked (Spray *et al.*, 1996), could be substantially reduced by treatment with GA (Hedden and Croker, 1992). These results are consistent with increased rates of GA20ox activity in the dwarf lines, indicating that this enzyme is under feedback regulation. This was confirmed by GA metabolism experiments in GA-deficient mutants of pea, the results from which indicated that both GA20ox and GA3ox activities were regulated by GA (Martin *et al.*, 1996).

On the basis of transcript abundance, it is has been shown in many species that some *GA20ox* and *GA3ox* genes are transcriptionally regulated by GA signalling: in *Arabidopsis* three of the five *GA20ox* genes and the *GA3ox1* gene are down-regulated by GA treatment (Chiang *et al.*, 1995; Phillips *et al.*, 1995; Mitchum *et al.*, 2006; Rieu *et al.*, 2008), while earlier genes in the pathway are not influenced by GA signalling. This feedback regulation is part of a broader homoeostatic mechanism, which includes GA-induced up-regulation of some *GA2ox* genes, encoding GA-inactivating enzymes, and down-regulation of the *GID1* GA-receptor genes (Thomas *et al.*, 1999; Griffiths *et al.*, 2006). As described in Chapter 6, GAs function by initiating the ubiquitination and degradation of DELLA proteins, which act to modify gene expression. In terms of growth, DELLAs function primarily as repressors, although they activate at least as many genes as they repress. There is now substantial evidence showing that expression of those *GA20ox* and *GA3ox* genes that are subject to feedback regulation is promoted by DELLA proteins, and the speed of this response indicates that the effect may be direct (reviewed in Hedden and Thomas, 2012). Attempts to understand the molecular mechanism underlying feedback regulation have focussed on identifying the relevant transcription factors. Current information is summarised in Figure 2.5. Several potential mechanisms have been proposed, some such as those involving the b-ZIP transcription factor REPRESSION OF SHOOT GROWTH (RSG) in tobacco (Fukazawa *et al.*, 2010), the C_2C_2 zinc finger protein OsYABBY1 in rice (Dai *et al.*, 2007) and the *Arabidopsis* AT hook protein AGF1 (Matsushita *et al.*, 2007), are gene specific. While GA suppressed the nuclear localisation of RSG, an activator of *NtGA20ox1* expression, and enhanced expression of OsYABBY1, which suppresses *OsGA3ox2* expression, the involvement of DELLA proteins in these processes is unclear. AGF1 was shown to be involved in the GA-induced down-regulation of *AtGA3ox1*, although the mechanism by which GA regulates this process is unknown. The conclusion from these observations was that feedback regulation of *GA20ox* and *GA3ox* genes operates by different mechanisms. It is also possible that several mechanisms exist for the feedback regulation of single genes, for example to allow interaction with developmental and environmental regulation. AGF1 was found to be necessary, but not sufficient for feedback regulation of *AtGA3ox1* (Matsushita *et al.*, 2007), supporting the requirement for multiple factors in this process.

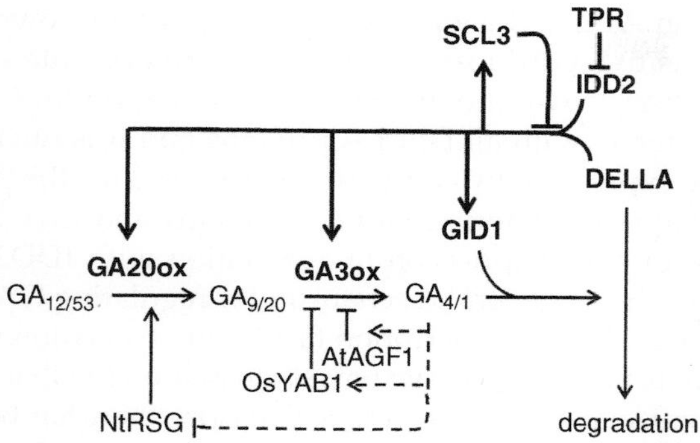

Figure 2.5 Feedback regulation of GA biosynthesis. Binding of the biologically active GA_1 or GA_4 to the receptor GIBBERELLIN INSENSITIVE DWARF 1 (GID1) initiates the degradation of DELLA proteins. In *Arabidopsis*, the DELLA protein GA INSENSITIVE (GAI) in association with the INDETERMINATE DOMAIN transcription factor GAI-ASSOCIATED FACTOR1 (GAF1; IDD2) was shown to promote expression of certain *GA20ox* and *GA3ox* biosynthetic genes as well as *GID1*. Interaction of GAF1 with TOPLESS RELATED (TPR) interfered with the GAI-GAF1 association and repressed feedback regulation. Thus GAI and TPR act antagonistically to regulate IDD2-induced promotion of GA biosynthesis. The DELLA-IDD complex also promotes expression of the GRAS protein SCARECROW-LIKE3 (SCL3), which, by binding to DELLA, interferes with this promotion. The transcription factors REPRESSION OF SHOOT GROWTH (NtRSG) in tobacco, YABBY1 (OsYAB1) in rice and AT hook protein (AGF1) in *Arabidopsis* also modify expression of GA-biosynthesis genes and are regulated by GA, but through incompletely understood mechanisms (dotted arrows).

The *Arabidopsis* feedback-regulated genes *AtGA20ox1, -2, -3* and *AtGA3ox1* were shown to be up-regulated in a mutant lacking SCARECROW-LIKE3 (SCL3), indicating that this protein suppresses expression of these genes (Zhang *et al.*, 2011b). SCL3, like DELLAs, is a GRAS protein, although unlike DELLAs it lacks the N-terminal domain that is required for GA-induced degradation. Expression of *SCL3* is positively regulated by DELLA in association with INDETERMINATE DOMAIN (IDD) transcription factors (Yoshida *et al.*, 2014). SCL3 interacts with both DELLAs and IDDs, attenuating its own gene expression by competing with DELLA for association with IDD (Yoshida *et al.*, 2014). Suppression of expression of the GA-biosynthetic genes by SCL3 may have a similar mechanism since it has been shown recently that the DELLA protein GAI interacts with the IDD protein GAF1, also known as IDD2, to promote expression of *AtGA20ox2*, *AtGA3ox1* and the receptor gene *GID1B* (Fukazawa *et al.*, 2014). GAF1 promotes expression of the genes in association with GAI, while a WD-repeat protein TOPLESS-RELATED (TPR), binds GAF1 to suppress expression, such that expression of *AtGA20ox2* and presumably also of other feedback-regulated genes is regulated by a balance between DELLA and TPR. Interestingly, ectopic expression of a closely related IDD, ENHYDROUS or IDD1, in seedlings from an ABA-inducible promoter resulted in down-regulation of the feedback-sensitive GA-biosynthesis genes, the opposite of what would be anticipated for an activator of their expression (Feurtado *et al.*, 2011). IDD1 was shown to interact with all DELLA proteins and it is unclear whether it suppresses gene expression by competing with IDD2 for the promoter binding site, or perhaps sequestering DELLA proteins and thereby attenuating their promotion of gene expression in association with IDD2. These recent findings point to a more general mechanism for regulating GA homoeostasis through the action of IDD transcription factors in association with DELLAs.

The experiments described above have focussed on feedback regulation at the transcriptional level, while for technical reasons there has been little attention paid to effects on enzyme protein content. However, Lee and Zeevaart (2007) could show by Western blotting that the spinach enzyme SoGA20ox1 is elevated in petioles and shoot tips after treatment with inhibitors of GA biosynthesis, but they could find no effect of this treatment on transcript levels. This observation indicates the existence of additional mechanisms in GA homoeostasis through regulation of translation or protein turnover.

2.6.3 Regulation by other hormones

As discussed in Chapter 8, there is increasing evidence for cross-talk between the GA signalling pathway and those of other plant hormones, in many cases mediated by DELLA proteins. Support for interaction at the level of hormone metabolism is less strong (Ross *et al.*, 2011), with the exception of auxin regulation of GA biosynthesis. It has been demonstrated in several species that auxin (indole-3-acetic acid; IAA) promotes GA biosynthesis in stems, including those of pea (Ross *et al.*, 2000), tobacco (Wolbang and Ross, 2001), barley (Wolbang *et al.*, 2004) and *Arabidopsis* (O'Neill *et al.*, 2010). Gibberellin production and elongation growth are stimulated by application of IAA to stems that have been depleted of auxin by decapitation or dissection, indicating that GA at least partly mediates growth enhancement by auxin in these tissues. Auxin has been shown to promote GA biosynthesis also in pea roots (Weston *et al.*, 2009) and ovaries; in the latter case the active agent is 4-chloroIAA, rather than IAA (Ozga and Reinecke, 1999; Ngo *et al.*, 2002). Since the presence of seeds promotes GA biosynthesis and growth of the ovary in a similar manner to 4-chloroIAA, it is suggested that this auxin acts as the seed-derived messenger that, through induction of GA biosynthesis, promotes the growth of the ovary after fertilisation (Ozga *et al.*, 2009). Similarly, auxin-induced parthenocarpic fruit development in tomato is partly mediated by enhanced GA biosynthesis (Serrani *et al.*, 2008). Auxin promotes GA biosynthesis by up-regulating expression of *GA20ox* and/or *GA3ox* genes, as well as down-regulating certain *GA2ox* genes, the primary targets depending on species. There is genetic evidence for direct involvement of the AUX/IAA-ARF signalling system in the regulation of GA-biosynthetic gene expression by auxin (Frigerio *et al.*, 2006; Mai *et al.*,

2011), and it is independent of DELLA proteins (O'Neill *et al.*, 2010), although auxin regulation may be masked to some extent by the DELLA-mediated feedback mechanism (Desgagne-Penix and Sponsel, 2008).

Imbibed seeds of *Arabidopsis* mutants that are deficient in abscisic acid (ABA) content were found to have enhanced GA formation, indicating regulation of GA biosynthesis by ABA in germinating seeds (Seo *et al.*, 2006). Thermoinhibition of seed germination is associated with increased ABA and decreased GA production, with reduced levels of expression of *AtGA20ox2*, *AtGA20ox3*, *AtGA3ox1* and *AtGA3ox2* (Toh *et al.*, 2008). Expression of the *GA3ox* genes was insensitive to high temperature in the ABA-deficient mutant *aba2-2*, while there was some recovery of *GA20ox* expression, indicating that the temperature inhibition of GA biosynthesis was at least partly mediated by ABA. It was suggested that ABA acts through the CHOTTO1 transcription factor which is regulated by this hormone in imbibed seeds, in which *chotto1* mutants had higher levels of GA_4 and of transcripts for *AtGA3ox1* and *AtGA3ox2* than the Columbia wild type (Yano *et al.*, 2009).

Despite considerable evidence indicating that GA biosynthesis is independent of brassinosteroid signalling (reviewed in Ross *et al.*, 2011 and Chapter 8), there are recent reports to the contrary. Applied brassinosteroid promoted expression of *GA20ox* genes in *Arabidopsis* (Lilley *et al.*, 2013) and of *OsGA3ox2* in rice (Tong *et al.*, 2014), while mutants deficient in brassinosteroid content or signalling contained reduced GA levels (Tong *et al.*, 2014). However, the relationship is complex and dose-dependent since application of high brassinosteroid concentrations ($>10^{-7}$ M) suppressed *OsGA20ox2* and promoted expression of the GA-deactivating *OsGA2ox3* gene, resulting in reduced GA biosynthesis. There is also evidence for inhibition of GA biosynthesis by jasmonate (JA). Wild tobacco plants (*Nicotiana attenuate*) that were silenced in two calcium-dependant kinases contained elevated levels of JA and exhibited strong dwarfism that was rescued by GA application (Heinrich *et al.*, 2013). The plants were found to contain reduced levels of GA_1 and 13-hydroxy GA precursors, consistent with the observed reduction in *GA20ox* transcript.

2.6.4 Regulation by environmental factors

Developmental responses to changes in the environment are mediated to a major extent by the GA signalling pathway, with GA metabolism being particularly sensitive to environmental stimuli, including light, temperature and different stress factors. Environmental regulation of GA metabolism involves primarily the ODD-encoding genes, with biosynthesis and GA 2-oxidase-mediated deactivation responding oppositely to the stimuli. Regulation of *GA2ox* expression is discussed in Chapter 3. As pointed out above for auxin regulation of GA biosynthesis, interpretation of the observed changes in gene expression in response to environmental factors is complicated by feedback regulation for some genes.

Light

The pathways by which GA biosynthesis is regulated by light are summarized for *Arabidopsis* in Figure 2.6. Germination of photoblastic seeds, such as those of *Arabidopsis*, in response to red light involves induction of GA biosynthesis and ABA inactivation (Seo *et al.*, 2009 and Chapter 9). In the presence of red light, germination of *Arabidopsis* seeds is promoted through the phytochrome-mediated degradation of the PIL5 transcription factor, which in the absence of light inhibits GA production by suppressing expression of *AtGA3ox1* and *AtGA3ox2*, as well as promoting *GA2ox* expression (Oh *et al.*, 2006). The CCCH-type zinc finger protein SOMNUS also suppresses *AtGA3ox1* and *AtGA3ox2* expression and since it is a direct transcriptional target of PIL5, it is suggested to be an intermediary in the PIL5 response (Kim *et al.*, 2008). A second potential intermediary, the Dof1 transcription factor DAG1, is also up-regulated by PIL5, although indirectly, and acts as a direct repressor of *AtGA3ox1* expression, but not of the other GA-metabolism genes (Gabriele *et al.*, 2010). Germination of dormant cereal seeds, in contrast, is inhibited by light, in this case blue light, which slows the after-ripening process (Gubler *et al.*, 2008). In imbibed barley seeds, blue light, detected by cryptochrome, reduces expression of the GA-biosynthetic gene *HvGA3ox2* as well as enhancing *GA2ox* expression and promoting accumulation of ABA (Gubler *et al.*, 2008; Hai Ha *et al.*, 2014).

Photomorphogenesis on exposure of dark-grown seedlings to light is associated with a rapid reduction of GA content through the action of both phytochrome and cryptochrome light receptors. In pea, red and blue light irradiation results in down-regulation of *PsGA3ox1* expression as well as up-regulation of *PsGA2ox2* (Reid *et al.*, 2002; Foo *et al.*, 2006), while in *Arabidopsis* hypocotyls, cryptochrome mediates suppression of *AtGA20ox1* and *AtGA3ox1* expression by blue light (Zhao *et al.*, 2007). Similarly, blue light reduces GA_1 content in dark-grown rice seedlings through phytochrome-mediated suppression of *OsGA20ox2*, *OsGA20ox4* and *OsGA3ox2*, while up-regulation of several *GA2ox* genes by blue light is mediated by cryptochromes (Hirose *et al.*, 2012).

GA biosynthesis is influenced by photoperiod in a number of long-day plants, in which GA has a role in the induction of flowering and bolting (reviewed in Mutasa-Gottgens and Hedden, 2009). Exposure to long days has been shown to promote GA production in several species, mainly through enhanced expression of *GA20ox* genes, particularly in leaves. Examples include spinach (Lee and Zeevaart, 2007), *Arabidopsis* (Hisamatsu *et al.*, 2005), barley (Boden *et al.*, 2014) and *Lolium* species (MacMillan *et al.*, 2005; King *et al.*, 2006). GAs produced in the leaves may act to promote flowering after translocation to the shoot apex or by induction of the protein messenger, FLOWERING LOCUS T (FT) in leaves (King *et al.*, 2006; Hisamatsu & King, 2008; Porri *et al.*, 2012). However, there is evidence for regulation of GA metabolism at the shoot apex in response to long days. The floral inhibitor SHORT VEGETATIVE PHASE (SVP) inhibits expression of *AtGA20ox2* at

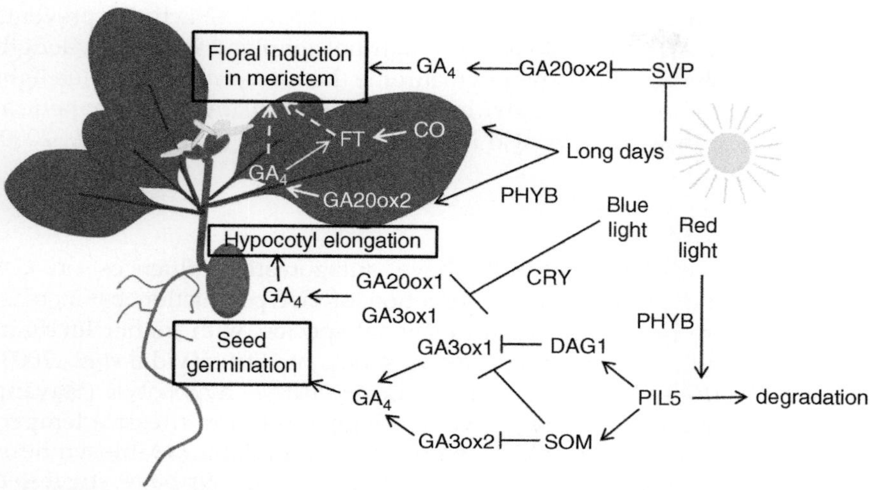

Figure 2.6 Light-induced development in *Arabidopsis* shown to be mediated by modification of GA biosynthesis. Red-light-induced seed germination involves GA_4 biosynthesis through up-regulation of *GA3ox* expression, which is otherwise repressed by the bHLH transcription factor PIL5 via SOMNUS (SOM) and DOF AFFECTING GERMINATION 1 (DAG1). In red light PIL5 is degraded through the action of PHYTOCHROME B (PHYB). Inhibition of hypocotyl elongation by blue light involves CRYTOCHROME (CRY)-mediated repression of *AtGA20ox1* and *AtGA3ox1* expression. Promotion of flowering in long days includes up-regulation of *AtGA20ox2* expression in leaves mediated by PHYB. Flowering is induced by transport of the resulting GA_4 to the shoot apex or by induction of the mobile floral integrator FLOWERING LOCUS T (FT), which is also strongly up-regulated by CONSTANS (CO). GA production in the shoot apex is also promoted in long days by suppression of the MADS box transcription factor SHORT VEGETATIVE PHASE (SVP), which otherwise represses expression of *AtGA20ox2*.

the *Arabidopsis* shoot apex (Andres *et al.*, 2014) so that a drop in expression of *SVP* under floral inductive photoperiods would promote GA production. Exposure to long days was shown to increase expression of several *GA20ox* genes at the shoot apex in wheat with evidence that photoperiod acts through leaf-derived FT to promote GA production and spike formation (Pearce *et al.*, 2013).

Photoperiod-sensitive species of potato form tubers in short days, a process that is inhibited by GAs, which promote vegetative growth. Exposure of the *andigena* sub-species of *Solanum tuberosum* to short day photoperiods reduced *StGA3ox2* expression in stolons, but promoted expression of this gene in vegetative tissues, consistent with a redistribution of GAs under this condition (Bou-Torrent *et al.*, 2011). Over-expression of *StGA3ox2* in leaves of transgenic *S. tuberosum* plants promoted both shoot growth and tuberisation, while expression of this gene in stolons delayed tuber formation. It was suggested by Bou-Torrent *et al.* (2011) that the enzyme product GA_1 is less mobile than its precursor GA_{20} such that *StGA3ox2* activity prevents movement of GA to the stolons. Although not involved in the day-length response, which is mediated by phytochrome (Jackson *et al.*, 1998), blue light promotes expression of *StGA20ox1* in photoperiod-sensitive and day-neutral potato species, in which GA also inhibits tuberisation (Jackson *et al.*, 2000; Fixen *et al.*, 2012).

Temperature

Temperature and light generally have antagonistic influences on GA biosynthesis and plant growth. Promotion of GA production by increasing temperature has been shown in several species, with higher levels of expression of a *GA20ox* gene reported in shoots of Citrus (Vidal *et al.*, 2003) and of both *AtGA20ox1* and *AtGA3ox1* in *Arabidopsis* hypocotyls (Stavang *et al.*, 2009) when plants were grown at a higher, but non-stressful temperature. The mechanism by which temperature stimulates GA-biosynthesis gene expression is not known, but Stavang *et al.* (2009) have suggested the involvement of heat-shock proteins. The importance of GA signalling to temperature-induced growth promotion was demonstrated in the GA-insensitive *Rht-B1c* (*Rht3*) dwarf mutant of wheat, leaf growth in which was similar to that of the tall (*Rht-B1a*) line when plants were grown at 10 °C, but was severely restricted relative to the tall line at 20 °C (Tonkinson *et al.*, 1997). A mechanism for temperature-enhanced hypocotyl extension is suggested by the observation that it requires the action of the PIF4 transcription factor (Stavang *et al.*, 2009), whose activity is regulated by GA signalling through interaction with DELLA proteins (see Chapter 7 and 8). Light and temperature also interact in the induction of seed germination: dormant *Arabidopsis* seeds can be induced to germinate by cold treatment (stratification), which promotes GA biosynthesis through up-regulation of *AtGA3ox1* (Yamauchi *et al.*, 2004). Expression of this gene was found to be repressed by the bHLH transcription factor SPATULA (SPT), which has been proposed to act together with PIL5 to repress germination of dormant *Arabidopsis* seeds (Penfield *et al.*, 2005). While PIL5 is degraded in the light, SPT is light-stable and is thought to play a role in the cold response, although the mechanism is unclear.

Stress

A major mechanism for the reduced GA content observed in plants subjected to abiotic stress is increased expression of the *GA2ox* genes (see Chapter 3). However, meta-analyses of gene expression data indicate that stress also results in down-regulation of GA-biosynthetic genes, including members of the *GA20ox* and *GA3ox* families (Shan *et al.*, 2013; Colebrook *et al.*, 2014). While stress-responsive AP2/ERF family transcription factors have been found to mediate the effects of stress on expression of *GA2ox* and DELLA genes, their regulation of GA biosynthesis is less well documented. An interesting exception is the observation that dwarfism and reduced GA_1

content after expression of the *Arabidopsis* gene *AtCBF1* in tobacco (*N. tabacum*) was associated with reduced expression of a *GGPPS* gene (Niu *et al.*, 2014). It was reported recently that the zinc-finger domain transcription factor Cm-BBX24, which was up-regulated in *Chrysanthemum morifolium* by dehydration and cold treatments, modifies GA biosynthesis by acting on *GA20ox* and *GA3ox* gene expression (Yang *et al.*, 2014). Over-expression of *Cm-BBX24* enhanced stress tolerance as well as delayed flowering, while silencing of this gene had the opposite effects, these traits being associated with the altered GA content.

2.7 Concluding remarks

The demonstration that the *CYP714B* genes from rice encoded GA_{12} 13-hydroxylases (Magome *et al.*, 2013) completed the identification of all the genes/enzymes in the pathway from GGDP to GA_1 in this plant species. However, this is unlikely to be the whole story. Plants lacking both CYP714B paralogues still produced small amounts of 13-hydroxyGAs. The CYP714B sub-family may be restricted to monocots, while the *Arabidopsis* enzyme CYP714A2 catalyses the 13-hydroxylation of *ent*-kaurenoic acid to steviol, but acts as a 12α-hydroxylase on GA_{12} (Nomura *et al.*, 2013). Furthermore, the CYP714 family is absent from gymnosperms (Nelson and Werck-Reichhart, 2011), which are nevertheless capable of producing GA_1 (MacMillan, 2002). It is therefore clear that further GA 13-hydroxylase classes remain to be identified, perhaps belonging to other CYP clades, or to other enzyme types, such as the ODDs. The physiological function of 13-hydroxylation is uncertain: in *Arabidopsis* and rice it results in a slight loss of biological activity and may have a regulatory role (Magome *et al.*, 2013). In most species for which information is available, the 13-hydroxylation pathway predominates and would be assumed to have an important function.

Considerable progress has been made towards understanding the mechanisms by which GA biosynthesis is regulated by developmental and environmental factors. Of particular note is the recent finding that the transcription factor GAF1/IDD2 activates expression of feedback regulated GA-biosynthetic genes only in association with a DELLA protein (Fukazawa *et al.*, 2014). This has provided a mechanism for DELLA-mediated GA homoeostasis and perhaps more generally for gene activation by DELLA proteins. Whereas a number of other transcription factors have been implicated in feedback regulation of GA biosynthesis, there is no clear involvement of DELLA proteins in these cases, despite the importance of DELLAs for this process.

In order to understand how GA concentrations are regulated it is important to map the cellular sites of GA biosynthesis, which has been addressed at the level of gene expression, but should ideally include measurement of the bioactive end-products of GA biosynthesis and their precursors. Physicochemical analysis of GA content is still not achievable at the cellular level, although imaging methods, such as the use of fluorescent probes or FRET, have sufficient sensitivity, if not the same level of quantitative accuracy. As highlighted in Chapter 4, there is renewed interest in GA transporters, which may act on GAs or precursors, indicating the importance of GA transport as part of the regulatory mechanisms that determine GA concentration at its site of action.

The advances in our understanding of the genetics of GA biosynthesis have opened up new opportunities for controlling plant growth. Members of the ODD gene families, although showing some redundancy of function, have sufficient specificity to be exploited in the introduction of beneficial traits into crop species. The prime examples are *ga20ox2* mutants of rice (*sd1*) and barley (*sdw1/denso*), and the *ga3ox1* (*le*) mutant of pea, which are incorporated into crop varieties to reduce stature without affecting fertility and yield (see Chapter 13). The increasing availability of genome sequences for important species is simplifying the identification of useful genes that could be targeted for the production of potentially useful mutants.

Acknowledgements

The author receives support from the 20:20 Wheat® Institute Strategic Programme at Rothamsted Research funded by the Biotechnology and Biological Sciences Research Council of the United Kingdom.

References

Aach, H., Böse, G. and Graebe, J.E. (1995). *ent*-Kaurene biosynthesis in a cell-free system from wheat (*Triticum aestivum* L.) seedlings and the localization of *ent*-kaurene synthetase in plastids of 3 species. *Planta* **197**, 333–342.

Aach, H., Bode, H., Robinson, D.G. and Graebe, J.E. (1997). *ent*-Kaurene synthase is located in proplastids of meristematic shoot tissues. *Planta* **202**, 211–219.

Abe, A., Takagi, H., Fujibe, T. *et al.* (2012). *OsGA20ox1*, a candidate gene for a major QTL controlling seedling vigor in rice. *Theoretical and Applied Genetics* **125**, 647–657.

Albone, K.S., Gaskin, P., Macmillan, J. *et al.* (1990). Biosynthetic origin of gibberellin A_3 and gibberellin A_7 in cell-free preparations from seeds of *Marah macrocarpus* and *Malus domestica*. *Plant Physiology* **94**, 132–142.

Andres, F., Porri, A., Torti, S. *et al.* (2014). SHORT VEGETATIVE PHASE reduces gibberellin biosynthesis at the *Arabidopsis* shoot apex to regulate the floral transition. *Proceedings of the National Academy of Sciences USA* **111**, E2760–E2769.

Appleford, N.E.J. and Lenton, J.R. (1991). Gibberellins and leaf expansion in near-isogenic wheat lines containing *Rht1* and *Rht3* dwarfing alleles. *Planta* **183**, 229–236.

Appleford, N.E.J., Evans, D.J., Lenton, J.R. *et al.* (2006). Function and transcript analysis of gibberellin-biosynthetic enzymes in wheat. *Planta* **223**, 568–582.

Arnaud, N., Girin, T., Sorefan, K. *et al.* (2010). Gibberellins control fruit patterning in *Arabidopsis thaliana*. *Genes & Development* **24**, 2127–2132.

Barboza, L., Effgen, S., Alonso-Blanco, C. *et al.* (2013). *Arabidopsis* semidwarfs evolved from independent mutations in GA20ox1, ortholog to green revolution dwarf alleles in rice and barley. *Proceedings of the National Academy of Sciences USA* **110**, 15818–15823.

Bearder, J.R., Macmillan, J. and Phinney, B.O. (1976). Origin of oxygen atoms in the lactone bridge of C_{19}-gibberellins. *Journal of the Chemical Society, Chemical Communications*, 834–835.

Beck, G., Coman, D., Herren, E. *et al.* (2013). Characterization of the GGPP synthase gene family in *Arabidopsis thaliana*. *Plant Molecular Biology* **82**, 393–416.

Bensen, R.J., Johal, G.S., Crane, V.C. *et al.* (1995). Cloning and characterization of the maize *An1* gene. *The Plant Cell* **7**, 75–84.

Bhattacharya, A., Ward, D.A., Hedden, P. *et al.* (2012). Engineering gibberellin metabolism in *Solanum nigrum* L. by ectopic expression of gibberellin oxidase genes. *Plant Cell Reports* **31**, 945–953.

Blechschmidt, S., Castel, U., Gaskin, P. *et al.* (1984). GC-MS analysis of the plant hormones in seeds of *Cucurbita maxima*. *Phytochemistry* **23**, 553–558.

Boden, S.A., Weiss, D., Ross, J.J. *et al.* (2014). EARLY FLOWERING3 regulates flowering in spring barley by mediating gibberellin production and *FLOWERING LOCUS T* expression. *The Plant Cell* **26**, 1557–1569.

Bou-Torrent, J., Martinez-Garcia, J.F., Luis Garcia-Martinez, J. and Prat, S. (2011). Gibberellin A_1 metabolism contributes to the control of photoperiod-mediated tuberization in potato. *PloS ONE* **6**, e24458.

Carrera, E., Bou, J., GarciaMartinez, J.L. and Prat, S. (2000). Changes in GA 20-oxidase gene expression strongly affect stem length, tuber induction and tuber yield of potato plants. *The Plant Journal* **22**, 247–256.

Castellaro, S.J., Dolan, S.C., Hedden, P. *et al.* (1990). Stereochemistry of the metabolic steps from kaurenoic acids to kaurenolides and gibberellins. *Phytochemistry* **29**, 1833–1839.

Chen, H., Banerjee, A.K. and Hannapel, D.J. (2004). The tandem complex of BEL and KNOX partners is required for transcriptional repression of *ga20ox1*. *The Plant Journal* **38**, 276–284.

Chen, Y., Hou, M., Liu, L. *et al.* (2014). The maize *DWARF1* encodes a gibberellin 3-oxidase and is dual localized to the nucleus and cytosol. *Plant Physiology* **166**, 2028-U1267.

Chiang, H.H., Hwang, I. and Goodman, H.M. (1995). Isolation of the *Arabidopsis GA4* locus. *The Plant Cell* **7**, 195–201.

Christianson, D.W. (2006). Structural biology and chemistry of the terpenoid cyclases. *Chemical Reviews* **106**, 3412–3442.

Colebrook, E.H., Thomas, S.G., Phillips, A.L. and Hedden, P. (2014). The role of gibberellin signalling in plant responses to abiotic stress. *Journal of Experimental Biology* **217**, 67–75.

Coles, J.P., Phillips, A.L., Croker, S.J. *et al.* (1999). Modification of gibberellin production and plant development in *Arabidopsis* by sense and antisense expression of gibberellin 20-oxidase genes. *The Plant Journal* **17**, 547–556.

Cross, B.E., Hanson, J.R. and Galt, R.H.B. (1964). The biosynthesis of gibberellins. I. (–)-Kaurene as precursor of gibberellic acid. *Journal of the Chemical Society*, 295–300.
Curaba, J., Moritz, T., Blervaque, R. *et al.* (2004). *AtGA3ox2*, a key gene responsible for bioactive gibberellin biosynthesis, is regulated during embryogenesis by LEAFY COTYLEDON2 and FUSCA3 in *Arabidopsis*. *Plant Physiology* **136**, 3660–3669.
Dai, M., Zhao, Y., Ma, Q. *et al.* (2007). The rice *YABBY1* gene is involved in the feedback regulation of gibberellin metabolism. *Plant Physiology* **144**, 121–133.
Davidson, S.E., Elliott, R.C., Helliwell, C.A. *et al.* (2003). The pea gene *NA* encodes *ent*-kaurenoic acid oxidase. *Plant Physiology* **131**, 335–344.
Dayan, J., Voronin, N., Gong, F. *et al.* (2012). Leaf-induced gibberellin signaling is essential for internode elongation, cambial activity, and fiber differentiation in tobacco stems. *The Plant Cell* **24**, 66–79.
Desgagne-Penix, I. and Sponsel, V.M. (2008). Expression of gibberellin 20-oxidase1 (*AtGA20ox1*) in *Arabidopsis* seedlings with altered auxin status is regulated at multiple levels. *Journal of Experimental Botany* **59**, 2057–2070.
Dugardeyn, J., Vandenbussche, F. and Van der Straeten, D. (2008). To grow or not to grow: what can we learn on ethylene-gibberellin cross-talk by in silico gene expression analysis? *Journal of Experimental Botany* **59**, 1–16.
Duncan, J.D. and West, C.A. (1981). Properties of kaurene synthetase from *Marah macrocarpus* endosperm – evidence for the participation of separate but interacting enzymes. *Plant Physiology* **68**, 1128–1134.
Eriksson, S., Bohlenius, H., Moritz, T. and Nilsson, O. (2006). GA_4 is the active gibberellin in the regulation of *LEAFY* transcription and *Arabidopsis* floral initiation. *The Plant Cell* **18**, 2172–2181.
Fambrini, M., Mariotti, L., Parlanti, S. *et al.* (2011). The extreme dwarf phenotype of the GA-sensitive mutant of sunflower, *dwarf2*, is generated by a deletion in the *ent*-kaurenoic acid oxidase1 (*HaKAO1*) gene sequence. *Plant Molecular Biology* **75**, 431–450.
Feurtado, J.A., Huang, D., Wicki-Stordeur, L. *et al.* (2011). The *Arabidopsis* C2H2 zinc finger INDETERMINATE DOMAIN1/ENHYDROUS promotes the transition to germination by regulating light and hormonal signaling during seed maturation. *The Plant Cell* **23**, 1772–1794.
Fixen, K.R., Thomas, S.C. and Tong, C.B.S. (2012). Blue light inhibition of tuberization in a day-neutral potato. *Journal of Plant Growth Regulation* **31**, 342–350.
Fleet, C.M., Yamaguchi, S., Hanada, A. *et al.* (2003). Overexpression of *AtCPS* and *AtKS* in *Arabidopsis* confers increased *ent*-kaurene production but no increase in bioactive gibberellins. *Plant Physiology* **132**, 830–839.
Foo, E., Platten, J.D., Weller, J.L. and Reid, J.B. (2006). PhyA and Cry1 act redundantly to regulate gibberellin levels during de-etiolation in blue light. *Physiologia Plantarum* **127**, 149–156.
Frigerio, M., Alabadi, D., Perez-Gomez, J. *et al.* (2006). Transcriptional regulation of gibberellin metabolism genes by auxin signaling in Arabidopsis. *Plant Physiology* **142**, 553–563.
Fujioka, S., Yamane, H., Spray, C.R. *et al.* (1988a). Qualitative and quantitative-analyses of gibberellins in vegetative shoots of normal, *dwarf-1*, *dwarf-2*, *dwarf-3*, and *dwarf-5* seedlings of *Zea mays* L. *Plant Physiology* **88**, 1367–1372.
Fujioka, S., Yamane, H., Spray, C.R. *et al.* (1988b). The dominant non-gibberellin-responding dwarf mutant (*d8*) of maize accumulates native gibberellins. *Proceedings of the National Academy of Sciences USA* **85**, 9031–9035.

Fukazawa, J., Nakata, M., Ito, T. *et al.* (2010). The transcription factor RSG regulates negative feedback of *NtGA20ox1* encoding GA 20-oxidase. *The Plant Journal* **62**, 1035–1045.

Fukazawa, J., Teramura, H., Murakoshi, S. *et al.* (2014). DELLAs function as coactivators of GAI-ASSOCIATED FACTOR1 in regulation of gibberellin homeostasis and signaling in *Arabidopsis*. *The Plant Cell* **26**, 2920–2938.

Gabriele, S., Rizza, A., Martone, J. *et al.* (2010). The Dof protein DAG1 mediates PIL5 activity on seed germination by negatively regulating GA biosynthetic gene *AtGA3ox1*. *The Plant Journal* **61**, 312–323.

Garcia-Martinez, J.L. and Rappaport, L. (1982). Contribution of leaves to the gibberellin-induced epicotyl elongation in cowpea. *Journal of Plant Growth Regulation* **1**, 129–137.

Gilmour, S.J., Bleecker, A.B. and Zeevaart, J.A.D. (1987). Partial purification of gibberellin oxidases from spinach leaves. *Plant Physiology* **85**, 87–90.

Gilmour, S.J., Zeevaart, J.A.D., Schwenen, L. and Graebe, J.E. (1986). Gibberellin metabolism in cell-free extracts from spinach leaves in relation to photoperiod. *Plant Physiology* **82**, 190–195.

Gomez-Mena, C., de Folter, S., Costa, M.M.R. *et al.* (2005). Transcriptional program controlled by the floral homeotic gene *AGAMOUS* during early organogenesis. *Development* **132**, 429–438.

Graebe, J.E. (1987) Gibberellin biosynthesis and control. *Annual Review of Plant Physiology and Plant Molecular Biology* **38**, 419–465.

Graebe, J.E., Dennis, D.T., Upper, C.D. and West, C.A. (1965). Biosynthesis of gibberellins. I. Biosynthesis of (–)-kaurene, (–)-kauren-19-ol and *trans*-geranylgeraniol in endosperm nucellus of *Echinocystis macrocarpa* Greene. *Journal of Biological Chemistry* **240**, 1847–1854.

Graebe, J.E., Bowen, D.H. and MacMillan, J. (1972). Conversion of mevalonic acid into gibberellin A_{12}-aldehyde in a cell-free system from *Cucurbita pepo*. *Planta* **102**, 261–271.

Griffiths, J., Murase, K., Rieu, I. *et al.* (2006). Genetic characterization and functional analysis of the GID1 gibberellin receptors in *Arabidopsis*. *The Plant Cell* **18**, 3399–3414.

Grosselindemann, E., Lewis, M.J., Hedden, P. and Graebe, J.E. (1992). Gibberellin biosynthesis from gibberellin A_{12}-aldehyde in a cell-free system from germinating barley (*Hordeum vulgare* L., cv Himalaya) embryos. *Planta* **188**, 252–257.

Gubler, F., Hughes, T., Waterhouse, P. and Jacobsen, J. (2008). Regulation of dormancy in barley by blue light and after-ripening: Effects on abscisic acid and gibberellin metabolism. *Plant Physiology* **147**, 886–896.

Hai Ha, H., Sechet, J., Bailly, C. *et al.* (2014). Inhibition of germination of dormant barley (*Hordeum vulgare* L.) grains by blue light as related to oxygen and hormonal regulation. *Plant Cell and Environment* **37**, 1393–1403.

Hay, A., Kaur, H., Phillips, A. *et al.* (2002). The gibberellin pathway mediates KNOTTED1-type homeobox function in plants with different body plans. *Current Biology* **12**, 1557–1565.

Hedden, P. and Croker, S.J. (1992). Regulation of gibberellin biosynthesis in maize seedlings. In *Progress in Plant Growth Regulation: Proceedings of the 14th International Conference on Plant Growth Substances* (eds. C.M. Karssen, L.C. Van Loon and D. Vreugdenhil). Kluwer, Dordrecht, pp. 534–544.

Hedden, P. and Graebe, J.E. (1981). Kaurenolide biosynthesis in a cell-free system from *Cucurbita maxima* seeds. *Phytochemistry* **20**, 1011–1015.

Hedden, P. and Phinney, B.O. (1979). Comparison of *ent*-kaurene and *ent*-isokaurene synthesis in cell-free systems from etiolated shoots of normal and dwarf-5 maize seedlings. *Phytochemistry* **18**, 1475–1479.

Hedden, P. and Thomas, S.G. (2012). Gibberellin biosynthesis and its regulation. *Biochemical Journal* **444**, 11–25.

Hedden, P., Phillips, A.L., Rojas, M.C. *et al.* (2002). Gibberellin biosynthesis in plants and fungi: A case of convergent evolution? *Journal of Plant Growth Regulation* **20**, 319–331.

Heinrich, M., Hettenhausen, C., Lange, T. *et al.* (2013). High levels of jasmonic acid antagonize the biosynthesis of gibberellins and inhibit the growth of *Nicotiana attenuata* stems. *The Plant Journal* **73**, 591–606.

Helliwell, C.A., Chandler, P.M., Poole, A. *et al.* (2001a). The CYP88A cytochrome P450, *ent*-kaurenoic acid oxidase, catalyzes three steps of the gibberellin biosynthesis pathway. *Proceedings of the National Academy of Sciences USA* **98**, 2065–2070.

Helliwell, C.A., Sullivan, J.A., Mould, R.M. *et al.* (2001b). A plastid envelope location of *Arabidopsis ent*-kaurene oxidase links the plastid and endoplasmic reticulum steps of the gibberellin biosynthesis pathway. *The Plant Journal* **28**, 201–208.

Hirose, F., Inagaki, N., Hanada, A. *et al.* (2012). Cryptochrome and phytochrome cooperatively but independently reduce active gibberellin content in rice seedlings under light irradiation. *Plant and Cell Physiology* **53**, 1570–1582.

Hisamatsu, T. and King, R.W. (2008). The nature of floral signals in *Arabidopsis*. II. Roles for FLOWERING LOCUS T (FT) and gibberellin. *Journal of Experimental Botany* **59**, 3821–3829.

Hisamatsu, T., King, R.W., Helliwell, C.A. and Koshioka, M. (2005). The involvement of gibberellin 20-oxidase genes in phytochrome-regulated petiole elongation of *Arabidopsis*. *Plant Physiology* **138**, 1106–1116.

Hu, J.H., Mitchum, M.G., Barnaby, N. *et al.* (2008). Potential sites of bioactive gibberellin production during reproductive growth in *Arabidopsis*. *The Plant Cell* **20**, 320–336.

Huang, S.S., Raman, A.S., Ream, J.E. *et al.* (1998). Overexpression of 20-oxidase confers a gibberellin-overproduction phenotype in *Arabidopsis*. *Plant Physiology* **118**, 773–781.

Itoh, H., Ueguchi-Tanaka, M., Sentoku, N. *et al.* (2001). Cloning and functional analysis of two gibberellin 3β-hydroxylase genes that are differently expressed during the growth of rice. *Proceedings of the National Academy of Sciences USA* **98**, 8909–8914.

Itoh, H., Tatsumi, T., Sakamoto, T. *et al.* (2004). A rice semi-dwarf gene, Tan-Ginbozu (D35), encodes the gibberellin biosynthesis enzyme, *ent*-kaurene oxidase. *Plant Molecular Biology* **54**, 533–547.

Jackson, S.D., James, P., Prat, S. and Thomas, B. (1998). Phytochrome B affects the levels of a graft-transmissible signal involved in tuberization. *Plant Physiology* **117**, 29–32.

Jackson, S.D., James, P.E., Carrera, E. *et al.* (2000). Regulation of transcript levels of a potato gibberellin 20-oxidase gene by light and phytochrome B. *Plant Physiology* **124**, 423–430.

Kamiya, Y. and Graebe, J.E. (1983). The biosynthesis of all major pea gibberellins in a cell-free system from *Pisum sativum*. *Phytochemistry* **22**, 681–689.

Kamiya, Y., Takahashi, N. and Graebe, J.E. (1986). The loss of carbon-20 in C_{19}-gibberellin biosynthesis in a cell-free system from *Pisum sativum* L. *Planta* **169**, 524–528.

Kaneko, M., Itoh, H., Inukai, Y. *et al.* (2003). Where do gibberellin biosynthesis and gibberellin signaling occur in rice plants? *The Plant Journal* **35**, 104–115.

Kasahara, H., Hanada, A., Kuzuyama, T. *et al.* (2002). Contribution of the mevalonate and methylerythritol phosphate pathways to the biosynthesis of gibberellins in *Arabidopsis*. *Journal of Biological Chemistry* **277**, 45188–45194.

Katsumi, M., Phinney, B.O., Henrick, C.A. and Jefferies, P. (1964). Growth response of the d-5 and an-1 mutants of maize to some kaurene derivatives. *Science* **144**, 849–850.

Kawai, Y., Ono, E. and Mizutani, M. (2014). Evolution and diversity of the 2-oxoglutarate-dependent dioxygenase superfamily in plants. *The Plant Journal* **78**, 328–343.

Kim, D.H., Yamaguchi, S., Lim, S. *et al.* (2008). SOMNUS, a CCCH-type zinc finger protein in *Arabidopsis*, negatively regulates light-dependent seed germination downstream of PIL5. *The Plant Cell* **20**, 1260–1277.

King, R.W., Moritz, T., Evans, L.T. *et al.* (2001). Long-day induction of flowering in *Lolium temulentum* involves sequential increases in specific gibberellins at the shoot apex. *Plant Physiology* **127**, 624–632.

King, R.W., Moritz, T., Evans, L.T. *et al.* (2006). Regulation of flowering in the long-day grass *Lolium temulentum* by gibberellins and the *FLOWERING LOCUS T* gene. *Plant Physiology* **141**, 498–507.

Koornneef, M. and van der Veen, J.H. (1980). Induction and analysis of gibberellin sensitive mutants in *Arabidopsis thaliana* (L) Heynh. *Theoretical and Applied Genetics* **58**, 257–263.

Lange, M.J.P., Liebrandt, A., Arnold, L. *et al.* (2013). Functional characterization of gibberellin oxidases from cucumber, *Cucumis sativus* L. *Phytochemistry* **90**, 62–69.

Lange, T. (1994). Purification and partial amino-acid sequence of gibberellin 20-oxidase from *Cucurbita maxima* L endosperm. *Planta* **195**, 108–115.

Lange, T. (1997). Cloning gibberellin dioxygenase genes from pumpkin endosperm by heterologous expression off enzyme activities in *Escherichia coli*. *Proceedings of the National Academy of Sciences USA* **94**, 6553–6558.

Lange, T. and Graebe, J.E. (1993). Enzymes of gibberellin biosynthesis. In *Methods in Plant Biochemistry: Volume 9. Enzymes of Secondary Metabolism* (ed. P.J. Lea). Academic Press, London, pp. 403–430.

Lange, T., Hedden, P. and Graebe, J.E. (1993). Biosynthesis of 12α-hydroxylated and 13-hydroxylated gibberellins in a cell-free system from *Cucurbita maxima* endosperm and the identification of new endogenous gibberellins. *Planta* **189**, 340–349.

Lange, T., Hedden, P. and Graebe, J.E. (1994a). Expression cloning of a gibberellin 20-oxidase, a multifunctional enzyme involved in gibberellin biosynthesis. *Proceedings of the National Academy of Sciences USA* **91**, 8552–8556.

Lange, T., Schweimer, A., Ward, D.A. *et al.* (1994b). Separation and characterization of three 2-oxoglutarate-dependent dioxygenases from *Cucurbita maxima* L. endosperm involved in gibberellin biosynthesis. *Planta* **195**, 98–107.

Lee, D.J. and Zeevaart, J.A.D. (2007). Regulation of *gibberellin 20-oxidase1* expression in spinach by photoperiod. *Planta* **226**, 35–44.

Lester, D.R., Ross, J.J., Davies, P.J. and Reid, J.B. (1997). Mendel's stem length gene (*Le*) encodes a gibberellin 3β-hydroxylase. *The Plant Cell* **9**, 1435–1443.

Lilley, J.L.S., Gan, Y., Graham, I.A. and Nemhauser, J.L. (2013). The effects of DELLAs on growth change with developmental stage and brassinosteroid levels. *The Plant Journal* **76**, 165–173.
MacMillan, C.P., Blundell, C.A. and King, R.W. (2005). Flowering of the grass *Lolium perenne*. Effects of vernalization and long days on gibberellin biosynthesis and signaling. *Plant Physiology* **138**, 1794–1806.
MacMillan, J. (2002). Occurrence of gibberellins in vascular plants, fungi and bacteria. *Journal of Plant Growth Regulation* **20**, 387–442.
MacMillan, J. and Gaskin, P. (1996). Gibberellins in endosperm and embryos of *Marah macrocarpus*. *Phytochemistry* **42**, 1263–1266.
Magome, H., Nomura, T., Hanada, A. *et al.* (2013). CYP714B1 and CYP714B2 encode gibberellin 13-oxidases that reduce gibberellin activity in rice. *Proceedings of the National Academy of Sciences USA* **110**, 1947–1952.
Mai, Y.-X., Wang, L. and Yang, H.-Q. (2011). A gain-of-function mutation in IAA7/AXR2 confers late flowering under short-day light in *Arabidopsis*. *Journal of Integrative Plant Biology* **53**, 480–492.
Martin, D.N., Proebsting, W.M., Parks, T.D. *et al.* (1996). Feedback regulation of gibberellin biosynthesis and gene expression in *Pisum sativum* L. *Planta* **200**, 159–166.
Martin, D.N., Proebsting, W.M. and Hedden, P. (1997). Mendel's dwarfing gene: cDNAs from the *Le* alleles and function of the expressed proteins. *Proceedings of the National Academy of Sciences USA* **94**, 8907–8911.
Matsushita, A., Furumoto, T., Ishida, S. and Takahashi, Y. (2007). AGF1, an AT-hook protein, is necessary for the negative feedback of *AtGA3ox1* encoding GA 3-oxidase. *Plant Physiology* **143**, 1152–1162.
Mehrshahi, P., Johnny, C. and DellaPenna, D. (2014). Redefining the metabolic continuity of chloroplasts and ER. *Trends in Plant Science* **19**, 501–507.
Mitchum, M.G., Yamaguchi, S., Hanada, A. *et al.* (2006). Distinct and overlapping roles of two gibberellin 3-oxidases in *Arabidopsis* development. *The Plant Journal* **45**, 804–818.
Moore, T.C. and Coolbaugh, R.C. (1976). Conversion of geranylgeranyl pyrophosphate to *ent*-kaurene in enzyme extracts of sonicated chloroplasts. *Phytochemistry* **15**, 1241–1247.
Morrone, D., Chen, X., Coates, R.M. and Peters, R.J. (2010). Characterization of the kaurene oxidase CYP701A3, a multifunctional cytochrome P450 from gibberellin biosynthesis. *Biochemical Journal* **431**, 337–344.
Mutasa-Gottgens, E. and Hedden, P. (2009). Gibberellin as a factor in floral regulatory networks. *Journal of Experimental Botany* **60**, 1979–1989.
Nelson, D. and Werck-Reichhart, D. (2011). A P450-centric view of plant evolution. *The Plant Journal* **66**, 194–211.
Ngo, P., Ozga, J.A. and Reinecke, D.M. (2002). Specificity of auxin regulation of gibberellin 20-oxidase gene expression in pea pericarp. *Plant Molecular Biology* **49**, 439–448.
Niu, S., Gao, Q., Li, Z. *et al.* (2014). The role of gibberellin in the CBF1-mediated stress-response pathway. *Plant Molecular Biology Reporter* **32**, 852–863.
Nomura, T., Magome, H., Hanada, A. *et al.* (2013). Functional analysis of *Arabidopsis* CYP714A1 and CYP714A2 reveals that they are distinct gibberellin modification enzymes. *Plant and Cell Physiology* **54**, 1837–1851.
O'Neill, D.P., Davidson, S.E., Clarke, V.C. *et al.* (2010). Regulation of the gibberellin pathway by auxin and DELLA proteins. *Planta* **232**, 1141–1149.

Oh, E., Yamaguchi, S., Kamiya, Y. *et al.* (2006). Light activates the degradation of PIL5 protein to promote seed germination through gibberellin in *Arabidopsis*. *The Plant Journal* **47**, 124–139.

Okada, K., Saito, T., Nakagawa, T. *et al.* (2000). Five geranylgeranyl diphosphate synthases expressed in different organs are localized into three subcellular compartments in *Arabidopsis*. *Plant Physiology* **122**, 1045–1056.

Okada, K., Kasahara, H., Yamaguchi, S. *et al.* (2008). Genetic evidence for the role of isopentenyl diphosphate isomerases in the mevalonate pathway and plant development in *Arabidopsis*. *Plant and Cell Physiology* **49**, 604–616.

Otomo, K., Kenmoku, H., Oikawa, H. *et al.* (2004). Biological functions of *ent*- and *syn*-copalyl diphosphate synthases in rice: key enzymes for the branch point of gibberellin and phytoalexin biosynthesis. *The Plant Journal* **39**, 886–893.

Otsuka, M., Kenmoku, H., Ogawa, M. *et al.* (2004). Emission of *ent*-kaurene, a diterpenoid hydrocarbon precursor for gibberellins, into the headspace from plants. *Plant and Cell Physiology* **45**, 1129–1138.

Ozga, J.A. and Reinecke, D.M. (1999). Interaction of 4-chloroindole-3-acetic acid and gibberellins in early pea fruit development. *Plant Growth Regulation* **27**, 33–38.

Ozga, J.A., Reinecke, D.M., Ayele, B.T. *et al.* (2009). Developmental and hormonal regulation of gibberellin biosynthesis and catabolism in pea fruit. *Plant Physiology* **150**, 448–462.

Pearce, S., Vanzetti, L.S. and Dubcovsky, J. (2013). Exogenous gibberellins induce wheat spike development under short days only in the presence of VERNALIZATION1. *Plant Physiology* **163**, 1433–1445.

Pearce, S.P., Huttly, A.K., Prosser, I.P. *et al.* (2015). Heterologous expression and transcript analysis of gibberellin biosynthetic genes of grasses reveals novel functionality in the *GA3ox* family. *BMC Plant Biology* **15**, 130.

Penfield, S., Josse, E.M., Kannangara, R. *et al.* (2005). Cold and light control seed germination through the bHLH transcription factor SPATULA. *Current Biology* **15**, 1998–2006.

Peters, R.J. (2006). Uncovering the complex metabolic network underlying diterpenoid phytoalexin biosynthesis in rice and other cereal crop plants. *Phytochemistry* **67**, 2307–2317.

Phillips, A.L., Ward, D.A., Uknes, S. *et al.* (1995). Isolation and expression of three gibberellin 20-oxidase cDNA clones from Arabidopsis. *Plant Physiology* **108**, 1049–1057.

Phillips, M.A., D'Auria, J.C., Gershenzon, J. and Pichersky, E. (2008). The *Arabidopsis thaliana* type I isopentenyl diphosphate isomerases are targeted to multiple subcellular compartments and have overlapping functions in isoprenoid biosynthesis. *The Plant Cell* **20**, 677–696.

Plackett, A.R.G., Powers, S.J., Fernandez-Garcia, N. *et al.* (2012). Analysis of the developmental roles of the *Arabidopsis* gibberellin 20-oxidases demonstrates that GA20ox1, -2, and -3 are the dominant paralogs. *The Plant Cell* **24**, 941–960.

Porri, A., Torti, S., Romera-Branchat, M. and Coupland, G. (2012). Spatially distinct regulatory roles for gibberellins in the promotion of flowering of *Arabidopsis* under long photoperiods. *Development* **139**, 2198–2209.

Prisic, S. and Peters, R.J. (2007). Synergistic substrate inhibition of *ent*-copalyl diphosphate synthase: A potential feed-forward inhibition mechanism limiting gibberellin metabolism. *Plant Physiology* **144**, 445–454.

Prisic, S., Xu, J.J., Coates, R.M. and Peters, R.J. (2007). Probing the role of the DXDD motif in class II diterpene cyclases. *Chembiochem* **8**, 869–874.

Radi, A., Lange, T., Niki, T. *et al.* (2006). Ectopic expression of pumpkin gibberellin oxidases alters gibberellin biosynthesis and development of transgenic *Arabidopsis* plants. *Plant Physiology* **140**, 528–536.

Radley, M. (1970). Comparison of endogenous gibberellins and response to applied gibberellin of some dwarf and tall wheat cultivars. *Planta* **92**, 292–300.

Railton, I.D., Fellows, B. and West, C.A. (1984). *ent*-Kaurene synthesis in chloroplasts from higher plants. *Phytochemistry* **23**, 1261–1267.

Regnault, T., Daviere, J.-M., Heintz, D. *et al.* (2014). The gibberellin biosynthetic genes *AtKAO1* and *AtKAO2* have overlapping roles throughout *Arabidopsis* development. *The Plant Journal* **80**, 462–474.

Reid, J.B., Botwright, N.A., Smith, J.J. *et al.* (2002). Control of gibberellin levels and gene expression during de-etiolation in pea. *Plant Physiology* **128**, 734–741.

Reinecke, D.M., Wickramarathna, A.D., Ozga, J.A. *et al.* (2013). Gibberellin 3-oxidase gene expression patterns influence gibberellin biosynthesis, growth, and development in pea. *Plant Physiology* **163**, 929–945.

Rieu, I., Ruiz-Rivero, O., Fernandez-Garcia, N. *et al.* (2008). The gibberellin biosynthetic genes *AtGA20ox1* and *AtGA20ox2* act, partially redundantly, to promote growth and development throughout the *Arabidopsis* life cycle. *The Plant Journal* **53**, 488–504.

Rojas, M.C., Urrutia, O., Cruz, C. *et al.* (2004). Kaurenolides and fujenoic acids are side products of the gibberellin P450-1 monooxygenase in *Gibberella fujikuroi*. *Phytochemistry* **65**, 821–830.

Ross, J.J., O'Neill, D.P., Smith, J.J. *et al.* (2000). Evidence that auxin promotes gibberellin A_1 biosynthesis in pea. *The Plant Journal* **21**, 547–552.

Ross, J.J., Davidson, S.E., Wolbang, C.M. *et al.* (2003). Developmental regulation of the gibberellin pathway in pea shoots. *Functional Plant Biology* **30**, 83–89.

Ross, J.J., Weston, D.E., Davidson, S.E. and Reid, J.B. (2011). Plant hormone interactions: how complex are they? *Physiologia Plantarum* **141**, 299–309.

Sakamoto, T., Kamiya, N., Ueguchi-Tanaka, M. *et al.* (2001). KNOX homeodomain protein directly suppresses the expression of a gibberellin biosynthetic gene in the tobacco shoot apical meristem. *Genes & Development* **15**, 581–590.

Sakamoto, T., Miura, K., Itoh, H. *et al.* (2004). An overview of gibberellin metabolism enzyme genes and their related mutants in rice. *Plant Physiology* **134**, 1642–1653.

Seo, M., Hanada, A., Kuwahara, A. *et al.* (2006). Regulation of hormone metabolism in *Arabidopsis* seeds: phytochrome regulation of abscisic acid metabolism and abscisic acid regulation of gibberellin metabolism. *The Plant Journal* **48**, 354–366.

Seo, M., Nambara, E., Choi, G. and Yamaguchi, S. (2009). Interaction of light and hormone signals in germinating seeds. *Plant Molecular Biology* **69**, 463–472.

Serrani, J.C., Ruiz-Rivero, O., Fos, M. and Garcia-Martinez, J.L. (2008). Auxin-induced fruit-set in tomato is mediated in part by gibberellins. *The Plant Journal* **56**, 922–934.

Shan, X., Li, Y., Jiang, Y. *et al.* (2013). Transcriptome profile analysis of maize seedlings in response to high-salinity, drought and cold stresses by deep sequencing. *Plant Molecular Biology Reporter* **31**, 1485–1491.

Sherwin, P.F. and Coates, R.M. (1982). Stereospecificity of the oxidation of *ent*-kauren-19-ol to *ent*-kaurenal by a microsomal-enzyme preparation from *Marah macrocarpus*. *Journal of the Chemical Society-Chemical Communications*, 1013–1014.

Silverstone, A.L., Chang, C.W., Krol, E. and Sun, T.P. (1997). Developmental regulation of the gibberellin biosynthetic gene *GA1* in *Arabidopsis thaliana*. *The Plant Journal* **12**, 9–19.

Simcox, P.D., Dennis, D.T. and West, C.A. (1975). Kaurene synthetase from plastids of developing plant tissues. *Biochemical and Biophysical Research Communications* **66**, 166–172.

Smith, M.W., Yamaguchi, S., Ait-Ali, T. and Kamiya, Y. (1998). The first step of gibberellin biosynthesis in pumpkin is catalyzed by at least two copalyl diphosphate synthases encoded by differentially regulated genes. *Plant Physiology* **118**, 1411–1419.

Smith, V.A., Knatt, C.J., Gaskin, P. and Reid, J.B. (1992). The distribution of gibberellins in vegetative tissues of *Pisum-sativum* L.1. Biological and biochemical consequences of the *le* mutation. *Plant Physiology* **99**, 368–371.

Spray, C.R., Kobayashi, M., Suzuki, Y. *et al.* (1996). The dwarf-1 (d1) mutant of *Zea mays* blocks three steps in the gibberellin-biosynthetic pathway. *Proceedings of the National Academy of Sciences USA* **93**, 10515–10518.

Stavang, J.A., Gallego-Bartolome, J., Gomez, M.D. *et al.* (2009). Hormonal regulation of temperature-induced growth in *Arabidopsis*. *The Plant Journal* **60**, 589–601.

Sun, T.P. and Kamiya, Y. (1994). The *Arabidopsis GA1* locus encodes the cyclase *ent*-kaurene synthetase A of gibberellin biosynthesis. *The Plant Cell* **6**, 1509–1518.

Sun, T.P. and Kamiya, Y. (1997). Regulation and cellular localization of *ent*-kaurene synthesis. *Physiologia Plantarum* **101**, 701–708.

Talon, M., Koornneef, M. and Zeevaart, J.A.D. (1990). Accumulation of C_{19}-gibberellins in the gibberellin-insensitive dwarf mutant *GAI* of *Arabidopsis thaliana* (L) Heynh. *Planta* **182**, 501–505.

Teng, F., Zhai, L., Liu, R. *et al.* (2013). *ZmGA3ox2*, a candidate gene for a major QTL, qPH3.1, for plant height in maize. *The Plant Journal* **73**, 405–416.

Thomas, S.G., Phillips, A.L. and Hedden, P. (1999). Molecular cloning and functional expression of gibberellin 2-oxidases, multifunctional enzymes involved in gibberellin deactivation. *Proceedings of the National Academy of Sciences USA* **96**, 4698–4703.

Toh, S., Imamura, A., Watanabe, A. *et al.* (2008). High temperature-induced abscisic acid biosynthesis and its role in the inhibition of gibberellin action in *Arabidopsis* seeds. *Plant Physiology* **146**, 1368–1385.

Tong, H., Xiao, Y., Liu, D. *et al.* (2014). Brassinosteroid regulates cell elongation by modulating gibberellin metabolism in rice. *The Plant Cell* **26**, 4376–4393.

Tonkinson, C.L., Lyndon, R.F., Arnold, G.M. and Lenton, J.R. (1997). The effects of temperature and the *Rht3* dwarfing gene on growth, cell extension, and gibberellin content and responsiveness in the wheat leaf. *Journal of Experimental Botany* **48**, 963–970.

Toyomasu, T., Kagahara, T., Hirose, Y. *et al.* (2009). Cloning and characterization of cDNAs encoding *ent*-copalyl diphosphate synthases in wheat: Insight into the evolution of rice phytoalexin biosynthetic genes. *Bioscience Biotechnology and Biochemistry* **73**, 772–775.

Tritsch, D., Hemmerlin, A., Bach, T.J. and Rohmer, M. (2010). Plant isoprenoid biosynthesis via the MEP pathway: In vivo IPP/DMAPP ratio produced by (E)-4-hydroxy-3-methylbut-2-enyl diphosphate reductase in tobacco BY-2 cell cultures. *FEBS Letters* **584**, 129–134.

Ueguchi-Tanaka, M., Ashikari, M., Nakajima, M. *et al.* (2005). GIBBERELLIN INSENSITIVE DWARF1 encodes a soluble receptor for gibberellin. *Nature* **437**, 693–698.
van Schie, C.C.N., Ament, K., Schmidt, A. *et al.* (2007). Geranyl diphosphate synthase is required for biosynthesis of gibberellins. *The Plant Journal* **52**, 752–762.
Vidal, A.M., Gisbert, C., Talon, M. *et al.* (2001). The ectopic overexpression of a citrus gibberellin 20-oxidase enhances the non-13-hydroxylation pathway of gibberellin biosynthesis and induces an extremely elongated phenotype in tobacco. *Physiologia Plantarum* **112**, 251–260.
Vidal, A.M., Ben-Cheikh, W., Talon, M. and Garcia-Martinez, J.L. (2003). Regulation of gibberellin 20-oxidase gene expression and gibberellin content in citrus by temperature and citrus exocortis viroid. *Planta* **217**, 442–448.
Vranova, E., Coman, D. and Gruissem, W. (2013). Network analysis of the MVA and MEP pathways for isoprenoid synthesis. *Annual Review of Plant Biology* **64**, 665–700.
Wang, Q., Hillwig, M.L., Wu, Y. and Peters, R.J. (2012). CYP701A8: A rice *ent*-kaurene oxidase paralog diverted to more specialized diterpenoid metabolism. *Plant Physiology* **158**, 1418–1425.
Ward, D.A., MacMillan, J., Gong, F. *et al.* (2010). Gibberellin 3-oxidases in developing embryos of the southern wild cucumber, *Marah macrocarpus*. *Phytochemistry* **71**, 2010–2018.
Ward, J.L., Jackson, G.J., Beale, M.H. *et al.* (1997). Stereochemistry of the oxidation of gibberellin 20-alcohols, GA_{15} and GA_{44}, to 20-aldehydes by gibberellin 20-oxidases. *Chemical Communications*, 13–14.
Ward, J.L., Gaskin, P., Brown, R.G.S. *et al.* (2002). Probing the mechanism of loss of carbon-20 in gibberellin biosynthesis. Synthesis of gibberellin 3α,20-hemiacetal and 19,20-lactol analogues and their metabolism by a recombinant GA 20-oxidase. *Journal of the Chemical Society, Perkin Transactions* **1**, 232–241.
Weiss, D. and Halevy, A.H. (1989). Stamens and gibberellin in the regulation of corolla pigmentation and growth in *Petunia hybrida*. *Planta* **179**, 89–96.
Weston, D.E., Reid, J.B. and Ross, J.J. (2009). Auxin regulation of gibberellin biosynthesis in the roots of pea (*Pisum sativum*). *Functional Plant Biology* **36**, 362–369.
Williams, J., Phillips, A.L., Gaskin, P. and Hedden, P. (1998). Function and substrate specificity of the gibberellin 3β-hydroxylase encoded by the Arabidopsis *GA4* gene. *Plant Physiology* **117**, 559–563.
Wolbang, C.M. and Ross, J.J. (2001). Auxin promotes gibberellin biosynthesis in decapitated tobacco plants. *Planta* **214**, 153–157.
Wolbang, C.M., Chandler, P.M., Smith, J.J. and Ross, J.J. (2004). Auxin from the developing inflorescence is required for the biosynthesis of active gibberellins in barley stems. *Plant Physiology* **134**, 769–776.
Wu, Y., Zhou, K., Toyomasu, T. *et al.* (2012). Functional characterization of wheat copalyl diphosphate synthases sheds light on the early evolution of labdane-related diterpenoid metabolism in the cereals. *Phytochemistry* **84**, 40–46.
Xu, M.M., Wilderman, P.R., Morrone, D. *et al.* (2007). Functional characterization of the rice kaurene synthase-like gene family. *Phytochemistry* **68**, 312–326.
Xu, M.M., Galhano, R., Wiemann, P. *et al.* (2012). Genetic evidence for natural product-mediated plant-plant allelopathy in rice (*Oryza sativa*). *New Phytologist* **193**, 570–575.
Yamaguchi, S. (2008) Gibberellin metabolism and its regulation. *Annual Review of Plant Biology* **59**, 225–251.

Yamaguchi, S., Saito, T., Abe, H. *et al.* (1996). Molecular cloning and characterization of a cDNA encoding the gibberellin biosynthetic enzyme *ent*-kaurene synthase B from pumpkin (*Cucurbita maxima* L.). *The Plant Journal* **10**, 203–213.

Yamaguchi, S., Kamiya, Y. and Sun, T.P. (2001). Distinct cell-specific expression patterns of early and late gibberellin biosynthetic genes during *Arabidopsis* seed germination. *The Plant Journal* **28**, 443–453.

Yamauchi, Y., Ogawa, M., Kuwahara, A. *et al.* (2004). Activation of gibberellin biosynthesis and response pathways by low temperature during imbibition of *Arabidopsis thaliana* seeds. *The Plant Cell* **16**, 367–378.

Yang, Y., Ma, C., Xu, Y. *et al.* (2014). A zinc finger protein regulates flowering time and abiotic stress tolerance in chrysanthemum by modulating gibberellin biosynthesis. *The Plant Cell* **26**, 2038–2054.

Yano, R., Kanno, Y., Jikumaru, Y. *et al.* (2009). CHOTTO1, a putative double APETALA2 repeat transcription factor, is involved in abscisic acid-mediated repression of gibberellin biosynthesis during seed germination in *Arabidopsis*. *Plant Physiology* **151**, 641–654.

Yeung, E.C. and Meinke, D.W. (1993). Embryogenesis in angiosperms - development of the suspensor. *The Plant Cell* **5**, 1371–1381.

Yoshida, H., Hirano, K., Sato, T. *et al.* (2014). DELLA protein functions as a transcriptional activator through the DNA binding of the INDETERMINATE DOMAIN family proteins. *Proceedings of the National Academy of Sciences USA* **111**, 7861–7866.

Zhang, Y., Zhang, B., Yan, D. *et al.* (2011a). Two *Arabidopsis* cytochrome P450 monooxygenases, CYP714A1 and CYP714A2, function redundantly in plant development through gibberellin deactivation. *The Plant Journal* **67**, 342–353.

Zhang, Z.-L., Ogawa, M., Fleet, C.M. *et al.* (2011b). SCARECROW-LIKE 3 promotes gibberellin signaling by antagonizing master growth repressor DELLA in *Arabidopsis*. *Proceedings of the National Academy of Sciences USA* **108**, 2160–2165.

Zhao, X., Yu, X., Foo, E. *et al.* (2007). A study of gibberellin homeostasis and cryptochrome-mediated blue light inhibition of hypocotyl elongation. *Plant Physiology* **145**, 106–118.

Zhou, K., Xu, M., Tiernan, M. *et al.* (2012). Functional characterization of wheat *ent*-kaurene(-like) synthases indicates continuing evolution of labdane-related diterpenoid metabolism in the cereals. *Phytochemistry* **84**, 47–55.

Zhu, X.F., Suzuki, K., Saito, T. *et al.* (1997). Geranylgeranyl pyrophosphate synthase encoded by the newly isolated gene GGPS6 from *Arabidopsis thaliana* is localized in mitochondria. *Plant Molecular Biology* **35**, 331–341.

Zhu, Y.Y., Nomura, T., Xu, Y.H. *et al.* (2006). *ELONGATED UPPERMOST INTERNODE* encodes a cytochrome P450 monooxygenase that epoxidizes gibberellins in a novel deactivation reaction in rice. *The Plant Cell* **18**, 442–456.

Annual Plant Reviews (2016) **49**, 73–94
doi: 10.1002/9781119210436.ch3
http://onlinelibrary.wiley.com

Chapter 3

INACTIVATION PROCESSES

Hiroshi Magome[1] and Yuji Kamiya

RIKEN Center for Sustainable Resources, Japan
[1]Current address: Japan Tobacco Inc., Leaf Tobacco Research Center, Japan

Abstract: Gibberellins (GAs) play a role in many aspects of growth and development in plants. Endogenous levels of bioactive GAs are controlled by both biosynthetic and inactivation processes. There is accumulating evidence that inactivation is important for control of endogenous GA levels as well as its biosynthesis. Recent studies demonstrate that there are multiple inactivation mechanisms, and some of them are controlled by developmental and environmental cues, the latter including factors, such as light, temperature and even various stresses. In this chapter we describe our current understanding of GA inactivation and its control mechanisms in flowering plants through the action of various enzymes, including GA 13-oxidase, which is responsible for the biosynthesis of the bioactive GA_1. Recent findings following the isolation of GA 13-oxidase genes suggest that their function is to partially decrease GA bioactivity.

Keywords: Inactivation, 2-oxoglutarate-dependent dioxygenase, methyltransferase, cytochrome P450 monooxygenase, glycosylation

3.1 Introduction

The plant hormones gibberellins (GAs) are growth promoters involved in a wide range of processes from seed germination to flowering and seed production. Of the more than 130 known GA compounds in nature, GA_1, GA_3, GA_4 and GA_7 are the major bioactive GAs in flowering plants. GAs are synthesised from geranylgeranyl diphosphate via a multi-step process (Figure 3.1; the GA biosynthesis pathway is described in detail in Chapter 2). Because GA_1 has been identified in a variety of plant species such as rice (*Orysa sativa*), maize (*Zea mays*) and pea (*Pisum sativum*), this compound is thought to be the most common bioactive form. Meanwhile, GA_4 is also found in most species along with GA_1, and is the major bioactive form in *Arabidopsis*

Annual Plant Reviews, Volume 49: The Gibberellins, First Edition.
Edited by Peter Hedden and Stephen G. Thomas.

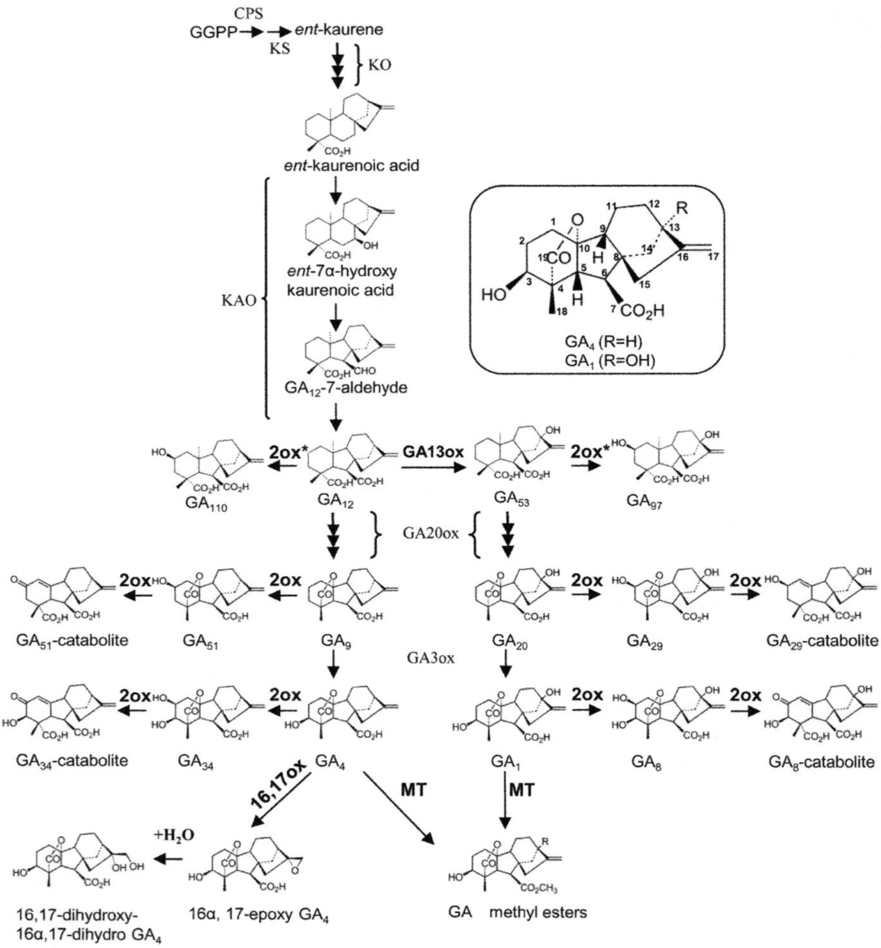

Figure 3.1 Gibberellin biosynthesis and inactivation pathways in flowering plants. Three classes of GA inactivation pathways catalysed by GA 2-oxidase (2ox, C_{19}-GA 2-oxidase; 2ox*, C_{20}-GA 2-oxidase), GA methyltransferase (MT) and GA 16,17-oxidase (16,17ox) are shown. Note that MTs catalyse the methylation of both 13-hydroxy and non-13-hydroxy GA precursors and bioactive GAs into GA-methyl esters; 16,17ox catalyses 16,17-oxidation of non-13-hydroxy GAs, including GA_4. Black solid arrows indicate GA biosynthesis pathway. GGDP, geranylgeranyl diphosphate; CPS, *ent*-copalyl diphosphate synthase; KS, *ent*-kaurene synthase; KO, *ent*-kaurene oxidase; KAO, *ent*-kaurenoic acid oxidase; GA20ox, GA 20-oxidase; GA3ox, GA 3-oxidase; GA13ox, GA 13-oxidase.

thaliana (*Arabidopsis*) and some Cucurbitaceae species. It is necessary to regulate GA levels precisely in plant tissues in order to achieve proper growth under changing developmental and environmental conditions; not only GA biosynthesis, but also GA inactivation is important for this regulation. GA inactivation occurs by the modification of precursors or bioactive GAs to lower the binding affinity of resultant end products to the receptor GIBERELLIN INSENSITIVE 1 (GID1). Recent studies revealed that there are multiple GA inactivation pathways. In this chapter, we focus on our current knowledge of these pathways, their regulations and physiological roles.

3.2 Gibberellin inactivation

Recent molecular genetic and biochemical studies have contributed toward the identification of GA inactivation genes. So far three classes of GA inactivation enzymes have been identified: GA 2-oxidase, GA methyltransferase and GA 16,17-oxidase. In addition to these, we also describe GA 13-/12α-oxidase in this section because recent findings suggest that these enzymes have the potential to decrease GA activity. Table 3.1 summarises genes encoding GA inactivation and potential GA inactivation enzymes in *Arabidopsis* and rice.

3.2.1 Gibberellin 2-oxidase

GA 2-oxidase (GA2ox), which encodes a soluble 2-oxoglutarate-dependent dioxygenase (2-ODD), converts the bioactive and intermediate forms of GAs to inactive forms by 2β-hydroxylation (Figure 3.1). The GA2ox was the first recognised inactivation enzyme, and has been researched most extensively. Given the presence of orthologous *GA2ox* genes in a number of plant species, such as pea (Lester *et al.*, 1999; Martin *et al.*, 1999), *Arabidopsis* (Thomas *et al.*, 1999), rice (Sakamoto *et al.*, 2001b) and hybrid poplar (*Populus tremula* × *Populus alba*) (Busov *et al.*, 2003), this inactivation is most common in flowering plants. The GA2oxs can be divided into two groups, C_{19}-GA2oxs and C_{20}-GA2oxs on the basis of substrate specificities. The C_{19}-GA2oxs use bioactive C_{19}-GAs (GA_1 and GA_4) and their immediate precursors (GA_9 and GA_{20}) as substrates (Figure 3.1; Lester *et al.*, 1999; Thomas *et al.*, 1999), and they are further divided into class I or II on the basis of amino acid sequences. Some C_{19}-GA2oxs are multi-functional enzymes that are capable of oxidising the 2β-hydroxy group further to the ketone, with the resulting production of GA-catabolites (Figure 3.1; Thomas *et al.*, 1999). The 2-keto GA-catabolites are dicarboxylic acids in which the lactone has opened, although this is thought to be an artifact during analysis (Thomas *et al.*, 1999). It has been reported that the multi-functional activity might be a characteristic of class I (Serrani *et al.*, 2007). However, *CsGA2ox2*, a cucumber GA2ox that belongs to class II has been shown to be multi-functional, indicating that functional differences

Table 3.1 Genes encoding GA inactivation enzyme or GA 13-oxidase in *Arabidopsis* and rice.

Enzyme	Type	Name	Locus (AGI / RAP)	Reference
Arabidopsis				
GA 2-oxidase (I)	2-ODD	*AtGA2ox1*	At1g78440	Thomas *et al.*, 1999
GA 2-oxidase (I)	2-ODD	*AtGA2ox2*	At1g30040	Thomas *et al.*, 1999
GA 2-oxidase (I)	2-ODD	*AtGA2ox3*	At2g34555	Thomas *et al.*, 1999
GA 2-oxidase (II)	2-ODD	*AtGA2ox4*	At1g47990	Hedden and Phillips, 2000
GA 2-oxidase (II)	2-ODD	*AtGA2ox6*	At1g02400	Hedden and Phillips, 2000
GA 2-oxidase (III)	2-ODD	*AtGA2ox7*	At1g50960	Schomberg *et al.*, 2003
GA 2-oxidase (III)	2-ODD	*AtGA2ox8*	At4g21200	Schomberg *et al.*, 2003
GA methyltransferase	Methyltransferase	*GAMT1*	At4g26420	Varbanova *et al.*, 2007
GA methyltransferase	Methyltransferase	*GAMT2*	At5g56300	Varbanova *et al.*, 2007
GA 16,17-oxidase	Cytochrome P450	*CYP714A1*	At5g24910	Zhang *et al.*, 2011; Nomura *et al.*, 2013
GA 13-/12α-oxidase	Cytochrome P450	*CYP714A2*	At5g24900	Zhang *et al.*, 2011; Nomura *et al.*, 2013
Rice				
GA 2-oxidase (I)	2-ODD	*OsGA2ox3*	Os01g0757200	Sakai *et al.*, 2003
GA 2-oxidase (I)	2-ODD	*OsGA2ox4*	Os05g0514600	Sakamoto *et al.*, 2004
GA 2-oxidase (I)	2-ODD	*OsGA2ox7*	Os01g0209700	Lo *et al.*, 2008
GA 2-oxidase (I)	2-ODD	*OsGA2ox8*	Os05g0560900	Lo *et al.*, 2008
GA 2-oxidase (I)	2-ODD	*OsGA2ox10*	Os05g0208550	Lo *et al.*, 2008
GA 2-oxidase (II)	2-ODD	*OsGA2ox1*	Os05g0158600	Sakamoto *et al.*, 2001
GA 2-oxidase (II)	2-ODD	*OsGA2ox2*	Os01g0332300	Sakai *et al.*, 2003
GA 2-oxidase (III)	2-ODD	*OsGA2ox5*	Os07g0103500	Lee and Zeevaart, 2005; Lo *et al.*, 2008
GA 2-oxidase (III)	2-ODD	*OsGA2ox6*	Os04g0522500	Lee and Zeevaart, 2005; Lo *et al.*, 2008
GA 2-oxidase (III)	2-ODD	*OsGA2ox9*	Os02g0630300	Lo *et al.*, 2008
GA 13-oxidase	Cytochrome P450	*CYP714B1*	Os07g0681300	Magome *et al.*, 2013
GA 13-oxidase	Cytochrome P450	*CYP714B2*	Os03g0332000	Magome *et al.*, 2013
GA 16,17-oxidase	Cytochrome P450	*CYP714D1*	Os05g0482400	Zhu *et al.*, 2006

between classes I and II are unclear (Lange *et al.*, 2013). Another group, the C_{20}-GA2oxs that catalyse the conversion of C_{20}-GAs, GA_{12} and GA_{53} to GA_{110} and GA_{97}, respectively, is categorised into class III on the basis of amino acid sequences (Figure 3.1; Schomburg *et al.*, 2003; Lee and Zeevaart, 2005; Lo *et al.*, 2008). However, there is also an exception with the cucumber CsGA2ox4, which is phylogenetically close to class I, but has both C_{19}- and C_{20}- GA2ox activity (Lange *et al.*, 2013). Similar to the GA20ox and GA3ox, GA2ox proteins are encoded by small gene families. Thus far, seven GA2ox (candidate) genes have been reported in *Arabidopsis* and 10 genes in rice (Table 3.1).

In spite of its major role in GA inactivation process, only a small number of loss of function phenotypes of *ga2ox* mutants are known. Presumably, this is because of the high redundancy of *GA2ox* genes and homeostatic regulation of GA biosynthesis genes (see below) that help to compensate for the effect of the mutation (Schomburg *et al.*, 2003; Rieu *et al.*, 2008a). An exception is the pea *slender* mutant, which has a defective *PsGA2ox1/SLENDER (SLN)* gene. The *sln* mutation leads to an accumulation of GA_{20} in mature seeds, resulting in the hyper-accumulation of bioactive GA_1 by 3β-hydroxylation during germination, and a lengthened seedling phenotype (Reid *et al.*, 1992; Ross *et al.*, 1995; Lester *et al.*, 1999; Martin *et al.*, 1999). The loss of function *atga2ox8* and *atga2ox7 atga2ox8* mutants show early flowering in short days, are resistance to the GA biosynthesis inhibitor ancymidol on germination, and have elongated hypocotyls (Schomburg *et al.*, 2003). The *atga2ox* quintuple mutant (disruption of five *C_{19}-GA2oxs*: *AtGA2ox1, 2, 3, 4* and *6*) shows many GA over-dose phenotypes, including an elongated stem, early flowering and reduced fertility, which is most likely due to the unequal elongation of the pistil and stamens (Rieu *et al.*, 2008a).

3.2.2 Gibberellin methyltransferase

The GA methyltransferases GAMT1 and GAMT2, members of the SABATH methyltransferase family, have been identified in *Arabidopsis* (Varbanova *et al.*, 2007; Xing *et al.*, 2007). Both enzymes catalyse methylation of the C-6 carboxy group of GAs using S-adenosyl-L-methionine (SAM) as a methyl donor. *In vitro*, GAMT1 and GAMT2 use a wide range of GAs, including precursor and bioactive GAs, as substrates. Over-expression of *GAMTs* in *Arabidopsis* resulted in dwarfism with reduced levels of bioactive GAs. These genes are mainly expressed in developing and germinating seeds. The *gamt1* and *gamt2* double mutant exhibited elevated levels of bioactive GAs in siliques, and showed resistance to a GA biosynthesis inhibitor, ancymidol, during germination. Since GA methyltransferase has been reported only in *Arabidopsis*, it is unclear how common methylation of GA is in other plant species.

3.2.3 Gibberellin 16,17-oxidase

The ELONGATED UPPERMOST INTERNODE (EUI) / CYP714D1 was first identified as a cytochrome P450 mono-oxygenase-type GA inactivation enzyme in rice. This enzyme catalyses 16α, 17-epoxidation of non-13-hydroxy GAs (GA_{12}, GA_9 and GA_4) *in vitro* (Zhu *et al.*, 2006) (Figures 3.1 and 3.2). The recessive *eui* mutant plants are morphologically normal until the heading stage when there is abnormally rapid elongation of the final internode. GA analysis showed that exceptionally large amounts of bioactive GAs, GA_4 and GA_1, are accumulated in the uppermost internode of the *eui* mutant (Zhu *et al.*, 2006). In contrast, over-expression of the *EUI* gene in rice resulted in severe dwarfism due to GA deficiency (Luo *et al.*, 2006; Zhu *et al.*, 2006). Accumulation of 16α,17-dihydrodiols, but not epoxides, are detected in the rice over-expressor plants, suggesting that the epoxides are hydrated either *in vivo* or during extraction and purification (Figure 3.1). *Arabidopsis* has two CYP714 members, one of which, *CYP714A1*, produced a severe dwarf phenotype when over-expressed in *Arabidopsis* (Zhang *et al.*, 2011; Nomura *et al.*, 2013). Functional analysis of CYP714A1 showed that this enzyme catalyses the conversion of GA_{12} to 16-carboxylated GA_{12} (Figure 3.2; Nomura *et al.*, 2013), probably produced via a multi-step oxidation. Moreover, like CYP714D1, CYP714A1 showed a trend of substrate preference for non-13-hydroxy GAs rather than 13-hydroxy GAs (Nomura *et al.*, 2013). GA 16,17-dihydrodiols have been found in other plant species, such as pea fruits (Santes *et al.*, 1995), apple (*Malus domestica*) developing seeds (Hedden *et al.*, 1993), sporophytes of a tree fern *Cybotium glaucum* (Yamane *et al.*, 1988) and capsules of *Populus trichocarpa* (Pearce *et al.*, 2002). These results support the notion that C-16,17 oxidation of GAs is a general inactivation mechanism in many plant species.

3.2.4 Gibberellin 13-oxidase and 12α-oxidase

GA 13-oxidase is a key enzyme of the early GA 13-hydroxylation pathway, and its activity determines the ratio of GA_4 and GA_1 (Figure 3.1): GA_1 is synthesised through the early GA 13-hydroxylation pathway, while GA_4 is synthesised via the non-13-hydroxylation pathway. The occurrence of GA 13-hydroxylation activity was first found in a microsomal fraction from pea embryos (Ropers *et al.*, 1978; Kamiya and Graebe, 1983). A recent work demonstrated that rice *CYP714B1* and *CYP714B2*, which are in the same family as *EUI/CYP714D1*, encode GA 13-oxidase (GA13ox), catalysing the conversion of GA_{12} to GA_{53} *in vitro* (Figures 3.1 and 3.2; Magome *et al.*, 2013). The *cyp714b1 cyp714b2* double mutants showed decreased levels of 13-hydroxy GAs, including GA_1, but increased levels of non-13-hydroxy GAs, including GA_4. Like *eui* plants, the double mutant plants had elongated uppermost internodes at the heading stage (Magome *et al.*, 2013). In contrast, *Arabidopsis* plants over-expressing *CYP714B* gene show semi-dwarfism

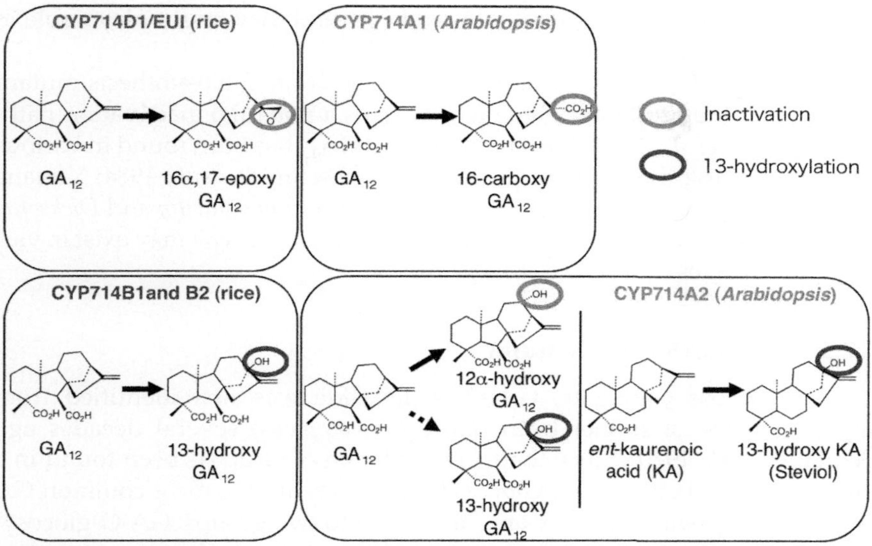

Figure 3.2 Function of CYP714 members of *Arabidopsis* and rice. Black arrows indicate strong (solid) or weak (dotted) enzyme activities of CYP714 members.

probably due to enhanced early GA 13-hydroxylation pathway. These results are likely because the bioactivity of GA_4 (13-H GA) is stronger than that of GA_1 (13-OH GA) in *Arabidopsis* (Talon *et al.*, 1990; Yang *et al.*, 1995) and because the selectivity of GID1 showed that GA_4 is the most effective form in rice (Ueguchi-Tanaka *et al.*, 2007) As described above, 13-OH GAs, including GA_1, are not substrates of EUI/CYP714D1 (Zhu *et al.*, 2006). These results suggest that the role of GA 13-hydroxylation is to synthesise less active, but EUI/CYP714D1-resistant GAs. Thus a combination of strong (GA_4) and weak (GA_1) GAs might allow plants to achieve fine and flexible tuning of GA-dependent growth (Magome *et al.*, 2013). In *Arabidopsis*, functional analysis showed that the CYP714A2 protein catalysed the conversion of *ent*-kaurenoic acid (which is a more upstream precursor than GA_{12}) to 13-hydroxy *ent*-kaurenoic acid (steviol); when GA_{12} was a substrate for CYP714A2, 12α-hydroxy GA_{12} (GA_{111}) was produced as a major product and 13-OH GA_{12} (GA_{53}) as a minor product (Figure 3.2; Nomura *et al.*, 2013). An *ent*-kaurenoic acid 13-oxidase (CYP716D) that belongs to another P450 family was found in stevia (*Stevia rebaudiana*) (Brandle and Telmer, 2007). In addition, 2-ODD type 13-hydroxylase has been found in a soluble fraction of spinach (*Spinacia oleracea*) leaves (Gilmour *et al.*, 1986). It was also reported that TaGA3ox2 protein (2-ODD of wheat, *Triticum aestivum*) and MmGA3ox2 protein (2-ODD of the southern wild cucumber, *Marah macrocarpus*) exhibited some GA 13-oxidase activity, besides GA 3-oxidase activity *in vitro* (Appleford *et al.*, 2006; Ward *et al.*, 2010). These results suggest that there are other classes of GA 13-oxidases, and they might play a role in plant development.

The biological activity study using *Arabidopsis ga1-3* biosynthesis mutant showed that the 12α-hydroxylation of GA_{12} is a potential inactivation pathway (Nomura *et al.*, 2013). 12α-Hydroxylated GA_{12} has been found in various plants, including the seeds of pumpkin (Blechschmidt *et al.*, 1984; Yamane *et al.*, 1988) and sporophytes of the tree ferns *Cybotium glaucum* and *Dicksonia antarctica* (Yamane *et al.*, 1988), suggesting that this pathway may exist in various plant species.

3.2.5 Conjugation with sugar

The GA-glucose conjugate, GA_8-2-O-glucoside was first identified from maturing fruits of runner bean (*Phaseolus coccineus*) several decades ago (Schreiber *et al.*, 1970). Since then, glycosylation of GAs has been found in a wide range of plant species. Glucosyl conjugates are the most common GA conjugates in plants, and they are classified into two groups: GA-O-glucosyl ethers and GA-glucosyl esters. In GA-O-glucosyl ethers, the glucose moiety can be linked to the hydroxy group at C-2, C-3, C-11, C-13 or C-17, while in GA-glucosyl esters the glucose moiety is linked to the 6-carboxy group (Schneider and Schliemann, 1994). It remains unclear what the physiological role of GA conjugations is, because no GA-glycosyl transferase gene has yet been identified. Conjugation with glucose might be an inactivation reaction. Moreover, given their polar character and their reversible formation (Schneider *et al.*, 1992), the glucosyl conjugates may play a role in the transport or subcellular compartmentalisation for storage (Schneider and Schliemann, 1994). A recent paper showed that expression of a fungal β-glucosidase in chloroplast of transplastomic tobacco plants resulted in elevated levels of phytohormones including GA_4 and GA_1 with an increase of biomass (Jin *et al.*, 2011). This suggests that certain quantities of hormones that can be released by β-glucosidase from their conjugates are stored in chloroplasts.

3.3 Regulation of gibberellin inactivation

In order to understand the whole picture of the GA inactivation process, it is essential to clarify where and how this metabolism occurs. As described above, there are multiple classes of GA inactivation enzymes, and some of the classes consist of multiple isozymes. Gene expression atlas data show that the genes encoding these GA inactivation enzymes are differentially expressed in a variety of tissues, but some of them have overlapping expression patterns, indicating the complexity of this regulation. Generally, high levels of bioactive GAs can be observed in actively growing tissues in which not only GA biosynthesis, but also inactivation genes show high expression, indicating that both GA biosynthesis and inactivation occur at or close to the site of action, and this combination is likely important for the control of bioactive GA levels. Indeed, recent high resolution hormone analysis shows that a narrow peak of bioactive GAs in the maize leaf might be developmentally controlled by the interplay between GA biosynthesis and inactivation (Nelissen *et al.*, 2012). Accumulating evidence has shown also that GA inactivation genes are regulated by hormonal and environmental cues. Several transcription factors directly or indirectly involved in GA inactivation have also been identified.

3.3.1 Developmental regulation

Shoot apical meristem

The shoot apical meristem (SAM) provides the structural architecture of above-ground organs. It has been reported in several plant species that the KNOTTED-like homoeobox protein (KNOX) transcription factor protein is involved in the reduction of GA levels in the SAM. Thus maintaining low GA levels is probably important for SAM function (Tanaka-Ueguchi *et al.*, 1998; Hay *et al.*, 2002). In *Arabidopsis*, a KNOX protein STM (SHOOTMERISTEMLESS) up-regulates the expression of *AtGA2ox2* and *AtGA2ox4* at the base of the SAM (Figure 3.3; Jasinski *et al.*, 2005). This might in part be mediated by

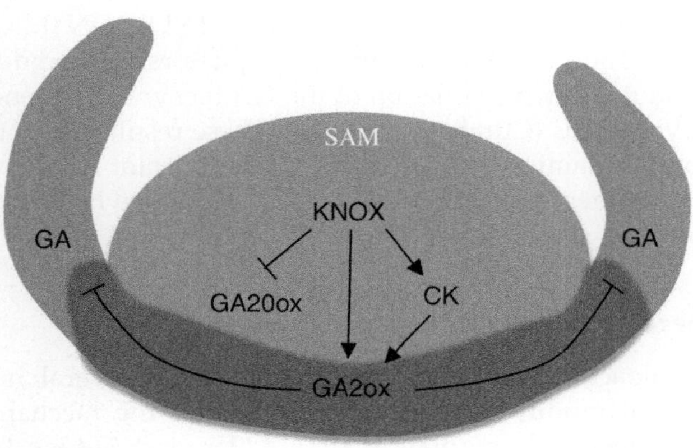

Figure 3.3 Involvement of KNOX transcription factors for reducing GA levels in the SAM. Summary of the role of KNOX proteins derived from the analysis of several plant species is shown. KNOX proteins suppress the expression of the GA biosynthesis gene, *GA20ox*. KNOX proteins also up-regulate the expression of *GA2ox* at the base of the SAM. This is in part likely to be mediated by activation of the cytokinin (CK) pathway. Maize KNOX proteins directly up-regulate *GA2ox* expression. The expression of *GA2ox* at the base of the SAM is thought to contribute to the low GA levels by preventing GA synthesised in the developing leaves from entering the SAM. (See insert for colour representation of this figure.)

cytokinin (CK), as STM also up-regulates CK levels through the induction of the *IPT* (isopentenyltransferase) genes, and the resultant CK up-regulates *AtGA2ox2* expression. The expression pattern of *AtGA2ox2* and *AtGA2ox4* is similar to that of *OsGA2ox1* in the SAM of rice (Sakamoto *et al.*, 2001b). On the other hand, Bolduc and Hake (2009) reported that maize KNOX protein directly up-regulated the *ZmGA2ox1* gene by binding its intron sequence. In addition, the tobacco KNOX protein NTH15 suppresses the expression of a *GA20ox* gene through binding to a sequence within its first intron (Sakamoto *et al.*, 2001a). These results indicate that the low GA levels in the SAM brought about by regulation of GA biosynthesis and inactivation is a common mechanism in plant species.

Seed development

It is widely known that high levels of various GAs, which includes inactive and intermediate as well as bioactive forms, are accumulated in developing seeds of a variety of species. GAs are required for embryo growth and seed development (Swain *et al.*, 1997) and GA biosynthesis is highly active in developing seeds (Kanno *et al.*, 2010). Meanwhile, a number of GA inactivation genes, *AtGA2oxs* (1, 2, 3, 4 and 6), *GAMTs* (1 and 2) and *CYP714A1* are also highly expressed in developmental seeds and siliques (Varbanova *et al.*, 2007; Rieu *et al.*, 2008a; Nomura *et al.*, 2013). Among these genes, *AtGA2ox6* has been reported to be directly regulated by a MADS [MCM1 (MINICHROMOSOME MAINTENANCE 1), AGAMOUS, DEFICIENS, and the SRF (serum-response factor)]-box transcription factor, AGAMOUS-Like 15 (Wang *et al.*, 2004). In addition, the LEAFY COTYLEDON (LEC) transcription factors LEC2 and FUS3, which are key regulators for seed development, are involved in the down-regulation of the GA biosynthesis gene *AtGA3ox2* during embryogenesis (Curaba *et al.*, 2004). These results indicate that strict spatiotemporal dynamics of bioactive GAs is important for embryo growth and seed development, and that dynamics are regulated by GA biosynthesis and inactivation.

3.3.2 Gibberellin homoeostasis

Endogenous bioactive GA levels are regulated by several homoeostatic mechanisms that modify GA metabolism. One of the mechanisms is the transcriptional feedback regulation of *GA20ox* and *GA3ox* biosynthesis genes. In *Arabidopsis*, for example, transcript levels of *AtGA20ox1, 2* and *3* and *AtGA3ox1* are increased when endogenous GA levels are low, whereas they are decreased when bioactive GA was applied exogenously (Chiang *et al.*, 1995; Phillips *et al.*, 1995; Rieu *et al.* 2008b). GA inactivation is also subject to homoeostatic regulation; transcript levels of several *GA2ox* genes have been shown to be controlled by feed-forward regulation (Thomas *et al.*, 1999; Elliott *et al.*, 2001). In *Arabidopsis*, five out of seven *GA2ox* genes (C_{19}-GA2ox: *AtGA2ox1, 2, 4* and *6*; C_{20}-GA2ox: *AtGA2ox8*) are transcriptionally up-regulated by exogenously applied bioactive GA (Rieu *et al.*, 2008a). DELLA proteins play a pivotal role in the negative regulation of GA signalling as transcriptional regulators (Peng *et al.*, 1997; Silverstone *et al.*, 1998; Sun, 2011), and also play a central role in homoeostatic regulation of GA metabolism. Zentella *et al.* (2007) showed that *AtGA20ox2* and *AtGA3ox1* are probably direct DELLA targets. In pea, a study of the loss-of-function DELLA mutants revealed that the feed-forward regulation of *PsGA2ox1* and *PsGA2ox2* is controlled by two DELLAs, LA and CRY (Weston *et al.*, 2008). It has been reported that *EUI/CYP714D1*, the GA 16,17-oxidase gene is also controlled by feedback regulation in rice seedlings (Zhang *et al.*, 2008; Magome *et al.*, 2013). Homoeostatic regulation is also observed for the GA13-oxidase genes, *CYP714B1* and *CYP714B2*, in rice seedlings, suggesting that GA signals are regulated not only through the quantity, but also the nature of GAs (Magome *et al.*, 2013).

3.3.3 Regulation by other hormones

Plant hormone interactions occur in many aspects of plant growth and development. It has been known that endogenous GA levels are affected by other phytohormones, some of which have been attributed to changing GA inactivation status. The regulation of endogenous GA levels by auxin has been attributed to changes in the expression level of the GA biosynthesis and/or inactivation enzymes. Decapitation and indole-3-acetic acid (IAA) treatment studies in pea demonstrated that IAA derived from the apical bud enhances the levels of *PsGA3ox1* transcript and suppresses those of *PsGA2ox1* transcript, resulting in the promotion of GA biosynthesis in the stem (O'Neill and Ross, 2002). Similarly, a study on barley demonstrated that IAA from the inflorescence plays a role in stem elongation by the up-regulation of GA 3β-hydroxylation and down regulation of GA 2β-hydroxylation (Wolbang *et al.*, 2004). In *Arabidopsis*, it has been shown that exogenous application of auxin up-regulates several GA metabolism genes including *AtGA20oxs* and *AtGA2oxs* (Frigerio *et al.*, 2006). Analyses of auxin-signalling mutants revealed that these expression changes are regulated by the Aux/IAA and ARF (pivotal components of auxin signalling) pathway.

It has long been known that GA and ABA (abscisic acid) play antagonistic roles, with GA acting as a stimulant in seed germination and ABA as an inhibitor. Two reports indicate the involvement of ABA in GA inactivation process: seeds of the ABA-deficient mutant *aba2-2* were shown to contain increased levels of *AtGA3oxs* and *AtGA20oxs* transcripts and decreased levels of *AtGA2ox6*, suggesting that ABA suppresses GA biosynthesis gene expression and promotes GA inactivation during seed maturation (Seo *et al.*, 2006). Another study reported that exogenously applied ABA up-regulated transcript levels of *EUI/CYP714D1* in rice (Yaish *et al.*, 2010). However, the molecular mechanisms for the regulation of these GA inactivation enzyme genes by ABA remain to be clarified.

3.3.4 Environmental regulation

Regulation by light

Plants perceive the quality and quantity of light, as well as day length, and optimise their photosynthesis, growth and development accordingly. In seed germination of *Arabidopsis*, phytochromes mediate light inhibition of ABA biosynthesis and the activation of GA biosynthesis via transcriptional regulation of ABA and GA metabolic genes (Ogawa *et al.*, 2003; Seo *et al.* 2006). Two factors involved in this process have been identified; PIF1 (phytochrome interacting factor 1)/PIL5 (PIF3-LIKE5), encoding a basic helix-loop-helix (bHLH) transcription factor, inhibits seed germination by decreasing the levels of bioactive GA in the absence of light through repressing *AtGA3ox1* and *2* expression and promoting that of *AtGA2ox2*. In the light this transcription factor is degraded by the 26S proteasome in a phytochrome-mediated process (Oh *et al.*, 2006). Besides this, PIF1/PIL5 directly up-regulates DELLAs (GAI and RGA) and also indirectly modulates ABA metabolism (Oh *et al.*, 2007). A follow-up paper showed that SOM (SOMNUS), a C3H-type zinc-finger protein, is a key negative regulator of seed germination that acts downstream of PIF1/PIL5 by repressing the expression of *AtGA3ox1* and *2* and promoting the expression of *AtGA2ox2* (Kim *et al.*, 2008). A rapid decrease in endogenous GA levels mediated by phytochrome and cryptochrome by suppression of biosynthesis and promotion of inactivation is implicated in the de-etiolation of dark-grown pea seedlings (Reid *et al.*, 2002; Foo *et al.*, 2006). Similar changes have been inferred to occur in *Arabidopsis* (Zhao *et al.*, 2007; Symons *et al.*, 2008). Analysis of the *long1* pea mutant revealed that the bZIP transcription factor LONG1 plays a role in up-regulation of *PsGA2ox2* expression and the rapid drop of GA_1 levels on pea de-etiolation (Weller *et al.*, 2009). LONG1 is an orthologue of *Arabidopsis* HY5 (ELONGATED HYPOCOTYL 5), whose stability is likely regulated by the ubiquitin ligase COP1 (CONSTITUTIVELY PHOTOMORPHOGENIC 1) orthologue LIP1 (LIGHT-INDEPENDENT PHOTOMORPHOGENESIS 1), as is the case with HY5 and COP1 in *Arabidopsis* (Ang *et al.*, 1998).

Responses to photoperiodic signals are partly mediated by GA signalling (reviewed in Jackson, 2009). In rice, a short-day (SD) plant, *OsGA2ox1* is highly expressed below the shoot apical meristem, and the expression is greatly decreased after the phase transition from vegetative to reproductive growth (Sakamoto *et al.*, 2001b). In the grass *Lolium temulentum*, a long-day (LD) plant, it is suggested that GA_5, a bioactive GA is produced in the leaf under LD conditions, and is transported to the shoot apex for flower induction (King *et al.*, 2006). Just like in rice, *GA2ox* genes are expressed below the shoot apex in this grass species as well (King *et al.*, 2008). These GA2ox enzymes likely inactivate GA_1 and GA_4 transported from leaves to the shoot apex. However, GA_5 is resistant to inactivation by GA2ox due to the presence of a C-2,3 double bond, and is therefore transported to the SAM where it promotes flowering. Potato (*Solanum tuberosum*) tuberisation is generally promoted by shifting LD to SD. Analyses of *StGA2ox1*, an SD-inducible gene in the stolon, and transgenic plants in which the gene is over-expressed or silenced revealed that this inactivation gene is up-regulated prior to tuber development within the sub-apical stolon region, and likely facilitates tuber development and growth (Kloosterman *et al.*, 2007).

Regulation by temperature

In addition to light signals as described above, it has been widely known that exposure of seeds to low temperatures (typically 2–5 °C) immediately after imbibition (stratification) promotes germination in many plant species. Microarray experiments with *Arabidopsis* seeds revealed that cold treatment up-regulates the expression of *AtGA3ox1* and *AtGA20ox2*, whereas it down-regulates *AtGA2ox2* expression (Yamauchi *et al.*, 2004; Yamauchi *et al.*, 2007). During seed maturation, low temperature conditions result in increased dormancy of mature *Arabidopsis* seeds (Donohue *et al.*, 2008). It was reported that increased expression of *AtGA2ox6* mediated by DELAY OF GERMINATION1 (DOG1) protein is involved in this process (Kendall *et al.*, 2011). Intriguingly, CBF (C-REPEAT-BINDING FACTOR)/DREB (DEHYDRATION RESPONSE ELEMENT BINDING FACTOR) proteins, which are known as key AP2 (APETALA2)/ERF (ETHYLENE RESPONSE FACTOR) transcription factors for cold or dehydration stress tolerance (see below), are necessary for the *AtGA2ox6* and *DOG1* expression, although *CBF/DREBs* are not likely transcriptionally regulated in seeds, unlike the case of their stress response at the seedling stage. Stavang *et al.* (2009) reported that the promotion of hypocotyl growth by increasing ambient temperature requires modulation of the auxin, brassinosteroids and GA pathways with the down-regulation of *AtGA2ox1* and up-regulation of *AtGA20ox1* and *AtGA3ox1* expression. In pea, it has been shown that a moderate drop in ambient temperature reduces stem growth, which is associated with a reduction of GA_1 content via up-regulation of *PsGA2ox2* expression (Stavang *et al.*, 2005).

Abiotic stress regulation

In the past decade, involvement of GA in abiotic stresses including salinity, cold and drought have been acknowledged (reviewed in Colebrook *et al.*, 2014). Growth arrest is the primary response to stress in plants. Recent studies have demonstrated that the GA signalling repressor DELLAs and GA metabolism are regulated by various stresses, and the resultant growth arrest and physiological changes confer stress adaptation (Achard *et al.*, 2006; Achard *et al.*, 2007; Achard *et al.*, 2008b). In this section, we focus on the current knowledge about regulation of GA inactivation in response to abiotic stress. In *Arabidopsis*, high-salinity stress induces transcription of six *GA2ox* genes, among which *AtGA2ox7* was directly up-regulated by the salinity responsive CBF/DREB1 transcription factor, DDF1 (DWARF AND DELAYED FLOWERING1) (Figure 3.4; Magome *et al.*, 2008). Indeed, bioactive GA levels were decreased in response to high-salinity stress, leading

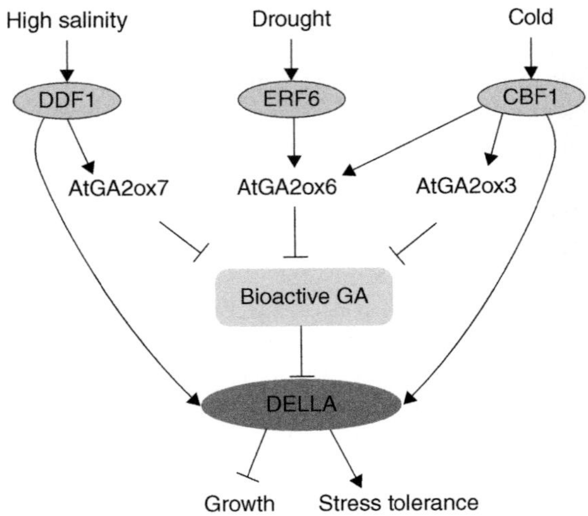

Figure 3.4 Stress-responsive AP2/ERF transcription factors up-regulate expression of GA 2-oxidase genes under individual abiotic stress. DDF1 (in response to high-salinity), CBF1/DREB1B (in response to cold) and ERF6 (in response to mild drought) up-regulate expression of *GA2ox* gene(s), and the reduction of bioactive GAs results in DELLA stabilisation. DELLA plays a role in growth arrest and stress tolerance. DDF1 and CBF1/DREB1B also up-regulate expression of the *RGL3* gene encoding a DELLA protein. Note that these AP2/ERF transcription factors also up-regulate transcription of a number of other stress-response genes that confer stress tolerance.

to the stabilisation of DELLA proteins, and the resultant accumulation of these proteins restrains growth (Achard *et al.*, 2006). Dwarf mutants defective in GA biosynthesis or signalling exhibit salt tolerance (Magome *et al.*, 2004, Achard *et al.*, 2006). Similarly, cold stress also induces expression of three *GA2ox* genes in *Arabidopsis*, among which *AtGA2ox3* and *AtGA2ox6* are indirectly up-regulated by the cold responsive transcription factor, CBF1/DREB1B (Figure 3.4; Achard *et al.*, 2008a). Moreover, these studies also showed that DDF1 and CBF1/DREB1B up-regulated the expression of the *RGL3* gene, which encodes a DELLA protein, suggesting another route to the accumulation of these proteins. Dubois *et al.* (2013) reported that expression of *AtGA2ox6* was up-regulated by ERF6, another class of AP2/ERF transcription factor, under mild drought stress (Figure 3.4). Touch also leads to morphological alterations such as growth retardation and delayed flowering, called thigmomorphogenesis. Recently it was shown that *AtGA2ox7* is a touch-responsive gene, and that its loss-of-function mutant plants did not respond to this stimulus (Lange and Lange, 2015). Thus, this gene is likely a key regulator of thigmomorphogenesis in *Arabidopsis*.

Biotic stress regulation

It has been widely known that hormones, such as jasmonic acid (JA), salicylic acid (SA) and ethylene, play a role in plant disease response. Recent studies revealed that other hormones including GA are also implicated in the response to this biotic stress (Grant and Jones, 2009). A study using the *Arabidopsis* quadruple-*della* loss-of-function and dominant *della* gain-of-function mutants indicated that DELLAs promote necrotrophic pathogen resistance via potentiating JA signalling, but increase susceptibility to biotrophic pathogens by repressing SA signalling (Navarro *et al.*, 2008). It has been shown that DELLAs up-regulate expression of genes encoding reactive oxygen species (ROS) detoxification enzymes thus decreasing ROS levels and promoting tolerance to abiotic and biotic stress (Achard *et al.*, 2008b). Yang *et al.* (2008) reported that transgenic plants over-expressing *EUI/CYP714D1* showed increased resistance, whereas their loss-of-function counterparts showed compromised resistance to bacterial blight (*Xanthomonas orysae* pv. *orysae*) and rice blast (*Magnaporthe orysae*), suggesting that a change in endogenous GA levels affects plant disease resistance. Interestingly, it has been shown that up-regulation of *OsGA2ox3* and down-regulation of *OsGA20ox3* expression (and thus accumulation of DELLA) via promotion of brassinosteroid signalling in the fungus *Pythium graminicola* attenuate resistance to this infection in rice (De Vleesschauwer *et al.*, 2012).

3.4 Concluding remarks

Studies over the past decade have revealed the complexity of GA inactivation; three classes of GA inactivation enzymes, GA 2-oxidase, GA methyltransferase and GA 16,17-oxidase with different properties are implicated (at least in *Arabidopsis*). However, to understand the whole process of GA inactivation, it would be necessary to identify GA-glycosyltransferases. The discovery of rice GA13-oxidase offers a glimpse of its physiological role, and identification of GA 13-oxidase of many other plant species will shed more light on the essential physiological function of this enzyme. We have highlighted transcriptional regulation of these GA inactivation enzymes in many aspects of the plant life cycle, but in many cases, the molecular mechanisms controlling these transcriptional changes are still largely unknown. Research on GA transport has come under the spotlight in the past several years (Chapter 4), and this is another major factor that controls hormone concentrations in certain tissues. Future integrated research, including biosynthesis, inactivation and transport will allow us to better understand how GA levels are regulated in plants.

References

Achard, P., Cheng, H., De Grauwe, L. *et al.* (2006). Integration of plant responses to environmentally activated phytohormonal signals. *Science* **311**, 91–94.

Achard, P., Baghour, M., Chapple, A. *et al.* (2007). The plant stress hormone ethylene controls floral transition via DELLA-dependent regulation of floral meristem-identity genes. *Proceedings of the National Academy of Sciences USA* **104**, 6484–6489.

Achard, P., Gong, F., Cheminant, S. *et al.* (2008a). The cold-inducible CBF1 factor-dependent signaling pathway modulates the accumulation of the growth-repressing DELLA proteins via its effect on gibberellin metabolism. *The Plant Cell* **20**, 2117–2129.

Achard, P., Renou, J.P., Berthome, R. *et al.* (2008b). Plant DELLAs restrain growth and promote survival of adversity by reducing the levels of reactive oxygen species. *Current Biology* **18**, 656–660.

Ang, L.H., Chattopadhyay, S., Wei, N. *et al.* (1998). Molecular interaction between COP1 and HY5 defines a regulatory switch for light control of *Arabidopsis* development. *Molecular Cell* **1**, 213–222.

Appleford, N.E., Evans, D.J., Lenton, J.R. *et al.* (2006). Function and transcript analysis of gibberellin-biosynthetic enzymes in wheat. *Planta* **223**, 568–582.

Blechschmidt, S., Castel, U., Gaskin, P. *et al.* (1984). GC/MS analysis of the plant hormones in seeds of *Cucurbita maxima*. *Phytochemistry* **23**, 553–558.

Bolduc, N. and Hake, S. (2009). The maize transcription factor KNOTTED1 directly regulates the gibberellin catabolism gene *ga2ox1*. *The Plant Cell* **21**, 1647–1658.

Brandle, J.E. and Telmer, P.G. (2007). Steviol glycoside biosynthesis. *Phytochemistry* **68**, 1855–1863.

Busov, V.B., Meilan, R., Pearce, D.W. *et al.* (2003). Activation tagging of a dominant gibberellin catabolism gene (*GA 2-oxidase*) from poplar that regulates tree stature. *Plant Physiology* **132**, 1283–1291.

Chiang, H.H., Hwang, I. and Goodman, H.M. (1995). Isolation of the *Arabidopsis GA4* locus. *The Plant Cell* **7**, 195–201.

Colebrook, E.H., Thomas, S.G., Phillips, A.L. and Hedden, P. (2014). The role of gibberellin signalling in plant responses to abiotic stress. *The Journal of Experimental Biology* **217**, 67–75.

Curaba, J., Moritz, T., Blervaque, R. *et al.* (2004). *AtGA3ox2*, a key gene responsible for bioactive gibberellin biosynthesis, is regulated during embryogenesis by LEAFY COTYLEDON2 and FUSCA3 in *Arabidopsis*. *Plant Physiology* **136**, 3660–3669.

De Vleesschauwer, D., Van Buyten, E., Satoh, K. *et al.* (2012). Brassinosteroids antagonize gibberellin- and salicylate-mediated root immunity in rice. *Plant Physiology* **158**, 1833–1846.

Donohue, K., Heschel, M.S., Butler, C.M. *et al.* (2008). Diversification of phytochrome contributions to germination as a function of seed–maturation environment. *New Phytologist* **177**, 367–379.

Dubois, M., Skirycz, A., Claeys, H. *et al.* (2013). ETHYLENE RESPONSE FACTOR6 acts as a central regulator of leaf growth under water-limiting conditions in *Arabidopsis*. *Plant Physiology* **162**, 319–332.

Elliott, R.C., Ross, J.J., Smith, J.L. *et al.* (2001). Feed-forward regulation of gibberellin deactivation in pea. *Journal of Plant Growth Regulation* **20**, 87–94.

Foo, E., Platten, J.D., Weller, J.L. and Reid, J.B. (2006). PhyA and cry1 act redundantly to regulate gibberellin levels during de-etiolation in blue light. *Physiologia Plantarum* **127**, 149–156.

Frigerio, M., Alabadi, D., Perez-Gomez, J. *et al.* (2006). Transcriptional regulation of gibberellin metabolism genes by auxin signaling in *Arabidopsis*. *Plant Physiology* **142**, 553–563.

Gilmour, S.J., Zeevaart, J.A.D., Schwenen, L. and Graebe, J.E. (1986). Gibberellin metabolism in cell-free-extracts from spinach leaves in relation to photoperiod. *Plant Physiology* **82**, 190–195.

Grant, M.R. and Jones, J.D. (2009). Hormone (dis)harmony moulds plant health and disease. *Science* **324**, 750–752.

Hay, A., Kaur, H., Phillips, A. *et al.* (2002). The gibberellin pathway mediates KNOTTED1-type homeobox function in plants with different body plans. *Current Biology* **12**, 1557–1565.

Hedden, P. and Phillips, A.L. (2000). Gibberellin metabolism: new insights revealed by the genes. *Trends in Plant Science* **5**, 523–530.

Hedden, P., Hoad, G.V., Gaskin, P. *et al.* (1993). Kaurenoids and gibberellins, including the newly characterized gibberellin A_{88}, in developing apple seeds. *Phytochemistry* **32**, 231–237.

Jackson, S.D. (2009). Plant responses to photoperiod. *New Phytologist* **181**, 517–531.

Jasinski, S., Piazza, P., Craft, J. *et al.* (2005). KNOX action in *Arabidopsis* is mediated by coordinate regulation of cytokinin and gibberellin activities. *Current Biology* **15**, 1560–1565.

Jin, S., Kanagaraj, A., Verma, D. *et al.* (2011). Release of hormones from conjugates: chloroplast expression of β-glucosidase results in elevated phytohormone levels associated with significant increase in biomass and protection from aphids or whiteflies conferred by sucrose esters. *Plant Physiology* **155**, 222–235.

Kamiya, Y. and Graebe, J.E. (1983). The biosynthesis of all major pea gibberellins in a cell-free system from *Pisum sativum*. *Phytochemistry* **22**, 681–689.

Kanno, Y., Jikumaru, Y., Hanada, A. *et al.* (2010). Comprehensive hormone profiling in developing *Arabidopsis* seeds: examination of the site of ABA biosynthesis, ABA transport and hormone interactions. *Plant and Cell Physiology* **51**, 1988–2001.

Kendall, S.L., Hellwege, A., Marriot, P. *et al.* (2011). Induction of dormancy in *Arabidopsis* summer annuals requires parallel regulation of DOG1 and hormone metabolism by low temperature and CBF transcription factors. *The Plant Cell* **23**, 2568–2580.

Kim, D.H., Yamaguchi, S., Lim, S. *et al.* (2008). SOMNUS, a CCCH-type zinc finger protein in *Arabidopsis*, negatively regulates light-dependent seed germination downstream of PIL5. *The Plant Cell* **20**, 1260–1277.

King, R.W., Moritz, T., Evans, L.T. *et al.* (2006). Regulation of flowering in the long-day grass *Lolium temulentum* by gibberellins and the FLOWERING LOCUS T gene. *Plant Physiology* **141**, 498–507.

King, R.W., Mander, L.N., Asp, T. *et al.* (2008). Selective deactivation of gibberellins below the shoot apex is critical to flowering but not to stem elongation of *Lolium*. *Molecular Plant* **1**, 295–307.

Kloosterman, B., Navarro, C., Bijsterbosch, G. *et al.* (2007). *StGA2ox1* is induced prior to stolon swelling and controls GA levels during potato tuber development. *The Plant Journal* **52**, 362–373.

Lange, M.J.P. and Lange, T. (2015). Touch-induced changes in *Arabidopsis* morphology dependent on gibberellin breakdown. *Nature Plants* **1**:14025.

Lange, M.J.P., Liebrandt, A., Arnold, L. *et al.* (2013). Functional characterization of gibberellin oxidases from cucumber, *Cucumis sativus* L. *Phytochemistry* **90**, 62–69.
Lee, D.J. and Zeevaart, J.A. (2005). Molecular cloning of *GA 2-oxidase3* from spinach and its ectopic expression in *Nicotiana sylvestris*. *Plant Physiology* **138**, 243–254.
Lester, D.R., Ross, J.J., Smith, J.J. *et al.* (1999). Gibberellin 2-oxidation and the *SLN* gene of *Pisum sativum*. *The Plant Journal* **19**, 65–73.
Lo, S.F., Yang, S.Y., Chen, K.T. *et al.* (2008). A novel class of gibberellin 2-oxidases control semidwarfism, tillering, and root development in rice. *The Plant Cell* **20**, 2603–2618.
Luo, A.D., Qian, Q., Yin, H.F. *et al.* (2006). *EUI1*, encoding a putative cytochrome P450 monooxygenase, regulates internode elongation by modulating gibberellin responses in rice. *Plant and Cell Physiology* **47**, 181–191.
Magome, H., Yamaguchi, S., Hanada, A. *et al.* (2004). *dwarf and delayed-flowering 1*, a novel *Arabidopsis* mutant deficient in gibberellin biosynthesis because of overexpression of a putative AP2 transcription factor. *The Plant Journal* **37**, 720–729.
Magome, H., Yamaguchi, S., Hanada, A. *et al.* (2008). The DDF1 transcriptional activator upregulates expression of a gibberellin-deactivating gene, *GA2ox7*, under high-salinity stress in *Arabidopsis*. *The Plant Journal* **56**, 613–626.
Magome, H., Nomura, T., Hanada, A. *et al.* (2013). *CYP714B1* and *CYP714B2* encode gibberellin 13-oxidases that reduce gibberellin activity in rice. *Proceedings of the National Academy of Sciences USA* **110**, 1947–1952.
Martin, D.N., Proebsting, W.M. and Hedden, P. (1999). The *SLENDER* gene of pea encodes a gibberellin 2-oxidase. *Plant Physiology* **121**, 775–781.
Navarro, L., Bari, R., Achard, P. *et al.* (2008). DELLAs control plant immune responses by modulating the balance and salicylic acid signaling. *Current Biology* **18**, 650–655.
Nelissen, H., Rymen, B., Jikumaru, Y. *et al.* (2012). A local maximum in gibberellin levels regulates maize leaf growth by spatial control of cell division. *Current Biology* **22**, 1183–1187.
Nomura, T., Magome, H., Hanada, A. *et al.* (2013). Functional analysis of *Arabidopsis* CYP714A1 and CYP714A2 reveals that they are distinct gibberellin modification enzymes. *Plant and Cell Physiology* **54**, 1837–1851.
O'Neill, D.P. and Ross, J.J. (2002). Auxin regulation of the gibberellin pathway in pea. *Plant Physiology* **130**, 1974–1982.
Ogawa, M., Hanada, A., Yamauchi, Y. *et al.* (2003). Gibberellin biosynthesis and response during *Arabidopsis* seed germination. *The Plant Cell* **15**, 1591–1604.
Oh, E., Yamaguchi, S., Kamiya, Y. *et al.* (2006). Light activates the degradation of PIL5 protein to promote seed germination through gibberellin in *Arabidopsis*. *The Plant Journal* **47**, 124–139.
Oh, E., Yamaguchi, S., Hu, J. *et al.* (2007). PIL5, a phytochrome-interacting bHLH protein, regulates gibberellin responsiveness by binding directly to the *GAI* and *RGA* promoters in *Arabidopsis* seeds. *The Plant Cell* **19**, 1192–1208.
Pearce, D.W., Hutt, O.E., Rood, S.B. and Mander, L.N. (2002). Gibberellins in shoots and developing capsules of *Populus* species. *Phytochemistry* **59**, 679–687.
Peng, J., Carol, P., Richards, D.E. *et al.* (1997). The *Arabidopsis GAI* gene defines a signaling pathway that negatively regulates gibberellin responses. *Genes & Development* **11**, 3194–3205.
Phillips, A.L., Ward, D.A., Uknes, S. *et al.* (1995). Isolation and expression of three gibberellin 20-oxidase cDNA clones from Arabidopsis. *Plant Physiology* **108**, 1049–1057.

Reid, J.B., Ross, J.J. and Swain, S.M. (1992). Internode length in *Pisum* : A new, slender mutant with elevated levels of C_{19} gibberellins. *Planta* **188**, 462–467.

Reid, J.B., Botwright, N.A., Smith, J.J. *et al.* (2002). Control of gibberellin levels and gene expression during de-etiolation in pea. *Plant Physiology* **128**, 734–741.

Rieu, I., Eriksson, S., Powers, S.J. *et al.* (2008a). Genetic analysis reveals that C_{19}-GA 2-oxidation is a major gibberellin inactivation pathway in *Arabidopsis*. *The Plant Cell* **20**, 2420–2436.

Rieu, I., Ruiz-Rivero, O., Fernandez-Garcia, N. *et al.* (2008b). The gibberellin biosynthetic genes *AtGA20ox1* and *AtGA20ox2* act, partially redundantly, to promote growth and development throughout the *Arabidopsis* life cycle. *The Plant Journal* **53**, 488–504.

Ropers, H.J., Graebe, J.E., Gaskin, P. and Macmillan, J. (1978). Gibberellin biosynthesis in a cell-free system from immature seeds of *Pisum sativum*. *Biochemical and Biophysical Research Communications* **80**, 690–697.

Ross, J.J., Reid, J.B., Swain, S.M. *et al.* (1995). Genetic regulation of gibberellin deactivation in *Pisum*. *The Plant Journal* **7**, 513–523.

Sakai, M., Sakamoto, T., Saito, T. *et al.* (2003). Expression of novel rice gibberellin 2-oxidase gene is under homeostatic regulation by biologically active gibberellins. *Journal of Plant Research* **116**, 161–164.

Sakamoto, T., Kamiya, N., Ueguchi-Tanaka, M. *et al.* (2001a). KNOX homeodomain protein directly suppresses the expression of a gibberellin biosynthetic gene in the tobacco shoot apical meristem. *Genes & Development* **15**, 581–590.

Sakamoto, T., Kobayashi, M., Itoh, H. *et al.* (2001b). Expression of a gibberellin 2-oxidase gene around the shoot apex is related to phase transition in rice. *Plant Physiology* **125**, 1508–1516.

Sakamoto, T., Miura, K., Itoh, H. *et al.* (2004). An overview of gibberellin metabolism enzyme genes and their related mutants in rice. *Plant Physiology* **134**, 1642–1653.

Santes, C.M., Hedden, P., Gaskin, P. and Garcia-Martinez, J.L. (1995). Gibberellins and related compounds in young gruits of pea and their relationship to fruit-set. *Phytochemistry* **40**, 1347–1355.

Schneider, G. and Schliemann, W. (1994). Gibberellin conjugates – an overview. *Plant Growth Regulation* **15**, 247–260.

Schneider, G., Jensen, E., Spray, C.R. and Phinney, B.O. (1992). Hydrolysis and reconjugation of gibberellin A_{20} glucosyl ester by seedlings of *Zea mays* L. *Proceedings of the National Academy of Sciences USA* **89**, 8045–8048.

Schomburg, F.M., Bizzell, C.M., Lee, D.J. *et al.* (2003). Overexpression of a novel class of gibberellin 2-oxidases decreases gibberellin levels and creates dwarf plants. *The Plant Cell* **15**, 151–163.

Schreiber, K., Weiland, J. and Sembdner, G. (1970). Isolierung von gibberellin-A8-O(3)-β-d-glucopyranosid aus früchten von *Phaseolus coccineus*. *Phytochemistry* **9**, 189–198.

Seo, M., Hanada, A., Kuwahara, A. *et al.* (2006). Regulation of hormone metabolism in *Arabidopsis* seeds: phytochrome regulation of abscisic acid metabolism and abscisic acid regulation of gibberellin metabolism. *The Plant Journal* **48**, 354–366.

Serrani, J.C., Sanjuan, R., Ruiz-Rivero, O. *et al.* (2007). Gibberellin regulation of fruit set and growth in tomato. *Plant Physiology* **145**, 246–257.

Silverstone, A.L., Ciampaglio, C.N. and Sun, T. (1998). The *Arabidopsis RGA* gene encodes a transcriptional regulator repressing the gibberellin signal transduction pathway. *The Plant Cell* **10**, 155–169.

Stavang, J.A., Lindgard, B., Erntsen, A. *et al.* (2005). Thermoperiodic stem elongation involves transcriptional regulation of gibberellin deactivation in pea. *Plant Physiology* **138**, 2344–2353.

Stavang, J.A., Gallego-Bartolome, J., Gomez, M.D. *et al.* (2009). Hormonal regulation of temperature-induced growth in *Arabidopsis*. *The Plant Journal* **60**, 589–601.

Sun, T.P. (2011). The molecular mechanism and evolution of the GA-GID1-DELLA signaling module in plants. *Current Biology* **21**, R338–345.

Swain, S.M., Reid, J.B. and Kamiya, Y. (1997). Gibberellins are required for embryo growth and seed development in pea. *The Plant Journal* **12**, 1329–1338.

Symons, G.M., Smith, J.J., Nomura, T. *et al.* (2008). The hormonal regulation of de-etiolation. *Planta* **227**, 1115–1125.

Talon, M., Koornneef, M. and Zeevaart, J.A. (1990). Endogenous gibberellins in *Arabidopsis thaliana* and possible steps blocked in the biosynthetic pathways of the semidwarf *ga4* and *ga5* mutants. *Proceedings of the National Academy of Sciences USA* **87**, 7983–7987.

Tanaka-Ueguchi, M., Itoh, H., Oyama, N. *et al.* (1998). Over-expression of a tobacco homeobox gene, NTH15, decreases the expression of a gibberellin biosynthetic gene encoding GA 20-oxidase. *The Plant Journal* **15**, 391–400.

Thomas, S.G., Phillips, A.L. and Hedden, P. (1999). Molecular cloning and functional expression of gibberellin 2- oxidases, multifunctional enzymes involved in gibberellin deactivation. *Proceedings of the National Academy of Sciences USA* **96**, 4698–4703.

Ueguchi-Tanaka, M., Nakajima, M., Katoh, E. *et al.* (2007). Molecular interactions of a soluble gibberellin receptor, GID1, with a rice DELLA protein, SLR1, and gibberellin. *The Plant Cell* **19**, 2140–2155.

Varbanova, M., Yamaguchi, S., Yang, Y. *et al.* (2007). Methylation of gibberellins by *Arabidopsis* GAMT1 and GAMT2. *The Plant Cell* **19**, 32–45.

Wang, H., Caruso, L.V., Downie, A.B. and Perry, S.E. (2004). The embryo MADS domain protein AGAMOUS-Like 15 directly regulates expression of a gene encoding an enzyme involved in gibberellin metabolism. *The Plant Cell* **16**, 1206–1219.

Ward, D.A., MacMillan, J., Gong, F. *et al.* (2010). Gibberellin 3-oxidases in developing embryos of the southern wild cucumber, *Marah macrocarpus*. *Phytochemistry* **71**, 2010–2018.

Weller, J.L., Hecht, V., Vander Schoor, J.K. *et al.* (2009). Light regulation of gibberellin biosynthesis in pea is mediated through the COP1/HY5 pathway. *The Plant Cell* **21**, 800–813.

Weston, D.E., Elliott, R.C., Lester, D.R. *et al.* (2008). The pea DELLA proteins LA and CRY are important regulators of gibberellin synthesis and root growth. *Plant Physiology* **147**, 199–205.

Wolbang, C.M., Chandler, P.M., Smith, J.J. and Ross, J.J. (2004). Auxin from the developing inflorescence is required for the biosynthesis of active gibberellins in barley stems. *Plant Physiology* **134**, 769–776.

Xing, S.F., Qin, G.J., Shi, Y. *et al.* (2007). *GAMT2* encodes a methyltransferase of gibberellic acid that is involved in seed maturation and germination in *Arabidopsis*. *Journal of Integrative Plant Biology* **49**, 368–381.

Yaish, M.W., El-kereamy, A., Zhu, T. *et al.* (2010). The APETALA-2-Like transcription factor OsAP2-39 controls key interactions between abscisic acid and gibberellin in rice. *PLoS Genetics* **6**, e1001098.

Yamane, H., Fujioka, S., Spray, C.R. *et al.* (1988). Endogenous gibberellins from sporophytes of two tree ferns, *Cibotium glaucum* and *Dicksonia antarctica*. *Plant Physiology* **86**, 857–862.

Yamauchi, Y., Ogawa, M., Kuwahara, A. *et al.* (2004). Activation of gibberellin biosynthesis and response pathways by low temperature during imbibition of *Arabidopsis thaliana* seeds. *The Plant Cell* **16**, 367–378.

Yamauchi, Y., Takeda-Kamiya, N., Hanada, A. *et al.* (2007). Contribution of gibberellin deactivation by AtGA2ox2 to the suppression of germination of dark-imbibed *Arabidopsis thaliana* seeds. *Plant and Cell Physiology* **48**, 555–561.

Yang, D.L., Li, Q., Deng, Y.W. *et al.* (2008). Altered disease development in the *eui* mutants and *Eui* overexpressors indicates that gibberellins negatively regulate rice basal disease resistance. *Molecular Plant* **1**, 528–537.

Yang, Y.Y., Nagatani, A., Zhao, Y.J. *et al.* (1995). Effects of gibberellins on seed germination of phytochrome-deficient mutants of *Arabidopsis thaliana*. *Plant and Cell Physiology* **36**, 1205–1211.

Zentella, R., Zhang, Z.L., Park, M. *et al.* (2007). Global analysis of DELLA direct targets in early gibberellin signaling in *Arabidopsis*. *The Plant Cell* **19**, 3037–3057.

Zhang, Y., Zhu, Y., Peng, Y. *et al.* (2008). Gibberellin homeostasis and plant height control by EUI and a role for gibberellin in root gravity responses in rice. *Cell Research* **18**, 412–421.

Zhang, Y.Y., Zhang, B.C., Yan, D.W. *et al.* (2011). Two *Arabidopsis* cytochrome P450 monooxygenases, CYP714A1 and CYP714A2, function redundantly in plant development through gibberellin deactivation. *The Plant Journal* **67**, 342–353.

Zhao, X.Y., Yu, X.H., Foo, E. *et al.* (2007). A study of gibberellin homeostasis and cryptochrome-mediated blue light inhibition of hypocotyl elongation. *Plant Physiology* **145**, 106–118.

Zhu, Y., Nomura, T., Xu, Y. *et al.* (2006). *ELONGATED UPPERMOST INTERNODE* encodes a cytochrome P450 monooxygenase that epoxidizes gibberellins in a novel deactivation reaction in rice. *The Plant Cell* **18**, 442–456.

Annual Plant Reviews (2016) **49**, 95–120
doi: 10.1002/9781119210436.ch4 http://onlinelibrary.wiley.com

Chapter 4

GIBBERELLIN TRANSPORT

Jonathan Dayan

Department of Biology, Duke University, USA

Abstract: As reviewed in the accompanying chapters, significant progress has been achieved in elucidating the gibberellin (GA) signal transduction pathway. Yet, the hormonal nature of GA with respect to its source, transport, and sink sites remains debatable. However, compelling evidence concerning the significance of GA transport is beginning to emerge. Since plants are immobile and lack a central nervous system they use translocation of signalling molecules, mainly hormones, to maintain synchronised growth and response to environmental cues. GA belongs to this group of hormones. While addressing concerns about its local signalling, the chapter reviews the known molecular and physiological experiments that elucidate the significance of GA transport. On the basis of these findings a transport map can be proposed: maturing leaves are a potential origin of the mobile signal in the shoot, with stems and petioles, and stems and roots as routes for polar and non-polar transport, respectively. Recent findings identify GA_{12}, a non-active GA precursor, as the major mobile GA signal for long-distance transport in *Arabidopsis*. The continuous cambium and fibre system, as well as unfolded leaves are important GA-signalling sinks. The chapter concludes with a discussion of the advances from GA labelling and *in vivo* monitoring that would facilitate future research on the GA transport mechanism and its components.

Keywords: gibberellin transport, grafting, fibre, cambium, labelled GA, fluorescein, reporter gibberellin, gibberellin nanosensors, GID1 DELLA yeast three-hybrid, gibberellin-induced FRET

4.1 Introduction

In principal, hormones are defined as substances that are synthesised at low concentrations in one part of an organism, facilitating processes such as growth, metabolism and reproduction in other organs (Davies, 2004). As for the autocrine and paracrine systems in animals, the characterisation

Annual Plant Reviews, Volume 49: The Gibberellins, First Edition.
Edited by Peter Hedden and Stephen G. Thomas.

of hormones in plants is broader than the conventional endocrine system (Gaspar *et al.*, 2003). In plants, auxin and abscisic acid represent classical endocrine-like hormones, being produced in one tissue or organ, while regulating signalling at distant locations throughout the plant (Teale *et al.*, 2006; Kuromori *et al.*, 2014). On the other hand, other hormones such as brassinosteroids and gibberellins (GAs) have been shown to induce signalling at their production sites (Kaneko *et al.*, 2003; Symons and Reid, 2004). Although brassinosteroids (BR) are perceived by a cell surface receptor, BRI1, they are presumed to be produced in the cells where their signalling is needed (Symons and Reid, 2004; Clouse, 2011). These conclusions on local action and lack of transport rely on expression profiles of BR biosynthetic genes in conjunction with the inability to rescue phenotypes associated with BR deficient mutants by grafting. In contrast to these examples of transport or lack thereof, research on GAs shows evidence for local production, paracrine-like signalling, and long-distance transport capable of complementing GA-deficient mutants through grafting. The following chapter reviews these topics on GA transport and discusses the advances achieved in GA labelling and intracellular monitoring techniques, and how these tools may be further exploited to facilitate the identification of components of a GA transport mechanism.

4.2 Gibberellins can be translocated along plant bodies

In the 1920s, the discovery of GA itself was the first indication that the bioactive substance is able to be translocated along plants, inducing physiological processes far from its point of contact. GA was found to be secreted from the fungus *Gibberella fujikuroi* causing a rice disease that made the plants grow faster, exhibiting a spindly and pale phenotype (Brian *et al.*, 1954). External applications of the hormone had substantiated its ability to be transported and affect growth responses throughout the plant, inducing enhanced, faster growth and the development of longer and more fibres (Stant, 1961; Stant, 1963; Digby and Wareing, 1966; Aloni, 1979; Suttle and Hultstrand, 1987; Singh *et al.*, 2002). These responses were observed regardless of whether GA was applied to the roots or the shoots (Dayan *et al.*, 2012). The translocation of GAs within organs and through different plant organs was confirmed by radiolabelled and fluorescein-tagged GA studies (Jacobs and Kaldewey, 1970; Jacobs and Pruett, 1973; Drake, 1979; Kwiatkowska, 1991; Rinne *et al.*, 2001; Eriksson *et al.*, 2006).

Otsuka *et.al.* (2004) have reported that *ent*-kaurene, a diterpene GA precursor, is a naturally occurring volatile in different plant species. When produced it is released into the plant's headspace in addition to its accumulation in the plant body. They showed that in a closed environment, airborne *ent*-kaurene released from over-producer transgenic plants was able to fully complement the dwarf phenotype of *ga1-3* and *ga2-1* mutants (defective in *ent*-kaurene synthesis), but not that of the *ga3-1* mutant (defective in *ent*-kaurene oxidase). Therefore, they proposed that *ent*-kaurene may be efficiently metabolised into bioactive GAs in *Arabidopsis* when supplied as a volatile. These results suggest a mechanism for GA-mediated interaction between plants or between organs within the same plant (Otsuka *et al.*, 2004).

Additionally, the physical properties of GAs suggest they may be transported. Similar to abscisic acid (ABA) and auxin (indole-3-acetic acid; IAA), GAs are weak acids and therefore subject to the ion-trapping mechanism that would concentrate them in the neutral cytoplasm. This implies that GAs would require an efflux carrier to transport the negatively charged GA molecule from the cytosol, similar to the transport of ABA and auxin (Vieten *et al.*, 2007; Kuromori *et al.*, 2010). On the other hand, in the acidic apoplast these hormones are protonated and could potentially be membrane-permeable. Although the uncharged hormones can, potentially, diffuse through membranes, it has been shown that cells use specific carriers for auxin and ABA uptake (Vieten *et al.*, 2007; Kang *et al.*, 2010), and the same is predicted to be the case for GA. Furthermore, it was shown by electron spin resonance of GA-treated liposomes that GA perturbs phospholipid membranes and associates with its surface rather than penetrating into the interior of the bilayer, thereby potentially restricting the diffusion of GA into the cell (Pauls *et al.*, 1982). Therefore, it is hypothesised that influx carriers should exist for GA uptake, and to overcome the ion-trapping mechanism efflux carriers are needed as well.

Support for the requirement for membrane carriers for GA transport stems from a report on theoretical decay lengths of the weak-acid hormones. This study calculated the distances each of the hormones can remain in the apoplast before diffusing into cells (Kramer, 2006). In this model system, assuming the absence of influx or efflux carriers, Kramer shows that the 13-hydroxylated GAs, GA_1 (bioactive) and GA_{20} (immediate precursor) have longer decay lengths compared to the non-13-hydroxylated GA_4 (bioactive) and GA_9 (immediate precursor). The long decay lengths are comparable with those of ABA and GA_3, while the shorter ones compare to auxin. The very short decay lengths translate to substances that are readily membrane-permeable or have an influx transporter that would keep them out of the extracellular space. Interestingly, in application experiments both GA_4 and GA_3 are capable of translocation along the whole plant body. Combining these assessments suggests that carriers would be needed to eliminate ion-trapping along the flow, especially in the case of GA_4. Supporting the notion of the existence of GA transporters, it was found that AIT3 (ABA import transporter), a membrane-localised transporter, is an ABA transporter that can also transport GA_3, but not IAA or jasmonic acid (JA) (Kanno *et al.*, 2012). Likewise, GTR1, a transporter of the biologically active JA-isoleucine conjugate, has been shown to transport GA_3 in oocytes (Saito *et al.*, 2015). However, the expression of the transporter in oocytes did not induce preferential uptake for the bioactive GA_1 and GA_4 hormones or their precursors. Further research is needed to verify this activity *in planta*. Nevertheless, Saito *et al.* (2015) have substantiated their finding by reporting, for the first time, the physiological significance of a non-specific GA transporter *in planta*. They showed that exogenous GA, but not JA can rescue retarded stamen development phenotypes exhibited in *gtr1* knockout mutants.

Although the aforementioned studies address the possibility of movement through the apoplast, GA has long been thought to be transported through the symplast (Drake and Carr, 1979; Kwiatkowska, 1991; Rinne *et al.*, 2001). Using radiolabelled GA applied to oat coleoptile cells, it was demonstrated that GA_1 transport is non-polar and proceeds at two distinct velocities. The authors attributed each velocity pool to different cell-specific transport routes or, alternatively, GA conjugation. They also detected that GA_8, the inactive derivative of GA_1 oxidation, is immobile and remains in the area of its production (Drake, 1979). The studies were able to show that the transport proceeded through plasmodesmata (Drake and Carr, 1979). Further support for these conclusions was obtained by analysis of GA transport in algae (*Chara vulgaris* L). It was shown that induced, as well as spontaneously occurring symplastic isolation of the antheridium from the thallus, drastically reduces labelled GA levels in antheridium cells (Kwiatkowska, 1991), suggesting that GA is transported from the thallus through the symplast. In the late 1990s, the first fluorescent labelled GA was synthesised (Pulici *et al.*, 1996) and shortly after, Rinne *et al.* (2001) used fluorescein-labelled GA_4 (FL-GA_4; Figure 4.1a) to verify plasmodesmata-dependent GA transport in the apical meristem, allowing these authors to correlate potential FL-GA transport with the release of bud dormancy. In dormant buds, FL-GA_4 injected into one cell does not spread to adjacent cells, similar to the control plasmodesmata reporter molecule. However, subsequent to chilling-induced release from dormancy, fluorescence was detected within seconds in adjacent cells following FL-GA injection to a single apical meristem cell. Astonishingly, although the value of *in vivo* hormone labelling and monitoring is indisputable, it took more than a decade to produce similar results. Shani *et al.* (2013) produced FL-GA_4 (Figure 4.1b) and showed translocation of the externally applied GA, which accumulated at the elongation zone within endodermal cells. Moreover, since accumulation in the endodermal layer was affected by competition by unlabelled GA, temperature alterations, and ATP inhibitors, they suggest that GA uptake is a saturable active process that results in regulated accumulation in the root elongation zone. Interestingly, although Pulici *et al.* (1996) have introduced a mercaptalkylthio group spacer to C-16 of GA_4 (Figure 4.1a), having assumed that the hydroxyl group on C-3 and carboxyl group on C-6 are important for activity, Shani *et al.* (2013) have found that modifying the C-6 carboxyl group, both in GA_3 and GA_4 (Figure 4.1b, c), did not deactivate the hormone. Both groups concluded that there is a positive correlation between longer linker lengths and bioactivity (four carbons compared to

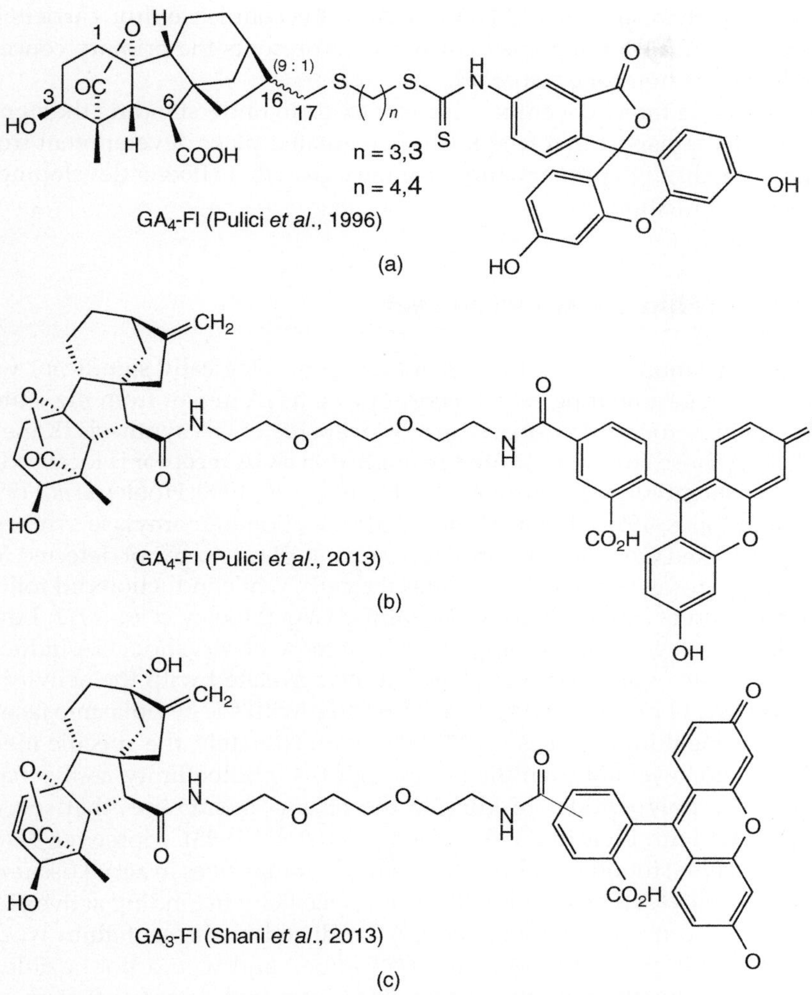

Figure 4.1 Molecular structures of fluorescein (FL) labelled GA_4 and GA_3 derivatives. a. GA_4-Fl labelling modification of C-16. Fluorescein label introduced into GA_4 and GA_3 at C-6, b and c, respectively.

three). These valuable findings should be used to enhance labelled-GA reporting systems and further the elucidation of GA transport in plants.

Establishing the notion that administered GAs, by pathogens or by artificial application, are transported raises questions about the significance of its endogenous transport, specifically at physiological concentration and originating from dedicated synthesising cells. External application of GA relies on uptake from the plant's surroundings and may utilise alternative transport mechanisms that do not necessitate cellular efflux carriers to overcome the ion-trapping mechanism that represents the primary concern for endogenous hormone transport.

The following topics cover the advances made in understanding the significance of GA transport that facilitate GA-regulated plant development from germination, through primary and secondary growth, to flower development and seed germination.

4.3 Gibberellin transport in seeds

The first indication that GA translocation is physiologically significant was obtained with germinating cereal seeds, in which GA moves from the scutellum epithelium to the aleurone, which is incapable of GA synthesis (Kaneko *et al.*, 2003). Those studies indicated strongly that a GA receptor is localised at the plasma membrane of aleurone cells (Hooley *et al.*, 1990; Hooley *et al.*, 1991; Gilroy and Jones, 1994). It was shown that induction of α-amylase synthesis and other related GA responses in aleurone protoplasts were not detected following high-dosage microinjection into the cells, while induction did follow the application of immobilised extra-cellular GAs (Hooley *et al.*, 1991; Lovegrove and Hooley, 2000). In compliance with these observation, GA induced α-amylase gene expression was shown to be correlated with the activity of heterotrimeric G proteins, which are associated with the cytoplasmic face of the plasma membrane (Jones *et al.*, 1998). Unfortunately, the specific membrane receptor was not identified, although GA photoaffinity assays have revealed two polypeptides residing in the plasma membranes that specifically interact with bioactive GAs (Lovegrove *et al.*, 1998). Moreover, Lovegrove *et al.* (1998) found membrane proteins of similar sizes in aerial tissues of pea and *Arabidopsis* that exhibited the same specificity in binding active GAs. With respect to transport, since the GA produced in the scutellum would be membrane impermeable it would be trapped and would not be able to translocate to the aleurone membrane had it not had a carrier that would facilitate its efflux. It would be very interesting to repeat these experiments with current bioinformatics and genetic tools which could shed light on the identity of the carriers or a membrane-localised receptor.

It should be noted that Yano *et al.* (2015) have recently published a comprehensive microarray analysis of GA-regulated genes in four GA signalling rice mutants and have elegantly demonstrated that GA perception, in all cases relating to GA-regulated transcription, proceeds through its nuclear perception by the GID1 receptor. This could contest the validity of the aforementioned findings, which imply that GA may be perceived on the plasma membrane and as a consequence, refute its implied transport. However, in their work Yano *et al.* (2015) could not rule out the need for the perception of an earlier GA signal on the plasma membrane since GA was administered by incubating the seeds in GA_3-containing media. Hence, all the GA responses that were found to proceed through GID1 perception resulted from GA that was in contact with the cell membrane beforehand. It would be interesting to resolve these perception mechanisms further.

4.4 Pattern of gibberellin biosynthesis in transport analysis

While labelled GA feeding studies substantiate its potential for transport, in most cases, expression studies on GA biosynthesis genes revealed that GAs are produced in sites that are actively involved in its signalling and perception.

As diterpenoids, GAs are synthesised from geranylgeranyl diphosphate (GGPP) (Hedden and Kamiya, 1997). GA biosynthesis proceeds through passage of its precursors between intracellular organelles and the cytosol. The first committed steps in GA biosynthesis occur in the plastid, leading to the synthesis of the tetracyclic hydrocarbon, *ent*-kaurene, which is oxidised on the outer side of the plastid envelope (Helliwell *et al.*, 2001). The second part of the pathway continues in the endoplasmic reticulum (ER) where *ent*-kaurenoic acid is sequentially oxidised to yield the first-formed GAs, GA_{12} and its 13-hydroxylated analogue GA_{53}. The last oxidations in which GA_{12} and GA_{53} are converted to other C_{20}-GAs and C_{19}-GAs proceed in the cytosol through the actions of the 2-oxoglutarate-dependent dioxygenases GA 20-oxidase and GA 3-oxidase, the latter catalysing the formation of the bioactive GAs (Hedden and Kamiya, 1997; Hedden and Proebsting, 1999; Kobayashi *et al.*, 2000; see Chapter 2). To maintain an efficiently regulated biosynthesis pathway it is therefore hypothesised that the intracellular localisation of precursors, as well as movement between the organelles, may require an intracellular transport mechanism. However, as discussed earlier, the volatility of *ent*-kaurene may facilitate its escape from the plastid and cells, while ER- and cytosol-localised precursors may be transported to other cells through the symplast to complete biosynthesis. Moreover, it cannot be ruled out that KAO may be associated with protein complexes containing dioxygenases on the surface of the ER, similar to a reported mechanism in the phenylpropanoid pathway (Bassard *et al.*, 2012), thereby bypassing the necessity for dedicated intracellular transporters. It would be interesting to observe where and how the biosynthetic enzymes localise in their respective organelles.

GA-deficient mutants are valuable tools that have been elegantly used in combination with grafting experiments to analyse the localisation of GA biosynthesis and to determine the physiological significance of endogenous GA transport. Mutated in GA biosynthesis genes, these mutants have reduced levels of active GAs in comparison to wild-type plants, resulting in a dwarfed phenotype that is completely rescued by GA application (Wilson *et al.*, 1992; Hedden and Proebsting, 1999; Koornneef *et al.*, 2002).

It has been reported that local GA production is coupled with its signalling. In rice, co-localisation of *GA 20-oxidase* and *GA 3-oxidase* gene expression with that of GA signalling genes indicated that bioactive GAs are synthesised at the sites of GA signalling, namely the rapidly elongating and dividing tissues or organs, with the exception of the aleurone cells, which are not capable of GA biosynthesis (Lovegrove and Hooley, 2000; Kaneko *et al.*, 2003). Accordingly, in tobacco, bioactive GAs are produced in dividing and elongating cells through the temporal and spatial control of *GA 3-oxidase* expression (MacMillan, 1997; Itoh *et al.*, 1999). Nevertheless, translocation of the hormone was not ruled out for the unexplained expansion of epidermal cells that do not express this gene (Itoh *et al.*, 1999). Other expression studies on *ent*-kaurene synthase revealed that its activity has been localised to the vascular tissues of some non-growing organs, such as expanded leaves, suggesting that these leaves may be a site of GA synthesis for transport to other organs (Silverstone *et al.*, 1997).

Using GA biosynthesis gene expression profiles, it was also shown that organ-specific GA transport is essential to facilitate normal floral development and maturation in *Arabidopsis* (Hu *et al.*, 2008). It was suggested that *de novo* synthesis of active GAs is necessary for stamen development in early flowers and that bioactive GAs made in the stamens and/or flower receptacles are transported to petals to promote their growth. In developing siliques, *GA 3-oxidases* are expressed in the replums, funiculi and the silique receptacles, as well as in the developing seeds. The immediate up-regulation of *GA 3-oxidase* genes after anthesis suggests that pollination and/or fertilisation is a prerequisite for *de novo* GA biosynthesis in fruit, which in turn promotes initial elongation of the silique. In this respect it was also shown that excessive GA signalling has a negative effect on plant fertility, which results in male sterility (Plackett *et al.*, 2014). Therefore, following the potency of local transport, it is proposed that post-anthesis GA levels would be tightly regulated by inactivation to maintain pollen development and viability. Since catabolism of biological active GAs mainly proceeds through irreversible hydroxylation by GA 2-*oxidases* (Lester *et al.*, 1999; Martin *et al.*, 1999; Thomas *et al.*, 1999; Sponsel and Hedden, 2004; Rieu *et al.*, 2008), it would be interesting to monitor and compare the expression profiles of GA 2-*oxidases* with those of the late GA biosynthesis genes to elucidate the spatiotemporal control of GA action. When coupled with local GA concentration measurements these would have the potential to clarify the temporal and spatial requirements for GA signalling. To this end, Dugardeyn *et al.* (2008) have used various gene expression databases along with numerous publications on expression profiles to establish a comprehensive picture of the points of GA synthesis and action. In roots, they propose the possibility of a lateral GA gradient rising from the epidermis to the endodermis. Additionally, they found that in the root zone, where fast elongation and cell specialisation takes place, *GA 2-oxidase* activity is relatively high and that GA synthesis genes are more prevalent in the meristematic zone. Given that GA directly regulates its synthesis through feedback inhibition of its biosynthesis and induction of its inactivation, it would be hard to conclude, without additional data, if these expression profiles result from feedback regulation, or the actual temporal developmental need for the hormone. In the perspective of the whole plant, they concluded that the expression of both the GA receptors and the DELLA proteins occurs throughout the plant and that the spatiotemporal expression along with other regulatory mechanisms such as hormone cross-talk coordinate the GA responses. Although the authors stipulate that tissue-specific GA response may primarily be regulated by the specificity of GA biosynthesis, the fact that they noticed an overlap between the expression of DELLA repressors and relatively high GA biosynthesis gene expression in the meristematic zone may indicate higher levels of complexity, such that GA transport is required to facilitate elongation in other parts of the root. This, in turn, would explain the higher GA 2-oxidase activity induced by higher bioactive GA concentrations in sink tissues.

4.5 Grafting experiments

Further substantiation for endogenous GA transport was obtained from grafting studies revealing that GA signals are transported between graft fusions (Katsumi *et al.*, 1983; Reid *et al.*, 1983; Proebsting *et al.*, 1992; Ragni *et al.*, 2011; Regnault *et al.*, 2015).

In a detailed study on GA movement along grafts in pea, Proebsting and colleagues had found that GA_{20} is the major transported GA (Proebsting *et al.*, 1992). Although these studies involved externally applied GAs, the use of precursors necessitates intracellular conversion to active GAs and therefore can be regarded as physiologically representative and meaningful. In dwarf *Zea mays* seedlings, GA_3 added to one member of the graft resulted in leaf-sheath elongation of both members (Katsumi *et al.*, 1983). Additionally, it was shown that GAs are found in the phloem, suggesting that they originate from vegetative tissues (Hoad *et al.*, 1993). Accordingly, in an elaborated grafting study, utilising natural genetic *Arabidopsis* ecotypes, it was shown that GA, originating from the shoot, is necessary for hypocotyl xylem expansion. This conclusion was evident since the graft transmitted signal rescued *ga1-3* hypocotyl expansion while GA 3-*oxidase* activity was not detected in the same tissue (Ragni *et al.*, 2011). A recent study by Regnault and colleagues (2015) has used an elaborated system of GA biosynthesis mutants to identify a specific mobile GA. Remarkably, although wild-type rootstocks (used as GA donor) were able to restore development to grafted *ga1-3* and *kao1 kao2* mutant scions (GA recipient tissues, mutated in the early stages of GA biosynthesis), wild-type rootstocks had no effect on the growth and flowering of *ga20ox1 ga20ox2 ga20ox3* and *ga3ox1 ga3ox2* mutant scions, which were statistically indistinguishable from those of *ga20ox1 ga20ox2 ga20ox3* and *ga3ox1 ga3ox2* self-grafts. Since GA_{12} is the immediate GA precursor product catalysed by KAO and substrate for GA 20-oxidase activity, Regnault and colleagues concluded that GA_{12} is the major mobile GA signal in *Arabidopsis*. Using the *kao* mutant eliminates any doubts that the rescued phenotype is a result of *ent*-kaurene volatility. These results were also observed on reciprocal grafting procedures, confirming the existence of both acro- and basipetal mobile GA signalling that stimulate shoot growth and root growth, respectively.

4.6 Significance for secondary growth

Morphological development associated with GA signalling is an important tool to assay the physiological significance of GA transport. It seems that not all the related morphological processes are always evaluated when concluding that a plant hormone is acting in the same locations as its production. In this respect, with the advancement of molecular biology, for the most part, the traditional physiological and anatomical perspectives had been neglected. Most of the correlation to biosynthesis was done on stem elongation and leaf expansion, the two most evident GA-related phenotypes that could be seen and easily quantified. But since GAs are produced in these tissues they have been regarded as local signals. Although the importance of the molecular data comprising the biosynthesis and the molecular perception of GAs, these pillar studies reveal only a fraction of the mechanisms accountable for GA physiology, including stem elongation, secondary growth, the expansion and shape of leaves, flowering, and seed germination (Yamaguchi, 2008). In the past few years, it is noticeable that methodologies are shifting back to emphasising the action of the hormone on a wide range of physiological processes. While grafting experiments that predominantly relied on stem and leaf elongation indicate the possibility of GA transport, secondary growth processes controlled by GA emphasise the significance of GA transport.

Early in its research, GA was found to regulate growth in unfolded primary leaves. Humphries and Wheeler (1963) had found that removal of cotyledons of *Phaseolus vulgaris* before the primary leaves have unfolded retarded young leaf expansion and lowered their GA content, while later removal did not have any effect. Moreover, removal of the apical bud induced expansion and increased GA levels in the primary leaves. They concluded that GAs and nutrients from the cotyledon are transported to unfolded primary leaves to induce their expansion (Humphries and Wheeler, 1963; Wheeler, 1966). These results were further corroborated by experiments showing that GA can rescue decreased cell division in the hypocotyl (Asahina *et al.*, 2002) as well as internode elongation, subsequent to cotyledon removal (Shininge, 1972). Apart from cotyledons, leaves are suggested to act as a source of GA signalling, only after the leaves have begun to elongate and before they completely mature (Wang *et al.*, 1997; Dayan *et al.*, 2012). In tobacco, it was shown that removal of maturing leaves while maintaining the apical primordial leaves (and very young leaves) eliminate almost all traces of bioactive GA from the plant (Dayan *et al.*, 2012). Although stem elongation does not proceed in the absence of maturing leaves, growth is restored by application of C_{19}-GAs at physiological concentrations below the elongating internodes, confirming leaves are required for GA signalling in stems. Moreover, these also support previous findings showing that GA_{20} may be the major transported form (Proebsting *et al.*, 1992). In contrast, Renault *et al.* (2015) have convincingly found that it is the earlier precursor, GA_{12} (a C_{20}-GA), that is the predominant mobile GA in *Arabidopsis.* Interestingly, analysis of the GA content data published in the tobacco decapitation study (Dayan *et al.*, 2012) indicates that GA_{53} (a 13-hydroxylated GA_{12}) is abundant in young internodes that do not elongate in decapitated plants. Although the authors did not give this observation special consideration, these results, in conjunction with those reported for the grafting studies, may indicate that the role of elongating leaves in the production of the bioactive signal is to regulate *GA 20-oxidase* activity. This regulation may take place in the leaf since the products of GA 20-oxidase, the C_{19}-GAs, have the capability of translocation (Dayan *et al.*, 2012). This would also explain the presence of GA_{20} in the phloem and between graft members of pea (Proebsting *et al.*, 1992; Hoad *et al.*, 1993) and its ability to restore growth to de-leafed tobacco plants away from point of application (Dayan *et al.*, 2012). It should be noted that concentrations of GA_{19} (an intermediate GA 20-oxidase product) in de-leafed young internodes were similar to normal, although all other products of GA 20- and 3-oxidase were barely or not detected (may be a result of the inability of GA_{19} to be metabolised or translocated out of the cells before decapitation). In essence, these results indicate that GA_{12} is a long-distance mobile GA signal that may require elongating leaves for its further metabolism into products of GA 20-oxidase and bioactive GAs that may be further transported within the shoot or root. However, their potential for translocation between these tissues remains debatable.

The identification of expanding leaves and cotyledons as sources for bioactive GA signalling enabled manipulation of GA content, which facilitated physiological experiments that unveiled the role of GA in secondary growth. Accordingly, removal of the source leaves affirmed the regulatory roles of GA in xylogenesis, controlling cambial activity and fibre formation (Hess and Sachs, 1972; Aloni, 1979; Dayan *et al.*, 2012). It was found that in the absence of the expanding leaves the vascular cambium remains inactivated and that this phenotype can be rescued by reintroducing GA to the plant. The same hormone function was evident in the maintenance of the interfasicular cambium between the vascular bundles (Dayan *et al.*, 2012). Although the induction of fibre differentiation is controlled by various plant hormones including auxin (Aloni *et al.*, 2003: Aloni *et al.*, 2004) and GA (Aloni, 1979) from leaves and cytokinin produced in roots (Aloni, 1982; Aloni *et al.*, 2004; Aloni *et al.*, 2006), it has been shown that GA signalling from maturing leaves is required for the induction of fibres in both the xylem and phloem (Aloni, 1987; Mauriat and Moritz, 2009; Dayan *et al.*, 2012).

Ragni and colleagues (2011) found that flowering-dependent hypocotyl xylem expansion is a general feature of herbaceous plants with a rosette growth habit. They found that flowering induction is sufficient to trigger xylem expansion in *Arabidopsis*. By contrast, neither flower formation nor elongation of the main inflorescence is required. Xylem production also does not depend on any particular flowering time pathway or absolute age. Through analyses of natural genetic variation, they reveal that ERECTA acts locally to restrict xylem expansion downstream of the GA pathway. Analysis of mutant and transgenic plants indicate that GA and its signalling pathway are both necessary and sufficient to directly trigger enhanced xylogenesis. Impaired GA signalling did not affect xylem expansion systemically, suggesting that it acts downstream of the mobile cue. By contrast, the GA effect was graft transmissible, suggesting that GA itself is the mobile shoot-derived signal (Ragni *et al.*, 2011).

In tobacco, stem development ends with the maturation of its inflorescence. In accordance with reports of *de novo* GA biosynthesis in flowers (Gallego-Giraldo *et al.*, 2007, Hu *et al.*, 2008), during flower development there is a burst of GA responses that are evident in fibre formation in the flower stalks (Aloni and Dayan, unpublished data). Cross sections in the upper internodes of the peduncle and pedicels revealed two distinguished developmental stages that were correlated with the developmental stages of the flowers. As the inflorescence develops, none of the sepal leaves open and the cambium in the stalks exhibits high production of unspecified cells (Figure 4.2). In such, even the application of GA to the bottom of the stems does not induce their differentiation to fibres, similar to the phenotype observed in the uppermost vegetative internodes. Once the flowers mature and open, these cells undergo swift specialisation, forming a highly fibre-rich, thick xylem (Figure 4.2). The few vessels in this xylem localise to its earlier development along its inner layers. The separation of the xylem into two distinct areas, vessels dominating the area closer to the pith and late-formed fibres closer to the cambium, may signify different regulation by GA and auxin (Aloni, 2013). Although it would be interesting to analyse this tissue for GA biosynthesis gene activity, it is evident that the flower exerts the GA signalling to provide its physical support.

Both the cambium and vascular fibres are continuous tissues; they depend on this continuum for their fundamental function in nutrient transport and structural support. The continuity of the cambium, vasculature and their supportive fibres require that their differentiating signal be spatially and temporally synchronised. Therefore it would be rational to assume that hormonal transport would facilitate the developmental synchronisation of the vascular tissue continuum spanning along the body. Similar to the continuous vessel differentiation dependence on auxin transport, the recent findings suggest

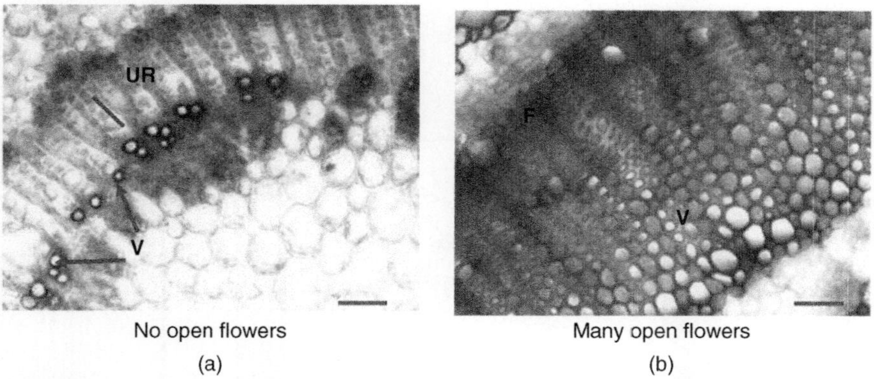

Figure 4.2 Flowers induce cambial activity and fibre formation. Cross sections in inflorescence stalks. (a) Depiction of the vasculature before flower maturation. A large undifferentiated cell region (UR) develops under young flowers. (b) Once the flowers in the inflorescence mature and their petals open, xylem cell differentiation occurs swiftly. In the inner xylem (first inside cells to differentiate) only vessels develop (V), while the later cells, closer to the cambium, form fibres (F). This depicts the gradient of hormonal effects from the developing flower. Beginning with high auxin to GA ratios (inducing vessels), and ending with high GA to auxin ratios, thereby inducing fibre formation. (Roni Aloni and Jonathan Dayan, unpublished.) (See insert for colour representation of this figure.)

that GA-specific control of fibre formation depends on its transport (Ragni *et al.*, 2011; Dayan *et al.*, 2012).

4.7 Orientation of gibberellin signal flow: source and sink tissues

A map detailing the orientation of GA signal flow emerges from integrating reports on biosynthetic and GA response gene expression along with grafting studies and physiological responses (Figure 4.3a). Since the later data incorporate results on secondary growth it is primarily attributed to GA flow in dicots.

As reviewed, GA_{12} is the earliest mobile GA (and predominant form in *Arabidopsis*), capable of long-distance transport throughout both the aerial and root tissues (Regnault *et al.*, 2015). In the shoot, its further oxidation to C_{19}-GAs is regulated by the development of maturing leaves, which thus act as a source for bioactive GA signalling, GA signal transport within the blade being non-polar (Drake, 1979). This has been substantiated by expression analysis of reporter gene fusions to GA-inducible promoters and GA-directed DELLA protein degradation using incisions in *Arabidopsis* leaf blades (Dayan *et al.*, 2012). Similar techniques of GA labelling and reporter gene activity, as well as GFP-RGA accumulation in petioles after blade excision, show

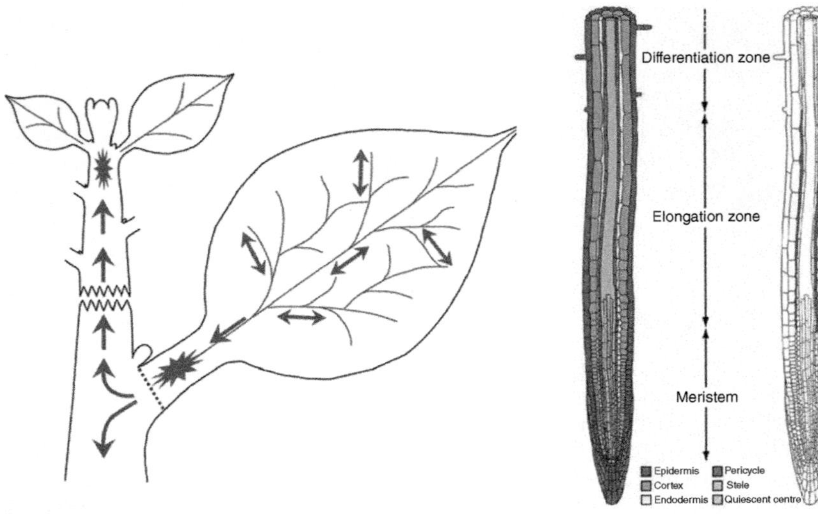

Figure 4.3 The source and translocation of the gibberellin signal. (a) An illustration modelling the translocation of a leaf-derived GA signal. The signal originates in developing leaves. Its flow is non-polar in the leaf blade and becomes polar only in the lower mid-vein towards the stem (arrows mark flow orientation). The unique anatomy at the base of the petiole (dotted line) potentially retards the flow which induces a local maximum (star), thereby acting as the leaf's elongation driving force. Although the signal flows in both directions along the stem, its upward movement from developing leaves reaches the young internodes (star) and induces stem elongation at the shoot apex. Throughout the flow along the stem, the signal results in bioactive GA accumulation that controls cambial activity and fibre differentiation. This illustration was originally published in Dayan *et al.* (2012). (b) Schematic illustrations of (left) the tissue organisation and zones within the *Arabidopsis* primary root and (right) the cellular distribution of the hormone signal GA. The concentric layers of epidermal, cortical, endodermal, pericycle and stele (vascular) tissues are colour coded (see key); the GA distribution is denoted in red. This illustration was originally published in Band and Bennett (2013). (See insert for colour representation of this figure.)

that in the petioles transport becomes polar oriented in the direction of the stem (Jacobs and Kaldewey, 1970, Dayan *et al.*, 2012). It should be noted that this accumulation is specifically attributed to GA after ruling out adverse effects of other hormones or wounding. Interestingly, elevated levels of *GUS* expression were detected in the petiole–stem junctions similar to the expression of the GA-inducible gene *LEAFY* (Blazquez *et al.*, 1997). These junctions have a distinct anatomy, characterised by short cells in both the ground and vascular tissues, which later participate in leaf abscission. It was hypothesised that this accumulation is caused by a local decreased cell size that slows the flow of the mobile signal, resembling a bottleneck. Incisions in the leaf blade and petioles were made to produce artificial barriers that successfully mimicked the hypothesised bottleneck. Furthermore, the observed patterns of GUS staining indicated that the non-polar flow in the blade is associated with the vascular veins, primarily in the phloem and bundle sheath cell, consistent with other findings that localised the primary GA responsive tissue in *Arabidopsis* roots to the endodermis (Ubeda-Tomas *et al.*, 2008; Shani *et al.*, 2013) (both the bundle sheath, in leaves, and the endodermis, in roots, surround the vascular tissues). It was proposed that the enhanced *GUS* expression in the petiole may be functionally related to the fact that the majority of leaf elongation occurs at its bottom third (Poethig and Sussex, 1985), However, comparison of bioactive GA content in the upper and lower halves of the petiole did not confirm an accumulation of GA at the petiole base, but these may need finer resolution to detect such an accumulation. Taken together, these results all conform to the notion of production of a mobile signal in developing leaves (Figure 4.3a).

In the stem, GA movement is non-polar, inducing the formation of the fibres and regulating the activity of the vascular cambium, regardless of the orientation from which it originates (Ragni *et al.*, 2011; Dayan *et al.*, 2012). Additional support for the locations of GA maxima and flow along the vasculature was given by analysis of the expression profile of a synthetic GA-responsive promoter. The expression of the promoter, constructed of conserved *cis*-motifs encompassed in GA-responsive genes, corresponds to the GA-regulated tissues and cells, particularly, around fibres in the vascular bundle and interfasicular region (Dayan *et al.*, 2012).

Notwithstanding the findings of Regnault *et al.* (2015) that GA_{12} is graft transmissible between aerial and root tissues and that it can rescue growth in *kao* and *ga1* mutants, much less is known on GA transport within roots. Although it is obvious that roots have the capacity for GA uptake and that such uptake can rescue GA-deficient phenotypes in the shoot, published data on endogenous GA transport is confined to the aerial tissues. However, elaborate studies in *Arabidopsis* roots demonstrate that the endodermis represents the primary GA-responsive tissue regulating organ growth and emphasise the importance of this regulation by showing that endodermal cell expansion is rate-limiting for elongation of other tissues and therefore of the root as a whole (Ubeda-Tomas *et al.*, 2008). Accordingly, as mentioned, it was found that the active uptake of FL-GA directs the accumulation of the signal in the endodermis (Shani *et al.*, 2013; Figure 4.3b). It would be interesting to see how and where the labelled-GA is further transported along the root, perhaps by labelling the known transported forms GA_{12}, GA_9, GA_4 and their 13-hydroxylated analogues.

The physiological significance of roots supplying the shoot with endogenous GA is unclear. GA deficiency results from de-leafing, although roots are maintained intact (Dayan *et al.*, 2012), and roots respond to significantly lower GA concentrations compared to shoot (1 nM compared to 1 uM GA_3, (Tanimoto, 2012)). These results may suggest that the root and shoot tissues have separate GA transport systems, although some connectivity exists, presumably at the GA_{12} precursor level, or for root-applied GA that can rescue GA-deficient phenotypes in the shoot.

4.8 Monitoring intra- and intercellular gibberellin concentration

Tools for detecting and monitoring GA with high spatial and temporal resolution in intact organs are invaluable for the elucidation of GA transport mechanisms and further understanding GA-regulated processes. As discussed, different labelling techniques were used to characterise GA transport, mainly by fluorescence and radio-labelling. Although the current data on specific routes or the proteins involved in GA transport are sparse, these labelling techniques have proven efficient, especially FL-GA, which was used to show accumulation of exogenous GA in the root endodermal layer and symplastic movement in buds (Rinne *et al.*, 2001; Shani *et al.*, 2013). However, direct GA labelling cannot overcome issues associated with external hormone application, such as selection of the correct application sites that correspond with GA source cells or application at physiological concentrations, i.e. in contrast to the potency of externally applied GA_4 to rescue GA deficient phenotypes of early GA biosynthesis mutants, in *Arabidopsis*, the endogenously synthesised GA_4 is not graft-transmissible. These caveats may be bypassed by the recent development of intracellular GA biosensors that can detect the presence, and, potentially, indicate concentrations of bioactive GAs (Miyamoto *et al.*, 2012). The biosensors rely on GA perception by its GID1A receptor. In the presence of bioactive GAs, GID1A and RGA, specifically interact, bringing about the proteasomal degradation of RGA (Ueguchi-Tanaka *et al.*, 2005; Griffiths *et al.*, 2006; Nakajima *et al.*, 2006; Ueguchi-Tanaka *et al.*, 2007; Murase *et al.*, 2008). The biosensor exploits this mechanism by expressing GID1A and RGA fusions to respective fluorophore FRET-pairs. The proteins only interact in the presence of bioactive GA and FRET is achieved when the distance between the fluorophores is between 1 and 10 nm. Only the N-terminal DELLA domain is used for the fusion, to eliminate GA-induced RGA degradation (Murase *et al.*, 2008). Originally, this reporter system was intended for producing a chemically inducible dimerisation system in mammalian cells. As described by the authors, the specificity of the system is assuring and its implementation *in planta* promising.

Another optional technique may be achieved by cloning GID1 gene in between a FRET pair that would respond to the conformational change of the receptor following GA interaction, similar to sugar nanosensors (Lager *et al.*, 2006). The added temporal advantage of such a biosensor stems from its ability to instantaneously report the recognition of cytosol-localised GA that does not depend on the expression and binding of a second protein.

Additionally, the GID1A-RGA interaction could be utilised for yeast-based assays to identify and test putative GA transporters, similar to the use of the

ABA receptor complex that successfully identified *AIT1/NRT1.2* as an ABA importer (Kanno *et al.*, 2012).

4.9 Conclusion: new aspects for gibberellin transport

In summary, there is compelling evidence for GA transport in plants, primarily dicots. The data suggest that the early GA precursor dedicated to GA biosynthesis, GA_{12}, has the potential to be transported throughout the whole plant. Furthermore, the bioactive GA accumulation in the apical internodes is regulated by developing leaves after leaf primordia development, and from petals and floral receptacles during flowering. The GA signalling originating from these organs regulates stem elongation, secondary growth, reproduction, and specifically induces continuous fibre formation.

The recent grafting data indicating that GA_{12} translocates through the transition zone is the first report on continuous GA transport between roots and shoots explicating the formation of the cambial and fibre continuum. Further research is needed to elucidate the specific transport route, its regulation and physiological significance within roots and shoots, and between the two.

4.9.1 Potential transporters

Acknowledging the significance of GA transport for plant development necessitates the further elucidation of its mechanism and characterisation of its components. The recent findings showing active GA transport and the possibility of non-specific ABA transporters being able to transport GAs (Kanno *et al.*, 2012; Boursiac *et al.*, 2013) as well as the non-specific GTR1 glucosinolate/JA-Ile/GA transporter (Saito *et al.*, 2015), are interesting, and may unfold into intriguing mechanisms of competitive transport as a means for hormone cross-talk. In this regard, other members of the ABC family have been identified as transporters of hormones including ABA, stringolactones, and auxin (Geisler *et al.*, 2005; Kang *et al.*, 2010; Kuromori *et al.*, 2010; Kretzschmar *et al.*, 2012). ABC transporters were first characterised as terpene and multi-drug transporters in mammalian cells and require ATP for function. As ATP-dependent transporters these proteins are relevant, primarily since GA uptake in roots was found to be reduced by ATP inhibitors (Shani *et al.*, 2013). It should be noted that when reported as a specific ABA importer, ABCG40/PDR12 tested negative for sclareol inhibition of ABA uptake, although it was previously shown to transport diterpenoids (Kang *et al.*, 2010). Given the non-specific transport of GA and ABA by AIT3, it would be interesting to test the capacity of PDR12 to transport GA as well.

Additionally, the inability of aleurone cells to locally produce GAs, supported by the lack of GA biosynthesis gene expression, and their reliance on GA produced in the scutellum and elongating shoots, indicates that GA transport is needed for post-germination growth. With the advancements in genomic and bioinformatical data analysis, it would be interesting to re-analyse the putative aleurone membrane receptors, as well as test for other possible components of the GA transport mechanism in the seeds.

4.9.2 Analysis through perception

The elucidation of molecular GA perception represents an opportunity to characterise GA sink tissues and further advance our understanding of downstream signalling events leading to specific GA-induced developmental processes. In this regard, most of the current molecular studies related to GA localisation are based on localisation of its biosynthesis and expression profiles of the DELLA repressors. Since specific GA transport routes are becoming evident, it would be crucial to correlate the latter with the expression of the GID1 receptors, assuming they relay all GA-promoted growth responses (Willige *et al.*, 2007). To this end, only a few reports have been published on the localisation of the GID1 receptors, primarily using GUS fusions (Dugardeyn *et al.*, 2008; Suzuki *et al.*, 2009; Voegele *et al.*, 2011). These results may be crucial for identifying GA perception with higher cellular resolution. The resulting physiological processes that are controlled by GA can further help with interpretation.

With respect to transport, the use of fluorescent-labelled GA probes to show symplastic movement in apical meristems and accumulation of exogenously applied GA in root endodermis (Pulici *et al.*, 1996; Rinne *et al.*, 2001; Shani *et al.*, 2013) should be further exploited for more thorough transport analyses throughout the plant, including leaf and stem tissues.

Elucidation of GA transport mechanisms as the characterisation of its source and sink tissues are fundamental objectives for future GA research which, obviously, may contribute to applied agriculture and forestry.

4.9.3 Links to sugar transport

The origin of GA signalling from leaves and the accompanying phenotype of its signal diffusion around the leaf junctions (Figure 4.4) and vascular bundles (Dayan *et al.*, 2012) may indicate coupling with sugar signalling. Hypothetically, such coupling, which may be physical, would facilitate the simultaneous supply of building blocks, along with the hormone that signals growth promotion. This mechanism could explain GA-related phenotypes observed in sucrose transport inhibition (Paparelli *et al.*, 2013) and is substantiated by research indicating GA phloem loading (Hoad *et al.*, 1993), as well as the presence of the major mobile GA signal in *Arabidopsis*, GA_{12}, in the phloem (Regnault *et al.*, 2015).

It should be noted that GAs can be converted into sugar conjugates in plants (Rood *et al.*, 1983; Schneider and Schliemann, 1994). However, the roles of these conjugates remain unclear. The possibility that sugar

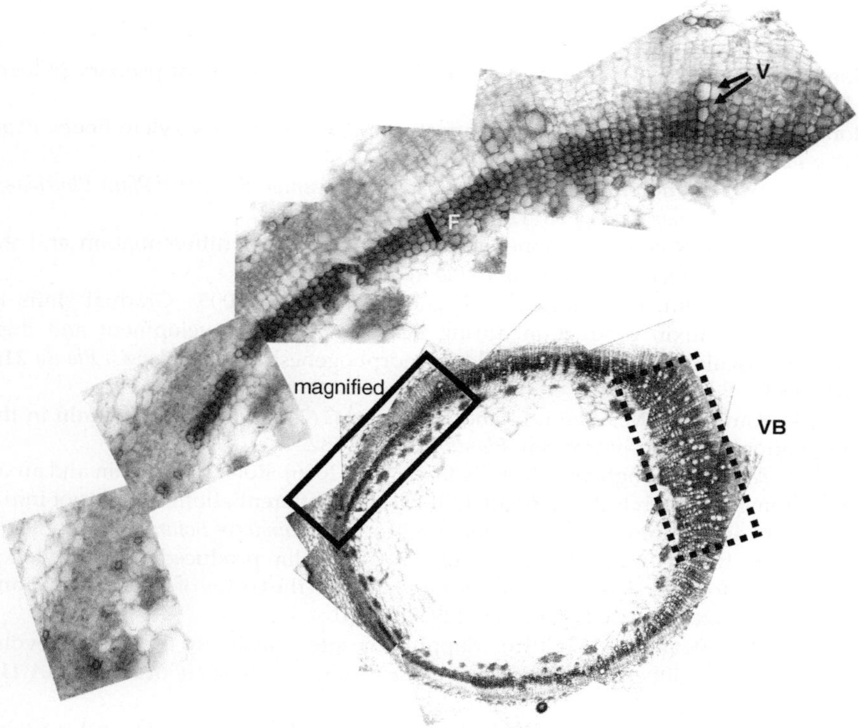

Figure 4.4 Diffusion of gibberellin signalling from a developing leaf. Effects of one leaf in otherwise de-leafed plants. Apical leaves were excised before becoming mature. A young leaf kept intact for 3.5 weeks, was cross sectioned in the internode beneath the leaf. Vessels form immediately beneath the leaf at the vascular bundle (VB). A fibre-cell layer is produced around the bundle and diminishes as the distance from the bundle grows. This layer resembles interfasicular fibers (here only one fasicular bundle has developed, indicated VB in the view of the whole cross section). (Roni Aloni and Jonathan Dayan, unpublished.) (See insert for colour representation of this figure.)

conjugation contributes to GA transport, similar to the non-polar movement of conjugated or bound auxin (Cohen and Bandurski, 1982; Lewis *et al.*, 2007), deserves further investigation.

Acknowledgements

I thank Professor Roni Aloni for proofreading this chapter and providing insightful remarks and guidance. Acknowledgement is also made to the support provided by BARD, the United States – Israel Binational Agricultural Research and Development Fund, Vaadia-BARD (FI-451-2011).

References

Aloni, R. (1979). Role of auxin and gibberellin in differentiation of primary phloem fibers. *Plant Physiology* **63**, 609–614.

Aloni, R. (1982). Role of cytokinin in differentiation of secondary xylem fibers. *Plant Physiology* **70**, 1631–1633.

Aloni, R. (1987). Differentiation of vascular tissues. *Annual Review of Plant Physiology and Plant Molecular Biology* **38**, 179–204.

Aloni, R. (2013). Role of hormones in controlling vascular differentiation and the mechanism of lateral root initiation. *Planta* **238**, 819–830.

Aloni, R., Schwalm, K., Langhans, M. and Ullrich, C.I. (2003). Gradual shifts in sites of free-auxin production during leaf-primordium development and their role in vascular differentiation and leaf morphogenesis in *Arabidopsis*. *Planta* **216**, 841–853.

Aloni, R., Langhans, M., Aloni, E. and Ullrich, C.I. (2004). Role of cytokinin in the regulation of root gravitropism. *Planta* **220**, 177–182.

Aloni, R., Aloni, E., Langhans, M. and Ullrich, C.I. (2006). Role of cytokinin and auxin in shaping root architecture: Regulating vascular differentiation, lateral root initiation, root apical dominance and root gravitropism. *Annals of Botany* **97**, 883–893.

Asahina, M., Iwai, H., Kikuchi, A. *et al.* (2002). Gibberellin produced in the cotyledon is required for cell division during tissue reunion in the cortex of cut cucumber and tomato hypocotyls. *Plant Physiology* **129**, 201–210.

Band, L.R. and Bennett, M.J. (2013). Mapping the site of action of the Green Revolution hormone gibberellin. *Proceedings of the National Academy of Sciences USA* **110**, 4443–4444.

Bassard, J.-E., Richert, L., Geerinck, J. *et al.* (2012). Protein–protein and protein–membrane associations in the lignin pathway. *The Plant Cell* **24**, 4465–4482.

Blazquez, M.A., Soowal, L.N., Lee, I. and Weigel, D. (1997). LEAFY expression and flower initiation in *Arabidopsis*. *Development* **124**, 3835–3844.

Boursiac, Y., Leran, S., Corratge-Faillie, C. *et al.* (2013). ABA transport and transporters. *Trends in Plant Science* **18**, 325–333.

Brian, P.W., Elson, G.W., Hemming, H.G. and Radley, M. (1954). The plant-growth-promoting properties of gibberellic acid, a metabolic product of the fungus *Gibberella fujikuroi*. *Journal of the Science of Food and Agriculture* **5**, 602–612.

Clouse, S.D. (2011). Brassinosteroid signal transduction: from receptor kinase activation to transcriptional networks regulating plant development. *The Plant Cell* **23**, 1219–1230.

Cohen, J.D. and Bandurski, R.S. (1982). Chemistry and physiology of the bound auxins. *Annual Review of Plant Physiology and Plant Molecular Biology*, **33**, 403–430.

Davies, P.J. (ed.) 2004. *Plant Hormones: Biosynthesis, Signal Transduction, Action!*, Dordrecht, Kluwer Academic.

Dayan, J., Voronin, N., Gong, F. *et al.* (2012). Leaf-induced gibberellin signaling is essential for internode elongation, cambial activity, and fiber differentiation in tobacco stems. *The Plant Cell* **24**, 66–79.

Digby, J. and Wareing, P.F. (1966). Effect of applied growth hormones on cambial division and differentiation of cambial derivatives. *Annals of Botany* **30**, 539–549.

Drake, G.A. (1979). Flux studies and compartmentation analysis of gibberellin A_1 in oat coleoptiles. *Journal of Experimental Botany* **30**, 429–437.

Drake, G.A. and Carr, D.J. (1979). Symplastic transport of gibberellins - evidence from flux and inhibitor studies. *Journal of Experimental Botany* **30**, 439–447.

Dugardeyn, J., Vandenbussche, F. and Van Der Straeten, D. (2008). To grow or not to grow: what can we learn on ethylene–gibberellin cross-talk by in silico gene expression analysis? *Journal of Experimental Botany* **59**, 1–16.

Eriksson, S., Bohlenius, H., Moritz, T. and Nilsson, O. (2006). GA_4 is the active gibberellin in the regulation of *LEAFY* transcription and *Arabidopsis* floral initiation. *The Plant Cell* **18**, 2172–2181.

Gallego-Giraldo, L., Garcia-Martinez, J.L., Moritz, T. and Lopez-Diaz, I. (2007). Flowering in tobacco needs gibberellins but is not promoted by the levels of active GA_1 and GA_4 in the apical shoot. *Plant and Cell Physiology* **48**, 615–625.

Gaspar, T., Kevers, C., Faivre-Rampant, O. *et al.* (2003). Changing concepts in plant hormone action. *In Vitro Cellular & Developmental Biology-Plant* **39**, 85–106.

Geisler, M., Blakeslee, J.J., Bouchard, R. *et al.* (2005). Cellular efflux of auxin catalyzed by the *Arabidopsis* MDR/PGP transporter AtPGP1. *The Plant Journal* **44**, 179–194.

Gilroy, S. and Jones, R.L. (1994). Perception of gibberellin and abscisic acid at the external face of the plasma membrane of barley (*Hordeum vulgare* L.) aleurone protoplasts. *Plant Physiology* **104**, 1185–1192.

Griffiths, J., Murase, K., Rieu, I. *et al.* (2006). Genetic characterization and functional analysis of the GID1 gibberellin receptors in *Arabidopsis*. *The Plant Cell* **18**, 3399–3414.

Hedden, P. and Kamiya, Y. (1997). Gibberellin biosynthesis: enzymes, genes and their regulation. *Annual Review of Plant Physiology and Plant Molecular Biology* **48**, 431–460.

Hedden, P. and Proebsting, W.M. (1999). Genetic analysis of gibberellin biosynthesis. *Plant Physiology* **119**, 365–370.

Helliwell, C.A., Sullivan, J.A., Mould, R.M. *et al.* (2001). A plastid envelope location of *Arabidopsis ent*-kaurene oxidase links the plastid and endoplasmic reticulum steps of the gibberellin biosynthesis pathway. *The Plant Journal* **28**, 201–208.

Hess, T. and Sachs, T. (1972). Influence of a mature leaf on xylem differentiation. *New Phytologist* **71**, 903–914.

Hoad, G.V., Retamales, J.A., Whiteside, R.J. and Lewis, M.J. (1993). Phloem translocation of gibberellins in three species of higher plants. *Journal of Plant Growth Regulation* **13**, 85–88.

Hooley, R., Beale, M.H. and Smith, S.J. (1990). Gibberellin perception in the *Avena fatua* aleurone. *Symposia of the Society for Experimental Biology* **44**, 79–86.

Hooley, R., Beale, M. and Smith, S. (1991). Gibberellin perception at the plasma membrane of *Avena fatua* aleurone protoplasts. *Planta* **183**, 274–280.

Hu, J., Mitchum, M.G., Barnaby, N. *et al.* (2008). Potential sites of bioactive gibberellin production during reproductive growth in *Arabidopsis*. *The Plant Cell* **20**, 320–336.

Humphries, E.C. and Wheeler, A.W. (1963). Physiology of leaf growth. *Annual Review of Plant Physiology and Plant Molecular Biology* **14**, 385–410.

Itoh, H., Tanaka-Ueguchi, M., Kawaide, H. *et al.* (1999). The gene encoding tobacco gibberellin 3 beta-hydroxylase is expressed at the site of GA action during stem elongation and flower organ development. *The Plant Journal* **20**, 15–24.

Jacobs, W.P. and Kaldewey, H. (1970). Polar movement of gibberellic acid through young *Coleus* petioles. *Plant Physiology* **45**, 539–541.

Jacobs, W.P. and Pruett, P.E. (1973). Time-course of polar movement of gibberellin through zea roots. *American Journal of Botany*, **60**, 896–900.

Jones, H.D., Smith, S.J., Desikan, R. *et al.* (1998). Heterotrimeric G proteins are implicated in gibberellin induction of a-amylase gene expression in wild oat aleurone. *The Plant Cell* **10**, 245–254.
Kaneko, M., Itoh, H., Inukai, Y. *et al.* (2003). Where do gibberellin biosynthesis and gibberellin signaling occur in rice plants? *The Plant Journal* **35**, 104–115.
Kang, J., Hwang, J.U., Lee, M. *et al.* (2010). PDR-type ABC transporter mediates cellular uptake of the phytohormone abscisic acid. *Proceedings of the National Academy of Sciences USA* **107**, 2355–2360.
Kanno, Y., Hanada, A., Chiba, Y. *et al.* (2012). Identification of an abscisic acid transporter by functional screening using the receptor complex as a sensor. *Proceedings of the National Academy of Sciences USA* **109**, 9653–9658.
Katsumi, M., Foard, D.E. and Phinney, B.O. (1983). Evidence for the translocation of gibberellin A_3 and gibberellin-like substances in grafts between normal, dwarf$_1$ and dwarf$_5$ seedlings of *Zea Mays* L. *Plant and Cell Physiology* **24**, 379–388.
Kobayashi, M., Macmillan, J., Phinney, B. *et al.* (2000). Gibberellin biosynthesis: metabolic evidence for three steps in the early 13-hydroxylation pathway of rice. *Phytochemistry* **55**, 317–321.
Koornneef, M., Bentsink, L. and Hilhorst, H. (2002). Seed dormancy and germination. *Current Opinion in Plant Biology* **5**, 33–36.
Kramer, E.M. (2006). How far can a molecule of weak acid travel in the apoplast or xylem? *Plant Physiology* **141**, 1233–1236.
Kretzschmar, T., Kohlen, W., Sasse, J. *et al.* (2012). A petunia ABC protein controls strigolactone-dependent symbiotic signalling and branching. *Nature* **483**, 341–344.
Kuromori, T., Miyaji, T., Yabuuchi, H. *et al.* (2010). ABC transporter AtABCG25 is involved in abscisic acid transport and responses. *Proceedings of the National Academy of Sciences USA* **107**, 2361–2366.
Kuromori, T., Sugimoto, E. and Shinozaki, K. (2014). Inter-tissue signal transfer of abscisic acid from vascular cells to guard cells. *Plant Physiology* **164**, 1587–1592.
Kwiatkowska, M. (1991). Autoradiographic studies on the role of plasmodesmata in the transport of gibberellin. *Planta* **183**, 294–299.
Lager, I., Looger, L.L., Hilpert, M. *et al.* (2006). Conversion of a putative agrobacterium sugar-binding protein into a FRET sensor with high selectivity for sucrose. *Journal of Biological Chemistry* **281**, 30875–30883.
Lester, D.R., Ross, J.J., Smith, J.J. *et al.* (1999). Gibberellin 2-oxidation and the *SLN* gene of *Pisum sativum*. *The Plant Journal* **19**, 65–73.
Lewis, D.R., Miller, N.D., Splitt, B.L. *et al.* (2007). Separating the roles of acropetal and basipetal auxin transport on gravitropism with mutations in two *Arabidopsis* multidrug resistance-like ABC transporter genes. *The Plant Cell* **19**, 1838–1850.
Lovegrove, A. and Hooley, R. (2000). Gibberellin and abscisic acid signalling in aleurone. *Trends in Plant Sciences* **5**, 102–110.
Lovegrove, A., Barratt, D.H., Beale, M.H. and Hooley, R. (1998). Gibberellin-photoaffinity labelling of two polypeptides in plant plasma membranes. *The Plant Journal* **15**, 311–320.
Macmillan, J. (1997). Biosynthesis of the gibberellin plant hormones. *Natural Product Reports* **14**, 221–243.
Martin, D.N., Proebsting, W.M. and Hedden, P. (1999). The *SLENDER* gene of pea encodes a gibberellin 2-oxidase. *Plant Physiology* **121**, 775–781.
Mauriat, M. and Moritz, T. (2009). Analyses of GA20ox- and GID1-over-expressing aspen suggest that gibberellins play two distinct roles in wood formation. *The Plant Journal* **58**, 989–1003.

Miyamoto, T., Derose, R., Suarez, A. *et al.* (2012). Rapid and orthogonal logic gating with a gibberellin-induced dimerization system. *Nature Chemical Biology* **8**, 465–470.

Murase, K., Hirano, Y., Sun, T.-P. and Hakoshima, T. (2008). Gibberellin-induced DELLA recognition by the gibberellin receptor GID1. *Nature* **456**, 459–463.

Nakajima, M., Shimada, A., Takashi, Y. *et al.* (2006). Identification and characterization of *Arabidopsis* gibberellin receptors. *The Plant Journal* **46**, 880–889.

Otsuka, M., Kenmoku, H., Ogawa, M. *et al.* (2004). Emission of *ent*-kaurene, a diterpenoid hydrocarbon precursor for gibberellins, into the headspace from plants. *Plant and Cell Physiology* **45**, 1129–1138.

Paparelli, E., Parlanti, S., Gonzali, S. *et al.* (2013). Nighttime sugar starvation orchestrates gibberellin biosynthesis and plant growth in *Arabidopsis*. *The Plant Cell* **25**, 3760–3769.

Pauls, K.P., Chambers, J.A., Dumbroff, E.B. and Thompson, J.E. (1982). Perturbation of phospholipid-membranes by gibberellins. *New Phytololgist* **91**, 1–17.

Plackett, A.R., Ferguson, A.C., Powers, S.J. *et al.* (2014). DELLA activity is required for successful pollen development in the Columbia ecotype of *Arabidopsis*. *New Phytologist* **201**, 825–836.

Poethig R.S. and Sussex I.M. (1985). The developmental morphology and growth dynamics of the tobacco leaf. *Planta* **165**, 158–169.

Proebsting, W.M., Hedden, P., Lewis, M.J. *et al.* (1992). Gibberellin concentration and transport in genetic lines of pea : Effects of grafting. *Plant Physiology* **100**, 1354–1360.

Pulici, M., Asami, T., Robertson, M. *et al.* (1996). Amylase induction activity of fluorescein labeled gibberellin in barley aleurone protoplasts. *Bioorganic & Medicinal Chemistry Letters* **6**, 2549–2552.

Ragni, L., Nieminen, K., Pacheco-Villalobos, D. *et al.* (2011). Mobile gibberellin directly stimulates *Arabidopsis* hypocotyl xylem expansion. *The Plant Cell*, **23**, 1322–1336.

Regnault, T., Davière, J.-M., Wild, M. *et al.* (2015). The gibberellin precursor GA_{12} acts as a long-distance growth signal in *Arabidopsis*. *Nature Plants* **15**:15073.

Reid, J.B., Murfet, I.C. and Potts, W.C. (1983). Internode length in Pisum .2. Additional information on the relationship and action of loci Le, La, Cry, Na and Lm. *Journal of Experimental Botany* **34**, 349–364.

Rieu, I., Eriksson, S., Powers, S.J. *et al.* (2008). Genetic analysis reveals that C_{19}-GA 2-oxidation is a major gibberellin inactivation pathway in *Arabidopsis*. *The Plant Cell* **20**, 2420–2436.

Rinne, P.L.H., Kaikuranta, P.M. and Van Der Schoot, C. (2001). The shoot apical meristem restores its symplasmic organization during chilling-induced release from dormancy. *The Plant Journal* **26**, 249–264.

Rood, S.B., Pharis, R.P. and Koshioka, M. (1983). Reversible conjugation of gibberellins *in situ* in maize. *Plant Physiology* **73**, 340–346.

Saito, H., Oikawa, T., Hamamoto, S. *et al.* (2015). The jasmonate-responsive GTR1 transporter is required for gibberellin-mediated stamen development in *Arabidopsis*. *Nature Communications* **6**:6095.

Schneider, G. and Schliemann, W. (1994). Gibberellin conjugates – an overview. *Plant Growth Regulation* **15**, 247–260.

Shani, E., Weinstain, R., Zhang, Y. *et al.* (2013). Gibberellins accumulate in the elongating endodermal cells of *Arabidopsis* root. *Proceedings of the National Academy of Sciences USA* **110**, 4834–4839.

Shininge, Tl (1972). Gibberellin substitution for requirement of cotyledons in stem elongation in *Pisum Sativum* seedlings. *Plant Physiology* **49**, 341–344.

Silverstone, A.L., Chang, C., Krol, E. and Sun, T.P. (1997). Developmental regulation of the gibberellin biosynthetic gene *GA1* in *Arabidopsis thaliana*. *The Plant Journal* **12**, 9–19.

Singh, D.P., Jermakow, A.M. and Swain, S.M. (2002). Gibberellins are required for seed development and pollen tube growth in *Arabidopsis*. *The Plant Cell* **14**, 3133–3147.

Sponsel, V.M. and Hedden, P. (2004). Gibberellin biosynthesis and inactivation. In: *Plant Hormones. Biosynthesis, Signal transduction, Action!* (ed. P.J. Davies). New York, United States, Kluwer Academic Publishers.

Stant, M.Y. (1961). The effect of gibberellic acid on fibre-cell length. *Annals of Botany* **25**, 453–462.

Stant, M.Y. (1963). The effect of gibberellic acid on cell width and the cell-wall of some phloem fibres. *Annals of Botany*, **27**, 185–196.

Suttle, J.C. and Hultstrand, J.F. (1987). Physiological studies of a synthetic gibberellin-like bioregulator: II. Effect of site of application on biological activity. *Plant Physiology* **84**, 1068–1073.

Suzuki, H., Park, S.H., Okubo, K. *et al.* (2009). Differential expression and affinities of *Arabidopsis* gibberellin receptors can explain variation in phenotypes of multiple knock-out mutants. *The Plant Journal* **60**, 48–55.

Symons, G.M. and Reid, J.B. (2004). Brassinosteroids do not undergo long-distance transport in pea. Implications for the regulation of endogenous brassinosteroid levels. *Plant Physiology* **135**, 2196–2206.

Tanimoto, E. (2012). Tall or short? Slender or thick? A plant strategy for regulating elongation growth of roots by low concentrations of gibberellin. *Annals of Botany* **110**, 373–381.

Teale, W.D., Paponov, I.A. and Palme, K. (2006). Auxin in action: signalling, transport and the control of plant growth and development. *Nature Reviews Molecular Cell Biology* **7**, 847–859.

Thomas, S.G., Phillips, A.L. and Hedden, P. (1999). Molecular cloning and functional expression of gibberellin 2-oxidases, multifunctional enzymes involved in gibberellin deactivation. *Proceedings of the National Academy of Sciences USA* **96**, 4698–4703.

Ubeda-Tomas, S., Swarup, R., Coates, J. *et al.* (2008). Root growth in *Arabidopsis* requires gibberellin/DELLA signalling in the endodermis. *Nature Cell Biology*, **10**, 625–628.

Ueguchi-Tanaka, M., Ashikari, M., Nakajima, M. *et al.* (2005). *GIBBERELLIN INSENSITIVE DWARF1* encodes a soluble receptor for gibberellin. *Nature* **437**, 693–698.

Ueguchi-Tanaka, M., Nakajima, M., Katoh, E. *et al.* (2007). Molecular interactions of a soluble gibberellin receptor, GID1, with a rice DELLA protein, SLR1, and gibberellin. *The Plant Cell* **19**, 2140–2155.

Vieten, A., Sauer, M., Brewer, P.B. and Friml, J. (2007). Molecular and cellular aspects of auxin-transport-mediated development. *Trends in Plant Science* **12**, 160–168.

Voegele, A., Linkies, A., Muller, K. and Leubner-Metzger, G. (2011). Members of the gibberellin receptor gene family *GID1 (GIBBERELLIN INSENSITIVE DWARF1)* play distinct roles during *Lepidium sativum* and *Arabidopsis thaliana* seed germination. *Journal of Experimental Botany* **62**, 5131–5147.

Wang, Q., Little, C.H.A. and Oden, P.C. (1997). Control of longitudinal and cambial growth by gibberellins and indole-3-acetic acid in current-year shoots of Pinus sylvestris. *Tree Physiology* **17**, 715–721.

Wheeler, A.W. (1966). Effect of removing cotyledons apical growing region or trifoliate leaves on growth and growth-substance content of dwarf french bean (*Phaseolus Vulgaris*). *Journal of Experimental Botany* **17**, 621–626.
Willige, B.C., Ghosh, S., Nill, C. *et al.* (2007). The DELLA domain of GA INSENSITIVE mediates the interaction with the GA INSENSITIVE DWARF1A gibberellin receptor of *Arabidopsis*. *The Plant Cell* **19**, 1209–1220.
Wilson, R.N., Heckman, J.W. and Somerville, C.R. (1992). Gibberellin is required for flowering in *Arabidopsis thaliana* under short days. *Plant Physiology*, **100**, 403–408.
Yamaguchi, S. (2008). Gibberellin metabolism and its regulation. *Annual Reviews of Plant Biology* **59**, 225–251.
Yano, K., Aya, K., Hirano, K. *et al.* (2015). Comprehensive gene expression analysis of rice aleurone cells: probing the existence of an alternative gibberellin receptor. *Plant Physiology* **167**, 531–544.

Annual Plant Reviews (2016) **49**, 121–152
doi: 10.1002/9781119210436.ch5

Chapter 5

GIBBERELLINS IN FUNGI, BACTERIA AND LOWER PLANTS: BIOSYNTHESIS, FUNCTION AND EVOLUTION

Bettina Tudzynski[1], Lena Studt[1] and María Cecilia Rojas[2]

[1] *Westfälische Wilhelms-Universität Münster, Institut für Biologie und Biotechnologie der Pflanzen, Germany*
[2] *Laboratorio de Bioorgánica, Departamento de Química, Facultad de Ciencias, Universidad de Chile, Chile*

Abstract: The rice pathogen *Fusarium fujikuroi*, as well as two distantly related fungi, *Sphaceloma manihoticola* and *Phaeosphaeriae* sp., contains clusters of gibberellin (GA) biosynthesis genes and produces GAs. Fungal GAs are structurally identical to those synthesised by higher plants, although the respective biosynthetic pathways, genes and enzymes differ. Besides fungi, some bacteria synthesise GAs. An operon of GA biosynthesis genes was found in the symbiotic rhizobacterium *Bradyrhizobium japonicum* and in other *Rhizobium* species. This operon encodes the enzymes of GA_9 biosynthesis and includes a reductase or alcohol dehydrogenase gene not present in fungal clusters. Differences between genes and enzymes indicate convergent evolution of the GA biosynthesis pathway in higher plants, fungi and bacteria. In contrast to higher plants, GAs are not produced by all lower plants. Mosses or liverworts synthesise *ent*-kaurenoids, but not GAs, while lycophytes and ferns synthesise GAs or GA-related products that participate in the regulation of reproductive development.

Keywords: Gibberellins, fungi, *Fusarium*, *Sphaceloma*, rhizobacteria, *Bradyrhizobium*, lower plants, mosses, lycophyte, ferns, function, evolution

Annual Plant Reviews, Volume 49: The Gibberellins, First Edition.
Edited by Peter Hedden and Stephen G. Thomas.

5.1 Introduction

Although gibberellins (GAs) are plant hormones, they were first identified in a fungus belonging to the genus *Fusarium*. This fungus causes 'bakanae' (foolish seedling) disease of rice which was already described more than 100 years ago by the Japanese scientist S. Hori (Hori, 1898). Later it has been shown that the bakanae fungus secretes a 'stimulus' that leads to the elongation of rice internodes (Sawada, 1912). The chemical structure of this metabolite was achieved when Yabuta and his colleague Sumiki succeeded in crystallising the active principle (gibberellin A) and confirmed its biological properties as stimulation of seedling elongation not only in rice, but also in a number of economically important crop plants (Yabuta and Sumiki, 1938). Only in the mid-1950s, Margaret Radley at ICI demonstrated the presence of gibberellin-like substances in higher plants, indicating that GAs were naturally occurring plant hormones (Radley, 1956) (for details see Chapter 1).

Beside *Fusarium* (*Gibberella*) *fujikuroi*, two distantly related fungi were also shown to produce GAs: *Sphaceloma manihoticola* (Rademacher and Graebe, 1979; Zeigler *et al.*, 1980) and *Phaeosphaeria sp.* (Sassa *et al.*, 1989). Currently, there are 136 GAs known, which are termed GA_1 to GA_{136}, and 25 of these GAs are produced by fungi (MacMillan, 2002; Sponsel and Hedden, 2004).

Only recently has it been shown that bacteria, and mainly plant growth promoting rhizobacteria, are also able to synthesise GAs. The best known example is the legume-associated root-colonising bacterium *Bradyrhizobium japonicum*, which contains genes encoding GA biosynthesis enzymes (Morrone *et al.*, 2009; Méndez *et al.* 2014). In addition, GA_1 and GA_3 were found to be produced by *Azospirillum* spp. (Bottini *et al.*, 1989; Piccoli and Bottini, 1994) and GA_4 by *Rhizobium phaseoli* and *Bacillus* spp. (Atzorn *et al.*, 1988; Gutierrez-Mañero *et al.*, 2001).

While GAs are ubiquitous in seed plants, they are not produced by all lower plants. Thus, the lycopod *Selaginella moellendorffii* biosynthesises GAs, but no GAs have been found in the moss *Physcomitrella patens*. However, *P. patens* produces *ent*-kaurene and *ent*-kaurenoic acid, both common precursors for GAs (Hayashi *et al.*, 2010).

This review summarises the progress that was made during the last decades in studying GA production in fungi, bacteria and lower plants, with specific emphasis on the evolution of GA biosynthetic genes in these organisms.

5.2 Gibberellin biosynthesis in fungi

5.2.1 The biosynthetic pathway in *F. fujikuroi*: genes and enzymes

Fungal GAs are synthesised via the mevalonic acid (MVA) pathway, starting from acetyl-CoA via hydroxymethylglutaryl (HMG) coenzyme A, MVA, the basic isoprenoid unit isopentenyl diphosphate (IPP), geranyldiphosphate (GDP), farnesyl diphosphate (FDP) and geranylgeranyl diphosphate (GGDP) (Homann *et al.*, 1996; Mende *et al.*, 1997; Woitek *et al.*, 1997).

In the 1960s and early 1970s, the general biosynthetic pathway to gibberellic acid (GA_3), the end-product of GA biosynthesis in *F. fujikuroi*, was determined by feeding ^{14}C-labelled acetate and mevalonate to mutant strains in which specific steps of the pathway were blocked. Subsequently, labelled intermediates were isolated and structurally elucidated (reviewed in Bearder, 1983).

A milestone in research on GA biosynthesis was the identification of genes encoding GA biosynthesis enzymes from higher plants (Lange *et al.*, 1994; Sun and Kamiya, 1994; Phillips *et al.*, 1995; Hedden and Phillips, 2000) on the one hand, and from *F. fujikuroi* (and two other GA-producing fungi) on the other (Kawaide *et al.*, 1997; Tudzynski *et al.*, 1998; Tudzynski and Hölter, 1998; Bömke *et al.*, 2008a). Only then was it possible to discover the fundamental differences in GA biosynthetic pathways between these organisms at the chemical, enzymatic, and genetic levels (Hedden *et al.*, 2002; Yamaguchi, 2008; Bömke and Tudzynski, 2009). This contradicted earlier speculation that fungi acquired the pathway genes from higher plants *via* horizontal gene transfer (Chapman and Regan, 1980).

In filamentous fungi, clustering of genes involved in secondary metabolism is a common feature. In 1997, a GA biosynthetic gene , *P450-1*, was first cloned from *F. fujikuroi* by differential cDNA screening, comparing the expression pattern of cDNA clones under nitrogen-limiting and nitrogen-sufficient conditions (Tudzynski *et al.*, 1998; Tudzynski and Hölter, 1998). A subsequent chromosome walking approach led to the identification of seven GA biosynthetic genes altogether in *F. fujikuroi* that form the GA gene cluster (Tudzynski and Hölter, 1998; Linnemannstöns *et al.*, 1999; Tudzynski *et al.*, 2001; 2002; 2003; Rojas *et al.*, 2001; Figure 5.1). Surprisingly, the cluster contains a second geranylgeranyl diphosphate synthase gene (*GGS2*) in addition to *GGS1* (Mende *et al.*, 1997). GGS2 probably provides GGDP exclusively for GA biosynthesis as deletion of the *GGS2* gene resulted in total loss of GA biosynthesis (B. Tudzynski, unpublished results).

Adjacent to *GGS2* is the diterpene cyclase-encoding gene *CPS/KS*, which is involved in formation of *ent*-kaurene from GGDP via *ent*-copalyl diphosphate. In plants, the two cyclisation steps from GGDP to *ent*-kaurene are catalysed by two separate terpene cyclases (*ent*-copalyl diphosphate synthase CPS and *ent*-kaurene diphosphate synthase KS). In addition, the gene cluster contains four P450 mono-oxygenase-encoding genes (*P450-1* to *P450-4*), and a desaturase gene (*DES*) (reviewed in Tudzynski, 2005).

The function of these seven genes was determined by targeted gene replacement and subsequent identification of accumulated intermediates. Furthermore, each single P450 mono-oxygenase-encoding gene was individually expressed in a *F. fujikuroi* mutant lacking the entire gene cluster. Feeding of labelled precursors to these mutants and elucidation of products allowed the determination of the multi-functional character of the

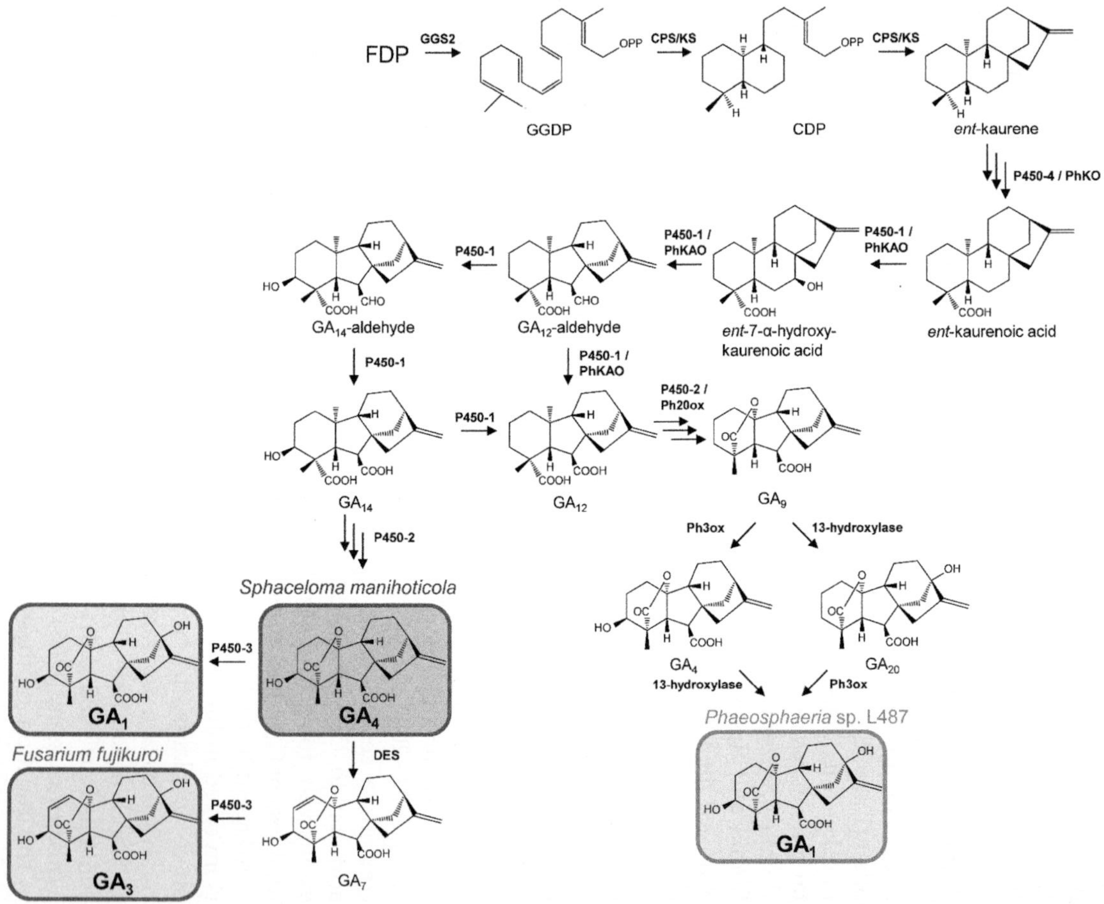

Figure 5.1 The gibberellin biosynthesis pathways in *F. fujikuroi*, *S. manihoticola* and *Phaeosphaeria* sp. L487. The final products are shown in shaded boxes.

ent-kaurene oxidase P450-4 and *ent*-kaurenoic acid oxidase P450-1 (Rojas *et al.*, 2001; Tudzynski *et al.*, 2001; 2002; 2003). P450-4 catalyses the three oxidation steps from *ent*-kaurene to *ent*-kaurenoic acid via *ent*-kaurenol and *ent*-kaurenal (Tudzynski *et al.*, 2001) (Figure 5.1). Remarkably, P450-1 shows *ent*-kaurenoic acid oxidase and 3β-hydroxylase activity. Cultures of mutants containing only P450-1 converted *ent*- [^{14}C]kaurenoic acid efficiently into [^{14}C]GA_{14}, indicating that P450-1 catalyses four sequential steps in the GA-biosynthetic pathway: 7β-hydroxylation, contraction of ring B by oxidation at C-6, 3β-hydroxylation, and oxidation at C-7. The GA precursors *ent*-7α-hydroxy[^{14}C]kaurenoic acid, [^{14}C]GA_{12}-aldehyde, and [^{14}C]GA_{12} were also converted to [^{14}C]GA_{14}, demonstrating the remarkable multi-functionality of this enzyme (Rojas *et al.*, 2001).

In higher plants, two families of P450 mono-oxygenases, the *ent*-kaurene oxidases (KO; CYP701As) and *ent*-kaurenoic acid oxidases (KAO; CYP88As) catalyse the oxidation of *ent*-kaurene to *ent*-kaurenoic acid, and of *ent*-kaurenoic acid to GA_{12}, respectively (see Chapter 2). However, despite their similar catalytic activities, the fungal P450s are not closely related to the plant enzymes and have only about 10–15% amino acid identity (Hedden *et al.*, 2002).

Thus, the first steps of the pathway to GA_{12}-aldehyde are identical in the fungus and in higher plants. After GA_{12}-aldehyde, the pathways in higher plants and *F. fujikuroi* differ. In the fungus, GA_{12}-aldehyde is first 3β-hydroxylated to GA_{14}-aldehyde, which is then oxidised at C-7 to form GA_{14} (Hedden *et al.* 1974; Urrutia *et al.* 2001). In contrast to the fungus, where 3β-hydroxylation is catalysed by P450-1 at an early step, 3β-hydroxylation is the final step of the GA biosynthetic pathway in plants and is catalysed by 2-oxoglutarate-dependent dioxygenases (Hedden and Kamiya, 1997).

In *F. fujikuroi*, GA_{14} is then converted to the 19-carbon gibberellin GA_4 by another P450 mono-oxygenase, the 20-oxidase P450-2, which converts GA_{14} to GA_4 (Tudzynski *et al.*, 2002). The formation of bioactive C_{19} GA-skeletons requires successive oxidation of C-20 from the methyl group of GA_{14} in the fungus, and GA_{12} or GA_{53} in plants. At the end, C-20 is lost as CO_2. Interestingly, the removal of C-20 by GA 20-oxidases is catalysed by soluble 2-oxoglutarate-dependent dioxygenases in plants (Phillips *et al.*, 1995) in contrast to a P450 in the fungus.

GA_4 is then converted to GA_7 by an unusual GA_4 desaturase (DES) that introduces the 1,2-double bond (Tudzynski *et al.*, 2003). Only recently, the nature of this oxidase has been identified by heterologous expression of the fungal gene in *Escherichia coli*. The enzyme has the characteristics of a 2-oxoglutarate-dependent dioxygenase, such as putative iron- and 2-oxoglutarate-binding domains. However, the amino acid sequence homology to known 2-oxoglutarate-dependent dioxygenases is very low (Bhattacharya *et al.*, 2012). Interestingly, expression of *DES* from the cauliflower mosaic virus 35S promoter in the plant species *Solanum nigrum*, *Solanum dulcamara*, and *Nicotiana sylvestris* resulted in substantial growth stimulation, with a 3-fold increase in height in *S. dulcamara* compared with controls (Bhattacharya *et al.*, 2012).

The final step in the fungal pathway is the 13-hydroxylation of GA_7 to produce GA_3, the final product of the GA pathway in *F. fujikuroi*. This reaction is catalysed by P450-3, an enzyme which also catalyses 13-hydroxylation of GA_4 to produce the minor product GA_1 (Tudzynski *et al.*, 2003).

In summary, higher plants and *F. fujikuroi* produce structurally identical GAs. The early steps from GGDP to the formation of GA_{12}-aldehyde are identical in both systems. However, after GA_{12}-aldehyde, there are important differences in the pathways and enzymes involved. 3β-hydroxylation occurs early in the fungal pathway and is catalysed by a cytochrome P450 monooxygenase, whereas it is usually the final step in plants and is catalysed by 2-oxoglutarate-dependent dioxygenases (reviewed in Hedden *et al.*, 2002). Even for 20-oxidation different enzymes are used: dioxygenases in plants and a cytochrome P450 in the fungus. These profound differences indicate that higher plants and fungi have evolved their complex biosynthetic pathways to GAs independently and not by horizontal gene transfer.

5.2.2 Gibberellin production in distantly related fungi

For a long time the bakanae fungus *F. fujikuroi* was thought to be the only fungus that is able to produce GAs. However, Rademacher and Graebe (1979) as well as Zeigler and co-workers (1980) independently discovered that a distantly related fungus, *Sphaceloma manihoticola* (teleomorph: *Elsinoë brasiliensis*), produces considerable amounts of GA_4, but not GA_1, GA_7 or GA_3. The fungus causes superelongation disease in cassava (*Manihot esculenta*), a starchy root crop grown widely in the tropics. However, it took almost 30 years until the GA biosynthetic gene cluster was identified in *S. manihoticola* (Bömke *et al.*, 2008a). This gene cluster consists of five genes, lacking the left and right border genes of the *F. fujikuroi* gene cluster that encodes GA_4 desaturase DES and the 13-hydroxylase P450-3 (Figure 5.2). As in *F. fujikuroi*, *CPS/KS* and the pathway-specific GGDP synthase gene *GGS2*, as well as two of the cytochrome P450 mono-oxygenase genes, *Sm-P450-1* and *Sm-P450-4*, form transcriptional units by sharing the same promoter. Compared with the situation in *F. fujikuroi*, the *Sm-P450-1/Sm-P450-4* complex as well as the *Sm-P450-*2 gene are inverted in the *Sphaceloma* genome.

The *S. manihoticola* cluster genes were functionally characterised by complementing the corresponding *F. fujikuroi* deletion mutants (Bömke *et al.*, 2008a). For instance, *SmCPS/KS* fully restored GA production in the *F. fujikuroi CPS/KS* mutant. Similarly, *SmP450-1*, *SmP450-2* and *SmP450-4* restored GA_3 biosynthesis in the *Fusarium* mutants, confirming that the new gene cluster is indeed responsible for GA biosynthesis. Heterologous expression of *Sm*P450-1 in the *F. fujikuroi* SG139 mutant lacking the gene cluster demonstrated that it is a multi-functional enzyme involved in GA_{14} synthesis from *ent*-kaurenoic acid. As for FfP450-1, ^{14}C-labelled precursors

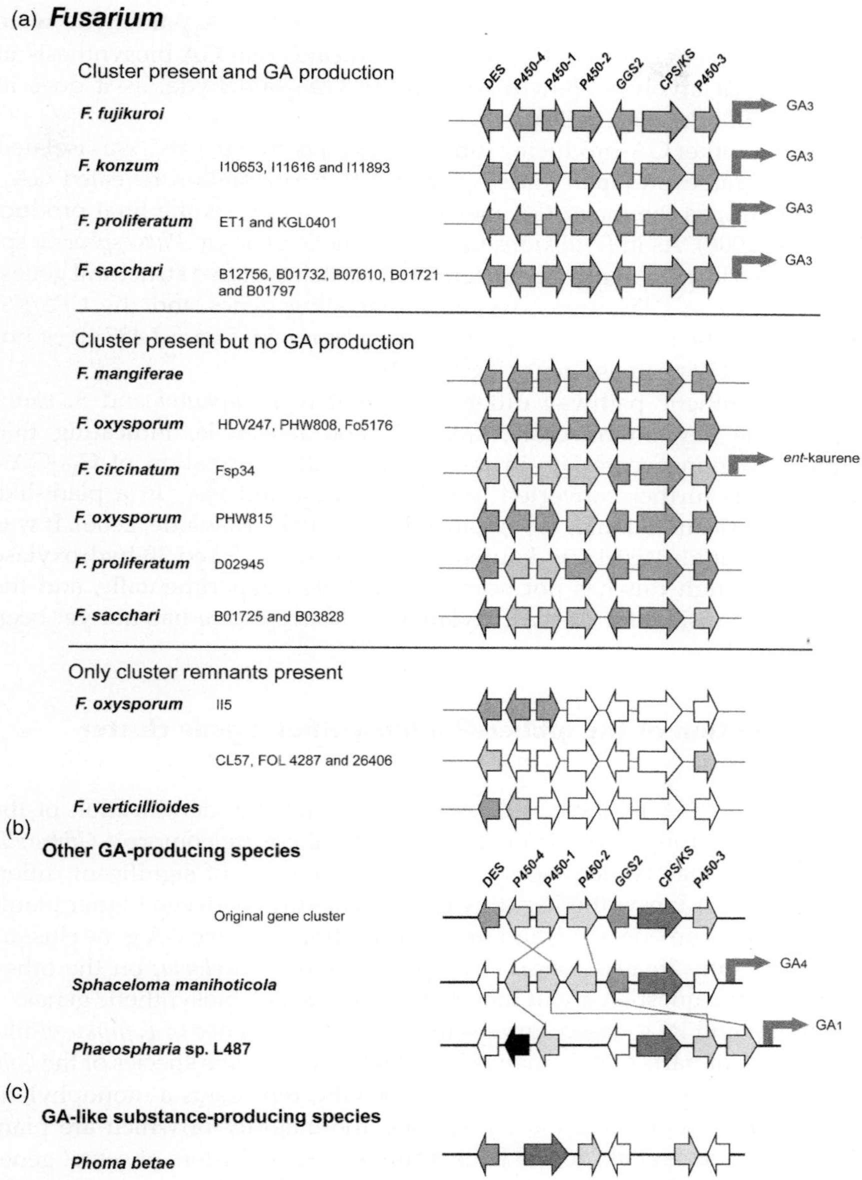

Figure 5.2 Comparison of GA biosynthetic gene clusters in fungal genomes. (a) Presence of GA gene clusters or parts of it in the genomes of *Fusarium* species. Green arrows represent functional genes or genes without obvious mutations. Yellow arrows represent genes with frame shift mutations, early stop codons or amino acid substitutions leading to non-functional enzymes. White arrows represent genes missing in the genome of this *Fusarium* isolate. (b) Functional GA gene clusters in the genomes of distantly related fungal species. (c) Aphidicolin biosynthetic gene cluster in *Phoma betae*. The cluster contains a similar gene pair consisting of a GGS2-like gene and a CPS/KS-like diterpene cyclase gene, both transcribed from the same bidirectional promoter. (See insert for colour representation of this figure.)

such as *ent*-7α-hydroxy-[^{14}C]kaurenoic acid and [^{14}C]GA_{12}-aldehyde were efficiently converted into [^{14}C]GA_{14}. This indicates that GA biosynthesis in *S. manihoticola* involves 3β-hydroxylation of GA_{12}-aldehyde, as it does in *F. fujikuroi* (Bömke *et al.*, 2008a).

In 1989, another GA-producing fungus, *Phaeosphaeria* sp. L487, was isolated from reed grasses in Japan (Sassa *et al.*, 1989). Fermentation revealed GA_1, GA_4 and GA_9 as the major GA metabolites, with GA_1 as the final product (Kawaide, 2006). As in *F. fujikuroi*, GA biosynthetic genes in *Phaeosphaeria* sp. L487 are also located in a gene cluster, which consists of five structural genes: four cytochrome P450 mono-oxygenase-encoding genes and the *CPS/KS*. Interestingly, the GA gene cluster in *Phaeosphaeria* sp. strain L487 does not contain a GGDP synthase gene (Figure 5.2).

The biosynthetic pathway differs from that in *F. fujikuroi* and *S. manihoticola*. The key intermediate, GA_{14}, is not detectable, indicating that 3β-hydroxylation occurs late in the pathway after formation of C_{19}-GAs. Thus, GA_9 is further converted to GA_4 or GA_{20} and GA_1 in a plant-like manner (Sassa and Suzuki, 1990; Kawaide *et al.*, 1995; Kawaide, 2006). It was suggested that PhP450-1 (*ent*-kaurenoic acid oxidase) lacked 3β-hydroxylase activity although this has not been demonstrated experimentally, and the enzyme responsible for 3β-hydroxylation in *Phaeosphaeria* has not yet been characterised.

5.2.3 Evolution of the gibberellin biosynthetic gene cluster in fungi

The history of GA research is closely linked with the identification of the GA-producing fungus, *Fusarium fujikuroi* Nirenberg (teleomorph *Gibberella fujikuroi* (Sawada) Wollenweber). After the discovery of significant differences in the GA biosynthetic pathways and enzymes between higher plants and the fungus on one hand, and the identification of active GA gene clusters in two distantly related fungi, *Phaeosphaeria* and *Sphaceloma*, on the other, there arose the question about the origin of fungal GA biosynthetic genes.

Southern and PCR-based surveys indicated the presence of *F. fujikuroi*-like GA clusters, or parts of it, in many closely related *Fusarium* species of the *Gibberella fujikuroi* species complex (GFC). The GFC represents a monophyletic assemblage of about 50 *Fusarium* species, the majority of which are plant pathogens (Fourie *et al.*, 2013). Thus, complete gene clusters of seven genes have been identified, e.g. in *F. proliferatum*, *F. sacchari*, *F. konzum*, *F. foetens*, *F. napiforme*, *F. miscanthi*, and *F. nisikadoi* (Malonek *et al.*, 2005a) (Figure 5.2). However, production of GAs has only been detected in some isolates of *F. sacchari* and *F. konzum* as well as the *F. proliferatum* isolates ET1 and KGL0401 (Rim *et al.*, 2005; Tsavkelova *et al.*, 2008; Troncoso *et al.*, 2010). The cluster is absent in the genomes of the more distantly related species *F. graminearum* and *F. solani*, which do not belong to the GFC (Malonek *et al.*, 2005a).

The recent genome sequencing of several *Fusarium* species, such as those of *F. fujikuroi* (Wiemann *et al.*, 2013), *F. verticillioides* (Broad Institute), *F. circinatum* (Wingfield *et al.*, 2012), *F. mangiferae* (Wiemann *et al.*, 2013), and 11 additional isolates of *F. oxysporum* (Broad Institute), but also of other ascomycetous fungi, such as *Magnaporthe oryzae* (Dean *et al.*, 2005) or *Claviceps purpurea* (Schardl *et al.*, 2012), provides an opportunity for a more comprehensive analysis of the origin of fungal GA biosynthetic genes than has been possible previously. This genome-based phylogenetic analysis of *Fusarium* species revealed the presence of homologues of the entire *F. fujikuroi* GA cluster in all the members of the GFC sequenced so far: *F. circinatum*, *F. mangiferae*, and *F. proliferatum* (Wingfield *et al.*, 2012; Wiemann *et al.*, 2013; B. Tudzynski and U. Güldener, unpublished). The only exception is the maize pathogen *F. verticillioides*, that contains only one gene (*DES*) of the GA gene cluster (Bömke *et al.*, 2008b). Surprisingly, five isolates of *F. oxysporum*, a species that is closely related, but does not belong to the GFC, also contain the entire gene cluster, but they do not produce any GAs under *in vitro* conditions. The seven other *F. oxysporum* genome sequences have only partial GA gene clusters (Figure 5.2) (Wiemann *et al.*, 2013).

All intact clusters share the same gene order and orientation as the previously described clusters in *F. fujikuroi* and *F. proliferatum* (Malonek *et al.*, 2005a, 2005b). However, some genes of certain isolates seem not to encode functional proteins, e.g. *GGS2* and *CPS/KS* in *F. proliferatum* isolates D02945 and D00502 (Malonek *et al.*, 2005c), and *P450-2* and *P450-3* in *F. oxysporum* isolates PHW815 and FOSC 3-a, whose coding regions are interrupted by premature stop codons (Wiemann *et al.*, 2013) (Figure 5.2). The most closely related *Fusarium* species, such as *F. proliferatum* and *F. mangiferae*, share considerable synteny in content and arrangement of flanking genes compared to *F. fujikuroi*. The *F. circinatum* flanking region has a 38-kb insert and the GA cluster itself is inverted relative to *F. fujikuroi* (Figure 5.2). The syntenic arrangement of GA cluster genes in all species or single isolates with a complete GA cluster suggests that an ancient GA cluster was present in the ancestral *Fusarium* genome before divergence of GFC and *F. oxysporum*.

Surprisingly, almost the same structure of the GA gene cluster can be found in *S. manihoticola*, except for the missing left and right border genes, *DES* and *P450-3*, and two gene inversions (Bömke *et al.*, 2008a). It remains unclear if the GA gene cluster of *Sphaceloma* ancestors has lost the flanking *DES* and *P450-3* homologous genes, perhaps by rearrangements, or if an ancient *Fusarium* species acquired this possibly basic GA gene cluster via horizontal gene transfer and it was coincidentally inserted between *DES* and *P450-3*, which could contribute to GA biosynthesis. It is also possible that *S. manihoticola* gained a part of the cluster by horizontal gene transfer from *Fusarium* or an ancestor.

Interestingly, the genome of the rice blast fungus *M. oryzae* (formerly *M. grisea*) contains two diterpene cyclase genes (MGG_14722 and MGG_01949) with high levels (65%) of sequence identity with the *F. fujikuroi CPS/KS*

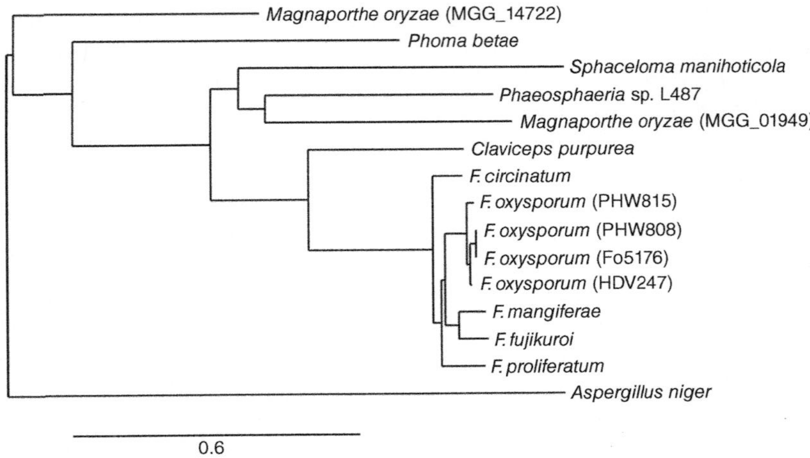

Figure 5.3 Maximum likelihood phylogenetic tree of Cps/Ks and related diterpene synthases from multiple *Fusarium* species and the distantly related fungi *Sphaceloma manihoticola* (AC CAP07655), *Phaeosphaeria* sp. L487 (AC O13284), *Phoma betae* (diterpene cyclase from the aphidicolin gene cluster; AC BAD29971), *Magnaporthe oryzae* (AC MGG_14722 and MGG_01949), *Claviceps purpurea*, (CPUR_08907; kindly provided by Prof. Paul Tudzynski), and *Aspergillus niger* (AC XP001398730) (modified after Wiemann *et al.*, 2013). Sequences for *F. proliferatum*, and non-*Fusarium* genera other than *Claviceps* were obtained from the NCBI database via BLASTp analysis with *F. fujikuroi* Cps/Ks. *F. oxysporum* sequences were obtained from the Broad Institute's *Fusarium* Comparative Database. Sequences for *F. mangiferae* and *F. circinatum* are from Wiemann *et al.* (2013) and unpublished data.

gene (http://www.broad.mit.edu/ annotation/fungi/fgi/) (Figure 5.3). One of them, MGG_14722, is part of a hypothetical gene cluster, also containing a putative pathway-specific GGDP synthase-encoding gene, MGG_00026, and a cytochrome P450 monooxygenase-encoding gene MGG_00024 with homology to the 20-oxidase *P450-2* genes from *F. fujikuroi* and *S. manihoticola*. Its function has not yet been studied experimentally. However, gene MGG_14722 was shown to encode a functional CPS/KS as it is able to restore GA production in the *CPS/KS* deletion mutant of *F. fujikuroi* (Bömke and Tudzynski, 2009). Full functionality has also been shown by the same complementation approach for a *CPS/KS* homologue in the ergot fungus *C. purpurea* (J. Schürmann and P. Tudzynski, personal communication). It remains unclear whether ancestors of these two phytopathogenic fungi once produced GAs and contained an entire GA biosynthetic gene cluster, which was lost during evolution.

A typical feature for all GA biosynthetic gene clusters (except for that from *Phaeosphaeria*) is the physical link of the *CPS/KS* gene with a pathway-specific GGDP synthase gene (*GGS2*) (Figure 5.2). A similar situation has been found in other fungal diterpene gene clusters, e. g. in the aphidicolin gene cluster in *Phoma betae* (Toyomasu *et al.*, 2004). This characteristic feature supports the hypothesis that the presence of a dedicated GGDP synthase-encoding gene might be a kind of molecular signature that is conserved among diterpene biosynthetic gene clusters (Zhang *et al.*, 2004) (Figures 5.2 and 5.3). An explanation might be a cellular compartmentation of the respective biosynthetic pathway with separate substrate pools.

Phylogenetic analysis of the currently identified fungal *CPS/KS*-like diterpene synthase genes resolved the *Fusarium* genes into a single clade that is distinct from *CPS/KS* genes in other fungi. This result indicates that the GA cluster in the analysed *Fusarium* species has likely evolved by vertical inheritance from a common ancestor (Figure 5.3). In addition, previously reported PCR and Southern data, as well as recent genome sequence data, indicate that partial GA clusters are derived via similar patterns of gene loss in three phylogenetically distinct lineages of the genus *Fusarium*: species of the GFC, the *F. oxysporum* species complex and *F. minscanthi/F. nisikadoi*. In all three lineages, partial GA clusters always lack the same four genes: *P450-2, GGS2, CPS/KS* and *P450-3* (Wiemann *et al.*, 2013). The similar patterns of gene loss in different *Fusarium* lineages indicate that degeneration of the cluster is not random.

5.2.4 The role of gibberellins in plant infection

Despite the broad distribution of the GA gene cluster among the genus *Fusarium*, the ability to produce GAs is restricted mainly to the rice pathogen *F. fujikuroi*. Some isolates of the highly related species *F. proliferatum, F. konzum* and *F. sacchari* also produce GAs, although in low amounts (Tsavkelova *et al.*, 2008; Troncoso *et al.*, 2010).

Although 'bakanae' disease was described more than 100 years ago, the role of GAs in pathogenesis of *F. fujikuroi* on rice is not well understood. Recently, extended rice infection assays were performed with a DsRed-labelled *F. fujikuroi* wild-type strain and the GA-non-producing mutant SG139, which lacks the entire GA gene cluster. Microscopic analysis of infected rice roots revealed that the two strains can equally penetrate the rice root epidermis. Both strains also show the same apoplastic growth behavior within the parenchyma cells of the epidermis and the cortex. However, the total number of successfully invaded symplasts per rice root differed significantly indicating that GAs contribute to the ability of *F. fujikuroi* to grow invasively in symplasts of parenchyma cells of rice epidermis and cortex rather than play a role in initial root colonisation (Wiemann *et al.*, 2013).

It is well known that jasmonates (JAs), lipid-derived small-molecule phytohormones, regulate multiple plant growth responses, such as the adaptation to abiotic stresses and the defense against pathogens and insects. Recent studies have demonstrated that GAs and DELLA proteins, a family of plant growth repressors, have an important function in mediating JA signalling.

DELLAs were shown to mediate elevated resistance of plants to necrotrophs via potentiating JA signalling. GAs modulate JA signalling by promoting the degradation of DELLAs and therefore reduce the plant's resistance to pathogens (Navarro *et al.*, 2008). Studying the molecular details of this phytohormone interaction, Hou and colleagues showed that the JAZ1 protein, a key repressor of JA signalling, interacts *in vivo* with DELLA proteins. The latter prevent the interaction of JAZ1 with a key transcriptional activator of JA responses, MYC2. As a consequence, MYC2 can activate the expression of its target genes. However, the presence of GAs triggers the degradation of DELLAs, allowing JAZ1 to bind MYC2 and suppress MYC2-dependent JA-signalling outputs (Hou *et al.*, 2010).

In conclusion, recent genome-based studies revealed the presence of intact or partial GA gene clusters in multiple *Fusarium* species/isolates. However, the ability to produce high levels of GAs and express GA biosynthetic genes is limited to the species *F. fujikuroi* and probably provides a selective advantage during infection of the preferred host plant, rice. Although the presence of a functional GA gene cluster is not essential for *F. fujikuroi* to colonise rice root cells, secretion of GAs by the pathogen contributes to further invasion in rice tissue, probably by interfering with the JA-mediated defence mechanisms.

5.2.5 Strain improvement

GAs have applications in agriculture and horticulture to enhance crop yields, fruit size and quality, and to induce earlier flowering (see Chapter 11). The development of high-producing fungal strains is a prerequisite for the efficient biotechnological production of GAs. In the past all strain improvement attempts were based on classic genetic approaches, such as random mutagenesis and selection. Identification of the GA biosynthetic genes, as well as early isoprenoid genes such as *FPPS* and *HMGR* encoding the farnesyl diphosphate synthase and the hydroxymethylglutaryl coenzyme A reductase, respectively (Homann *et al.*, 1996; Woitek *et al.*, 1997) allow molecular approaches to directly affect GA biosynthesis. However, over-expression of *FPPS* and *HMGR* resulted in reduced production levels, probably due to negative feedback regulation of HMGR activity by precursors of ergosterol biosynthesis. All eukaryotic HMGR enzymes described so far consist of two distinct domains: a hydrophobic N-terminal domain with eight membrane-spanning regions containing a sterol-sensing domain, and a hydrophilic C-terminal domain containing the catalytic domain (DeBose-Boyd, 2008). To circumvent the postulated feedback inhibition of HMGR by intermediates of the isoprenoid pathway, the deduced N-terminal transmembrane domain of the *F. fujikuroi* HMGR protein was deleted, and the remaining catalytic domain was over-expressed resulting in a 2.5-fold increased GA yield compared to the wild type (Albermann *et al.*, 2013a). Also the over-expression of the first GA-specific pathway gene *GGS2* by a strong promoter, or integration of additional *GGS2* and *CPS/KS* gene copies resulted in an enhanced production level of 150% (Albermann *et al.*, 2013a).

Another possibility to increase GA yield is the deletion or down-regulation of competing pathways, e.g. those for the biosynthesis of polyketide synthase (PKSs) and non-ribosomal peptide synthase (NRPSs)-derived secondary metabolites. To achieve the immediate loss of all PKS- and NRPS-derived products, the 4′- phosphopantheteinyl-transferase-encoding gene *PPT1* has been recently deleted in *F. fujikuroi.* This enzyme is essential for post-translational modification of acyl carrier protein modules that are present in all PKS and NRPS enzymes. The *PPT1* deletion mutant indeed produced significantly increased GA amounts, at least in a low-producing wild-type strain (Wiemann *et al.*, 2012).

Besides the quantity, the spectrum of the produced GAs can also be directly affected by molecular approaches. Thus, deletion of the 13-hydroxylase-encoding gene *P450-3* led to production of GA_7 instead of GA_3 and GA_1. Deletion of the GA_4 desaturase gene *DES* resulted in formation of GA_1 as the major product, while double deletion of both genes led to accumulation of GA_4 (Albermann *et al.*, 2013b).

5.3 Gibberellin biosynthesis in bacteria

5.3.1 Free-living rhizobacteria

Some bacterial systems, mainly rhizobacteria that improve plant growth, are known to produce GAs, which have been detected in low amounts in the culture media. Thus, bioactive GAs, GA_1, GA_3 and/or GA_4, have been identified by gas chromatography-mass spectrometry (GC-MS) in liquid cultures of several free-living rhizobacteria isolated from roots or from the rhizosphere: these include *Azospirillum lipoferum* (Bottini *et al.*, 1989), *Azospirillum brasilense* (Janzen *et al.*, 1992), *Acetobacter diazotrophicus* (Bastian *et al.*, 1998), *Bacillus licheniformis, Bacillus pumilus* (Gutierrez-Mañero *et al.*, 2001) and others (Piccoli *et al.*, 2011; Kang *et al.*, 2012; 2013). These micro-organisms improve plant growth and yield of different crops through complex mechanisms that may include nitrogen fixation, nitrate reductase activity, siderophore production and/or phytohormone biosynthesis, such as GAs, which would be released into the host plant (Bottini *et al.*, 2004).

Under laboratory conditions, GA_1 and GA_3 were detected in media from cultures of *A. lipoferum* and *A. brasilense*, two endophytic bacteria that colonise the roots of cereals including wheat or rice (Bottini *et al.*, 1989; Janzen *et al.*, 1992; Piccoli *et al.*, 1996; 1997). The C_{19} products GA_9 and GA_{20} as well as the 13-hydroxylated C_{20} precursor GA_{19} were also found in *A. lipoferum* cultures (Piccoli *et al.*, 1996; 1997). GA biosynthesis by *Azospirillum* spp. was also demonstrated *in vivo*, i.e. in association with higher plants. Both *A. lipoferum* and *A. brasilense* were able to restore wild-type-like growth to GA-deficient maize or rice dwarf mutants blocked at GA 3-oxidase or *ent*-kaurene oxidase, respectively (Lucangeli and Bottini, 1996), or to reverse the inhibitory growth effect of uniconazole in maize seedlings (Lucangeli and Bottini, 1997). In association with the rice GA-deficient mutant *dy* lacking 3β-hydroxylase, *A. lipoferum* hydrolysed $[^2H_2]GA_{20}$ glucosyl conjugates and converted the corresponding aglycone into $[^2H_2]GA_1$, promoting sheath leaf elongation (Cassán *et al.*, 2001a; b).

Besides producing GAs, *A. lipoferum* cultures metabolised exogenously added $[^2H_2]GA_{20}$ or $[^2H_2]GA_9$ (Piccoli and Bottini, 1994; Piccoli *et al.*, 1996). $[^2H_2]GA_{20}$ was converted into $[^2H_2]GA_1$, identified from its mass spectrum (Piccoli and Bottini, 1994), while $[^2H_2]GA_9$ was converted exclusively into $[^2H_2]GA_3$ (Piccoli *et al.*, 1996). These data suggest the existence of two branches for GA biosynthesis in *A. lipoferum*: an early 13-hydroxylation pathway that would give GA_1 through GA_{20}, and an early non-hydroxylation branch that forms GA_9, which would be further converted to GA_3 (Bottini *et al.*, 2004). However, neither the intermediates or earlier precursors involved in these reactions, nor the corresponding genes or enzymes have been identified so far.

5.3.2 Symbiotic rhizobacteria: genes and reactions of the gibberellin biosynthetic pathway

In contrast to the situation for free-living rhizobacteria, genes and enzymes involved in GA biosynthesis have been described for several nitrogen-fixing symbiotic rhizobacteria, belonging to the four major rhizobia genera, associated with their host legumes (Tully *et al.*, 1998; Keister *et al.*, 1999; Morrone *et al.*, 2009; Hershey *et al.*, 2014; Méndez *et al.*, 2014). Initially, Tully *et al.* (1998) identified and sequenced an operon of eight genes proposed to be involved in diterpenoid biosynthesis in the soybean (*Glycine max*) symbiont *B. japonicum*. This operon consists of genes putatively encoding three cytochrome P450 monooxygenases (CYP112, CYP114 and CYP117), a [3Fe-4S]ferredoxin (Fd), a short chain alcohol dehydrogenase or reductase (SDR), an isoprenyl diphosphate synthase and two diterpene synthases. A similar gene cluster was found in other *Rhizobium* species (Keister *et al.*, 1999; Freiberg *et al.*, 1997; Hershey *et al.*, 2014). By heterologous expression in *Escherichia coli* complemented with a plant geranylgeranyl diphosphate synthase (GGPS), the two putative diterpene synthases BjCPS and BjKS from *B. japonicum* were biochemically characterised and shown to have *ent*-copalyl diphosphate synthase (CPS) and *ent*-kaurene synthase (KS) activity, respectively (Morrone *et al.*, 2009). Coexpression of BjCPS and BjKS in *E. coli* containing GGPS produced *ent*-kaurene, the first committed intermediate of GA biosynthesis, detected in hexane extracts of liquid cultures by GC-MS. Moreover, BjCPS or BjKS could functionally substitute for *ent*-CDP-specific CPS or KS from *Arabidopsis thaliana* when coexpressed in *E. coli* together with GGPS, leading to *ent*-kaurene production. The catalytic functions of *B. japonicum* diterpene cyclases, as well as their stereochemistry confirmed that the diterpenoid gene operon may be involved in GA biosynthesis (Morrone *et al.*, 2009).

Besides *B. japonicum*, *CPS* and *KS* homologues have been identified in different species of the four main genera of rhizobia, i.e. *Rhizobium*, *Sinorhizobium*, *Mezorhizobium* and *Bradyrhizobium* (Hershey *et al.*, 2014). In all cases coexpression of *CPS* and *KS* homologues in *E. coli* complemented with a plant GGPS led to production of *ent*-kaurene. Examination of the genomic context for each of the characterised *CPS* and *KS* genes revealed homologues of all the other genes from the *B. japonicum* diterpenoid operon in the same relative order, which indicates that the gene cluster is conserved and has the same phylogenetic origin. In *R. etli* and *M. loti* some genes are fused (*CYP114* and *Fd* or *Fd* and *SDR* respectively), but still present homology to the separate genes from *B. japonicum*, while in *R. etli* the isoprenyl diphosphate synthase gene is disrupted with a second isoprenyl diphosphate synthase gene in close proximity to the diterpenoid operon (Hershey *et al.*, 2014).

The only other gene of the rhizobial operon that has been functionally characterised is the isoprenyl diphosphate synthase gene (putative *GGPS*) adjacent to *CPS* (Hershey *et al.*, 2014). In this case two fragments of the *Sinorhizobium fredii* operon containing *GGPS-CPS-KS* or *CPS-KS* were over-expressed in *Sinorhizobium meliloti*, a closely related bacterium that lacks the rhizobial diterpenoid operon. Expression of *GGPS-CPS-KS* resulted in *ent*-kaurene production, while expression of *CPS-KS* did not, confirming that the isoprenyl diphosphate synthase from *S. fredii* is indeed a GGDP synthase and confirming the capability of the rhizobial diterpenoid biosynthesis operon at least to produce *ent*-kaurene (Hershey *et al.*, 2014).

Oxidative reactions of the GA biosynthesis pathway after *ent*-kaurene have been recently demonstrated in *B. japonicum* bacteroids by administrating labelled GA precursors and identification of the metabolic products by GC-MS (Méndez *et al.*, 2014). Symbiotic bacteroids isolated from soybean root nodules efficiently converted $[^{14}C_1]GA_{12}$ into several C-20 oxidation products, identified by GC-MS as $[^{14}C_1]GA_{24}$ (C-20 aldehyde), $[^{14}C_1]GA_{15}$ (lactonic C-20 alcohol product) and $[^{14}C_1]GA_9$ (C_{19}-GA formed by elimination of C-20 as CO_2). The GA_9-17-nor-16-one, formed by oxidative cleavage of the 16-17 double bond, was additionally detected as a major product. The early GA precursors, *ent*-$[^{14}C_1]$kaurenoic acid and $[^{14}C_4]GA_{12}$ aldehyde, were also converted into ^{14}C-GA_9 plus intermediates partially oxidised at C-20 and the GA_9-17-nor-16-one (found as a labelled product from $[^{14}C_4]GA_{12}$ aldehyde). Finally *ent*-$[^{14}C_4]$kaurene was converted into $[^{14}C_4]GA_9$, although with a low efficiency, probably due to difficult access by this hydrophobic precursor to the bacteroid GA oxidases. Interestingly, no 3β- or 13-hydroxylated GAs were formed as products from any of the labelled precursors indicating that only the non-hydroxylation pathway of GA biosynthesis is present in *B. japonicum* (Figure 5.4). This was confirmed by the lack of conversion of $[^{14}C_1]GA_4$ (3β-hydroxylated C_{19}-GA) or $[^{14}C_1]GA_{20}$ (13-hydroxylated C_{19}-GA) by the bacteroids. The bacterial 20-oxidase utilises

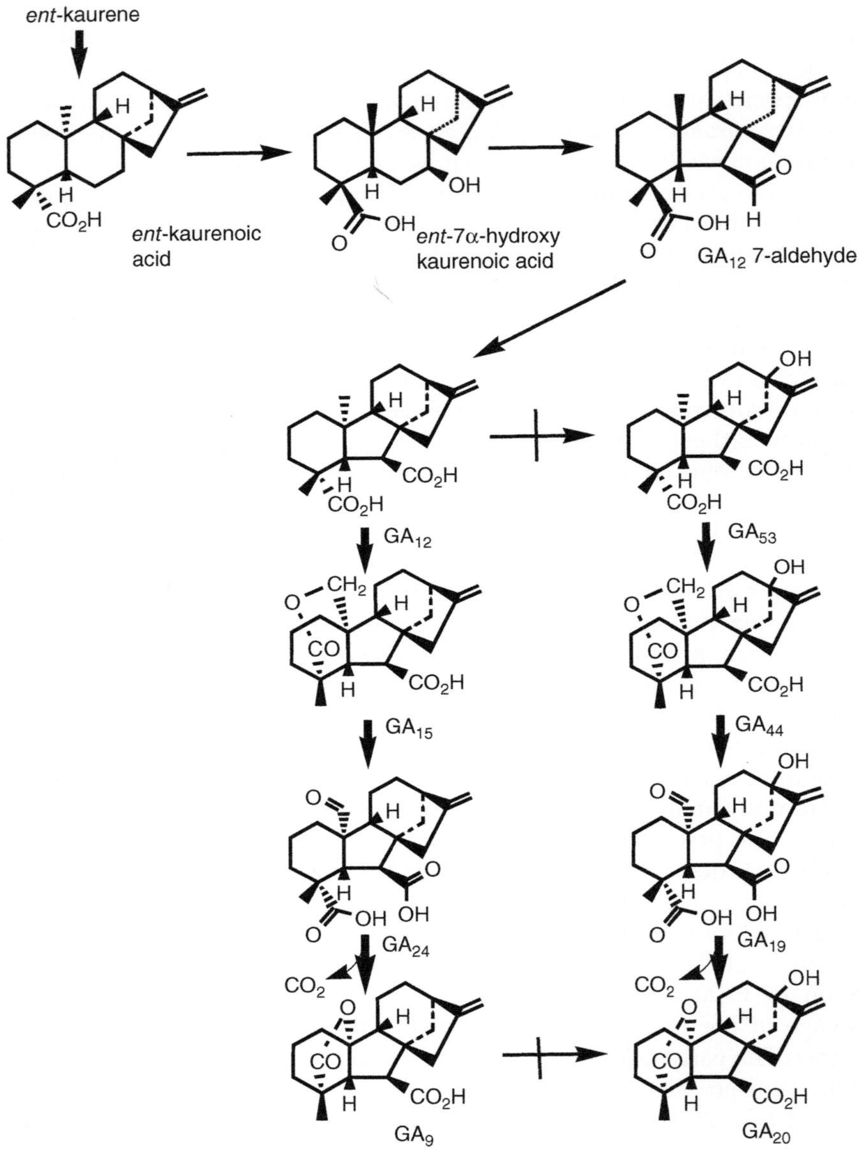

Figure 5.4 Reactions of GA biosynthesis demonstrated in *B. japonicum* bacteroids.

exogenous 13-hydroxylated GAs besides [$^{14}C_1$]GA_{12}, [$^{14}C_1$]GA_{53} (C-20 methyl), [$^{14}C_1$]GA_{44} (lactonic C-20 alcohol) and [$^{14}C_1$]GA_{19} (C-20 aldehyde), which were converted into the C_{19} final product [$^{14}C_1$]GA_{20}. No further products were formed, which confirmed the absence of 3β-hydroxylase activity in *B. japonicum* (Méndez *et al.*, 2014). Thus, the products of the GA biosynthetic pathway in this symbiotic rhizobacteria differ from those in plants or fungi, which synthesise mainly 3β and/or 3β,13-hydroxylated GAs (Hedden *et al.*, 2002).

Oxidative reactions of GA biosynthesis demonstrated in *B. japonicum* correspond to those catalysed by *ent*-kaurene oxidase, *ent*-kaurenoic acid oxidase and 20-oxidase that would likely be multi-functional P450 mono-oxygenases corresponding to those encoded from the bacterial diterpenoid gene operon that contains three P450 genes: *CYP112*, *CYP114* and *CYP117* (Tully *et al.*, 1998, Hershey *et al.*, 2014). In addition, an alcohol dehydrogenase or reductase could participate in some of the oxidative steps, as suggested by the presence of a gene with homology to short chain alcohol dehydrogenases or reductases in the operon. The finding of strong GA oxidase activities in *B. japonicum* bacteroids together with demonstration of *ent*-kaurene synthesis by heterologously expressed *BjCPS* and *BjKS* would be indicative of the ability to produce GAs by *Rhizobium* species that contain the diterpenoid biosynthesis gene operon. Even when GA biosynthesis by isolated bacteroids has not been demonstrated, endogenous GA_9 and GA_{24}, two products formed from GA precursors by *B. japonicum* bacteroids, were identified by GC-MS in methanolic extracts of soybean symbiotic root nodules, while GA_9 was detected in the plant cytosol nodule fraction (Méndez *et al.*, 2014). This suggests GA_9 synthesis by the bacteroids, which would be secreted into the plant tissue and further transformed into bioactive GAs by the plant 3β and/or 13-hydroxylases.

5.3.3 Function and evolution

GAs produced by plant growth-promoting rhizobacteria are secondary metabolites whose function may be related to the interaction they maintain with their host plants. In *Rhizobium* spp. there is evidence for a possible role of GAs in symbiosis since the diterpenoid biosynthesis gene operon is located in the symbiotic module, plasmid or symbiotic island (Hershey *et al.*, 2014). Moreover, expression of the diterpenoid biosynthesis genes as well as activity of GA oxidases have been demonstrated in the differentiated bacterial form present in symbiotic root nodules (bacteroids), but not in free-living *Rhizobium* cultures (Hauser *et al.*, 2007; Pessi *et al.*, 2007; Salazar *et al.*, 2010; Méndez *et al.*, 2014). Transcripts of *CYP112* and *CYP114* were detected in *B. japonicum* bacteroids in contrast to aerobic liquid cultures (Pessi *et al.*, 2007) while *CPS* and *KS* transcripts were found in symbiotic bacteroids from soybean plants around the time of flowering (Méndez *et al.*, 2014). GA 20-oxidase activity was also present in *B. japonicum* bacteroids in contrast to aerobic liquid cultures that were unable to convert GA oxidase substrates. All these findings are consistent with known induction of symbiotically relevant genes by low oxygen tension plus other factors that may be present in legume root nodules (Pessi *et al.*, 2007). In fact, the expression of *CYP112* and *CYP114* in *B. japonicum* as well as expression of the diterpenoid biosynthesis genes in *R. etli* have been shown to depend on the symbiosis-specific sigma factor RpoN and associated oxygen responsive transcription factor NifA (Hauser *et al.*, 2007; Salazar *et al.*, 2010) while *M. loti* and *S. fredii* contain sequences upstream of the diterpenoid biosynthesis operon that also suggest this kind of regulation (Hershey *et al.*, 2014).

The composition of the rhizobial diterpene biosynthesis operon together with production of the non-hydroxylated C_{19} product GA_9, but not of 3β- or 13-hydroxylated GAs, by the GA oxidases from *B. japonicum* indicate that rhizobial GA biosynthesis would represent an independently assembled pathway from that in plants or fungi. Although sequence comparisons between diterpene cyclases involved in *ent*-kaurene biosynthesis suggest a potential common origin for these enzymes in rhizobia, plants or fungi (Morrone *et al.*, 2009), there are important differences in further oxidative steps of the GA biosynthetic pathway. Plants use P450 mono-oxygenases and several 2-oxoglutarate-dependent dioxygenases to produce bioactive GAs (Hedden and Thomas, 2012; MacMillan, 1997), while fungi use only P450 mono-oxygenases (Bömke and Tudzynski, 2009) and, for *Fusarium* species, a desaturase that has been recently demonstrated to be a 2-oxoglutarate-dependent dioxygenase (Batthacharya *et al.*, 2012). In contrast, there are three P450 genes, with no significant homology to plant P450s, in the rhizobial operon and no 2-oxoglutarate-dependent dioxygenase genes, and also a gene with homology to short-chain alcohol dehydrogenases or reductases that could participate in some oxidative steps. Thus, GA biosynthesis by rhizobia would represent a third convergent evolutionary solution for production of these diterpene phytohormones (Morrone *et al.*, 2009). Interestingly the free-living rhizobacterium *A. lipoferum* has been reported to synthesise 3β-and 13-hydroxylated GA products, GA_1 and GA_3 (Bottini *et al.*, 2004), which indicates diversity in the molecular organisation of GA biosynthesis by soil bacteria.

The diterpenoid biosynthesis gene operon is not present in all *Rhizobium* spp., but has a scattered distribution within the plasmid or symbiotic island of the four main rhizobia genera with functional conservation of at least the ability to synthesise *ent*-kaurene (Hershey *et al.*, 2014). Although it has been described that polar disruption of the *CYP112* gene in *B. japonicum,* resulting in the loss of both CYP112 and CYP114, has no significant effect on soybean nodulation or nitrogen fixation (Tully and Keister, 1993), the conservation of the rhizobial operon in the different rhizobial genera together with its location in the symbiotic module and expression in bacteroids strongly suggests that bacterial GAs may be synthesised during symbiosis and would provide a selective advantage by modulating interaction of the symbiont with the host plant (Hershey *et al.*, 2014). Further functional characterisation of the operon genes, as well as identification of the reactions and final products of the GA pathway in other *Rhizobium* spp. will give a better understanding of rhizobial GA biosynthesis and its possible role in the symbiotic relationship with legume plants.

5.4 Gibberellin biosynthesis and signalling components in lower plants

GA-like substances have been reported in algae (Mowat, 1965), lichens and mosses (Ergün *et al.*, 2002), although unequivocal identification of GAs in lower plants by reliable chemical methods has only been described in the lycophyte *S. moellendorffii*, that produces GA_4 (Hirano *et al.*, 2007), and in tree ferns that produce GAs and/or GA methyl esters (Yamane *et al*, 1988; Yamauchi *et al.*, 1996; Menéndez *et al.*, 2006a; b; Abul *et al.*, 2010). In contrast, GAs have never been found in bryophytes, but GA-related *ent*-kaurenoid diterpenes have been identified in liverwort *Jungermannia* species (Nagashima *et al.*, 2003; Nozaki *et al.*, 2007), as well as in certain mosses.

The model moss *Physcomitrella patens* synthesises *ent*-kaurene, 16α-hydroxy-*ent*-kaurane and *ent*-kaurenoic acid, but not GAs (Hayashi *et al.*, 2006; 2010; Miyazaki *et al.*, 2011) while 16α-hydroxy-*ent*-kaurane has been identified in the moss *Saelania glaucescens* (von Schwartzenberg *et al.*, 2004).

Due to the availability of genome information (Rensing *et al.*, 2008), the moss *P. patens* has become a model organism for studying the phylogeny and evolution of GA biosynthesis and signalling in land plants. The genome of this primitive non-vascular land plant contains several putative GA biosynthetic genes, including a bifunctional *ent*-kaurene synthase gene (*PpCPS/KS*) and a cytochrome P450 gene encoding an *ent*-kaurene oxidase (CYP701B1) (Anterola and Shanle, 2008). Both genes have been cloned and functionally characterised (Hayashi *et al.*, 2006; Miyazaki *et al.*, 2011). As in fungi, the *P. patens ent*-kaurene synthase is a bifunctional enzyme (PpCPS/KS) containing both *ent*-copalyl diphosphate synthase and *ent*-kaurene synthase activities (CPS and KS, respectively), as demonstrated in a cell-free protein expression system (Hayashi *et al.*, 2006). Upon incubation with GGDP and $MgCl_2$ two products were formed, identified by GC-MS as *ent*-kaurene and 16α-hydroxy-*ent*-kaurane. The latter is found at high levels in *P. patens* tissue and released into the air (von Schwartzenberg *et al.*, 2004). The 101-kDa polypeptide encoded by *PpCPS/KS* has two aspartate-rich motifs, DVDD in the N-terminal domain, and DDYFD at the C-terminal domain, that correspond to conserved motifs in CPS and KS plant enzymes (Hayashi *et al.*, 2006). Mutational analysis of conserved motifs demonstrated that DVDD is responsible for CPS activity in PpCPS/KS (protonation-initiated cyclisation of GGDP to *ent*-CDP) while DDYFD is essential for KS activity (diphosphate ionisation-initiated cyclisation of *ent*-CDP to *ent*-kaurene or alternatively to 16α-hydroxy-*ent*-kaurane).

A bifunctional *ent*-kaurene synthase was also identified in *Jungermannia subulata* (*JsCPS/KS*), a liverwort that accumulates *ent*-kaurene but not 16α-hydroxy-*ent*-kaurane (Kawaide *et al.*, 2011). JsCPS/KS catalyses the cyclisation reaction of GGDP via *ent*-CDP to *ent*-kaurene but not to 16α-hydroxy-*ent*-kaurane, even though PpCPS/KS and JsCPS/KS share 60% identity. The analysis of chimeric JsCPS/KS and PpCPS/KS proteins together with mutational analysis demonstrated that the hydrophobicity and size of a single amino acid residue of PpCPS/KS, Ala710, is responsible for the formation of 16α-hydroxy-*ent*-kaurane by addition of a water molecule to the *ent*-kauranyl cation in the final step of GGDP cyclisation (Kawaide *et al.*, 2011). The *ent*-kaurene synthase from *B. japonicum* that forms *ent*-kaurene but not 16α-OH-*ent*-kaurane from *ent*-CDP, also has a larger residue at this position (Leu140) which together with Tyr and Phe residues at the active site prevent water from quenching the carbocation formed on ionisation (Liu *et al.*, 2014). Thus, *P. patens* and *J. subulata ent*-kaurene synthases are similar to fungal *ent*-kaurene synthases, which directly synthesise *ent*-kaurene from GGDP, and differ from *ent*-kaurene synthesis by higher plants or rhizobia that involve separate *ent*-copalyl synthases and *ent*-kaurene synthases (reviewed in Bömke and Tudzynski, 2009).

Besides *ent*-kaurene and 16α-hydroxy-*ent*-kaurane, *P. patens* produces *ent*-kaurenoic acid, which is formed by the cytochrome P450 enzyme encoded by *CYP701B1* (Miyazaki *et al.*, 2011). The *P. patens ent*-kaurene oxidase (CYP701B1, PpKO) was functionally characterised by heterologous expression in yeast and demonstrated to catalyse sequential oxidation at C-19 of *ent*-kaurene to *ent*-kaurenoic acid (Miyazaki *et al.*, 2011). In contrast to *ent*-kaurene oxidase from higher plants, PpKO is highly resistant to uniconazole-P. No further GA biosynthesis genes have been demonstrated in *P. patens*, although several genes with homology to GA 20-oxidases or to GA 3-oxidases are present in the *P. patens* genome. However, the encoded enzymes were shown not to encode active GA oxidases, in agreement with the absence of endogenous GAs in mosses (Hirano *et al.*, 2007).

A question arose as to whether *ent*-kaurene has a biological function in moss. A role in spore germination was assigned to *ent*-kaurene produced by *P. patens* since AMO-1618, an inhibitor of *ent*-kaurene synthesis, suppressed spore germination (Anterola *et al.* 2009). However, the *PpCPS/KS* disruption mutant was found to have the same spore germination rate as the wild-type, indicating that AMO1618 inhibits through a pathway different from *ent*-kaurene biosynthesis (Hayashi *et al.*, 2010). Phenotypic analysis of the *PpCPS/KS* disruption mutant showed that the mutant lines had a defect in protonemal differentiation of chloronemata (the protonemal cells produced from spore germination) to caulonemata (faster growing cells that produce gametophores) (Hayashi *et al.*, 2010). The phenotypic defect was recovered by applications of *ent*-kaurene or *ent*-kaurenoic acid, in contrast to GA_3 or GA_4, which had no effect. 16α-hydroxy-*ent*-kaurane was also inefficient in reversing the decrease in growth of caulonema filaments, indicating that this kaurenoid is a by-product of PpCPS/KS activity with no physiological role in *P. patens*. Interestingly, GA_9 methyl ester, a fern antheridiogen, recovered the protonemal mutant although with less efficiency than *ent*-kaurenoic acid. *ent*-Kaurene diterpene metabolites would be also involved in auxin- and light-regulated protonemal differentiation, since disruption of *ent*-kaurene production confers auxin resistance in protonemal development and affects photomorphogenesis in *P. patens* (Hayashi *et al.*, 2010). All these findings indicate that protonemal differentiation in *P. patens* is regulated by *ent*-kaurene-derived diterpenes distinct from active GAs found in vascular plants.

Since externally added GAs do not induce specific developmental processes or growth responses in mosses it was proposed that the GA signalling pathway developed later in land plant evolution, after the bryophyte divergency. However, as such pathways do not appear completely *de novo*, precursors for GA perception may be present in mosses (Hirano *et al.*, 2007; Vandenbussche *et al.*, 2007; Yasumura *et al.*, 2007). The *P. patens* genome contains two *GID1*-Like (*PpGID1L*) and two *DELLA*-Like (*PpDELLAL*) genes, but no evidence was found for functional conservation in the corresponding encoded proteins. *In vitro* GA binding assays demonstrated that PpGID1Ls have no or very low affinity to GA_4 in the presence or absence of DELLAs (Hirano *et al.*, 2007), while no binding to GA_9, GA_{12} or epi-GA_4 to GID1L was found. Moreover, PpGID1L and PpGID2L did not interact with any DELLA protein in the presence of GA_4 in a yeast two-hybrid assay (Hirano *et al.*, 2007; Yasumura *et al.*, 2007) and similarly, the PpDELLAL proteins did not interact with any GID1 protein (Hirano *et al.*, 2007). In addition, interactions between PpGID1L and PpDELLAL proteins were not observed in the presence of any added GA, while a PpDELLAL-deficient mutant did not show any growth defect (Yasumura *et al.*, 2007). This indicates that *P. patens* does not contain a functional GA perception mechanism mediated by GID1-DELLA.

In contrast to *P. patens*, the more recently diverging lycophyte *Selaginella*, a member of one of the first lineages of vascular plants to diverge from flowering plants, contains a functional orthologue of each of the elements of the GA signalling pathway (Hirano *et al.*, 2007; Yasumura *et al.*, 2007; reviewed by Hirano *et al.*, 2008). The proteins encoded by *Selaginella GID1* and *DELLA* homologues function as GA receptors or negative modulators of GA signalling, respectively (Hirano *et al.*, 2007; Yasumura *et al.*, 2007). GID1 and DELLA proteins from *Selaginella kraussiana* or *S. moellendorffii* interact and form a complex in the presence of GA_3 or GA_4 (Yasumura *et al.*, 2007). Furthermore, SmDELLAs are degraded upon treatment with GA_4, consistent with the presence of SLY1/GID2 homologues in *S. moellendorffii*. Also, down-regulation of the *SmGID1b* gene by GA_4 was found, as is the case, for example, in rice or *Arabidopsis*. Both DELLA and GID1 proteins from *S. moellendorffii* can functionally substitute for SLR1 and GID1 in rice (Hirano *et al.*, 2007) while the *S. kraussiana* DELLA repressed growth in *Arabidopsis* (Yasumura *et al.*, 2007). Added exogenous GAs, however, did not affect growth at the vegetative stage in *Selaginella*. GA_3 did not promote growth of *S. kraussiana* (Yasumura *et al.*, 2007), and GA_4 had no significant effect on stem growth or leaf elongation in *S. moellendorffii* (Hirano *et al.*, 2007). GA_4 treatment, however, promotes the outer wall development of microspores in *S. moellendorffii* (Aya *et al.*, 2011). Thus homologues of GID1 and DELLA in *Selaginella* spp. function as in flowering plants, but the GA signalling pathway does not regulate growth and rather has a role in spore wall formation in the lycophyte (Aya *et al.*, 2011).

Besides the components of the GA signalling pathway, *S. moellendorffii* contain functional homologues of GA 20-oxidase and GA 3-oxidase genes, which were characterised by heterologous expression in *E. coli* (Hirano *et al.*, 2007). The recombinant SmGA 20-oxidase catalyses the conversion of GA_{12} into GA_{15}, GA_{24} and GA_9 while SmGA 3-oxidase converts GA_9 into GA_4 but not GA_{20} into GA_1. Interestingly, GA_4 and low levels of GA_{24} have been detected as endogenous GAs in young shoots of *S. moellendorffii* by LC/MS/MS, while no GA_1 was found (Hirano *et al.*, 2007; reviewed in Sun, 2011). GA_4 and GA_9 were also identified in *S. moellendorffii* microsporangia, while GA_4 was detected to a lesser extent in young stems (Aya *et al.*, 2011). These results indicate that *S. moellendorffii* has a functional GA biosynthesis pathway similar to the non-13 hydroxylation pathway in angiosperms, synthesising GA_4 as a final product.

Finally, GAs and GA-related compounds have been identified in different fern species (Yamane *et al.*, 1988; Yamauchi *et al.*, 1996; Kurumatani *et al.*, 2001; Menéndez *et al.*, 2006a; b; Abul *et al.*, 2010). The *Schizaceous Lygodium* ferns synthesise GA_{73} methyl ester ($GA_{73}Me$), GA_9 methyl ester (GA_9Me) and GA_{20} methyl ester as antheridiogens, regulators of antheridium development in prothalia. The main antheridiogen appears to be $GA_{73}Me$, a 9,11-ene derivative of GA_9Me, which induces male and/or represses female gametophyte development (Yamauchi *et al.*, 1996; Kurumatani *et al.*, 2001). 12α- or 12β-hydroxylated derivatives of $GA_{73}Me$ have also been identified in *Lygodium microphylus* or *Lygodium reticulatum*. Hydroxylation at 12α- or 12β- positions decreases antheridium-inducing activity of $GA_{73}Me$ and may be the main catabolic reactions of this antheridiogen in *Lygodium* prothalia leading to its inactivation (Kurumatani *et al.*, 2001). Besides antheridiogens, several GAs have been identified in ferns. GA_1, GA_4, GA_9 and/or GA_{20} have been demonstrated by GC-MS in sporophytes of *Cibotium glaucum* (Yamane *et al.*, 1988), in gametophytes of *Blechnum spicant* (Menéndez *et al.*, 2006a) or in the rhizome and aerial shoots of the fern-like vascular plant *Psilotum nudum* (Abul *et al.*, 2010). The physiological function of GAs in ferns, however, is not clear. In *B. spicant* GAs do not appear to have a significant effect in inducing antheridia or archegonia (Abul *et al.*, 2010), while in *Dyopteris affinis* it has been suggested that GAs would participate in the induction and differentiation of apogamous embryo development (Menéndez *et al.*, 2006b). In *Lygodium japonicum* it has been found that GA_4 triggers antheridium formation in prothallia through a mechanism that involves splitting of the GA biosynthesis pathway between individuals at different developmental stages (Tanaka *et al.*, 2014). First, GA_9 methyl ester is produced by early-maturing prothallia, and secreted and transmitted to late-maturing neighbouring prothallia, which convert it into GA_4, the active component which is perceived by GID1-DELLA, inducing antheridia and suppressing archegonia formation. This mechanism would favor cross-fertilisation and genetic variation in *L. japonicum* by controlling the sex ratio among prothallia in a colony (Tanaka *et al.*, 2014).

In summary, GAs are not present in mosses or liverworts but are produced by lycophyte *Selaginella* species and ferns. This suggests that GA biosynthesis may have evolved in the vascular plant lineage after the bryophyte divergence at the same time or just after the plant vascular system, and before plants acquired the capacity for seed reproduction. *ent*-Kaurene diterpenes are, however, produced by mosses and liverworts and may have a role in growth and development. The GID1-DELLA-mediated GA signalling pathway also developed after bryophyte divergence in land plant evolution, as indicated by the presence of a functional GA signalling pathway in *Selaginella spp.*, in contrast to *P. patens* that contains non-functional precursors of GID1 and DELLA (Vandenbussche *et al.*, 2007; Hartweck, 2008). In *Selaginella* GAs would participate mainly in spore wall formation rather than in growth, whereas in ferns GAs and/or their methyl esters serve as antheridiogens, which suggests that after the establishment of GA biosynthesis in land plant evolution, its first function would be related to spore development or sexual organ formation.

5.5 Concluding remarks

Gibberellins are *ent*-kaurene-derived diterpenoid phytohormones that are not only produced by higher plants, but also by some fungi and bacteria, as well as some lower plants. In fact, they were first isolated as 'toxic' metabolites of the rice pathogenic fungus *F. fujikuroi*, and it took more than 50 years until the discovery that GAs are natural plant hormones. A major milestone in research on GA biosynthesis was the identification of the GA biosynthetic genes in *A. thaliana* on the one hand, and *F. fujikuroi* and two other fungi, *Phaeosphaeria* sp. L487 and *S. manihoticola*, on the other. Recently, the first *ent*-copalyl diphosphate synthase and *ent*-kaurene synthase-encoding genes have been identified from the bacterium *B. japonicum*. In contrast, GAs are not present in mosses or liverworts, but are produced by lycophyte *Selaginella* species and ferns. The fast development of technologies to inexpensively sequence entire genomes provides the opportunity to search for

GA biosynthetic genes and whole gene clusters and operons in more fungal, bacterial and lower plant genomes.

Comparison of genes and enzymes involved in GA biosynthesis revealed remarkable differences between higher plants, fungi and bacteria. While plants and bacteria typically use two separate diterpene synthases to form *ent*-kaurene, fungi use only a single bifunctional diterpene synthase. On the other hand, GA biosynthetic genes are scattered through the plant genomes, while they are organised in gene clusters and operons in fungi and bacteria. The different nature of the enzymes utilised for GA biosynthesis in the three classes of organisms indicate that the pathways in plants, fungi and bacteria evolved independently from each other although there is a potential homology between diterpene synthases from different origins. In addition, profound differences were found in enzymes catalysing downstream oxidation steps: while plants use P450s and dioxygenases to produce bioactive GAs, *F. fujikuroi* uses four P450s and one unusual 2-oxoglutarate-dependent dioxygenase, DES, which has only low amino acid homology to known 2-oxoglutarate-dependent dioxygenases (Bhattacharya *et al.*, 2012). In contrast, *S. manihoticola* lacks the genes *P450-3* and *DES*, which encode the 13-hydroxylase and the GA_4 desaturase, respectively, and therefore uses only P450s. The CPS- and KS-containing operons in *Bradyrhizobium* contain three P450s and a short-chain alcohol dehydrogenase. However, there is less than 15% identity at the amino acid level between the fungal and *B. japonicum* P450s on the one hand (Morrone *et al.*, 2009), and only about 10% identity between the fungal and plant P450s (e.g. between the *ent*-kaurene oxidases), on the other (Hedden *et al.*, 2002). In addition, the short-chain alcohol dehydrogenase conserved in the bacterial operons is not present in any fungal gene cluster.

In conclusion, there are profound differences regarding the GA biosynthetic genes and enzymes, their regulation, and the biological function of GAs between higher and lower plants, fungi and bacteria indicating convergent evolution of GA biosynthetic pathways in these organisms. In vascular plants it appears that the GA pathway evolved to regulate reproductive development (still an important function in seed plants), but then broadened its function to control vegetative growth. This is probably related to the interaction of DELLA proteins with different transcription factors.

References

Abul, Y., Menéndez, V., Gómez-Campo, C. *et al.* (2010). Occurrence of plant growth regulators in *Psilotum nudum*. *Journal of Plant Physiology* **167**, 1211–1213.

Albermann, S., Linnemannstöns, P. and Tudzynski, B. (2013a). Strategies for strain improvement in *Fusarium fujikuroi*: overexpression and localization of key enzymes of the isoprenoid pathway and their impact on gibberellin biosynthesis. *Applied Microbiology and Biotechnology* **97**, 2979–2995.

Albermann, S., Elter, T., Teubner, A. *et al.* (2013b). Characterization of novel mutants with an altered gibberellin spectrum in comparison to different wild-type strains of *Fusarium fujikuroi. Applied Microbiology and Biotechnology* **97**, 7779–7790.

Anterola, A. and Shanle, E.K. (2008). Genomic insights in moss gibberellin biosynthesis. *Bryologist* **111**, 218–230.

Anterola, A., Shanle, E.K., Mansouri, K. *et al.* (2009). Gibberellin precursor is involved in spore germination in the moss *Physcomitrella patens*. *Planta* **229**, 1003–1007.

Atzorn, R., Crozier, A., Wheeler, C. and Sandberg, G. (1988). Production of gibberellins and indole 3-acetic acid by *Rhizobium phaseoli* in relation to nodulation of *Phaseolus vulgaris* roots. *Planta* **175**, 532–538.

Aya, K., Hiwatashi, Y., Kojima, M. *et al.* (2011). The gibberellin perception system evolved to regulate a pre-existing GAMYB-mediated system during land plant evolution. *Nature Communications* **2**, 544.

Bastian, F., Cohen, A., Piccoli, P. *et al.* (1998). Production of indole-3-acetic acid and gibberellins A_1 and A_3 by *Acetobacter diazotrophicus* and *Herbaspirillum seropedicae* in chemically-defined culture media. *Plant Growth Regulation* **24**, 7–11.

Bearder, J.R. (1983). *In vivo* diterpenoid biosynthesis in *Gibberella fujikuroi*: the pathway after *ent*-kaurene. In *The Biochemistry and Physiology of Gibberellins* (ed. A. Crozier), Praeger Publishers, New York, pp. 251–387.

Bhattacharya, A., Kourmpetli, S., Ward, D.A. *et al.* (2012). Characterization of the fungal gibberellin desaturase as a 2-oxoglutarate-dependent dioxygenase and its utilization for enhancing plant growth. *Plant Physiology* **160**, 837–845.

Bömke, C. and Tudzynski, B. (2009). Diversity, regulation and evolution of the gibberellin biosynthetic pathway in fungi compared to plants and bacteria. *Phytochemistry* **70**, 1876–1893.

Bömke, C., Rojas, M.C., Gong, F. *et al.* (2008a). Isolation and characterization of the gibberellin biosynthetic gene cluster in *Sphaceloma manihoticola. Applied and Environmental Microbiology* **74**, 5325–5339.

Bömke, C., Rojas, M.C., Hedden, P. and Tudzynski, B. (2008b). Loss of gibberellin production is due to a deletion in the gibberellic acid gene cluster in *Fusarium verticillioides* (*G. Fujikuroi* MP-A) *Applied and Environmental Microbiology* **74**, 7790–7801.

Bottini, R., Fulchieri, M., Pearce, D. and Pharis, R.P. (1989). Identification of gibberellins A_1, A_3, and isoA_3 in cultures of *Azospirillum lipoferum*. *Plant Physiology* **90**, 45–47.

Bottini, R., Cassán, F. and Piccoli, P. (2004). Gibberellin production by bacteria and its involvement in plant growth promotion and yield increase. *Applied Microbiology and Biotechnology* **65**, 497–503.

Cassán, F., Lucangeli, C.D., Bottini, R. and Piccoli, P. (2001a). *Azospirillum spp.* metabolize [17,17-$^2H_2$]gibberellin A_{20} to [17,17-$^2H_2$]gibberellin A_1 *in vivo dy* rice mutant seedlings. *Plant Cell Physiology* **42**, 763–767.

Cassán, F., Bottini, R., Schneider, G. and Piccoli, P. (2001b). *Azospirillum brasilense* and *Azospirillum lipoferum* hydrolyze conjugates of GA_{20} and metabolize the resultant aglicones to GA_1 in seedlings of rice dwarf mutants. *Plant Physiology* **125**, 2053–2058.

Chapman, M.A. and Regan M.A. (1980). Evolution of a biochemical pathway: evidence from comparative biochemistry. *Annual Review of Plant Physiology* **31**, 639–645.

Dean, R.A., Talbot, N.J., Ebbole, D.J. *et al.* (2005). The genome sequence of the rice blast fungus *Magnaporthe grisea*. *Nature* **434**, 980–986.

DeBose-Boyd, R.A. (2008). Feedback regulation of cholesterol synthesis: sterol-accelerated ubiquitination and degradation of HMG CoA reductase. *Cell Research* **18**, 609–621.

Ergün, N., Topcuoglu, S.F. and Yildiz, A. (2002). Auxin (indole-3-acetic acid), gibberellic acid (GA_3), abscisic acid (ABA) and cytokinin (zeatin) production by some species of mosses and lichens. *Turkish Journal of Botany* **26**, 13–18.

Fourie, G., van der Merwe, N.A., Wingfield, B.D. *et al.* (2013). Evidence for inter-specific recombination among the mitochondrial genomes of *Fusarium* species in the *Gibberella fujikuroi* complex. *BMC Genomics* **14**, 605–619.

Freiberg, C., Fellay, R., Bairoch, A. *et al.* (1997). Molecular basis of symbiosis between *Rhizobium* and legumes. *Nature* **387**, 394–401.

Gutierrez-Mañero, F., Ramos-Solano, B., Probanza, A. *et al.* (2001). The plant growth-promoting rhizobacteria *Bacillus pumilus* and *Bacillus licheniformis* produce high amounts of physiologically active gibberellins. *Physiologia Plantarum* **111**, 206–211.

Hartweck, L.M. (2008). Gibberellin signaling. *Planta* **229**, 1–13.

Hauser, F., Pessi, G., Friberg, M. *et al.* (2007). Dissection of the *Bradyrhizobium japonicum* NifA+σ^{54} regulon, and identification of a ferredoxin gene (*fdxN*) for symbiotic nitrogen fixation. *Molecular Genetics and Genomics* **278**, 255–271.

Hayashi, K., Kawaide, H., Notomi , *et al.* (2006). Identification and functional analysis of bifunctional *ent*-kaurene synthase from the moss *Physcomitrella patens*. *FEBS Letters* **580**, 6175–6181.

Hayashi, K., Horie, K., Hiwatashi, Y. *et al.* (2010). Endogenous diterpenes derived from *ent*-kaurene, a common gibberellin precursor, regulate protonema differentiation of the moss *Physcomitrella patens*. *Plant Physiology* **153**, 1085–1097.

Hedden, P. and Kamiya, Y. (1997). Gibberellin biosynthesis: enzymes, genes and their regulation. *Annual Review of Plant Physiology and Plant Molecular Biology* **48**, 431–460.

Hedden, P. and Phillips, A.L. (2000). Manipulation of hormone biosynthetic genes in transgenic plants. *Current Opinion in Biotechnology* **11**, 130–137.

Hedden, P. and Thomas, S.G. (2012). Gibberellin biosynthesis and its regulation. *Biochemical Journal* **444**, 11–25.

Hedden, P., MacMillan, J. and Phinney, B.O. (1974). Fungal products. Part XII. Gibberellin A_{14}-aldehyde, an intermediate in gibberellin biosynthesis in *Gibberella fujikuroi*. *Journal of the Chemical Society, Perkin Transactions* **1**, 587–592.

Hedden, P., Phillips, A.L., Rojas, M.C., *et al.* (2002). Gibberellin biosynthesis in plants and fungi: a case of convergent evolution? *Journal of Plant Growth Regulation* **20**, 319–331.

Hershey, D.M., Lu, X., Zi, J. and Peters, R.J. (2014). Functional conservation of the capacity for *ent*-kaurene biosynthesis and an associated operon in certain rhizobia. *Journal of Bacteriology* **196**, 100–106.

Hirano, K., Nakajima, M., Asano, K. *et al.* (2007). The GID1-mediated gibberellin perception mechanism is conserved in the lycophyte *Selaginella moellendorffii* but not in the bryophyte *Physcomitrella patens*. *The Plant Cell* **19**, 3058–3079.

Hirano, K., Ueguchi-Tanaka, M. and Matsuoka, M. (2008). GID1-mediated gibberellin signaling in plants. *Trends in Plant Science* **13**, 192–199.

Homann, V., Mende, K., Arntz, C. *et al.* (1996). The isoprenoid pathway: cloning and characterization of fungal *FPPS* genes. *Current Genetics* **30**, 232–239.

Hori, S. (1898). Some observations on 'Bakanae' disease of the rice plant. *Memoirs of Agricultural Research (Tokyo)* **12**, 110–119.

Hou, X., Lee, L.Y., Xia, K. *et al.* (2010). DELLAs modulate jasmonate signaling via competitive binding to JAZs. *Developmental Cell* **19**, 884–894.

Janzen, R., Rood, S., Dormar, J. and McGill, W. (1992). *Azospirillum brasilense* produces gibberellins in pure cultures, chemically-defined medium and in co-culture on straw. *Soil Biology and Biochemistry* **24**, 1061–1064.

Kang, S-M., Khan, A.L., Hamayun, M. *et al.* (2012). Gibberellin-producing *Promicromonospora* sp. SE188 improves *Solanum lycopersicum* plant growth and influences endogenous plant hormones. *Journal of Microbiology* **50**, 902–909.

Kang, S-M., Khan, A.L., You, Y-H., *et al.* (2013). Gibberellin production by newly isolated strain of *Leifsonia soli* SE134 and its potential for plant growth promotion. *Journal of Microbiology and Biotechnology* **24**, 106–112.

Kawaide, H., Sassa, T. and Kamiya, Y. (1995). Plant-like biosynthesis of gibberellin A_1 in the fungus *Phaeosphaeria* sp. L487. *Phytochemistry* **39**, 305–310.

Kawaide, H., Imai, R., Sassa, T. and Kamiya, Y. (1997). *ent*-Kaurene synthase from the fungus *Phaeosphaeria* sp. L487: cDNA isolation, characterization, and bacterial expression of a bifunctional diterpene cyclase in fungal gibberellin biosynthesis. *Journal of Biological Chemistry* **272**, 21706–21712.

Kawaide, H. (2006). Biochemical and molecular analyses of gibberellin biosynthesis in fungi. *Bioscience, Biotechnology, and Biochemistry* **70**, 583–590.

Kawaide, H., Hayashi, K., Kawanabe, R. *et al.* (2011). Identification of the single amino acid involved in quenching the *ent*-kauranyl cation by a water molecule in *ent*-kaurene synthase of *Physcomitrella patens*. *FEBS Journal* **278**, 123–133.

Keister, D.L., Tully, R.E. and van Berkum, P. (1999). A cytochrome P450 gene cluster in the *Rhizobiaceae*. *The Journal of General and Applied Microbiology* **45**, 301–303.

Kurumatani, M., Yagi, K., Murata, T. *et al.* (2001). Isolation and identification of antheridiogens in the ferns *Lygodium microphyllum* and *Lygodium reticulatum*. *Bioscience Biotechnology and Biochemistry* **65**, 2311–2314.

Lange, T., Hedden, P. and Graebe, J.E. (1994). Expression cloning of a gibberellin 20-oxidase, a multifunctional enzyme involved in gibberellin biosynthesis. *Proceedings of the National Academy of Sciences USA* **91**, 8552–8556.

Linnemannstöns, P., Voß, T., Hedden, P. *et al.* (1999). Deletions in the gibberellin biosynthesis gene cluster of *Gibberella fujikuroi* by restriction enzyme-mediated mutagenesis. *Applied and Environmental Microbiology* **65**, 2558–2564.

Liu, W., Feng, X., Zheng, Y. *et al.*, (2014). Structure, function and inhibition of *ent*-kaurene synthase from *Bradyrhizobium japonicum*. *Scientific Reports* **4**, 6214.

Lucangeli, C. and Bottini, R. (1996). Reversion of dwarfism in *dwarf-1* maize (*Zea mays* L.) and *dwarf-x* rice (*Oryza sativa* L.) mutants by endophytic *Azospirillum spp. Biocell* **20**, 223–228.

Lucangeli, C. and Bottini, R. (1997). Effects of *Azospirillum spp.* on endogenous gibberellins content and growth of maize (*Zea mays* L.) treated with uniconazole. *Symbiosis* **23**, 63–72.

MacMillan, J. (1997). Biosynthesis of the gibberellin plant hormones. *Natural Product Reports* **14**, 221–243.

MacMillan, J. (2002). Occurrence of gibberellins in vascular plants, fungi and bacteria. *Journal of Plant Growth Regulation* **20**, 387–442.

Malonek, S., Bömke, C., Bornberg-Bauer, E. *et al.* (2005a). Distribution of gibberellin biosynthetic genes and gibberellin production in the *Gibberella fujikuroi* species complex. *Phytochemistry* **66**, 1296–1311.

Malonek, S., Rojas, M.C., Hedden, P., *et al.*, (2005b). Functional characterization of two cytochrome P450 monooxygenase genes, P450–1 and P450–4, of the gibberellic acid gene cluster in *Fusarium proliferatum* (*Gibberella fujikuroi* MP-D). *Applied and Environmental Microbiology* **71**, 1462–1472.

Malonek, S., Rojas, M.C., Hedden, P., *et al.*, (2005c). Restoration of gibberellin production in *Fusarium proliferatum* by functional complementation of enzymatic blocks. *Applied and Environmental Microbiology* **71**, 6014–6025.

Mende, K., Homann, V. and Tudzynski, B. (1997). Molecular characterization of the geranylgeranyl diphosphate synthase gene of *Gibberella fujikuroi*. *Molecular and General Genetics* **255**, 96–105.

Méndez, C., Baginsky, C., Hedden, P. *et al.* (2014). Gibberellin oxidase activities in *Bradyrhizobium japonicum* bacteroids. *Phytochemistry* **98**, 101–109.

Menéndez, V., Revilla, M.A., Bernard, P., *et al.* (2006a). Gibberellins and antheridiogen on sex in *Blechnum spicant* L. *Plant Cell Reports* **25**, 1104–1110.

Menéndez, V., Villacorta, N.F., Revilla, M.A., *et al.* (2006b). Exogenous and endogenous growth regulators on apogamy in *Dryopteris affinis* (Lowe) Fraser-Jenkins sp. *affinis*. *Plant Cell Reports* **25**, 85–91.

Miyazaki, S., Katsumata, T., Natsume, M. and Kawaide, H. (2011). The CYP701B1 of *Physcomitrella patens* is an *ent*-kaurene oxidase that resists inhibition by uniconazole-P. *FEBS Letters* **585**, 1879–1883.

Morrone, D., Chambers, J., Lowry, L. *et al.* (2009). Gibberellin biosynthesis in bacteria: separate *ent*-copalyl diphosphate and *ent*-kaurene synthases in *Bradyrhizobium japonicum*. *FEBS Letters* **583**, 475–480.

Mowat, J.A. (1965). A survey of results on the occurrence of auxins and gibberellins in algae. *Botanica Marina* **8**, 149–155.

Nagashima, F., Kasai, W., Kondoh, M., *et al.* (2003). New *ent*-kaurene-type diterpenoids possessing cytotoxicity from the New Zealand liverwort *Jungermannia* species. *Chemical and Pharmaceutical Bulletin* **51**, 1189–1192.

Navarro, L, Bari, R., Achard, P. *et al.* (2008). DELLAs control plant immune responses by modulating the balance of jasmonic acid and salicylic acid signaling. *Current Biology* **18**, 650–655.

Nozaki, H., Hayashi, K., Okuda, K., *et al.* (2007). *ent*-Kaurene-type diterpenoids from a suspension culture of the liverwort *Jungermannia subulata*. *Planta Medica* **73**, 689–695.

Pessi, G., Ahrens, C.H., Rehrauer, H. *et al.* (2007). Genome-wide transcript analysis of *Bradyrhizobium japonicum* bacteroids in soybean root nodules. *Molecular Plant–Microbe Interactions* **20**, 1353–1363.

Phillips, A.L., Ward, D.A., Uknes, S. *et al.* (1995). Isolation and expression of three gibberellin 20-oxidase cDNA clones from *Arabidopsis*. *Plant Physiology* **108**, 1049–1057.

Piccoli, P. and Bottini, R. (1994). Metabolism of 17,17-[^{2}H]gibberellin A_{20} to 17,17-[^{2}H]gibberellin A_1 by *Azospirillum lipoferum* cultures. *AgriScientia* **9**, 13–15.

Piccoli, P., Masciarelli, O. and Bottini, R. (1996). Metabolism of 17,17[$^2H_2$]-gibberellins A_4, A_9 and A_{20} by *Azospirillum lipoferum* in chemically-defined culture medium. *Symbiosis* **21**, 263–274.

Piccoli, P., Lucangelli, D., Schneider, G. and Bottini, R. (1997). Hydrolysis of [17, 17-$^2H_2$] gibberellin A_{20}-glucosyl ester by *Azospirillum lipoferum* cultured in a nitrogen-free biotin-based chemically-defined medium. *Plant Growth Regulation* **23**, 179–182.

Piccoli, P., Travaglia, C., Cohen, A. *et al.* (2011). An endophytic bacterium isolated from roots of the halophyte *Prosopis strombulifera* produces ABA, IAA, gibberellins A_1 and A_3 and jasmonic acid in chemically-defined culture medium. *Plant Growth Regulation* **64**, 207–210.

Rademacher, W. and Graebe, J.E. (1979). Gibberellin A_4 produced by *Sphaceloma manihoticola*, the cause of the superelongation disease of cassava (*Manihot esculenta*). *Biochemical and Biophysical Research Communications* **91**, 35–40.

Radley, M. (1956). Occurrence of substances similar to gibberellic acid in higher plants. *Nature* **178**, 1070–1071.

Rensing, S.A., Lang, D., Zimmer, A.D. *et al.* (2008). The *Physcomitrella patens* genome reveals evolutionary insights into the conquest of land by plants. *Science* **319**, 64–69.

Rim, S.-O., Lee, J.-H., Choi, W.-Y., *et al.* (2005). *Fusarium proliferatum* KGL0401 as a new gibberellin-producing fungus. *Journal of Microbiology and Biotechnology* **15**, 809–814.

Rojas, M.C., Hedden, P., Gaskin, P. and Tudzynski, B. (2001). The *P450-1* gene of *Gibberella fujikuroi* encodes a multifunctional enzyme in gibberellin biosynthesis. *Proceedings of the National Academy of Sciences USA* **98**, 5838–5843.

Salazar, E., Díaz-Mejía, J., Moreno-Hagelsieb, G. *et al.* (2010). Characterization of the NifA-RpoN Regulon in *Rhizobium etli* in free life and in symbiosis with *Phaseolus vulgaris*. *Applied and Environmental Microbiology* **76**, 4510–4520.

Sassa, T. and Suzuki, K. (1990). Metabolism of gibberellin A_9 to gibberellin A_4 in a new gibberellin-producing fungus, *Phaeosphaeria* sp. L487. *Agricultural and Biological Chemistry* **54**, 3373–3375.

Sassa, T., Suzuki, K. and Haruki, E. (1989). Isolation and identification of gibberellins A_4 and A_9 from a fungus *Phaeosphaeria* sp. *Agricultural and Biological Chemistry* **53**, 303–304.

Sawada, K. (1912). Diseases of agricultural products in Japan. *Formosan Agricultural Reviews* **63**, 10–16.

Schardl, C.L., Young, C.A., Hesse, U., *et al.* (2012). Plant-symbiotic fungi as chemical engineers: multi-genome analysis of the *Clavicipitaceae* reveals dynamics of alkaloid loci. *PLoS Genetics* **9**, e1003323.

Sponsel, V.M. and Hedden, P. (2004). Gibberellin biosynthesis and inactivation. In *Plant Hormones. Biosynthesis, Signal transduction, Action!* (ed. P.J. Davies.), Kluwer Academic Publishers, Dordrecht, the Netherlands, pp. 63–94.

Sun, T-P. (2011) The molecular mechanism and evolution of the GA-GID1-DELLA signaling module in plants. *Current Biology* **21**, R338-R345.

Sun, T.-P. and Kamiya, Y. (1994). The *Arabidopsis GA1* locus encodes the cyclase *ent*-kaurene synthase A of gibberellin biosynthesis. *The Plant Cell* **6**, 1509–1518.

Tanaka, J., Yano, K., Aya, K., *et al.* (2014). Antheridiogen determines sex in ferns via spatiotemporally split gibberellin synthesis pathway. *Science* **346**, 469–473.

Toyomasu, T., Nakaminami, K., Toshima, H. *et al.* (2004). Cloning of a gene cluster responsible for the biosynthesis of diterpene aphidicolin, a specific inhibitor of DNA polymerase α. *Bioscience, Biotechnology, and Biochemistry* **68**, 146–152.

Troncoso, C., González, X., Bömke, C. *et al.* (2010). Gibberellin biosynthesis and gibberellin oxidase activities in *Fusarium sacchari, Fusarium konzum* and *Fusarium subglutinans* strains. *Phytochemistry* **71**, 1322–1331.

Tsavkelova, E.A., Bömke, C., Netrusov, A.I., *et al.* (2008). Production of gibberellic acids by an orchid-associated *Fusarium proliferatum* strain. *Fungal Genetics and Biology* **45**, 1393–1403.

Tudzynski, B. (2005). Gibberellin biosynthesis in fungi: genes, enzymes and impact on biotechnology. *Applied Microbiology and Biotechnology* **66**, 597–611.

Tudzynski, B. and Hölter, K. (1998). Gibberellin biosynthetic pathway in *Gibberella fujikuroi*: evidence for a gene cluster. *Fungal Genetics and Biology* **25**, 157–170.

Tudzynski, B., Kawaide, H. and Kamiya, Y. (1998). Gibberellin biosynthesis in *Gibberella fujikuroi*: cloning and characterization of the copalyl diphosphate synthase gene. *Current Genetics* **34**, 234–240.

Tudzynski, B., Hedden, P., Carrera, E. and Gaskin P. (2001). The *P450-4* gene of *Gibberella fujikuroi* encodes *ent*-kaurene oxidase in the gibberellin biosynthesis pathway. *Applied and Environmental Microbiology* **67**, 3514–3522.

Tudzynski, B., Rojas, M.C., Gaskin, P. and Hedden, P. (2002). The gibberellin 20-oxidase of *Gibberella fujikuroi* is a multifunctional monooxygenase. *Journal of Biological Chemistry* **277**, 21246–21253.

Tudzynski, B., Mihlan, M., Rojas, M.C., *et al.* (2003). Characterization of the final two genes of the gibberellin biosynthesis gene cluster of *Gibberella fujikuroi*: *des* and *P450-3* encode GA_4 desaturase and the 13-hydroxylase, respectively. *Journal of Biological Chemistry* **278**, 28635–28643.

Tully, R.E. and Keister, D.L. (1993). Cloning and mutagenesis of a cytochrome P-450 locus from *Bradyrhizobium japonicum* that is expressed anaerobically and symbiotically. *Applied and Environmental Microbiology* **59**, 4136–4142.

Tully, R.E., van Berkum, P., Lovins, *et al.* (1998). Identification and sequencing of a cytochrome P450 gene cluster from *Bradyrhizobium japonicum*. *Biochimica et Biophysica Acta* **1398**, 243–255.

Urrutia, O., Hedden, P., Rojas, M.C., *et al.* (2001). Monooxygenases involved in GA_{12} and GA_{14} synthesis in *Gibberella fujikuroi*. *Phytochemistry* **56**, 5005–5011.

Vandenbussche, F., Fierro, A.C., Wiedemann, G., *et al.* (2007). Evolutionary conservation of plant gibberellin signalling pathway components. *BMC Plant Biology* **7**, 1–17.

von Schwartzenberg, K., Schulte, W. and Kassner, H. (2004). The moss *Physcomitrella patens* releases a tetracyclic diterpene. *Plant Cell Reports* **22**, 780–786.

Wiemann, P., Albermann, S., Niehaus, E. *et al.* (2012). The Sfp-type 4′-phosphopantetheinyl transferase Ppt1 of *Fusarium fujikuroi* controls development, secondary metabolism and pathogenicity. *PLoS One* **7**, e37519.

Wiemann, P., Sieber, C.M.-K., von Bargen, K.W. *et al.* (2013). Deciphering the cryptic genome: genome-wide analyses of the rice pathogen *Fusarium fujikuroi* reveal complex regulation of secondary metabolism and novel metabolites. *PLoS Pathogens* **9**, e1003475.

Wingfield, B.D., Steenkamp. E.T., Santana Q. *et al.* (2012). First fungal genome sequence from Africa: Consequences scientific and regional. *South African Journal of Science* **108**, 1–9.

Woitek, S., Unkles, S.E., Kinghorn, J.R. and Tudzynski, B. (1997). 3-Hydroxy-3-methylglutaryl-CoA reductase gene of *Gibberella fujikuroi*: isolation and characterization. *Current Genetics* **31**, 38–47.

Yabuta, T. and Sumiki, Y. (1938). On the crystal of gibberellin, a substance to promote plant growth. *Journal of the Agricultural Chemical Society of Japan* **14**, 1526.

Yamane, H., Fujioka, S., Spray, C.R. *et al.* (1988). Endogenous gibberellins from sporophytes of two tree ferns, *Cibotium glaucum* and *Dicksonia antarctica*. *Plant Physiology* **86**, 857–862.

Yamauchi, T., Oyama, N., Yamane, H. *et al.* (1996). Identification of antheridiogens in *Lygodium circinnatum* and *Lygodium flexuosum*. *Plant Physiology* **111**, 741–745.
Yamaguchi, S. (2008). Gibberellin metabolism and its regulation. *Annual Review of Plant Biology* **59**, 225–251.
Yasumura, Y., Crumpton-Taylor, M., Fuentes, S. and Harberd, N.P. (2007). Step-by-step acquisition of the gibberellin-DELLA growth-regulatory mechanism during land-plant evolution. *Current Biology* **17**, 1225–1230.
Zhang, S., Monahan, B.J., Tkacz, J.S. and Scott, B. (2004). Indole-diterpene gene cluster from *Aspergillus flavus*. *Applied and Environmental Microbiology* **70**, 6875–6883.
Zeigler, R.S., Powell, L.E. and Thurston, H.D. (1980). Gibberellin A_4 production by *Sphaceloma manihoticola*, causal agent of cassava superelongation disease. *Phytopathology* **70**, 589–593.

Annual Plant Reviews (2016) **49**, 153–188
doi: 10.1002/9781119210436.ch6

Chapter 6

GIBBERELLIN HORMONE SIGNAL PERCEPTION: DOWN-REGULATING DELLA REPRESSORS OF PLANT GROWTH AND DEVELOPMENT

Sven K. Nelson[1,4] and Camille M. Steber[1,2,3]

[1] *Molecular Plant Sciences Program, Washington State University, USA*
[2] *USDA-ARS, Wheat Health, Genetics, and Quality Unit, Pullman, WA, USA*
[3] *Department of Crop and Soil Science, Washington State University, USA*
[4] *Current address: USDA-ARS, Plant Genetic Research Unit, University of Missouri, USA*

Abstract: The gibberellin (GA) hormone signal is perceived by a receptor with homology to hormone-sensitive lipases, GID1 (GA-INSENSITIVE DWARF1). This leads to GA-stimulated responses, including stem elongation, seed germination and the transition to flowering. GA-binding enables GID1 to interact with and block the function of the DELLA repressors of GA responses. DELLA repression can be blocked both by proteolytic and non-proteolytic mechanisms triggered by the formation of a GID1-GA-DELLA complex. DELLA is down-regulated by the SLEEPY1/GID2 F-box proteins via the ubiquitin-proteasome pathway, and can be regulated by other post-translational modifications. This chapter reviews the structural requirements for GA-binding by GID1 and for GID1-GA-DELLA protein complex formation, and reviews the current understanding of the mechanisms regulating DELLA repressors.

Keywords: GID1, DELLA, SLY1, GID2, ubiquitin, proteasome, gibberellin, signalling, EL1, SPY

Annual Plant Reviews, Volume 49: The Gibberellins, First Edition.
Edited by Peter Hedden and Stephen G. Thomas.

6.1 Introduction

Gibberellins (GAs) are tetracyclic diterpenoid plant hormones that stimulate seed germination, stem elongation, the transition to flowering and fertility in diverse plant species (see Chapter 1; reviewed in Sun and Gubler, 2004; Ueguchi-Tanaka *et al.*, 2007b; Yamaguchi, 2008). While 136 GAs have been identified in plants, fungi and bacteria, only a small sub-set of these are biologically active. The predominant bioactive GAs are GA_1 and GA_4. The DELLA (Asp-Glu-Leu-Leu-Ala) domain family of proteins act as repressors of GA responses through effects on gene transcription (see Chapter 7; Silverstone *et al.*, 1998; Itoh *et al.*, 2002; Zentella *et al.*, 2007). This chapter reviews the mechanisms by which the GA hormone signal is perceived and transduced to release DELLA repression of GA responses based mainly on evidence from rice (*Orzya sativa*) and *Arabidopsis thaliana* (*Arabidopsis*). In the 'DELLA destruction' model of GA signalling, GA binding allows the GA receptor GID1 (GA-INSENSITIVE DWARF1) to interact with DELLA repressors, thereby triggering DELLA destruction through the ubiquitin-proteasome pathway (Figure 6.1a, b). Alternative mechanisms for GA signalling in which the GA signal is transduced without DELLA destruction, or in which DELLA repressors are regulated by post-translational modification or in a GA-independent manner are also described. GA signalling mechanisms downstream of DELLA are described in Chapter 7.

6.2 DELLA proteins are repressors of gibberellin responses

DELLA proteins are nuclear-localised negative regulators of GA signalling defined by the presence of an N-terminal DELLA regulatory domain, and a C-terminal GRAS (GAI, RGA, and SCARECROW) functional domain (Figure 6.2; Silverstone *et al.*, 1998; Ikeda *et al.*, 2001; Itoh *et al.*, 2002). Mutations in the GRAS functional domain result in loss of DELLA repressor function, leading to a tall or 'slender' plant growth phenotype. This recessive phenotype is observed in the GRAS domain mutations in *SLN1* (*SLENDER1*) and *SLR1* (*SLENDER RICE1*), the sole DELLA genes in barley (*Hordeum vulgare*) and rice, respectively (Ikeda *et al.*, 2001; Itoh *et al.*, 2002; Chandler *et al.*, 2002). GRAS domain genes are a large family of transcriptional regulators unique to plants, and conserved in mosses, rice and *Arabidopsis* (Engstrom, 2011). The C-terminal GRAS domain contains a nuclear localisation sequence (NLS), two leucine heptad repeat motifs (LHR1 and LHR2) that flank the VHIID amino acid motif, and the PFYRE and SAW motifs (Figure 6.2; Richards *et al.*, 2000; Levy and Darnell, 2002; Bolle, 2004). The C-terminal PFYRE and SAW motifs have some homology to mammalian STAT (Signal Transducers and Activators of Transcription) transcription factors. Thus far, only one GRAS protein has been demonstrated to directly bind to DNA, a legume protein called NSP1 (NODULATION SIGNALING

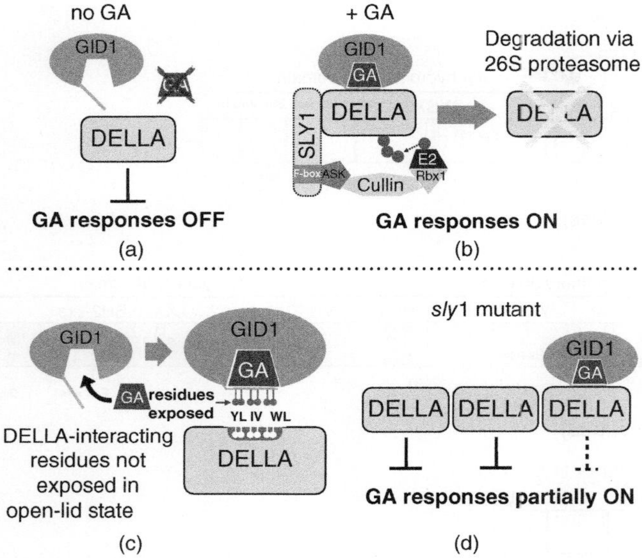

Figure 6.1 Models of proteolysis-dependent and -independent GA signalling. (a, b) The DELLA destruction model: (a) In the absence of GA, DELLA proteins are stable and repress GA responses. (b) GID1 binding to GA allows formation of the GID1-GA-DELLA complex, which in turn allows the SLY1/GID2 F-box protein to bind and polyubiquitinate DELLA, thereby targeting DELLA for destruction by the 26S proteasome. This lifts DELLA repression of GA responses. The SCF^{SLY1} E3 ubiquitin ligase complex consists of the Skp1 homologue ASK1, Cullin, the SLY1 F-box protein and RBX1. The SCF E3 catalyses transfer of ubiquitin (dark grey circles) from E2 to DELLA. (c) GID1 lid closure model: Without GA, the GID1 lid is believed to be open and unable to bind DELLA. When GA is bound, the GID1 lid closes exposing the hydrophobic residues (L, W, V, I, L and Y) needed to interact with DELLA protein. (d) Non-proteolytic DELLA down-regulation: In the *sly1* mutant, DELLA cannot be targeted for degradation, and DELLA over-accumulation represses GA responses. Formation of the GID1-GA-DELLA complex down-regulates some DELLA, partially relieving repression of GA responses.

PATHWAY1), suggesting that most GRAS proteins may indirectly regulate gene transcription (Hirsch *et al.*, 2009).

The N-terminal DELLA regulatory domain contains the DELLA, VHYNP (also called TVHYNP), and poly S/T/V motifs (Figure 6.2). Deletions in these motifs result in increased DELLA repression due to an inability to respond to GA hormone, leading to a semi-dominant semi-dwarf phenotype (Itoh *et al.*, 2002). The first DELLA mutant, *gai-1* (*GA-insensitive-1*), was isolated as a GA-insensitive semi-dominant semi-dwarf in *Arabidopsis* resulting from a 17-amino acid deletion within the DELLA/LExLE motif (Koornneef *et al.*, 1985; Peng *et al.*, 1997). The cloning of the two *Arabidopsis* DELLA genes, *GAI* and *RGA*, led to the cloning of similar GA-insensitive semi-dominant semi-dwarf DELLA mutants in wheat and maize (Silverstone *et al.*, 1997b;

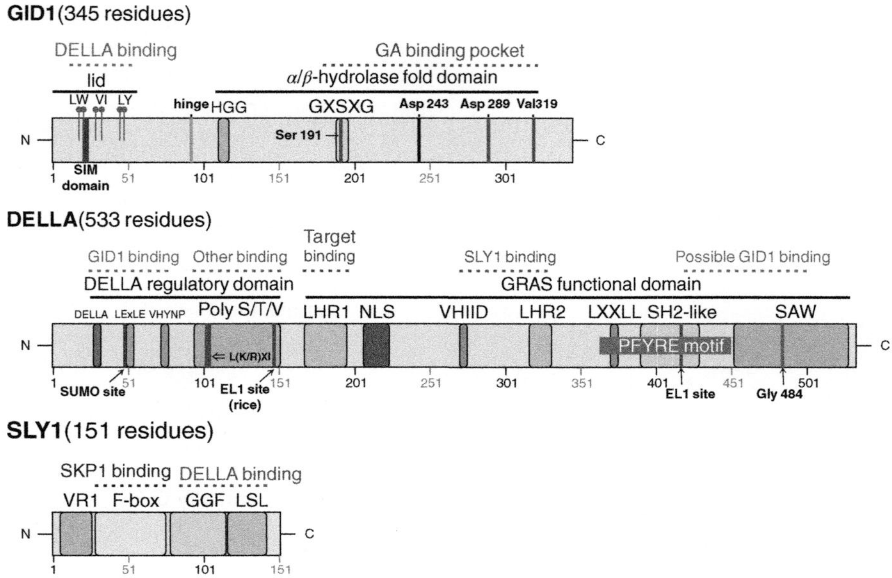

Figure 6.2 Diagrams of the GID1, SLY1 and DELLA domains, motifs and key amino acid residues. Proteins are drawn to scale based on the amino acid sequences of *Arabidopsis* GID1a, DELLA GAI and SLY1. Regions involved in specific protein–protein or protein–ligand interactions are indicated with dotted bars above the diagram. GID1: Two major domains marked by black bars above diagram are the GID1 lid and the α/β hydrolase fold domain including the core GA-binding pocket. Key motifs and significant amino acid residues include: (1) the hinge residue (orange bar for GID1a Pro 92, OsGID1 P99), (2) the catalytic triad (Ser 191, Asp 289, Val 319) involved in GA binding, (3) the six lid hydrophobic residues involved in DELLA-binding (lollipops for L, W, V, I, L, Y), (4) the SUMO-Interaction-Motif (SIM) domain (WVLI) and (5) the HGG and GXSXG motifs characteristic of hormone-sensitive lipases. DELLA: The major DELLA regulatory domain and the GRAS functional domain are marked by black bars above the diagram. Within the DELLA regulatory domain the DELLA, LExLE and VHYNP motifs are involved in GID1 binding (shaded boxes), and the poly S/T/V motif contains the L(K/R)XI motif likely involved in binding an undetermined 'other' GA signalling component. The GRAS functional domain contains: two leucine heptad repeats (LHR1 and LHR2), a nuclear localisation signal (NLS), the VHIID, the PFYRE, LXXLL, SH2-like, and SAW motifs (shaded boxes). Significant residues are marked by full-height bars. SLY1: SLY1 and GID2 contain the F-box domain that binds SKP1, the GGF and LSL motifs involved in DELLA-binding, and a variable region (VR1). (See insert for colour representation of this figure.)

Peng *et al.*, 1997; Silverstone *et al.*, 1998; Peng *et al.*, 1999). The shorter, thicker stems of semi-dwarf DELLA mutants of maize and wheat enabled yield increase that is now called the 'Green Revolution' by allowing farmers to use modern fertilisers to increase yield without causing the plants to fall over or lodge (Allan, 1986).

DELLA proteins appear to act in complex with transcription factors as coactivators or corepressors (reviewed in Chapter 7; Hauvermale *et al.*, 2012). DELLA proteins have been shown to interact with a wide range of transcriptional regulators including PIF3 (PHYTOCHROME INTERACTING FACTOR3), PIF4, PIF1/PIL5 (for PIF3-LIKE5), PIL2, JAZ1 (JASMONATE ZIM DOMAIN1), ALC (ALCATRAZ), SPT (SPATULA), BZR1 (BRASSINOZALE-RESISTANT1) and the GRAS protein SCL3 (SCARECROW-LIKE3) (Zentella *et al.*, 2007; de Lucas *et al.*, 2008; Feng *et al.*, 2008; Gallego-Bartolomé *et al.*, 2010; Arnaud *et al.*, 2010; Hou *et al.*, 2010; Heo *et al.*, 2011; Zhang *et al.*, 2011; Hirano *et al.*, 2012; Bai *et al.*, 2012; Gallego-Bartolomé *et al.*, 2012). DELLA proteins also interact with the chromatin remodelling factor SWI3C (SWITCH3C) (Sarnowska *et al.*, 2013). JAZ1, PIF4, and BZR1 interact with DELLA proteins via the LHR1 motif (de Lucas *et al.*, 2008; Hou *et al.*, 2010; Gallego-Bartolomé *et al.*, 2012). It has been proposed that DELLA proteins function: (1) as coactivators of genes that negatively regulate GA signalling, (2) as repressors of transcriptional activators by blocking the ability of a transcription factor to bind its promoter and (3) as factors that recruit chromatin remodelling complexes to promoter elements.

6.3 Gibberellin signalling lifts DELLA repression of gibberellin responses

The partly overlapping roles of the five *Arabidopsis* DELLA repressors were defined based on the ability of DELLA loss-of-function alleles to rescue the phenotypes of the strong GA biosynthesis mutant, *ga1-3* (King *et al.*, 2001; Dill and Sun 2001; Cheng *et al.*, 2004; Cao *et al.*, 2005). This 10-kb deletion of the *GA1* gene encoding *ent*-copalyl diphosphate synthase (CPS) results in failure to germinate, extreme dwarfism, inability to transition to flowering under short days, and under-developed flowers (Koornneef and van der Veen 1980; Wilson *et al.*, 1992; Sun and Kamiya 1994; Silverstone *et al.*, 1997a). These phenotypes are rescued by GA hormone application, or by combinations of DELLA loss-of-function mutations. Thus, DELLAs act downstream of *GA1* to repress GA responses. GA stimulates GA responses by lifting DELLA repression. The five *Arabidopsis* DELLA genes encode proteins with 55.2–73.9% amino acid identity, and are named *GAI* (*GA-INSENSITIVE*), *RGA* (*REPRESSOR OF GA*), *RGL1*, *RGL2* and *RGL3* (*RGA-LIKE*) (Figure 6.3). The DELLAs GAI and RGA are the main repressors, and DELLA RGL1 a minor repressor, of stem elongation (King *et al.*, 2001; Dill and Sun, 2001; Wen *et al.*, 2002). However, DELLAs RGA, GAI, RGL1 and RGL2 all repress stem elongation under high temperature stress (Stavang *et al.*, 2009). The DELLA RGL2 is the main repressor of seed germination, since *ga1-3 rgl2-1* is the only *ga1-3 della* double mutant that can germinate without GA application in the light, but not in the dark (Lee *et al.*, 2002; Tyler *et al.*, 2004; Cao *et al.*, 2005). Since the *ga1-3 rgl2-1 gai-t6 rga-t2* mutant can germinate in the dark as well as the light, DELLAs GAI and RGA can also repress seed germination. The DELLA

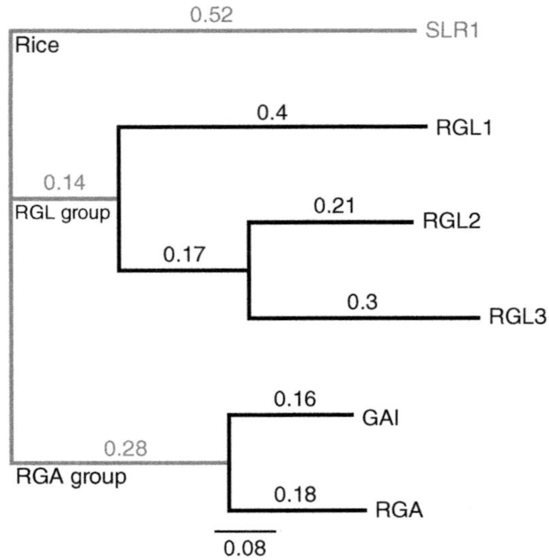

Figure 6.3 Phylogenetic analysis of DELLA protein homologues in *Arabidopsis* and rice based on predicted amino acid sequence analysis using ClustalΩ (Sievers *et al.*, 2011). A maximum likelihood tree was produced based on the JTT model (Jones *et al.*, 1992) and bootstrapping was performed with 1000 bootstrap replicates. Length of horizontal branches are proportional to the estimated number of amino acid substitutions per residue, which is indicated above each branch. The proposed RGL and RGA groups are indicated at their respective branching points.

RGL3 has been implicated in jasmonate (JA) signalling and in endosperm rupture during seed germination (Piskurewicz and Lopez-Molina, 2009; Wild *et al.*, 2012). RGL2 and RGA are the main DELLA repressors, and RGL1 a minor repressor, of the transition to flowering (Cheng *et al.*, 2004). GA also stimulates floral development by inducing the expression of floral homeotic transcripts, *APETALA3*, *PISTILLATA* and *AGAMOUS* (Yu *et al.*, 2004; see Chapter 11). RGL1 and RGA are the main DELLA repressors of floral development, whereas RGL2 plays a minor role. The defects in *ga1-3* floral development were partly rescued in *ga1-3 rgl1-1* and *ga1-3 rga-t2* double mutants, and were almost completely rescued in the *ga1-3 rgl1-1 rga-t2 rgl2-1* and *ga1-3 gai-t6 rga-t2 rgl1-1 rgl2-1* multiple mutants. The knockout of the four DELLAs *gai-t6 rga-t2 rgl1-1 rgl2-1* in the *ga1-3* background was referred to as 'penta' because it contains five mutations, although it is not a knock-out of all five DELLA genes (Cao *et al.*, 2005). A knockout of all five DELLA genes in the Landsberg *erecta* background was published by Feng *et al.* (2008).

A promoter swap experiment was used to examine whether the functional specialisation of the five DELLA genes was due to gene expression pattern or to differences in protein sequence (Gallego-Bartolomé *et al.*, 2010). DELLA RGL2 normally represses seed germination, but not stem elongation. DELLA

RGA normally represses stem elongation, but has a minor role in seed germination. When *GFP-RGA* and *GFP-RGL2* translational fusions were expressed on a 2-kb *RGA* promoter element, both genes served equally well to partly repress the growth of *ga1-3 gai-t6 rga-24*, and to partly restore feedback regulation of *GA20ox* GA biosynthesis gene expression. Conversely, *GFP-RGA* could function similarly to *GFP-RGL2* in repressing seed germination when expressed on the *RGL2* promoter. This suggests that some of the differences in the roles of *RGA* and *RGL2* are due not to differences in protein functionality, but due to differences in the timing and location of promoter expression. While both DELLA RGA and RGL2 proteins were able to interact with bHLH (basic-Helix-Loop-Helix) transcription factors PIF4 and PIF1/PIL5 in a yeast 2-hybrid assay, we cannot rule out that these DELLAs have different affinities for other DELLA-interacting proteins. For example, different DELLA proteins show different affinities for the three *Arabidopsis* GA receptors (Suzuki *et al.*, 2009), suggesting that differences in DELLA protein structure may lead to differential regulation by the three *Arabidopsis* GA receptors.

6.4 The gibberellin receptor GID1 (GA-INSENSITIVE DWARF1)

DELLA proteins function as negative regulators of GA responses that are down-regulated as a result of GA-stimulated protein–protein interaction with the GA-receptor, GID1 (GA-INSENSITIVE DWARF1) (Ueguchi-Tanaka *et al.*, 2005). The GA receptor was first identified by map-based cloning of the severely dwarfed mutant in rice, *gid1*. Loss of GID1 function in rice causes failure to respond to GA stimulation of leaf and cell elongation, flowering and fertility, and α-amylase expression during seed germination. Furthermore, these mutants accumulate bioactive GA at much higher levels than wild type. This increase in endogenous GA levels is likely due to up-regulation of GA biosynthesis genes as a feedback response to reduced GA signalling. While there is a single *GID1* GA receptor gene in rice, there are three GA receptor genes in *Arabidopsis*, *GID1a*, *GID1b*, and *GID1c* (Nakajima *et al.*, 2006; Yano *et al.*, 2015). The *Arabidopsis gid1* triple mutant exhibits severe GA-insensitive phenotypes including: failure to germinate unless the seed coat is cut, severe dwarfism and complete infertility (Griffiths *et al.*, 2006; Willige *et al.*, 2007; Iuchi *et al.*, 2007). Interestingly, the first *GID1* alleles were identified in barley based on reduced GA sensitivity during leaf elongation (Chandler and Robertson, 1999; Chandler *et al.*, 2008). After the cloning of the rice *GID1* gene, these semi-dwarf *gse1* (*GA sensitivity1*) mutants were found to be missense alleles of barley *GID1*.

The three *Arabidopsis GID1* genes have partially overlapping roles in GA signalling. No single *GID1* T-DNA insertion allele shows a strong GA-insensitive phenotype. However, double and triple mutants show varying degrees of GA-insensitive phenotypes that provide clues to the specialisation of *GID1* gene function in *Arabidopsis* (Griffiths *et al.*, 2006; Willige *et al.*, 2007; Iuchi *et al.*, 2007). The *gid1a gid1c* mutant has a stronger dwarf phenotype than *gid1a gid1b* or *gid1b gid1c*, suggesting that *GID1a* and *GID1c* play a stronger role in stem elongation. The *gid1a gid1b gid1c* triple mutant is far more severely dwarfed than any *gid1* double mutant, suggesting that *GID1b* also plays a role in stem elongation. Consistent with this, both *GID1a* and *GID1c* mRNAs and GID1-GUS translational fusions are more strongly expressed in inflorescence stems than *GID1b* (Griffiths *et al.*, 2006; Suzuki *et al.*, 2009). The *gid1a gid1b* double mutant had the strongest decrease in silique length and fertility, and the *gid1a gid1c* double mutant showed the most severe reduction in germination efficiency (4% germination) (Griffiths *et al.*, 2006; Voegele *et al.*, 2011). However, a *gid1b* allele in the Nossen ecotype showed a strong decrease in GA sensitivity during seed germination, suggesting that *GID1b* also stimulates seed germination. *GID1a* and *GID1b* stimulate floral bud formation, as the *gid1a gid1b* double mutant has lower fertility associated with shorter stamens (Griffiths *et al.*, 2006; Willige *et al.*, 2007; Iuchi *et al.*, 2007). While one research group published that the *Arabidopsis gid1a gid1b gid1c* triple mutant fails to flower under long day conditions, another published that the same triple mutant flowered under their light conditions (Willige *et al.*, 2007; Plackett *et al.*, 2014). It appears that the requirement for *GID1* genes during *Arabidopsis* flowering may depend on as yet uncharacterised environmental conditions, such as temperature, humidity, light quality or intensity.

The functional specialisation of the three *Arabidopsis GID1* genes may result in part from differences in their ability to regulate different DELLA proteins, given that the five *Arabidopsis* DELLA proteins have partly specialised functions. The strength of the DELLA-GID1 interaction was examined in the presence of GA_4 using both competitive yeast 3-hybrid and *in vitro* QCM (quartz crystal microbalance) assays for each of the three *Arabidopsis* GID1 proteins with each of the five DELLA proteins (Suzuki *et al.*, 2009). Since DELLAs RGA and GAI are the main repressors of stem elongation, we would expect them to interact more strongly with GID1a and GID1c, the main GA receptors regulating stem elongation. Instead, RGA and GAI exhibited the strongest preference for GID1b-binding. DELLA RGL2 and GID1a play strong roles in regulating seed germination. Consistent with this, RGL2 had the highest affinity for GID1a, followed by its affinity for GID1b. RGL1 and RGL3 had the strongest affinity for GID1a and the lowest affinity for GID1b. No DELLA protein had a strong preference for GID1c. Thus, the *Arabidopsis* DELLA proteins can be placed into two groups based on GID1 preference: (1) the RGA group with higher affinity for GID1b including GAI and RGA and (2) the RGL group with higher affinity for GID1a including RGL1, RGL2 and RGL3. While this grouping does not fully explain functional differences, it does coincide with the two phylogenetic groups based on overall DELLA amino acid sequence homology (Figure 6.3; Hirano *et al.*, 2007).

6.5 The structural requirements for gibberellin binding by GID1

The GID1 protein is a soluble GA receptor that localises to both the nucleus and the cytoplasm of rice and *Arabidopsis* cells (Ueguchi-Tanaka *et al.*, 2005; Willige *et al.*, 2007). Early work in barley showed that GA could be perceived both at the cell membrane and in the cytoplasm of barley aleurone cells (Hooley *et al.*, 1991; Gilroy and Jones 1994). While it has been postulated that both membrane-bound and cytosolic GA receptors may exist, no membrane-bound GA receptor has yet been identified (Nakajima *et al.*, 1997; Park *et al.*, 2005; Nakajima *et al.*, 2006). The nuclear localisation of GID1 is consistent with the fact that rice GID1 is the only GA receptor controlling DELLA-regulated gene expression in the nucleus (Yano *et al.*, 2015). GA hormone has also been shown to regulate calcium-dependent protein kinase function, calcium signalling and α-amylase secretion in the cytoplasm of barley aleurone cells (McCubbin *et al.*, 2004). It may be that GID1 also functions in cytoplasmic GA signalling, given that *Arabidopsis* GID1a fused to GFP and a nuclear exclusion signal was able to partially rescue germination and growth phenotypes of the *gid1a gid1c* double mutant (Livne and Weiss, 2014). Future work will need to better characterise GID1 function in cytoplasmic GA signalling.

GID1 is a homologue of the mammalian family of HSLs (hormone sensitive lipases) where the lipid-binding domain has become a GA hormone-binding domain (Østerlund, 2001; Ueguchi-Tanaka *et al.*, 2005; Nakajima *et al.*, 2006; Ueguchi-Tanaka *et al.*, 2007a; Hirano *et al.*, 2008; Murase *et al.*, 2008). GID1 proteins lack hydrolase activity, likely because either Val or Ile replace the His amino acid residue of the Ser-His-Asp catalytic triad (Figures 6.2 and 6.5; Nakajima *et al.*, 2006). Instead, this site forms the binding core for bioactive GAs, including GA_1, GA_3, GA_4 and GA_7 (Murase *et al.*, 2008; Shimada *et al.*, 2008). Both rice and *Arabidopsis* GID1 proteins have the highest affinity for GA_4, although GA_1 is the predominant bioactive GA in vegetative tissues of monocots (Ueguchi-Tanaka *et al.*, 2005; Nakajima *et al.*, 2006; Ueguchi-Tanaka *et al.*, 2007a). Bioactive GAs contain a γ-lactone ring between C-4 and C-10, a carboxyl group at C-6 and are hydroxylated at C-3. The Val/Ile residue of the catalytic triad plays a key role in GA binding through a non-polar interaction with the γ-lactone ring of bioactive GA molecules (Shimada *et al.*, 2008; Ueguchi-Tanaka and Matsuoka, 2010). The crystal structures of the rice OsGID1 (*Oryza sativa* GID1) and the *Arabidopsis* GID1a proteins bound to GA_4 as well as GA_3 have been solved (Murase *et al.*, 2008; Shimada *et al.*, 2008). GID1 resembles hormone-sensitive lipases in that it is composed of a C-terminal core with an N-terminal extension referred to as the lid (Figures 6.2 and 6.4). The core, also referred to as the α/β hydrolase fold domain, is composed of an α/β hydrolase fold surrounded by an eight-stranded β-sheet with α-helices packing the sides. The core contains the conserved HGG and GXSXG motifs characteristic of hormone-sensitive lipases and other carboxylesterases (Figure 6.5). The catalytic triad within the GID1 core forms a GA-binding pocket. There are six water molecules at the bottom of the binding pocket that form a hydrogen-bonding network with the polar side of GA. GID1 has lower affinity for GA_3 and GA_1 because, unlike GA_4, these GAs contain a 13-hydroxyl group that is inserted close to a negatively charged Asp residue (Asp 243 in GID1a) in the binding pocket (Nakajima *et al.*, 2006; Murase *et al.*, 2008; Shimada *et al.*, 2008). The N-terminal extension of GID1 consists of a loop and three α-helices (αa, αb and αc) that form a flat lid domain that covers both GA and the GA-binding pocket (Figure 6.4). The hydrophobic side of GA interacts with the GID1-lid to induce a stable conformational change. Hydrophobic interactions between the GA molecule and the lid are likely involved in pulling the lid closed. The GA-stimulated folding of the N-terminal lid creates a binding domain for DELLA protein on the outer face of GID1 (Figure 6.1c). There are no direct interactions between GA and DELLA, such that the N-terminal lid serves as 'molecular glue' between the GA-binding core on one face and the DELLA protein on the other face (Murase *et al.*, 2008).

6.6 The structural requirements for the GID1-DELLA protein–protein interaction

The current model of GID1 binding proposes that in its unbound form the exposed surface of the open GID1 lid is hydrophilic, but undergoes a conformational change upon GA binding to expose hydrophobic DELLA-interacting residues (Figure 6.1c; Shimada *et al.*, 2008; Murase *et al.*, 2008; Ueguchi-Tanaka and Matsuoka, 2010). This model is based solely on the structure of the GA-bound form of GID1, since the structure of the unbound form has not been solved. Thus, it is not known whether or not the unbound form is stably 'open' (Hao *et al.*, 2013). GID1 in its GA-bound form has a number of hydrophobic amino acid side chains that protrude from the outer surface of the N-terminal lid, providing a binding domain for DELLA proteins. These hydrophobic residues are Leu-18, Trp-21, Leu- or Val-29, and Ile-33 in α-helix αb, and Leu-45 and Tyr-48 in α-helix αc of OsGID1 and GID1a (Figures 6.2 and 6.5; Shimada *et al.*, 2008; Murase *et al.*, 2008). Alanine scanning has shown that these hydrophobic residues in rice GID1 are required for protein–protein interaction with the DELLA SLR1, but not for GA binding (Shimada *et al.*, 2008). Moreover, the crystal structure of the *Arabidopsis* GID1a-GA-DELLA complex showed that these residues are the major sites of interactions between the closed lid of GID1a and DELLA GAI (Murase *et al.*, 2008). Hydrophobicity is a major force in protein folding, given that native protein structure in an aqueous environment generally does not allow exposure of hydrophobic side chains (Rose *et al.*, 1985; Dill, 1990; Huang *et al.*, 1995). GID1's GA-binding activity is stronger in the presence than in the absence of DELLA (Nakajima *et al.*, 2006; Ueguchi-Tanaka *et al.*,

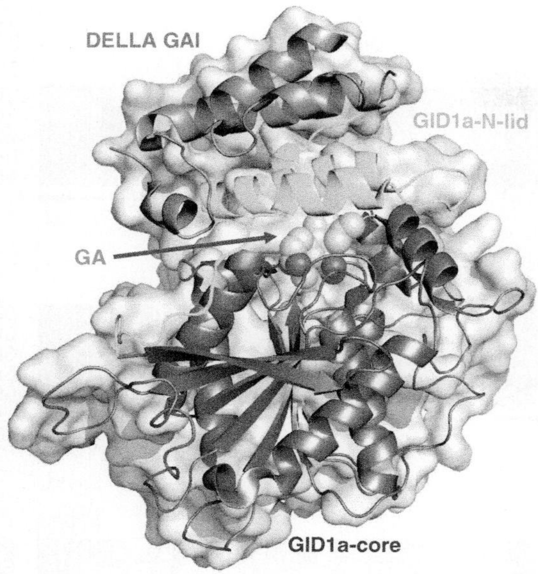

Figure 6.4 The GID1a-GA_3-DELLA complex based on the 1.8 angstrom crystal structure (Murase *et al.*, 2008). Ribbon representation of GID1a in complex with DELLA GAI and GA_3. The N-terminal GAI DELLA domain residues 11–113 (pink) is shown in complex with GID1a residues 1–344. The GID1a N-terminal extension or lid domain (GID1a-N-lid) is shown in blue and the GID1a α/β core domain in purple (GID1a-core). The GA_3 molecule (arrow) is shown in its binding pocket as a space-filling model where carbon is grey and oxygen red. (Figure was kindly provided by Toshio Hakoshima.) (See insert for colour representation of this figure.)

2007a). It may be that the presence of DELLA facilitates the exposure of the hydrophobic GID1 lid residues needed for stable DELLA protein binding.

Major DELLA protein motifs involved in the GID1-DELLA protein–protein interaction are located in the DELLA regulatory domain. The two neighbouring motifs, DELLA and LExLE (sometimes collectively referred to as the DELLA motif), are required for DELLA interaction with GID1 protein (Murase *et al.*, 2008). The crystal structure showed direct GID1 binding to the DELLA motif at the residues DeLLaΦLxYxV and MAxVAxxLExLExΦ, where capitalised residues represent sites of direct interactions, Φ represents a non-polar residue, and 'x' can represent any residue. Mutation analysis demonstrated that the DELLA motif is essential for GID1 binding, whereas mutations in the LExLE motif only resulted in decreased affinity for GA-bound GID1 (Figure 6.2: Peng *et al.*, 1997; Dill and Sun, 2001; Itoh *et al.*, 2002; Murase *et al.*, 2008). The VHYNP motif also plays a role in GID1-binding by stabilising the GID1-DELLA interaction via the residues TVhynPxxLxxWxxxM.

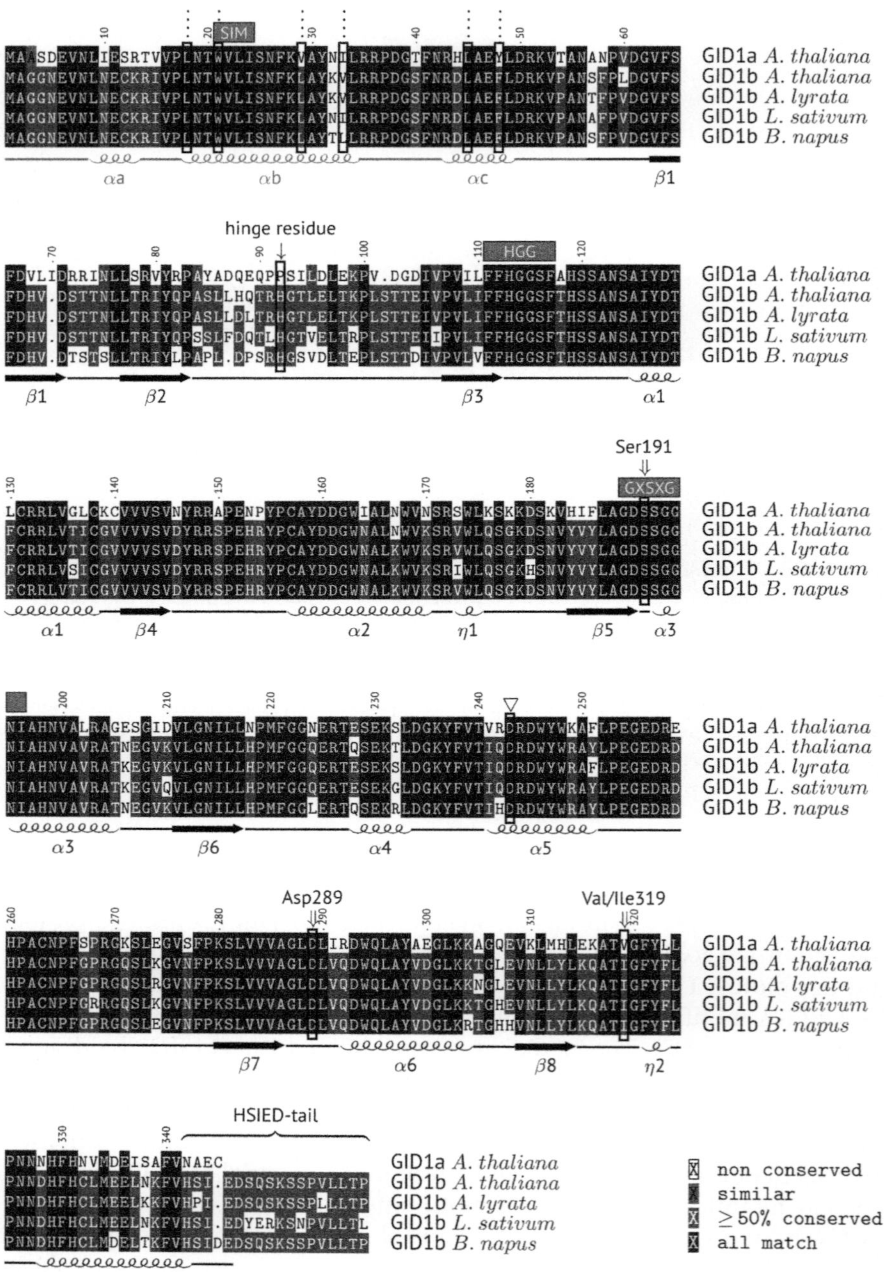
SIM
HGG
hinge residue
Ser191
GXSXG
Asp289
Val/Ile319
HSIED-tail
GID1a A. thaliana
GID1b A. thaliana
GID1b A. lyrata
GID1b L. sativum
GID1b B. napus
non conserved
similar
≥50% conserved
all match

The GID1-GA-DELLA complex has a highly ordered structure, but investigation of DELLA in its unbound state revealed that the N-terminal region of DELLA proteins is intrinsically unstructured (Murase *et al.*, 2008; Sun *et al.*, 2010; Sheerin *et al.*, 2011). About 70% of signalling proteins are predicted to be intrinsically unstructured proteins (IUPs), containing long disordered regions believed to play a role in molecular recognition (Dunker *et al.*, 2000; Iakoucheva *et al.*, 2002; Oldfield *et al.*, 2005). Such IUPs can contain short regions of relative order within their unstructured regions called molecular recognition features (or MoRFs) (Oldfield *et al.*, 2005; Mohan *et al.*, 2006). MoRFs undergo a disorder-to-order conformational change upon recognition of their binding partner, and are designated as α, β or ι based on their preferred bound conformational state of α-helix, β-strand, or irregular structure, respectively (Fuxreiter *et al.*, 2004; Mohan *et al.*, 2006). Research suggested that the DELLA N-terminal region is almost entirely disordered, with the exception of the predicted α-MoRFs, DELLA/LExLE and L(K/R)XI, and the ι-MoRF VHYNP (Figure 6.2; Uversky 2002; Sun *et al.*, 2010). The C-terminal GRAS domain was predicted to be mostly ordered. The predicted MoRFs represent the key binding sites for the DELLA interaction with GID1, but for the small L(K/R)KI motif located within the poly S/T/V domain. It is possible that this motif is involved in DELLA binding with another component of the signalling pathway (Figure 6.2).

The ι-MoRF region in the VHYNP motif is an irregular loop (VHYNPSD loop) involved in binding-induced folding of the RGA, but not the RGL group of DELLAs (Figure 6.3; Sun *et al.*, 2010). The VHYNPSD loop of the RGA group, consisting of RGA, GAI and OsSLR1, undergoes a conformational change upon GID1 binding. The RGL group, consisting of RGL1, RGL2 and RGL3, does not appear to undergo a conformational change. One explanation for this difference is that the N-terminal DELLA domain of the RGA group

Figure 6.5 Predicted amino acid sequence alignment of *Arabidopsis thaliana* GID1a and GID1b with GID1b-type homologues from *Arabidopsis lyrata, Lepidum sativum* and *Brassica napus* showing GID1b-type specific regions of homology using ClustalΩ for alignment and TeXshade package in LaTeX (Beitz 2000; Sievers *et al.*, 2011). Amino acid residue numbers are based on GID1a. Significant residues (boxed) include the DELLA interacting residues in the lid (four dots mark these six residues), the 'hinge residue' in a loop of the lid that differs between GID1ac- and GID1b-type receptors, the catalytic triad involved in binding GA (Ser 191, Asp 289, and Val/Ile 319), and the negatively charged Asp 243 that likely reduces affinity for GA_1 and GA_3 compared to GA_4 (downward facing triangle). Significant motifs are the SUMO-Interaction-Motif (SIM), the HGG motif and the GXSXG motif (grey box). Predicted secondary structures are presented below the aligned sequences as a solid line (loop), spiral (α indicates α-helix, η indicates 3_{10}-helix) and block arrow (β-strand); where the lid containing α-helices αa, αb and αc is grey and the core domain is black. Some regions are conserved only among GID1b-type receptors, such as hinge residue and the C-terminal HSIED-tail (bracket). (See insert for colour representation of this figure.)

may be more structured and less flexible than the DELLA domain of the RGL group. Interestingly, *Arabidopsis* DELLAs can be divided into the same two groups based on amino acid sequence homology and preference for GID1a, GID1b or GID1c as a binding partner (Suzuki *et al.*, 2009). It is possible that the structural difference described above helps to determine the preference for GID1-binding partner.

Some amino acid residues in the GRAS functional domain also appear to participate in GID1 binding (Figure 6.2; Hirano *et al.*, 2010; Sato *et al.*, 2014). The semi-dominant mutation in the rice DELLA, SLR1^{G576V}, resulted in reduced GID1 binding in yeast 2-hybrid studies, suggesting that the SAW motif may participate in the GID1-DELLA protein–protein interaction (Hirano *et al.*, 2010). Alanine scanning also detected some decrease in GID1 binding due to changes in the VHIID motif. Another study demonstrated that the SLR1 GRAS domain bound to GID1 with much lower affinity than the DELLA domain, using pull-down assays, NMR spectroscopy and surface plasmon resonance (SPR) analysis (Sato *et al.*, 2014). This suggested that, under physiological conditions, GID1 binding to the GRAS domain likely occurs after GID1 binding to the DELLA domain. Based on mutation analysis, Gly-576 of the SAW motif appears to be a key residue in the GRAS-GID1 interaction. A rice homologue of DELLA SLR1, SLRL1 (SLR1-Like1), was able to function like SLR1 to repress growth when over-expressed (Itoh *et al.*, 2005b). SLRL1 lacks a DELLA domain, but was apparently able to bind to GID1 via the GRAS domain alone. Because it has no DELLA domain, SLRL1 did not undergo GA-stimulated protein destruction. Future work will need to examine whether non-DELLA GRAS proteins can function via protein–protein interaction with GID1. GID1 binding results in a C-terminal conformational change in DELLA that likely promotes binding to the SLEEPY1 (SLY1) F-box protein via the VHIID and LHR2 motifs (Sasaki *et al.*, 2003; Murase *et al.*, 2008; Shimada *et al.*, 2008).

6.7 The DELLA destruction model: negative regulation of DELLA repressors by SLY1/GID2 and the ubiquitin-proteasome pathway

GA signalling down-regulates DELLA repressors of GA responses by targeting them for destruction via the ubiquitin-proteasome pathway (Griffiths *et al.*, 2006; Nakajima *et al.*, 2006; Ariizumi *et al.*, 2008; Wang *et al.*, 2009). The 'DELLA destruction model' for GA signalling originated with the observation that GA rescue of GA biosynthesis mutants was associated with the rapid disappearance of the DELLA protein RGA (Figure 6.1a, b; Silverstone *et al.*, 2001). All of the DELLA proteins of *Arabidopsis* and other plants characterised thus far degrade as quickly as 5 to 60 minutes after GA treatment (Itoh *et al.*, 2002; Fu *et al.*, 2002; 2004; Tyler *et al.*, 2004; Ariizumi and Steber, 2007; Wang *et al.*, 2009; Zhang *et al.*, 2010). Thus, it is widely accepted that GA lifts DELLA repression of seed germination, stem elongation, and flowering and fertility via DELLA protein proteolysis.

DELLA is ubiquitinated and targeted for destruction by an SCF (Skp1, Cullin, F-box) E3 ubiquitin ligase (Figure 6.1b; Sasaki *et al.*, 2003; McGinnis *et al.*, 2003; Dill *et al.*, 2004; Gomi *et al.*, 2004; Fu *et al.*, 2004; Hussain *et al.*, 2005; Ariizumi *et al.*, 2011). The *Arabidopsis* SLY1 and rice GID2 proteins are the F-box sub-units of the SCF complex that specifically binds to DELLA proteins, leading to their polyubiquitination. Mutations in the F-box genes *Arabidopsis SLY1* (*SLEEPY1*) and rice *GID2* (*GA-INSENSITIVE DWARF2*) block GA-induced DELLA proteolysis, leading to GA-insensitive phenotypes, including dwarfism, infertility and increased seed dormancy in *sly1* (Steber *et al.*, 1998; Sasaki *et al.*, 2003; McGinnis *et al.*, 2003). Thus, DELLA over-accumulation is associated with decreased GA signalling (Figure 6.1d).

Protein ubiquitination occurs via a multi-stage process that concludes with the covalent linkage of the 76-amino-acid ubiquitin peptide to the target protein (reviewed by Smalle and Vierstra, 2004; Wang and Deng, 2011). The E1 ubiquitin activating enzyme catalyses the formation of a thio-ester bond between the C-terminal glycine of ubiquitin and an E1 cysteine residue. The activated ubiquitin is transferred to a cysteine residue of the ubiquitin conjugating enzyme E2 by transesterification. The E2 ubiquitin conjugating enzyme transfers ubiquitin to a lysine residue on the target protein. An E3 ubiquitin ligase like $SCF^{SLY1/GID2}$ can catalyse the transfer of ubiquitin to a specific target by bringing the E2 and the target protein together in a single complex. Addition of a polyubiquitin chain containing four ubiquitin moieties targets a protein for destruction by the 26S proteasome. In a cell-free system, DELLA was polyubiquitinated predominantly by a ubiquitin chain with Lys-29 linkages between ubiquitin moieties, rather than the usual Lys48 linkages (Wang *et al.*, 2009). The Lys residue(s) modified by ubiquitination of DELLA have not yet been identified and may be a good avenue for future investigation.

As shown in Figure 6.1b, the SCF E3 ubiquitin ligase of GA signalling is apparently comprised of: (1) the SLY1/GID2 F-box protein that binds the DELLA target at its C-terminus and binds an ASK (*Arabidopsis* SKP1 homologue) via the F-box motif, (2) the ASK protein that binds CUL1 (CULLIN1), (3) CUL1, the backbone of the complex that binds ASK at its N-terminus and an RBX1 (RING BOX1 protein) at its C-terminus and (4) an RBX1 homologue that binds an E2 ubiquitin conjugating enzyme (Figure 6.1b; Gagne *et al.*, 2002; Gray *et al.*, 2002; Risseeuw *et al.*, 2003; Fu *et al.*, 2004; Wang *et al.*, 2009; Ariizumi *et al.*, 2011). *Arabidopsis* SLY1 and rice GID2 are small proteins of 151 and 212 amino acids, respectively, that show 36.8% amino acid identity and 56% similarity (Sasaki *et al.*, 2003; McGinnis *et al.*, 2003; Itoh *et al.*, 2003). They contain an F-box motif and a C-terminal domain required for interaction with DELLA proteins (Figure 6.2; Dill *et al.*, 2004; Fu *et al.*, 2004; Hirano *et al.*, 2010). The F-box motif, conserved in yeast, mammals and plants, binds to SKP1 homologues, allowing SCF complex formation (Schulman *et al.*, 2000). SLY1 and GID2 both interact with SKP1 homologues in yeast 2-hybrid assays (Gagne *et al.*, 2002; Sasaki *et al.*, 2003; Fu *et al.*, 2004). SLY1 coimmunoprecipitation with CUL1 depends on the presence of an intact F-box motif, indicating that SLY1 forms an SCF complex *in planta* via the F-box motif (Ariizumi *et al.*, 2011). The conserved GGF and LSL amino acid motifs in the SLY1/GID2 C-terminus are required for interaction with the DELLA protein VHIID and LHR2 motifs (Figure 6.2; Hirano *et al.*, 2010; Ariizumi *et al.*, 2011). Rice *GID2* mutants carrying a 19- or 31-bp deletion in the F-box motif resulted in a GA-insensitive phenotype associated with dwarfism, complete infertility and failure to induce the GA-induced enzyme α-amylase during seed germination (Sasaki *et al.*, 2003). *Arabidopsis sly1* mutants also result in dwarfism associated with partial, rather than complete infertility (Steber *et al.*, 1998; Steber and McCourt, 2001). The *sly1* mutants also show increased seed dormancy, consistent with the role of GA signalling in seed germination (Ariizumi and Steber, 2007). Thus, *SLY1* and *GID2* are required for normal GA responses as well as for DELLA protein destruction.

The formation of the GID1-GA-DELLA complex is the signal that causes $SCF^{SLY1/GID2}$ to polyubiquitinate DELLA, thereby targeting DELLA for destruction by the 26S proteasome (Figure 6.1a, b). While some interaction was initially detected between SLY1 and DELLA protein by yeast 2-hybrid, later work showed the SLY1/GID2 affinity for DELLA is greatly enhanced when DELLA is in the GID1-GA-DELLA complex (Sasaki *et al.*, 2003; Fu *et al.*, 2004; Griffiths *et al.*, 2006; Willige *et al.*, 2007; Ariizumi *et al.* 2011; Hirano *et al.*, 2010). Thus, both the SLY1-DELLA protein–protein interaction and DELLA destruction are stimulated by GA hormone perception. As demonstrated *in vivo* and in cell-free extracts, both *GID1* and *SLY1* are necessary for efficient DELLA proteolysis in response to GA (Sasaki *et al.*, 2003; McGinnis *et al.*, 2003; Ueguchi-Tanaka *et al.*, 2005; Willige *et al.*, 2007; Wang *et al.*, 2009). It is widely accepted that the 26S proteasome is responsible for DELLA proteolysis because GA-stimulated DELLA destruction is blocked by 26S proteasome inhibitors, leading to accumulation of ubiquitinated DELLA protein (Fu *et al.*, 2002; Sasaki *et al.*, 2003; Hussain *et al.*, 2005; Wang *et al.*, 2009).

Originally, it was assumed that the *sly1* loss-of-function mutants have a GA-insensitive phenotype because there is only one copy of the *SLY1* gene in *Arabidopsis*. However, the *sly1* mutant phenotypes are not as severe as those of the GA biosynthesis mutant *ga1-3* or the *gid1a gid1b gid1c* triple mutant. A possible explanation for this was a predicted homologue of *SLY*1 in *Arabidopsis* with 23.7% amino acid identity (Itoh *et al.*, 2003). A screen for genes that suppress *sly1* phenotypes when over-expressed on the 35S promoter identified this *SLY1* homologue, referred to as *SNE* (*SNEEZY*) or as *SLY2* in *Arabidopsis* (Fu *et al.*, 2004; Strader *et al.*, 2004). *SNE* over-expression partly rescues *sly1* mutations and results in decreased DELLA protein levels, suggesting that the SNE F-box protein can functionally replace SLY1. An HA:SNE fusion protein coimmunoprecipitated with DELLA RGA, but not with RGL2, whereas HA:SLY1 coimmunoprecipitated with both RGA and RGL2 (Ariizumi *et al.*, 2011). Thus, it appears that SNE can only down-regulate a sub-set of DELLA repressors. If SNE normally functions in GA signalling, then *sne* mutants should show *sly1*-like GA-insensitive phenotypes. However, T-DNA insertion alleles of *SNE* showed no phenotype (Ariizumi and Steber, 2011). Moreover, the *sly1 sne* double mutant only showed a slight increase in seed dormancy and slight decrease in plant height, suggesting that loss of *SNE* does not eliminate GA signalling in *sly1* mutants. Thus, SLY1 is the major F-box protein directing DELLA degradation. Future work will need to determine whether there are environmental conditions under which the SNE F-box protein plays a stronger role in GA signalling, and determine whether SNE regulates proteins other than DELLA repressors. For example, *SLY1* and *SNE* are expressed in different root cells types, and the *sne-1* mutant exhibited a shortened root phenotype under dry conditions (Cui and Benfey, 2009). This suggests that *SNE* may have a unique function in stimulating root growth.

6.8 Regulation of DELLA by phosphorylation and *O*-GlcNAc modification

The discovery that DELLA proteins can be phosphorylated followed fast on the heels of the discovery that DELLAs are regulated by the ubiquitin-proteasome pathway (Sasaki *et al.*, 2003; Fu *et al.*, 2004). However, the functional significance of DELLA phosphorylation has been elusive. Many proteins regulated by the ubiquitin-proteasome pathway are ubiquitinated and targeted for destruction in response to phosphorylation (reviewed by Willems *et al.*, 1999; Nguyen *et al.*, 2013). For example, phosphorylation of yeast CYCLIN2 stimulates the interaction of the cyclin with the F-box protein GRR1 (GLUCOSE REPRESSION-RESISTANT1), leading to cyclin ubiquitination and destruction by the 26S proteasome. Thus, early models of GA signalling hypothesised that DELLA phosphorylation was the signal for DELLA ubiquitination and destruction (Sasaki *et al.*, 2003; Gomi *et al.*, 2004; Fu *et al.*, 2004). In this model DELLA phosphorylation would stimulate SLY1/GID2 binding to DELLA, thereby targeting DELLA for GA-stimulated destruction. This model was disproved when it was found that the DELLA SLR1 phosphorylation detected in the TVHYNP and poly S/T/V motifs was GA-independent, and that both the phosphorylated and unphosphorylated forms of DELLA SLR1 interacted with the F-box GID2 (Itoh *et al.*, 2005a). Subsequent research proved that the interaction of the F-box protein with DELLA depended, not upon DELLA phosphorylation, but on the formation of the DELLA-GA-GID1 complex (Ueguchi-Tanaka *et al.*, 2005; Griffiths *et al.*, 2006; Nakajima *et al.*, 2006). Moreover, protein phosphatase inhibitors appeared to block degradation of barley DELLA SLN1 and *Arabidopsis* DELLAs RGA and RGL2 (Fu *et al.*, 2002; Hussain *et al.*, 2005; Wang *et al.*,

2009). This would suggest that DELLA phosphorylation stabilises, rather than targets DELLA protein for destruction.

Results of genetic studies of the rice Ser/Thr casein kinase I *EL1* (*EARLY FLOWERING1*) are consistent with the idea that phosphorylation positively regulates DELLA repression of GA signalling (Figure 6.6a; Dai and Xue, 2010). The *el1* loss-of-function mutant flowered early and enhanced GA-mediated DELLA degradation. The *el1* mutant has other phenotypes consistent with increased GA sensitivity, including a small increase in stem elongation and an ABA-insensitive increase in α-amylase expression during seed germination. Over-expression of *EL1* resulted in dwarfism. Thus, *EL1* behaves like a negative regulator of GA signalling. Several lines of evidence suggest that EL1 regulates DELLA SLR1 by phosphorylation: (1) EL1 can phosphorylate DELLA SLR1 *in vitro,* (2) loss of the predicted SLR1 phosphorylation sites, in S196A and S510A mutants, leads to reduced accumulation of DELLA activated transcripts and (3) phosphomimic mutations, S196D and S510D, lead to increased accumulation of DELLA-activated transcripts. Moreover, the rice *el1* mutation suppresses the dwarf phenotype associated with SLR1 over-expression, indicating that EL1 is directly or indirectly required for DELLA SLR1 repression of stem elongation. If *EL1* acts primarily through DELLA SLR1 phosphorylation *in planta,* then this suggests that DELLA repressors are positively regulated by EL1-mediated phosphorylation (Figure 6.6a). This is consistent with phosphatase inhibitor studies suggesting that phosphorylation stabilises DELLA repressors in *Arabidopsis* and barley (Fu *et al.*, 2002; Hussain *et al.*, 2005; Wang *et al.*, 2009). Based on an amino acid alignment, the SLR1 Ser-510 residue is conserved in *Arabidopsis* DELLAs RGA, GAI and RGL1, corresponding to Ser-417 in GAI (Figure 6.2). The DELLAs RGL2 and RGL3 have an Ala residue in place of the Ser. The SLR1 Ser-196 residue does not appear to be conserved in *Arabidopsis* DELLAs. Future work will need to examine whether EL1-mediated DELLA phosphorylation occurs *in planta,* is conserved in other plant species and whether there is a direct connection between *el1* phenotypes and DELLA phosphorylation state. It will also be interesting to learn what effects DELLA phosophorylation at Ser-196 in the poly S/T/V motif and/or at Ser-510 in the PFYRE motif may have on DELLA function and protein–protein interactions. This is the first phenotypic evidence suggesting that phosphorylation may stabilise DELLA protein and promote DELLA repression of GA signalling.

Investigation of the *Arabidopsis* protein phosphatase, TOPP4 (TYPE ONE PROTEIN PHOSPHATASE4), provided further evidence that phosphorylation may positively regulate and dephosphorylation negatively regulate DELLA repression of GA signalling (Qin *et al.*, 2014). The *Arabidopsis* TOPP family contains nine members implicated in regulation of plant growth and development (Smith and Walker, 1993; Lin *et al.*, 1998). The dominant negative *topp4-1* mutation results in GA-insensitive phenotypes, including dwarfism, poor fertility, delayed flowering and failure to induce GA-responsive gene expression (Qin *et al.*, 2014). The *topp4-1* phenotypes

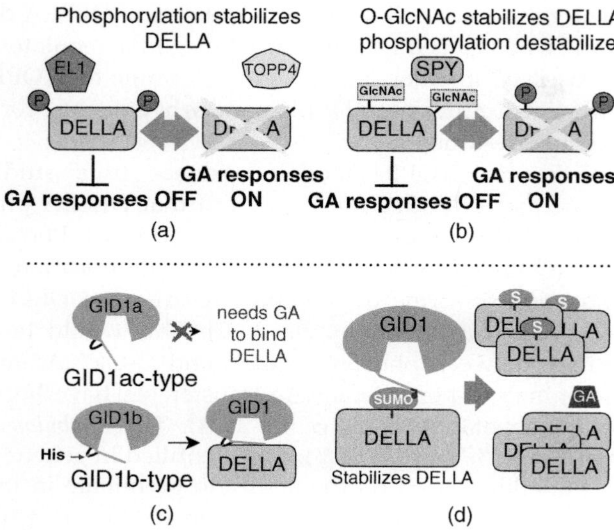

Figure 6.6 Alternative models for DELLA regulation. (a) The EL1/TOPP4 model. EL1-mediated phosphorylation of DELLA proteins stabilises DELLA protein, thereby increasing DELLA repression of GA responses. TOPP4-mediated dephosphorylation destabilises DELLA, thereby stimulating GA responses. (b) The SPY model. Phosphorylation *destabilises* DELLA thereby lifting DELLA repression of GA responses (opposite of EL1 model). SPY directs *O*-GlcNAc modification of DELLA at the same Ser/Thr residues subject to phosphorylation, leading to DELLA stabilisation and repression of GA responses. The effect of phosphorylation on DELLA activity may differ based on the location of the phosphorylation site or changes in binding partners. (c) The GID1b-type lid model. The GID1ac-type receptors can only bind DELLA when GA stimulates lid closure. The GID1b-type hinge has His 91 in place of the Pro in GID1ac-type receptors. This residue causes the lid domain to be partly closed without GA-binding leading to a low level of GA-independent DELLA binding and GA signalling. (d) The SUMO model. DELLA is SUMOylated at a residue directly before the LExLE motif involved in GID1 binding. SUMOylation of DELLA allows GA-independent binding to the SIM domain in the GID1 lid. GID1-binding by SUMOylated DELLA sequesters GID1 away from non-SUMOylated DELLA, thereby preventing DELLA ubiquitination and destruction, resulting in a build-up of both SUMOylated and non-SUMOylated forms of DELLA.

were associated with increased DELLA accumulation and delayed DELLA degradation following GA application. *In vitro* phosphatase assays suggested that wild-type TOPP4, but not topp4-1 protein, can dephosphorylate DELLAs GAI and RGA. Thus, it appears that *TOPP4* is a positive regulator of GA signalling that may negatively regulate DELLA by dephosphorylation. GA treatment stimulated the accumulation of *TOPP4* mRNA and protein. Thus, the proposed model is: (1) phosphorylation by EL1 or other kinases stabilises DELLA proteins, thereby repressing GA responses and (2) GA stimulates TOPP4 accumulation thereby destabilising DELLA via dephosphorylation and proteasomal degradation, thus stimulating GA responses (Figure 6.6a).

Future work will need to establish the *in vivo* effects of DELLA dephosphorylation, examine whether TOPP4 function as a negative regulator of DELLA repressors is conserved in other species, and determine the TOPP4 dephosphorylated amino acid residues in DELLA and whether they correspond to EL1-phosphorylated residues.

Contrary to the EL1/TOPP4 model, evidence from studies of the *O*-GlcNAc transferase, *SPY* (*SPINDLY*) suggest that DELLA phosphorylation can negatively regulate GA signalling (Figure 6.6b). Phosphorylation and *O*-GlcNAc (*O*-linked *N*-acetylglucosamine) modification may compete for modification of serine or threonine residues on DELLA proteins (Shimada *et al.*, 2006; Silverstone *et al.*, 2007). This would be analogous to the competition between phosphorylation and *O*-GlcNAc modification observed in mammals, where *O*-GlcNAc transferases have been found in complex with phosphatases (Wells *et al.*, 2004). The *Arabidopsis* putative *O*-GlcNAc transferase, *SPY* (*SPINDLY*), was identified in genetic screens for increased GA signalling based on the ability to germinate in the presence of a GA biosynthesis inhibitor and suppression of the *ga1-3* biosynthesis mutant (Jacobsen and Olszewski, 1993; Silverstone *et al.*, 1997b). Direct protein–protein interaction between SPY and DELLA has not been observed and specific sites of modification have yet to be proposed. *SPY* is defined as a negative regulator of GA signalling in *Arabidopsis*, barley and rice because loss-of-function results in increased GA signalling associated with increased stem elongation and lack of seed dormancy (Robertson *et al.*, 1998; Swain *et al.*, 2001; Shimada *et al.*, 2006; Filardo *et al.*, 2009). Silencing of rice *SPY* suppresses the GA-insensitive dwarfism of the *gid1* GA receptor and the *gid2* F-box mutants without any change in DELLA protein levels, suggesting that the increase in GA signalling in *spy* requires neither the GA receptor nor DELLA destruction (Shimada *et al.*, 2006). The *Arabidopsis spy* mutant also suppresses the GA-insensitive dwarfism of DELLA gain-of-function mutation *rga-Δ17* (Silverstone *et al.*, 2007). The *spy* suppression of *rga-Δ17* and *spy* silencing in rice was associated with an apparent increase in DELLA phosphorylation. Thus, the current model is that SPY activates the DELLA repressor via *O*-GlcNAc modification and that increased phosphorylation in *spy* mutants *inactivates* the DELLA repressor leading to increased GA signalling (Figure 6.6b). This disagrees with the *EL1/TOPP4* model in which phosphorylation *activates* DELLA repression of GA signalling. Thus, the role of phosphorylation in controlling DELLA protein function may be more complex than turning DELLA repression on or off. Future work will need to examine whether DELLA phosphorylation at different Ser or Thr residues has different functions, serving either to stimulate or block DELLA repression. One important consideration for future investigations will be to clearly ascertain whether changes in DELLA electrophoretic mobility are due to phosphorylation, *O*-GlcNAc modification or the newly discovered SUMOylation of DELLA protein (Conti *et al.*, 2014).

6.9 Evidence for gibberellin-independent DELLA regulation

Arabidopsis GID1b and some GID1b-type homologues such as soybean GID1b-2 have the ability to interact with DELLA proteins to some degree even in the absence of GA, suggesting that GID1b-type receptors may serve to 'prime the pump' of GA signalling under conditions when GA levels are low (Figure 6.6c; Griffiths *et al.*, 2006; Nakajima *et al.* 2006; Yamamoto *et al.*, 2010). While GA binding stimulates the interaction of GID1b with DELLA, the fact that GID1b can bind DELLA in the absence of GA suggests that GID1b-type receptors can initiate GA-independent GA signalling via DELLA destruction. However, no one has yet demonstrated that a GA-independent GID1b-DELLA protein interaction can stimulate interaction of DELLA with SLY1 leading to DELLA proteolysis. Sequence alignment between *Arabidopsis* GID1 protein sequences shows 85% amino acid identity between GID1a and GID1c, but only 66% and 67% identity of GID1a with GID1b and of GID1c with GID1b, respectively (Figure 6.5). Based on amino acid homology, higher plant GID1 homologues can be divided into two groups: GID1ac-type and GID1b-type receptors (Yamamoto *et al.*, 2010; Voegele *et al.*, 2011). Like the GID1ac-type receptors, monocot GID1 proteins such as OsGID1 show only GA-dependent interaction with DELLA proteins.

Mutation analysis of rice GID1 provided clues to the structural basis for the GA-independent GID1b-DELLA protein–protein interaction, and led to a model to explain this interaction (Yamamoto *et al.*, 2010). A missense mutation causing a P99S amino acid substitution in a loop between the N-terminal lid domain and the body of rice GID1 (in the hinge of the lid) resulted in a GID1b-mimic phenotype, allowing OsGID1^{P99S} to bind DELLA in the absence of GA and suppressing the GA-insensitive phenotype of the *gid1-8* loss-of-function mutation. The same Pro residue is present in the loop region of *Arabidopsis* GID1a at Pro-92 and GID1c at Pro-91, but is replaced by His-91 in GID1b (Figure 6.2 and 6.5). Site-directed mutation analysis showed that P99I, P99V and P99A amino acid substitutions in OsGID1 resulted in GA-independent DELLA-binding, and that a H91P substitution in *Arabidopsis* GID1b reduced DELLA-binding in the absence of GA. In the model proposed, *Arabidopsis* GID1a Pro-92 (OsGID1 Pro-99) is needed to prevent DELLA-binding when GA is not present to 'pull the lid closed' on the GID1 receptor. In this model, having His instead of Pro in the GID1b loop/hinge region causes the GID1b lid domain to remain partly closed, allowing the lid to bind DELLA in the absence of GA (Figure 6.5 and 6.6c). Note that the partially closed lid does not bind DELLA as well as a fully closed lid, so that GA binding greatly increases the affinity of GID1b-type receptors for DELLA. Interestingly, GID1b homologues in *Brassica* and soybean also showed GA-independent DELLA-binding activity, suggesting that multiple plant species have evolved GA-independent DELLA signalling.

In addition to having greater affinity for DELLA in the absence of GA, GID1b-type receptors also have higher affinity for GA. *Arabidopsis* GID1b has higher affinity for GA_4 (K_d = 4.8 x 10^{-7} M) than either GID1a or GID1c (K_d = *ca.* 2 x 10^{-6} M) (Nakajima *et al.*, 2006). Kinetic studies revealed that GID1b association with GA_4 occurs at about the same rate as GID1a, but that GA_4 dissociation from GID1b is about 17 times slower than from GID1a (Yamamoto *et al.*, 2010). The P99A amino acid substitution in OsGID1 resulted in GA_4 binding kinetics that more closely resembled those of GID1b. GID1b also shows optimal function over a narrower range of pH conditions (optimal pH 6.8) than GID1a or GID1c, which exhibited a consistently high level of binding activity between pH 6.4 and pH 8.3 (Nakajima *et al.*, 2006). Thus, sequence differences between GID1b- and GID1ac-type GA receptors may impact multiple functions. Basal GA signalling by *Arabidopsis* GID1b may explain why the *ga1-3* mutant that produces little or no GA hormone is not as extremely dwarfed as the *gid1a gid1b gid1c* triple mutant (Griffiths *et al.*, 2006).

Comparison of the *Arabidopsis* GID1b predicted amino acid sequence with three other GID1b-type receptors from other eudicot species, *Arabidopsis lyrata, Lepidum sativum* and *Brassica napus*, indicated that there is a higher degree of homology between predicted GID1b-type receptors of these four species than between GID1a and GID1b of *Arabidopsis* (Figure 6.5). There are many additional regions with conserved predicted amino acid sequence, in addition to the conserved His91 in the GID1b hinge. For example, there is a region with high homology at the C-terminal end of the four GID1b-type receptors, where the *Arabidopsis* GID1b amino acid sequence is HSIEDSQSKSSPVLLTP. Predicting GID1b structure based on the crystal structure of GID1a, it is possible that this C-terminal HSIED-tail motif of GID1b-type receptors might be oriented such that it could play a role in lid closure or GA binding (Murase *et al.*, 2008; Shimada *et al.*, 2008). Thus, future work will need to examine whether additional amino acid regions participate in the unique properties of GID1b-type receptors. Taken together, this information suggests that the GID1b-type receptors of eudicot plant species may have evolved for a unique and as yet undefined purpose in plant growth and development.

Recent evidence in *Arabidopsis* has suggested another method of GA-independent signalling through an increase in DELLA repression due to SUMO (Small Ubiquitin-like Modifier) modification of DELLAs (Conti *et al.*, 2014). Like ubiquitin, SUMO is a short peptide that can be covalently linked to a protein sequence (reviewed by Vierstra, 2012). SUMOylation of DELLA RGA protein was found within the DELLA regulatory domain at a conserved lysine residue (Lys-65 in RGA, Lys-49 of GAI) immediately before the LExLE motif involved in GID1 binding (Figure 6.2; Murase *et al.*, 2008; Conti *et al.*, 2014). GID1a protein was shown to bind SUMOylated DELLA RGA in the absence of GA via a SUMO-Interaction-Motif (SIM) with the sequence WVLI (residues 21–24 in GAI) (Figure 6.2 and 6.5). This SIM domain includes the Trp-21 residue in the GID1 lid known to directly interact with DELLA protein. A double mutant in the SUMO proteases, *OTS1* and *OTS2* (*OVERLY TOLERANT TO SALT1* and 2), resulted in increased accumulation of both SUMOylated and non-SUMOylated forms of DELLA RGA and GAI protein associated with shorter roots, which was further enhanced under salt stress. This short root phenotype was suppressed by a *rga* mutation, suggesting that this phenotype resulted from DELLA RGA repression of root growth. Interestingly, *ots1 ots2* exhibited early flowering, which was enhanced by an *rga* mutation, suggesting that both DELLA and OTS negatively regulate the transition to flowering. RGA was shown to be deSUMOylated by OTS1 *in vitro*, suggesting that DELLAs are direct targets of OTS1. The short root phenotype and enhanced response to the GA biosynthesis inhibitor paclobutrazol during seed germination suggested that the *ots1 ots2* mutant results in decreased GA sensitivity. However, *ots1 ots2* showed no significant change in endogenous GA levels suggesting that these phenotypes result from altered signalling. *OTS1* over-expression suppressed the dwarfism of the partially GA deficient *ga1-5* mutant and resulted in decreased DELLA protein accumulation. Thus, it appears that lack of DELLA deSUMOylation results in increased DELLA repression of plant growth, and increased DELLA deSUMOylation results in decreased DELLA repression of plant growth. The proposed model is that SUMOylation of DELLA, such as in response to salt stress, results in a GA-independent interaction of SUMOylated-DELLA with GID1 protein (Figure 6.6d). The GID1 interaction with SUMOylated-DELLA reduces the amount of GID1 available for GA-dependent interaction with non-SUMOylated DELLA, leading to decreased DELLA ubiquitination/destruction and increased DELLA repressor protein levels. Higher DELLA accumulation under high salt represses root growth, presumably preventing damage due to salt stress. Further validation of this model will require experiments to examine whether SUMOylated-DELLA does one of the following: blocks GA-binding by GID1, blocks the GA-dependent GID1-DELLA protein–protein interaction, or blocks SLY1-binding to DELLA. Future work will need to determine whether SUMOylation of DELLA protein occurs in other plant species or in response to other forms of environmental stress. Increased DELLA protein accumulation and repression of plant growth has been observed in response to stress hormones ABA and ethylene, and in response to environmental stresses including salt, cold and submergence (Achard *et al.*, 2003; 2006; 2008; Fukao and Bailey-Serres, 2008). The notion that stress-induced DELLA SUMOylation leads to increased DELLA repression of plant growth offers an attractive model to explain these observations.

6.10 Evidence for gibberellin signalling without DELLA destruction

The GA receptor GID1 can transmit the GA hormone signal without DELLA proteolysis, referred to as 'non-proteolytic GA signalling'. Based on the DELLA destruction model, the level of GA signalling should negatively correlate with the level of DELLA repressor protein accumulation (Figure 6.1a, b). In other words, mutants with higher DELLA protein levels should be shorter than mutants with lower DELLA protein levels. Paradoxically, the F-box mutants, *Arabidopsis sly1* and rice *gid2*, accumulate higher levels of DELLA protein than GA biosynthesis mutants or *GID1* null lines, but exhibit less severe GA-insensitive phenotypes (McGinnis *et al.*, 2003; Willige *et al.*, 2007; Ariizumi *et al.*, 2008; Ueguchi-Tanaka *et al.*, 2008). For example, the *Arabidopsis ga1-3* biosynthesis mutant and the *gid1a gid1b gid1c* triple mutants cannot germinate unaided, are severely dwarfed and are completely infertile. The *sly1-2* mutant has dormant seeds that eventually after-ripen, is a semi-dwarf and is only partly infertile. Thus, the *sly1* and *gid2* mutants appear capable of a low level of GA signalling. In fact, these mutants are not completely GA-insensitive, since GA treatment resulted in some increase in stem elongation (Ariizumi *et al.*, 2008; Ueguchi-Tanaka *et al.*, 2008).

The non-proteolytic GA signalling in *sly1* and *gid2* depends on GA and GID1 (Ariizumi *et al.*, 2008; Ueguchi-Tanaka *et al.*, 2008; Ariizumi *et al.*, 2013). For example, the *ga1-3 sly1-10* double mutant is more strongly dwarfed and infertile, and accumulates less DELLA protein than the *sly1-10* single mutant. Moreover, *gid1* mutations exacerbated the GA-insensitive phenotypes of *sly1* and *gid2*, while at the same time reducing DELLA accumulation. Based on the DELLA destruction model, reduced DELLA accumulation should be associated with decreased rather than increased severity of GA-insensitive phenotypes. These results indicate that GA and GID1 are needed both for non-proteolytic GA signalling and for the high level of DELLA protein accumulation observed in *sly1* and *gid2* mutants.

It appears that GID1 can mediate GA signalling without DELLA destruction. *GID1* over-expression partly rescued the GA-insensitive mutant phenotypes of *sly1* and *gid2* mutants without causing a decrease in DELLA protein levels (Ariizumi *et al.*, 2008; Ueguchi-Tanaka *et al.*, 2008; Ariizumi *et al.*, 2013). Thus, GID1 and GA can down-regulate DELLA repressors in F-box mutants that cannot destroy DELLA repressors via the 26S proteasome. Moreover, rescue by *GID1* over-expression was blocked by deletion of the DELLA motif required for GID1-DELLA interaction, suggesting that GID1-GA-DELLA complex formation is required. Higher levels of HA:GID1 protein expression were associated with increased coimmunoprecipation of DELLA and with better rescue of seed germination and stem elongation in *Arabidopsis sly1* (Ariizumi *et al.*, 2013). Thus, the proposed model is that formation of the GID1-GA-DELLA complex decreases the ability of DELLA to repress GA responses leading to increased GA response without DELLA destruction (Figure 6.1d). A final proof of this model would require a direct assay for DELLA function, so that the notion that GID1-GA-DELLA complex formation results in decreased DELLA function could be tested directly. Non-proteolytic DELLA down-regulation is not exclusive to *sly1/gid2* mutants, because loss of *SPY* resulted in increased GA signalling without any apparent decrease in DELLA protein accumulation (Shimada *et al.*,

2006). Thus, future work should examine whether SPY-directed *O*-GlcNAc modification or other DELLA post-translational modifications play a role in non-proteolytic GA signalling. Such work will need to examine whether non-proteolytic GA signalling is important under environmental conditions that reduce DELLA destruction, such as drought and salt stress (Achard *et al.*, 2003; 2006; 2008).

Genetic analysis suggests that the relative roles of the *GID1a*, *GID1b* and *GID1c* genes in non-proteolytic GA signalling in *sly1-2* mutants differed somewhat from their roles in proteolytic GA signalling in the wild-type *SLY1* background (Griffiths *et al.*, 2006; Willige *et al.*, 2007; Ariizumi *et al.*, 2008; 2013; Hauvermale *et al.*, 2014). While *gid1c-2* has an apparently stronger effect on seed germination during proteolytic GA signalling, *gid1a-1* had a stronger effect than *gid1c-2* in the *sly1-2* mutant, interfering with the ability of *sly1-2* seeds to lose dormancy through a long (20 month) period of dry after-ripening (Voegele *et al.*, 2011; Ariizumi *et al.*, 2013). The *sly1-2 gid1b-1* double mutant seed also failed to germinate. *GID1a* appeared to play the strongest role in controlling plant height in both proteolytic and non-proteolytic GA signalling (Griffiths *et al.*, 2006; Willige *et al.*, 2007; Ariizumi *et al.*, 2013; Hauvermale *et al.*, 2014). Whereas *GID1c* had the strongest secondary effect on plant height in proteolytic GA signalling, *GID1b* had the strongest secondary effect in non-proteolytic GA signalling. For fertility, *GID1a* had the primary and *GID1b* the secondary role in proteolytic GA signalling, whereas *GID1b* had the primary and *GID1a* the secondary role in non-proteolytic GA signalling. The *sly1-2 gid1b-1* double mutant had a much stronger infertility phenotype than *sly1-2 gid1a-1* or *sly1-2 gid1c-1*, indicating that *GID1b* plays the major role in stimulating fertility during non-proteolytic GA signalling.

6.11 Concluding remarks

Our understanding of the mechanisms of GA signalling has come a long way since the first mutations in *Arabidopsis* GA biosynthesis genes were identified in 1980 (Figure 6.7; Koornneef and van der Veen, 1980). Genetic studies in rice and in *Arabidopsis* have identified components of and elucidated mechanisms in the GA signalling pathway (Koornneef *et al.*, 1985; Silverstone *et al.*, 1997b; Steber *et al.*, 1998; Steber and McCourt 2001; Griffiths *et al.*, 2006; Nakajima *et al.*, 2006; Willige *et al.*, 2007; Iuchi *et al.*, 2007). The canonical DELLA destruction model was based on: (1) the observation that DELLA repressors disappear after GA treatment and (2) the identification of the SLY1 and GID2 F-box proteins as major positive regulators of GA signalling (Silverstone *et al.*, 2001; Itoh *et al.*, 2002; Sasaki *et al.*, 2003; McGinnis *et al.*, 2003; Gomi *et al.*, 2004; Fu *et al.*, 2004). The cloning of the GA receptor GID1 led to an understanding of how GID1-GA-DELLA complex formation stimulates DELLA destruction (Ueguchi-Tanaka *et al.*, 2005). Biochemical

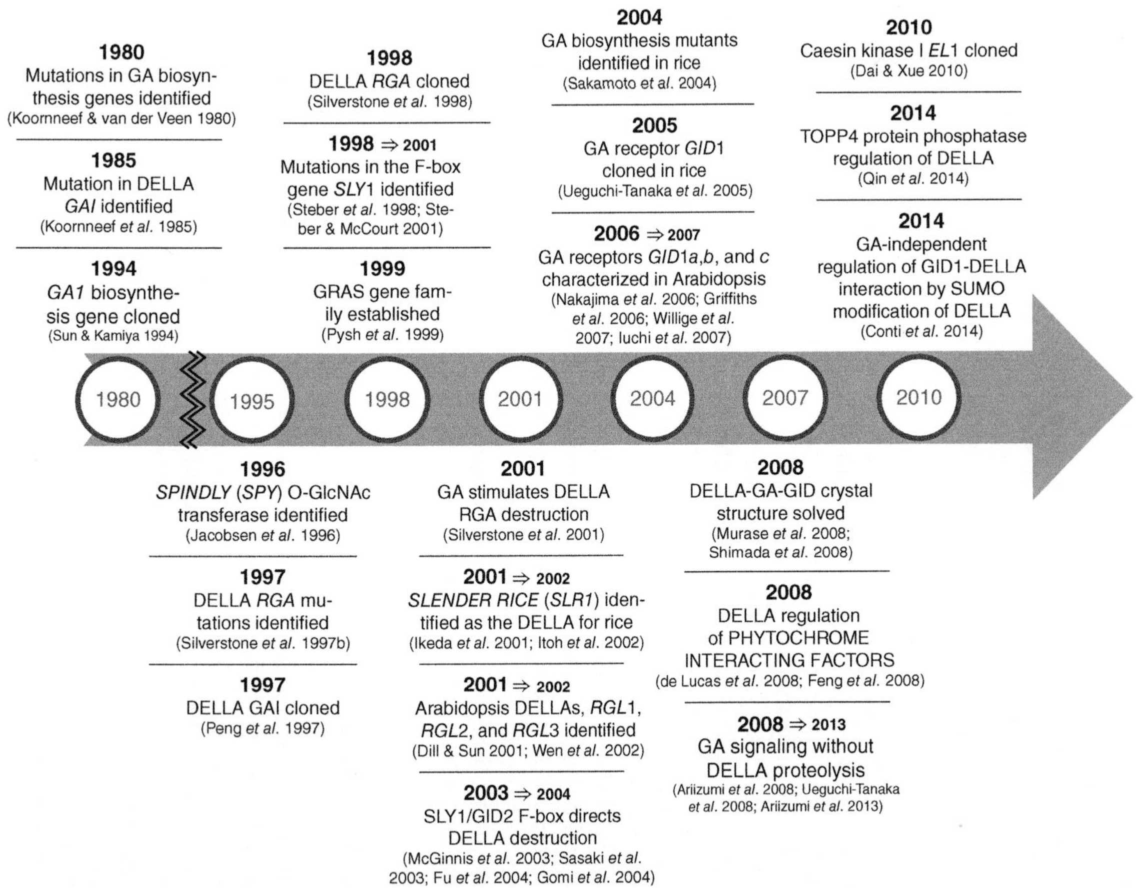

Figure 6.7 A timeline of significant advances in the understanding of GA signalling from 1980 to the present day.

studies and the crystal structure of the OsGID1-GA and GID1a-GA-DELLA has provided a clear understanding of the amino acid motifs involved in this complex interaction (Murase *et al.*, 2008; Shimada *et al.*, 2008). Recent studies have investigated DELLA-targets, post-translational modification of DELLA proteins, and alternative mechanisms of GA signalling in the absence of DELLA-proteolysis or without GID1-GA interaction (de Lucas *et al.*, 2008; Feng *et al.*, 2008; Ariizumi *et al.*, 2008; Ueguchi-Tanaka *et al.*, 2008; Dai and Xue, 2010; Ariizumi *et al.*, 2013; Conti *et al.*, 2014). The new knowledge gained has raised as many interesting new questions as it has answered. Thus, this chapter should be viewed as a starting point rather than as the finished story of GA signalling.

Acknowledgements

The authors would like to acknowledge support from the National Science Foundation (Award no. 0850981 to C.M.S.) and the USDA-Agricultural Research Service. The authors wish to thank Miyako Ueguchi-Tanaka for her comments on the manuscript, and Toshio Hakoshima for providing the crystal structure image in Figure. 6.4.

References

Achard, P., Vriezen, W.H., Van Der Straeten, D. and Harberd, N.P. (2003). Ethylene regulates *Arabidopsis* development via the modulation of DELLA protein growth repressor function. *The Plant Cell* **15**, 2816–2825.

Achard, P., Cheng, H., De Grauwe, L., *et al.* (2006). Integration of plant responses to environmentally activated phytohormonal signals. *Science* **311**, 91–94.

Achard, P., Gong, F., Cheminant, S., *et al.* (2008). The cold-inducible CBF1 factor-dependent signaling pathway modulates the accumulation of the growth-repressing DELLA proteins via its effect on gibberellin metabolism. *The Plant Cell* **20**, 2117–2129.

Allan, R.E. (1986). Agronomic comparisons among wheat lines nearly isogenic for three reduced-height genes. *Crop Science* **26**, 707–710.

Ariizumi, T. and Steber, C.M. (2007). Seed germination of GA-insensitive *sleepy1* mutants does not require RGL2 protein disappearance in *Arabidopsis. The Plant Cell* **19**, 791–804.

Ariizumi, T. and Steber, C.M. (2011). Mutations in the F-box gene *SNEEZY* result in decreased Arabidopsis GA signaling. *Plant Signaling and Behavior* **6**, 831–833.

Ariizumi, T., Murase, K., Sun, T.-P. and Steber, C.M. (2008). Proteolysis-independent downregulation of DELLA repression in *Arabidopsis* by the gibberellin receptor GIBBERELLIN INSENSITIVE DWARF1. *The Plant Cell* **20**, 2447–2459.

Ariizumi, T., Lawrence, P.K. and Steber, C.M. (2011). The role of two F-box proteins, *SLEEPY1* and *SNEEZY*, in Arabidopsis gibberellin signaling. *Plant Physiology* **155**, 765–775.

Ariizumi, T., Hauvermale, A.L., Nelson, S.K., *et al.* (2013). Lifting DELLA repression of Arabidopsis seed germination by nonproteolytic gibberellin signaling. *Plant Physiology* **162**, 2125–2139.

Arnaud, N., Girin, T., Sorefan, K., *et al.* (2010). Gibberellins control fruit patterning in *Arabidopsis* thaliana. *Genes and Development* **24**, 2127–2132.

Bai, M.-Y., Shang, J.-X., Oh, E., *et al.* (2012). Brassinosteroid, gibberellin and phytochrome impinge on a common transcription module in *Arabidopsis*. *Nature Cell Biology* **14**, 810–817.

Beitz, E. (2000). TEXshade: shading and labeling of multiple sequence alignments using LATEX2 epsilon. *Bioinformatics* **16**, 135–139.

Bolle, C. (2004). The role of GRAS proteins in plant signal transduction and development. *Planta* **218**, 683–692.

Cao, D., Hussain, A., Cheng, H. and Peng, J. (2005). Loss of function of four DELLA genes leads to light- and gibberellin-independent seed germination in *Arabidopsis*. *Planta* **223**, 105–113.

Chandler, P. and Robertson, M. (1999). Gibberellin dose-response curves and the characterization of dwarf mutants of barley. *Plant Physiology* **120**, 623–632.

Chandler, P.M., Marion-Poll, A., Ellis, M. and Gubler, F. (2002). Mutants at the *Slender1* locus of barley cv Himalaya. molecular and physiological characterization. *Plant Physiology* **129**, 181–190.

Chandler, P.M., Harding, C.A., Ashton, A.R. *et al.* (2008). Characterization of gibberellin receptor mutants of barley (Hordeum vulgare L.). *Molecular Plant* **1**, 285–294.

Cheng, H., Qin, L., Lee, S., *et al.* (2004). Gibberellin regulates *Arabidopsis* floral development via suppression of DELLA protein function. *Development* **131**, 1055–1064.

Conti, L., Conti, L., Nelis, S., *et al.* (2014). Small Ubiquitin-like Modifier Protein SUMO enables plants to control growth independently of the phytohormone gibberellin. *Developmental Cell* **28**, 102–110.

Cui, H. and Benfey, P.N. (2009). Interplay between SCARECROW, GA and LIKE HETEROCHROMATIN PROTEIN 1 in ground tissue patterning in the *Arabidopsis* root. *The Plant Journal* **58**, 1016–1027.

Dai, C. and Xue, H.-W. (2010). Rice *early flowering1*, a CKI, phosphorylates DELLA protein SLR1 to negatively regulate gibberellin signalling. *The EMBO Journal* **29**, 1916–1927.

de Lucas, M., Davière, J.-M., Rodríguez-Falcón, M., *et al.* (2008). A molecular framework for light and gibberellin control of cell elongation. *Nature* **451**, 480–484.

Dill, A. and Sun, T-p. (2001). Synergistic derepression of gibberellin signaling by removing RGA and GAI function in *Arabidopsis thaliana*. *Genetics* **159**, 777–785.

Dill, A., Thomas, S.G., Hu, J., Steber, C.M. and Sun, T.-P. (2004). The *Arabidopsis* F-box protein SLEEPY1 targets gibberellin signaling repressors for gibberellin-induced degradation. *The Plant Cell* **16**, 1392–1405.

Dill, K.A. (1990). The meaning of hydrophobicity. *Science* **250**, 297–298.

Dunker, A.K., Obradovic, Z., Romero, P., Garner, E.C. and Brown, C.J. (2000). Intrinsic protein disorder in complete genomes. *Genome Informatics* **11**, 161–171.

Engstrom, E.M. (2011). Phylogenetic analysis of GRAS proteins from moss, lycophyte and vascular plant lineages reveals that GRAS genes arose and underwent substantial diversification in the ancestral lineage common to bryophytes and vascular plants. *Plant Signaling and Behavior* **6**, 850–854.

Feng, S., Martinez, C., Gusmaroli, G., *et al.* (2008). Coordinated regulation of *Arabidopsis thaliana* development by light and gibberellins. *Nature* **451**, 475–479.

Filardo, F., Robertson, M., Singh, D.P., Parish, R.W. and Swain, S.M. (2009). Functional analysis of HvSPY, a negative regulator of GA response, in barley aleurone cells and *Arabidopsis*. *Planta* **229**, 523–537.

Fu, X., Richards, D.E., Ait-ali, T., *et al.* (2002). Gibberellin-mediated proteasome-dependent degradation of the barley DELLA protein SLN1 repressor. *The Plant Cell* **14**, 3191–3200.

Fu, X., Richards, D.E., Fleck, B., *et al.* (2004). The *Arabidopsis* mutant sleepy1$^{gar2\text{-}1}$ protein promotes plant growth by increasing the affinity of the SCF^{SLY1} E3 ubiquitin ligase for DELLA protein substrates. *The Plant Cell* **16**, 1406–1418.

Fukao, T. and Bailey-Serres, J. (2008). Submergence tolerance conferred by *Sub1A* is mediated by SLR1 and SLRL1 restriction of gibberellin responses in rice. *Proceedings of the National Academy of Sciences USA* **105**, 16814–16819.

Fuxreiter, M., Simon, I., Friedrich, P. and Tompa, P. (2004). Preformed structural elements feature in partner recognition by intrinsically unstructured proteins. *Journal of Molecular Biology* **338**, 1015–1026.

Gagne, J.M., Downes, B.P., Shiu, S.H., Durski, A.M. and Vierstra, R.D. (2002). The F-box subunit of the SCF E3 complex is encoded by a diverse superfamily of genes in *Arabidopsis*. *Proceedings of the National Academy of Sciences USA* **99**, 11519–11524.

Gallego-Bartolomé, J., Minguet, E.G., Marín, J.A., *et al.* (2010). Transcriptional diversification and functional conservation between DELLA proteins in *Arabidopsis*. *Molecular Biology and Evolution* **27**, 1247–1256.

Gallego-Bartolomé, J., Minguet, E.G., Grau-Enguix, F., *et al.* (2012). Molecular mechanism for the interaction between gibberellin and brassinosteroid signaling pathways in *Arabidopsis*. *Proceedings of the National Academy of Sciences USA* **109**, 13446–13451.

Gilroy, S. and Jones, R.L. (1994). Perception of gibberellin and abscisic acid at the external face of the plasma membrane of barley (*Hordeum vulgare* L.) aleurone protoplasts. *Plant Physiology* **104**, 1185–1192.

Gomi, K., Sasaki, A., Itoh, H., *et al.* (2004). GID2, an F-box subunit of the SCF E3 complex, specifically interacts with phosphorylated SLR1 protein and regulates the gibberellin-dependent degradation of SLR1 in rice. *The Plant Journal* **37**, 626–634.

Gray, W.M., Hellmann, H., Dharmasiri, S. and Estelle, M. (2002). Role of the *Arabidopsis* RING-H2 protein RBX1 in RUB modification and SCF function. *The Plant Cell* **14**, 2137–2144.

Griffiths, J., Murase, K., Rieu, I., *et al.* (2006). Genetic characterization and functional analysis of the GID1 gibberellin receptors in *Arabidopsis*. *The Plant Cell* **18**, 3399–3414.

Hao, G.-F., Yang, S.-G., Yang, G.-F. and Zhan, C.-G. (2013). Computational gibberellin-binding channel discovery unraveling the unexpected perception mechanism of hormone signal by gibberellin receptor. *Journal of Computational Chemistry* **34**, 2055–2064.

Hauvermale, A.L., Ariizumi, T. and Steber, C.M. (2012). Gibberellin signaling: a theme and variations on DELLA repression. *Plant Physiology* **160**, 83–92.

Hauvermale, A.L., Ariizumi, T. and Steber, C.M. (2014). The roles of the GA receptors *GID1a, GID1b,* and *GID1c* in *sly1*-independent GA signaling. *Plant Signaling and Behavior* **9**, e28030.

Heo, J.-O., Chang, K.S., Kim, I.A., *et al.* (2011). Funneling of gibberellin signaling by the GRAS transcription regulator SCARECROW-LIKE 3 in the *Arabidopsis* root. *Proceedings of the National Academy of Sciences USA* **108**, 2166–2171.

Hirano, K., Nakajima, M., Asano, K., *et al.* (2007). The GID1-mediated gibberellin perception mechanism is conserved in the Lycophyte *Selaginella moellendorffii* but not in the Bryophyte *Physcomitrella patens*. *The Plant Cell* **19**, 3058–3079.

Hirano, K., Ueguchi-Tanaka, M. and Matsuoka, M. (2008). GID1-mediated gibberellin signaling in plants. *Trends in Plant Science* **13**, 192–199.

Hirano, K., Asano, K., Tsuji, H., *et al.* (2010). Characterization of the molecular mechanism underlying gibberellin perception complex formation in rice. *The Plant Cell* **22**, 2680–2696.

Hirano, K., Kouketu, E., Katoh, H., *et al.* (2012). The suppressive function of the rice DELLA protein SLR1 is dependent on its transcriptional activation activity. *The Plant Journal* **71**, 443–453.

Hirsch, S., Hirsch, S., Kim, J., *et al.* (2009). GRAS proteins form a DNA binding complex to induce gene expression during nodulation signaling in *Medicago truncatula*. *The Plant Cell* **21**, 545–557.

Hooley, R., Beale, M.H. and Smith, S.J. (1991). Gibberellin perception at the plasma membrane of *Avena fatua* aleurone protoplasts. *Planta* **183**, 274–280.

Hou, X., Lee, L.Y.C., Xia, K., Yan, Y. and Yu, H. (2010). DELLAs modulate jasmonate signaling via competitive binding to JAZs. *Developmental Cell* **19**, 884–894.

Huang, E.S., Subbiah, S. and Levitt, M. (1995). Recognizing native folds by the arrangement of hydrophobic and polar residues. *Journal of Molecular Biology* **252**, 709–720.

Hussain, A., Cao, D., Cheng, H., Wen, Z. and Peng, J. (2005). Identification of the conserved serine/threonine residues important for gibberellin-sensitivity of *Arabidopsis* RGL2 protein. *The Plant Journal* **44**, 88–99.

Iakoucheva, L.M., Brown, C.J., Lawson, J.D., Obradović, Z. and Dunker, A.K. (2002). Intrinsic disorder in cell-signaling and cancer-associated proteins. *Journal of Molecular Biology* **323**, 573–584.

Ikeda, A., Ikeda, A., Ueguchi-Tanaka, M., *et al.* (2001). *slender* rice, a constitutive gibberellin response mutant, is caused by a null mutation of the *SLR1* gene, an ortholog of the height-regulating gene *GAI/RGA/RHT/D8*. *The Plant Cell* **13**, 999–1010.

Itoh, H., Ueguchi-Tanaka, M., Sato, Y., Ashikari, M. and Matsuoka, M. (2002). The gibberellin signaling pathway is regulated by the appearance and disappearance of SLENDER RICE1 in nuclei. *The Plant Cell* **14**, 57–70.

Itoh, H., Matsuoka, M. and Steber, C.M. (2003). A role for the ubiquitin-26S-proteasome pathway in gibberellin signaling. *Trends in Plant Science* **8**, 492–497.

Itoh, H., Sasaki, A., Ueguchi-Tanaka, M., *et al.* (2005a). Dissection of the phosphorylation of rice DELLA protein, SLENDER RICE1. *Plant and Cell Physiology* **46**, 1392–1399.

Itoh, H., Shimada, A., Ueguchi-Tanaka, M., *et al.* (2005b). Overexpression of a GRAS protein lacking the DELLA domain confers altered gibberellin responses in rice. *The Plant Journal* **44**, 669–679.

Iuchi, S., Suzuki, H., Kim, Y.-C., *et al.* (2007). Multiple loss-of-function of *Arabidopsis* gibberellin receptor AtGID1s completely shuts down a gibberellin signal. *The Plant Journal* **50**, 958–966.

Jacobsen, S.E. and Olszewski, N.E. (1993). Mutations at the *SPINDLY* locus of *Arabidopsis* alter gibberellin signal transduction. *The Plant Cell* **5**, 887–896.

Jones, D.T., Taylor, W.R. and Thornton, J.M. (1992). The rapid generation of mutation data matrices from protein sequences. *Computer Applications in the Biosciences* **8**, 275–282.

King, K.E., Moritz, T. and Harberd, N.P. (2001). Gibberellins are not required for normal stem growth in *Arabidopsis thaliana* in the absence of GAI and RGA. *Genetics* **159**, 767–776.

Koornneef, M. and van der Veen, J.H. (1980). Induction and analysis of gibberellin sensitive mutants in *Arabidopsis thaliana* (L.) heynh. *Theoretical and Applied Genetics* **58**, 257–263.

Koornneef, M., Elgersma, A., Hanhart, C.J., *et al.* (1985). A gibberellin insensitive mutant of *Arabidopsis thaliana*. *Physiologia Plantarum* **65**, 33–39.

Lee, S., Cheng, H., King, K.E., *et al.* (2002). Gibberellin regulates *Arabidopsis* seed germination via *RGL2*, a *GAI/RGA*-like gene whose expression is up-regulated following imbibition. *Genes and Development* **16**, 646–658.

Levy, D.E. and Darnell, J.E. (2002). Stats: transcriptional control and biological impact. *Nature Reviews Molecular Cell Biology* **3**, 651–662.

Lin, Q., Li, J., Smith, R.D. and Walker, J.C. (1998). Molecular cloning and chromosomal mapping of type one serine/threonine protein phosphatases in *Arabidopsis thaliana*. *Plant Molecular Biology* **37**, 471–481.

Livne, S., and Weiss, D. (2014). Cytosolic activity of the gibberellin receptor GIBBERELLIN INSENSITIVE DWARF1a. *Plant and Cell Physiology* **55**, 1727–1733.

McCubbin, A.G., Ritchie, S.M., Swanson, S.J. and Gilroy, S. (2004). The calcium-dependent protein kinase HvCDPK1 mediates the gibberellic acid response of the barley aleurone through regulation of vacuolar function. *The Plant Journal* **39**, 206–218.

McGinnis, K.M., Thomas, S.G., Soule, J.D., *et al.* (2003). The *Arabidopsis SLEEPY1* gene encodes a putative F-box subunit of an SCF E3 ubiquitin ligase. *The Plant Cell* **15**, 1120–1130.

Mohan, A., Oldfield, C.J., Radivojac, P., *et al.* (2006). Analysis of molecular recognition features (MoRFs). *Journal of Molecular Biology* **362**, 1043–1059.

Murase, K., Hirano, Y., Sun, T.-P. and Hakoshima, T. (2008). Gibberellin-induced DELLA recognition by the gibberellin receptor GID1. *Nature* **456**, 459–463.

Nakajima, M., Takita, K., Wada, H., *et al.* (1997). Partial purification and characterization of a gibberellin-binding protein from seedlings of *Azukia angularis*. *Biochemical and Biophysical Research Communications* **241**, 782–786.

Nakajima, M., Shimada, A., Takashi, Y., *et al.* (2006). Identification and characterization of *Arabidopsis* gibberellin receptors. *The Plant Journal* **46**, 880–889.

Nguyen, L.K., Kolch, W. and Kholodenko, B.N. (2013). When ubiquitination meets phosphorylation: a systems biology perspective of EGFR/MAPK signalling. *Cell Communication and Signaling* **11**:52.

Oldfield, C.J., Cheng, Y., Cortese, M.S., *et al.* (2005). Coupled folding and binding with alpha-helix-forming molecular recognition elements. *Biochemistry* **44**, 12454–12470.

Østerlund, T. (2001). Structure-function relationships of hormone-sensitive lipase. *European Journal of Biochemistry* **268**, 1899–1907.

Park, S.-H., Nakajima, M., Hasegawa, M. and Yamaguchi, I. (2005). Similarities and differences between the characteristics of gibberellin-binding protein and gibberellin 2-oxidases in adzuki bean (*Vigna angularis*) seedlings. *Bioscience, Biotechnology, and Biochemistry* **69**, 1508–1514.

Peng, J., Carol, P., Richards, D.E., *et al.* (1997). The *Arabidopsis GAI* gene defines a signaling pathway that negatively regulates gibberellin responses. *Genes and Development* **11**, 3194–3205.

Peng, J., Richards, D.E., Hartley, N.M., *et al.* (1999). 'Green revolution' genes encode mutant gibberellin response modulators. *Nature* **400**, 256–261.

Piskurewicz, U. and Lopez-Molina, L. (2009). The GA-signaling repressor RGL3 represses testa rupture in response to changes in GA and ABA levels. *Plant Signaling and Behavior* **4**, 63–65.

Plackett, A.R.G., Ferguson, A.C., Powers, S.J., *et al.* (2014). DELLA activity is required for successful pollen development in the Columbia ecotype of *Arabidopsis*. *The New Phytologist* **201**, 825–836.

Qin, Q., Wang, W., Guo, X., *et al.* (2014). *Arabidopsis* DELLA protein degradation is controlled by a type-one protein phosphatase, TOPP4. *PLoS Genetics* **10**, e1004464.

Richards, D.E., Peng, J. and Harberd, N.P. (2000). Plant GRAS and metazoan STATs: one family? *BioEssays* **22**, 573–577.

Risseeuw, E.P., Daskalchuk, T.E., Banks, T.W., *et al.* (2003). Protein interaction analysis of SCF ubiquitin E3 ligase subunits from *Arabidopsis*. *The Plant Journal* **34**, 753–767.

Robertson, M., Swain, S.M., Chandler, P.M. and Olszewski, N.E. (1998). Identification of a negative regulator of gibberellin action, HvSPY, in barley. *The Plant Cell* **10**, 995–1007.

Rose, G.D., Geselowitz, A.R., Lesser, G.J., Lee, R.H. and Zehfus, M.H. (1985). Hydrophobicity of amino acid residues in globular proteins. *Science* **229**, 834–838.

Sarnowska, E.A., Sarnowska, E.A., Rolicka, A.T., *et al.* (2013). DELLA-interacting SWI3C core subunit of switch/sucrose nonfermenting chromatin remodeling complex modulates gibberellin responses and hormonal cross talk in Arabidopsis. *Plant Physiology* **163**, 305–317.

Sasaki, A., Itoh, H., Gomi, K., *et al.* (2003). Accumulation of phosphorylated repressor for gibberellin signaling in an F-box mutant. *Science* **299**, 1896–1898.

Sato, T., Miyanoiri, Y., Takeda, M., *et al.* (2014). Expression and purification of a GRAS domain of SLR1, the rice DELLA protein. *Protein Expression and Purification* **95**, 248–258.

Schulman, B.A., Carrano, A.C., Jeffrey, P.D., *et al.* (2000). Insights into SCF ubiquitin ligases from the structure of the Skp1-Skp2 complex. *Nature* **408**, 381–386.

Sheerin, D.J., Buchanan, J., Kirk, C., *et al.* (2011). Inter- and intra-molecular interactions of *Arabidopsis thaliana* DELLA protein RGL1. *The Biochemical Journal* **435**, 629–639.

Shimada, A., Ueguchi-Tanaka, M., Sakamoto, T., *et al.* (2006). The rice *SPINDLY* gene functions as a negative regulator of gibberellin signaling by controlling the suppressive function of the DELLA protein, SLR1, and modulating brassinosteroid synthesis. *The Plant Journal* **48**, 390–402.

Shimada, A., Ueguchi-Tanaka, M., Nakatsu, T., *et al.* (2008). Structural basis for gibberellin recognition by its receptor GID1. *Nature* **456**, 520–523.

Sievers, F., Wilm, A., Dineen, D., *et al.* (2011). Fast, scalable generation of high-quality protein multiple sequence alignments using Clustal Omega. *Molecular Systems Biology* **7**:539.

Silverstone, A.L., Chang, C., Krol, E. and Sun, T.-P. (1997a). Developmental regulation of the gibberellin biosynthetic gene GA1 in *Arabidopsis thaliana*. *The Plant Journal* **12**, 9–19.

Silverstone, A.L., Mak, P.Y., Martínez, E.C. and Sun, T.-P. (1997b). The new *RGA* locus encodes a negative regulator of gibberellin response in *Arabidopsis thaliana*. *Genetics* **146**, 1087–1099.

Silverstone, A.L., Ciampaglio, C.N. and Sun, T. (1998). The *Arabidopsis RGA* gene encodes a transcriptional regulator repressing the gibberellin signal transduction pathway. *The Plant Cell* **10**, 155–169.

Silverstone, A.L., Jung, H.S., Dill, A., *et al.* (2001). Repressing a repressor: gibberellin-induced rapid reduction of the RGA protein in *Arabidopsis*. *The Plant Cell* **13**, 1555–1566.

Silverstone, A.L., Tseng, T.-S., Swain, S.M., *et al.* (2007). Functional analysis of SPINDLY in gibberellin signaling in Arabidopsis. *Plant Physiology* **143**, 987–1000.

Smalle, J. and Vierstra, R.D. (2004). The ubiquitin 26S proteasome proteolytic pathway. *Annual Review of Plant Biology* **55**, 555–590.

Smith, R.D. and Walker, J.C. (1993). Expression of multiple type 1 phosphoprotein phosphatases in *Arabidopsis thaliana*. *Plant Molecular Biology* **21**, 307–316.

Stavang, J.A., Gallego-Bartolomé, J., Gómez, M.D., *et al.* (2009). Hormonal regulation of temperature-induced growth in *Arabidopsis*. *The Plant Journal* **60**, 589–601.

Steber, C.M. and McCourt, P. (2001). A role for brassinosteroids in germination in Arabidopsis. *Plant Physiology* **125**, 763–769.

Steber, C.M., Cooney, S.E. and McCourt, P. (1998). Isolation of the GA-response mutant *sly1* as a suppressor of *ABI1-1* in *Arabidopsis thaliana*. *Genetics* **149**, 509–521.

Strader, L.C., Ritchie, S., Soule, J.D., McGinnis, K.M. and Steber, C.M. (2004). Recessive-interfering mutations in the gibberellin signaling gene *SLEEPY1* are rescued by overexpression of its homologue, *SNEEZY*. *Proceedings of the National Academy of Sciences USA* **101**, 12771–12776.

Sun, T.-P. and Gubler, F. (2004). Molecular mechanism of gibberellin signaling in plants. *Annual Review of Plant Biology* **55**, 197–223.

Sun, T.-P. and Kamiya, Y. (1994). The *Arabidopsis GA1* locus encodes the cyclase *ent*-kaurene synthetase A of gibberellin biosynthesis. *The Plant Cell* **6**, 1509–1518.

Sun, X., Jones, W.T., Harvey, D., *et al.* (2010). N-terminal domains of DELLA proteins are intrinsically unstructured in the absence of interaction with GID1/gibberellic acid receptors. *Journal of Biological Chemistry* **285**, 11557–11571.

Suzuki, H., Park, S.-H., Okubo, K., *et al.* (2009). Differential expression and affinities of *Arabidopsis* gibberellin receptors can explain variation in phenotypes of multiple knock-out mutants. *The Plant Journal* **60**, 48–55.

Swain, S.M., Tseng, T.S. and Olszewski, N.E. (2001). Altered expression of *SPINDLY* affects gibberellin response and plant development. *Plant Physiology* **126**, 1174–1185.

Tyler, L., Thomas, S.G., Hu, J., *et al.* (2004). DELLA proteins and gibberellin-regulated seed germination and floral development in Arabidopsis. *Plant Physiology* **135**, 1008–1019.

Ueguchi-Tanaka, M. and Matsuoka, M. (2010). The perception of gibberellins: clues from receptor structure. *Current Opinion in Plant Biology* **13**, 503–508.

Ueguchi-Tanaka, M., Ashikari, M., Nakajima, M., *et al.* (2005). *GIBBERELLIN INSENSITIVE DWARF1* encodes a soluble receptor for gibberellin. *Nature* **437**, 693–698.

Ueguchi-Tanaka, M., Nakajima, M., Katoh, E., *et al.* (2007a). Molecular interactions of a soluble gibberellin receptor, GID1, with a rice DELLA protein, SLR1, and gibberellin. *The Plant Cell* **19**, 2140–2155.

Ueguchi-Tanaka, M., Nakajima, M., Motoyuki, A. and Matsuoka, M. (2007b). Gibberellin receptor and its role in gibberellin signaling in plants. *Annual Review of Plant Biology* **58**, 183–198.

Ueguchi-Tanaka, M., Hirano, K., Hasegawa, Y., Kitano, H. and Matsuoka, M. (2008). Release of the repressive activity of rice DELLA protein SLR1 by gibberellin does not require SLR1 degradation in the *gid2* mutant. *The Plant Cell* **20**, 2437–2446.

Uversky, V.N. (2002). What does it mean to be natively unfolded? *European Journal of Biochemistry* **269**, 2–12.

Vierstra, R.D. (2012). The expanding universe of ubiquitin and ubiquitin-like modifiers. *Plant Physiology* **160**, 2–14.

Voegele, A., Linkies, A., Müller, K. and Leubner-Metzger, G. (2011). Members of the gibberellin receptor gene family *GID1* (*GIBBERELLIN INSENSITIVE DWARF1*) play distinct roles during *Lepidium sativum* and *Arabidopsis thaliana* seed germination. *Journal of Experimental Botany* **62**, 5131–5147.

Wang, F. and Deng, X.W. (2011). Plant ubiquitin-proteasome pathway and its role in gibberellin signaling. *Cell Research* **21**, 1286–1294.

Wang, F., Zhu, D., Huang, X., *et al.* (2009). Biochemical insights on degradation of *Arabidopsis* DELLA proteins gained from a cell-free assay system. *The Plant Cell* **21**, 2378–2390.

Wells, L., Kreppel, L.K., Comer, F.I., Wadzinski, B.E. and Hart, G.W. (2004). *O*-GlcNAc transferase is in a functional complex with protein phosphatase 1 catalytic subunits. *Journal of Biological Chemistry* **279**, 38466–38470.

Wen, C.-K., Wen, C.-K., Chang, C. and Chang, C. (2002). *Arabidopsis RGL1* encodes a negative regulator of gibberellin responses. *The Plant Cell* **14**, 87–100.

Wild, M., Daviere, J.M., Cheminant, S. *et al.*, (2012). The *Arabidopsis* DELLA *RGA-LIKE3* is a direct target of MYC2 and modulates jasmonate signaling responses. *The Plant Cell* **24**, 3307–19.

Willems, A.R., Goh, T., Taylor, L., *et al.* (1999). SCF ubiquitin protein ligases and phosphorylation-dependent proteolysis. *Philosophical Transactions B* **354**, 1533–1550.

Willige, B.C., Ghosh, S., Nill, C., *et al.* (2007). The DELLA domain of GA INSENSITIVE mediates the interaction with the GA INSENSITIVE DWARF1A gibberellin receptor of *Arabidopsis*. *The Plant Cell* **19**, 1209–1220.

Wilson, R.N., Heckman, J.W. and Somerville, C.R. (1992). Gibberellin is required for flowering in *Arabidopsis thaliana* under short days. *Plant Physiology* **100**, 403–408.

Yamaguchi, S. (2008). Gibberellin metabolism and its regulation. *Annual Review of Plant Biology* **59**, 225–251.

Yamamoto, Y., Hirai, T., Yamamoto, E., *et al.* (2010). A rice *gid1* suppressor mutant reveals that gibberellin is not always required for interaction between its receptor, GID1, and DELLA proteins. *The Plant Cell* **22**, 3589–3602.

Yano, K., Aya, K., Hirano, K., *et al.* (2015). Comprehensive gene expression analysis of rice aleurone cells: Probing the existence of an alternative gibberellin receptor. *Plant Physiology* **167**, 531–544.

Yu, H., Ito, T., Zhao, Y., *et al.* (2004). Floral homeotic genes are targets of gibberellin signaling in flower development. *Proceedings of the National Academy of Sciences USA* **101**, 7827–7832.

Zentella, R., Zhang, Z.-L., Park, M., *et al.* (2007). Global analysis of DELLA direct targets in early gibberellin signaling in *Arabidopsis*. *The Plant Cell* **19**, 3037–3057.

Zhang, Y., Liu, Z., Wang, L., *et al.* (2010). Sucrose-induced hypocotyl elongation of *Arabidopsis* seedlings in darkness depends on the presence of gibberellins. *Journal of Plant Physiology* **167**, 1130–1136.

Zhang, Z.-L., Ogawa, M., Fleet, C.M., *et al.* (2011). SCARECROW-LIKE 3 promotes gibberellin signaling by antagonizing master growth repressor DELLA in *Arabidopsis*. *Proceedings of the National Academy of Sciences USA* **108**, 2160–2165.

Annual Plant Reviews (2016) **49**, 189–228
doi: 10.1002/9781119210436.ch7 http://onlinelibrary.wiley.com

Chapter 7

DELLA PROTEINS: MASTER REGULATORS OF GIBBERELLIN-RESPONSIVE GROWTH AND DEVELOPMENT

Stephen G. Thomas[1], Miguel A. Blázquez[2] and David Alabadí[2]

[1] *Plant Biology and Crop Science Department, Rothamsted Research, UK*
[2] *Instituto de Biología Molecular y Celular de Plantas (CSIC-UPV), Spain*

Abstract: In flowering plants, DELLA proteins (DELLAs) have emerged as the master transcriptional regulators that are responsible for repressing all aspects of gibberellin (GA)-dependent growth and development. Although we now have a good understanding of how GAs relieve the growth repression exerted by DELLAs, our understanding of the mechanisms through which these proteins act has only started to emerge in the last few years. It has become apparent that DELLAs physically associate with a myriad of proteins, including transcription factors, altering their activities and promoting the transcriptional changes that control GA-responsive growth and development. Fascinatingly, the role of DELLAs does not appear to revolve exclusively around transcriptional control, with a recent report demonstrating that they are directly involved in the reorientation of microtubules in the *Arabidopsis* hypocotyl. These recent observations add a further dimension to a remarkable class of regulatory proteins.

Keywords: DELLA proteins, GA, GAMYB, gibberellin, GRAS proteins, hub proteins, microtubule orientation, protein–protein interactions, transcription factors, transcriptional regulation

Annual Plant Reviews, Volume 49: The Gibberellins, First Edition.
Edited by Peter Hedden and Stephen G. Thomas.

7.1 Introduction

Since the discovery that gibberellins (GAs) act as phytohormones controlling diverse aspects of plant growth and development there have been considerable efforts to understand the underlying molecular mechanisms. However, the multitude of cellular processes that are regulated by this phytohormone presented a potential problem for these early investigations, with the likelihood that they are controlled by different mechanisms. The discovery that GAs are synthesized and secreted from the germinating cereal grain embryo and then act specifically on the aleurone cells, to trigger the secretion of hydrolytic enzymes that mobilise endosperm reserves required for seedling establishment, provided an important early experimental system for these studies (reviewed in Sun and Gubler, 2004). These studies have provided some important insights into the GA-mediated transcriptional regulation of genes encoding hydrolytic enzymes, most notably the demonstration that the transcription factor (TF) GAMYB plays a central role in this process. Where the aleurone system has failed to shed further light is in the identification of upstream GA signalling components and the mechanisms through which they regulate these transcriptional responses. It was the use of molecular genetics approaches using the model plant *Arabidopsis* that eventually provided the keys that would unlock this field of research. A pioneering screen by Marten Koornneef and colleagues led to the identification of the *gibberellin-insensitive* (*gai*; later renamed *gai-1*) mutant, which exhibited a classical GA-related dwarf phenotype, but which was insensitive to exogenous application of GA (Koornneef *et al.*, 1985). This phenotype suggested that this dominant gain-of-function mutation affected a gene involved in the GA signal transduction pathway. Several years later, a screen performed by Silverstone and colleagues led to the identification of *rga* (for repressor of *ga1-3*) mutant alleles which rescued the severe phenotypic defects of an *Arabidopsis* GA-deficient mutant, *ga1-3* (Silverstone *et al.*, 1997). Genetic analysis and the subsequent cloning of the *GAI* and *RGA* genes demonstrated that they encoded homologous proteins that are negative regulators of GA signalling (now commonly known as DELLA proteins or DELLAs), acting as potential transcriptional regulators (Peng *et al.*, 1997; Silverstone *et al.*, 1997; Silverstone *et al.*, 1998). Importantly, bioactive GAs were later shown to relieve the repression exerted by RGA through its targeted degradation (Silverstone *et al.*, 2001; reviewed in Chapter 6). In this chapter, we discuss these and other early studies that were influential in identifying and characterising the role of DELLAs in regulating GA-responsive growth. However, the main focus of this review is to discuss extensive recent work that has led to an understanding of the mechanisms through which DELLAs control a myriad of physiological processes.

7.2 DELLAs regulate downstream gibberellin signalling

More than 50 years ago, it was proposed that GAs promote plant growth and development by acting as 'inhibitors of inhibitors' (Brian, 1957). This hypothesis was based on the observation that the garden pea (*Pisum sativum* L.) mutant *la cry* displays a GA overdose (slender) phenotype that is not dependent on the levels of GAs in the plant (Figure 7.1; Brian, 1957; Potts *et al.*, 1985; Weston *et al.*, 2008). Essentially, this model proposed by Brian (1957) predicted that GAs act to overcome the growth inhibition exerted by the *LA* and *CRY* gene products. The identification of mutants in other plant species that displayed a similar slender phenotype supported a common mechanism for the control of growth by GAs (Foster, 1977; Peng *et al.*, 1997; Silverstone *et al.*, 1997). Whilst the mutations that underlie the slender growth habit are recessive in nature, another class of mutations resulting in a contrasting dwarf phenotype was also identified in several plant species, including maize (Phinney, 1956), wheat (Gale and Marshall, 1973), *Arabidopsis* (Koornneef *et al.*, 1985), barley (Chandler *et al.*, 2002) and rice (Asano *et al.*, 2009). These mutations are partially dominant and give rise to a

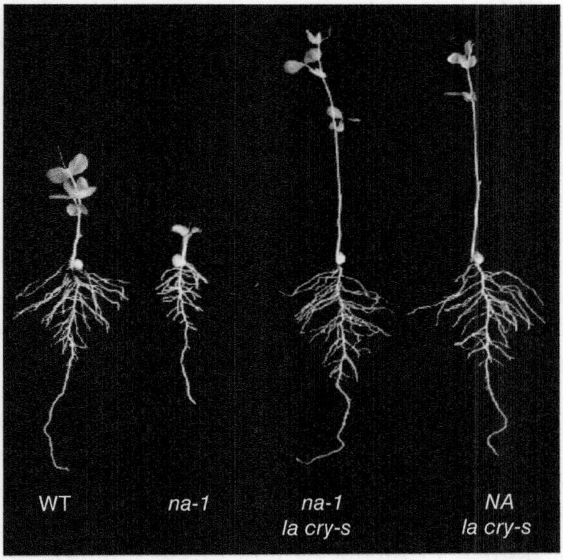

Figure 7.1 Shoot and root phenotypes of pea genotypes. The *la cry-s* alleles completely suppress the phenotype of the GA-deficient mutant *na-1*. (Weston *et al.* (2008). Reproduced with permission from American Society of Plant Biologists.)

GA-insensitive dwarf phenotype. It was the *Arabidopsis* mutant belonging to this class, *gai-1*, which led to further supporting evidence for the hypothesis proposed by Brian (1957). Using an elegant genetic approach, Peng and colleagues (1997) demonstrated that transposon-mediated disruption at the *GAI* locus reverted *gai-1* to a phenotype that was almost identical to wild-type plants. Importantly, it also resulted in partially GA-independent growth based on a resistance to the GA biosynthesis inhibitor, paclobutrazol. These observations indicated that *GAI* encodes a GA-inactivated repressor (similar to *LA* and *CRY*) and that the gain-of-function *gai-1* mutation acts to prevent this process, leading to constitutive repression of GA growth responses (Peng *et al.*, 1997; Harberd *et al.*, 1998). Notably, this work led to the cloning of the *GAI* gene, allowing its further functional and genetic analysis. A homologous *Arabidopsis* gene, *RGA*, was then identified as encoding another negative regulator of GA signalling in an elegant screen and cloning strategy employed by Silverstone and colleagues (1997; 1998).

The GAI and RGA proteins belong to the GRAS family of putative transcriptional regulators (named after the founding members: GAI, RGA and SCR), which in *Arabidopsis* includes 32 members (Pysh *et al.*, 1999; Engstrom, 2011). This family of proteins is characterised by a highly conserved C-terminal GRAS domain that is important for functionality. In the cases of RGA and GAI they also possess two highly conserved N-terminal domains named after amino acid residues found within them: the DELLA and TVHYNP domains. The DELLA domain has now been adopted as a name for this group of repressors of GA signalling (Wen and Chang, 2002). A potential role for RGA and GAI in regulating transcriptional responses that control GA-responsive growth was supported by the presence of bipartite nuclear localisation signals in the predicted protein sequences and the subsequent demonstration that a GFP-RGA fusion protein was targeted to the nucleus in transgenic plants (Peng *et al.*, 1997; Silverstone *et al.*, 1998; Silverstone *et al.*, 2001). Furthermore, epistasis studies have established the *DELLA* genes as encoding downstream components within the GA signalling cascade (McGinnis *et al.*, 2003; Griffiths *et al.*, 2006). The identification of these genes in *Arabidopsis* quickly led to the cloning of *DELLA* orthologues in other plant species (Peng *et al.*, 1999; Ogawa *et al.*, 2000). In rice, barley and tomato DELLA loss-of-function alleles were demonstrated to exhibit a classical GA-insensitive slender phenotype (Ikeda *et al.*, 2001; Chandler *et al.*, 2002; Jasinski *et al.*, 2008). This highlighted a conserved role in repressing GA-responsive growth in monocots and dicots. Importantly, the *LA* and *CRY* genes in the garden pea were also found to encode DELLA proteins, identifying them as the growth inhibitors that Brian originally proposed are inhibited by the action of GA (Brian, 1957; Weston *et al.*, 2008).

7.3 Gibberellins relieve DELLA-growth repression by targeting their degradation

The cloning of the *DELLA* genes did not immediately provide any clear insights into how GAs inactivate these growth repressors. However, some clues were provided by the observations that the partially dominant GA-insensitive dwarfing mutations in *Arabidopsis*, wheat, maize, barley and rice affected residues within the N-terminal DELLA and TVHYNP domains (Peng *et al.*, 1997; Peng *et al.*, 1999; Chandler *et al.*, 2002; Asano *et al.*, 2009). The phenotype of these plants suggested that these mutations result in constitutively active DELLA growth repressors that cannot be inactivated by GAs and also highlighted the importance of these domains in controlling this process. The breakthrough in understanding how GAs inactivate the DELLAs was provided by the demonstration that a GFP-RGA fusion protein was rapidly degraded *in planta* following treatment with bioactive GAs (Silverstone *et al.*, 2001). The endogenous DELLA proteins in *Arabidopsis* and other plant species were found to respond in a similar fashion, thus demonstrating that targeted degradation of DELLAs is a common mechanism through which GA relieves growth repression exerted by these proteins (Silverstone *et al.*, 2001; Chandler *et al.*, 2002; Itoh *et al.*, 2002; Dill *et al.*, 2004). In the case of gain-of-function DELLA mutations responsible for conferring GA-insensitivity, it was found that their effect is to abolish or reduce GA-mediated degradation of the encoded proteins (Dill *et al.*, 2001; Chandler *et al.*, 2002; Dill *et al.*, 2004; Asano *et al.*, 2009). This demonstrated that the GA-insensitive phenotype was caused by the presence of mutant DELLAs that constitutively repressed GA responses because they are unable to be inactivated through GA-mediated degradation. It also highlighted the importance of the DELLA and TVHYNP domains in controlling this regulatory process. It is now known that GA-mediated degradation is achieved through the action of a soluble GA receptor, GIBBERELLIN INSENSITIVE DWARF1 (GID1), which is localised within the nucleus and cytoplasm (Ueguchi-Tanaka *et al.*, 2005; reviewed in Chapter 6). The binding of bioactive GAs to GID1 promotes the formation of a GA-GID1-DELLA complex, which ultimately results in the targeted degradation of DELLA via an ubiquitin-26S proteasome pathway (see Chapter 6). In a seminal study, Murase and colleagues (2008) resolved the crystal structure of the ternary complex of the *Arabidopsis* GA receptor GID1A bound to bioactive GA and the N-terminus of GAI, thereby defining the DELLA domains that were required for binding to the GID1-GA complex. It highlighted the importance of the DELLA, LExLE and TVHYNP domains within which conserved residues are in direct contact with GID1. Mutations within these conserved domains cause GA-insensitive dwarfism because they prevent binding of

DELLAs to the GID1-GA complex, thereby blocking GA-mediated degradation and causing constitutive growth repression (Griffiths *et al.*, 2006; Willige *et al.*, 2007; Asano *et al.*, 2009; Pearce *et al.*, 2011). See also Chapter 6 for a more detailed description of the processes controlling GA-mediated DELLA degradation.

7.4 Functional diversification of *DELLA* genes

A major difference between the GA signalling components in monocots and dicots is the level of genetic redundancy. In monocots, including rice and barley, genetic evidence and genome sequence data supports the existence of a single *DELLA* gene in these plants (Ikeda *et al.*, 2001; Chandler *et al.*, 2002). The rice and barley DELLA loss-of-function mutants, *slr1* and *sln1*, respectively, display extremely slender phenotypes that mimic constitutively active GA signalling (Foster, 1977; Ikeda *et al.*, 2001; Chandler *et al.*, 2002). These defects are consistent with SLR1 and SLN1 acting to repress all aspects of GA-responsive growth in these cereals. However, interrogation of the rice genome sequence has identified two *SLR1-like* genes encoding SLRL1 and SLRL2 that have high homology to the GRAS domain of SLR1, but which lack the N-terminal DELLA and TVHYNP domains required for GID1-mediated degradation (Itoh *et al.*, 2005). Interestingly, the existence of SLRL-type GRAS proteins appears to be restricted to monocots. When over-expressed in transgenic plants, SLRL1 was able to suppress the slender phenotype of the loss-of-function *slr1* mutant, supporting its role in controlling GA-responsive growth. As SLRL1 lacks the GID1 recognition domains, this raises the prospect that it is able to repress GA-responsive growth in the presence of bioactive GAs. In fact, a recent study of submergence-tolerant rice has implicated SLRL1 and SLR1 in repressing GA-dependent stem elongation following submergence (Fukao and Bailey-Serres, 2008). To confirm this role for SLRL1 it will be important to identify and characterise the submergence response in an *slrl1* loss-of-function mutant.

In contrast to rice and barley, the genome of the model dicot *Arabidopsis* contains five *DELLA* genes: *RGA*, *GAI*, *RGA-LIKE1* (*RGL1*), *RGL2* and *RGL3* (Dill and Sun, 2001; Tyler *et al.*, 2004). A phylogenetic analysis of DELLA proteins from evolutionarily diverse flowering plants supports their expansion occurring after the monocot-dicot split (Figure 7.2; Dill and Sun, 2001; Gallego-Bartolome *et al.*, 2010). Although two main clades have been identified in dicots, species within the Asterids genera only possess DELLAs belonging to one of these. For example, genetic analysis and interrogation of genome sequence data supports the existence of a single *DELLA* gene in *Solanum lycopersicum* (Jasinski *et al.*, 2008; Fleishon *et al.*, 2011). In contrast, species within the Rosids possess DELLAs that group within both of the two main clades (Gallego-Bartolome *et al.*, 2010). It has therefore been suggested

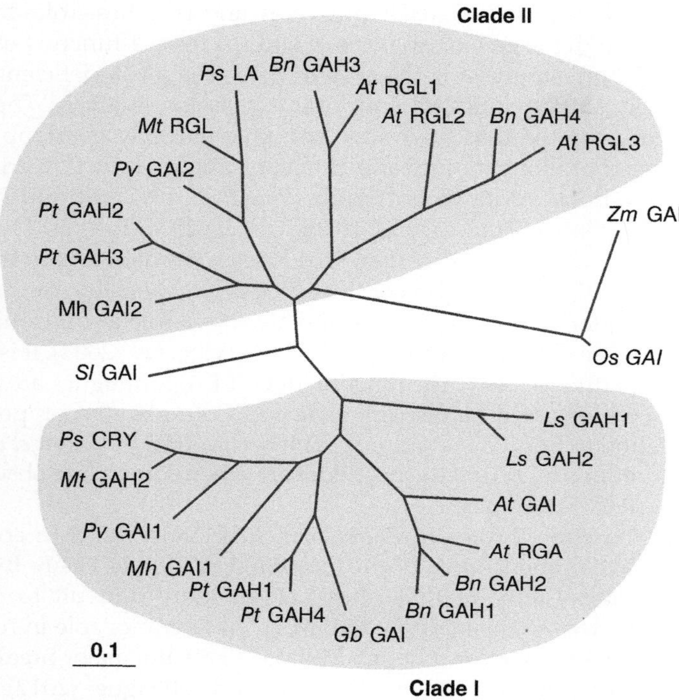

Figure 7.2 Phylogenetic history of *DELLA* genes. The image depicts the phylogenetic relationships between the multiple *DELLA* genes arisen from successive duplications in dicotyledoneous plants. (Adapted from Gallego-Bartolomé *et al.* 2010. Reproduced with permission from Oxford University Press.) *At, Arabidopsis thaliana; Bn, Brassica napus; Gb, Gossypium barbadense; Ls, Lactuca sativa; Mh, Malus hupehensis; Mt, Medicago truncatula; Os, Oryza sativa; Ps, Pisum sativum; Pt, Populus trichocarpa; Pv, Phaseolus vulgaris; Sl, Solanum lycopersicum; Zm, Zea mays.*

that this diversification in DELLAs occurred through the duplication of an ancestral gene within the Rosids. Further duplication events have occurred in the Brassicaceae family, which have led to the presence of the five paralogues in *Arabidopsis*. The order of these duplication events is supported by the presence of these genes within syntenic regions of the *Arabidopsis* genome (Dill and Sun, 2001; Gallego-Bartolome *et al.*, 2010).

Extensive studies to establish the roles of the five *Arabidopsis DELLA* genes in controlling GA-responsive growth have been performed (Silverstone *et al.*, 1997; Dill and Sun, 2001; King *et al.*, 2001; Lee *et al.*, 2002; Wen and Chang, 2002; Cheng *et al.*, 2004; Tyler *et al.*, 2004; Wild *et al.*, 2012). In addition to confirming that all of them encode repressors of GA signalling, these studies have demonstrated that they have both distinct and overlapping roles in controlling *Arabidopsis* growth (a more detailed description of these roles controlling seed germination, vegetative and reproductive development and how they were elucidated are provided in Chapters 9, 10 and 11). These roles have been elucidated through the genetic analysis of DELLA loss-of-function mutations and their effect on suppressing the growth defects of a GA-deficient mutant, *ga1-3*, coupled with spatial and temporal expression analysis. These studies have demonstrated that *RGA* and *GAI* have partially overlapping roles in controlling root elongation, leaf expansion, initiation of flowering, stem elongation and floral development (Silverstone *et al.*, 1997; Dill and Sun, 2001; King *et al.*, 2001; Fu and Harberd, 2003; Cheng *et al.*, 2004; Tyler *et al.*, 2004; Cao *et al.*, 2006). However, it appears that *RGA* has a dominant role in the majority of these growth processes. In contrast, *RGL2* is responsible for repressing germination under certain conditions, and also has a role in controlling floral development (Lee *et al.*, 2002; Cheng *et al.*, 2004; Tyler *et al.*, 2004). It is interesting to note that differences in the function of DELLA paralogues are apparent between the widely used Landsberg *erecta* and Columbia-0 ecotypes, specifically in relation to floral development (Tyler *et al.*, 2004; Plackett *et al.*, 2014). These observations highlight the need for care when comparing observations between these two ecotypes.

While the functional roles of *RGA*, *GAI*, *RGL1* and *RGL2* in controlling *Arabidopsis* development have been uncovered relatively easily by genetic analysis, the importance of *RGL3* has proven more difficult to establish. Although there was some evidence that *RGL3* has a minor role in repressing germination (Piskurewicz and Lopez-Molina, 2009), the major breakthrough in our understanding was uncovered by Wild and colleagues (2012). By interrogating publically available microarray data, they established that *RGL3* is an early target of jasmonate (JA) signalling that is induced by this pathway. The impact of elevated RGL3 levels is to enhance the output of JA signalling through a direct repression of the JASMONATE ZIM-DOMAIN (JAZ) JA response repressors (the mechanisms involved are discussed in more detail in Section 7.7.3; see also Chapter 8). To establish the biological significance of this response, Wild and colleagues (2012) investigated the role of RGL3 in the JA-dependent resistance to the necrotrophic fungus *Botrytis cinerea*. Consistent with a role in modulating JA signalling, the absence of RGL3 was found to enhance susceptibility to *Botrytis cinerea*. Furthermore, they also established that JA-mediated susceptibility to the hemibiotroph *Pseudomonas syringae* was reduced in the absence of RGL3. Because RGL3 is degraded in response to GA signalling (Feng *et al.*, 2008; Wild *et al.*, 2012), the status of GAs within the plant will determine the modulation of the JA response. This suggests that the transcriptional regulation of *RGL3* by JA signalling provides a mechanism to integrate growth status of the plant with its ability to respond to pathogens. It is interesting to note that *RGL3* is also regulated by other abiotic stresses, including cold and drought (Achard *et al.*, 2008; Colebrook *et al.*, 2014), suggesting that this is a conserved mechanism for controlling growth of the plant in response to multiple environmental factors.

The presence of multiple *DELLA* genes with functionally divergent roles in controlling GA-responsive growth in dicots raises the important question as to how this functional specificity is conferred. Characterisation of the tissue-specific expression patterns of the five *DELLA* paralogues in *Arabidopsis* has highlighted clear differences that potentially underlie their functional specificity (Tyler *et al.*, 2004), although this study could not rule out other factors, such as differential activities regulating downstream targets. This question was addressed by Gallego-Bartolome and colleagues (2010) who used promoter-swapping experiments to investigate whether *RGA* and *RGL2* could complement each other's function when expressed in the respective *rgl2* or *rga* loss-of-function mutant background. These two DELLAs were most suitable for this study because they control distinct developmental processes that could be easily monitored (RGA and RGL2 control vegetative growth and seed germination, respectively). Characterisation of the transgenic lines that were generated in this study demonstrated that RGA and RGL2 were in fact capable of functional complementation. These findings have led to the conclusion that the functional diversification of *Arabidopsis DELLA* genes can result from differences in their spatial and temporal expression patterns. In view of the importance of GA signalling in controlling plant growth in response to changes in environmental conditions, it is tempting to speculate that the expansion and functional diversification of *DELLA* genes in some plant species provides a fitness benefit by allowing fine tuning of environmental responses.

7.5 DELLA activity invokes rapid changes in the transcriptome

Genomic approaches have shown that GAs induce rapid changes in the transcriptome, likely mediated by DELLA activity (Ogawa *et al.*, 2003; Cao *et al.*, 2006; Willige *et al.*, 2007). In agreement with the idea that DELLAs are transcriptional regulators, GAI and RGA are able to alter gene expression very rapidly (Zentella *et al.*, 2007; Gallego-Bartolomé *et al.*, 2011a). In these two studies, the *Arabidopsis* transcriptome was interrogated in lines expressing the dominant versions rga-Δ17 and gai-1 under the control of inducible promoters, and microarray analyses identified 475 and 148 genes whose expression was altered at least 1.5-fold within the first four hours after the induction, respectively. Consistently, more genes were up- than down-regulated in response to DELLA activity, whereas the opposite situation is observed following treatment with GA (Willige *et al.*, 2007). From a mechanistic point of view, it is important to note that both DELLAs were able to affect gene expression in a direct manner, as expected for bona fide transcriptional regulators. This was supported by two lines of evidence. First, chromatin immunoprecipitation (ChIP) analyses showed that *in vivo* RGA associates with the promoters of some of its target genes (Zentella *et al.*, 2007). This study defined a core of 14 *bona fide* RGA- and GA-responsive genes, of which RGA was associated with the promoters of a sub-set of eight genes, which were all DELLA up-regulated. Nevertheless, RGA also associates with the promoters of genes that are down-regulated by DELLAs (Park *et al.*, 2013), consistent with the finding that the expression of other RGA targets is reduced in response to RGA activity (Zentella *et al.*, 2007). This was a remarkable result, since it indicated that DELLAs can act as *cis*-acting transcriptional regulators on target genes. The lack of a recognizable DNA binding motif within DELLAs suggests that they bind chromatin through an interaction with other *trans*-acting factors (see below). Second, gai-1 was able to regulate the transcription of target genes in the absence of protein synthesis (Gallego-Bartolomé *et al.*, 2011a, b, c). This was demonstrated by using a transgenic line that expresses a translational fusion between gai-1 and the receptor domain of the rat glucocorticoid receptor (GR) enabling the chimeric gai-1-GR to move from the cytosol, where it accumulates, to the nucleus after treatment with the synthetic steroid, dexamethasone (Gallego-Bartolomé *et al.*, 2011c). This technology allows the activation of gai-1-GR, i.e. relocation into the nucleus, without the need for protein synthesis. Thus, the combination of dexamethasone and cycloheximide treatments demonstrated protein-synthesis-independent changes in gene expression for many GAI targets, indicating that GAI can directly regulate gene expression. The regulation of direct targets is compatible with DELLAs associating directly with the promoter of their target genes, but does not exclude alternative mechanisms, such as sequestration of TFs (see below). *In silico* analysis of the promoter regions of DELLA-regulated genes identified several *cis* elements that were over-represented (Gallego-Bartolomé *et al.*, 2011a). Some of these elements provided important clues about the classes of TFs that might bind the promoter of these genes, mediating regulation by DELLAs (see below). For instance, the presence of G (CACGTG) and E (CACATG) boxes, the CArG-like, and the AAAG and NGATT sequences suggests that bHLHs, bZIPs, MADSs, Dofs and type-b response regulators, respectively, are likely involved in regulating these genes (see below).

7.6 DELLA proteins activate transcription

Soon after their discovery, it was apparent that DELLAs were able to activate transcription. This was first demonstrated for the rice DELLA SLR1 in trans-activation assays performed in spinach leaves (Ogawa *et al.*, 2000). SLR1 activated transcription of a reporter gene when fused to the DNA-binding domain of the yeast GAL4 TF. Deletion analyses identified the N-terminal DELLA domain of SLR1 as responsible for this transcriptional activation ability. A more comprehensive study showed that both the DELLA and TVHYNP motifs residing in the N-terminal region are important for this activity (Hirano *et al.*, 2012). Several loss-of-function mutations in DELLA proteins reside in the GRAS domain, which, however, was dispensable for transactivation activity. The introduction of these mutations in a chimeric protein lacking the N-terminal region, but fused to the strong transcriptional activator VP16 lowered its ability to exert the repressive function in transgenic rice plants, suggesting that the GRAS domain, despite not being necessary for transactivation, is required to manifest the full DELLA activity *in planta* (Hirano *et al.*, 2012). It is becoming apparent that this activity of the GRAS domain is most likely mediated through its interaction with TFs that direct DELLAs to specific target promoters (Fukazawa *et al.*, 2014; Yoshida *et al.*, 2014; see below). Based on the association between the growth phenotype and transcriptional activity of *SLR1* deletion alleles, it appears that DELLAs control plant growth at least partly through transcriptional activation (Hirano *et al.*, 2012). The transcriptional activation ability is manifested in heterologous systems as well, such as yeast (Hirano *et al.*, 2012), indicating that DELLAs might interact with conserved elements of the basal transcriptional machinery. This ability was inhibited upon interaction with GID1, both in rice callus and in yeast (Hirano *et al.*, 2012), which is consistent with the capacity of the receptor to inactivate the DELLA protein through a physical association (Ariizumi *et al.*, 2008; Ueguchi-Tanaka *et al.*, 2008). It is somewhat surprising that no reduced or loss-of-function DELLA mutations affecting residues within the N-terminal transactivation domains have been identified. However, it is possible that the impact of such mutations could be masked by them also affecting GA-mediated degradation of DELLA. The identification of additional components involved in DELLA-mediated transactivation should shed further light on the mechanisms responsible.

7.7 DELLAs regulate transcription by physical interaction with transcriptional regulators

The lines of evidence presented in the previous sections strongly suggest that DELLAs exert their activity, at least in part, by modulating the transcriptional rate of their target genes. In brief: (1) there are motifs in their primary structure typical of transcriptional regulators, although none of them resemble known DNA-binding domains (Gubler *et al.*, 2002), (2) DELLAs possess an intrinsic transcriptional activation ability (Ogawa *et al.*, 2000; Hirano *et al.*, 2012), (3) they are able to rapidly alter transcription of target genes (Zentella *et al.*, 2007; Gallego-Bartolomé *et al.*, 2011a) and (4) they associate with several target gene promoters (Zentella *et al.*, 2007; Lim *et al.*, 2013; Park *et al.*, 2013; Fukazawa *et al.*, 2014; Yamaguchi *et al.*, 2014). Nonetheless, an outstanding question still remained to be answered: how do DELLAs regulate transcription? The key to the answer came from the identification of DELLA protein partners, as demonstrated by pioneering studies from Salome Prat's and Xing-Wang Deng's laboratories (de Lucas *et al.*, 2008; Feng *et al.*, 2008).

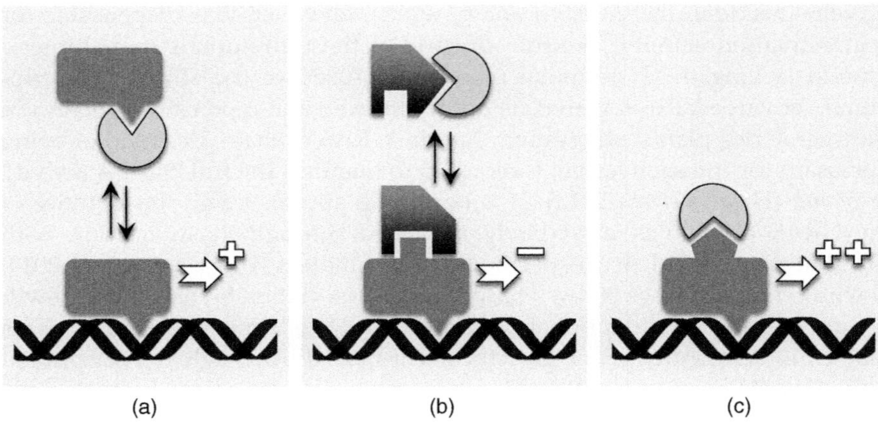

Figure 7.3 Molecular mechanisms of DELLA activity. (a) DELLA proteins prevent the interaction between a TF and its target DNA sequence. (b) DELLA proteins facilitate the activity of a TF by releasing negative regulation imposed by a repressor. (c) DELLA proteins promote transcriptional activation of a TF at the target loci.

Subsequently, several classes of TFs and transcriptional regulators have been identified as bona fide DELLA interactors, mainly in *Arabidopsis* (see below). Remarkably, the elucidation of the molecular events that are a consequence of these interactions has defined several different mechanisms by which DELLAs regulate transcription (Figure 7.3). In addition and equally important, the identification of DELLA interactors provides us with the possibility to answer, from a mechanistic point of view, two other relevant questions in the field: (1) how GAs regulate so many different processes during the life cycle of the plant and (2) how the GA pathway interacts with other signalling pathways to exert this regulation.

7.7.1 DELLAs sequester *bona fide* TFs by physical interaction

The first TFs identified as DELLA interactors were PHYTOCHROME INTERACTING FACTORS (PIFs), PIF3 (Feng *et al.*, 2008) and PIF4 (de Lucas *et al.*, 2008). These TFs belong to the bHLH superfamily (Toledo-Ortiz *et al.*, 2003) and were formerly identified as growth-promoting, negative regulators of phytochrome-regulated responses (Huq and Quail, 2002; Kim *et al.*, 2003). Importantly, it was shown by ChIP and transcriptional assays that the ability of these TFs to bind to the promoter of several of their target genes was inhibited upon interaction with the DELLA protein (de Lucas *et al.*, 2008; Feng *et al.*, 2008). Thus, a series of biochemical assays unambiguously showed that DELLAs inhibit PIF3 and PIF4 upon interaction with the conserved bHLH DNA-binding domain, providing the first functional evidence of a mechanism by which DELLAs regulate transcription, i.e. by sequestering TFs (Figure 7.3a).

The next important question was to establish the physiological relevance of these interactions. These authors demonstrated that the interactions are important for regulating hypocotyl elongation in response to two antagonistic signals, GAs and light (de Lucas *et al.*, 2008; Feng *et al.*, 2008). More specifically, light represses hypocotyl growth by negatively acting on PIF3 and PIF4 by two means, on the one hand it induces their rapid phosphorylation and degradation by the 26S proteasome upon interaction with photoactivated phytochromes (Al-Sady *et al.*, 2006; de Lucas *et al.*, 2008), and on the other, it enhances accumulation of DELLAs (Achard *et al.*, 2007) that can interact with and inactivate the remaining pool of PIF proteins. Thus, the identification of the DELLA-PIF3/4 interaction not only provided the first mechanistic understanding of how DELLAs regulate transcription, but also a molecular mechanism for the interaction between GAs and light in controlling photomorphogenic growth in plants.

The activity of these and other related bHLH TFs (mainly PIF1 and PIF5) is required for the correct progression of skotomorphogenesis that takes place just after germination in darkness (Leivar *et al.*, 2008; Shin *et al.*, 2009). The basic features of the skotomorphogenic growth are that hypocotyls grow fast and form an apical hook to protect the shoot apical meristem and cotyledons, which remain small and folded. In addition, the gene expression programmes associated with photomorphogenesis remain off. Accordingly, this developmental program also requires low DELLA levels in order to progress (Alabadí *et al.*, 2004; Achard *et al.*, 2007). Despite their levels being low, DELLAs inactivate PIF5 by physical interaction, and this is relevant to the control of ethylene production and auxin transport during apical hook development (Gallego-Bartolomé *et al.*, 2011b; Willige *et al.*, 2012). Similarly, low DELLA levels seem to be important to guarantee the activity of the PIFs controlling hypocotyl elongation and cotyledon opening (Alabadí *et al.*, 2008), and also to regulate the expression of many genes involved in photosynthesis before and during the greening process triggered upon illumination, either directly, such as *CONDITIONAL CHLORINA* (*CHLH*), *NADPH:PROTOCHLOROPHYLLIDE OXIDOREDUCTASE C* (*PORC*) and *CHLOROPHYLL A OXYGENASE* (*CAO*) (Cheminant *et al.*, 2011), or indirectly, through the TFs GATA, NITRATE-INDUCIBLE, CARBON-METABOLISM INVOLVED (GNC) and GNC-LIKE (GNL) (Richter *et al.*, 2010). The repression of *PORC* in skotomorphogenic seedlings is also exerted by SCARECROW-LIKE 27 (SCL27), a GRAS TF, by binding to GT elements in its promoter (Ma *et al.*, 2014). This repression is alleviated by a dual mechanism upon illumination: (1) accumulating DELLAs interact and sequester SCL27 and (2) light induces the expression of the miR171 that targets *SCL27* mRNA for degradation.

The interplay between GAs and light in the control of growth responses might be subject to daily fluctuation, since seedlings are naturally growing under photoperiod. The coordinated activity of light signalling through the phytochromes and the circadian clock leaves a daily window of high PIF activity towards the end of the night that causes maximum hypocotyl growth rates (Nozue *et al.*, 2007; Nusinow *et al.*, 2011; Soy *et al.*, 2012). Remarkably, the circadian clock gates GA signalling by preventing the accumulation of DELLAs at the end of the night as a way to allow hypocotyl growth (Arana *et al.*, 2011). Therefore, the regulation of PIFs by DELLAs might also be part of the mechanism that generates the daily rhythm in hypocotyl elongation.

The activities of both GAs and PIFs are intimately related in the control of other developmental responses involving elongation growth. For instance, petiole and hypocotyl elongation in response to either shade or increased ambient temperature (Djakovic-Petrovic *et al.*, 2007; Lorrain *et al.*, 2008; Koini *et al.*, 2009; Stavang *et al.*, 2009). Both environmental cues seem to act primarily on the GA pathway by promoting the degradation of DELLA proteins (Djakovic-Petrovic *et al.*, 2007; Stavang *et al.*, 2009; Bou-Torrent *et al.*, 2014). The enhancement of PIF activity occurs at different levels in response to these environmental cues: increasing ambient temperature acts at both the transcriptional and post-transcriptional levels, promoting *PIF4* and *PIF5* expression (Koini *et al.*, 2009; Stavang *et al.*, 2009) and stabilizing the protein (Foreman *et al.*, 2011), while only the latter occurs in response to shade (Lorrain *et al.*, 2008). In both situations, it is reasonable to think that the interaction between DELLA and PIFs performs a regulatory role.

DELLAs also interact with other bHLH TFs belonging to the same clade as PIFs, including ALCATRAZ (ALC), PIF3-LIKE 2 (PIL2) and SPATULA (SPT) (Arnaud *et al.*, 2010; Gallego-Bartolomé *et al.*, 2010). The interaction between DELLAs and the developmental regulator ALC was the first example highlighting how the regulatory module formed by a DELLA-bHLH TF association has been recruited to control processes other than elongation growth and also revealed a previously unidentified role for GAs in fruit patterning (Arnaud *et al.*, 2010). In particular, DELLA accumulation in the valve margins of the *Arabidopsis* fruit inactivates ALC, thus preventing the differentiation of the separation layer needed for fruit opening and seed dispersal. As fruit development progresses, the upstream regulator INDEHISCENT (IND) promotes GA biosynthesis in the valve margins through induction of *AtGA3ox1* gene expression, causing DELLA degradation and the concomitant activation of ALC, which in turn directs the differentiation of the separation layer. However, no role has been identified for the interactions between DELLAs and the other members of the PIF clade, PIL2 and SPT. It is interesting to note that both DELLAs and SPT act as repressors of cotyledon expansion (Josse *et al.*, 2011). They exert this effect as part of a regulatory circuit in which SPT levels, but not activity, are negatively regulated by DELLAs in response to the light quality of the environment as a way to self-regulate the total growth repressive activity of both proteins.

DELLAs are able to interact with and inactivate bHLH TFs that are members of other clades within this family, such as MYC2, GLABRA 3 (GL3) and ENHANCER OF GLABRA 3 (EGL3) (Hong *et al.*, 2012a; Qi *et al.*, 2014). The interaction of DELLAs with MYC2 regulates volatile production in

Arabidopsis flowers, uncovering a role for GAs in this process (Hong *et al.*, 2012a), whereas the interaction with GL3 and EGL3 provides a molecular mechanism that explains the long-standing role for GAs promoting trichome development (Chien and Sussex, 1996; Qi *et al.*, 2014). In the latter case, the DELLA interactions extend to the inhibition of the MYB GLABRA1 (GL1) component of the WD-repeat/bHLH/MYB regulatory complex, which is also needed to trigger trichome development. These DELLA partners are also inactivated by the negative regulators of JA signalling, the JAZ proteins (Chini *et al.*, 2007; Thines *et al.*, 2007; Qi *et al.*, 2011). In both processes, GAs and JA act synergistically to enhance the expression of the genes involved in volatile production or in trichome development through releasing the repression imposed by DELLAs and JAZs. Therefore, MYC2 and GL1/GL3/EGL3 integrate both GA and JA signalling, and the molecular interaction described represents only one of the mechanisms for the cross-regulation between both hormone pathways (see below).

Remarkably, the cell-type determination mechanism represented by the interaction between DELLAs and the epidermis-specific TFs, GL1, GL3 and EGL3 provides clues about how GAs regulate other processes in a tissue-specific manner, such as the regulation of cell elongation and cell division in roots (Úbeda-Tomás *et al.*, 2008; Úbeda-Tomás *et al.*, 2009; Shani *et al.*, 2013), anther development (Plackett *et al.*, 2014), or flowering time (Galvao *et al.*, 2012; Porri *et al.*, 2012), i.e. by DELLA interaction with tissue-specific partners. Indeed, the dual role of DELLAs on the activity of SQUAMOSA PROMOTER BINDING PROTEIN-LIKE 9 (SPL9), first inactivating it to delay flowering (Yu *et al.*, 2012), and later on acting together as a coactivator to promote *APETALA 1* (*AP1*) expression and thus flower formation (Yamaguchi *et al.*, 2014), or the inactivation of the HD-ZIP TFs MERISTEM LAYER 1 (ML1) and PROTODERMAL FACTOR 2 (PDF2) upon interaction to regulate seed germination (Rombola-Caldentey *et al.*, 2014), represent other examples of how tissue- or organ-specific TFs provide spatial specificity for certain roles of GAs.

The interaction and inactivation of BRASSINAZOLE-RESISTANT 1 (BZR1)/BRI1-EMS-SUPPRESSOR 1 (BES1) (Bai *et al.*, 2012; Gallego-Bartolomé *et al.*, 2012; Li *et al.*, 2012), ETHYLENE INSENSITIVE 3 (EIN3) (An *et al.*, 2012) and RELATED TO AP2 3 (RAP2.3) (Marin-de la Rosa *et al.*, 2014), belonging to different families of TFs, represent cross-regulatory points between the GA and the brassinosteroid (BR) and ethylene signalling pathways, respectively (see also Chapter 8). The interaction with BZR1/BES1 explains the synergistic effect of GAs and BRs promoting hypocotyl elongation, both in seedlings undergoing skoto- and photomorphogenesis. In contrast, the interaction of DELLA with EIN3 and RAP2.3 is important for regulating apical hook development in etiolated seedlings. In this situation, DELLAs repress the expression of the *HOOKLESS 1* (*HLS1*) gene leading to hook opening through inactivation of the positive regulator EIN3 (An *et al.*, 2012), whereas the inactivation of RAP2.3 affects a discrete transcriptional branch downstream of EIN3 also leading to hook opening (Marin-de la Rosa *et al.*, 2014). It seems likely therefore that GAs control several aspects of skotomorphogenesis by controlling a transcriptional network through the interaction between DELLAs and several TFs – PIFs, BZR1/BES1, EIN3 and RAP2.3 – that promote various aspects of this developmental program. PIF4 and BZR1 also interact physically, acting together to promote elongation growth both in light grown and in etiolated seedlings and defining a complex transcriptional module negatively regulated by DELLAs that also integrates BR and light information (Bai *et al.*, 2012). Interestingly, a recent report has broadened this module with the addition of two AUXIN RESPONSE FACTORs (ARFs) that mediate transcriptional regulation in response to auxin, ARF6 and ARF8, forming the so-called BAP module (Oh *et al.*, 2014), and providing a molecular mechanism for the observation of the synergism between many BR and auxin-regulated responses (Nemhauser *et al.*, 2004). Oh and colleagues have demonstrated the existence of large sets of genes coregulated by ARF6/8-PIF4 and ARF6/8-BZR1 complexes, in addition to those regulated by BZR1-PIF4 (Bai *et al.*, 2012), and that are relevant to regulate growth by cell elongation, both in seedlings growing in darkness and in the light (Oh *et al.*, 2014). Importantly, DELLAs are able to prevent the activity of all these pairs of TFs through a physical association.

Plant growth and form is a consequence of coordinated cell elongation and cell division. Interestingly, two recent studies report on the role that DELLAs exert in mediating the effects of GAs in promoting cell division, both at the shoot and the root apical meristems (Daviere *et al.*, 2014; Resentini *et al.*, 2015), in addition to their well-established role in regulating cell elongation. Both works identify the class I TCP TFs as targets for inactivation by DELLAs upon interaction. A series of biochemical assays demonstrated that the DNA binding ability of these TFs, and thus their transcriptional activity, was compromised by DELLAs. This effect results in reduced expression of core cell-cycle genes, which lowers the cell division capacity at the shoot apex (Daviere *et al.*, 2014). Similarly, it also provides a likely explanation for the reduced cell division that occurs at the root apex when DELLA levels increase (Achard *et al.*, 2009; Úbeda-Tomás *et al.*, 2009; Resentini *et al.*, 2015).

7.7.2 DELLAs interact with TFs in the context of promoters

All of the DELLA-TF interactions discussed above result in the inactivation of the particular TF by sequestering it away from its target genes. Nonetheless, several reports show that DELLAs can associate with specific target gene promoters (Zentella *et al.*, 2007; Lim *et al.*, 2013; Park *et al.*, 2013; Fukazawa *et al.*, 2014; Yamaguchi *et al.*, 2014; Figure 7.3c). Remarkably, ChIP followed by massive sequencing has demonstrated the binding of RGA to the regulatory regions (2500 bp up- or 500 bp downstream of a gene or within introns or UTRs) of 421 genes, indicating that transcriptional regulation by DELLAs in the context of the chromatin might be a common mechanism as well

(Marín-de la Rosa *et al.*, 2015). In this study, a series of *cis* elements enriched in the peak regions were also identified through an *in silico* approach that led to predict possible TFs responsible for the binding of RGA. The list included sequence elements bound by the INDETERMINATE DOMAIN (IDD) and bZIP families, some of their members being known interactors of DELLA proteins thus demonstrating the strength of the approach (see below; Lim *et al.*, 2013; Fukazawa *et al.*, 2014; Marin-de la Rosa *et al.*, 2014; Yoshida *et al.*, 2014). Interestingly, *cis* elements bound by the GARP-*ARABIDOPSIS* RESPONSE REGULATOR (ARR) family were also found. Type-b GARP-ARRs ARR1, ARR2 and ARR14 are positive regulators within the cytokinin (CK) signal transduction pathway and interact with both GAI and RGA (Marin-de la Rosa *et al.*, 2014, 2015). A series of biochemical, molecular genetics and physiological assays showed that these two DELLAs enhance the activation ability of ARR1 over a variety of target genes and that this effect occurs in the vicinity of their promoters and controls at least two physiological responses known to be antagonistically regulated by GAs and CKs: root meristem size and de-etiolation (Marín-de la Rosa *et al.*, 2015). Furthermore, Lim and colleagues (2013) have demonstrated that DELLAs are targeted to the promoter of *SOMNUS* (*SOM*), a negative regulator of germination, through an interaction with ABSCISIC ACID INSENSITIVE3 (ABI3) and the bZIP ABSCISIC ACID INSENSITIVE5 (ABI5). This interaction leads to an increase in *SOM* expression when ambient temperatures are high resulting in the inhibition of germination (Lim *et al.*, 2013). Importantly, these two recent studies provide molecular mechanisms for the interaction between the GA and the ABA and CK signalling pathways during the control of these particular processes. However, it is highly probable that these mechanisms of cross-talk could also operate at other stages of development influenced by these two hormones antagonistic to GAs. Similarly, and as mentioned above, DELLAs promote flower formation thanks to their ability to activate the expression of the floral identity gene *AP1* by acting in the context of its promoter through interaction with SPL9 (Yamaguchi *et al.*, 2014). The role that DELLAs perform in these transcriptional complexes has not been explicitly investigated. However, it is tempting to speculate that they are acting as transcriptional coactivators that are targeted to the *SOM* and *AP1* promoters by the TFs, as demonstrated for DELLA-ARR1 (Marín-de la Rosa *et al.*, 2015). This has also been demonstrated for the interaction between DELLAs and several members of the INDETERMINATE DOMAIN (IDD) family in *Arabidopsis* (Fukazawa *et al.*, 2014; Yoshida *et al.*, 2014). These two studies demonstrate that DELLAs participate in a mechanism in which the interchange of transcriptional coactivators (DELLAs) and corepressors (SCL3 and TOPLESS-RELATED (TPR)) bound to IDDs activate or repress gene targets in a GA-dependent manner (Figure 7.4a). Despite their similarities, the mechanisms are not identical. For instance, IDD3, 4, 5, 9 and 10 activate transcription of target genes upon interaction with DELLAs (Yoshida *et al.*, 2014), including *SCL3*, which encodes a coactivator of GA signalling (Heo *et al.*, 2011; Zhang *et al.*, 2011). Previous studies had indicated that SCL3 promotes GA responses, including promotion of *SCL3* expression itself, by antagonizing DELLA function through a direct protein–protein interaction (Zhang *et al.*, 2011). In light of these findings, Yoshida and colleagues (2014) proposed a regulatory transcriptional feedback loop controlled by SCL3 and DELLA levels that buffers GA signalling by providing a homeostatic control mechanism for this pathway, the important factor being that *SCL3* is a transcriptional target of both the activator DELLA-IDD and the repressor SCL3-IDD complexes. Further insights into the importance of the DELLA-IDD transcriptional module in controlling GA responses have been gained from a recent study with *Arabidopsis* characterising an interaction between GAF1 (GAI-ASSOCIATED FACTOR1)/IDD2 and the DELLA, GAI (Fukazawa *et al.*, 2014). GAF1/IDD2 activates transcription of its target genes when it interacts with DELLAs, but represses transcription when it interacts with the TPR1 and TPR4 corepressors. Importantly, this transcriptional system provides a molecular mechanism for the feedback regulation that operates to control GA levels and perception: GAF1/IDD2 binds to the promoters of the *GA20ox2*, *GA3ox1* and *GID1b* genes to activate their expression through interaction with DELLAs when GA levels are low, but repress their expression through interaction with TPR1 and TPR4 when GA levels are high (see also Chapter 2). In essence, the levels of expression of these GA-feedback regulated genes are controlled by a balance between DELLA and TPR levels, which is determined by the GA concentration. These interactions define two key mechanisms that provide homeostasis to the GA pathway, including biosynthesis, perception and signalling. The closest homologue to GAF1/IDD2 is ENHYDROUS (ENY)/IDD1, which has also been demonstrated to physically interact with DELLAs (Feurtado *et al.*, 2011). ENY/IDD1 appears to have a positive effect on GA responsiveness during seed development, and although the mechanism of action is not known, it is likely that it acts similarly to GAF1/IDD2.

7.7.3 DELLAs interact with other transcriptional regulators

DELLAs not only control transcription by binding to bona fide TFs, they also interact with other transcriptional regulators that function in a similar manner to DELLAs by regulating transcription without binding directly to DNA (Figure 7.3b). This ability likely amplifies the effect of DELLAs on transcription, given that each transcriptional regulator might influence the activity of multiple TFs. One of the best known examples is the interaction between DELLAs and JAZ proteins (Hou *et al.*, 2010; Yang *et al.*, 2012). Hou and colleagues (2010) demonstrated that this interaction mediates the antagonism between GAs and JA in the control of root growth in *Arabidopsis*. In this context, DELLA accumulation inactivates the JAZ1 protein, thus allowing MYC2, and likely other JAZ-regulated TFs, to prevent root growth and enhance the expression of JA-responsive genes. This regulatory module

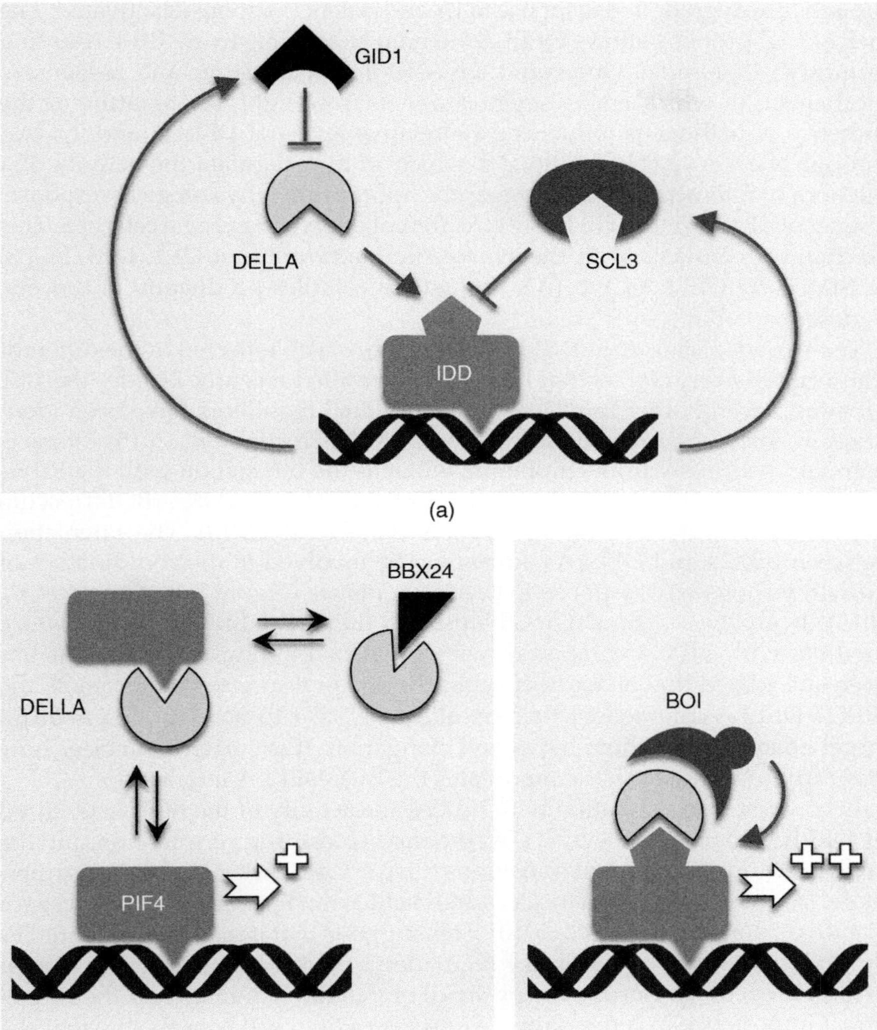

(a) (b) (c)

Figure 7.4 Modes of regulation of DELLA activity. (a) SCL3 and other transcriptional repressors compete with DELLAs to prevent DELLA-mediated transcriptional activation through IDDs. (b) The inhibition of PIF4 by DELLAs is relieved by BBX24 in the context of shade avoidance. (c) BOIs promote DELLA-mediated transcriptional activation.

involves direct competition between DELLAs and MYC2 for binding to the JAZ protein. A similar mechanism operates in other physiological contexts, including coordination of JA-regulated defence and growth responses (Yang *et al.*, 2012). It is competition between JAZ9 and PIF3 for DELLA binding that provides the mechanism for these responses. When defence is a priority, JAZ proteins are degraded in response to JA and defence responses activated. The lack of JAZ proteins allows DELLAs to bind to and inactivate PIFs, resulting in growth repression. One could envision a very dynamic and responsive mechanism in which environmental signals from light, temperature or the presence of pathogens influences the relative levels of DELLA and/or JAZ through the GA and JA pathways, which in turn regulate the activity of a plethora of interacting TFs to trigger the appropriate physiological response. The set of TFs targeted will likely be different depending on the cell type. This mechanism contrasts with the synergistic inactivation of GL1/GL3/EGL3, or MYC2, by DELLAs and JAZ to regulate volatiles production in flowers, as described above.

The variety of classes of DELLA interactors is also reflected in the different transcriptional regulators that have been identified recently. Besides the JAZ proteins, several other unrelated transcriptional regulators have been identified in *Arabidopsis*. The double B-box BBX24 (Crocco *et al.*, 2015) interacts with DELLAs away from chromatin, whereas the interaction with the RING finger proteins BOTRYTIS SUSCEPTIBLE1 INTERACTORs (BOIs) occurs within the vicinity of target promoters (Park *et al.*, 2013). The interaction between BBX24 and DELLAs appears to be involved in the coordination of growth responses in response to shade avoidance (Figure 7.4b; Crocco *et al.*, 2015). It involves a similar mechanism to the JAZ-DELLA-PIF regulatory module, with BBX24 competing with PIF4 for binding to DELLA. It has been speculated that an early effect of the shade response is to promote the BBX24-DELLA interaction, thereby allowing PIF4 to activate downstream target genes that promote hypocotyl elongation. It is currently unclear how this early response to shade modulates the BBX-DELLA interaction.

In contrast to the situation with BBX24, the activity of the BOIs is required by DELLAs to regulate several GA responses, including germination, and the juvenile-to-adult and floral transitions (Figure 7.4c; Park *et al.*, 2013). Furthermore, BOI has been shown to act as an E3 ubiquitin ligase due to the presence of a RING domain (Luo *et al.*, 2010). This suggests that it may ubiquitinate TFs or other proteins involved in the regulation of GA-responsive genes to suppress their activity. Despite the presence of the RING domain and their functional importance in repressing GA responses, it is still conceivable that the role of BOIs in regulating transcription is unrelated to their E3 ligase activity. For example, DELLAs and BOIs might interact and together recruit the TFs or coregulators of transcription, allowing their association with chromatin and initiating downstream transcriptional events that promote growth responses.

7.7.4 DELLAs regulate chromatin dynamics

The activity of TFs must ultimately allow the assembly of the basal transcriptional machinery to initiate transcription of the target genes. This is usually achieved by targeting chromatin remodelling and/or histone modifying complexes to the promoter through interaction with TFs (Clapier and Cairns,

2009). It is reasonable to think that all DELLA-TF interactions described above result indirectly in modifications of the epigenetic landscape at the target promoters, whether DELLAs sequester the TF or act bound to the chromatin, ultimately changing gene expression. Remarkably, a recent study in *Arabidopsis* demonstrates that DELLAs can directly regulate epigenetic mark deposition through their physical association with PICKLE (PKL), a chromatin remodelling protein of the CHD3 type (Zhang *et al.*, 2014). In this study, it is demonstrated that PKL is targeted to genes promoting cell elongation, such as *IAA19* and *PRE1*, through an interaction with PIF3, BZR1 or both, in etiolated seedlings. This targeting of PKL reduces trimethylation of lysine 27 of histone 3 (H3K27me3) at these promoters, which would normally act as a repressive mark. The effect of the DELLA interaction with PKL is therefore to increase the H3K27me3 mark in the promoters of *IAA19* and *PACLOBUTRAZOL RESISTANCE 1* (*PRE1*) and reduce cell elongation. Together with the direct regulation of PIF3 and BZR1 through their association with DELLAs, this mechanism increases the negative effect that DELLAs exert on cell elongation. Thus, the DELLA-PKL interaction represents an extra layer of regulation coupled to the PIF-BZR1 module (Bai *et al.*, 2012). It seems likely that this form of epigenetic regulation will also affect the activity of the BAP module (Oh *et al.*, 2014). Whether the DELLA-PKL interaction occurs in physiological contexts distinct from the regulation of skotomorphogenesis requires further investigation. It has recently been reported that RGL2 and RGL3 interact with the catalytic sub-unit SWI3C of the SWITCH/SUCROSE NONFERMENTING (SWI/SNF) chromatin remodeller, suggesting another potential mechanism for epigenetic regulation by DELLAs (Sarnowska *et al.*, 2013). However, the physiological significance of these interactions has yet to be established.

7.8 A non-genomic response regulated by DELLAs

All DELLA interactions described above are related to their role as transcriptional regulators. However, the identification of prefoldin 5 (PFD5) and PFD3 as partners of DELLA provided hints of a non-genomic role for DELLAs (Locascio *et al.*, 2013). These proteins are part of the heterohexameric PFD complex that is formed by sub-units PFD1 to 6. This complex is conserved from yeast to humans and functions as a cochaperonin in the cytosol, presenting nascent polypeptides of actins and tubulins to the molecular chaperone CCT (Vainberg *et al.*, 1998). The activity of the PFD complex controls the formation of the α/β-tubulin heterodimers that are the forming units of the microtubules (MTs). Indeed, several mutants of *Arabidopsis* affecting sub-units of the complex have defects in MT organization (Gu *et al.*, 2008; Rodriguez-Milla and Salinas, 2009). It is generally accepted that the MT organization determines the cell growth direction, in such a way that transverse MTs allow anisotropic cell elongation, whereas random or longitudinal MT arrays prevent it (Lloyd, 2011). It has been known for several years that GAs are able to promote the transverse organization of MTs in epidermal cells (Wenzel *et al.*, 2000). The mechanism for this role of GAs has remained, however, unknown. The DELLA-PFD interaction has provided an explanation (Locascio *et al.*, 2013). The PFD localizes to the nucleus upon interaction with DELLA proteins and this has immediate consequences for the cytosolic function of the complex. The amount of properly folded α/β-tubulins heterodimers drops, likely causing the disorganization of MTs that prevents anisotropic growth. Thus, the MT organization is indirectly regulated by GA levels through the DELLA-PFD interaction. The regulation of PFD by DELLAs seems to fluctuate daily, thus contributing to restricting the maximum growth rate of seedlings towards the end of the night (Arana *et al.*, 2011; Locascio *et al.*, 2013). Interestingly, a role for the yeast PFD complex in the nucleus has been recently described, showing that it participates in transcription elongation of long genes (Millán-Zambrano *et al.*, 2013). Given the conservation of the PFD in different organisms, a similar role for the plant counterpart could be envisaged.

7.9 Analysis of DELLA protein structure-function

An understanding of the structure-function relationship for DELLA proteins is slowly emerging through a combination of genetic analysis, protein structure-function studies and the identification of interacting downstream components. As discussed earlier, DELLAs belong to one of ten predicted sub-groups of the GRAS family of plant transcriptional regulators (Sun *et al.*, 2012). Members within these different sub-groups have been demonstrated to have diverse roles in controlling plant growth and development, including root radial patterning and elongation, shoot apical meristem maintenance, formation of nodules in legumes, axillary meristem development and light signalling. Proteins belonging to this family have a highly conserved C-terminal GRAS domain that is important for functionality and suggests a common mode of action. Within the GRAS domain there are five highly conserved motifs: LHR1, VHIID, LHR2, PFYRE and SAW (Figure 7.5). The sub-groups of the GRAS family share common N-terminal motifs that are predicted to occur within intrinsically disordered regions of the protein (Sun *et al.*, 2012). In the case of the DELLA sub-group the common N-terminal motif includes the DELLA, LEXLE and TVHYNP domains that are required for binding to the GID1-GA complex (Murase *et al.*, 2008). Extensive structure-function studies of the GA signalling components SLR1, GID1 and GID2 in rice have further extended our understanding of the mechanisms controlling GA-induced DELLA degradation (Hirano *et al.*, 2010; see also Chapter 6). Based on these studies, a model has emerged in which binding of GID1-GA to the N-terminal domains within SLR1 is predicted to induce a conformational change that allows stabilisation of the SLR1-GID1-GA

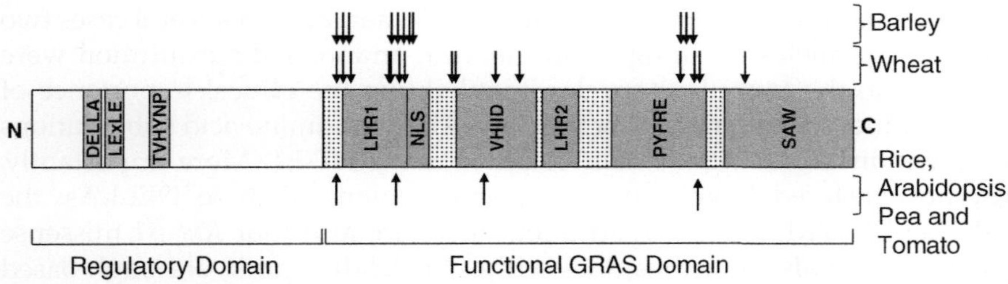

Figure 7.5 Conserved domains in DELLA proteins and sites affecting functionality. The position of mutation sites of reduced- and loss-of-function mutants are indicated by arrows. The plant species that are the source of the mutations are indicated on the right. See also Figure 6.2 for information on the regions within DELLAs that are necessary for their post-translational modification. (Adapted from Chandler and Harding, 2013).

complex through an additional interaction between GID1 and the GRAS domain of SLR1. This secondary interaction with GID1 involves the VHIID, PYFRE and SAW domains within the C-terminus of SLR1. The stabilised SLR1-GID1-GA complex is then recognised and bound by the F-Box component, GID2, presumably facilitating ubiquitination through the associated E3 Ubiquitin ligase activity (Hirano *et al.*, 2010). Multiple regions within the GRAS domain are required for SLR1 binding to GID2, but the VHIID and LHR2 domains appear to have a major role. A less well characterised function of the GRAS domain is to facilitate dimerization. Although this has yet to be established *in planta*, self-interaction studies using SLR1 or RGA in yeast have demonstrated that homodimerisation occurs through the LHR1 domain (Itoh *et al.*, 2002; Bai *et al.*, 2012). Given the importance of this domain in facilitating the interaction with downstream components (see below) it will be intriguing to establish whether this represents another mechanism for regulating DELLA activity.

The characterisation of reduced or loss-of-function *della* mutants has provided some important insights into the domains that are required for conferring the ability to repress downstream GA responses. It is apparent that an almost complete GRAS domain is necessary for this functionality, with mutations that lead to very short C-terminal truncations causing complete loss of activity (the *slr1-4* allele encodes a protein lacking six residues from the C-terminus; Ikeda *et al.*, 2001). Whilst much of our knowledge of the GA signalling pathway has come from the study of the model plants rice and *Arabidopsis*, an extremely extensive collection of *della* mutants have recently been identified from more genetically challenging sources. Using suppressor screens of barley and wheat GA biosynthesis and signalling mutants, Chandler and Harding (2013) identified over-growth mutants that exhibited increased leaf and stem elongation. In the majority of alleles, the mutations were demonstrated to occur in the barley and wheat *DELLA* genes, *SLN1* and *Rht-B1*, respectively. A total of 26 independent *della* missense mutations were identified in these screens (Figure 7.5). Remarkably, in several cases two independent alleles resulting in an identical amino acid substitution were identified in the two plant species, highlighting the critical importance of these residues in conferring functionality. All of the amino acid substitutions occur within the GRAS domain of SLN1 and RHT-B1. More importantly, they are clustered largely in two specific regions of these DELLAs: the LHR1 and PFYRE motifs. In addition, there are also four *Rht-B1* missense mutations that also affect residues within the VHIID motif, although based on the suppression of stem elongation in these mutants their impact on reducing DELLA functionality appears to be less. Other *della* missense mutations in *Arabidopsis* (*rga-2*; Silverstone *et al.*, 1998), pea (*cry-c*; Weston *et al.*, 2008), rice (*slr1-6*; (Hirano *et al.*, 2012) and tomato (*procera*; Bassel *et al.*, 2008) have also been demonstrated to affect residues within the LHR1, PFYRE and VHIID motifs. In addition to these domains identified from studying *della* mutants, an alanine-scanning approach in transgenic rice has also implicated the LHR2 motif in controlling the repressive activity of SLR1 (Hirano *et al.*, 2010).

In many of the cases in which DELLAs have been demonstrated to modulate the activity of transcriptional regulators through direct protein–protein contact, these findings have been based on yeast two-hybrid screens that have uncovered these interactions (For examples see: de Lucas *et al.*, 2008; Hou *et al.*, 2010; Gallego-Bartolomé *et al.*, 2012; Fukazawa *et al.*, 2014; Yoshida *et al.*, 2014). These assays have subsequently allowed the DELLA-interaction domains to be interrogated by deletion analysis. In general, these studies have highlighted the importance of the GRAS domain in mediating the interactions. Only in the case of the interaction with JAZ1 is there any evidence that the N-terminus of DELLA has any role (Hou *et al.*, 2010). However, this is not supported by similar interaction studies between JAZ9 and GAI, which identify only the GRAS domain as being required (Yang *et al.*, 2012). Further investigations are necessary to clarify the nature of DELLA-JAZ interaction domains. The finding that the GRAS domain is largely responsible for the interaction with multiple transcriptional regulators is consistent with the genetic evidence that supports its functional role in regulating downstream GA responses. What is not currently clear is whether all of the characterised DELLA interactors bind to a conserved binding site within the GRAS domain. Deletion of the LHR1 domain has been shown to abolish the interaction with BZR1, MYC2, JAZ1 and EIN3 in yeast (Hou *et al.*, 2010; An *et al.*, 2012; Gallego-Bartolomé *et al.*, 2012; Hong *et al.*, 2012a). In addition, C-terminal deletion of the SAW and/or PFYRE domains also abolished interactions with BZR1, MYC2 and EIN3 (An *et al.*, 2012; Gallego-Bartolomé *et al.*, 2012; Hong *et al.*, 2012a). An exception was the interaction between JAZ1 and RGA, which did not appear to require 180 residues of the C-terminal GRAS domain (Hou *et al.*, 2010). Because DELLAs have been shown to interact with IDD proteins to directly activate the expression of GA signalling components to control homoeostasis of this pathway, it raised the prospect that this association might shed some light on the proposed transactivation role of the N-terminus (Fukazawa *et al.*, 2014; Yoshida *et al.*, 2014b). However, interaction studies in yeast have also demonstrated that it is only the GRAS domain, including the LHR1 and PFYRE/SAW domains, that is required for this association. It therefore seems likely that the components and mechanisms responsible for controlling the transactivation activity of the N-terminal domain of DELLAs are yet to be determined. Within the GRAS domain it is clear that multiple regions are important for the interactions with downstream components, but there is still considerable uncertainty about the precise identity of these regions. By investigating how amino acid substitutions that alter DELLA function affect these interactions will further refine the extent of these domains. There is the possibility that different regions of the GRAS domain are important for interaction with specific partners that control different growth responses. At present, the genetic and biochemical evidence for this is limited. However, characterisation of leaf elongation and α-amylase induction in the aleurone in barley and wheat over-growth mutants did provide some evidence that the regions of the GRAS domain controlling these GA responses are different (Chandler and Harding, 2013). The identification of downstream components responsible for controlling these GA responses should provide us with further clues regarding the specificity of the GRAS domain. The ultimate goal will be the elucidation of the 3-D crystal structures of these binary complexes, which will establish the precise structural requirements that specify these interactions. The recent expression and purification of recombinant SLR1 provides an important step towards this goal (Sato *et al.*, 2014).

7.10 GAMYB: A transcriptional regulator of gibberellin responses during cereal grain germination and pollen development

During the germination of a cereal grain, GAs, which are released from the imbibed embryo, act on aleurone cells to promote the secretion of hydrolytic enzymes into the endosperm (Jacobsen *et al.*, 1995). This produces the nutrients and other components required to sustain heterotrophic growth of the germinating seedling until it reaches a stage and environment that allows auxotrophic growth. The cereal aleurone cells provided some important advantages over other systems for studying GA response: (1) these cells do not synthesize GAs that could interfere with response assays, (2) substantial increases in the expression of genes encoding hydrolytic enzymes, including α-amylases, which breakdown starch reserves within the endosperm, are rapidly induced by GA providing convenient reporters for monitoring responses, (3) aleurone cells can be easily separated from other grain tissues in the form of protoplasts or de-embryonated half seeds, allowing GA response to be monitored in a single cell type and (4) transient transformations can be performed on aleurone cells allowing the dissection of the signalling pathways responsible for the expression of hydrolytic enzymes. These factors have led to a good understanding of the mechanisms controlling GA-regulated expression of α-amylase genes, although there are some important gaps in this knowledge. In light of the extensive research carried out in this area, we are unable to provide a comprehensive review of the literature. The reader is therefore directed to some excellent articles, which provide a more comprehensive review of earlier studies (Jacobsen *et al.*, 1995; Bethke *et al.*, 1997; Sun and Gubler, 2004). In this section we will mainly focus on recent developments in the field.

7.10.1 GAMYB positively regulates gene expression in cereal aleurone cells

In cereal aleurone cells the GA induction of α-amylase activity has been studied extensively over the last 50 years (reviewed in Jacobsen *et al.*, 1995 and Sun and Gubler, 2004). Up-regulation of α-amylase gene expression is observed after around 3 to 6 hours following a GA treatment to barley aleurone cells (Chandler *et al.*, 1984; Gubler *et al.*, 2002). Analysis of the α-*Amy1* promoter has identified several *cis*-elements that are required for the transcriptional control by GA and ultimately led to the identification of multiple transcriptional regulators that have a regulatory role (Sun and Gubler, 2004). Most notably, the R2R3 MYB TF GAMYB has been demonstrated to bind to the GA response element (GARE) of the α-amylase promoter and activate expression in barley (Gubler *et al.*, 1995; Gubler *et al.*, 1999). In rice, a role for GAMYB in this process has been confirmed by the absence of GA-induced α-amylase expression in a *gamyb* loss-of-function mutant (Kaneko *et al.*, 2004). Other *cis*-elements that have been identified to have a role in the GA-regulated expression of the α-*Amy1* gene include the pyrimidine box and the TATCCA (TA) box (reviewed in Sun and Gubler, 2004). In addition to being regulated by hormonal signals, α-amylase genes within the aleurone are also regulated by sugar and nutrient starvation signals, allowing coordination of endosperm reserve mobilisation to facilitate optimal seedling establishment. The TA box has been demonstrated to be an important element for the induction of α-amylase expression in response to sugar starvation. Activation of α-amylase expression is induced under sugar starvation by another MYB TF, MYBS1. Identified in rice, MYBS1 is sugar repressible and activates α-amylase expression through binding to the TA-box element (Hong *et al.*, 2012b). The importance of the TA box in regulating both GA and sugar starvation responses has highlighted an important mechanism of cross-talk between these signalling pathways (see below).

The mechanisms through which GA signalling activates GAMYB to promote the expression of direct downstream response genes in the aleurone are still not entirely clear. It has been demonstrated in barley and rice that

DELLAs act as upstream repressors of this response (Ikeda *et al.*, 2001; Gubler *et al.*, 2002). In barley aleurone cells, GA-mediated degradation of SLN1 protein is complete within 10 minutes of a GA treatment (Gubler *et al.*, 2002). A subsequent increase in *GAMYB* transcript levels is then observed around 1 to 2 hours later and is thought to be responsible for the induction in α-amylase gene expression. However, recent studies have suggested that GA signalling also causes changes in the sub-cellular localisation of GAMYB (Hong *et al.*, 2012b). Based on the study of a GAMYB-GFP reporter in transiently transformed barley aleurone cells, a 24-hour treatment with GA was demonstrated to promote GAMYB-GFP redistribution from the cytoplasm to the nucleus. If this sub-cellular redistribution is required for the activation of downstream GA-response genes it will be important to establish the kinetics of GA-mediated GAMYB nuclear redistribution, to determine whether it correlates with activation of α-amylase gene expression. Interestingly, the same study by Hong and colleagues (2012b) demonstrated that nucleocytoplasmic partitioning of MYBS1 is also important for mediating aleurone transcriptional responses to both nutrient starvation and GA signalling. When nutrients, including glucose, phosphate and nitrogen, are available, MYBS1 is located predominantly in the cytoplasm through active nuclear export. Under nutrient starvation a partial redistribution of MYBS1 to the nucleus occurs, allowing it to bind to the TA-box elements within the promoters of target genes resulting in their activation for resource mobilisation. Acting through GAMYB, GAs enhance the activation of MYBS1 target genes in both the presence and absence of nutrients. An increase in GAMYB levels in response to GA signalling facilitates the nuclear import of a GAMYB-MYBS1 complex. Within the nucleus, binding of GAMYB and MYBS1 to the GARE and TA box, respectively, results in the strong activation of target gene expression that is necessary for resource mobilisation and seedling establishment. It has been hypothesized that this cross-talk between nutrient and GA signalling, mediated by these two classes of MYB TFs, is important for the environmental regulation of the transition between the dormant seed and actively growing seedling (Hong *et al.*, 2012b). However, further studies to confirm the physiological importance of this cross-talk mechanism are still required.

A major gap in our knowledge of GA signalling within the aleurone cells is an understanding of how DELLA proteins act to repress GAMYB. In barley, the lag between SLN1 degradation and an increase in GAMYB transcript levels suggests that the regulation of transcription is unlikely to be direct (Gubler *et al.*, 2002). It is conceivable that SLN1 represses an activator of GAMYB expression. Alternatively, it is tempting to speculate that SLN1 may regulate the nucleocytoplasmic distribution of GAMYB, by inhibiting its nuclear import. Although there is currently limited evidence for such a form of regulation, if established, it could provide an explanation as to why the presence of GAMYB transcripts in the absence of GA signalling does not result in expression of downstream target genes (Gubler *et al.*, 1995;

Gubler *et al.*, 2002). It is likely that DELLAs have multiple targets to regulate different aleurone responses.

The recent demonstration that DELLAs interact with the ABA signalling components ABI3 and ABI5 in *Arabidopsis* (Lim *et al.*, 2013; see above) provides a potential mechanism for the well-characterised GA and ABA cross-talk within this cell type. In cereals, ABI5- and ABI3-related TFs have been identified and demonstrated to regulate aleurone responses (Casaretto and Ho, 2003; Johnson *et al.*, 2008; Harris *et al.*, 2013). Understanding whether these components represent direct activation targets for DELLAs could provide some important insights into the mechanisms controlling ABA and GA cross-talk during this response.

7.10.2 GAMYB regulates gibberellin-dependent anther development

The importance of GAMYB in controlling downstream GA responses is not restricted to the control of endosperm mobilisation in cereal grains. As discussed in more detail in Chapter 11, GAMYB also functions as a TF that is essential for promoting GA-dependent anther development required for the formation of viable pollen in angiosperms (Murray *et al.*, 2003; Kaneko *et al.*, 2004; Millar and Gubler, 2005; Aya *et al.*, 2009). This role, which appears restricted to the tapetum cell layer, involves the promotion of both tapetal programmed cell death (PCD) and pollen wall biogenesis, processes that are essential for pollen development (Aya *et al.*, 2009). Potential genes that control these developmental processes have been identified in rice anthers and demonstrated to be acting downstream of the GA-GID1-DELLA-GAMYB regulatory system. These include genes encoding proteases or protease inhibitors and putative sporopollenin biosynthetic genes that have been implicated in controlling tapetal PCD and pollen exine production, respectively. Notably, many of the genes that are positively regulated by GAMYB contain GAMYB binding motifs within their promoters, suggesting that they are direct transcriptional targets. Aya and colleagues (2009) have provided strong evidence supporting this model by demonstrating that GAMYB directly regulates the sporopollenin biosynthesis genes, CYP703A3 and KAR, through its interaction with these promoter binding sites. In view of the fact that aleurone and tapetal cells undergo similar developmental programmes involving targeted secretion/reserve mobilisation and ultimately PCD in response to GA signalling, it is tempting to speculate that DELLA inhibition of GAMYB occurs through a similar mechanism. Interestingly, recent work suggests that the GA-GID1-DELLA regulatory system evolved in early vascular plants to regulate aspects of reproductive development through its control of GAMYB (Aya *et al.*, 2011). Therefore, understanding how DELLAs regulate GAMYB is of key significance for understanding how the GA signalling pathway evolved to have such an important role controlling plant development.

7.11 Concluding remarks

During the last eight years we have gained a considerable understanding of the molecular mechanisms by which DELLA proteins act to control a multitude of cellular responses. As well as providing important insights into how GAs control plant growth and development, these studies have also uncovered some of the mechanisms responsible for controlling cross-talk with other phytohormone and environmental signalling pathways. It is perhaps not surprising that these recent studies have established that DELLAs control transcription through their association with other transcriptional regulators; however the diversity of protein classes which these include and the different transcriptional regulatory mechanisms employed is truly remarkable (Figure 7.6). DELLAs have therefore emerged as important signalling hubs, true connectors modulating the activity of the interacting transcriptional regulators and therefore the signalling pathways in which they participate. It is likely that the extent of these connections will expand substantially as our knowledge of these growth repressors improves. Therefore, to fully understand the implications on plant growth and development it will be necessary to employ systems-biology-based approaches. Furthermore, because DELLAs are essentially a reporter for the level of bioactive GAs, it will also be important to improve our understanding of

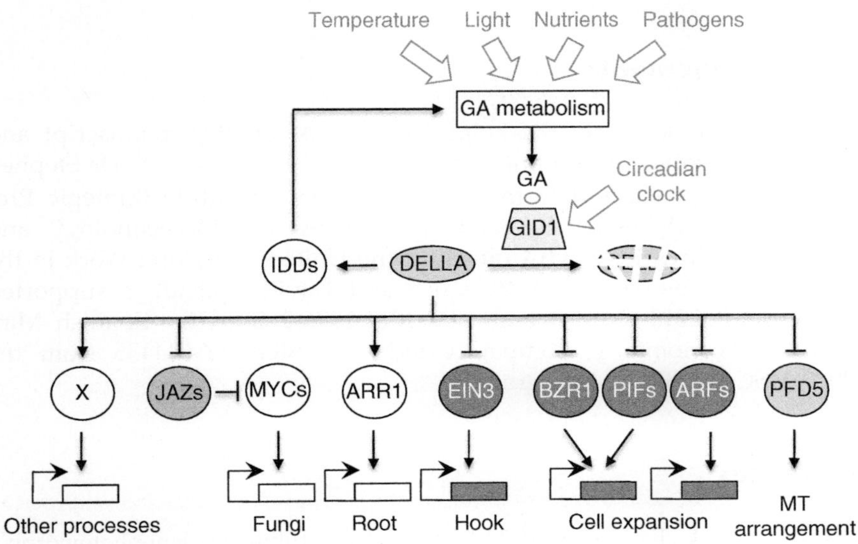

Figure 7.6 DELLA proteins act as hubs. Different environmental cues regulate GA metabolism, eventually affecting DELLA stability. DELLA levels thus integrate environmental conditions and relay this information to prewired transcriptional networks and other cellular processes, facilitating the most appropriate adaptive response.

how developmental and environmental signals control GA accumulation at the cellular level.

Perhaps the most surprising recent finding is that DELLAs have modes of action that are distinct from controlling transcription. In addition to controlling microtubule reorientation, a recent study in rice has provided evidence that SLR1 interacts with the strigolactone receptor, D14, in a strigolactone-dependent manner (Nakamura *et al.*, 2013). At present the physiological significance of this interaction is yet to be established. Nevertheless, it highlights the need to keep an open mind when investigating DELLA function.

Not only have recent insights into DELLA functionality helped to improve our understanding of how GAs control plant growth and development, they also have important implications for studying how other GRAS proteins function. In light of the sequence homology within their functional GRAS domains, it is likely that all members of this family function as hub proteins that interact with a myriad of downstream components. Arguably, our knowledge of how DELLAs act has progressed more rapidly than for other GRAS proteins because we know that they are subjected to receptor-mediated degradation by a process involving their N-terminal domain (see above; see also Chapter 6). As most other GRAS proteins contain N-terminal domains that are predicted to serve as interaction domains (Sun *et al.*, 2011; Sun *et al.*, 2012), it will be fascinating to see whether receptor-mediated degradation is a common mechanism through which they are regulated.

Acknowledgements

We thank Camille Steber for helpful comments on this manuscript and John Ross for kindly providing the image used in Figure 7.1. Dr Stephen Thomas receives support from the 20:20 Wheat® Institute Strategic Programme at Rothamsted Research funded by the Biotechnology and Biological Sciences Research Council of the United Kingdom. Work in the laboratory of Drs Miguel A Blázquez and David Alabadí is supported by grants BIO2013-43184-P and RTC-2014-2876-2 from the Spanish Ministerio de Economía y Competitividad, and SIGNAT-644435 from the H2020-MSCA-RISE-2014 program.

References

Achard, P., Liao, L., Jiang, C. *et al.* (2007). DELLAs contribute to plant photomorphogenesis. *Plant Physiology* **143**, 1163–1172.

Achard, P., Gong, F., Cheminant, S. *et al.* (2008). The cold-inducible CBF1 factor-dependent signaling pathway modulates the accumulation of the growth-repressing DELLA proteins via its effect on gibberellin metabolism. *The Plant Cell* **20**, 2117–2129.

Achard, P., Gusti, A., Cheminant, S. *et al.* (2009). Gibberellin signaling controls cell proliferation rate in *Arabidopsis*. *Current Biology* **19**, 1188–1193.

Al-Sady, B., Ni, W., Kircher, S. *et al.* (2006). Photoactivated phytochrome induces rapid PIF3 phosphorylation prior to proteasome-mediated degradation. *Molecular Cell* **23**, 439–446.

Alabadí, D., Gil, J., Blazquez, M.A. and Garcia-Martinez, J.L. (2004). Gibberellins repress photomorphogenesis in darkness. *Plant Physiology* **134**, 1050–1057.

Alabadí, D., Gallego-Bartolomé, J., Orlando, L. *et al.* (2008). Gibberellins modulate light signaling pathways to prevent *Arabidopsis* seedling de-etiolation in darkness. *The Plant Journal* **53**, 324–335.

An, F., Zhang, X., Zhu, Z. *et al.* (2012). Coordinated regulation of apical hook development by gibberellins and ethylene in etiolated *Arabidopsis* seedlings. *Cell Research* **22**, 915–927.

Arana, M.V., Marín-de la Rosa, N., Maloof, J.N. *et al.* (2011). Circadian oscillation of gibberellin signaling in *Arabidopsis*. *Proceedings of the National Academy of Sciences USA* **108**, 9292–9297.

Ariizumi, T., Murase, K., Sun, T.P. and Steber, C.M. (2008). Proteolysis-independent downregulation of DELLA repression in *Arabidopsis* by the gibberellin receptor GIBBERELLIN INSENSITIVE DWARF1. *The Plant Cell* **20**, 2447–2459.

Arnaud, N., Girin, T., Sorefan, K. *et al.* (2010). Gibberellins control fruit patterning in *Arabidopsis* thaliana. *Genes and Development* **24**, 2127–2132.

Asano, K., Hirano, K., Ueguchi-Tanaka, M. *et al.* (2009). Isolation and characterization of dominant dwarf mutants, *Slr1-d*, in rice. *Molecular Genetics and Genomics* **281**, 223–231.

Aya, K., Ueguchi-Tanaka, M., Kondo, M. *et al.* (2009). Gibberellin modulates anther development in rice via the transcriptional regulation of GAMYB. *The Plant Cell* **21**, 1453–1472.

Aya, K., Hiwatashi, Y., Kojima, M. *et al.* (2011). The Gibberellin perception system evolved to regulate a pre-existing GAMYB-mediated system during land plant evolution. *Nature Communications* **2**, 544.

Bai, M.Y., Shang, J.X., Oh, E. *et al.* (2012). Brassinosteroid, gibberellin and phytochrome impinge on a common transcription module in *Arabidopsis*. *Nature Cell Biology* **14**, 810–817.

Bassel, G.W., Mullen, R.T. and Bewley, J.D. (2008). *procera* is a putative DELLA mutant in tomato (*Solanum lycopersicum*): effects on the seed and vegetative plant. *Journal of Experimental Botany* **59**, 585–593.

Bethke, P.C., Schuurink, R. and Jones, R.L. (1997). Hormonal signalling in cereal aleurone. *Journal of Experimental Botany* **48**, 1337–1356.

Bou-Torrent, J., Galstyan, A., Gallemi, M. *et al.* (2014). Plant proximity perception dynamically modulates hormone levels and sensitivity in *Arabidopsis*. *Journal of Experimental Botany*, **65**, 2937–2947.

Brian, P.W. (1957). The effects of some microbial metabolic products on plant growth. *Symposia of the Society for Experimental Biology* **11**, 166–182.

Cao, D.N., Cheng, H., Wu, W. *et al.* (2006). Gibberellin mobilizes distinct DELLA-dependent transcriptomes to regulate seed germination and floral development in *Arabidopsis*. *Plant Physiology* **142**, 509–525.

Casaretto, J. and Ho, T.H. (2003). The transcription factors HvABI5 and HvVP1 are required for the abscisic acid induction of gene expression in barley aleurone cells. *The Plant Cell* **15**, 271–284.

Chandler, P.M. and Harding, C.A. (2013). 'Overgrowth' mutants in barley and wheat: new alleles and phenotypes of the 'Green Revolution' *Della* gene. *Journal of Experimental Botany* **64**, 1603–1613.
Chandler, P.M., Zwar, J.A., Jacobsen, J.V. *et al.* (1984). The effects of gibberellic acid and abscisic acid on alpha-amylase mRNA levels in barley aleurone layers studies using an alpha-amylase cDNA clone. *Plant Molecular Biology* **3**, 407–418.
Chandler, P.M., Marion-Poll, A., Ellis, M. and Gubler, F. (2002). Mutants at the *Slender1* locus of barley cv Himalaya. Molecular and physiological characterization. *Plant Physiology* **129**, 181–190.
Cheminant, S., Wild, M., Bouvier, F. *et al.* (2011). DELLAs regulate chlorophyll and carotenoid biosynthesis to prevent photooxidative damage during seedling deetiolation in *Arabidopsis*. *The Plant Cell* **23**, 1849–1860.
Cheng, H., Qin, L., Lee, S. *et al.* (2004). Gibberellin regulates *Arabidopsis* floral development via suppression of DELLA protein function. *Development* **131**, 1055–1064.
Chien, J.C. and Sussex, I.M. (1996). Differential regulation of trichome formation on the adaxial and abaxial leaf surfaces by gibberellins and photoperiod in *Arabidopsis thaliana* (L.) Heynh. *Plant Physiology* **111**, 1321–1328.
Chini, A., Fonseca, S., Fernandez, G. *et al.* (2007). The JAZ family of repressors is the missing link in jasmonate signalling. *Nature* **448**, 666-671.
Clapier, C.R. and Cairns, B.R. (2009). The biology of chromatin remodeling complexes. *Annual Review of Biochemistry* **78**, 273–304.
Colebrook, E.H., Thomas, S.G., Phillips, A.L. and Hedden, P. (2014). The role of gibberellin signalling in plant responses to abiotic stress. *Journal of Experimental Biology* **217**, 67–75.
Crocco, C.D., Locascio, A., Escudero, C.M. *et al.* (2015). The transcriptional regulator BBX24 impairs DELLA activity to promote shade avoidance in *Arabidopsis* thaliana. *Nature Communications* **6**, 6202.
Daviere, J.M., Wild, M., Regnault, T. *et al.* (2014). Class I TCP-DELLA interactions in inflorescence shoot apex determine plant height. *Current Biology* **24**, 1923–1928.
de Lucas, M., Daviere, J.M., Rodriguez-Falcon, M. *et al.* (2008). A molecular framework for light and gibberellin control of cell elongation. *Nature* **451**, 480–484.
Dill, A. and Sun, T. (2001). Synergistic derepression of gibberellin signaling by removing RGA and GAI function in *Arabidopsis thaliana*. *Genetics* **159**, 777–785.
Dill, A., Jung, H.S. and Sun, T.P. (2001). The DELLA motif is essential for gibberellin-induced degradation of RGA. *Proceedings of the National Academy of Sciences USA* **98**, 14162–14167.
Dill, A., Thomas, S.G., Hu, J. *et al.* (2004). The *Arabidopsis* F-box protein SLEEPY1 targets gibberellin signaling repressors for gibberellin-induced degradation. *The Plant Cell* **16**, 1392–1405.
Djakovic-Petrovic, T., de Wit, M., Voesenek, L.A. and Pierik, R. (2007). DELLA protein function in growth responses to canopy signals. *The Plant Journal* **51**, 117–126.
Engstrom, E.M. (2011). Phylogenetic analysis of GRAS proteins from moss, lycophyte and vascular plant lineages reveals that GRAS genes arose and underwent substantial diversification in the ancestral lineage common to bryophytes and vascular plants. *Plant Signaling and Behavior* **6**, 850–854.
Feng, S., Martinez, C., Gusmaroli, G. *et al.* (2008). Coordinated regulation of *Arabidopsis thaliana* development by light and gibberellins. *Nature* **451**, 475–479.
Feurtado, J.A., Huang, D., Wicki-Stordeur, L. *et al.* (2011). The *Arabidopsis* C2H2 zinc finger INDETERMINATE DOMAIN1/ENHYDROUS promotes the transition to

germination by regulating light and hormonal signaling during seed maturation. *The Plant Cell* **23**, 1772–1794.

Fleishon, S., Shani, E., Ori, N. and Weiss, D. (2011). Negative reciprocal interactions between gibberellin and cytokinin in tomato. *New Phytologist* **190**, 609–617.

Foreman, J., Johansson, H., Hornitschek, P. *et al.* (2011). Light receptor action is critical for maintaining plant biomass at warm ambient temperatures. *The Plant Journal* **65**, 441–452.

Foster, C.A. (1977). Slender: an accelerated extension growth mutant of barley. *Barley Genetics Newsletter* **7**, 24–27.

Fu, X. and Harberd, N.P. (2003). Auxin promotes *Arabidopsis* root growth by modulating gibberellin response. *Nature* **421**, 740–743.

Fukao, T. and Bailey-Serres, J. (2008). Submergence tolerance conferred by *Sub1A* is mediated by SLR1 and SLRL1 restriction of gibberellin responses in rice. *Proceedings of the National Academy of Sciences USA* **105**, 16814–16819.

Fukazawa, J., Teramura, H., Murakoshi, S. *et al.* (2014). DELLAs function as coactivators of GAI-ASSOCIATED FACTOR1 in regulation of gibberellin homeostasis and signaling in *Arabidopsis*. *The Plant Cell* **26**, 2920–2938.

Gale, M.D. and Marshall, G.A. (1973). Insensitivity to gibberellin in dwarf wheats. *Annals of Botany* **37**, 729–735.

Gallego-Bartolomé, J., Alabadí, D. and Blázquez, M.A. (2011a). DELLA-induced early transcriptional changes during etiolated development in *Arabidopsis thaliana*. *PLoS ONE* **6**, e23918.

Gallego-Bartolomé, J., Arana, M.V., Vandenbussche, F. *et al.* (2011b). Hierarchy of hormone action controlling apical hook development in *Arabidopsis*. *The Plant Journal* **67**, 622–634.

Gallego-Bartolomé, J., Kami, C., Fankhauser, C. *et al.* (2011c). A hormonal regulatory module that provides flexibility to tropic responses. *Plant Physiology* **156**, 1819–1825.

Gallego-Bartolomé, J., Minguet, E.G., Grau-Enguix, F. *et al.* (2012). Molecular mechanism for the interaction between gibberellin and brassinosteroid signaling pathways in *Arabidopsis*. *Proceedings of the National Academy of Sciences USA* **109**, 13446–13451.

Gallego-Bartolome, J., Minguet, E.G., Marin, J.A. *et al.* (2010). Transcriptional diversification and functional conservation between DELLA proteins in *Arabidopsis*. *Molecular Biology and Evolution* **27**, 1247–1256.

Galvao, V.C., Horrer, D., Kuttner, F. and Schmid, M. (2012). Spatial control of flowering by DELLA proteins in *Arabidopsis thaliana*. *Development* **139**, 4072–4082.

Griffiths, J., Murase, K., Rieu, I. *et al.* (2006). Genetic characterization and functional analysis of the GID1 gibberellin receptors in *Arabidopsis*. *Plant Cell* **18**, 3399–3414.

Gu, Y., Deng, Z., Paredez, A.R. *et al.* (2008). Prefoldin 6 is required for normal microtubule dynamics and organization in *Arabidopsis*. *Proceedings of the National Academy of Sciences USA* **105**, 18064–18069.

Gubler, F., Kalla, R., Roberts, J.K. and Jacobsen, J.V. (1995). Gibberellin-regulated expression of a *myb* gene in barley aleurone cells: evidence for Myb transactivation of a high-pI alpha-amylase gene promoter. *The Plant Cell* **7**, 1879–1891.

Gubler, F., Raventos, D., Keys, M. *et al.* (1999). Target genes and regulatory domains of the GAMYB transcriptional activator in cereal aleurone. *The Plant Journal* **17**, 1–9.

Gubler, F., Chandler, P.M., White, R.G. *et al.* (2002). Gibberellin signaling in barley aleurone cells. Control of SLN1 and GAMYB expression. *Plant Physiology* **129**, 191–200.

Harberd, N.P., King, K.E., Carol, P. *et al.* (1998). Gibberellin: inhibitor of an inhibitor of… ? *Bioessays* **20**, 1001–1008.

Harris, L.J., Martinez, S.A., Keyser, B.R. *et al.* (2013). Functional analysis of TaABF1 during abscisic acid and gibberellin signalling in aleurone cells of cereal grains. *Seed Science Research* **23**, 89–98.

Heo, J.O., Chang, K.S., Kim, I.A. *et al.* (2011). Funneling of gibberellin signaling by the GRAS transcription regulator SCARECROW-LIKE 3 in the *Arabidopsis* root. *Proceedings of the National Academy of Sciences USA* **108**, 2166–2171.

Hirano, K., Asano, K., Tsuji, H. *et al.* (2010). Characterization of the Molecular Mechanism Underlying Gibberellin Perception Complex Formation in Rice. *The Plant Cell* **22**, 2680–2696.

Hirano, K., Kouketu, E., Katoh, H. *et al.* (2012). The suppressive function of the rice DELLA protein SLR1 is dependent on its transcriptional activation activity. *The Plant Journal* **71**, 443–453.

Hong, G.J., Xue, X.Y., Mao, Y.B. *et al.* (2012a). *Arabidopsis* MYC2 interacts with DELLA proteins in regulating sesquiterpene synthase gene expression. *The Plant Cell* **24**, 2635–2648.

Hong, Y.-F., Ho, T.-H.D., Wu, C.-F. *et al.* (2012b). Convergent starvation signals and hormone crosstalk in regulating nutrient mobilization upon germination in cereals. *The Plant Cell* **24**, 2857–2873.

Hou, X., Lee, L.Y., Xia, K. *et al.* (2010). DELLAs modulate jasmonate signaling via competitive binding to JAZs. *Developmental Cell* **19**, 884–894.

Huq, E. and Quail, P.H. (2002). PIF4, a phytochrome-interacting bHLH factor, functions as a negative regulator of phytochrome B signaling in *Arabidopsis*. *EMBO Journal* **21**, 2441–2450.

Ikeda, A., Ueguchi-Tanaka, M., Sonoda, Y. *et al.* (2001). *slender* rice, a constitutive gibberellin response mutant, is caused by a null mutation of the *SLR1* gene, an ortholog of the height-regulating gene *GAI/RGA/RHT/D8*. *The Plant Cell* **13**, 999–1010.

Itoh, H., Ueguchi-Tanaka, M., Sato, Y. *et al.* (2002). The gibberellin signaling pathway is regulated by the appearance and disappearance of SLENDER RICE1 in nuclei. *The Plant Cell* **14**, 57–70.

Itoh, H., Shimada, A., Ueguchi-Tanaka, M. *et al.* (2005). Overexpression of a GRAS protein lacking the DELLA domain confers altered gibberellin responses in rice. *The Plant Journal* **44**, 669–679.

Jacobsen, J.V., Gubler, F. and Chandler, P.M. (1995) Gibberellin action in germinated cereal grains. *Plant Hormones: Physiology, Biochemistry and Molecular Biology* (ed. P.J. Davies), Kluwer Academic Publishers, Netherlands, pp.246–271

Jasinski, S., Tattersall, A., Piazza, P. *et al.* (2008). *PROCERA* encodes a DELLA protein that mediates control of dissected leaf form in tomato. *The Plant Journal* **56**, 603–612.

Johnson, R.R., Shin, M. and Shen, J.Q. (2008). The wheat PKABA1-interacting factor TaABF1 mediates both abscisic acid-suppressed and abscisic acid-induced gene expression in bombarded aleurone cells. *Plant Molecular Biology* **68**, 93–103.

Josse, E.M., Gan, Y., Bou-Torrent, J. *et al.* (2011). A DELLA in disguise: SPATULA restrains the growth of the developing *Arabidopsis* seedling. *The Plant Cell* **23**, 1337–1351.

Kaneko, M., Inukai, Y., Ueguchi-Tanaka, M. *et al.* (2004). Loss-of-function mutations of the rice *GAMYB* gene impair α-amylase expression in aleurone and flower development. *The Plant Cell* **16**, 33–44.

Kim, J., Yi, H., Choi, G. *et al.* (2003). Functional characterization of phytochrome interacting factor 3 in phytochrome-mediated light signal transduction. *The Plant Cell* **15**, 2399–2407.

King, K.E., Moritz, T. and Harberd, N.P. (2001). Gibberellins are not required for normal stem growth in *Arabidopsis thaliana* in the absence of GAI and RGA. *Genetics* **159**, 767–776.

Koini, M.A., Alvey, L., Allen, T. *et al.* (2009). High temperature-mediated adaptations in plant architecture require the bHLH transcription factor PIF4. *Current Biology* **19**, 408–413.

Koornneef, M., Elgersma, A., Hanhart, C.J. *et al.* (1985). A gibberellin-insensitive mutant of *Arabidopsis thaliana*. *Physiologia Plantarum* **65**, 33–39.

Lee, S., Cheng, H., King, K.E. *et al.* (2002). Gibberellin regulates *Arabidopsis* seed germination via *RGL2*, a *GAI/RGA*-like gene whose expression is up-regulated following imbibition. *Genes and Development* **16**, 646–658.

Leivar, P., Monte, E., Oka, Y. *et al.* (2008). Multiple phytochrome-interacting bHLH transcription factors repress premature seedling photomorphogenesis in darkness. *Current Biology* **18**, 1815–1823.

Li, Q.F., Wang, C., Jiang, L. *et al.* (2012). An interaction between BZR1 and DELLAs mediates direct signaling crosstalk between brassinosteroids and gibberellins in *Arabidopsis*. *Science Signaling* **5**, ra72.

Lim, S., Park, J., Lee, N. *et al.* (2013). ABA-INSENSITIVE3, ABA-INSENSITIVE5, and DELLAs interact to activate the expression of *SOMNUS* and other high-temperature-inducible genes in imbibed seeds in *Arabidopsis*. *The Plant Cell* **25**, 4863–4878.

Lloyd, C. (2011). Dynamic microtubules and the texture of plant cell walls. *International Reviews in Cell and Molecular Biology* **287**, 287–329.

Locascio, A., Blazquez, M.A. and Alabadí, D. (2013). Dynamic regulation of cortical microtubule organization through prefoldin-DELLA interaction. *Current Biology* **23**, 804–809.

Lorrain, S., Allen, T., Duek, P.D. *et al.* (2008). Phytochrome-mediated inhibition of shade avoidance involves degradation of growth-promoting bHLH transcription factors. *The Plant Journal* **53**, 312–323.

Luo, H., Laluk, K., Lai, Z. *et al.* (2010). The *Arabidopsis* Botrytis Susceptible1 Interactor defines a subclass of RING E3 ligases that regulate pathogen and stress responses. *Plant Physiology* **154**, 1766–1782.

Ma, Z., Hu, X., Cai, W. *et al.* (2014). *Arabidopsis* miR171-targeted scarecrow-like proteins bind to GT *cis*-elements and mediate gibberellin-regulated chlorophyll biosynthesis under light conditions. *PLoS Genetics* **10**, e1004519.

Marin-de la Rosa, N., Sotillo, B., Miskolczi, P. *et al.* (2014). Large-scale identification of gibberellin-related transcription factors defines Group VII ERFs as functional DELLA partners. *Plant Physiology* **116**, 1022–1032.

Marín-de la Rosa N, Pfeiffer A, Hill K *et al.* (2015) Genome wide binding site analysis reveals transcriptional coactivation of cytokinin-responsive genes by DELLA proteins. *PLoS Genetics* **11**, e1005337.

McGinnis, K.M., Thomas, S.G., Soule, J.D. *et al.* (2003). The *Arabidopsis SLEEPY1* gene encodes a putative F-box subunit of an SCF E3 ubiquitin ligase. *The Plant Cell* **15**, 1120–1130.

Millán-Zambrano, G., Rodríguez-Gil, A., Penate, X. *et al.* (2013). The prefoldin complex regulates chromatin dynamics during transcription elongation. *PLoS Genetics* **9**, e1003776.

Millar, A.A. and Gubler, F. (2005). The *Arabidopsis GAMYB-like* genes, *MYB33* and *MYB65*, are MicroRNA-regulated genes that redundantly facilitate anther development. *The Plant Cell* **17**, 705–721.

Murase, K., Hirano, Y., Sun, T.P. and Hakoshima, T. (2008). Gibberellin-induced DELLA recognition by the gibberellin receptor GID1. *Nature* **456**, 459–464.

Murray, F., Kalla, R., Jacobsen, J. and Gubler, F. (2003). A role for HvGAMYB in anther development. *The Plant Journal* **33**, 481–491.

Nakamura, H., Xue, Y.L., Miyakawa, T. *et al.* (2013). Molecular mechanism of strigolactone perception by DWARF14. *Nature Communications* **4**, 2613.

Nemhauser, J.L., Mockler, T.C. and Chory, J. (2004). Interdependency of brassinosteroid and auxin signaling in *Arabidopsis*. *PLoS Biology* **2**, E258.

Nozue, K., Covington, M.F., Duek, P.D. *et al.* (2007). Rhythmic growth explained by coincidence between internal and external cues. *Nature* **448**, 358–361.

Nusinow, D.A., Helfer, A., Hamilton, E.E. *et al.* (2011). The ELF4-ELF3-LUX complex links the circadian clock to diurnal control of hypocotyl growth. *Nature* **475**, 398–402.

Ogawa, M., Kusano, T., Katsumi, M. and Sano, H. (2000). Rice gibberellin-insensitive gene homolog, *OsGAI*, encodes a nuclear-localized protein capable of gene activation at transcriptional level. *Gene* **245**, 21–29.

Ogawa, M., Hanada, A., Yamauchi, Y. *et al.* (2003). Gibberellin biosynthesis and response during *Arabidopsis* seed germination. *The Plant Cell* **15**, 1591–604.

Oh, E., Zhu, J.Y., Bai, M.Y. *et al.* (2014). Cell elongation is regulated through a central circuit of interacting transcription factors in the *Arabidopsis* hypocotyl. *Elife* **3**, e03031.

Park, J., Nguyen, K.T., Park, E. *et al.* (2013). DELLA proteins and their interacting RING finger proteins repress gibberellin responses by binding to the promoters of a subset of gibberellin-responsive genes in *Arabidopsis*. *The Plant Cell* **25**, 927–943.

Pearce, S., Saville, R., Vaughan, S.P. *et al.* (2011). Molecular characterization of *Rht-1* dwarfing genes in hexaploid wheat. *Plant Physiology* **157**, 1820–1831.

Peng, J. and Harberd, N.P. (1997). Gibberellin deficiency and response mutations suppress the stem elongation phenotype of phytochrome-deficient mutants of *Arabidopsis*. *Plant Physiology* **113**, 1051–1058.

Peng, J., Carol, P., Richards, D.E. *et al.* (1997). The *Arabidopsis GAI* gene defines a signaling pathway that negatively regulates gibberellin responses. *Genes and Development* **11**, 3194–3205.

Peng, J., Richards, D.E., Hartley, N.M. *et al.* (1999). 'Green revolution' genes encode mutant gibberellin response modulators. *Nature* **400**, 256–261.

Phinney, B.O. (1956). Growth response of single-gene dwarf mutants in maize to gibberellic acid. *Proceedings of the National Academy of Sciences USA* **42**, 185–189.

Piskurewicz, U. and Lopez-Molina, L. (2009). The GA-signaling repressor RGL3 represses testa rupture in response to changes in GA and ABA levels. *Plant Signaling and Behavior* **4**, 63-65.

Plackett, A.R.G., Ferguson, A.C., Powers, S.J. *et al.* (2014). DELLA activity is required for successful pollen development in the Columbia ecotype of *Arabidopsis*. *New Phytologist* **201**, 825–836.

Porri, A., Torti, S., Romera-Branchat, M. and Coupland, G. (2012). Spatially distinct regulatory roles for gibberellins in the promotion of flowering of *Arabidopsis* under long photoperiods. *Development* **139**, 2198–2209.

Potts, W.C., Reid, J.B. and Murfet, I.C. (1985). Internode length in *Pisum.* Gibberellins and the slender phenotype. *Physiologia Plantarum* **63**, 357–364.

Pysh, L.D., Wysocka-Diller, J.W., Camilleri, C. *et al.* (1999). The GRAS gene family in *Arabidopsis*: sequence characterization and basic expression analysis of the *SCARECROW-LIKE* genes. *The Plant Journal* **18**, 111–119.

Qi, T., Song, S., Ren, Q. *et al.* (2011). The Jasmonate-ZIM-domain proteins interact with the WD-Repeat/bHLH/MYB complexes to regulate Jasmonate-mediated anthocyanin accumulation and trichome initiation in *Arabidopsis thaliana. The Plant Cell* **23**, 1795–1814.

Qi, T., Huang, H., Wu, D. *et al.* (2014). *Arabidopsis* DELLA and JAZ proteins bind the WD-repeat/bHLH/MYB complex to modulate gibberellin and jasmonate signaling synergy. *The Plant Cell,* **26**, 1118–1133.

Resentini, F., Felipo-Benavent, A., Colombo, L. *et al.* (2015). TCP14 and TCP15 mediate the promotion of seed germination by gibberellins in *Arabidopsis thaliana. Molecular Plant* **8**, 482–485.

Richter, R., Behringer, C., Muller, I.K. and Schwechheimer, C. (2010). The GATA-type transcription factors GNC and GNL/CGA1 repress gibberellin signaling downstream from DELLA proteins and PHYTOCHROME-INTERACTING FACTORS. *Genes and Development* **24**, 2093–2104.

Rodriguez-Milla, M.A. and Salinas, J. (2009). Prefoldins 3 and 5 play an essential role in *Arabidopsis* tolerance to salt stress. *Molecular Plant* **2**, 526–534.

Rombola-Caldentey, B., Rueda-Romero, P., Iglesias-Fernandez, R. *et al.* (2014). *Arabidopsis* DELLA and two HD-ZIP transcription factors regulate GA signaling in the epidermis through the L1 box *cis*-element. *The Plant Cell* **26**, 2905–2919.

Sarnowska, E.A., Rolicka, A.T., Bucior, E. *et al.* (2013). DELLA-interacting SWI3C core subunit of switch/sucrose nonfermenting chromatin remodeling complex modulates gibberellin responses and hormonal cross talk in *Arabidopsis. Plant Physiology* **163**, 305–317.

Sato, T., Miyanoiri, Y., Takeda, M. *et al.* (2014). Expression and purification of a GRAS domain of SLR1, the rice DELLA protein. *Protein Expression and Purification* **95**, 248–258.

Shani, E., Weinstain, R., Zhang, Y. *et al.* (2013). Gibberellins accumulate in the elongating endodermal cells of *Arabidopsis* root. *Proceedings of the National Academy of Sciences USA* **110**, 4834–4839.

Shin, J., Kim, K., Kang, H. *et al.* (2009). Phytochromes promote seedling light responses by inhibiting four negatively-acting phytochrome-interacting factors. *Proceedings of the National Academy of Sciences USA* **106**, 7660–7665.

Silverstone, A.L., Mak, P.Y., Martinez, E.C. and Sun, T.P. (1997). The new *RGA* locus encodes a negative regulator of gibberellin response in *Arabidopsis thaliana. Genetics* **146**, 1087–1099.

Silverstone, A.L., Ciampaglio, C.N. and Sun, T. (1998). The *Arabidopsis RGA* gene encodes a transcriptional regulator repressing the gibberellin signal transduction pathway. *Plant Cell* **10**, 155–169.

Silverstone, A.L., Jung, H.S., Dill, A. *et al.* (2001). Repressing a repressor: gibberellin-induced rapid reduction of the RGA protein in *Arabidopsis. The Plant Cell* **13**, 1555–1566.

Soy, J., Leivar, P., Gonzalez-Schain, N. *et al.* (2012). Phytochrome-imposed oscillations in PIF3 protein abundance regulate hypocotyl growth under diurnal light/dark conditions in *Arabidopsis. The Plant Journal* **71**, 390–401.

Stavang, J.A., Gallego-Bartolomé, J., Gómez, M.D. *et al.* (2009). Hormonal regulation of temperature-induced growth in *Arabidopsis*. *The Plant Journal* **60**, 589–601.

Sun, T.P. and Gubler, F. (2004). Molecular mechanism of gibberellin signaling in plants. *Annual Review of Plant Biology* **55**, 197–223.

Sun, X., Xue, B., Jones, W.T. *et al.* (2011). A functionally required unfoldome from the plant kingdom: intrinsically disordered N-terminal domains of GRAS proteins are involved in molecular recognition during plant development. *Plant Molecular Biology* **77**, 205–223.

Sun, X., Jones, W.T. and Rikkerink, E.H.A. (2012). GRAS proteins: the versatile roles of intrinsically disordered proteins in plant signalling. *Biochemical Journal* **442**, 1–12.

Thines, B., Katsir, L., Melotto, M. *et al.* (2007). JAZ repressor proteins are targets of the SCF(COI1) complex during jasmonate signalling. *Nature* **448**, 661–665.

Toledo-Ortiz, G., Huq, E. and Quail, P.H. (2003). The *Arabidopsis* basic/helix-loop-helix transcription factor family. *The Plant Cell* **15**, 1749–1770.

Tyler, L., Thomas, S.G., Hu, J. *et al.* (2004). DELLA proteins and gibberellin-regulated seed germination and floral development in *Arabidopsis*. *Plant Physiology* **135**, 1008–1019.

Úbeda-Tomás, S., Swarup, R., Coates, J. *et al.* (2008). Root growth in *Arabidopsis* requires gibberellin/DELLA signalling in the endodermis. *Nature Cell Biology* **10**, 625–628.

Úbeda-Tomás, S., Federici, F., Casimiro, I. *et al.* (2009). Gibberellin signaling in the endodermis controls *Arabidopsis* root meristem size. *Current Biology* **19**, 1194–1199.

Ueguchi-Tanaka, M., Ashikari, M., Nakajima, M. *et al.* (2005). *GIBBERELLIN INSENSITIVE DWARF1* encodes a soluble receptor for gibberellin. *Nature* **437**, 693–698.

Ueguchi-Tanaka, M., Hirano, K., Hasegawa, Y. *et al.* (2008). Release of the repressive activity of rice DELLA protein SLR1 by gibberellin does not require SLR1 degradation in the *gid2* mutant. *The Plant Cell* **20**, 2437–2446.

Vainberg, I.E., Lewis, S.A., Rommelaere, H. *et al.* (1998). Prefoldin, a chaperone that delivers unfolded proteins to cytosolic chaperonin. *Cell* **93**, 863–873.

Wen, C.K. and Chang, C. (2002). *Arabidopsis RGL1* encodes a negative regulator of gibberellin responses. *The Plant Cell* **14**, 87–100.

Wenzel, C.L., Williamson, R.E. and Wasteneys, G.O. (2000). Gibberellin-induced changes in growth anisotropy precede gibberellin-dependent changes in cortical microtubule orientation in developing epidermal cells of barley leaves: Kinematic and cytological studies on a gibberellin-responsive dwarf mutant, M489. *Plant Physiology* **124**, 813–822.

Weston, D.E., Elliott, R.C., Lester, D.R. *et al.* (2008). The Pea DELLA proteins LA and CRY are important regulators of gibberellin synthesis and root growth. *Plant Physiology* **147**, 199–205.

Wild, M., Daviere, J.M., Cheminant, S. *et al.* (2012). The *Arabidopsis* DELLA *RGA-LIKE3* is a direct target of MYC2 and modulates jasmonate signaling responses. *The Plant Cell* **24**, 3307–3319.

Willige, B.C., Ghosh, S., Nill, C. *et al.* (2007). The DELLA domain of GA INSENSITIVE mediates the interaction with the GA INSENSITIVE DWARF1A gibberellin receptor of *Arabidopsis*. *The Plant Cell* **19**, 1209–1220.

Willige, B.C., Ogiso-Tanaka, E., Zourelidou, M. and Schwechheimer, C. (2012). WAG2 represses apical hook opening downstream from gibberellin and PHYTOCHROME INTERACTING FACTOR 5. *Development* **139**, 4020–4028.

Yamaguchi, N., Winter, C.M., Wu, M.F. *et al.* (2014). Gibberellin acts positively then negatively to control onset of flower formation in *Arabidopsis*. *Science* **344**, 638–641.
Yang, D.L., Yao, J., Mei, C.S. *et al.* (2012). Plant hormone jasmonate prioritizes defense over growth by interfering with gibberellin signaling cascade. *Proceedings of the National Academy of Sciences USA* **109**, 1192–1200.
Yoshida, H., Hirano, K., Sato, T. *et al.* (2014). DELLA protein functions as a transcriptional activator through the DNA binding of the INDETERMINATE DOMAIN family proteins. *Proceedings of the National Academy of Sciences USA* **111**, 7861–7866.
Yu, S., Galvao, V.C., Zhang, Y.C. *et al.* (2012). Gibberellin regulates the *Arabidopsis* floral transition through miR156-targeted *SQUAMOSA* promoter binding-like transcription factors. *The Plant Cell* **24**, 3320–3332.
Zentella, R., Zhang, Z.L., Park, M. *et al.* (2007). Global analysis of DELLA direct targets in early gibberellin signaling in *Arabidopsis*. *The Plant Cell* **19**, 3037–3057.
Zhang, D., Jing, Y., Jiang, Z. and Lin, R. (2014). The chromatin-remodeling factor PICKLE integrates brassinosteroid and gibberellin signaling during skotomorphogenic growth in *Arabidopsis*. *The Plant Cell* **26**, 2472–2485.
Zhang, Z.L., Ogawa, M., Fleet, C.M. *et al.* (2011). SCARECROW-LIKE 3 promotes gibberellin signaling by antagonizing master growth repressor DELLA in *Arabidopsis*. *Proceedings of the National Academy of Sciences USA* **108**, 2160–2165.

Annual Plant Reviews (2016) **49**, 229–252
doi: 10.1002/9781119210436.ch8

http://onlinelibrary.wiley.com

Chapter 8

INTERACTIONS BETWEEN GIBBERELLINS AND OTHER HORMONES

John J. Ross, Asemeh Miraghazadeh[1], Amelia H. Beckett, Laura J. Quittenden and Erin L. McAdam

School of Biological Sciences, University of Tasmania, Australia
[1]Current address: College of Medicine, Biology and Environment, Australian National University, Australia

Abstract: In this chapter we discuss interactions between the gibberellins (GAs) and other plant hormones. Auxin and ethylene exert strong effects on GA levels, and at least some of the physiological responses to auxin are mediated by increased GA content. Abscisic acid (ABA) and GA appear to negatively regulate each other's levels in seeds, consistent with their antagonistic effects on germination. It has been suggested that several non-GA hormones affect the stability of DELLAs, the key signalling proteins that are destabilised by GAs. However, it appears that the reported effects of these additional hormones on DELLA stability are either mediated by changes in GA content, or have not yet been confirmed. Recent papers provide evidence that DELLAs physically interact with transcriptional regulators from the signalling pathways of the brassinosteroids, ethylene, jasmonic acid and ABA. This may represent a physiologically significant way in which GAs interact with other signalling pathways, and helps to explain how DELLAs, which lack a conserved DNA binding domain, can regulate gene transcription.

Keywords: Gibberellin, auxin, ethylene, abscisic acid, brassinosteroid, jasmonic acid, strigolactone, cytokinin, interaction

8.1 Introduction

Interactions involving two plant hormones can be classified into two broad categories: those where the first hormone affects the synthesis, deactivation and/or transport of the second, thereby affecting its levels; and interactions

Annual Plant Reviews, Volume 49: The Gibberellins, First Edition.
Edited by Peter Hedden and Stephen G. Thomas.

where the first affects the signal transduction pathway of the second (Santner and Estelle, 2009; Vanstraelen and Benkov, 2012; Claeys *et al.*, 2014). For both types of interaction, mutual effects are possible. There are several well-documented examples of the first type of interaction, where the resulting change in hormone level is responsible for substantial effects on plant growth and development. In addition, recent papers document interactions of the second type, in which signalling components intersect, with important consequences for the plant. However, some interactions only slightly affect the levels or signalling of the second hormone, with the effects being quite indirect and of little physiological significance. Indeed, recent years have seen a proliferation of suggested interactions, not all of which are of functional importance in plant systems (Ross *et al.*, 2011).

Here we review recent advances in the study of the major interactions between gibberellins (GA) and other hormones, covering both main types of interaction mentioned above, and with a focus on their physiological significance.

8.2 Interactions involving effects of other hormones on gibberellin levels

Bioactive GAs are synthesised by a multi-step pathway, with the later steps catalysed by enzymes of the 2-oxoglutarate-dependent dioxygenase (2-ODD) class. The final, activation step in GA synthesis is catalysed by enzymes known as GA 3-oxidases, while the penultimate step is catalysed by GA 20-oxidases (Hedden and Thomas, 2012; see also Chapter 2). The deactivation of bioactive GAs (which in most cases are GA_4 and/or GA_1) is usually catalysed by another group of 2-ODDs, the GA 2-oxidases (see Chapter 3). The dramatic phenotypes caused by mutations in key GA synthesis and deactivation genes highlight the importance of both processes in regulating bioactive GA levels (Hedden and Phillips, 2000). Our thorough understanding of GA synthesis and deactivation pathways has facilitated the characterisation of interactions in which other hormones affect GA steady-state levels. These will be discussed below.

8.2.1 Auxin promotes gibberellin biosynthesis

Auxins and GAs were both discovered in the early 1900s, but whether or not they interacted was a major question that remained unresolved for decades. As late as the 1990s, Kende and Zeevaart (1997) noted, 'The relationship between auxin and GA action … awaits resolution'. However, around that time, important clues were reported. Van Huizen *et al.* (1997) found that a particular auxin, 4-chloro indole-3-acetic acid (4-Cl IAA), can stimulate a GA biosynthetic step in pea pods. This was followed by evidence that the classical auxin, indole-3-acetic acid (IAA), strongly promotes the synthesis of the bioactive GA, GA_1, in stems of pea (Ross *et al.*, 2000; O'Neill *et al.*, 2010). In fact, when auxin levels were low, as observed in decapitated pea stems, the production of GA_1 was virtually eliminated (Ross *et al.*, 2000). Subsequently, auxin has been shown to promote GA synthesis in a range of systems, including shoots of tobacco (Wolbang and Ross, 2001), barley (Wolbang *et al.*, 2004), *Arabidopsis* (Frigerio *et al.*, 2006), and rice (Yin *et al.*, 2007), as well as tomato fruits (Serrani *et al.*, 2008) and pea roots (Weston *et al.*, 2009). The effect of auxin is mediated by both the up-regulation of GA synthesis genes, in particular those encoding GA 20-oxidases and GA 3-oxidases, and by the down-regulation of genes encoding GA 2-oxidases (GA deactivating enzymes).

Little is known about the molecular links between auxin and the up-regulation of GA synthesis genes. However, Frigerio *et al.* (2006) investigated the possible involvement of Aux/IAA proteins in this process. These proteins, which are destabilised by auxin, are known to play a pivotal role in the regulation of other genes (Reed, 2001). Frigerio *et al.* (2006) utilized several gain-of-function mutations that stabilise the corresponding Aux/IAA. These mutations tended to reduce the capacity of auxin to up-regulate the two GA synthesis genes tested. The effect was particularly noticeable in the case of the 20-oxidase gene, *AtGA20ox1*. In wild-type (WT) plants this gene was strongly up-regulated by auxin, a result confirmed by O'Neill *et al.* (2010). However, in mutants affected in the Aux/IAA proteins IAA3, IAA7, IAA17, or IAA19, the up-regulation was considerably reduced or eliminated. This indicates that in WT plants, auxin destabilises Aux/IAA proteins, which in turn allows the expression of *AtGA20ox1* and other GA synthesis genes.

If auxin promotes elongation (at least in part) by increasing GA levels, we would expect that it is less effective when the capacity of the plant to synthesise GAs is impaired. Accordingly, Ross *et al.* (2003) showed that in excised pea stem segments, auxin was much less effective at promoting elongation of GA_1-deficient (Mendel's *le-1* dwarf mutant) segments, compared with those of the WT. More recently, Chapman *et al.* (2012) showed a similarly reduced capacity of auxin to promote hypocotyl elongation in mutants genetically impaired in GA synthesis, or in WT plants treated with paclobutrazol, an inhibitor of GA synthesis. Chapman *et al.* (2012) concluded that auxin acts in part by increasing GA content, but also that it acts independently of GA (Figure 8.1). The up-regulation of GA levels by auxin is important also for fruit development in pea (van Huizen *et al.*, 1997, Ozga *et al.*, 2009), tomato (Serrani *et al.*, 2008) and *Arabidopsis* (Dorcey *et al.*, 2009).

8.2.2 Ethylene inhibits gibberellin biosynthesis

There is strong evidence that increased ethylene signalling negatively regulates the synthesis of active GAs. Foo *et al.* (2006) reported that GA_1 levels

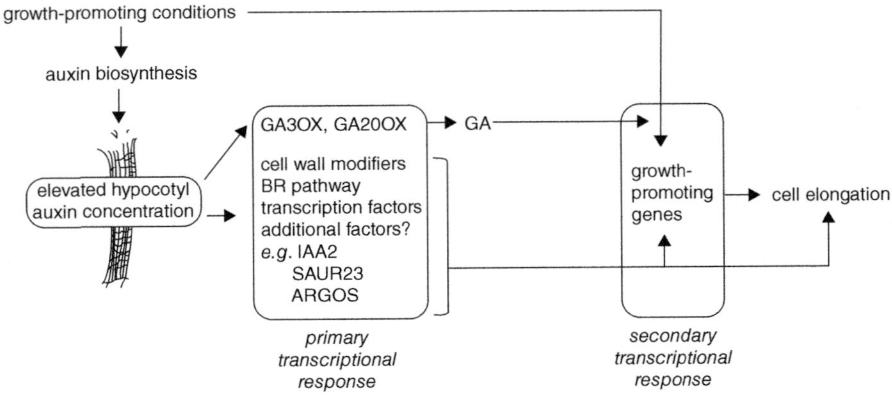

Figure 8.1 Auxin up-regulates GA biosynthesis in *Arabidopsis* hypocotyls, but can also promote elongation independently of GAs. According to this model, auxin does not directly destabilise DELLAs. (Figure adapted from Chapman *et al.* (2012) Hypocotyl transcriptome reveals auxin regulation of growth-promoting genes through GA-dependent and -independent pathways. © Chapman *et al.*)

were dramatically reduced (by up to 85%) in a mutant that over-produces ethylene (because of a deficiency of phytochromes), and that this effect was largely reversed by an inhibitor of ethylene synthesis. However, this inhibitor did not affect GA_1 levels in WT stems, indicating that ethylene levels might not limit GA synthesis under normal growing conditions. In *Arabidopsis*, Achard *et al.* (2007) showed that the constitutive ethylene response mutant *ctr1-1* contains substantially less GA_4 and GA_1 than WT plants in mature vegetative rosettes. The reduction in GA content appears to mediate the flower-delaying effect of *ctr1-1*, since this effect was reversed by GA application (Achard *et al.*, 2007).

There is also evidence for the reverse interaction – an inhibition of ethylene synthesis by GA (De Grauwe *et al.*, 2008). Consistent with that, Ferguson *et al.* (2011) reported that ethylene production was elevated in the GA-deficient mutant of pea, *na-1*. It was suggested that this increased ethylene production contributes to the reduced root nodule formation observed in *na-1* plants.

8.2.3 Do gibberellin and abscisic acid inhibit each other's synthesis?

Gibberellin and abscisic acid (ABA) often play antagonistic roles in plant growth and development, and it is frequently suggested that they also inhibit each other's biosynthesis. Seo *et al.* (2006) reported that in ABA-deficient seeds, GA levels were elevated, while Oh *et al.* (2007) found that in a GA-deficient mutant, seed ABA levels were elevated. Again, young seeds of the GA-deficient pea mutant, *lh-2*, contain low levels of bioactive GA and elevated levels of ABA (Batge *et al.*, 1999). Further data showing the antagonistic relationship in seeds were published by Piskurewicz *et al.* (2008). The two hormones play key roles in regulating seed dormancy in those species with dormant seeds, and the capacity of ABA and GA to mutually down-regulate levels is therefore of physiological significance.

However, the situation is not as clear in vegetative tissues. Ross *et al.* (2011) treated WT pea seedlings with ABA and did not find a reduction in GA levels, while shoots of the *lh-2* mutant (Batge *et al.*, 1999) and of two other GA-deficient pea mutants (Ross *et al.*, 2011) contained less, not more, ABA than the WT. Therefore, the GA-ABA antagonism, with regard to hormone content, might not be a general phenomenon in shoots and is possibly restricted to seeds. Deepwater rice, however, is a vegetative plant system in which ABA may well inhibit GA biosynthesis. Deepwater rice varieties can withstand the prolonged floods that occur annually in the river delta areas of Southeast Asia, West Africa and South America (Catling, 1992). When submerged, the internodes rapidly elongate, at up to 25 cm per day, to keep the apical leaves above the rising water surface (Catling, 1992). According to the prevailing model, internode elongation is initiated by increased levels of ethylene in the submerged internodes (rising from 0.02 µl/L to 1 µl/L) (Métraux and Kende, 1983), attributable to reduced diffusion of, and increased synthesis of, the gaseous hormone (Jackson, 1985). The accumulated ethylene is thought to rapidly reduce ABA levels in the internodes by up to 75% (Hoffmann-Benning and Kende, 1992). This is suggested to lead in turn to increased (approximately four-fold) levels of GA_1 in submerged internodes (Hoffmann-Benning and Kende, 1992). GA is considered to be the final effector that induces the rapid internode elongation (Bleecker *et al.*, 1986). However, it is not clear whether the GA-ABA link has been tested by directly applying ABA to the system and then monitoring GA levels. Clearly, the GA-ethylene interaction involved in deepwater rice differs from that discussed in the above section, since in deepwater rice elevated ethylene content enhances, not reduces, GA levels.

Lowland rice (*Oryza sativa*) tolerates short-term submergence, but unlike deepwater rice, these plants stop elongating until flood waters recede (Fukao *et al.*, 2006). Like deepwater rice, however, the response mechanism to submergence involves both ethylene and the GA signal transduction pathway (Fukao *et al.*, 2006; Xu *et al.*, 2006). The response mechanism is primarily controlled by the *SUBMERGENCE-1 (SUB-1)* genes: *SUB1A, SUB1B* and *SUB1C* (Xu *et al.*, 2006). In wild-type rice, the *SUB1A* gene is absent, but in the submergence-tolerant variety, *SUB1A* limits growth by adjusting ethylene levels and repressing GA signal transduction pathways (Fukao *et al.*, 2006; Xu *et al.*, 2006). The higher levels of ethylene that accumulate during flooding induce *SUB1A* mRNA accumulation, which ultimately suppresses ethylene production (Fukao *et al.*, 2006). *SUB1A* also stabilizes the DELLA proteins SLENDER RICE-1 (SLR1) and SLR-LIKE-1 (SLRL1), which are repressors of GA signalling (Fukao and Bailey-Serres, 2008; see Chapter 6). Both lowered ethylene levels and inhibited GA signalling result in halted plant growth (Fukao and Xiong, 2013). If the water subsides within two weeks, growth of the plant is renewed (Fukao *et al.*, 2006).

8.2.4 Do brassinosteroids act by affecting gibberellin levels?

An interaction between brassinosteroid (BR) and GA biosynthesis has been suggested (Bouquin *et al.*, 2001). The BR accumulation rice mutant *m107* has GA_1 levels elevated 5.7-fold, and rice seedlings treated with brassinolide (10^{-6} M) exhibited a 2-fold increase in GA_1 levels (Tong *et al.*, 2014). These data were presented alongside evidence for the induction of GA biosynthetic genes (*GA20ox2, GA3ox2*) and the repression of a GA deactivation gene (*GA2ox3*) by BR application and in BR over-accumulation (*m107*) or enhanced-signalling lines (*Do-2*), with the opposite observed in BR-deficient (*d2* and *d11*) or decreased-BR-signalling mutants (*d61-1, d61-2, Go-2, dlt*). Furthermore, application of GA_3 (10^{-7} M) rescued the phenotype of *d2* (a rice BR-deficient mutant) in terms of leaf sheath length. Taken together, the Tong *et al.* (2014) results suggest that BR promotes rice cell elongation by increasing GA biosynthesis.

However, in pea, sunflower and *Arabidopsis*, the BR status of the plant has been shown to have little effect on the actual content of bioactive GAs, with the level of the precursor, GA_{20}, instead being reduced on BR application (Jager *et al.*, 2005; Kurepin *et al.*, 2012). These observations do not support the idea that BRs promote growth via GAs in these species. Furthermore, GA application in pea did not restore BR deficient lines (*lka* and *lkb*) to the wild-type internode length phenotype (Reid and Ross, 1989). Similarly, *Arabidopsis* BR-deficient (*de-etiolated2*) and BR-insensitive (*bri1-119*) mutants were insensitive to GA in terms of hypocotyl elongation (Bai *et al.*, 2012; Gallego-Bartolomé *et al.*, 2012). It is possible that in different species, plant tissues and/or developmental stages, there are different interactions between BRs and GA (Tong *et al.*, 2014).

8.2.5 Possible effects of other hormones on gibberellin synthesis

Cytokinins also have been implicated in the regulation of GA levels, but only on the basis of cytokinin effects on GA deactivation (*GA 2-oxidase*) gene expression; GA levels were not measured (Jasinski *et al.*, 2005).

8.3 Interactions between hormone signal transduction pathways

Bioactive GAs exert their effects by initially interacting with a GA receptor (see Chapter 6). In 2005 this receptor was shown to be GID1, first identified in rice (Ueguchi-Tanaka *et al.*, 2005) and subsequently in other species, including *Arabidopsis* (Griffiths *et al.*, 2006). The next stage involves the proteins known as DELLAs, which are negative regulators of GA signalling. It appears that GID1, bioactive GA and DELLAs interact, resulting in the degradation of DELLAs. Therefore high GA levels result in low DELLA levels and GA signalling can proceed. In other words, GA is an 'inhibitor of an inhibitor' (Harberd *et al.*, 1998; see Chapters 6 and 7).

In reports on cross-talk between GA signalling and the signalling of other hormones, the DELLA proteins have received the most attention. During the 2000s, there was a trend to extend the role of DELLAs beyond GA signalling, with a series of papers suggesting that other hormones affect DELLA stability (Achard *et al.*, 2003; Fu and Harberd, 2003; Achard *et al.*, 2006). However, the evidence for this type of cross-talk is questionable. There is convincing evidence, however, for a second type of DELLA-related interaction, in which DELLAs physically interact with key proteins from the signal transduction pathways of other hormones. These two different themes are explored below, concentrating on auxin, brassinosteroids, ethylene, ABA, jasmonic acid and strigolactones.

8.3.1 Do other hormones affect DELLA stability?

Auxin and DELLA stability

Auxin was the first hormone, after GA itself, to be implicated in the regulation of DELLA stability. Fu and Harberd (2003) suggested that in *Arabidopsis* roots, auxin acts by facilitating the capacity of GAs to degrade DELLAs. According to their model, if auxin levels are low, GAs are less effective at destabilizing DELLAs and therefore at promoting growth. Fu and Harberd (2003) decapitated the shoots of young *Arabidopsis* seedlings, which appeared to result in auxin deficiency in the roots. To restore the root auxin content they applied auxin to the cut stump of the seedlings. They then examined the capacity of applied GA to reduce the signal from the fusion protein GFP-RGA (RGA is a DELLA protein). After four hours of GA treatment, there was more GFP-RGA in the decapitated plants than in intact plants or decapitated plants treated with auxin. But the decapitated plants started with more GFP-RGA signal than the other two cases. In fact, the rate of disappearance of GFP-RGA in response to GA was remarkably similar, and rapid, in all three cases (it is possible to count the coloured nuclei in the root elongation zones). This observation implies that the capacity of GA to destabilise DELLAs does not depend on auxin. However, Fu and Harberd (2003) drew the opposite conclusion – that the capacity of GAs to destabilise DELLAs *does* depend on auxin.

Fu and Harberd (2003) implied a generality for their model, on the basis of an earlier report (Brian and Hemming, 1958) that internodes of decapitated dwarf pea plants (Mendel's *le-1* mutant) do not elongate in response to GA, even though this mutant shows a very strong response when the plant is intact. However, Ross *et al.* (2002) demonstrated that decapitated, and therefore auxin-deficient, *le-1* internodes respond strongly to GA (applied in ethanol to an expanded leaf), showing a response 70% of that observed in intact plants and which could not be increased by applying auxin concurrently. The data presented in these studies (Ross *et al.*, 2002, Ross *et al.*, 2003) suggest that GA does not require auxin in order to degrade DELLAs and promote elongation.

Chapman *et al.* (2012) also tested whether auxin affects DELLA stability. They found that the rate of disappearance of DELLAs after auxin treatment was strongly reduced by the inhibitor of GA synthesis, paclobutrazol. Chapman *et al.* (2012) concluded that auxin degrades DELLAs indirectly, by up-regulating GA synthesis, rather than by affecting DELLAs directly (Figure 8.1). In summary, it appears that auxin does not directly influence DELLA stability, with the Fu and Harberd (2003) model receiving very little experimental support since its inception.

Since auxin is a key player in tropisms, the possible role of GA-auxin interactions in the bending response has received some attention. The classical Cholodny–Went theory stipulates that gravistimulation leads to an asymmetric distribution of auxin across a horizontally placed organ, which in turn leads to asymmetric growth and therefore bending (Trewavas, 1992). If auxin acts in gravistimulation by affecting either GA levels or response, then the response ought to be reduced in DELLA-deficient mutants, in which GA-stimulated elongation is already maximal on both sides of the gravistimulated organ. This prediction appears to apply in *Arabidopsis* roots (Löfke *et al.*, 2013). However, in barley pulvini, a DELLA-deficient mutant showed a *stronger* bending response than WT plants (Wolbang *et al.*, 2007). This latter observation is not consistent with the idea that the initial, response-triggering role of auxin in tropisms is mediated by changes in GA level or signal transduction. On the other hand, the effects of DELLA deficiency on the bending response do indicate a role for these proteins and by implication, for GA signalling, in tropic responses.

Ethylene and DELLA stability

Ethylene is another hormone that has been implicated in DELLA stability. In contrast to the GAs which destabilise DELLAs, it has been suggested that ethylene has a stabilising effect (Achard *et al.*, 2003). However, when 1-aminocyclopropane-1-carboxylic acid (ACC), a key ethylene precursor, was applied to GA-treated primary roots of pRGA:GFP-RGA seedlings of *Arabidopsis*, there was no increase in RGA protein level compared with controls, and with the possible exception of the root tip itself, there was a rapid GA-induced disappearance of RGA protein, even in the presence of ACC and therefore ethylene (Achard *et al.*, 2003). More recent papers attribute the reported effect of ethylene on DELLA stability to the capacity of the gaseous hormone to reduce GA content (Achard *et al.*, 2007; Vandenbussche *et al.*, 2007; De Grauwe *et al.*, 2008).

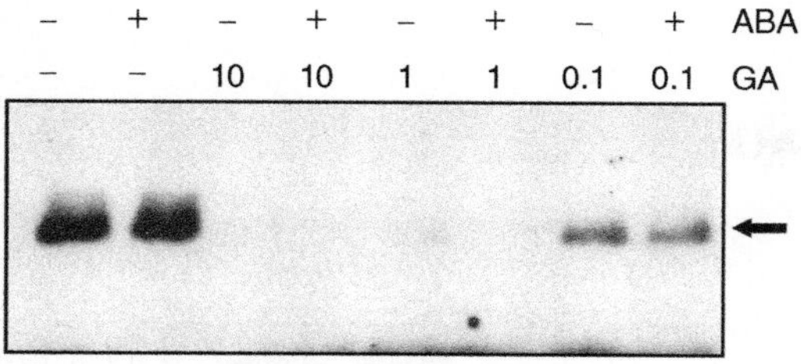

Figure 8.2 ABA does not prevent destabilisation of the DELLA protein RGA by GA. Protein extracts from whole 5-day-old *Arabidopsis* seedlings were immunoblotted with affinity-purified anti-RGA antibody. The seedlings (GA-deficient, genotype *ga1-3*) were pretreated with either water or 20 μM ABA for 2 h, followed by treatment with the indicated concentration (μM) of GA_3 for 1 h. The arrow indicates the RGA protein, with no RGA protein present in seedlings that were exposed to high levels of GA regardless of ABA application. (Adapted from Zentella *et al.* (2007). Reproduced with permission from the American Society of Plant Biologists.)

Abscisic acid and DELLA stability

ABA has also been suggested to stabilise DELLA proteins (Achard *et al.*, 2006). However, Gubler *et al.* (2002) had previously reported that ABA did not 'protect' the barley DELLA protein, SLN1, from GA-induced degradation. To resolve the issue, Zentella *et al.* (2007) tested whether ABA stabilises the *Arabidopsis* DELLA protein, RGA, in the presence of GA (Figure 8.2). They found no evidence that this was the case with the endogenous RGA protein, although there was some evidence of protection of the GFP-RGA fusion protein. Zentella *et al.* (2007) suggested that there could be a difference in behaviour between the endogenous and fusion proteins, and consistent with that, they reported that 'RGA responded more rapidly to GA than the GFP-RGA fusion protein.'

8.3.2 DELLAs interact with proteins from the signalling pathways of other hormones

Recently, several papers have reported that DELLAs physically interact with key proteins from the signalling pathways of other hormones. These protein–protein interactions are potentially physiologically relevant, given that DELLAs themselves lack a conserved DNA-binding domain and probably depend on such interactions in order to regulate gene expression.

Interactions between DELLAs and brassinosteroid signalling proteins

The signal transduction pathway of the growth-promoting hormones brassinosteroids (BRs) involves the protein BZR1, which is a positive regulator of the BR response (He *et al.*, 2005). BZR1 functions as a transcription factor that binds to the promoters of target genes. In 2012, three papers reported that BZR1 interacts with DELLA proteins, with DELLA inhibiting BZR1

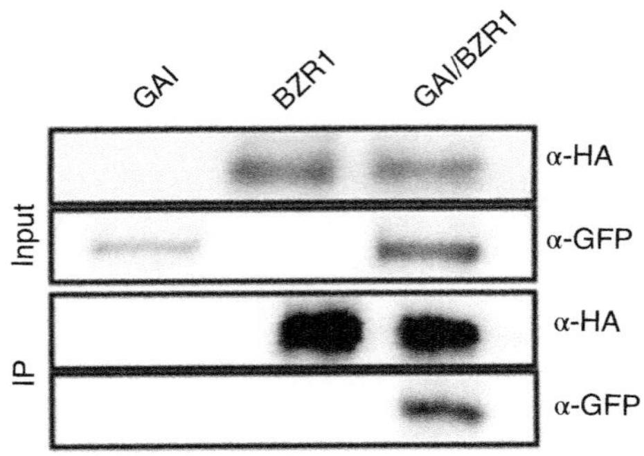

Figure 8.3 The DELLA protein GAI and the BR signalling protein BZR1 physically interact in the leaves of *Nicotiana benthamiana*, as shown by a coimmunoprecipitation assay. YFP-GAI and BZR1-HA were expressed alone or together; anti-GFP and anti-HA antibodies showed that GAI and BZR1 interact when expressed together, with YFP-GAI coimmunoprecipitating when HA-BZR1 was pulled down from leaf extracts. (Gallego-Bartolomé *et al.* (2012). Reproduced with permission from National Academy of Sciences [*Proceedings of the National Academy of Sciences USA*].)

function (Bai *et al.*, 2012, Gallego-Bartolomé *et al.*, 2012, Li *et al.*, 2012). The interaction was detected *in vitro* using yeast 2-hybrid assays, and *in vivo* by coimmunoprecipitation studies (Figure 8.3). Interestingly, deletion of the 'DELLA' domain of DELLA proteins had no effect on BZR1 binding, suggesting that BZR1 interacts with the DELLA 'GRAS' domain, rather than the DELLA domain (Bai *et al.*, 2012). The GRAS domain is conserved in all members of the DELLA family and BZR1 interacts with four of the five *Arabidopsis* DELLAs: RGA, GAI, RGL1 and RGL3. The DELLA proteins are thought to inhibit the transcriptional activity of BZR1 by blocking its DNA-binding activity (Bai *et al.*, 2012). Consistent with that, it has been shown that the level of BZR1 (Li *et al.*, 2012) and the binding of BZR1 to promoters of its target gene (Bai *et al.* 2012) increased after GA treatment, presumably through GA-dependent DELLA degradation.

Another transcription factor, PHYTOCHROME INTERACTING FACTOR 4 (PIF4), also plays a critical role in GA-BR signalling interactions. This protein, identified initially for its role in mediating light responses, interacts with DELLA and BZR1 proteins to form a three-way complex (Feng *et al.*, 2008, Oh *et al.*, 2012, 2014). PIF4, like BZR1, targets a set of growth-related genes. DELLA degradation by GAs presumably releases both BZR1 and PIFs, which then stimulate expression of their target genes (Figure 8.4).

The DELLA-BZR1 interaction appears to be important in hypocotyls

The DELLA-BZR1 interactions detailed above appear to be physiologically significant in *Arabidopsis* hypocotyls. In WT hypocotyls, BR application stimulates elongation, presumably by promoting the activity of BZR1 (Bai *et al.*, 2012). However, according to the recent model (Figure 8.4), when BR is

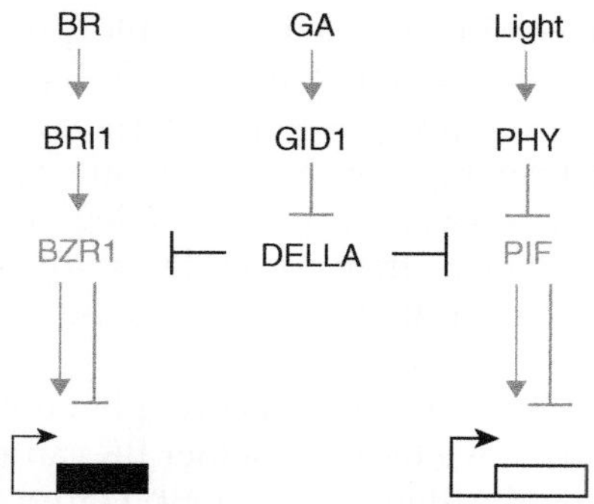

Figure 8.4 Model showing proposed interactions between DELLAs and other signalling proteins. By forming complexes with BZR1 and/or members of the PIF family, DELLAs inhibit the action of these regulatory proteins, preventing them from up- or down-regulating their target genes. (Gallego-Bartolomé *et al.* (2012). Reproduced with permission from National Academy of Sciences [*Proceedings of the National Academy of Sciences USA*].)

applied alone, it does not cause maximal elongation, since that is prevented by the presence of DELLA proteins. When GA is applied with BR, there is a synergistic effect (Tanaka *et al.*, 2003, Stewart Lilley *et al.*, 2013): the applied GA reduces DELLA content and removes the inhibition on BZR1 action. Consistent with that finding, BR dramatically promotes the hypocotyl elongation of DELLA-deficient mutant plants (Bai *et al.*, 2012); again, in this case there are no DELLAs present to reduce the capacity of BR-induced BZR1 to bind to the promoters of target genes.

As well as BR action being limited by DELLAs, the new models to some extent place BZR1 downstream of DELLAs, and consistent with that placement, BR application to the GA-deficient mutant *ga1-3* strongly stimulated hypocotyl length (Bai *et al.*, 2012). In fact, Bai *et al.* (2012) reported that in the light BR-treated *ga1-3* hypocotyls were 2–3 times longer than those of untreated WT hypocotyls. In other words, BR application rescued the hypocotyl phenotype of the GA-deficient mutant. This is a rare case in which a phenotypic aspect of a GA mutant was reversed by BR application. However, Steber and McCourt (2001) carried out a longer-term application experiment and found that epi-BL (an active BR) did not rescue dwarfism in the *ga1-3* mutant. They noted that epi-BL was more effective at promoting hypocotyl elongation than at the later stages of development.

Again consistent with the idea that BZR1 acts to some extent downstream of DELLAs, GA application is ineffective at stimulating *Arabidopsis* hypocotyl elongation when the level of active BZR1 is low (e.g. in the BR synthesis mutant *det2-1* or the receptor/response mutant *bri1-5*), but strongly promotes hypocotyl elongation in the BZR1-accumulating mutant *bzr1-1D* (Bai *et al.*, 2012).

The new model for GA-BR interactions would predict that many genes are regulated by both hormones and Bai *et al.* (2012) presented evidence that this is indeed the case. In *Arabidopsis* hypocotyls, 35% of genes with altered expression in a GA-deficient mutant were also affected in a BR-insensitive mutant. A similar overlap was reported by Gallego-Bartolomé *et al.* (2012). It should be noted, however, that an earlier study found an 'unexpectedly small' number of genes that are coregulated by multiple hormones (Nemhauser *et al.* 2006).

Since BR reportedly rescued the hypocotyl phenotype of a GA-deficient *Arabidopsis* mutant (*ga1-3*), we tested whether BR can rescue the phenotype of a pea (*Pisum sativum*) GA-deficient mutant. Figure 8.5 shows the results of applying epi-BL to internodes of the *ls-1* mutant. The *Arabidopsis ga1-3* and pea *ls-1* mutations both block the production of the key GA precursor, *ent*-kaurene (Hedden and Thomas, 2012), thereby strongly reducing the levels of bioactive GAs. For comparison, the BR synthesis mutant, *lkb*, was also included; this mutation blocks BR synthesis in pea just before the block imposed by *det2* in *Arabidopsis* (Hedden and Thomas, 2012). BR treatment only slightly promoted elongation in the *ls-1* mutant, although the effect was statistically significant (Figure 8.5, 10 ng: $P < 0.01$; 40 ng: $P < 0.05$). Importantly, internode length was not restored to that of the untreated WT: the response to BR was only 12% of the difference between *ls-1* and WT internodes (Figure 8.5). This result contrasts markedly with that reported for the GA-deficient *ga1-3 Arabidopsis* hypocotyls by Bai *et al.* (2012).

In this context it is relevant to note that the mutants affected in GA or BR biosynthesis in pea do not closely resemble each other; certainly they are not 'astonishingly similar' – a description applied to GA and BR mutants in *Arabidopsis* (Jaillais and Vert, 2012). In fact, BR-deficient pea mutants are characterised by stems with 'banding', thought to be due to unequal growth of the epidermis and internal tissues (Ross and Reid, 1986, Reid and Ross, 1989). In pea GA-deficient dwarves, in contrast, banding does not occur. In general terms, the tissue specificity of GA and BR action is an important consideration. Jaillais and Vert (2012) note: 'Brassinosteroids and gibberellin are (both) … involved in root growth, but have been shown to act in the epidermis and endodermis, respectively'. This localised and distinct site of action might imply that DELLA-BZR1 interactions do not play a significant role in regulating root growth.

Interactions between DELLAs and ethylene signalling proteins

DELLAs are thought to intersect with ethylene signalling, by physically interacting with, and repressing the function of, a key positive regulator from that pathway, the protein ETHYLENE INSENSITIVE 3 (EIN3) (Gallego-Bartolomé *et al.*, 2011; An *et al.*, 2012). An *et al.* (2012) performed yeast 2-hybrid assays, which showed that at least some DELLAs (GAI and RGA), interact with EIN3, and that the EIN3 region involved is the amino

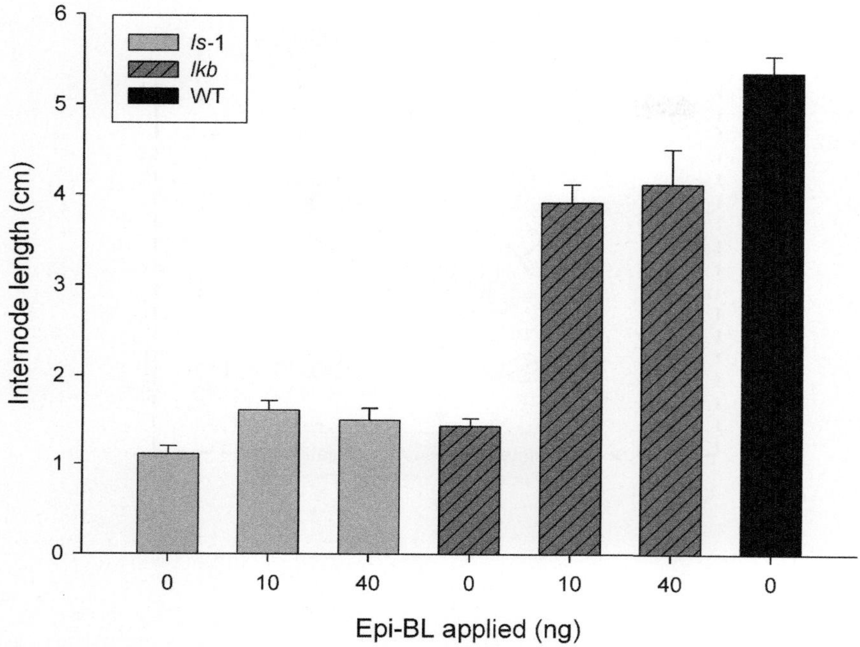

Figure 8.5 BR does not rescue the dwarf phenotype of a GA-deficient pea mutant. The bioactive BR, epi-BL, was applied directly (in 5 μL of ethanol containing 0.15% Tween 20) to the elongating internodes of the GA-deficient pea (*Pisum sativum*) mutant, *ls-1*, or of the BR synthesis pea mutant, *lkb*. Data are means ± standard errors (n = 10) of the final internode length between nodes 5 and 6 (1st scale leaf = node 1). The WT isoline was derived from cv. Torsdag.

acid sequence 200–500, thought to be responsible for DNA binding (Solano *et al.*, 1998; Chen *et al.*, 2009). These findings indicate that by reducing the DNA-binding ability of EIN3, DELLAs diminish the capacity of that protein to regulate its target genes. Similarly, recent evidence indicates that an *Arabidopsis* DELLA protein (GAI) impairs the ethylene-induced transcription factor RELATED TO APETALA2.3 on the target promoters (Marín-de la Rosa *et al.*, 2014).

This interaction appears to have important physiological implications, since the expression of *HOOKLESS1* (*HLS1*), a key EIN3-regulated gene, is repressed by DELLA proteins (An *et al.*, 2012). It is known that *HLS1* is required for apical hook formation, since *hls1* mutations reduce or abolish that process (Lehman *et al.*, 1996). Both GA and ethylene promote apical hook formation, and the model shown in Figure 8.6 describes how the effects of these two hormones are integrated at the level of EIN3. When GA levels are low and DELLAs are consequently abundant, the formation of the DELLA-EIN3 dimer is favoured, *HLS1* expression is down-regulated, and

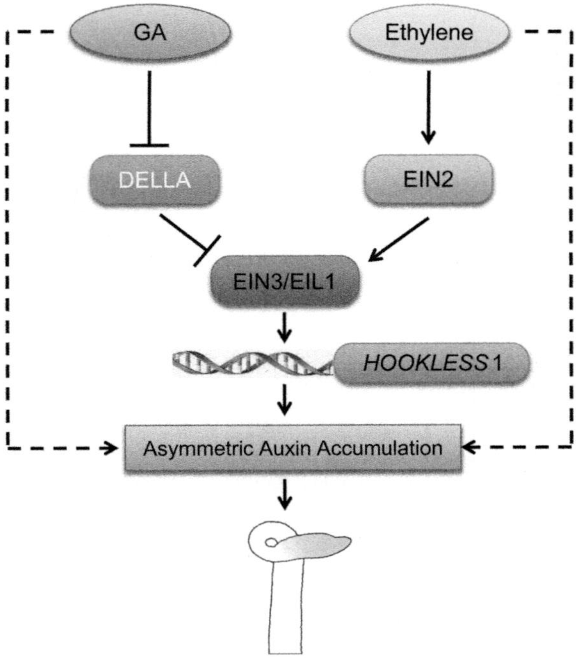

Figure 8.6 GAs and ethylene interact to regulate apical hook formation. Both ethylene and GAs induce *HOOKLESS1 (HLS1)* transcription through the activity of ETHYLENE INSENSITIVE 3/EIN3-LIKE 1 (EIN3/EIL1). Ethylene stabilizes EIN3/EIL1, while GAs remove the inhibitory effect of DELLA proteins on that protein. It has also been proposed (as indicated by the dotted lines) that GAs and ethylene also initiate *HLS1*-independent pathways to regulate hook curvature. (An *et al.* (2012). Reproduced with permission from Nature Publishing Group.) (See insert for colour representation of this figure.)

hook formation is inhibited (An *et al.*, 2012). It is unknown to what extent the proposed GA-ethylene interaction applies to other physiological aspects, although HLS1 is thought to affect a number of processes besides apical hook formation, including hypocotyl elongation (Lehman *et al.*, 1996) and regulation of light signalling (Li *et al.*, 2004).

Interactions between DELLAs and jasmonic acid signalling proteins

GA signalling also interacts, in an antagonistic way, with that of the defence hormone, jasmonic acid (JA). Plants often 'trade off' between growth on the one hand, and defence responses on the other. Usually, plant growth is reduced as a result of investment in defence against factors such as disease, herbivory, wounding or abiotic stresses (Hou *et al.*, 2013). Plants react to stress by increasing the level of JA (Zhang and Turner, 2008), which is perceived by the F-box protein COI1 (Katsir *et al.*, 2008). In the presence of JA, COI1 interacts with proteins known as JAZ proteins, triggering their degradation. Intact JAZ proteins form a complex with another protein, MYC2, with JAZ degradation releasing MYC2 proteins, which activate the expression of JA-responsive genes (Katsir *et al.*, 2008; Yan *et al.*, 2009; Hou *et al.*, 2013).

Recent studies have indicated that DELLAs compete with MYC2 for binding to JAZ proteins when GA levels drop (Figure 8.7; Hou *et al.*, 2010). This leaves unbound MYC2 to act in its role as the transcriptional activator of JA signalling. However, when GA levels increase, MYC2 binds to JAZ proteins with no competition from DELLAs (due to their degradation), and JA signalling is attenuated. Leone *et al.* (2014) used an *Arabidopsis* quintuple DELLA mutant, with impaired gene function, to demonstrate this reallocation of plant resources from defence to growth under varied light treatments. The stability of JAZ10 increased in a shade avoidance response after exposure to predominantly far-red light, reducing the response to JA. Other findings reveal that a number of JAZs stabilise under these light conditions, with a number of MYCs having a higher turnover, further suppressing JA response (Chico *et al.*, 2014)

Some pathogens appear to have evolved interference techniques targeting this GA/DELLA/JA interaction (Pieterse *et al.*, 2014). Navarro *et al.* (2008) found that DELLAs contribute to the intensity of JA signalling in response to pathogen stress. This explains why pathogens may use GA over-accumulation to promote virulence: increased GA levels reduce the DELLA content and consequently the MYC2 activation of JA-response genes (Navarro *et al.*, 2008; Grant and Jones, 2009).

Studies have also shown that JAs may modulate GA-mediated growth (Yang *et al.*, 2012). In the presence of JA, JAZ proteins degrade to release DELLAs from the JAZ-DELLA complex, thus leading to the suppression of GA-mediated growth (Figure 8.7; Hou *et al.*, 2010). In rice, studies on over-expression and knock-out mutants found that an interruption of the JA signalling pathway enhances the GA response, in terms of plant growth and seed germination (Yang *et al.*, 2012). Furthermore, Wild *et al.* (2012) reported that JA rapidly induces the expression of the *Arabidopsis* DELLA-encoding gene, *RGL3*, and that RGL3 is required to 'fully enhance' JA-mediated responses.

It seems possible that within different tissues, slightly different interactions occur between the hormones and signalling proteins involved in cell and organ growth. Recent evidence indicates that GA and JA degrade the DELLA and JAZ proteins, respectively, to cooperatively activate the WD-repeat/bHLH/MYB complex in order to induce trichome initiation in *Arabidopsis* (Qi *et al.*, 2014).

In summary, when JA levels are increased due to stress, GA signalling is diminished. Likewise, with decreased GA levels, JA response genes are triggered. Indeed, it is postulated that a signal cascade involving COI1-JAZ-DELLA-PIF proteins controls the back-and-forth trade off regarding plant defence versus growth (Yang *et al.*, 2012).

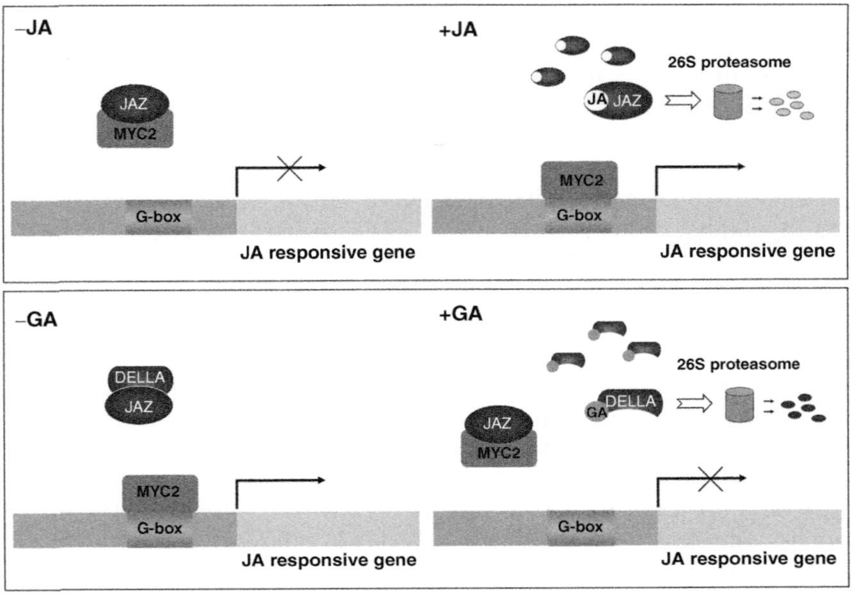

Figure 8.7 DELLAs compete with MYC2 for binding to JAZ, modulating JA signalling. When JA levels are low, JAZ proteins interact with MYC2, inhibiting its activity as a transcriptional activator. When JA levels are high, JAZ proteins are destabilised and MYC2 is released to activate the expression of JA-responsive genes. DELLA proteins compete with MYC2 for binding to JAZ; when GA levels are low, stabilised DELLAs bind to JAZ, leaving unbound MYC2 to activate JA-responsive genes. When GA is present, the DELLAs are degraded and MYC2 is able to bind to JAZs without competition, thus repressing the expression of JA-responsive genes. (Hou *et al.* (2010). Reproduced with permission from Elsevier.) (See insert for colour representation of this figure.)

Interactions between DELLAs and abscisic acid signalling proteins

A recent paper indicates that DELLAs interact with components of the ABA signalling pathway (Lim *et al.*, 2013). Again, yeast 2-hybrid and coimmunoprecipitation assays show that RGA physically interacts with ABA-INSENSITIVE 3 (ABI3) and ABI5. In *Arabidopsis* seeds, high temperatures are thought to promote the formation of an ABI3-ABI5-DELLA complex that targets the promoter of the germination-regulation gene, *SOM*. Activation of SOM, which is favoured by high ABA levels and low GA levels, then inhibits seed germination (Lim *et al.*, 2013).

Interactions between gibberellin and strigolactone signalling

The strigolactones (SLs) are the most recently discovered group of plant hormones, now known to be potent inhibitors of shoot branching. Interestingly, SL signalling is yet another pathway recently implicated with the DELLA proteins. In rice, Nakamura *et al.* (2013) reported that a strigolactone-related protein, D14, physically interacts with the DELLA SLR1, on the basis of yeast 2-hybrid assays. They report that D14 appears to play a pivotal role in SL signalling, by converting SLs into the active form (known as D-OH). Interestingly, in the SRL1 protein, the 'D14-interacting' domain overlaps with the domain responsible for forming the GA-GID1-SLR1 complex (Nakamura *et al.*, 2013). This might be expected since the D14 and GID1 proteins are related, derived from alpha/beta hydrolases. Whether or not the SLR1-D14 interaction occurs *in vivo* has not yet been determined.

On the other hand, there is strong recent evidence from studies on SL- and GA-related mutants that these two hormones act independently on internode length, at least in pea (de Saint Germain *et al.*, 2013). Importantly, de Saint Germain *et al.* (2013) also showed that the reduced stem elongation of SL-deficient mutants is not due to a general weakening of the main stem by the copious lateral branches of these mutants. This study therefore identified a role for SLs in stem elongation in addition to their role in branching.

8.4 Gibberellins and auxin transport

It has been suggested that GAs might play a role in tropisms by influencing auxin transport and consequently the establishment of auxin asymmetry after a tropic stimulus. Willige *et al.* (2011) showed a reduction in auxin transport in the inflorescences of *Arabidopsis* GA biosynthesis and signalling mutants; and this reduction was correlated with a reduction in PIN-FORMED (PIN) auxin efflux facilitators. Treatment of the GA-deficient mutants with GA restored the PIN protein level to that of WT plants. Willige *et al.* (2011) suggested that the gene required for PIN vacuolar targeting is regulated by DELLAs, and that PIN turnover is thereby modulated by GA levels. Löfke *et al.* (2013) also reported that GAs influence PIN stability via an effect on PIN vacuolar targeting. Willige *et al.* (2011) proposed that reduced auxin transport as a result of GA deficiency may also affect other processes, such as cotyledon differentiation, in addition to root gravitropic responses.

These recent reports linking GAs to auxin transport tend to revive an earlier theory, namely that GAs stimulate stem elongation, at least in part, by increasing auxin content (e.g. Law and Hamilton, 1984). Certainly, in the garden pea, there is evidence that GA-deficient mutants contain less IAA in stems than do WT plants (McKay *et al.*, 1994). Previously, this has been attributed in part to a reduced rate of IAA metabolism when GA levels are high (Law and Hamilton, 1984), but in view of the recent data (Björklund *et al.*, 2007; Willige *et al.*, 2011; Löfke *et al.*, 2013), it seems possible that there is also a reduction in IAA transport in GA-deficient mutants. Might the resulting two-fold drop in stem IAA content confer the dwarf habit? Evidence that this is not the case was presented by Ross *et al.* (2002). The strong elongation response of a decapitated GA-deficient mutant to applied GA (Ross *et al.*. 2002) indicates that GA does not promote stem elongation by increasing auxin content. For example, Ross *et al.* (2002) compared the elongation of decapitated stems treated with GA with that of stems from intact plants not treated with GA. The former internodes contained 15 times less IAA than the latter, but elongated more than twice as much. This indicates that GAs act downstream of auxin, not upstream, in regulating stem elongation.

8.5 Conclusion

This chapter has described two main types of GA-related hormone interactions that have received attention over the last 15 years: interactions involving changes in hormone synthesis, and those involving changes in hormone signal transduction. The latter group has been split into two categories: interactions where non-GA hormones affect DELLA stability and those where DELLAs form complexes with proteins from the signalling pathways of other hormones.

Certain interactions from the first group are well-established, and are clearly important for growth. Auxin and ethylene stimulate and inhibit GA biosynthesis, respectively; evidence for these interactions, especially the former, has been obtained from a number of species. However, the extent to which variations in endogenous ethylene content actually affect GA content is still unknown. In assessing interactions involving GAs, we should realise that the plant responds to alterations in bioactive GA content, not to changes in expression of GA synthesis or deactivation genes *per se*. These latter changes are physiologically meaningful only if they lead to changes in hormone level. Therefore, in this review, emphasis has been placed on reports that demonstrate not only changes in synthesis (or deactivation) gene expression, but actual alterations in hormone content as well.

During the 2000s, a series of papers suggested that several hormones, other than GA itself, affect DELLA stability. However, these effects now appear to be unlikely (auxin), attributable to a change in GA content (auxin, ethylene), or have not been repeated using a different approach (ABA). We should seriously question whether such cross-talk plays a physiologically functional role in the plant. Such questioning helps to avoid constructing overly complicated models of hormone interactions.

The second type of signal transduction interaction involves the capacity of DELLAs to physically interact with other key hormone signalling proteins. DELLAs appear to form complexes with BZR1 (from the BR pathway), EIN3 (ethylene), JAZ (jasmonates) and ABI3/5 (ABA). DELLAs also interact with the PIF family of transcription factors. These interactions have been detected by yeast 2-hybrid assays, with their *in planta* occurrence indicated by coimmunoprecipitation. Given that DELLA proteins exist in gene families, as do (typically) their partner proteins, there is potentially a bewildering array of protein complexes that might form. However, specific types of protein–protein interactions may well be limited to certain environmental conditions and/or plant parts, limiting the number of complexes formed at any one time.

It seems likely that the protein–protein complexes involving DELLAs are of considerable physiological significance, especially since DELLAs themselves do not have conserved DNA binding domains. Interacting with transcription factors provides a means for DELLAs to regulate gene expression. However, it should also be stressed that the physiological experiments that are required to test the relevance of the DELLA-containing complexes have, at this stage, been largely restricted to one experimental system, the *Arabidopsis* hypocotyl. We should be continually aware that conclusions from this system are not necessarily applicable to other stages of development, or to other species.

Acknowledgements

We thank the Australian Research Council for financial support, and Scott McAdam for comments on the manuscript.

References

Achard, P., Vriezen, W. H., Van Der Straeten, D. and Harberd, N. P. (2003). Ethylene regulates *Arabidopsis* development via the modulation of DELLA protein growth repressor function. *The Plant Cell* **15**, 2816–2825.

Achard, P., Cheng, H., De Grauwe, L., *et al.* (2006). Integration of plant responses to environmentally activated phytohormonal signals. *Science* **311**, 91–94.

Achard, P., Baghour, M., Chapple, A., *et al.* (2007). The plant stress hormone ethylene controls floral transition via DELLA-dependent regulation of floral meristem-identity genes. *Proceedings of the National Academy of Sciences USA* **104**, 6484–6489.

An, F., Zhang, X., Zhu, Z., *et al.* (2012). Coordinated regulation of apical hook development by gibberellins and ethylene in etiolated *Arabidopsis* seedlings. *Cell Research* **22**, 915–927.

Bai, M.-Y., Shang, J.-X., Oh, E., *et al.* (2012). Brassinosteroid, gibberellin and phytochrome impinge on a common transcription module in *Arabidopsis*. *Nature Cell Biology* **14**, 810–817.

Batge, S. L., Ross, J. J. and Reid, J. B. (1999). Abscisic acid levels in seeds of the gibberellin-deficient mutant *lh-2* of pea (*Pisum sativum*). *Physiologia Plantarum* **105**, 485–490.

Björklund, S., Antti, H., Uddestrand, I., *et al.* (2007). Cross-talk between gibberellin and auxin in development of *Populus* wood: gibberellin stimulates polar auxin transport and has a common transcriptome with auxin. *The Plant Journal* **52**, 499–511.

Bleecker, A. B., Schuette, J. L. and Kende, H. (1986). Anatomical analysis of growth and developmental patterns in the internode of deepwater rice. *Planta* **169**, 490–497.

Bouquin, T., Meier, C., Foster, R., *et al.* (2001). Control of specific gene expression by gibberellin and brassinosteroid. *Plant Physiology* **127**, 450–458.

Brian, P. W. and Hemming, M. G. (1958). Complementary action of gibberellic acid and auxins in pea internode extension. *Annals of Botany* **22**, 1–17.

Catling, D. (1992). *Rice in Deepwater*. Macmillan Press, London.

Chapman, E. J., Greenham, K., Castillejo, C., *et al.* (2012). Hypocotyl transcriptome reveals auxin regulation of growth-promoting genes through GA-dependent and -independent pathways. *PLoS ONE* **7**, e36210.

Chen, H., Xue, L., Chintamanani, S., *et al.* (2009). ETHYLENE INSENSITIVE3 and ETHYLENE INSENSITIVE3-LIKE1 repress *SALICYLIC ACID INDUCTION DEFICIENT2* expression to negatively regulate plant innate immunity in *Arabidopsis*. *The Plant Cell* **21**, 2527–2540.

Chico, J.-M., Fernández-Barbero, G., Chini, A., *et al.* (2014). Repression of jasmonate-dependent defenses by shade involves differential regulation of protein stability of MYC transcription factors and their JAZ repressors in *Arabidopsis*. *The Plant Cell* **26**, 1967–1980.

Claeys, H., De Bodt, S. and Inzé, D. (2014). Gibberellins and DELLAs: central nodes in growth regulatory networks. *Trends in Plant Science* **19**, 231–239.

De Grauwe, L., Dugardeyn, J. and Van Der Straeten, D. (2008). Novel mechanisms of ethylene-gibberellin crosstalk revealed by the *gai eto2-1* double mutant. *Plant Signaling and Behavior* **3**, 1113–1115.

de Saint Germain, A., Ligerot, Y., Dun, E. A., *et al.* (2013). Strigolactones stimulate internode elongation independently of gibberellins. *Plant Physiology* **163**, 1012–1025.

Dorcey, E., Urbez, C., Blázquez, M. A., *et al.* (2009). Fertilization-dependent auxin response in ovules triggers fruit development through the modulation of gibberellin metabolism in *Arabidopsis*. *The Plant Journal* **58**, 318–332.

Feng, S., Martinez, C., Gusmaroli, G., *et al.* (2008). Coordinated regulation of *Arabidopsis thaliana* development by light and gibberellins. *Nature* **451**, 475–479.

Ferguson, B. J., Foo, E., Ross, J. J. and Reid, J. B. (2011). Relationship between gibberellin, ethylene and nodulation in *Pisum sativum*. *New Phytologist* **189**, 829–842.

Foo, E., Platten, J. D., Weller, J. L. and Reid, J. B. (2006). PhyA and cry1 act redundantly to regulate gibberellin levels during de-etiolation in blue light. *Physiologia Plantarum* **127**, 149–156.

Frigerio, M., Alabadí, D., Pérez-Gómez, J., *et al.* (2006). Transcriptional regulation of gibberellin metabolism genes by auxin signaling in Arabidopsis. *Plant Physiology* **142**, 553–563.

Fu, X. and Harberd, N. P. (2003). Auxin promotes *Arabidopsis* root growth by modulating gibberellin response. *Nature* **421**, 740–743.

Fukao, T. and Bailey-Serres, J. (2008). Submergence tolerance conferred by *Sub1A* is mediated by SLR1 and SLRL1 restriction of gibberellin responses in rice. *Proceedings of the National Academy of Sciences USA* **105**, 16814–16819.

Fukao, T. and Xiong, L. (2013). Genetic mechanisms conferring adaptation to submergence and drought in rice: simple or complex? *Current Opinion in Plant Biology* **16**, 196–204.

Fukao, T., Xu, K., Ronald, P. C. and Bailey-Serres, J. (2006). A variable cluster of ethylene response factor-like genes regulates metabolic and developmental acclimation responses to subergence in rice. *The Plant Cell* **18**, 2021–2034.

Gallego-Bartolomé, J., Arana, M. V., Vandenbussche, F., *et al.* (2011). Hierarchy of hormone action controlling apical hook development in *Arabidopsis*. *The Plant Journal* **67**, 622–634.

Gallego-Bartolomé, J., Minguet, E. G., Grau-Enguix, F., *et al.* (2012). Molecular mechanism for the interaction between gibberellin and brassinosteroid signaling pathways in *Arabidopsis*. *Proceedings of the National Academy of Sciences USA* **109**, 13446–13451.

Grant, M. R. and Jones, J. D. G. (2009). Hormone (dis)harmony moulds plant health and disease. *Science* **324**, 750–752.

Griffiths, J., Murase, K., Rieu, I., *et al.* (2006). Genetic characterization and functional analysis of the GID1 gibberellin receptors in *Arabidopsis*. *The Plant Cell* **18**, 3399–3414.

Gubler, F., Chandler, P. M., White, R. G., *et al.* (2002). Gibberellin signaling in barley aleurone cells. Control of SLN1 and GAMYB expression. *Plant Physiology* **129**, 191–200.

Harberd, N. P., King, K. E., Carol, P., *et al.* (1998). Gibberellin: inhibitor of an inhibitor of . . . ? *BioEssays* **20**, 1001–1008.

He, J.-X., Gendron, J. M., Sun, Y., *et al.* (2005). BZR1 is a transcriptional repressor with dual roles in brassinosteroid homeostasis and growth responses. *Science* **307**, 1634–1638.

Hedden, P. and Phillips, A. L. (2000). Gibberellin metabolism: new insights revealed by the genes. *Trends in Plant Science* **5**, 523–530.

Hedden, P. and Thomas, S. G. (2012). Gibberellin biosynthesis and its regulation. *Biochemical Journal* **444**, 11–25.

Hoffmann-Benning, S. and Kende, H. (1992). On the role of abscisic acid and gibberellin in the regulation of growth in rice. *Plant Physiology* **99**, 1156–1161.

Hou, X., Lee, L. Y. C., Xia, K., *et al.* (2010). DELLAs modulate jasmonate signaling via competitive binding to JAZs. *Developmental Cell* **19**, 884–894.

Hou, X., Ding, L. and Yu, H. (2013). Crosstalk between GA and JA signaling mediates plant growth and defense. *Plant Cell Reports* **32**, 1067–1074.

Jackson, M. B. (1985). Ethylene and responses of plants to soil waterlogging and submergence. *Annual Review of Plant Physiology* **36**, 145–174.

Jager, C., Symons, G., Ross, J., *et al.* (2005). The brassinosteroid growth response in pea is not mediated by changes in gibberellin content. *Planta* **221**, 141–148.

Jaillais, Y. and Vert, G. (2012). Brassinosteroids, gibberellins and light-mediated signalling are the three-way controls of plant sprouting. *Nature Cell Biology* **14**, 788–790.

Jasinski, S., Piazza, P., Craft, J., *et al.* (2005). KNOX action in *Arabidopsis* is mediated by coordinate regulation of cytokinin and gibberellin activities. *Current Biology* **15**, 1560–1565.

Katsir, L., Schilmiller, A. L., Staswick, P. E., *et al.* (2008). COI1 is a critical component of a receptor for jasmonate and the bacterial virulence factor coronatine. *Proceedings of the National Academy of Sciences USA* **105**, 7100–7105.

Kende, H. and Zeevaart, J. A. D. (1997). The five 'classical' plant hormones. *The Plant Cell* **9**, 1197–1210.

Kurepin, L. V., Joo, S. H., Kim, S. K., *et al.* (2012). Interaction of brassinosteroids with light quality and plant hormones in regulating shoot growth of young sunflower and *Arabidopsis* seedlings. *Journal of Plant Growth Regulation* **31**, 156–164.

Law, D. M. and Hamilton, R. H. (1984). Effects of gibberellic acid on endogenous indole-3-acetic acid and indoleacetyl aspartic acid levels in a dwarf pea. *Plant Physiology* **75**, 255–256.

Lehman, A., Black, R. and Ecker, J. R. (1996). *HOOKLESS1* an ethylene response gene, is required for differential cell elongation in the *Arabidopsis* hypocotyl. *Cell* **85**, 183–194.

Leone, M., Keller, M. M., Cerrudo, I. and Ballaré, C. L. (2014). To grow or defend? Low red: far-red ratios reduce jasmonate sensitivity in Arabidopsis seedlings by promoting DELLA degradation and increasing JAZ10 stability. *New Phytologist* **204**, 355–367.

Li, H., Johnson, P., Stepanova, A., *et al.* (2004). Convergence of signaling pathways in the control of differential cell growth in *Arabidopsis*. *Developmental Cell* **7**, 193–204.

Li, Q.-F., Wang, C., Jiang, L., *et al.* (2012). An interaction between BZR1 and DELLAs mediates direct signaling crosstalk between brassinosteroids and gibberellins in *Arabidopsis*. *Science Signaling* **5**, ra72.

Lim, S., Park, J., Lee, N., *et al.* (2013). ABA-INSENSITIVE3, ABA-INSENSITIVE5, and DELLAs interact to activate the expression of *SOMNUS* and other high-temperature-inducible genes in imbibed seeds in *Arabidopsis*. *The Plant Cell* **25**, 4863–4878.

Löfke, C., Zwiewka, M., Heilmann, I., *et al.* (2013). Asymmetric gibberellin signaling regulates vacuolar trafficking of PIN auxin transporters during root gravitropism. *Proceedings of the National Academy of Sciences USA* **110**, 3627–3632.

Marín-de la Rosa, N., Sotillo, B., Miskolczi, P., *et al.* (2014). Large-scale identification of gibberellin-related transcription factors defines group VII ETHYLENE RESPONSE FACTORS as functional DELLA partners. *Plant Physiology* **166**, 1022–1032.

McKay, M. J., Ross, J. J., Lawrence, N. L., *et al.* (1994). Control of internode length in *Pisum sativum*. *Plant Physiology* **106**, 1521–1526.

Métraux, J.-P. and Kende, H. (1983). The role of ethylene in the growth response of submerged deep water rice. *Plant Physiology* **72**, 441–446.

Nakamura, H., Xue, Y.-L., Miyakawa, T., *et al.* (2013). Molecular mechanism of strigolactone perception by DWARF14. *Nature Communications* **4**:2613.

Navarro, L., Bari, R., Achard, P., *et al.* (2008). DELLAs control plant immune responses by modulating the balance of jasmonic acid and salicylic acid signaling. *Current Biology* **18**, 650–655.

Nemhauser, J. L., Hong, F. and Chory, J. (2006). Different plant hormones regulate similar processes through largely nonoverlapping transcriptional responses. *Cell* **126**, 467–475.

O'Neill, D., Davidson, S., Clarke, V., *et al.* (2010). Regulation of the gibberellin pathway by auxin and DELLA proteins. *Planta* **232**, 1141–1149.

Oh, E., Yamaguchi, S., Hu, J., *et al.* (2007). PIL5, a phytochrome-interacting bHLH protein, regulates gibberellin responsiveness by binding directly to the *GAI* and *RGA* promoters in *Arabidopsis* seeds. *The Plant Cell* **19**, 1192–1208.

Oh, E., Zhu, J. Y. and Wang, Z. Y. (2012). Interaction between BZR1 and PIF4 integrates brassinosteroid and environmental responses. *Nature Cell Biology* **14**, 802–809.

Oh, E., Zhu, J.-Y., Bai, M.-Y., *et al.* (2014). Cell elongation is regulated through a central circuit of interacting transcription factors in the Arabidopsis hypocotyl. *eLife* **3**, e03031.

Ozga, J. A., Reinecke, D. M., Ayele, B. T., *et al.* (2009). Developmental and hormonal regulation of gibberellin biosynthesis and catabolism in pea fruit. *Plant Physiology* **150**, 448–462.

Pieterse, C. M. J., Pierik, R. and Van Wees, S. C. M. (2014). Different shades of JAZ during plant growth and defense. *New Phytologist* **204**, 261–264.

Piskurewicz, U., Jikumaru, Y., Kinoshita, N., *et al.* (2008). The gibberellic acid signaling repressor RGL2 inhibits *Arabidopsis* seed germination by stimulating abscisic acid synthesis and ABI5 activity. *The Plant Cell* **20**, 2729–2745.

Qi, T., Huang, H., Wu, D., *et al.* (2014). *Arabidopsis* DELLA and JAZ proteins bind the WD-repeat/bHLH/MYB complex to modulate gibberellin and jasmonate signaling synergy. *The Plant Cell* **26**, 1118–1133.

Reed, J. W. (2001). Roles and activities of Aux/IAA proteins in *Arabidopsis*. *Trends in Plant Science* **6**, 420–425.

Reid, J. B. and Ross, J. J. (1989). Internode length in *Pisum*. Two further gibberellin-insensitivity genes, *lka* and *lkb*. *Physiologia Plantarum* **75**, 81–88.

Ross, J. J. and Reid, J. B. (1986). Internode length in *Pisum*. The involvement of ethylene with the gibberellin-insensitive erectoides phenotype. *Physiologia Plantarum* **67**, 673–679.

Ross, J. J., O'Neill, D. P., Smith, J. J., *et al.* (2000). Evidence that auxin promotes gibberellin A_1 biosynthesis in pea. *The Plant Journal* **21**, 547–552.

Ross, J. J., O'Neill, D. P., Wolbang, C. M., *et al.* (2002). Auxin-gibberellin interactions and their role in plant growth. *Journal of Plant Growth Regulation* **20**, 336–353.

Ross, J., O'Neill, D. and Rathbone, D. (2003). Auxin-gibberellin interactions in pea: integrating the old with the new. *Journal of Plant Growth Regulation* **22**, 99–108.

Ross, J. J., Weston, D. E., Davidson, S. E. and Reid, J. B. (2011). Plant hormone interactions: how complex are they? *Physiologia Plantarum* **141**, 299–309.

Santner, A. and Estelle, M. (2009). Recent advances and emerging trends in plant hormone signalling. *Nature* **459**, 1071–1078.

Seo, M., Hanada, A., Kuwahara, A., *et al.* (2006). Regulation of hormone metabolism in *Arabidopsis* seeds: phytochrome regulation of abscisic acid metabolism and abscisic acid regulation of gibberellin metabolism. *The Plant Journal* **48**, 354–366.

Serrani, J. C., Ruiz-Rivero, O., Fos, M. and García-Martínez, J. L. (2008). Auxin-induced fruit-set in tomato is mediated in part by gibberellins. *The Plant Journal* **56**, 922–934.

Solano, R., Stepanova, A., Chao, Q. and Ecker, J. R. (1998). Nuclear events in ethylene signaling: A transcriptional cascade mediated by ETHYLENE-INSENSITIVE3 and ETHYLENE-RESPONSE-FACTOR1. *Genes and Development* **12**, 3703–3714.

Steber, C. M. and McCourt, P. (2001). A role for brassinosteroids in germination in Arabidopsis. *Plant Physiology* **125**, 763–769.

Stewart Lilley, J. L., Gan, Y., Graham, I. A. and Nemhauser, J. L. (2013). The effects of DELLAs on growth change with developmental stage and brassinosteroid levels. *The Plant Journal* **76**, 165–173.

Tanaka, K., Nakamura, Y., Asami, T., *et al.* (2003). Physiological roles of brassinosteroids in early growth of *Arabidopsis*: Brassinosteroids have a synergistic relationship with gibberellin as well as auxin in light-grown hypocotyl elongation. *Journal of Plant Growth Regulation* **22**, 259–271.

Tong, H., Xiao, Y., Liu, D., *et al.* (2014). Brassinosteroid regulates cell elongation by modulating gibberellin metabolism in rice. *The Plant Cell* **26**, 4376–4393.

Trewavas, A. J. (1992). What remains of the Cholodny-Went theory? Forum. *Plant, Cell and Environment* **15**, 761.

Ueguchi-Tanaka, M., Ashikari, M., Nakajima, M., *et al.* (2005). *GIBBERELLIN INSENSITIVE DWARF1* encodes a soluble receptor for gibberellin. *Nature* **437**, 693–698.

van Huizen, R., Ozga, J. A. and Reinecke, D. M. (1997). Seed and hormonal regulation of gibberellin 20-oxidase expression in pea pericarp. *Plant Physiology* **115**, 123–128.

Vandenbussche, F., Vancompernolle, B., Rieu, I., *et al.* (2007). Ethylene-induced *Arabidopsis* hypocotyl elongation is dependent on but not mediated by gibberellins. *Journal of Experimental Botany* **58**, 4269–4281.

Vanstraelen, M. and Benkov, E. (2012). Hormonal interactions in the regulation of plant development. *Annual Review of Cell and Developmental Biology* **28**, 463–487.

Weston, D. E., Reid, J. B. and Ross, J. J. (2009). Auxin regulation of gibberellin biosynthesis in the roots of pea (*Pisum sativum*). *Functional Plant Biology* **36**, 362–369.

Wild, M., Davière, J.-M., Cheminant, S., *et al.* (2012). The *Arabidopsis* DELLA *RGA-LIKE3* is a direct target of MYC2 and modulates jasmonate signaling responses. *The Plant Cell* **24**, 3307–3319.

Willige, B. C., Isono, E., Richter, R., *et al.* (2011). Gibberellin regulates PIN-FORMED abundance and is required for auxin transport-dependent growth and development in *Arabidopsis thaliana*. *The Plant Cell* **23**, 2184–2195.

Wolbang, C. M. and Ross, J. J. (2001). Auxin promotes gibberellin biosynthesis in decapitated tobacco plants. *Planta* **214**, 153–157.

Wolbang, C. M., Chandler, P. M., Smith, J. J. and Ross, J. J. (2004). Auxin from the developing inflorescence is required for the biosynthesis of active gibberellins in barley stems. *Plant Physiology* **134**, 769–776.

Wolbang, C. M., Davies, N. W., Taylor, S. A. and Ross, J. J. (2007). Gravistimulation leads to asymmetry of both auxin and gibberellin levels in barley pulvini. *Physiologia Plantarum* **131**, 140–148.

Xu, K., Xu, X., Fukao, T., *et al.* (2006). *Sub1A* is an ethylene-response-factor-like gene that confers submergence tolerance to rice. *Nature* **442**, 705–708.

Yan, J., Zhang, C., Gu, M., *et al.* (2009). The *Arabidopsis* CORONATINE INSENSITIVE1 protein is a jasmonate receptor. *The Plant Cell* **21**, 2220–2236.

Yang, D. L., Yao, J., Mei, C. S., *et al.* (2012). Plant hormone jasmonate prioritizes defense over growth by interfering with gibberellin signaling cascade. *Proceedings of the National Academy of Sciences USA* **109**, E1192–E1200.

Yin, C., Gan, L., Ng, D., *et al.* (2007). Decreased panicle-derived indole-3-acetic acid reduces gibberellin A_1 level in the uppermost internode, causing panicle enclosure in male sterile rice Zhenshan 97A. *Journal of Experimental Botany* **58**, 2441–2449.

Zentella, R., Zhang, Z. L., Park, M., *et al.* (2007). Global analysis of DELLA direct targets in early gibberellin signaling in *Arabidopsis*. *The Plant Cell* **19**, 3037–3057.

Zhang, Y. and Turner, J. G. (2008). Wound-induced endogenous jasmonates stunt plant growth by inhibiting mitosis. *PLoS ONE* **3**, e3699.

Annual Plant Reviews (2016) **49**, 253–284 http://onlinelibrary.wiley.com
doi: 10.1002/9781119210436.ch9

Chapter 9

GIBBERELLINS AND SEED GERMINATION

Terezie Urbanova[1] and Gerhard Leubner-Metzger[1,2]

[1] *Laboratory of Growth Regulators, Faculty of Science, Palacký University and Institute of Experimental Botany AS CR, UP&IEB AVCR and Centre of the Region Haná for Agricultural and Biotechnological Research, Czech Republic*
[2] *Royal Holloway University of London, School of Biological Sciences, Plant Molecular Science and Centre for Systems and Synthetic Biology, UK; Web: www.seedbiology.eu*

Abstract: Gibberellins (GA) promote seed germination, but they are not simple 'Go Ahead' (GA) molecules as the insight gained into the molecular mechanisms underlying their role in seeds appears to be complex. This chapter covers their central role in mediating the environmental and developmental control of seed germination, and how this differs from vegetative growth processes. Spatiotemporal patterns of GA metabolism and GID1-type receptor signalling in the key seed compartments determine tissue interactions and germination timing in response to ambient environmental cues. Gibberellins are key players in seed temperature responses; during thermoinhibition they interact with other hormonal pathways. Allelochemicals such as myrigalone A inhibit seed germination by specific interference with GA metabolism and signalling. This reveals important ecophysiological roles for GAs in seeds and suggests that they are fundamental for studying species adaptation and interaction in natural and agricultural ecosystems upon climate change.

Keywords: Abiotic stress and thermoinhibition, allelochemical myrigalone A, *Arabidopsis thaliana*, coat dormancy release, Delay of Germination1 dormancy gene, embryo growth potential, endosperm weakening, GID1-type gibberellin receptor signalling, *Lepidium sativum*, seed gibberellin metabolism

Annual Plant Reviews, Volume 49: The Gibberellins, First Edition.
Edited by Peter Hedden and Stephen G. Thomas.

9.1 Introduction

The gibberellin (GA) requirement of seed germination and the importance of *de novo* biosynthesis of bioactive GAs in imbibed seeds were recognized already during the early phase of GA research (e.g. Hashimoto and Yamaki, 1959; Ikuma and Thimann, 1960; Yomo and Iinuma, 1966). The GA requirement for seed germination was also instrumental in screens at the dawn of *Arabidopsis thaliana* (*Arabidopsis*) mutant research: Koornneef and van der Veen (1980) distinguished between 'germinating GA-dwarfs' and 'non-germinating GA-dwarfs' to isolate GA-deficient and GA-insensitive *Arabidopsis* mutants such as *ga1* and *gai* (Koornneef and van der Veen, 1980; Koornneef *et al.*, 1985; Sun *et al.*, 1992; Peng *et al.*, 1997; Koornneef and Meinke, 2010). Treatment with bioactive GA induces the germination of the GA-deficient mutant seeds, and is also used to break seed dormancy and induce seed germination of *Arabidopsis* and other model species, as well as many horticultural species. In this chapter we focus on GA metabolism and signalling during seed germination, with the focus on the non-dormant seed state; for a recent review on the dormant seed state see Graeber *et al.* (2012).

9.2 Spatiotemporal expression of gibberellin metabolism during Brassicaceae seed germination

The mature seeds of most angiosperms consist of the embryo surrounded by the diploid maternal testa (seed coat) and the triploid endosperm as distinct covering layers (Finch-Savage and Leubner-Metzger, 2006; Holdsworth *et al.*, 2008; Linkies *et al.*, 2010). Seed germination of the Brassicacea *Arabidopsis*, *Lepidium sativum* and *Sisymbrium officinale* progresses from imbibition and swelling by water uptake to the successively visible rupture of the testa and the endosperm (Figure 9.1a) (Liu *et al.*, 2005; Müller *et al.*, 2006; Iglesias-Fernandez and Matilla, 2010). The endosperm rupture is associated with visible radicle protrusion considered as the completion of the seed germination process, and is followed by seedling growth and establishment. Endosperm rupture and radicle protrusion depend on the balance between the weakening of the micropylar endosperm (CAP) surrounding the radicle and the increase in the embryo growth potential required for the elongation of the lower hypocotyl-radicle axis (RAD) (Nonogaki, 2006; Holdsworth *et al.*, 2008; Sliwinska *et al.*, 2009; Linkies and Leubner-Metzger, 2012). The completion of germination by endosperm weakening required for endosperm rupture is promoted by GA, which interacts with other promoting, e.g. ethylene, or inhibiting, e.g. abscisic acid (ABA), hormones, and the weakening also requires proteolysis in the CAP (Linkies *et al.*, 2009, 2010; Morris *et al.*, 2011). Environmental cues, including light, temperature and allelochemicals, mediate their effects, at least in part, by tissue-specific

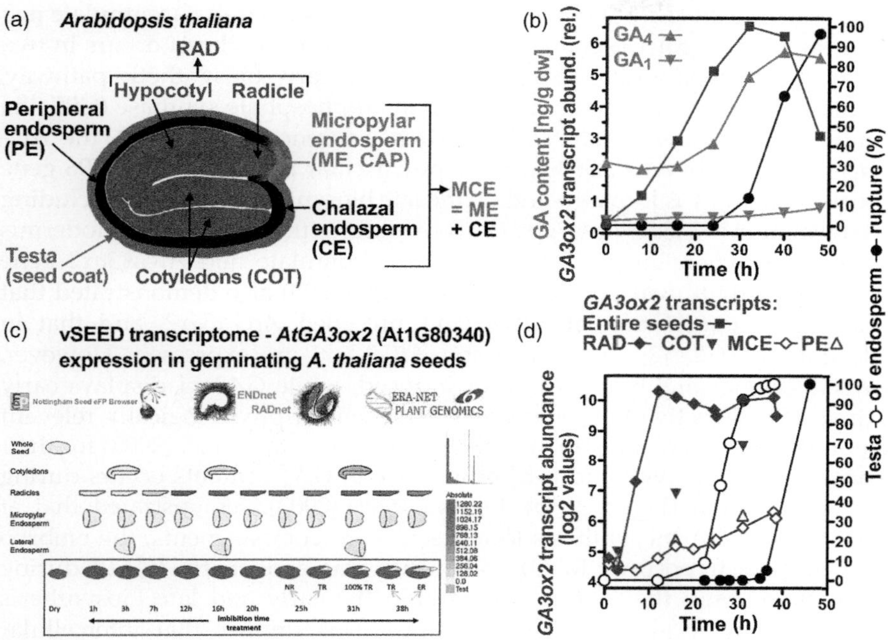

Figure 9.1 Spatio-temporal expression of the GA-biosynthetic gene *AtGA3ox2* during *Arabidopsis thaliana* seed germination. (a) Structure of a mature seed showing key seed compartments, including RAD (lower one-third of the hypocotyl/radicle axis, embryo growth zone) and CAP/ME (micropylar endosperm). (b) *AtGA3ox2* transcript abundance and bioactive GA_1 and GA_4 contents during seed germination. Note the increase in GA_4 content during late germination. (c) Seed compartment-specific transcriptome analysis during seed germination conducted by the ERA-NET Plant Genomics Consortium vSEED (eFP Browser results from vseed.nottingham.ac.uk). (d) Spatio-temporal *AtGA3ox2* expression during seed germination in relation to the kinetics of testa and endosperm rupture. (Results from (b) Ogawa *et al.*, 2003 and (d) Dekkers *et al.*, 2013.) (See insert for colour representation of this figure.)

alteration of hormone contents and responsiveness (e.g. Kucera *et al.*, 2005; Yamaguchi *et al.*, 2007; Preston *et al.*, 2009; Seo *et al.*, 2009; Weitbrecht *et al.*, 2011; Barua *et al.*, 2012; Oracz *et al.*, 2012). The importance of the tissue-specific nature of this regulation is, for example, evident from the differences in the GA metabolite contents in the *L. sativum* key seed compartments CAP and RAD (Oracz *et al.*, 2012) and from the distinct temporal and spatial pattern of GA biosynthesis and response gene expression in germinating *Arabidopsis* seeds (Ogawa *et al.*, 2003).

The temporal and spatial expression patterns of GA biosynthesis genes have been intensively studied during *Arabidopsis* seed germination (Yamaguchi *et al.*, 2001; Ogawa *et al.*, 2003; Yamauchi *et al.*, 2004; Rieu *et al.*, 2008; Preston *et al.*, 2009) and are summarised in several reviews (Kucera *et al.*,

2005; Yamaguchi *et al.*, 2007; Seo *et al.*, 2009). Bioactive GAs accumulate just prior to radicle protrusion and it seems that GA biosynthesis occurs in two separate locations within the embryo: (1) the early biosynthetic pathway, including the steps catalysed by *ent*-copalyl diphosphate synthase (CPS, the *Arabidopsis GA1* gene At4G02780) and *ent*-kaurene oxidase (KO, the *Arabidopsis GA3* gene At5G25900), in the provascular tissue where *AtCPS* gene promoter activity is localised and (2) the late biosynthetic pathway, including the formation of bioactive GA by GA 3-oxidase, in the cortex and endodermis of the root where *AtGA3ox2* transcripts accumulate and *AtGA3ox2* gene promoter activity was detected. Ogawa *et al.* (2003) also demonstrated that transcript accumulation of *AtGA20ox1* preceded *AtGA3ox2*, and that in addition *AtGA20ox2*, *AtGA20ox3* and *AtGA3ox1* are expressed. However, *GA20ox2* is not highly expressed or induced, while *GA3ox1* displays early expression. Bioactive GA_4 was already present in physiologically relevant amounts in the dry, after-ripened seeds used by Ogawa *et al.* (2003) for their transcriptome analysis and further increase in GA_4 contents occurs during late germination (Figure 9.1b). Ogawa *et al.* (2003) demonstrated that at least the late GA biosynthesis localises to both compartments, the embryo (radicle plus hypocotyl, RAD) and micropylar endosperm (CAP) during germination, and that within the embryo the early and late biosynthesis pathway may localise to distinct tissues. This implies that intercellular transport of an intermediate of the GA biosynthetic pathway (probably *ent*-kaurene) is required to produce bioactive GA (Kucera *et al.*, 2005).

A recent transcriptome analysis of *Arabidopsis* seed germination (Dekkers *et al.*, 2013) was conducted with a very high temporal and spatial resolution and in relation to the kinetics of testa and endosperm rupture (ER) (Figure 9.1c). The onset of testa rupture (TR) was at around 20 h in the population of imbibed seeds and the completion of TR was at *ca.* 31 h. The onset of endosperm rupture was around 31 h and the completion of ER was at *ca.* 45 h. RNA from defined seed compartments was extracted along the germination time course and, for the 25-h and 38-h time points, non-ruptured and ruptured seeds regarding TR and ER, respectively, were analysed separately. The transcriptome data set of Dekkers *et al.* (2013) reveals two transcriptional phases during germination that are separated by testa rupture. The first phase is marked by large transcriptome changes upon seed imbibition. The second transcriptional phase starts with testa rupture. At the 25-h transition time point (roughly 50% TR, Figure 9.1d) between the two phases these authors analysed the transcriptomes of seeds with and without TR separately. Seed compartment-specific transcriptome analysis was conducted from RAD (radicle plus hypocotyl) and MCE (micropylar endosperm (ME/CAP) plus chalazal endosperm (CE)) for all time points and in addition for COT (cotyledons) and PE (peripheral endosperm) at selected time points (Figure 9.1). The transcripts for the enzymes for *ent*-kaurene formation in the plastid, AtCPS and *ent*-kaurene synthase (AtKS, the *Arabidopsis GA2* gene At1G79460) were more abundant in the embryo compartments

(RAD, COT) compared to the endosperm (MCE, PE). Transcript abundance for genes encoding the enzymes AtKO and *ent*-kaurenoic acid oxidase (the *AtKAO1* and *AtKAO2* genes At1G05160 and At2G32440, respectively; KAO enzymes localise to the endoplasmatic reticulum) which catalyse the formation of GA_{12} (Hedden and Thomas, 2012), were higher in the RAD compared to the other three compartments (Dekkers *et al.*, 2013), supporting the hypothesis that the RAD is the major GA_{12} production site during *Arabidopsis* seed germination. This is in agreement with earlier work demonstrating that the RAD is a major site for the early GA biosynthesis pathway (Ogawa *et al.*, 2003; Kucera *et al.*, 2005; Seo *et al.*, 2009).

The GA_{12} metabolite marks the transition to the late GA biosynthesis pathway, which is localised in the cytosol (Hedden and Thomas, 2012). Of the five *Arabidopsis* GA 20-oxidase genes, three are expressed during seed germination (Yamaguchi *et al.*, 2007). Their transcript expression is induced very early upon seed imbibition and this gene expression differs between the seed compartments (Figure 9.2). *GA20ox1* is mainly expressed in the RAD and there is a striking decrease in the RAD *GA20ox1* transcript abundance associated with the transition between the two transcriptional phases upon testa rupture (see 25 h in Figure 9.2). This decrease in transcript abundance is also evident for *GA20ox2* and *GA20ox3* in the RAD, but is not evident for the MCE. *GA20ox2* transcript expression in the MCE peaked during early germination, and *GA20ox3* transcript expression is highest in the MCE and COT throughout germination (Figure 9.2). The late germination phase, after testa rupture and just prior to endosperm rupture, is therefore characterised by decreased GA 20-oxidase transcript abundances in the RAD, but considerable high *GA20x3* transcript abundance in the MCE. The GA 20-oxidases produce GA_9 and GA_{20}, the direct precursors of the bioactive GA_4 and GA_1, respectively (Figure 9.2). The distinct transcript expression patterns of GA 20-oxidases suggest that both the RAD and the MCE are able to produce GA_9 and GA_{20} during the early germination phase. Of the four *Arabidopsis* GA 3-oxidases, two are expressed during seed germination (Ogawa *et al.*, 2003; Seo *et al.*, 2009). This earlier finding is in agreement with the results obtained from the vSEED transcriptome published by Dekkers *et al.* (2013), but their spatial resolution provides new insight (Figure 9.2): During the early phase of germination *GA3ox1* and *GA3ox2* transcripts accumulate rapidly in the RAD, and in the late phase of germination in all seed compartments. For *GA3ox1* there is a striking increase in the RAD and MCE transcript abundances associated with the transition between the two transcriptional phases upon testa rupture (see 25 h in Figure 9.2). GA 2-oxidase genes are not expressed during germination, except for the very late phase in which *GA2ox6* is induced in the MCE and *GA2ox2* in the RAD. Taken together, this suggests that the RAD can convert direct precursors into bioactive GA_4 and GA_1 already during early germination, and that the further increase of bioactive GA is associated with enhanced expression of *GA3ox1* and *GA3ox2* transcript expression in the RAD and MCE upon testa rupture. Bioactive GA production by GA 3-oxidases in

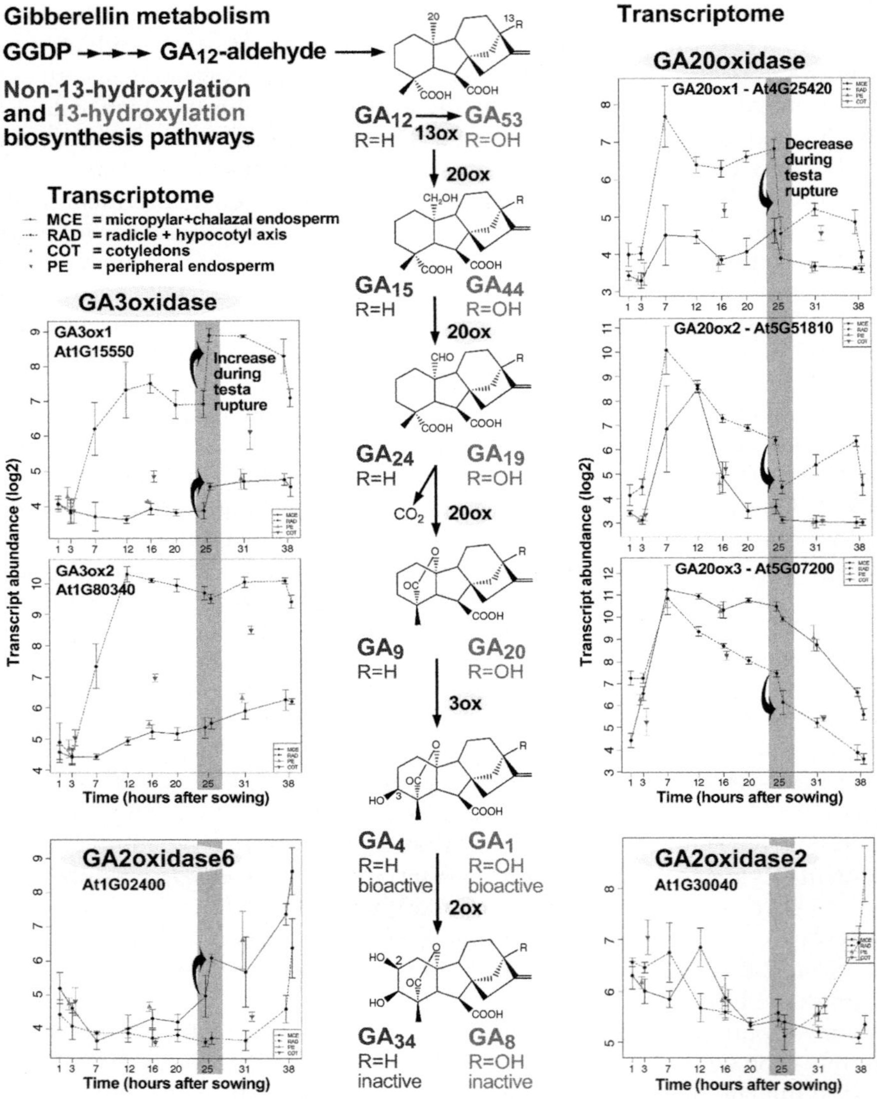

Figure 9.2 Spatiotemporal gene expression patterns of the *Arabidopsis thaliana* GA metabolic pathway during seed germination. Seed compartment-specific transcriptome results (vseed.nottingham.ac.uk) are presented for the GA 3-, 20-, and 2-oxidase genes, which show major regulation in imbibed seeds. Note that for *GA3ox1* the transcript abundance increases rapidly upon testa rupture (marked as grey area) in RAD and MCE (CAP+CE). The same is evident for *GA2ox6* in MCE, while for the three *GA20ox* genes there is a drastic decrease in transcript abundance in the RAD upon testa rupture. Transcript abundances are from the transcriptome of Dekkers *et al.* (2013) available at vseed.nottingham.ac.uk. The metabolites of the GA pathway are shown with the enzymatic steps indicated.

the late germination phase may therefore occur in all seed compartments. Testa rupture marks an important transition during germination and is associated with marked changes in *GA20ox* and *GA3ox* gene expression.

Two roles for bioactive GA during seed germination have been proposed (Kucera *et al.*, 2005; Nonogaki, 2006; Yamaguchi *et al.*, 2007; Linkies and Leubner-Metzger, 2012). Gibberellins increase the growth potential of the embryo to enable RAD elongation by cell growth and are necessary to overcome the mechanical restraint of seed covering layers by weakening of the tissues surrounding the radicle (testa, endosperm CAP). The early induction of GA biosynthesis gene transcripts in the *Arabidopsis* RAD is in concert with the hypothesis that an embryonic GA metabolite and/or bioactive GA itself (the base-level already evident, Figure 9.1b) can diffuse early on during imbibition to the CAP to make it competent for the subsequent weakening during the late germination phase. The transcript expression patterns for GA biosynthesis genes obtained by the vSEED transcriptome analysis (Figures 9.1 and 9.2) suggest that the conversion of inactive precursors into bioactive GAs is further enhanced upon testa rupture by the enhanced expression of *GA3ox* in the embryo (RAD, COT) and the endosperm (MCE, PE), leading to higher GA_4 contents (Figure 9.1b) required for endosperm CAP weakening. Indirect evidence for the timing of endosperm weakening of *Arabidopsis* seeds (Debeaujon and Koornneef, 2000; Bethke *et al.*, 2007) is in agreement with this.

Seed germination of GA-deficient mutants depends on the addition of GA to the medium during imbibition (Kucera *et al.*, 2005). *Arabidopsis* GA-deficient *ga1* and ABA-deficient *aba1* mutants, as well as several testa mutants have been studied regarding their GA requirement for dormancy release and seed germination (Debeaujon and Koornneef, 2000; Debeaujon *et al.*, 2000; North *et al.*, 2010). Many testa mutants exhibit reduced seed dormancy due to reduced testa tannin pigmentation. In the presence of GA biosynthesis inhibitors, or when transferred to a GA-deficient background, they are more sensitive to exogenous GA than wild-type. The germination capacity of the *ga1* mutant can be restored, without any contribution of exogenous GA, by removing the surrounding testa and endosperm, or by transferring it to a testa-mutant background. Debeaujon and Koornneef (2000) concluded that dormancy and germination are the consequence of a balance between many promoting and inhibiting factors, such as GA and ABA, which have the embryo and the testa as targets. Their results support the view that the GA requirement for dormancy release and germination is determined by ABA produced in the developing seeds and/or the state of dormancy set by ABA, as well as the amount of ABA produced upon imbibition, especially in dormant seeds. Furthermore, when the testa mutations weaken the restraint to radicle protrusion, the embryo growth potential threshold required for germination is decreased. Therefore, the testa characteristics, embryonic growth potential and embryonic ABA are the determining properties for the GA requirement of *Arabidopsis* seed germination (Kucera *et al.*, 2005).

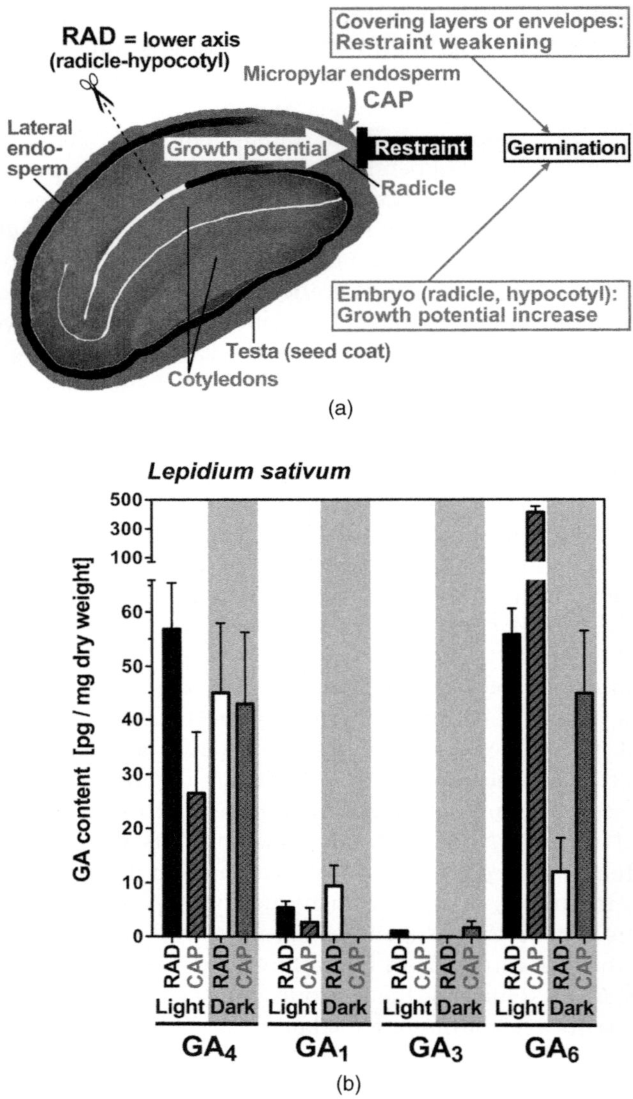

Figure 9.3 Opposing forces during seed germination and seed compartment-specific analysis of bioactive GAs in *Lepidium sativum*. (a) Seed germination is promoted by the growth potential of the embryonic axis (RAD: radicle plus lower hypocotyl) and inhibited by the restraint of the seed covering layers (testa, endosperm). The completion of germination by radicle emergence and endosperm rupture occurs when the increasing embryo growth potential overcomes the restraint of the micropylar endosperm (CAP). *L. sativum* seeds are a Brassicaceae model to study endosperm weakening as an important developmental process that precedes endosperm rupture. (From Graeber *et al.*, 2010.) (b) The contents of bioactive GAs GA_4, GA_1, GA_3 and GA_6, were quantified in both RAD and CAP during late germination (at 15 h, i.e. just prior to endosperm rupture) in light- and dark-imbibed seeds, respectively. (Results compiled from Oracz *et al.*, 2012 and Voegele *et al.*, 2012.)

Gibberellins are important during both the early and the late phases of germination and counteract ABA inhibition. Due to rapid ABA degradation, the ratio of GA/ABA increases *ca.* threefold during early germination and *ca.* 10-fold during late germination of non-dormant *Arabidopsis* seeds (Weitbrecht *et al.* 2011). While for the early germination phase Ogawa *et al.* (2003) did not find altered ABA contents upon treatment of GA-deficient *ga1-3 Arabidopsis* seeds with exogenous GA, Yano *et al.* (2009) found that GA_4 contents and *GA3ox1* transcript levels were decreased in ABA-over-producing *cyp707a2 Arabidopsis* seeds (Yano *et al.*, 2009). ABA therefore can inhibit GA biosynthesis during early seed germination. Transcript expression of specific *AtGA20ox*, *AtGA3ox* and *AtGA2ox* genes were demonstrated to be regulated by light (via phytochrome signalling) and temperature (cold-stratification and thermoinhibition) in imbibed *Arabidopsis* seeds (Yamaguchi *et al.*, 2007; Toh *et al.*, 2008; Seo *et al.*, 2006, 2009, 2011; Toh *et al.*, 2012a). *GA20ox* and *GA3ox* genes are induced by red light and cold-stratification. Moist cold-stratification of *Arabidopsis*, i.e. incubation of imbibed seeds at 4 °C in darkness for usually 1-4 days, is routinely used to break dormancy and promote subsequent germination in the light. Yamauchi *et al.* (2004) found that cold stratification is related to the accumulation of *GA20ox* and *GA3ox* transcripts and by increased contents of bioactive GAs (Figure 7c in Weitbrecht *et al.*, 2011). Furthermore, cold-stratification induced a spatial change in *GA3ox1* transcript expression in that it strongly accumulated in the CAP as well as in the RAD (Yamauchi *et al.*, 2004; Weitbrecht *et al.*, 2011). The interaction between the different key seed compartments is therefore of utmost importance for the control of seed germination by GA.

L. sativum seeds (Figure 9.3) are similar in structure and physiology to *Arabidopsis* seeds, for example with respect to the presence of a thin living endosperm layer surrounding the mature seed and a two-step germination process with visible testa and endosperm rupture (e.g. Müller *et al.*, 2006; Linkies *et al.*, 2009, 2010; Morris *et al.*, 2011; Linkies and Leubner-Metzger, 2012; Voegele *et al.*, 2012). *L. sativum* seeds, however, differ from *Arabidopsis* seeds in that they are larger, non-dormant, and do not require light for their germination. The larger seed size of *L. sativum* enables the direct biomechanical quantification of the endosperm CAP weakening (Müller *et al.*, 2006; Linkies *et al.*, 2010), as well as seed compartment-specific analysis of transcript and hormone contents, as demonstrated for the CAP and RAD (Linkies *et al.*, 2010; Voegele *et al.*, 2011; Oracz *et al.*, 2012; Voegele *et al.*, 2012). In *L. sativum* seeds the bioactive forms GA_4, GA_6, GA_1 and GA_3 were detected in both RAD and CAP during late germination (at 15 h, Figure 9.3b); it is therefore evident that both the 13-hydroxylation and non-13-hydroxylation pathways actively operate in *L. sativum* germinating seeds (Figure 9.4). GA_4 as the non-13-hydroxylated product was, on a dry-weight-basis, about twofold more abundant in the RAD compared to the CAP in the light; in dark-imbibed seeds the GA_4-contents in these tissues were equal (Figure 9.3b). Compared to GA_4, the GA_1 (corresponding

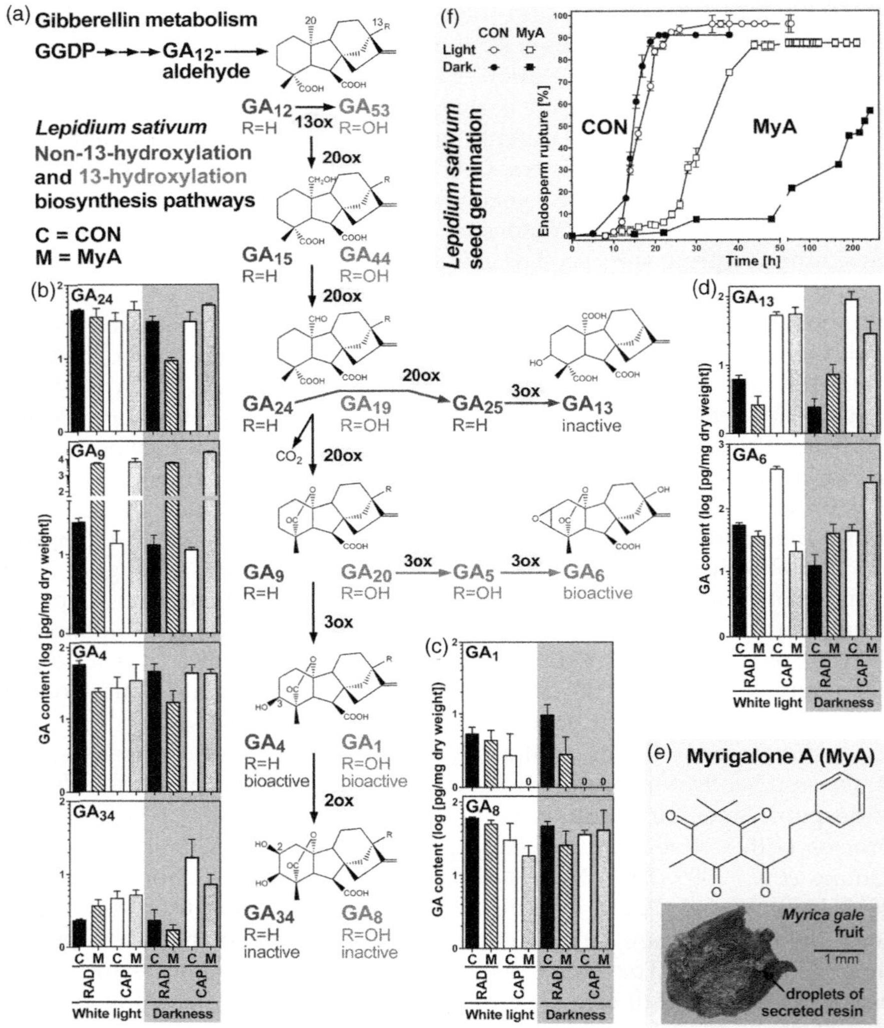

Figure 9.4 The compartment-specific (RAD, CAP) and light-dependent effect of myrigalone A (MyA) on GA metabolism during germination of *L. sativum* seeds. (a) The non-13-hydroxylation and 13-hydroxylation GA biosynthesis and inactivation pathways and important metabolites detected in *L. sativum* seeds. (b, c, d) Contents of bioactive GAs GA_1, GA_4 and GA_6, and precursors and inactive forms GA_{24}, GA_9, GA_{34}, GA_8, GA_{13} quantified in RAD and CAP excised from control (CON) and MyA-treated seeds incubated in continuous light or darkness for 15 h. (e) Chemical structure of MyA and *Myrica gale* fruit with resin droplets containing essential oils. (f) The inhibiting effect of 5×10^{-4} M MyA on the germination of *L. sativum* seeds imbibed in continuous light or in darkness. (Compiled results from Oracz *et al.*, 2012 and Voegele *et al.*, 2012.)

13-hydroxylated product) contents were low, and only traces were detected for GA_3. The 13-hydroxylated product GA_6 was detected at comparable levels to GA_4 in RAD and CAP, except for the CAP from light-imbibed seeds, where it accumulated 15-fold compared to GA_4.

Based on their work with *L. sativum* seeds, Oracz *et al.* (2012) proposed that in addition to GA_4, also GA_6 also plays an important role in seed germination. The unambiguous confirmation that the identified compound in *L. sativum* seeds is indeed GA_6 requires further analytical work in which verification with another method has to be obtained. Bioactivity of GA_6 has been proposed for the induction of flowering and stem extension growth in the grass *Lolium temutentum* (King *et al.*, 2003). GA 3-oxidases, including AtGA3ox1 expressed in *Arabidopsis* seeds (Figure 9.2), and the GA3ox from *Phaseolus* seeds and wheat grains, have been shown to have 2,3-epoxidase activity to produce GA_6 from GA_5 (Kwak *et al.*, 1988; Zhou *et al.*, 2004; Appleford *et al.*, 2006). The 2,3-epoxide group of GA_6 confers its resistance to inactivation by GA 2-oxidases. GA_6 may also serve for transport or accumulation as it is a stable bioactive GA (Pimenta Lange and Lange, 2006; Yamaguchi, 2008). As GA_6 accumulates specifically in the CAP of *L. sativum* seeds, it was proposed that it contributes, together with GA_4, to the endosperm CAP weakening (Oracz *et al.*, 2012; Voegele *et al.*, 2012; Graeber *et al.*, 2014). From these compartment-specific measurements (Figure 9.3b) the combined bioactive GA_4 and GA_6 concentrations can be estimated as *ca.* 20 nM in the RAD and *ca.* 125 nM in the CAP of *L. sativum* seeds imbibed in continuous light.

Figure 9.4 shows that GA_8 and GA_{13} were abundant inactive metabolites in *L. sativum* seeds (Oracz *et al.* 2012; Voegele *et al.* 2012). GA_8 is the 2β-hydroxylated inactivation product from GA_1 and its accumulation explains the low GA_1 contents. GA_{13} contents were 9- and 37-fold in CAP compared to RAD of seeds imbibed in continuous light and in darkness, respectively (Figure 9.4). It is known as an unusual product found in developing seeds requiring atypical GA20ox and GA3ox enzymes, its function is unknown and it is inactive (Pimenta Lange and Lange, 2006). Work on developing pumpkin seeds demonstrated that a GA 3-oxidase which converts GA_9 to GA_4, also converts GA_{25} to GA_{13} (Frisse *et al.* 2003). In support for this, GA_{13} contents were reduced in seedlings of the *A. thaliana ga3ox1* mutant (Talon *et al.*, 1990). As GA_{13} can bind to GA 2-oxidases, its accumulation in *L. sativum* CAP tissues may serve as GA2ox activity inhibitor and thereby prevent GA_4 inactivation. Recent work by Nomura *et al.* (2013) identified AtCYP714A1 (At5G24910) as an enzyme which converts GA_{12} to biologically inactive 16-carboxylated GA_{12}. During the early germination of *Arabidopsis* seeds, transcripts of AtCYP714A1 were highly abundant in the MCE and PE, but declined rapidly during the late phase and were low in the RAD and COT (vseed.nottingham.ac.uk, Dekkers *et al.* 2013). Epoxidation and hydroxylation, including those catalysed by GA 3-, GA 2- and GA 13-oxidases, mediate the fine-tuning of late GA metabolism (Hedden and Thomas, 2012; Magome *et al.*, 2013; Nomura *et al.*, 2013) and for germinating seeds this is also evident from the compartment-specific enzyme transcript and GA metabolite contents (Figures 9.1 to 9.4).

A recent publication by Graeber *et al.* (2014) shows that the dormancy gene *DOG1 (DELAY OF GERMINATION1)* controls dormancy by setting the optimal ambient temperature window for germination. This is achieved by temperature-dependent alteration of GA metabolism, while ABA metabolism is not appreciably affected. Over-expression of *AtDOG1* in transgenic *L. sativum* seeds leads to a generally enhanced GA biosynthesis by up-regulated *KAO* expression and an altered temperature-responsiveness of *GA20ox* expression. These findings suggest that DOG1 interferes with the negative feedback regulation of *GA20ox* in a temperature-dependent manner (Graeber *et al.*, 2014). The altered GA metabolism leads in turn to altered expression of genes required for the biomechanical weakening of the coats encasing the embryo. The over-expression of *DOG1* does not affect the embryo growth potential. Regulation of GA metabolism is therefore a key process in the DOG1-mediated conserved coat-dormancy mechanism of seeds.

9.3 Gibberellin signalling and seed germination

9.3.1 The GID1ac and GID1b pathways in seeds

The discovery that *GID1 (GIBBERELLIN INSENSITIVE DWARF 1)* encodes a soluble GA receptor in rice (Ueguchi-Tanaka *et al.*, 2005) and subsequent work on the GID1 receptors of other species (e.g. Griffiths *et al.*, 2006; Iuchi *et al.*, 2007; Willige *et al.*, 2007; Voegele *et al.*, 2011; Hauvermale *et al.*, 2012) has considerably advanced our understanding of the GA signal transduction cascade. The DELLA proteins are repressors of GA signalling and repress the GA responses, e.g. seedling growth and seed germination, by their interaction with transcription factors (see Chapter 7). Bioactive GAs promote DELLA repressor degradation in an environmentally and developmentally dependent manner. The GA-dependent alleviation of the DELLA-imposed repression is achieved by GID1 binding GA, which in turn enables GID1-DELLA complex formation resulting in DELLA recognition, ubiquitination, followed by 26S-proteasome-mediated DELLA degradation (see Chapter 6). Despite evidence, summarised in Chapter 1, indicating that GA-mediated induction of α-amylase genes in the aleurone of germinating cereal grains involves a plasma membrane receptor, recent work by Yano et al. (2015) strongly suggests that GA signalling in the rice aleurone utilises only OsGID1. GA signalling in the cereal aleurone has been reviewed by Sun and Gubler (2004) and is also discussed in Chapter 7.

Arabidopsis contains three GID1 receptors: AtGID1a, AtGID1b and AtGID1c. Figure 9.5 shows that environmental, hormonal and developmental cues affect expression of the three *AtGID1* genes. Gibberellins promote seedling growth and seed germination, while ethylene inhibits seedling growth, but promotes seed germination (Linkies *et al.*, 2009; Voegele *et al.*, 2011; Linkies and Leubner-Metzger, 2012). In germinating *Arabidopsis* seeds, treatment with GA or with ACC (1-aminocyclopropane-1-carboxylic acid, the direct biosynthetic precursor of ethylene) reduced *AtGID1* transcript expression (Figure 9.5a). In contrast to seeds, this GA- and ethylene-mediated negative regulation was not evident in growing seedlings (Figure 9.5b). This effect might have been caused by seedling saturation by GA (in contrast to seed). Furthermore, the response to treatment with ABA differed for AtGID1b between seeds and seedlings, in that ABA down-regulates *AtGID1b* transcript expression in seeds while it up-regulates it in seedlings. Different regulation of *AtGID1* transcripts in seeds and seedlings in response to ambient temperature is also evident (Figure 9.5c,d). For cold temperature this relationship is, however, complicated by the fact that *Arabidopsis* seeds required cold-stratification and light for dormancy release and/or germination (Figure 9.5c). An important point from these examples is that GA signalling in seeds is distinct from seedlings and that AtGID1 expression is part of this distinct regulation.

Voegele *et al.* (2011) showed by molecular phylogenetic analysis of the angiosperm GID1 receptor family individual clustering of GID1 proteins into three distinct groups: eudicot GID1ac, eudicot GID1b and monocot GID1 (Figure 9.5e). It has been demonstrated that the individual GID1 receptors of *Arabidopsis* (AtGID1a, AtGID1b, AtGID1c) display partial redundancy and have functional specificities for regulating the GA-responsiveness of different developmental processes (e.g. Griffiths *et al.* 2006; Iuchi *et al.* 2007; Voegele *et al.*, 2011). The distinction between two eudicot groups (GID1ac and GID1b) and the monocot group is also supported by biochemical evidence in studies with the three *Arabidopsis* GID1 receptors: Nakajima *et al.* (2006) showed that AtGID1a and AtGID1c bind GA_4 and GA_3 with lower affinity compared to AtGID1b. It should be mentioned, however, that less work has been carried out on monocot species seed germination so far, and that the degree of partial redundancy between the two eudicot pathways may differ, depending on the developmental process. The regulation of the GID1 pathways and its roles during seed germination differ from other developmental processes and may be restricted to the Brassicaceae family.

Voegele *et al.* (2011) analysed the *Arabidopsis* knockout mutants for the three AtGID1 receptors and showed that the AtGID1b receptor is not able to compensate for the seed germination phenotype of the *gid1agid1c* double mutant (Figure 9.5f). Thus GA signalling via the GID1ac receptors is required for seed germination. In contrast, the AtGID1a and AtGID1c receptors are partially redundant and can substitute for AtGID1b. Based on the seed germination phenotypes of the *Arabidopsis* knock-out mutants and the ABA-related *LesaGID1ac* and *LesaGID1b* transcript expression patterns in the micropylar endosperm (CAP) and the RAD of *L. sativum*, the GID1c receptors may have a major influence on seed germination. Transcript

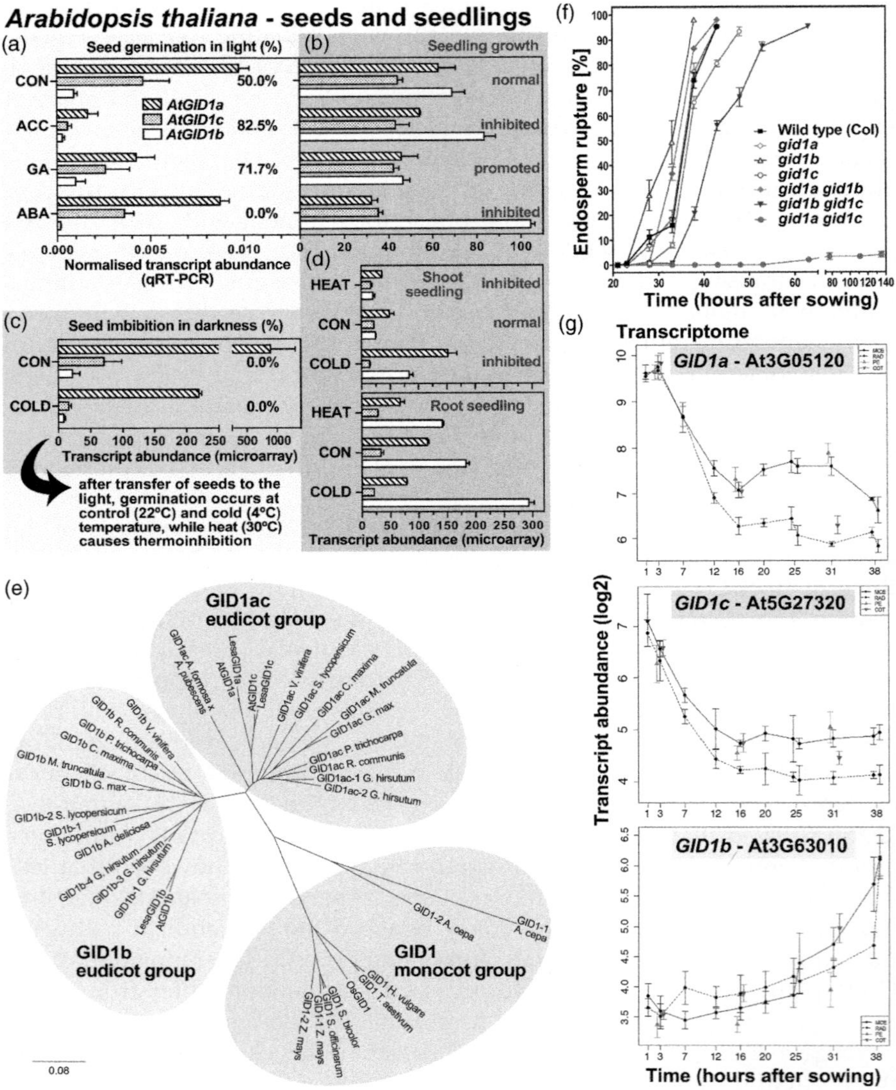
Arabidopsis thaliana - seeds and seedlings
(a)
Seed germination in light (%)
CON
ACC
GA
ABA
50.0%
82.5%
71.7%
0.0%
AtGID1a
AtGID1c
AtGID1b
0.000
0.005
0.010
Normalised transcript abundance (qRT-PCR)
(b)
Seedling growth
normal
inhibited
promoted
inhibited
0
20
40
60
80
100
(c)
Seed imbibition in darkness (%)
CON
COLD
0.0%
0.0%
0
50
100
150
200
250
500
1000
Transcript abundance (microarray)
after transfer of seeds to the light, germination occurs at control (22°C) and cold (4°C) temperature, while heat (30°C) causes thermoinhibition
(d)
HEAT
CON
COLD
Shoot seedling
inhibited
normal
inhibited
Root seedling
0
50
100
150
200
250
300
Transcript abundance (microarray)
(e)
GID1ac eudicot group
GID1b eudicot group
GID1 monocot group
GID1ac A. formosa x A. pubescens
AtGID1a
LesaGID1a
AtGID1c
LesaGID1c
GID1ac V. vinifera
GID1ac S. lycopersicum
GID1ac C. maxima
GID1ac M. truncatula
GID1ac G. max
GID1ac P. trichocarpa
GID1ac R. communis
GID1ac-1 G. hirsutum
GID1ac-2 G. hirsutum
GID1b V. vinifera
GID1b R. communis
GID1b P. trichocarpa
GID1b C. maxima
GID1b M. truncatula
GID1b G. max
GID1b-2 S. lycopersicum
GID1b-1 S. lycopersicum
GID1b A. deliciosa
GID1b-4 G. hirsutum
GID1b-3 G. hirsutum
GID1b-1 G. hirsutum
LesaGID1b
AtGID1b
GID1-2 A. cepa
GID1-1 A. cepa
GID1 H. vulgare
GID1 T. aestivum
OsGID1
GID1 S. bicolor
GID1 S. officinarum
GID1-1 Z. mays
GID1-2 Z. mays
0.08
(f)
Endosperm rupture [%]
Wild type (Col)
gid1a
gid1b
gid1c
gid1a gid1b
gid1b gid1c
gid1a gid1c
Time (hours after sowing)
(g)
Transcriptome
GID1a - At3G05120
GID1c - At5G27320
GID1b - At3G63010
Transcript abundance (log2)
Time (hours after sowing)

abundances in the order *AtGID1a* > *AtGID1c* > *AtGID1b* were evident in dry and imbibed seeds (Figure 9.5g). Their spatiotemporal transcript expression patterns during germination were similar for *AtGID1a* and *AtGID1c*, but distinct for *AtGID1b*: While the transcript abundance of *AtGID1b* increased during germination, those for *AtGID1a* and *AtGID1c* decreased (Figure 9.5g). The hypothesis of the different degrees of importance of the individual Brassicaceae *GID1* genes during seed germination, with *GID1b* being distinct from *GID1ac*, is also supported by the stronger GUS staining in seeds of *AtGID1ac*-promoter reporter gene lines compared to *AtGID1b* (Voegele *et al.*, 2011). This finding is further supported by the fact that Griffiths *et al.* (2006) found negative feedback-regulation by GA for all three *GID1* transcripts in *Arabidopsis* seedlings, while Voegele *et al.* (2011) demonstrated from transcript analyses and GUS reporter line staining results, combined with *in silico* analysis using the eFP browser, that down-regulation by GA during the germination of unstratified *Arabidopsis* seeds was evident only for the *AtGID1a* and *AtGID1c* transcripts, but not for the *AtGID1b* transcripts. An alternative interpretation might be a higher GA sensitivity

Figure 9.5 The effect of environmental, hormonal and developmental conditions on the expression of the three GID1 GA receptor genes in *Arabidopsis*. (a) *AtGID1a*, *AtGID1*b and *AtGID1c* transcript abundances were determined by qRT-PCR in whole seeds during germination. Seeds were incubated for 30 h in continuous white light without (CON) or with 1 mM ACC, 10 μM GA_{4+7} (GA) or 1 μM ABA added; ACC results (Voegele, unpublished results), and GA/ABA results (Voegele *et al.*, 2011) presented for comparison, are from the same experiment. Expression values relative to validated constitutive transcripts are presented as mean values ±SE of 4 × >1000 seeds. The percentage of endosperm rupture of the individual seed populations is indicated next to the bars. (b) GID1 transcript abundances from microarray data of *Arabidopsis* seedlings (Kilian *et al.* 2007). The effect of the treatment on seedling growth is indicated next to the bars. Microarray data for *Arabidopsis* were obtained from the BAR eFP-Browser (Winter *et al.*, 2007; Bassel *et al.*, 2008). (c) *GID1* transcript abundance from microarray data of *Arabidopsis* seeds imbibed in darkness at 22 °C (CON) and 4 °C (COLD) (Yamauchi *et al.*, 2004). Note that dormancy release by cold-stratification causes the indicated germination responses at different temperatures in the light. (d) *GID1* transcript abundance from microarray data from *Arabidopsis* seedling shoots and roots, grown at 30 °C (HEAT), 22 °C (CON), 4 °C (COLD) (Kilian *et al.* 2007). (e) The molecular phylogenetic analysis of the angiosperm GID1 receptor family individual clustering of GID1 proteins into three distinct groups: eudicot GID1ac, eudicot GID1b and monocot GID1; AtGID1 = *Arabidopsis thaliana* GID1, LesaGID1 = *Lepidium sativum* GID1, OsGID1 = *Oryza sativa* GID1 (from Voegele *et al.*, 2011). (f) Time-course analysis of endosperm rupture of after-ripened seeds of *Arabidopsis* single (*gid1a*, *gid1b*, *gid1c*) and double (*gid1agid1b*, *gid1bgid1c*, *gid1agid1c*) mutants imbibed at 24 °C in continuous light without preceding stratification (from Voegele *et al.*, 2011). (g) The spatiotemporal gene expression patterns of At*GID1a*, *AtGID1b* and *AtGID1c* during seed germination. Note that *GIDac* exhibit a similar pattern, which is distinct from *GID1b*. (Transcript abundances are from the transcriptome of Dekkers *et al.*, 2013 available at vseed.nottingham.ac.uk.)

of the negative feedback mechanism on *AtGID1b* transcript abundance compared to *AtGID1a* and *AtGID1c* as proposed by Iuchi *et al.* (2007). The endogenous GA content (Figure 9.1b) would then already be sufficient to decrease *AtGID1b* transcript abundance. In support of this proposal, the GA-triggered negative feedback loop on *AtGID1a* and *AtGID1c*, but not on *AtGID1b*, was also evident in imbibed *ga1-3 Ler* seeds (Ogawa *et al.*, 2003). Furthermore, during *L. sativum* seed germination, a GA-triggered negative feedback loop in the CAP and the RAD was only evident for the *LesaGID1a* and *LesaGID1c* transcripts (Voegele *et al.*, 2011). This strongly suggests that a GA-triggered negative feedback loop during seed germination exists for *GID1a* and *GID1c* in Brassicaceae seeds, while *GID1b*-type transcripts are not down-regulated. In addition, in both species expression patterns are similar regarding transcript abundance during early germination (8 h in *L. sativum* vs. 30 h *Arabidopsis*), in which the *GID1ac* transcript levels are usually higher compared to *GID1b*, and no significant regulation by ABA during early germination takes place. However, *GID1b* with its unique binding activity to GA_4 and its pH dependence is together with *GID1ac* required for Brassicaceae seed germination although their expression patterns differ spatially, temporally and hormonally. Taken together, the GID1ac and GID1b receptor groups are both important for proper seed germination, but both groups play distinct roles during this process.

9.3.2 DELLA proteins and seed germination

All three *Arabidopsis* GID1 receptors can interact with all five *Arabidopsis* DELLA repressor (GAI (GA-insensitive), RGA (repressor-of-ga1-3), RGL1 (RGA-like1), RGL2, RGL3) targets (Nakajima *et al.*, 2006; Willige *et al.*, 2007; Suzuki *et al.*, 2009). In an evolutionary context a separation between an GID1ac-type (interacting preferentially with the RGL1/RGL2/RGL3 group of DELLA repressors) and a GID1b-type (interacting preferentially with the GAI/RGA group of DELLA repressors) pathway (Suzuki *et al.*, 2009) hints to a greater specialization of eudicot GID1-mediated GA signalling, while such a partial functional separation has not occurred within the monocots. In monocot species, solely one group of GID1 receptors has been identified (Figure 9.5e). This might be associated with fewer DELLA proteins present in monocot: SLR1 in *Oryza sativa*, SLN1 in *Hordeum vulgare*, D8 and D9 in *Zea mays* (Peng *et al.*, 1999; Ikeda *et al.*, 2001; Gubler *et al.*, 2002; Weston *et al.*, 2008). Differential expression and distinct patterns of degradation of different DELLA repressors has been shown in seeds (Bassel *et al.*, 2004; Piskurewicz *et al.*, 2008, Piskurewicz and Lopez-Molina. 2009; Piskurewicz *et al.*, 2009; Voegele *et al.*, 2011; Ariizumi *et al.*, 2013; Chandler and Harding, 2013). It has been suggested that RGL1 possesses a more major role in seed germination than do GAI and RGA, but that RGL2 is the most crucial regulator of seed germination in *Arabidopsis* in response to GA (reviewed by Kucera *et al.*, 2005). Light- and GA- independent seed germination can be obtained by the loss of function of four *DELLA* genes (*RGL2, RGL1, RGA* and *GAI*). From former findings it has been proposed that RGA and GAI destabilization or inactivation in seeds might be triggered by GA. This fact supports the view that DELLA repressors integrate environmental and endogenous cues with the seed germination regulation (reviewed by Kucera *et al.*, 2005).

The role of light (via phytochrome) in the DELLA-mediated seed germination response has been elucidated in *Arabidopsis* (Oh *et al.*, 2006, 2007; Holdsworth *et al.*, 2008; Penfield and Hall, 2009). Red-light activated phytochrome (Pfr form) decreases PIL5 (phytochrome-interacting factor3-like5; bHLH protein) activity, which enhances germination. In darkness, PIL5 directly binds to the *RGA* and *GAI* gene promoters and stimulates DELLA repressor expression, as well as indirectly increasing *GA2ox* gene expression (GA inactivation) and expression of ABA biosynthetic genes. In addition, the GA biosynthetic genes *GA20ox* and *GA3ox* are down-regulated by PIL5. Penfield *et al.* (2005) found that the transcription factor SPATULA (SPT), together with PIF3 and PIL5, is involved in mediating the light and temperature responses of seed germination; and in dormant seeds this involves down-regulation of *GA3ox*. Moreover, PIL5 stimulates the expression of *SOMNUS* (*SOM*), encoding a CCHH-type zinc finger protein (Kim *et al.*, 2008; Park *et al.*, 2011). The role of PIL5 is significant, provided GA synthesis is sufficiently high; otherwise, high GAI and RGA protein levels persist to block germination. SOM represses germination by down-regulating and stimulating GA and ABA synthesis gene expression, respectively. ABA-INSENSITIVE3, ABA-INSENSITIVE5 and DELLAs interact to activate the expression of SOMNUS and other high-temperature-inducible genes in imbibed *Arabidopsis* seeds (Lim *et al.*, 2013). The GA-signalling repressors RGL2 and RGL3 inhibit *Arabidopsis* seed germination by stimulating ABA biosynthesis and ABI5 activity (Piskurewicz *et al.*, 2008; Piskurewicz and Lopez-Molina, 2009) and the inhibition of cell-wall remodelling protein gene expression (Morris *et al.*, 2011; Voegele *et al.*, 2011; Stamm *et al.*, 2012). RGL2 transcript and protein levels dominate relative to that of other DELLA factors such as RGA and GAI, whereas RGL3 transcripts could only be observed in the absence of RGL2 (as in an rgl2 mutant background) (Piskurewicz *et al.*, 2008, Piskurewicz and Lopez-Molina, 2009). Piskurewicz and Lopez-Molina (2009) reported that RGL2 represses testa rupture in response to changes in GA and ABA levels. The testa rupture in rgl2 mutants is insensitive to low GA or high ABA conditions. Far-red light inhibits germination through DELLA-dependent stimulation of ABA synthesis and the transcription factor ABA-INSENSITIVE 3 (ABI3) (Piskurewicz *et al.*, 2009). In white light, transgene expression of *GAI* and *RGA* driven by the *RGL2* promoter can substitute for *RGL2* to promote ABA synthesis and repress germination, consistent with the recent findings with RGL2. The three DELLA factors repress testa rupture, whereas ABI3 blocks endosperm rupture (Piskurewicz *et al.*, 2009). In the shade, endospermic ABA opposes phyA signalling through the transcription factor ABI5, which shares with the bHLH transcription factor PIF1 several target genes that negatively regulate germination in the embryo. ABI5 enhances the expression of phytochrome signalling genes *PIF1*, *SOMNUS*, *GAI* and *RGA*, but also of ABA and GA metabolic genes (Lee *et al.*, 2012). CHOTTO1, a putative double APETALA2 repeat transcription factor, acts downstream of ABA to repress GA biosynthesis during seed germination (Yano *et al.*, 2009). ABI4 is implicated in the regulation of the ABA-GA balance and reserve mobilisation in seeds (Penfield *et al.*, 2006; Cantoro *et al.*, 2013; Suh *et al.*, 2013). Park *et al.* (2013) isolated from seedling tissue DELLA-interacting RING domain proteins, BOI-RELATED GENE1 (BRG1), BRG2 and BRG3 (collectively referred to as BOIs). Single *Arabidopsis* mutants of each *BOI* gene did not significantly influence GA responses, but the *boi* quadruple mutant (*boiQ*) showed a higher seed germination frequency in the presence of paclobutrazol (GA inhibitor). They propose that the DELLA and BOI proteins inhibit GA responses by interacting with each other, binding to the same promoters of GA responsive genes, and down-regulating these genes.

9.4 Gibberellin and abiotic stress factors: thermoinhibition of seed germination

Thermoinhibition is the failure of seeds to germinate at high imbibition temperatures. The delayed or severely inhibited seed germination by thermoinhibition is mediated by complex alterations of hormone metabolism and signalling (Watt *et al.*, 2011; Linkies and Leubner-Metzger, 2012; Toh *et al.*, 2012a; Huo *et al.*, 2013). The research group of Kent Bradford studied the mechanisms of lettuce thermoinhibition by comparing the germination of *Lactuca sativa* cv. Salinas ('*L. sativa* Sal'), a cultivated variety exhibiting thermoinhibition, with *Lactuca serriola* UC96US23 ('*L. serriola* UC'), a wild progenitor accession exhibiting thermotolerance, at different temperatures (Argyris *et al.*, 2005, 2008, 2011; Schwember and Bradford, 2010a, 2010b; Huo *et al.*, 2013). While both lettuce genotypes germinated when imbibed in the light at 20 °C, only *L. serriola* UC was able to germinate at 35 °C (Figure 9.6a). *L. serriola* UC is thermotolerant until *ca.* 38 °C, while thermoinhibition of *L. sativa* Sal starts above *ca.* 25 °C (Argyris *et al.*, 2008; Huo *et al.*, 2013). The switch to thermoinhibition occurs within a narrow temperature window of 2–3 °C, and different lettuce genotypes differ in the temperature for this switch, while the hormonal mechanisms causing thermoinhibition above this temperature appear to be similar. The lettuce ABA biosynthesis gene *LsNCED4* (encoding a 9-*cis*-epoxycarotenoid dioxygenase) was identified as a major quantitative trait gene (QTG) by conducting quantitative trait loci (QTL) analysis of recombinant inbred line (RIL) populations derived from a cross between *L. sativa* Sal x *L. serriola* UC (Argyris *et al.*, 2008, 2011; Huo *et al.*, 2013). In agreement with a major role of the *LsNCED4* QTG during thermoinhibition, its expression was up-regulated and the seed ABA contents and sensitivities were increased at temperatures causing thermoinhibition. Argyris *et al.* (2008) conclude that the temperature sensitivity of *LsNCED4* expression may determine the upper temperature limit for lettuce seed germination. Figure 9.6c summarises their findings about how high temperature affects the expression of hormone-related genes and demonstrates that the up-regulation of ABA biosynthesis is associated with down-regulation of GA- and ethylene-related biosynthesis genes. Huo *et al.* (2013) demonstrated that silencing of the *LsNCED4* gene in transgenic lettuce seeds altered the expression of genes involved in ABA, GA and ethylene biosynthesis and signalling pathways. The hormonal cross-talk in lettuce seed thermoinhibition (Figure 9.6c) seems therefore to be mediated by indirect (elevated ABA contents due to up-regulated *LsNCED4* gene expression) and direct effects of the high temperature. These findings are in agreement with the knowledge that not an individual hormone, but the content ratios between promoting and inhibiting hormones (such as GA/ABA) combined with the seed sensitivities (interaction between the endogenous contents and the state of the corresponding signalling pathways) determine the seed responses (Kucera *et al.*, 2005; Holdsworth *et al.*, 2008; Linkies and Leubner-Metzger, 2012). Work with *Arabidopsis* supports the view that the hormonal interactions described for lettuce (Figure 9.6c) constitute a conserved mechanism for seed thermoinhibition (Gonai *et al.*, 2004; Toh *et al.*, 2008; Seo *et al.*, 2009; Toh *et al.*, 2012a; Toh *et al.*, 2012b). Regarding signalling components, increased expression of DELLA repressors, the ethylene-signalling component CTR1, as well as the ABA-related components ABI3, ABI4, ABI5, SNF4 (Figure 9.6c), and the transcription factor FUSCA3 have been proposed to be involved (Chiu *et al.* 2012).

In agreement with a major role for GAs in counteracting the ABA inhibition and in alleviating thermoinhibition, high temperature inhibited the expression of *GA3ox* genes in imbibed seeds of lettuce (Argyris *et al.*, 2008) and *Arabidopsis* (Toh *et al.*, 2008). In contrast to thermoinhibited *L. sativa* Sal seeds, *LsGA3ox1* expression was evident in the thermotolerant *L. serriola* UC seeds imbibed at 35 °C in continuous light (Figure 9.6a); similar results were obtained for *LsGA3ox2*. It was demonstrated for *Arabidopsis* that thermoinhibition is associated with decreased expression of *AtGA3ox2* and *AtGA3ox1* and of GA_4 and GA_1 contents, and, in agreement with a role of the decreased seed GA contents, can be alleviated by treatment with bioactive GA (Figure 9.6b).

Arabidopsis seed GA_4 and GA_1 contents are strongly influenced by imbibition temperature: compared to 22 °C they decreased in response to thermoinhibition (34 °C in the light) and, as described above, increased during cold-stratification (4 °C in the dark) (Yamauchi *et al.*, 2004; Toh *et al.*, 2008) (Figure 9.6b). This finding is consistent with the observed up-regulation of *AtGA3ox1* expression by low temperature. Expression of *AtGA3ox2*, the major *AtGA3ox* gene induced during seed germination, is, however, not

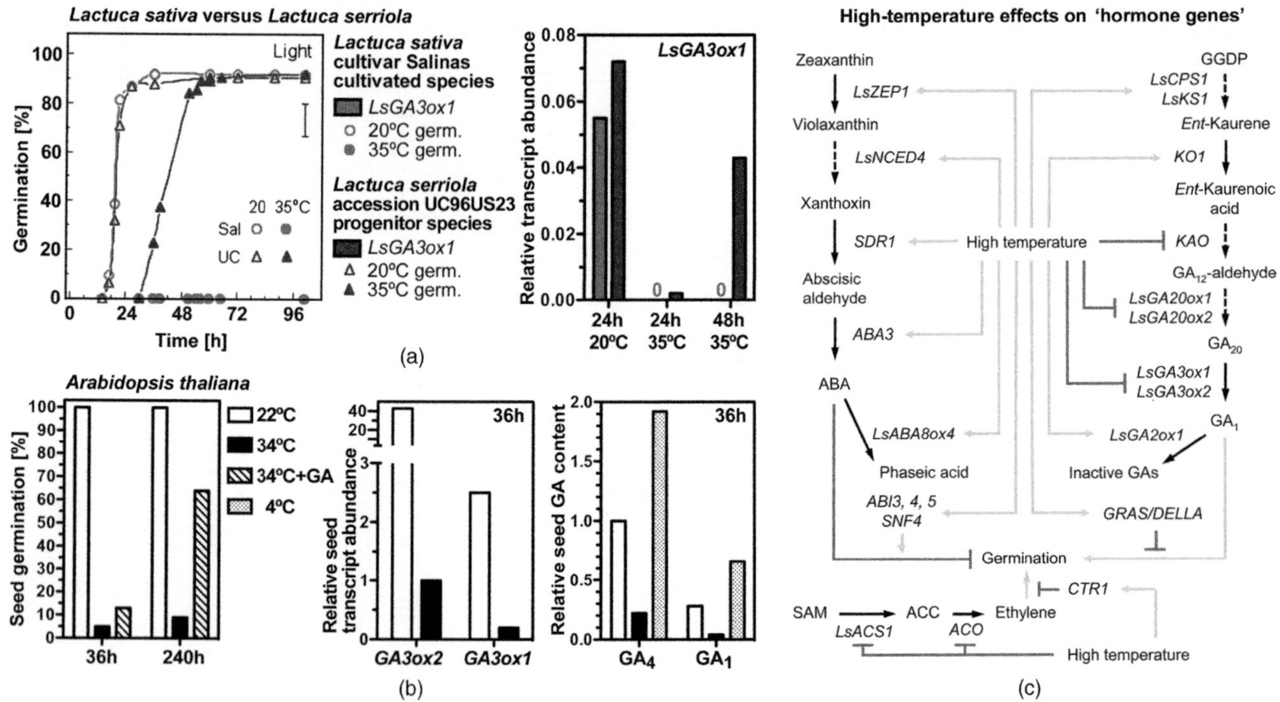

Figure 9.6 Hormonal aspects of thermoinhibition of lettuce and *Arabidopsis* seed germination. (a) The effect of supra-optimal temperature on germination (*left*) and *LsGA3ox1* gene expression (*right*) during germination of *Lactuca sativa* cv. Salinas (cultivated lettuce) and *Lactuca serriola* UC96US23 (thermotolerant lettuce). Note that compared to 20 °C, thermoinhibion of *L. sativa* occurs at 35 °C and is associated with inhibited *LsGA3ox1* expression, while the gene is expressed in thermotolerant *L. serriola* seeds (results from Argyris *et al.*, 2008).
(b) Thermoinhibition of *Arabidopsis* seeds imbibed in the light occurs at 34 °C and is alleviated by GA addition to the medium (*left*). The effect of temperature on the expression of the GA biosynthetic genes *AtGA3ox1* and *AtGA3ox2* and on the seed bioactive GA_1 and GA_4 contents is shown (*right*). Note that thermoinhibition is associated with a decrease, while cold-stratification (4 °C) is associated with an increase in the seed GA contents (results from Toh *et al.*, 2008 and from Yamauchi *et al.*, 2004). (c) Promoting (arrows) and inhibiting (bars) effects of high temperature on metabolism and signalling genes of hormones (GA, ABA and ethylene) involved in lettuce seed germination (from Argyris *et al.*, 2008). (See insert for colour representation of this figure.)

up-regulated during cold-stratification, while high temperature inhibited the expression of both *AtGA3ox2* and *AtGA3ox1* (Figure 9.6b, Toh *et al.*, 2008). Light (via phytochrome) as a required factor for *Arabidopsis* seed germination is also required for *AtGA3ox2* expression (Yamaguchi *et al.*, 2007). This provides a possible explanation for the finding that *AtGA3ox2* is not up-regulated as the cold-stratification is conducted in the dark. Light also promotes germination of lettuce seeds, in which *LsGA3ox2* expression is not induced in the dark at any temperature, while low-level expression of *LsGA3ox1* is induced at 20 °C in both genotypes, and at 35 °C only in thermotolerant *L. serriola* UC seeds, but not in thermoinhibited *L. sativa* Sal seeds (Argyris *et al.*, 2008).

Thermoinhibition also suppressed induction of *AtGA20ox2* and *AtGA20ox3* during the early imbibition of *Arabidopsis* seeds (Toh *et al.*, 2008), as it did for the *LsGA20ox1* and *LsGA20ox2* genes during the early imbibition of *L. sativa* Sal seeds (Argyris *et al.*, 2008). In contrast to the thermoinhibited *L. sativa* Sal seeds, *LsGA20ox1* and *LsGA20ox2* gene expression was evident at 35 °C in thermotolerant *L. serriola* UC seeds. Interestingly, except for *LsKAO*, early GA biosynthetic genes such as *LsCPS1*, *LsKO1*, and *LsKS1*, were expressed in thermoinhibited *L. sativa* Sal seeds. Also, there was no clear pattern for the regulation of *GA2ox* genes during *Arabidopsis* and lettuce thermoinhibition (Argyris *et al.*, 2008; Toh *et al.*, 2008), except that at 35 °C *LsGA2ox1* is more highly expressed in *L. sativa* Sal compared to *L. serriola* UC seeds. This is consistent with the finding that thermoinhibition causes elevated seed ABA contents, and that ABA promotes GA2ox-mediated GA inactivation (Seo *et al.*, 2006; Zentella *et al.*, 2007). Taken together, high temperature effects during seed thermoinhibition are caused by direct and indirect (via ABA) alteration of late GA metabolism which is mainly achieved by inhibited *GA20ox* and *GA3ox* gene expression (Figure 9.6c; Argyris *et al.*, 2008; Toh *et al.*, 2008).

9.5 Gibberellin and biotic stress factors: allelochemical interference of gibberellin biosynthesis during seed germination

Several plant- and microbe-derived phytotoxins have been proposed to be allelochemicals that affect seed germination and seedling growth of surrounding 'target' plants through leaching into the rhizosphere (Inderjit and Duke, 2003; Weston and Duke, 2003; Weir *et al.*, 2004). Examples of 'donor' plants with allelopathic phytotoxic potential include the juglon-producing walnut tree, as well as many invasive plant species. In many cases the mode of action of these allelochemicals is not known and, only in a few cases for seeds, was interference with hormone metabolism in the 'target' plant demonstrated (Bogatek and Gniazdowska, 2007; Oracz *et al.*, 2012; Voegele *et al.*, 2012). In addition to plant roots, microbes in the rhizosphere can also release phytotoxic compounds which interfere with GA-mediated plant growth. An example is the rhizobacterium *Bacillus subtilis* IJ-31, for which the culture extract, as well as its component hydrocinnamic acid (HCA) were shown to act as plant growth retardants for red pepper, ryegrass and *Arabidopsis* (Kim and Rhee, 2012). A dose-dependent inhibition effect of HCA on seedling root and shoot growth in *Arabidopsis*, as well as the down-regulation of a GA-inducible cell-wall remodelling gene by HCA or the culture extract was demonstrated. Most interestingly, Kim and Rhee (2012) showed using an *in vitro* enzyme activity assay with recombinant AtGA3ox1 protein that HCA and the *Bacillus subtilis* IJ-31 culture extract both inhibit GA3ox enzyme activity. The effects of these putative allelochemicals on seed germination has not been studied, but HCA has been shown to inhibit seedling growth in other species (Tang and Young, 1982; Williamson *et al.*, 1992; Chon *et al.*, 2002).

Myrigalone A (MyA), 3-(1-oxo-3-phenylpropyl)-1,1,5-trimethylcyclohexane-2,4,6-trione (Figure 9.4e), a phytotoxin produced by *Myrica gale* L. ('sweet gale', 'bog myrtle', Myricaceae) has been demonstrated to inhibit seed germination and seedling growth of 'target' species (Popovici *et al.*, 2011; Oracz *et al.*, 2012; Voegele *et al.*, 2012). *M. gale* is a deciduous shrub native to Northern and Western Europe and North America, adapted to flood-prone habitats (Skene *et al.*, 2000). It grows in acidic peat bogs and at the intertidal zone of lakes and rivers that are often flooded by frequent rise and fall in water level. *M. gale* fruits and leaves secrete resin droplets containing essential oils (Figure 9.4e). Fruit and leaf exudates of *M. gale* exhibit phytotoxic activity on seedling growth of invasive knotweed (*Fallopia x bohemica*) and other species (Popovici *et al.*, 2011) and inhibit the seed germination of *L. sativum* (Oracz *et al.*, 2012; Voegele *et al.*, 2012). *M. gale* fruit exudates contain rare flavonoids, with MyA being the major C-methylated dihydrochalcone (Anthonsen *et al.*, 1971; Mathiesen *et al.*, 1995; Popovici *et al.*, 2011). MyA inhibits shoot and root growth of etiolated eudicot (cress, mustard, knotweed) and monocot (sorghum) seedlings (Popovici *et al.*, 2011), and also inhibits the growth of cress seedlings in the light (Oracz *et al.*, 2012). MyA inhibits processes required for embryo elongation during seed germination, including endoreduplication and the formation of apoplastic reactive oxygen in the hypocotyl-radicle axis of *L. sativum* (Oracz *et al.*, 2012). This finding is in agreement with the known function of MyA as a radical scavenger (Mathiesen *et al.*, 1997). Several key weakening and growth processes during early and late seed germination of *L. sativum* were found to be targets for MyA (Voegele *et al.*, 2012): MyA enhanced testa permeability and water uptake (early germination phase), and also inhibited endosperm weakening and rupture, and embryo growth (late germination phase). The inhibitory effects of MyA on *L. sativum* seed were modulated by light conditions and ambient water potential, with the inhibition being stronger in darkness compared to the light (Figure 9.4f). The important point with regard to the topic of this article is that the inhibition of seed germination was mediated, at least in part, by interference of MyA with GA metabolism and signalling (Oracz *et al.*, 2012; Voegele *et al.*, 2012).

MyA specifically interferes with GA-regulated processes important for seed germination of *L. sativum*, inhibiting endosperm weakening, as well as embryo extension growth. As shown in Figure 9.4, MyA causes a *ca.* threefold decrease of the (bioactive) GA_4 contents in the RAD of seeds imbibed in continuous light or in darkness, but does not affect the GA_4 contents in the CAP (Oracz *et al.*, 2012; Voegele *et al.*, 2012). This suggests that the MyA inhibition of endosperm rupture is mediated, at least in part, by GA-promoted downstream mechanism(s) that affects the embryo growth potential. The MyA-mediated inhibition of GA_4 production in the RAD was accompanied by a 200-fold (light) or 400-fold (darkness) accumulation of its biosynthetic precursor GA_9 (Figure 9.4b), indicating a block of the GA3ox-catalysed conversion of GA_9 to GA_4. MyA inhibits GA 3β-hydroxylation in the RAD, but does not affect it in the CAP (the increased GA_9 contents in the CAP may be due to diffusion from the RAD), and does not affect the preceding GA 20-oxidase reactions (Oracz *et al.*, 2012).

In contrast to the inhibitory effects of MyA on the GA_4 contents in the RAD, it had no appreciable inhibitory effect on GA_1 in this tissue, but the GA_1 contents were very low compared to the GA_4 contents (Figure 9.4c). GA 3-oxidase activity is also required for the production of GA_6 and GA_{13} detected in RAD and CAP tissues (Figure 9.4d). In the light these steps are inhibited by MyA treatment in the RAD, but in the CAP MyA treatment caused a *ca.* 20-fold reduction in GA_6, but did not affect the contents of GA_{13} (Oracz *et al.*, 2012; Voegele *et al.*, 2012). In darkness these effects of MyA were reversed (Figure 9.4d), suggesting that light modulates the MyA inhibition of GA metabolism. These findings also show that the MyA-mediated inhibition of GA3ox is not a general effect on these enzymes, but has some specificity regarding RAD/CAP and substrate/product for which the molecular mechanisms remain unknown. Taken together, MyA acts as a GA3ox inhibitor in germinating *L. sativum* seeds with specificity for the catalytic step from GA_9 to bioactive GA_4 and this compromises proper endosperm weakening and embryo growth required for endosperm rupture.

GA signalling via the soluble GID1-type GA receptors mediates downstream processes that confer embryo extension growth and endosperm weakening (see Section 9.3). A GA-triggered negative feedback loop in the CAP and RAD of germinating *L. sativum* seeds was only evident for the *GID1ac* transcripts, but not for *GID1b*. The GID1b receptor proteins have a higher GA_4 binding affinity (K_D *ca.* 30 nM) than the GID1ac receptor proteins (K_D *ca.* 300 nM) (Nakajima *et al.*, 2006). If we therefore assume that GA signalling is mediated only by the GID1ac receptor proteins in the germinating seed, the low concentrations of bioactive GA_4 after MyA treatment would not allow optimal GA signalling. Taken together, these results support the view that GA signalling via both GID1-pathways is required for the endosperm rupture of Brassicaceae seeds, and that MyA acts by interfering with GA metabolism and signalling important for downstream cell-wall loosening mechanisms such as XTH/expansins and/or apoplastic ROS (Voegele *et al.*, 2011; Oracz *et al.*, 2012; Voegele *et al.*, 2012). MyA therefore acts as an inhibitor on important GA-regulated key processes of seed germination and seedling establishment as targets. Voegele *et al.* (2012) speculate that MyA is a soil seed bank-destroying allelochemical that secures the persistence of *M. gale* in its flood-prone environment.

9.6 Conclusions and perspectives

Seeds are diverse in structure and in a typical mature angiosperm seed the embryo is encased by a living endosperm layer (including the cereal grain aleurone) and a dead testa (seed coat) (Linkies *et al.*, 2010). The recent progress using novel approaches has deepened our knowledge about the roles of GA during seed germination. Important progress was obtained in seeds by moving away from whole-seed hormone and transcript quantification to seed tissues and key compartments (RAD, CAP), which for GA metabolites requires detection in minute amounts of tissue (Seo *et al.*, 2011; Oracz *et al.*, 2012; Urbanova *et al.*, 2013). The early induction of GA biosynthesis genes in the RAD is in accordance with the hypothesis that an embryonic GA metabolite and/or bioactive GA itself diffuses early during imbibition to the CAP to make it competent (release of coat dormancy) for the subsequent endosperm weakening during the late germination phase. Thus, the interaction between the key seed compartments is crucial for the control of seed germination by GA. GA signalling in seeds is distinct from seedlings, with expression of the GID1 receptor being part of this distinct regulation. Molecular phylogenetic analysis revealed that members of the angiosperm GID1 receptor family cluster into three distinct groups: eudicot GID1ac, eudicot GID1b and monocot GID1 (Voegele *et al.*, 2011). In Brassicaceae seeds the GID1ac receptors may be more important for the release of coat dormancy and the promotion of germination compared to the GID1b receptors, but proper seed germination requires both the GID1ac and the GID1b signalling pathways. All three types of GA receptor bind DELLA repressor proteins to target them for degradation which is important to promote seed germination. Environmental and hormonal, as well as developmental cues feed into these different variants of GA-signalling.

Gibberellin metabolism and signalling in seeds are involved in integrating environmental cues to control the timing of germination. Germination is influenced by abiotic stress factors, such as supra-optimal temperature (heat), which inhibits germination (Argyris *et al.*, 2008) and induction of *GA20ox* gene expression during early seed imbibition. Biotic stress factors such as the allelochemical myrigalone A (MyA) inhibit GA 3-oxidase enzyme activity and thereby the production of bioactive GAs required for seed germination (Oracz *et al.*, 2012; Voegele *et al.*, 2012). MyA targets different key processes, including the GA-induced endosperm weakening. These findings reinforce the importance of GA metabolism and signalling in mediating diverse seed-environment interactions, and more of these are yet to be discovered. The GA requirement for seed germination was instrumental in screens at the dawn of *Arabidopsis* mutant research (Koornneef and van der Veen, 1980, Koorneef *et al.*, 1985). Gibberellins are, however, not simple 'GO AHEAD' (GA) molecules, since there are distinct and specific GA-actions in key seed compartments and beyond GA and ABA there is a complex network of signalling molecules that interact to control the timing of seed germination (Linkies and Leubner-Metzger, 2012).

Acknowledgements

Support for positions and the research of the authors on gibberellins in seeds is gratefully acknowledged. This support was by the Deutsche Forschungsgemeinschaft (DFG LE720/6 and DFG LE720/7) to GLM, by Palacký University (Olomouc, CZ) (PrF_2013_023) and the National Program of Sustainability of the Czech Ministry of Education (LO1204) to TU.

References

Anthonsen, T., Falkenberg, I., Laake, M. *et al.* (1971). Some unusual flavonoids from *Myrica gale* L. *Acta Chemica Scandinavica* **25**, 1929–1930.

Appleford, N.E., Evans, D.J., Lenton, J.R. *et al.* (2006). Function and transcript analysis of gibberellin-biosynthetic enzymes in wheat. *Planta* **223**, 568–582.

Argyris, J., Truco, M.J., Ochoa, O. *et al.* (2005). Quantitative trait loci associated with seed and seedling traits in *Lactuca*. *Theoretical and Applied Genetics* **111**, 1365–1376.

Argyris, J., Dahal, P., Hayashi, E. *et al.* (2008). Genetic variation for lettuce seed thermoinhibition is associated with temperature-sensitive expression of abscisic acid, gibberellin, and ethylene biosynthesis, metabolism, and response genes. *Plant Physiology* **148**, 926–947.

Argyris, J., Truco, M.J., Ochoa, O. *et al.* (2011). A gene encoding an abscisic acid biosynthetic enzyme (LsNCED4) collocates with the high temperature germination locus Htg6.1 in lettuce (*Lactuca* sp.). *Theoretical and Applied Genetics* **122**, 95–108.

Ariizumi, T., Hauvermale, A.L., Nelson, S.K. *et al.* (2013). Lifting DELLA repression of *Arabidopsis* seed germination by nonproteolytic gibberellin signaling. *Plant Physiology* **162**, 2125–2139.

Barua, D., Butler, C., Tisdale, T.E. and Donohue, K. (2012). Natural variation in germination responses of *Arabidopsis* to seasonal cues and their associated physiological mechanisms. *Annals of Botany* **109**, 209–226.

Bassel, G.W., Zielinska, E., Mullen, R.T. and Bewley, J.D. (2004). Down-regulation of *DELLA* genes is not essential for germination of tomato, soybean, and Arabidopsis seeds. *Plant Physiology* **136**, 2782–2789.

Bethke, P.C., Libourel, I.G.L., Aoyama, N. *et al.* (2007). The Arabidopsis aleurone layer responds to nitric oxide, gibberellin, and abscisic acid and is sufficient and necessary for seed dormancy. *Plant Physiology* **143**, 1173–1188.

Bogatek, R. and Gniazdowska, A. (2007). ROS and phytohormones in plant-plant allelopathic interaction. *Plant Signaling and Behavior* **2**, 317–318.

Cantoro, R., Crocco, C.D., Benech-Arnold, R.L. and Rodriguez, M.V. (2013). In vitro binding of *Sorghum bicolor* transcription factors ABI4 and ABI5 to a conserved region of a GA 2-OXIDASE promoter: possible role of this interaction in the expression of seed dormancy. *Journal of Experimental Botany* **64**, 5721–5735.

Chandler, P.M. and Harding, C.A. (2013). 'Overgrowth' mutants in barley and wheat: new alleles and phenotypes of the 'Green Revolution' DELLA gene. *Journal of Experimental Botany* **64**, 1603–1613.

Chiu, R.S., Nahal, H., Provart, N.J. and Gazzarrini, S. (2012). The role of the Arabidopsis FUSCA3 transcription factor during inhibition of seed germination at high temperature. *BMC Plant Biology* **12**, 15–31.

Chon, S.-U., Choi, S.-K., Jung, S. *et al.* (2002). Effects of alfalfa leaf extracts and phenolic allelochemicals on early sedling growth and root morphology of alfalfa and barnyard grass. *Crop Protection* **21**, 1077–1082.

Debeaujon, I. and Koornneef, M. (2000). Gibberellin requirement for Arabidopsis seed germination is determined both by testa characteristics and embryonic abscisic acid. *Plant Physiology* **122**, 415–424.

Debeaujon, I., Léon-Kloosterziel, K.M. and Koornneef, M. (2000). Influence of the testa on seed dormancy, germination, and longevity in Arabidopsis. *Plant Physiology* **122**, 403–413.

Dekkers, B.J.W., Pearce, S., Van Bolderen-Veldkamp, R.P.M. *et al.* (2013). Transcriptional dynamics of two seed compartments with opposing roles in Arabidopsis seed germination. *Plant Physiology* **163**, 205–215.

Finch-Savage, W.E. and Leubner-Metzger, G. (2006). Seed dormancy and the control of germination. *New Phytologist* **171**, 501–523.

Frisse, A., Pimenta, M.J. and Lange, T. (2003). Expression studies of gibberellin oxidases in developing pumpkin seeds. *Plant Physiology* **131**, 1220–1227.

Gonai, T., Kawahara, S., Tougou, M. *et al.* (2004). Abscisic acid in the thermoinhibition of lettuce seed germination and enhancement of its catabolism by gibberellin. *Journal of Experimental Botany* **55**, 111–118.

Graeber, K., Linkies, A., Müller, K. *et al.* (2010). Cross-species approaches to seed dormancy and germination: Conservation and biodiversity of ABA-regulated mechanisms and the Brassicaceae *DOG1* genes. *Plant Molecular Biology* **73**, 67–87.

Graeber, K., Nakabayashi, K., Miatton, E. *et al.* (2012). Molecular mechanisms of seed dormancy. *Plant, Cell and Environment* **35**, 1769–1786.

Graeber, K., Linkies, A., Steinbrecher, T. *et al.* (2014). DELAY OF GERMINATION 1 mediates a conserved coat-dormancy mechanism for the temperature- and gibberellin-dependent control of seed germination. *Proceedings of the National Academy of Sciences USA* **111**, E3571–E3580.

Griffiths, J., Murase, K., Rieu, I. *et al.* (2006). Genetic characterization and functional analysis of the GID1 gibberellin receptors in *Arabidopsis*. *The Plant Cell* **18**, 3399–3414.

Gubler, F., Chandler, P.M., White, R.G. *et al.* (2002). Gibberellin signaling in barley aleurone cells. Control of SLN1 and GAMYB expression. *Plant Physiology* **129**, 191–200.

Hashimoto, T. and Yamaki, T. (1959). On the physiological effects of gibberellins A_1, A_2, A_3, and A_4. *Botanical Magazine (Tokyo)* **72**, 178–181.

Hauvermale, A.L., Ariizumi, T. and Steber, C.M. (2012). Gibberellin signaling: a theme and variations on DELLA repression. *Plant Physiology* **160**, 83–92.

Hedden, P. and Thomas, A.G. (2012). Gibberellin biosynthesis and its regulation. *Biochemical Journal* **444**, 11–25.

Holdsworth, M.J., Bentsink, L. and Soppe, W.J.J. (2008). Molecular networks regulating *Arabidopsis* seed maturation, after-ripening, dormancy and germination. *New Phytologist* **179**, 33–54.

Huo, H., Dahal, P., Kunusoth, K. *et al.* (2013). Expression of 9-cis-EPOXYCAROTENOID DIOXYGENASE4 is essential for thermoinhibition of lettuce seed germination but not for seed development or stress tolerance. *The Plant Cell* **25**, 884–900.

Iglesias-Fernandez, R. and Matilla, A.J. (2010). Genes involved in ethylene and gibberellins metabolism are required for endosperm-limited germination of *Sisymbrium officinale* L. seeds. *Planta* **231**, 653–664.

Ikeda, A., Ueguchi-Tanaka, M., Sonoda, Y. *et al.* (2001). Slender rice, a constitutive gibberellin response mutant, is caused by a null mutation of the *SLR1* gene, an ortholog of the height-regulating gene *GAI/RGA/RHT/D8*. *The Plant Cell* **13**, 999–1010.

Ikuma, H. and Thimann, K.V. (1960). Action of gibberellic acid on lettuce seed germination. *Plant Physiology* **35**, 557–566.

Inderjit and Duke, S.O. (2003). Ecophysiological aspects of allelopathy. *Planta* **217**, 529–539.

Iuchi, S., Suzuki, H., Kim, Y.-C. *et al.* (2007). Multiple loss-of-function of Arabidopsis gibberellin receptor AtGID1s completely shuts down a gibberellin signal. *The Plant Journal* **50**, 958–966.

Kilian, J., Whitehead, D., Horak, J. *et al.* (2007). The AtGenExpress global stress expression data set: protocols, evaluation and model data analysis of UV-B light, drought and cold stress responses. *The Plant Journal* **50**, 347–363.

Kim, W.C. and Rhee, I.K. (2012). Functional mechanism of plant growth retardation by Bacillus subtilis IJ-31 and its allelochemicals. *Journal of Microbiology and Biotechnology* **22**, 1375–1380.

Kim, D.H., Yamaguchi, S., Lim, S. *et al.* (2008). SOMNUS, a CCCH-type zinc finger protein in *Arabidopsis*, negatively regulates light-dependent seed germination downstream of PIL5. *The Plant Cell* **20**, 1260–1277.

King, R.W., Evans, L.T., Mander, L.N. *et al.* (2003). Synthesis of gibberellin GA_6 and its role in flowering of *Lolium temulentum*. *Phytochemistry* **62**, 77–82.

Koornneef, M. and Meinke, D. (2010). The development of Arabidopsis as a model plant. *The Plant Journal* **61**, 909–921.

Koornneef, M. and Van Der Veen, J.H. (1980). Induction and analysis of gibberellin sensitive mutants in *Arabidopsis thaliana* (L.) Heynh. *Theoretical and Applied Genetics* **58**, 257–263.

Koornneef, M., Elgersma, A., Hanhart, C.J. *et al.* (1985). A gibberellin insensitive mutant of *Arabidopsis thaliana*. *Physiologia Plantarum* **65**, 33–39.

Kucera, B., Cohn, M.A. and Leubner-Metzger, G. (2005). Plant hormone interactions during seed dormancy release and germination. *Seed Science Research* **15**, 281–307.

Kwak, S.-S., Kamiya, Y., Sakurai, A. *et al.* (1988). Partial purification and characterization of gibberellin 3β-hydroxylase from immature seeds of *Phaseolus vulgaris* L. *Plant and Cell Physiology* **29**, 935–943.

Lee, K.P., Piskurewicz, U., Tureckova, V. *et al.* (2012). Spatially and genetically distinct control of seed germination by phytochromes A and B. *Genes and Development* **26**, 1984–1996.

Lim, S., Park, J., Lee, N. *et al.* (2013). ABA-INSENSITIVE3, ABA-INSENSITIVE5, and DELLAs interact to activate the expression of SOMNUS and other high-temperature-inducible genes in imbibed seeds in *Arabidopsis*. *The Plant Cell* **25**, 4863–4878.

Linkies, A. and Leubner-Metzger, G. (2012). Beyond gibberellins and abscisic acid: how ethylene and jasmonates control seed germination. *Plant Cell Reports* **31**, 253–270.

Linkies, A., Müller, K., Morris, K. *et al.* (2009). Ethylene interacts with abscisic acid to regulate endosperm rupture during germination: A comparative approach using *Lepidium sativum* and *Arabidopsis thaliana*. *The Plant Cell* **21**, 3803–3822.

Linkies, A., Graeber, K., Knight, C. and Leubner-Metzger, G. (2010). The evolution of seeds. *New Phytologist* **186**, 817–831.

Liu, P.-P., Koizuka, N., Homrichhausen, T.M. *et al.* (2005). Large-scale screening of Arabidopsis enhancer-trap lines for seed germination-associated genes. *The Plant Journal* **41**, 936–944.

Magome, H., Nomura, T., Hanada, A. *et al.* (2013). CYP714B1 and CYP714B2 encode gibberellin 13-oxidases that reduce gibberellin activity in rice. *Proceedings of the National Academy of Sciences USA* **110**, 1947–1952.

Mathiesen, L., Malterud, K.E. and Sund, R.B. (1995). Antioxidant activity of fruit exudate and C-methylated dihydrochalcones from *Myrica gale*. *Planta Medica* **61**, 515–518.

Mathiesen, L., Malterud, K.E. and Sund, R.B. (1997). Hydrogen bond formation as basis for radical scavenging activity: a structure-activity study of C-methylated dihydrochalcones from *Myrica gale* and structurally related acetophenones. *Free Radical Biology and Medicine* **22**, 307–311.

Morris, K., Linkies, A., Müller, K. *et al.* (2011). Regulation of seed germination in the close Arabidopsis relative *Lepidium sativum*: a global tissue-specific transcript analysis. *Plant Physiology* **155**, 1851–1870.

Müller, K., Tintelnot, S. and Leubner-Metzger, G. (2006). Endosperm-limited Brassicaceae seed germination: Abscisic acid inhibits embryo-induced endosperm weakening of *Lepidium sativum* (cress) and endosperm rupture of cress and *Arabidopsis thaliana*. *Plant and Cell Physiology* **47**, 864–877.

Nakajima, M., Shimada, A., Takashi, Y. *et al.* (2006). Identification and characterization of Arabidopsis gibberellin receptors. *The Plant Journal* **46**, 880–889.

Nomura, T., Magome, H., Hanada, A. *et al.* (2013). Functional analysis of Arabidopsis CYP714A1 and CYP714A2 reveals that they are distinct gibberellin modification enzymes. *Plant and Cell Physiology* **54**, 1837–1851.

Nonogaki, H. (2006). Seed germination: The biochemical and molecular mechanisms. *Breeding Science* **56**, 93–105.

North, H., Baud, S., Debeaujon, I. *et al.* (2010). Arabidopsis seed secrets unravelled after a decade of genetic and omics-driven research. *The Plant Journal* **61**, 971–981.

Ogawa, M., Hanada, A., Yamauchi, Y. *et al.* (2003). Gibberellin biosynthesis and response during *Arabidopsis* seed germination. *The Plant Cell* **15**, 1591–1604.

Oh, E., Yamaguchi, S., Kamiya, Y. *et al.* (2006). Light activates the degradation of PIL5 protein to promote seed germination through gibberellin in Arabidopsis. *The Plant Journal* **47**, 124–139.

Oh, E., Yamaguchi, S., Hu, J. *et al.* (2007). PIL5, a phytochrome-interacting bHLH protein, regulates gibberellin responsiveness by binding directly to the GAI and RGA promoters in *Arabidopsis* seeds. *The Plant Cell* **19**, 1192–1208.

Oracz, K., Voegele, A., Tarkowska, D. *et al.* (2012). Myrigalone A inhibits *Lepidium sativum* seed germination by interference with gibberellin metabolism and apoplastic superoxide production required for embryo extension growth and endosperm rupture. *Plant and Cell Physiology* **53**, 81–95.

Park, J., Lee, N., Kim, W. *et al.* (2011). ABI3 and PIL5 collaboratively activate the expression of SOMNUS by directly binding to its promoter in imbibed *Arabidopsis* seeds. *The Plant Cell* **23**, 1404–1415.

Park, J., Nguyen, K.T., Park, E. *et al.* (2013). DELLA proteins and their interacting RING Finger proteins repress gibberellin responses by binding to the promoters of a subset of gibberellin-responsive genes in *Arabidopsis*. *The Plant Cell* **25**, 927–943.

Penfield, S. and Hall, A. (2009). A role for multiple circadian clock genes in the response to signals that break seed dormancy in *Arabidopsis*. *The Plant Cell* **21**, 1722–1732.

Penfield, S., Josse, E.-M., Kannangara, R. *et al.* (2005). Cold and light control seed germination through the bHLH transcription factor SPATULA. *Current Biology* **15**, 1998–2006.

Penfield, S., Li, Y., Gilday, A.D. *et al.* (2006). *Arabidopsis* ABA INSENSITIVE4 regulates lipid mobilization in the embryo and reveals repression of seed germination by the endosperm. *The Plant Cell* **18**, 1887–1899.

Peng, J., Carol, P., Richards, D.E. *et al.* (1997). The Arabidopsis *GAI* gene defines a signaling pathway that negatively regulates gibberellin responses. *Genes and Development* **11**, 3194–3205.

Peng, J., Richards, D.E., Hartley, N.M. *et al.* (1999). 'Green revolution' genes encode mutant gibberellin response modulators. *Nature* **400**, 256–261.

Pimenta Lange, M.J. and Lange, T. (2006). Gibberellin biosynthesis and the regulation of plant development. *Plant Biology* **8**, 281–290.

Piskurewicz, U. and Lopez-Molina, L. (2009). The GA-signaling repressor RGL3 represses testa rupture in response to changes in GA and ABA levels. *Plant Signaling and Behavior* **4**, 63–65.

Piskurewicz, U., Jikumaru, Y., Kinoshita, N. *et al.* (2008). The gibberellic acid signaling repressor RGL2 inhibits *Arabidopsis* seed germination by stimulating abscisic acid synthesis and ABI5 activity. *The Plant Cell* **20**, 2729–2745.

Piskurewicz, U., Tureckova, V., Lacombe, E. and Lopez-Molina, L. (2009). Far-red light inhibits germination through DELLA-dependent stimulation of ABA synthesis and ABI3 activity. *The EMBO Journal* **28**, 2259–2271.

Popovici, J., Bertrand, C., Jacquemoud, D. *et al.* (2011). An allelochemical from *Myrica gale* with strong phytotoxic activity against highly invasive *Fallopia x bohemica* taxa. *Molecules* **16**, 2323–2333.

Preston, J., Tatematsu, K., Kanno, Y. *et al.* (2009). Temporal expression patterns of hormone metabolism genes during imbibition of *Arabidopsis thaliana* seeds: a comparative study on dormant and non-dormant accessions. *Plant and Cell Physiology* **50**, 1786–1800.

Rieu, I., Ruiz-Rivero, O., Fernandez-Garcia, N. *et al.* (2008). The gibberellin biosynthetic genes *AtGA20ox1* and *AtGA20ox2* act, partially redundantly, to promote growth and development throughout the Arabidopsis life cycle. *The Plant Journal* **53**, 488–504.

Schwember, A.R. and Bradford, K.J. (2010a). A genetic locus and gene expression patterns associated with the priming effect on lettuce seed germination at elevated temperatures. *Plant Molecular Biology* **73**, 105–118.

Schwember, A.R. and Bradford, K.J. (2010b). Quantitative trait loci associated with longevity of lettuce seeds under conventional and controlled deterioration storage conditions. *Journal of Experimental Botany* **61**, 4423–4436.

Seo, M., Jikumaru, Y. and Kamiya, Y. (2011). Profiling of hormones and related metabolites in seed dormancy and germination studies. *Methods in Molecular Biology* **773**, 99–111.

Seo, M., Hanada, A., Kuwahara, A. *et al.* (2006). Regulation of hormone metabolism in Arabidopsis seeds: phytochrome regulation of abscisic acid metabolism and abscisic acid regulation of gibberellin metabolism. *The Plant Journal* **48**, 354–366.

Seo, M., Nambara, E., Choi, G. and Yamaguchi, S. (2009). Interaction of light and hormone signals in germinating seeds. *Plant Molecular Biology* **69**, 463–472.

Skene, K.R., Sprent, J.I., Raven, J.A. and Herdman, L. (2000). Myrica gale L. *Journal of Ecology* **88**, 1079–1094.

Sliwinska, E., Bassel, G.W. and Bewley, J.D. (2009). Germination of *Arabidopsis thaliana* seeds is not completed as a result of elongation of the radicle but of the adjacent transition zone and lower hypocotyl. *Journal of Experimental Botany* **60**, 3587–3594

Stamm, P., Ravindran, P., Mohanty, B. *et al.* (2012). Insights into the molecular mechanism of RGL2-mediated inhibition of seed germination in *Arabidopsis thaliana*. *BMC Plant Biology* **12**, 179–195.

Suh, K., Zhang, H., Wang, S. *et al.* (2013). ABI4 regulates primary seed dormancy by regulating the biogenesis of abscisic acid and gibberellins in Arabidopsis. *PLoS Genetics* **9**, e1003577–e1003591.

Sun, T.P. and Gubler, F. (2004). Molecular mechanism of gibberellin signaling in plants. *Annual Review of Plant Biology* **55**, 197–223.

Sun, T.P., Goodman, H.M. and Ausubel, F.M. (1992). Cloning the Arabidopsis *GA1* locus by genomic subtraction. *The Plant Cell* **4**, 119–128.

Suzuki, H., Park, S.H., Okubo, K. *et al.* (2009). Differential expression and affinities of Arabidopsis gibberellin receptors can explain variation in phenotypes of multiple knock-out mutants. *The Plant Journal* **60**, 48–55.

Talon, M., Koornneef, M. and Zeevaart, J.A. (1990). Endogenous gibberellins in *Arabidopsis thaliana* and possible steps blocked in the biosynthetic pathways of the semidwarf *ga4* and *ga5* mutants. *Proceedings of the National Academy of Sciences USA*, **87**, 7983–7987.

Tang, C.S. and Young, C.C. (1982). Collection and identification of allelopathic compounds from the undisturbed root system of bigalta limpograss (*Hemarthria altissima*). *Plant Physiology* **69**, 155–160.

Toh, S., Imamura, A., Watanabe, A. *et al.* (2008). High temperature-induced abscisic acid biosynthesis and its role in the inhibition of gibberellin action in Arabidopsis seeds. *Plant Physiology* **146**, 1368–1385.

Toh, S., Kamiya, Y., Kawakami, N. *et al.* (2012a). Thermoinhibition uncovers a role of strigolactones in Arabidopsis seed germination. *Plant and Cell Physiology* **53**, 107–117.

Toh, S., Mccourt, P. and Tsuchiya, Y. (2012b). HY5 is involved in strigolactone-dependent seed germination in Arabidopsis. *Plant Signaling and Behavior* **7**, 556–558.

Ueguchi-Tanaka, M., Ashikari, M., Nakajima, M. *et al.* (2005). *GIBBERELLIN INSENSITIVE DWARF1* encodes a soluble receptor for gibberellin. *Nature* **437**, 693–698.

Urbanova, T., Tarkowska, D., Novak, O. *et al.* (2013). Analysis of gibberellins as free acids by ultra performance liquid chromatography-tandem mass spectrometry. *Talanta* **112**, 85–94.

Voegele, A., Linkies, A., Müller, K. and Leubner-Metzger, G. (2011). Members of the gibberellin receptor gene family *GID1* (*GIBBERELLIN INSENSITIVE DWARF1*) play distinct roles during *Lepidium sativum* and *Arabidopsis thaliana* seed germination. *Journal of Experimental Botany* **62**, 5131–5147.

Voegele, A., Graeber, K., Oracz, K. *et al.* (2012). Embryo growth, testa permeability, and endosperm weakening are major targets for the environmentally regulated inhibition of *Lepidium sativum* seed germination by myrigalone A. *Journal of Experimental Botany* **63**, 5337–5350.

Watt, M.S., Bloomberg, M. and Finch-Savage, W.E. (2011). Development of a hydrothermal time model that accurately characterises how thermoinhibition regulates seed germination. *Plant, Cell and Environment* **34**, 870–876.

Weir, T.L., Park, S.W. and Vivanco, J.M. (2004). Biochemical and physiological mechanisms mediated by allelochemicals. *Current Opinions in Plant Biolology* **7**, 472–479.

Weitbrecht, K., Müller, K. and Leubner-Metzger, G. (2011). First off the mark: early seed germination. *Journal of Experimental Botany* **62**, 3289–3309.

Weston, D.E., Elliott, R.C., Lester, D.R. *et al.* (2008). The Pea DELLA proteins LA and CRY are important regulators of gibberellin synthesis and root growth. *Plant Physiology* **147**, 199–205.

Weston, L.A. and Duke, S.O. (2003). Weed and crop allelopathy. *Critical Reviews in Plant Sciences* **22**, 367–389.

Williamson, G.B., Obee, E.M. and Weidenhamer, J.D. (1992). Inhibition of *Schizachyrium scoparium* (Poaceae) by the allelochemical hydrocinnamic acid. *Journal of Chemical Ecology* **18**, 2095–2105.

Willige, B.C., Ghosh, S., Nill, C. *et al.* (2007). The DELLA Domain of GA INSENSITIVE Mediates the Interaction with the GA INSENSITIVE DWARF1A Gibberellin Receptor of *Arabidopsis*. *The Plant Cell* **19**, 1209–1220.

Winter, D., Vinegar, B., Nahal, H. *et al.* (2007). An "electronic fluorescent pictograph" browser for exploring and analyzing large–scale biological data sets. *PLoS ONE* **2**, e718–e732.

Yamaguchi, S. (2008). Gibberellin metabolism and its regulation. *Annual Review of Plant Biology* **59**, 225–251.

Yamaguchi, S., Kamiya, Y. and Sun, T.P. (2001). Distinct cell-specific expression patterns of early and late gibberellin biosynthetic genes during Arabidopsis seed germination. *The Plant Journal* **28**, 443–453.

Yamaguchi, S., Kamiya, Y. and Nambara, E. (2007). Regulation of ABA and GA levels during seed development and germination in *Arabidopsis*. *In:* Bradford, K. and Nonogaki, H. (eds) *Seed development, dormancy and germination*. Oxford, UK: Blackwell.

Yamauchi, Y., Ogawa, M., Kuwahara, A. *et al.* (2004). Activation of gibberellin biosynthesis and response pathways by low temperature during imbibition of *Arabidopsis thaliana* seeds. *The Plant Cell* **16**, 367–378.

Yano, K., Aya, K., Hirano, K. *et al.* (2015). Comprehensive gene expression analysis of rice aleurone cells: probing the existence of an alternative gibberellin receptor. *Plant Physiology* **167**, 531–544.

Yano, R., Kanno, Y., Jikumaru, Y. *et al.* (2009). CHOTTO1, a putative double APETALA2 repeat transcription factor, is involved in abscisic acid-mediated repression of gibberellin biosynthesis during seed germination in Arabidopsis. *Plant Physiology* **151**, 641–654.

Yomo, H. and Iinuma, H. (1966). Production of gibberellin-like substance in the embryo of barley during germination. *Planta* **71**, 113–118.

Zentella, R., Zhang, Z.L., Park, M. *et al.* (2007). Global analysis of DELLA direct targets in early gibberellin signaling in *Arabidopsis*. *The Plant Cell* **19**, 3037–3057.

Zhou, R., Yu, M. and Pharis, R.P. (2004). 16,17-dihydro gibberellin A_5 competitively inhibits a recombinant Arabidopsis GA 3β-hydroxylase encoded by the *GA4* gene. *Plant Physiology* **135**, 1000–1007.

Annual Plant Reviews (2016) **49**, 285–322
doi: 10.1002/9781119210436.ch10
http://onlinelibrary.wiley.com

Chapter 10

GIBBERELLINS AND PLANT VEGETATIVE GROWTH

Cristina Martínez, Ana Espinosa-Ruiz and Salomé Prat
Centro Nacional de Biotecnología (CNB-CSIC), Madrid, Spain

Abstract: Gibberellins (GAs) control multiple developmental processes throughout the plant life cycle. GA action relies on the degradation of the DELLA repressors, which modulate the activity of several families of transcription factors. In recent years, our knowledge of DELLAs mode of action has increased exponentially, and a novel transcriptional activation function for these repressors has also been reported. Control of cell elongation by these hormones has been dissected at the molecular level, and a role of GAs in cell division has been definitely established. At the physiological level, important progress has been made towards understanding how GAs modulate different developmental processes and extensive cross-talk interaction with other hormonal pathways, such as brassinosteroids, auxin, ethylene, or jasmonates could be unveiled. GA tissue-specific distribution has been demonstrated, and the first GA transporters have been identified. Here, we discuss recent findings concerning the function of GAs in the control of shoot and root growth, and responses to stress during plant vegetative growth.

Keywords: Aux/IAA, DELLA, endodermis, hormonal cross-talk, hypocotyl elongation, mycorrhiza, PIFs, polar auxin transport, root meristem, stress tolerance

10.1 Introduction

Gibberellins (GAs) are a class of tetracyclic diterpenoid hormones with a key role in plant growth and in multiple developmental processes, including seed germination, floral transition and fruit patterning. Plants affected in GA synthesis or response display dwarf phenotypes of different severity, due to defects in cell elongation and cell division (Figure 10.1). GAs are required for the rapid growth of the hypocotyl in dark germinated seedlings

Annual Plant Reviews, Volume 49: The Gibberellins, First Edition.
Edited by Peter Hedden and Stephen G. Thomas.

(Alabadí *et al.*, 2008; Feng *et al.*, 2008; de Lucas *et al.*, 2008), and the increased elongation of stem and petioles of plants grown under a shaded canopy (Djakovic-Petrovic *et al.*, 2007; Keuskamp *et al.*, 2010). Increased levels of these hormones in shaded plants suppress axillary meristem activity and inhibit shoot branching (Nakamura *et al.*, 2013). Moreover, GAs promote root growth by increasing both cell proliferation and expansion (Úbeda-Tomás *et al.*, 2008; 2009; Achard *et al.*, 2009), hence playing a crucial role in determining the overall plant shoot and root architecture.

Gibberellin signalling is also known to play a pivotal role in the trade-off between plant growth and adaptation to adverse environmental conditions (Achard *et al.*, 2006, 2008a; Magome *et al.*, 2008; Dubois *et al.*, 2013), and in the defence against pathogenic attack (Navarro *et al.*, 2008; Yang *et al.*, 2012), with relative activation of these opposing responses being critical to optimal plant survival in nature. Several of the high-yielding 'green revolution' wheat and rice semi-dwarf cultivars carry mutations in GA biosynthetic or the DELLA repressor genes (Hedden, 2003), hence highlighting the GA pathway as an important target for increasing the productivity of crops (see also Chapter 13).

GAs are synthesised from the diterpenoid precursor *trans*-geranylgeranyl diphosphate, via the sequential action of two plastidic terpene cyclases, *ent*-copalyl diphosphate synthase (CPS) and *ent*-kaurene synthase (KS). Outside the plastid, *ent*-kaurene is subjected to oxidation by the endoplasmic reticulum cytochrome P450 mono-oxygenase *ent*-kaurene oxidase (KO) and *ent*-kaurenoic acid oxidase (KAO) enzymes, to be further modified in the cytosol by soluble 2-oxoglutarate-dependent dioxygenases, leading to bioactive GA_1, GA_3, GA_4 and GA_7 (reviewed by Hedden and Thomas, 2012). *CPS* null mutations lead to a severe GA-deficient dwarf phenotype (Sun and Kamiya, 1994; Figure 10.1a). In contrast, GA 20-oxidases (GA20ox) and GA 3-oxidases (GA3ox) are encoded by small gene families, with mutations in these genes often leading to only a mild dwarf phenotype (Hu *et al.*, 2008; Rieu *et al.*, 2008; Figure 10.1a). A third class of dioxygenases, the GA 2-oxidases (GA2ox), catalyse the inactivation of GAs (see also Chapter 3). Genes encoding these different families of dioxygenases were shown to be the main sites of GA synthesis regulation in response to developmental and environmental signals (see also Chapter 2).

GAs promote growth by triggering the degradation of the DELLA proteins, a sub-group of the plant-specific GRAS family of regulators that function as repressors of GA signalling. Binding of bioactive GAs to GIBBERELLIN INSENSITIVE DWARF 1 (GID1) receptor facilitates GID1-DELLA interaction, and enables DELLA recognition by the F-box ubiquitin E3 ligase protein SLY1/GID2 (Hirano *et al.*, 2010). This results in polyubiquitination of DELLAs, marking them for degradation by the 26S proteasome (McGinnis *et al.*, 2003; Sasaki *et al.*, 2003; Dill *et al.*, 2004; Fu *et al.*, 2004; see also Chapter 6).

Although the biochemical function of DELLAs remained elusive for over a decade after their discovery, it is now well established that DELLAs control plant growth and development through direct interactions with

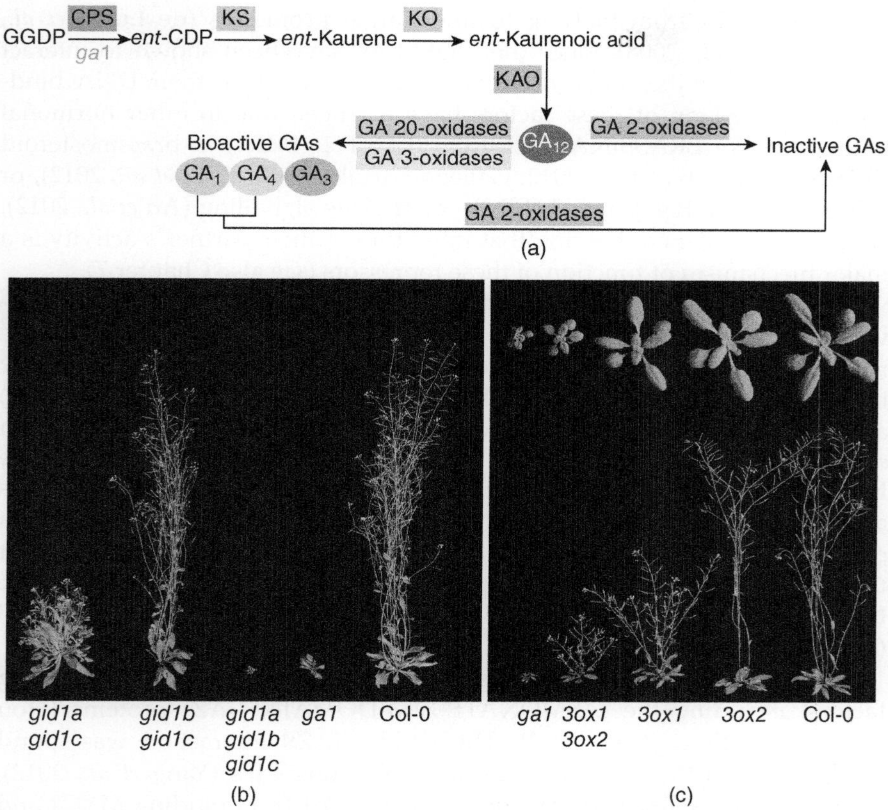

Figure 10.1 Phenotype of GA-deficient and insensitive mutants in *Arabidopsis thaliana*. (a) Schematic representation of the GA biosynthetic pathway and deactivation by GA 2-oxidase. GGDP, geranylgeranyl diphosphate; *ent*-CDP *ent*-copalyl diphosphate; CPS *ent*-copalyl diphosphate synthase; KS, *ent*-kaurene synthase; KO, *ent*-kaurene oxidase; KAO, *ent*-kaurenoic acid oxidase. (b) Phenotype of the *gid1* mutants. GID1, the gibberellin receptor, is encoded by three partially redundant genes in *Arabidopsis*. The semi-dwarf phenotype of *gid1a gid1c* double mutants reveals a predominant role of these two genes in stem elongation. Triple *gid1* mutant displays a more extreme dwarf phenotype than the biosynthetic *ga1* mutant. (Griffiths *et al.* (2006). Reproduced with permission of American Society of Plant Biologists.) (c) Additive phenotype of the *ga3ox* mutants. Among the four *ga3ox* genes in *Arabidopsis*, *GA3ox1* and *GA3ox2* control stem elongation and branching. (Mitchum *et al.* (2006). Reproduced with permission from John Wiley & Sons.) (See insert for colour representation of this figure.)

different families of transcriptional regulators, of which the bHLH PHYTOCHROME INTERACTING FACTORs (PIFs), PIF3 and PIF4 were the first to be identified (de Lucas *et al.*, 2008; Feng *et al.*, 2008). PIFs promote cell elongation by activating transcription of cell-wall-related enzymes and auxin signalling components; whereas DELLAs restrain growth by preventing PIFs from binding to their target promoters (de Lucas *et al.*, 2008; Feng *et al.*, 2008). Since then, DELLAs have been shown to interact with several families of transcription factors and block their DNA binding activity. Many of these factors have a crucial role in other hormonal pathways, like BRASSINAZOLE-RESISTANT 1 (BZR1) in brassinosteroid (BR) signalling (Bai *et al.*, 2012; Gallego-Bartolomé, 2012; Li *et al.*, 2012), or ETHYLENE INSENSITIVE 3 (EIN3) in ethylene signalling (An *et al.*, 2012), cumulative evidence showing that inhibition of their partner's activity is a major mechanism of function of these repressors (see also Chapter 7).

DELLAs have also been reported to associate with promoters to activate gene expression (Zentella *et al.*, 2007; Lim *et al.*, 2013). DELLAs were recently shown to inhibit seed germination at elevated temperatures, by interacting with the ABA-signalling factors ABA-INSENSITIVE3 (ABI3) and ABA-INSENSITIVE5 (ABI5) to activate transcription of *SOMNUS* (Lim *et al.*, 2013). Furthermore, DELLAs interact with several members of the INDETERMINATE DOMAIN (IDD) subfamily of C_2H_2 transcription factors and enhance their transcriptional activity (Feurtado *et al.*, 2011; Fukazawa *et al.*, 2014; Yoshida *et al.*, 2014). How this activity is achieved is at present unknown, although it might be conferred by chromatin remodelling, since DELLAs were found to interact with chromatin remodellers such as SWI3 (Sarnowska *et al.*, 2013) and PICKLE (Zhang *et al.*, 2014).

DELLAs have also been demonstrated to interact with non-DNA binding factors, including the JASMONATE-ZIM-DOMAIN (JAZ) proteins (Hou *et al.*, 2010; Yang *et al.*, 2012). The DELLA-JAZ9 interaction was found to prevent DELLAs from binding and inhibiting PIFs (Yang *et al.*, 2012), whereas DELLA–JAZ1 interaction prevents JAZ1 from binding MYC2 and inhibit JA-regulated gene expression (Hou *et al.*, 2010).

Hence, according to the current model, DELLAs function as modulators of their targets' activities. Unravelling the spatiotemporal distribution of DELLAs and their interactors, and the mechanism by which these repressors mediate activation of some genes, will be critical to fully understand the physiological role of GAs in different developmental processes (see also Chapter 7).

10.2 Gibberellins and shoot development

GAs have a well-established role in plant vegetative development, as exemplified by the dramatic reduction in leaf expansion and impaired stem and root elongation of the *Arabidopsis* GA-deficient *ga1-3* mutants, defective in CPS activity (Koornneef and van der Veen, 1980). Application of GAs rescues all growth defects of this severe mutant, underscoring an essential role of these hormones in growth promotion of both above- and below-ground organs.

10.2.1 Control of SAM function and leaf size

GAs stimulate, while DELLAs repress leaf growth through effects on both cell expansion and cell division (reviewed by González *et al.*, 2012). The molecular mechanisms underlying GA promotion of cell elongation are at present best understood, since GAs stimulate the expression of enzymes involved in cell wall loosening and promote microtubule rearrangement guiding directional cell expansion (Ogawa *et al.*, 2003; Achard *et al.*, 2009; Sambade *et al.*, 2012). Likewise, cross-talk between GAs and auxin positively modulates cell elongation, with GA biosynthesis and signalling shown to be required for efficient auxin transport and the appropriate accumulation of PIN-FORMED (PIN) auxin efflux transporters (Willige *et al.*, 2011). The notion that auxin controls cell expansion downstream of GA signalling is supported by microarray data showing that PIFs control both GA- and auxin-regulated gene sets (Gallego-Bartolomé *et al.*, 2011a; Nozue *et al.*, 2011).

Moreover, additional data indicate that these hormones control cell division. Plants in which DELLA proteins are stabilised, e.g. the *sly1* mutant, show a reduced leaf blade area partly due to a decreased cell division rate, whereas cell proliferation rate is higher in young leaves of the quadruple *della* mutant (Achard *et al.*, 2009). A positive regulator of cell division downstream of DELLAs is the receptor-like cytoplasmic kinase STUNTED, that stimulates cell proliferation by repressing the *SIAMESE* (*SIM*) and *SIAMESE RELATED1* (*SMR1*) cyclin-dependent kinase inhibitor genes (Lee *et al.*, 2012). Repression of STUNTED transcription by the DELLAs is lifted by GAs, although the molecular mechanism for DELLA-mediated control of this gene has not been described so far. Although the identification of GA targets among cell cycle regulators had remained elusive, DELLAs were recently shown to directly regulate the activity of the class I TCP factors, which bind the promoters of core cell-cycle genes (Davière *et al.*, 2014). DELLAs bind the DNA recognition domain of these factors and block their function, while GA-induced destruction of DELLAs antagonises such repression and allows cell-cycle gene activation.

A related GA function in cell proliferation was also described in maize leaves, which are an excellent model to leaf growth studies because of their linear organisation. GAs were shown to accumulate as a narrow peak at the leaf transition zone between cell division and elongation (Nelissen *et al.*, 2012). In *GA20ox* over-expressers, the number of proliferating cells at the leaf base is increased, while a reduction in these cells is observed in the *d8* mutant that accumulates a stable form of the maize DELLA (Lawit *et al.*, 2010; Nelissen *et al.*, 2012). Hence, these observations point to an essential role of DELLAs in signalling cell-cycle exit and the onset of cell differentiation in the leaves.

Likewise, GAs are required to promote cell proliferation in the root apical meristem (see below), where reduced GA levels or expression of the stable gai-1 protein lead to a substantial reduction in meristem size (Achard *et al.*,

2009; Úbeda-Tomás *et al.*, 2009). In contrast, GAs are excluded from the shoot apical meristem (SAM), highlighting a key differential function of GAs in the SAM and root meristems (Galinha *et al.*, 2009). In the shoot, an auxin maxima is associated with exit from the SAM and acquisition of lateral organ identity, while cells undergo more cell divisions than their root counterparts and still maintain a proliferative activity at the base of the leaf (Barkoulas *et al.*, 2007). The KNOTTED1 (KN1)-like homeobox (KNOX) factors promote SAM function by mediating high cytokinin (CK) activity (Jasinski *et al.*, 2005; Yanai *et al.*, 2005), whereas they negatively modulate GA accumulation by repressing *GA20ox* transcription (Hay *et al.*, 2002) (Figure 10.2). Loss-of-function *SHOOT MERISTEMLESS (stm)* mutants fail to maintain a meristem, whereas *KNOX* over-expression leads to ectopic meristem formation on the adaxial surface of the leaves, emphasising a central role of KNOX activity in meristem function. Over-expression of the CK biosynthetic *ISOPENTENYL TRANSFERASE* (*IPT*) genes partially rescues the meristem defects of *stm* mutants (Yanai *et al.*, 2005), while exogenous GA application or constitutive GA signalling suppress the KNOX gain-of-function phenotypes and enhance the meristem defects of weak *stm* mutants (Hay *et al.*, 2002; Jasinski *et al.*, 2005). Notably, increased GA activity is unable to mimic the meristem defects of *stm* mutants, although over-expression of the CYTOKININ OXIDASE 3 (CKX3) catabolic enzyme in the constitutive GA-signalling *spindly* (*spy*) mutant results in *stm*-like SAM defects (Jasinski *et al.*, 2005). This indicates that combined regulation of CK and GA activities, and antagonistic signalling by these hormones is required for normal SAM function (Galinha *et al.*, 2009). KNOX1, in concert with other not yet identified regulators, activates *GA2ox* transcription at the leaf-meristem boundary (Sakamoto *et al.*, 2001; Bolduc and Hake, 2009) (Figure 10.2). Hence, localised expression of this GA catabolic enzyme likely prevents GA transport from adjacent tissues into the stem cell niche, and aids the establishment of a high CK/low GA hormonal environment required for SAM function (Bolduc and Hake, 2009).

10.2.2 Elongation of the hypocotyl

Plants exhibit an enormous phenotypic plasticity, which in nature is critical for survival at the highly variable conditions of their immediate environment. This distinctive feature depends on their ability to sense changes in light quantity and quality, and in temperature, and on the integration of these informational cues into their own hormonal status to dynamically regulate growth of the different organs (Lau and Deng, 2010). Light-induced photomorphogenesis in *Arabidopsis* has proved to be an excellent model to study this complex interaction network, since GA- and BR-signalling were established to antagonise light effects on this developmental switch (Achard *et al.*, 2007; Alabadí *et al.*, 2008). Moreover, the main integrators in this process are conserved in the adult plant, where they play a prominent role in the control

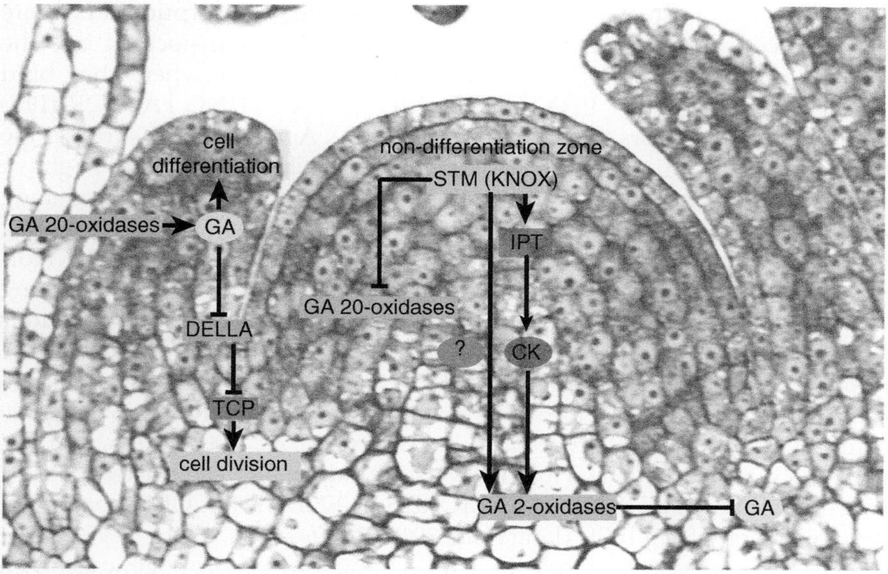

Figure 10.2 Model representing the role of GA in the shoot apical meristem. SHOOT MERISTEMLESS (STM), a member of the KNOTTED1-LIKE HOMEOBOX (KNOX) family of transcription factors, maintains the shoot apical meristem in an undifferentiated state increasing the content of cytokinins (CK), promoting the degradation of GA 20-oxidases and the activation of GA 2-oxidases through a CK-dependent and -independent manner, thus limiting the presence of GAs in the meristematic zone. Low GA levels promote DELLA accumulation in the shoot apex, where they actively restrict cell division through direct binding and inactivation of class I TCP transcription factors. However, in the emerging leaf primordium the expression of *KNOX* genes is absent, and the biosynthesis of GAs increases, promoting cell division and elongation. (The picture of the SAM was generously donated by Dr. Enrique Rojo.)

of increased stem and petiole elongation in response to vegetational shade or elevated temperatures, through a similar interaction network (Casal, 2013; Proveniers and van Zaten, 2013).

In the dark, the hypocotyl elongates very rapidly in an effort to reach the soil surface and enable the plant to start its photosynthetic life. The cotyledons remain unexpanded and the upper part of the hypocotyl folds into a hook-shaped structure that protects the meristem as the seedling pushes through the soil. In sunlight, growth of the hypocotyl ceases, the apical hook opens and the cotyledons expand and differentiate their chloroplasts, while the apical meristem is activated to start formation of true leaves (Kami *et al.*, 2010). Light triggers these developmental changes mostly through activation of the red light (RL) photoreceptor phytochromes (PHYs), of which PHYA prevails after early contact with light, whereas PHYB has a dominant role in suppressing hypocotyl elongation and inhibition of stem growth in the adult plant (Tepperman *et al.*, 2004, 2006). Upon light absorption, PHYs are photoconverted to their active Pfr form and rapidly translocated into the nucleus (Huq *et al.*, 2003; Fankhauser and Chen, 2008), where they bind several members of the PHYTOCHROME INTERACTING FACTOR (PIF) family of bHLH factors (Leivar and Quail, 2011). Activated PHYs trigger the rapid phosphorylation and degradation of PIFs, which function as negative regulators of light signalling by activating multiple cell-wall remodelling enzymes and auxin-related genes that control cell elongation, and repressing chloroplast biogenesis and photosynthetic genes (Nozue *et al.*, 2011; Hornitschek *et al.*, 2012; Li *et al.* 2012; Franklin *et al.*, 2011; Sun *et al.*, 2012; Zhang *et al.*, 2013). These bHLH factors share common and also distinct functions in several light-regulated processes (reviewed in Jeong and Choi, 2013). Loss-of-function mutations in the *PIF1, PIF3, PIF4* and *PIF5* genes, in the *pifq* mutant, lead to short hypocotyls and a constitutive photomorphogenic response in the dark (Leivar *et al.*, 2008; Shin *et al.*, 2009). PIFs, in turn, negatively regulate nuclear phyB levels, by facilitating the nuclear import and the CONSTITUTIVE PHOTOMORPHOGENESIS 1 (COP1)-dependent destabilisation of phyB Pfr (Jang *et al.*, 2010; Pfeiffer *et al.*, 2011).

GAs play an essential role in suppressing photomorphogenesis, with mutants defective in GA synthesis or signalling shown to display short hypocotyls, an open hook and activated expression of light-regulated genes when grown in darkness (Alabadí *et al.* 2004, 2008, Cheminant *et al.*, 2011). This de-etiolation response results from impaired DELLA degradation, these repressors found to directly interact with PIF3 and PIF4 and interfere with binding to the promoters of their gene targets (de Lucas *et al.*, 2008; Feng *et al.*, 2008). GA levels are high in dark-grown hypocotyls and decline during light-induced de-etiolation, which leads to DELLA stabilisation (Achard *et al.*, 2007; Alabadí *et al.*, 2008). Loss-of-function of DELLAs in the *global* mutant suppresses the de-etiolation defects of *ga1-3* seedlings or PAC-treated plants, with these mutants found to display tall hypocotyls and a partially etiolated response in the light (Feng *et al.*, 2008). Hence, GAs are required for full etiolated growth, to trigger degradation of DELLAs and allow activation of PIF-regulated genes.

DELLA-PIF interaction is also critical to suppress chlorophyll biosynthesis during etiolated growth. In contrast to the hypocotyls, relatively high levels of the GAI and RGA proteins were observed to accumulate in the dark in the cotyledons, and rapidly decrease following transfer to white light (Cheminant *et al.*, 2011, Josse *et al.*, 2011). Accumulation of DELLAs positively regulates the assembly of the photosynthetic machinery by de-repressing the chlorophyll and carotenoid biosynthetic pathways, and plays a role in protection against photo-oxidative damage, during initial light exposure. Accordingly, dark-grown *ga1-3* seedlings accumulate higher amounts of the phototoxic precursor *protochlorophyllide,* but are more resistant to photo-oxidative damage as they accumulate the photoprotective PROTOCHLOROPHYLLIDE OXIDOREDUCTASE (POR) enzyme in the dark (Cheminant *et al.*, 2011). DELLAs prevent PIF1 from repressing the *CONDITIONAL CHLORINA* (*CHLH*) gene, essential in *protochlorophyllide* biosynthesis, and *PHYTOENE SYNTHASE* (*PSY*), encoding the main rate-limiting enzyme in carotenoid biosynthesis (Toledo-Ortiz *et al.*, 2010, Cheminant *et al.*, 2011). Likewise, an interaction of DELLAs with SPATULA (SPT) modulates cotyledon expansion, with DELLAs proposed to negatively regulate the growth restraint imposed by SPT and prevent excessive growth inhibition by simultaneous SPT and DELLA activity (Josse *et al.*, 2011). Binding of DELLAs to PIF3 may also contribute to de-repress chlorophyll biosynthesis and photosynthetic gene expression, since PIF3 has been shown to interact with HISTONE DEACETYLASE15, the PIF3-HDA15 complex mediating decreased histone acetylation and repressing transcription of these genes by RNA polymerase II in etiolated seedlings (Liu *et al.*, 2013).

BRs exert also a major role in the control of etiolated growth (Depuydt and Hardtke, 2011), as indicated by the stronger dwarf phenotype and dark de-etiolated response of BR-related mutants as compared to GA mutants. A considerable advance in our understanding concerning how signalling by GAs and BRs is integrated, has come from the finding that BZR1, a transcription factor with a central role in BR-signalling, binds PIF4 and functions as a coactivator of this light-regulated factor (Oh *et al.*, 2012). Notably, PIF4 is unable to promote hypocotyl elongation in the absence of active BZR1, and conversely, the constitutive BR signalling phenotype of the *bzr1-D* mutant is lost in a *pifq* mutant background (Oh *et al.*, 2012). PIF4 and BZR1 were shown to share around 40% of their targets and bind as a complex to the promoters of several cell-wall-remodelling enzymes, auxin biosynthesis and transport, and auxin-signalling components like the *Aux/IAA* and *SAUR* genes, activating their expression (Oh *et al.*, 2012). In this way, coactivator function of these factors defines a 'core growth module' driving increased cell expansion (de Lucas and Prat, 2014). Importantly, the DELLA repressors negatively regulate this growth module, not only by inhibiting PIFs transcriptional activity, but also by binding BZR1 and its homologous BRI1-EMS-SUPPRESSOR1 (BES1), and suppress their activity via a similar mechanism as reported for PIFs (Gallego-Bartolomé *et al.*, 2012; Bai *et al.*, 2012; Li *et al.*, 2012). Up-regulation of auxin signalling is a major transcriptional output of the PIF4-BZR1 module, in agreement with a synergistic function of BRs and auxin in plant growth (Jaillais & Chory, 2010). Auxin signalling also up-regulates many of the PIF4-BZR1 gene targets (Jaillais and Chory, 2010; Nozue *et al.*, 2011), in addition to up-regulating GA synthesis (Frigerio *et al.*, 2006; O'Neill *et al.*, 2010; Chapman *et al.*, 2012), which leads to DELLA destabilisation and thus contributes to signal amplification through a positive feed-back loop. The current model for growth regulation by these hormones is shown in Figure 10.3.

Under diurnal conditions, the hypocotyl and petioles elongate in a time-of-day-specific manner, such that in short days (8 h light–16 h darkness) hypocotyl growth is fastest at dawn and suppressed during daytime (Nozue

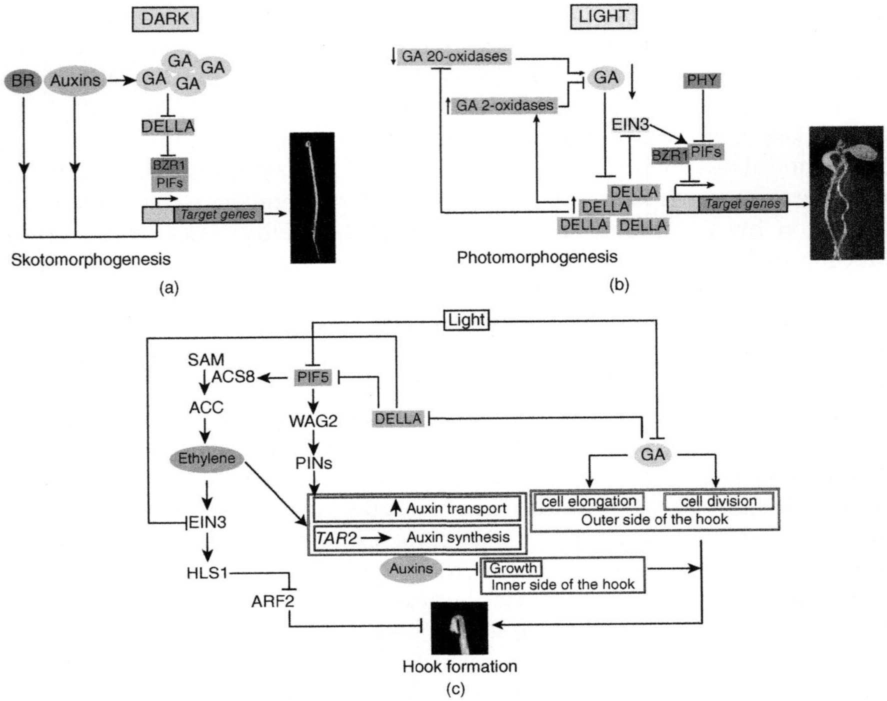

Figure 10.3 Schematic representation of light-dependent hypocotyl elongation. Circles represent hormones and squares represent proteins. Arrows and T-bars indicate positive and negative regulation, respectively. Green lines represent transcriptional regulation, and orange bars post-transcriptional regulation. (a) In dark-grown seedlings GA levels in the hypocotyl are high and DELLAs are degraded. Absence of Phy and DELLA repressor activities leads to PIF transcription factor accumulation. PIFs and BES1/BZR1 heterodimerise and activate their downstream targets, including cell wall remodelling enzymes and auxin response genes. Auxin response, in turn, increases GA levels inducing GA biosynthetic genes expression, in a positive feedback loop. (b) Under light conditions PIFs are destabilised and inactivated due to combined action of Phy signalling and DELLA accumulation. PIF transcriptional response is therefore blocked, ceasing hypocotyl growth. Downstream from ethylene, EIN3 has an opposite function, directly activating PIF3 transcription to promote hypocotyl elongation in the light. The ability of DELLAs to bind and inactivate EIN3 may provide a mechanism to fine-tune ethylene-mediated hypocotyl growth. (c) GA regulation of apical hook formation. Auxin biosynthesis and transport create an auxin gradient along the hypocotyl, which stops growth at the inner side of the hook, where an auxin maximum is reached. GAs, conversely, promote cell division and elongation in the outer side of the hook. Additionally, DELLAs counteract auxin signalling through auxin transport inhibition and direct inactivation of PIF5 and EIN3 transcription factors.

et al., 2007). This rhythmic growth pattern depends on the control of *PIF4* and *PIF5* transcription by the circadian clock and on the degradation of these proteins by light-activated phytochromes (Nozue *et al.*, 2007; Niwa *et al.*, 2009). *PIF4* and *PIF5* transcripts rise during late night and are repressed at dusk, dawn-specific hypocotyl elongation therefore being explained by the coincidence of high transcript levels and increased protein stability in the dark (Shin *et al.*, 2013). The inhibitory 'evening complex' formed by ELF3, ELF4 and the LUX/ARRYTHMO transcription factor, binds a conserved element in the *PIF4* and *PIF5* promoters, and represses transcription of these genes at dusk (Nusinow *et al.*, 2011). Mutations in the evening complex components actually lead to longer hypocotyls and up-regulated *PIF4* and *PIF5* transcript levels during early night, while *pif4 pif5* loss-of-function mutations suppress the phenotype of *elf3*, *elf4* and *lux* loss-of-function mutants (Nusinow *et al.*, 2011). The endogenous clock also gates GA signalling by regulating *GID1a* and *GID1b* expression, with levels of these transcripts observed to be elevated at night, but reduced during daytime (Arana *et al.*, 2011). Hence, GID1 protein oscillation further contributes to phase hypocotyl growth at dawn, with constitutive *GID1* expression found to expand the daily growth period, while loss of DELLAs leads to constitutive growth (Arana *et al.*, 2011).

PIFs also have a role in GA homeostasis, PIF1 being shown to activate *GAI* and *RGA* expression and indirectly repress the GA biosynthetic genes (Oh *et al.*, 2007). Furthermore, while the *PIF4* and *PIF5* transcripts display diurnal oscillation, *PIF3* is not regulated by the clock, although it was established that it plays a prominent role in promoting hypocotyl elongation at dawn (Soy *et al.*, 2012). Abundance of the PIF3 factor declines in the light due to photoactivated phyB-mediated degradation, but the protein accumulates during the night. Remarkably, while *pif3* mutants are impaired in ethylene-induced hypocotyl elongation in the light, *PIF3* over-expression rescues the short hypocotyl phenotype of *ein3 eil1* mutants, suggesting a function of PIF3 downstream of EIN3/EIL1 (Zhong *et al.*, 2012). In fact, *PIF3* is directly activated by EIN3, with another EIN3 target, the AP2-like ERF1 factor, found to suppress hypocotyl elongation and to be destabilised in the dark. Hence, the different light-dependent stabilities of the PIF3 and ERF1 proteins were proposed to mediate inverse ethylene effects in promoting hypocotyl elongation in the light, but suppress hypocotyl growth in darkness (Zhong *et al.*, 2012).

10.2.3 Apical hook formation

In dark-grown seedlings, GAs also play a major role in the control of apical hook formation, a 180° curvature of the hypocotyl below the cotyledons that protects the apical meristem and cotyledons as they push through the soil (Vriezen *et al.*, 2004; Li *et al.*, 2004). The apical hook is formed through asymmetric expansion of cells at both sides of the upper hypocotyl (Raz and Ecker, 1999; Vriezen *et al.*, 2004), and is coordinated by several plant hormones, including GAs, BRs, auxin and ethylene (reviewed by: Abbas *et al.*, 2013 and Mazzella *et al.*, 2014).

Accumulation of the DR5 auxin reporter at the inner hook side illustrates that differential growth of these cells depends on an asymmetrical distribution of auxin, which is progressively lost during hook opening (Zádníková *et al.* 2010; Gallego-Bartolomé *et al.*, 2011b). In seedlings treated with an auxin transport inhibitor, no auxin gradient is established and the hook is not formed. Formation of this auxin gradient requires the coordinated action of the PIN3, PIN4 and PIN7 efflux carriers (Zádníková *et al.* 2010; Vandenbussche *et al.*, 2010), and consequently *pin3 pin7* mutants lack an apical hook. Thus, increased auxin signalling marks the side with a lower growth rate, as seen for root gravitropism (Band *et al.*, 2012), in contrast to the shoot gravitropic and phototropic responses, where increased DR5 activity is associated with the side having a higher growth rate (Esmon *et al.*, 2006, Spalding, 2013).

Mutants with enhanced ethylene synthesis or signalling, such as the *eto1* and *eto2* over-producers or the *ctr1-1* constitutive signalling mutant, show an exaggerated hook curvature, while *etr1* mutants defective in ethylene perception and *ein2* mutants impaired in ethylene signalling, fail to form an apical hook (Vandenbussche *et al.*, 2010; Gallego-Bartolomé *et al.*, 2011b). Ethylene function is placed upstream of auxin, as auxin application compensates the hook defects of ethylene-insensitive mutants (Vandenbussche *et al.*, 2010), whereas treatment with the polar auxin transport inhibitor NPA, or mutations impairing auxin signalling or transport, suppress the exaggerated hook curvature of constitutive *ctr1-1* mutants (Zádníková *et al.*, 2010).

HOOKLESS1 (*HLS1*) was identified as an essential gene for hook formation, with *hls1* loss-of-function mutants failing to form both an apical hook and establish an auxin gradient in the hypocotyl (Lehman *et al.*, 1996; Li *et al.*, 2004). *HLS1* acts upstream of auxin signalling as mutations in the *AUXIN RESPONSE FACTOR 2* (*ARF2*) suppress the *hls1* phenotype, while HLS1 was observed to down-regulate *ARF2* expression (Li *et al.*, 2004). In response to ethylene, EIN3 directly binds the *HLS1* promoter and activates its expression (An *et al.*, 2012), indicating that HLS1 acts as a converging point between ethylene and auxin signalling (Figure 10.3c). However, HLS1 over-expressers (*HLS1ox*) still respond to ethylene by forming an exaggerated hook, which suggests that ethylene activates other HLS1-independent pathways. Ethylene in fact activates auxin biosynthesis in the inner side of the hook, by local up-regulation of *TAR2*, the tryptophan aminotransferase that catalyses the first step in IAA biosynthesis (Vandenbussche *et al.*, 2010). In addition, it modulates auxin transport by accelerating the turnover of AUX1 on the inner side of the hook (Vandenbussche *et al.*, 2010), and promoting PIN3 localisation to the lateral side of cortex cells at the outer hook side (Zádníková *et al.*, 2010).

GAs control both the speed and the degree of hook formation. Mutants impaired in GA synthesis or signalling do not form an apical hook, while *della* mutants display an exaggerated hook (Gallego-Bartolomé *et al.*, 2011b). Treatment with the GA-biosynthesis inhibitor paclobutrazol (PAC) abolishes differential DR5:GUS staining in the hypocotyl (Gallego-Bartolomé *et al.*, 2011b), indicating that GAs act in part by favouring generation of the auxin gradient.

GAs promote cell elongation and cell division at the outer side of the hook (Vriezen *et al.*, 2004). Therefore, hook formation involves both auxin-mediated growth inhibition at the inner hook side, and GA-mediated growth promotion at the outer side (Figure 10.3c).

A role for DELLAs in the control of a component of the ethylene signalling pathway has been suggested by the observation that PAC application suppresses the exaggerated hook of constitutive ethylene *ctr1-1* mutants (Vriezen *et al.*, 2004). Like ethylene, GAs induce *HLS1* expression (Gallego-Bartolomé *et al.*, 2011b; An *et al.*, 2012), and this activation is abolished in an *ein3 eil1* background. DELLA signalling requires HLS1, since the *hls1* hook phenotype is unaffected by GA treatment or DELLA loss-of-function mutations (An *et al.*, 2012). This response is indeed mediated by a direct interaction of DELLAs with EIN3 and EIL1 to block their transcriptional activity, with GAs activating *HLS1* expression by relieving the inhibitory effect of DELLAs. Over-expression of EIN3 restores hook formation of GA-deficient and PAC-treated seedlings and causes exaggerated hook bending in the presence of GAs, while *ein3 eil1* loss of function suppresses the exaggerated hook curvature of *della* mutants (An *et al.*, 2012). GAs and ethylene therefore act cooperatively to activate expression of *HLS1* (Figure 10.3c).

GAs are also required to maintain proper *PIN* gene expression and auxin transport from the cotyledons (Gallego-Bartolomé *et al.*, 2011b), with *pin3 pin7* mutants found to be resistant to GA treatment. On the other hand, expression of the stable gai-1 protein down-regulates the ethylene biosynthetic *ACC SYNTHASE8* (*ACS8*) and *ACS5/ETO2* genes (Gallego-Bartolomé *et al.*, 2011b). Remarkably, these genes were reported to be direct targets of PIF5, which binds a conserved G-box element in the *ACS8* promoter. Thus, these findings are consistent with a role of DELLAs in repressing *ACS8* expression by inhibiting PIF5 transcriptional activity (Gallego-Bartolomé *et al.*, 2011b). PIF5 also transcriptionally regulates WAG2, an AGC-type kinase shown to phosphorylate PINs (Willige *et al.*, 2012), pointing to a role of this factor in establishing the auxin gradient in the hook (Figure 10.3c). Expression of *HLS1* and *ACS5/ETO2* is in fact reduced in *pifq* seedlings (Leivar *et al.*, 2009; Shin *et al.*, 2009), which lack an apical hook and immediately enter into the opening phase. In cell-specific studies it was established that expression of the stable DELLA gai-1 protein in the endodermis is sufficient to impair hook formation, while no effect is observed when this protein is expressed in the epidermis (Gallego-Bartolomé *et al.*, 2011b). Notably, expression of *PIF1* in the endodermis of *pifq* seedlings is also able to restore hook formation (Kim *et al.*, 2011), thus supporting a function of PIFs downstream of GAs. Hence, suppression of PIF activity by both DELLA inhibition and phyB-mediated destabilisation is likely to play a relevant role in triggering apical hook opening in the light.

10.3 Gibberellin function in root development

Early studies in pea and lettuce suggested a role for GAs in the promotion of root growth (reviewed in Tanimoto, 2005). This function was confirmed in *Arabidopsis*, where the mechanisms of GA action in the control of root development are starting to be dissected. GA-deficient *ga1-3* mutants (Fu and Harberd, 2003) and wild-type seedlings treated with PAC (Achard *et al.*, 2009) display a severe reduction in root growth, which is fully recovered in the quadruple *della* mutant background, indicating that GA promotion of root growth is mediated by DELLA degradation (Fu and Harberd, 2003; Achard *et al.*, 2009).

During root growth, cell division and elongation are spatially separated, with cell division taking place in the root meristem and cell expansion in the root elongation-differentiation zone (Figure 10.4a). Consistent with the accepted role of GAs in promoting cell expansion, PAC application leads to a reduction in root cell length, indicating that DELLAs restrict root cell elongation (Achard *et al.*, 2009). Expression of the stable gai-1 protein in the root meristem leads to reduced cell elongation and shorter roots (Achard *et al.*, 2009), with both cell number and root meristem size observed to be greatly reduced in *ga1-3* mutants (Úbeda-Tomas *et al.*, 2009; Achard *et al.*, 2009). By using the *pCYCD:GUS* fusion as a reporter for cell division, it could be established that the *ga1-3* mutation or PAC treatments restrain cell division rate in the root meristem (Achard *et al.*, 2009). Similar results were obtained using a *pCYCB:GFP* fusion and anti-KNOLLE antibodies in roots of WT, *ga1-3* and *ga3ox1ga3ox2* double knock-out mutants (Úbeda-Tomás *et al.*, 2009). Hence, appropriate GA levels in the root are required to promote cell division and increase meristem size, in addition to controlling cell elongation.

In contrast, *ga1-3* or PAC-treated seedlings are not altered in mRNA, protein levels or the pattern of expression of genes involved in stem cell niche specification, such as *SHORT-ROOT* (*SHR*), *SCARECROW* (*SCR*) and *PLETHORA* (*PLT*) (Achard *et al.*, 2009), indicating that GAs control root meristem size by regulating proliferation of proximal meristem cells rather than affecting QC function.

10.3.1 Hormonal control of root growth

The development of the root meristem is regulated by a complex interaction of hormone signalling pathways (reviewed by Sozzani *et al.*, 2014; Pacifici *et al.*, 2015). The fundamental role of auxin in initiating, organising and maintaining the root apical meristem was soon established based on the strong phenotypes of mutants deficient in auxin biosynthesis or transport, which fail to initiate roots and show a premature arrest in meristem and stem cell function (Hamann *et al.*, 2002, Blilou *et al.*, 2005). Auxin accumulates within the distal stem cell region via both local biosynthesis (Brady *et al.*, 2007, Stepanova *et al.*, 2008) and polar transport from the shoot (Bliluo *et al.*, 2005),

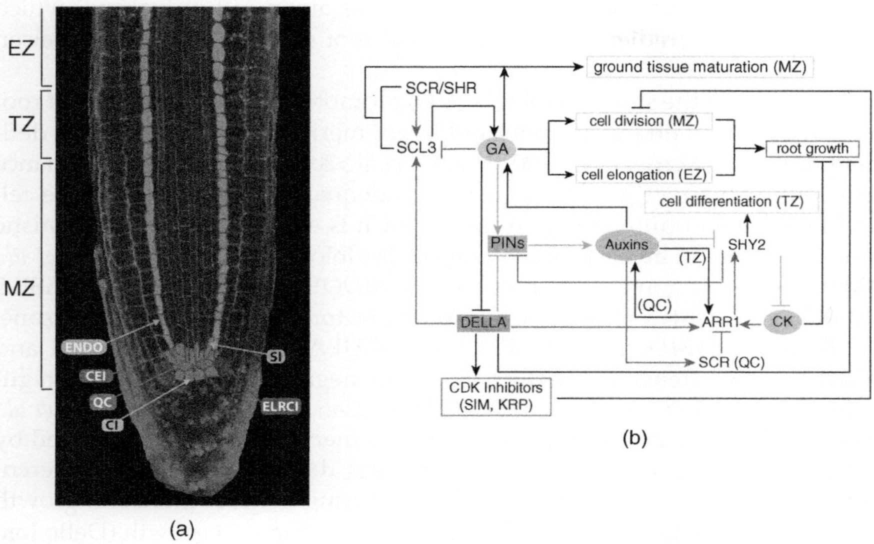

Figure 10.4 GA regulation of root development. (a) Longitudinal view of the *Arabidopsis* root tip. False-coloured confocal microscope image showing tissue layers and different developmental zones: MZ = meristematic zone, TZ = transition zone, EZ = elongation zone, QC = quiescent centre, CI = columela initials, CEI = cortex/endodermis initials, SI = stele initials, ELRCI = epidermis/lateral root cap initials, ENDO = endodermis. Gibberellins (GA) act in the endodermis of the meristematic and elongation zone (yellow cells), not affecting directly the stem cell niche, which includes the stem cells together with the quiescent centre (QC). (b) Schematic representation of the role of GAs in root growth. During the first 3 days post-germination (dpg) GA levels increase promoting cell division and elongation in detriment of cell differentiation and consequently, meristem growth. 5 dpg GA levels decrease and DELLA proteins increase in the TZ. As a consequence, DELLAs induce the expression of *ARR1* keeping SHY2 levels high, repressing auxin flow and thus balancing cell division with differentiation. Additionally, a DELLA-dependent accumulation of CDK inhibitors of the SIM and KRP families, promotes a reduction in cell division. Cytokinins (CK) in the TZ induce cell differentiation through up-regulation of SHY2, mediated by ARR1. On the other hand, auxins promote SHY2 degradation, thus sustaining PIN activity and cell division in the proximal meristem. Auxins also positively regulate GA biosynthesis. The resultant increased GA signalling induces polar auxin transport, and represses *ARR1*. In the QC, SCR represses *ARR1* expression and therefore, the synthesis of auxins. The auxins produced in the QC are transported through PIN action to the TZ where they induce *ARR1* expression promoting in this zone cell differentiation. In the endodermis of the MZ, SCL3, a positive regulator of GA signalling, and a DELLA-induced gene, is under the control of SCR and SHR, regulating together with GAs the ground tissue maturation. (The picture of the RAM was generously donated by Dr. Enrique Rojo.) (See insert for colour representation of this figure.)

through the activity of PIN-FORMED (PIN) auxin efflux carriers, which create an auxin gradient that peaks at the root stem cell niche (Grieneisen *et al.*, 2007).

In contrast to the shoot, cytokinin (CK) promotes the differentiation of root meristem cells, with longer roots and larger meristems observed in CK deficient mutants (Werner *et al.*, 2003, Riefler *et al.*, 2006). Meristem maintenance and continuous root growth requires an adequate balance between the cell proliferation and differentiation rates, which is ensured by the antagonistic function of auxin and CK signalling (Dello Ioio *et al.*, 2007; Ruzicka *et al.*, 2009). CK perception by the receptor *ARABIDOPSIS* HISTIDINE KINASE 3 (AHK3) activates two primary CK response factors in the root transition zone, the B-type *ARABIDOPSIS* RESPONSE REGULATORS1 and 12 (ARR1 and ARR12), which transcriptionally activate the negative auxin signalling regulator *SHORT HYPOCOTYL2* (*SHY2/IAA3*) (Tian *et al.*, 2002, Dello Ioio *et al.*, 2008). SHY2 has a crucial role in controlling meristem size, as illustrated by the enlarged root meristem of *shy2-31* mutants, due to a delay in cell differentiation. In contrast, increased SHY2 levels during the root meristem growth phase are sufficient to reduce meristem size and stop root growth (Dello Ioio *et al.*, 2008; Moubayidin *et al.*, 2010). SHY2 down-regulates *PIN* gene expression, whilst auxins signal the degradation of this Aux/IAA protein, and lead to increased PIN activity and sustained cell division (Dello Ioio *et al.*, 2008; Moubayidin *et al.*, 2010).

During the first days post-germination, the auxin/CK balance is shifted towards auxin signalling, to allow meristem growth. In *Arabidopsis*, the cell division and cell differentiation rates equilibrate by day 5 post-germination, and the size of the meristem becomes constant. This balance is largely mediated by *SHY2*, whose transcript levels are under tight temporal control and gradually increase following germination, to reach a plateau by 5 days after germination (Moubayidin *et al.*, 2010). Several lines of evidence indicate that temporal control of *SHY2* expression is set by GAs. First, GA levels in the root are elevated following germination, as evidenced by the high levels of expression of GA biosynthetic genes and the reduced accumulation of *pRGA*:RGA-GFP (Moubayidin *et al.*, 2010), then decrease by the time the meristem size becomes constant. Second, GA application decreases, whereas PAC treatment increases *ARR1* expression in the root, both *ARR1* and *SHY2* expression levels being reduced in *rga-24* mutants, which display a larger root meristem. Third, GA and PAC treatments do not affect root meristem size of *arr1* mutants (Moubayidin *et al.*, 2010), suggesting that RGA activates *ARR1* expression in a direct or indirect fashion. Hence, increased RGA abundance during the first days after germination promotes *SHY2* expression, which negatively regulates auxin signalling to balance the auxin/CK ratio and set the root meristem size (Figure 10.4b).

In *ga1-3* seedlings, an increase in transcription of the CYCLIN-DEPENDENT KINASE INHIBITORS (CKIs) of the SIAMESE (SIM) and KIP-RELATED PROTEIN (KRP) families has been observed, which is reverted by GA application (Achard *et al.*, 2009; Figure 10.4b). Accumulation of CKIs would explain decreased cell division in the root meristem, although it remains to be established whether activation of these genes is directly mediated by the DELLAs or results from increased SHY2 accumulation.

Intact auxin signalling is required for GA-promotion of root growth, as application of the auxin efflux inhibitor NPA inhibits GA-induced root growth of *ga1-3* mutants. Roots of decapitated *ga1-3* plants are also less responsive to GA than roots of intact *ga1-3* plants, while IAA restores this response (Fu and Harberd, 2003). GA-induced degradation of DELLA proteins is delayed in roots of the auxin insensitive *axr1-12* mutant and in seedlings pretreated with NPA, indicating that auxin increases DELLA sensitivity to GAs (Fu and Harberd, 2003). Auxins were reported to increase GA levels in *Arabidopsis*, pea and tobacco, mediated by an up-regulation of GA biosynthetic gene expression (Ross *et al.*, 2000; O'Neill and Ross, 2002; Wolbang *et al.*, 2004; Reid *et al.*, 2011; Chapman *et al.*, 2012; see also Chapter 8). In *Arabidopsis*, expression of the *GA20ox* gene is rapidly induced in response to auxin and a similar response is observed in the *rga-24 gai-t6* mutant, indicating that auxin-mediated activation of this gene is independent of DELLAs. The fast kinetics of induction suggests that short-lived Aux/IAA proteins may mediate this response, supported by the finding that cycloheximide, which potentiates Aux/IAA depletion by blocking *de novo* protein synthesis, mimics auxin effects, whereas MG132, which inhibits Aux/IAA destabilisation, prevents activation of this gene (Frigerio *et al.*, 2006).

In addition to their role in root growth, GAs also modulate root gravitropism, a response that largely depends on auxin transport. Gravity is perceived by the starch statoliths within the root columella cells which, upon gravistimulation, promote relocation of the auxin PIN transporters into the lower cell side and redirect auxin flow to this side of the root, causing the asymmetric growth of root cells and downward root bending (Ottenschläger *et al.*, 2003). GA signalling modulates this response as *ga1-3* and double *gid1* mutants display reduced gravitropism, while gravitropic curvature is fully recovered in the *ga1-3* quadruple *della* mutant (Willige *et al.*, 2011). During gravitropism, the PIN2 auxin transporter is degraded at the upper side of the root, but stabilised at the lower side, with *pin2* loss-of-function mutants showing a root agravitropic phenotype that is amplified by the GA biosynthesis inhibitor, uniconazole. In addition, application of this inhibitor or expression of the stable rgaΔ17 protein rescues the gravitropic defects of *35S::PIN1* over-expressers, suggesting a role for GAs in root gravitropism by regulating PINs abundance (Löfke *et al.*, 2013). PIN2 levels are in fact lower in *ga1-3* mutants, although GAs do not activate *PIN2* gene expression. Moreover, plasma membrane levels of all PIN-GFP fusions were observed to increase after GA treatment and be reduced by PAC application (Willige *et al.*, 2011; Li *et al.*, 2015), suggesting that GAs control their protein turnover. PIN levels are indeed regulated by the balance between recycling to the plasma membrane and traffic to the lytic vacuole. Inhibition of vacuolar transport in *ga1-3* mutant was found to increase PIN2 levels, pointing to a role of GAs in the control of PINs stability, by modulating their vacuolar targeting (Vieten *et al.*, 2007; Willige *et al.*, 2011). Remarkably, immunodetection with an antibody against GA revealed that root bending is also associated with an asymmetric distribution of GAs, with a maximum observed at the lower root side (Löfke *et al.*, 2013). Higher GA levels correlate with preferential degradation of the DELLAs in this region, and increased amounts of the PIN2 transporter at the plasma membrane, which help in consolidating asymmetric auxin distribution and differential cell growth. PINs were reported to be phosphorylated by the D6 PROTEIN KINASE (D6PK) sub-family of AGCVII kinases, which directly regulate PIN-mediated auxin transport (Willige *et al.*, 2013). Auxins promote D6PK-mediated phosphorylation of PINs, with this modification found to be crucial for activation of these efflux carriers (Zourelidou *et al.*, 2014). Given that *d6pk* mutants exhibit an impaired root gravitropic response as well as defects in polar auxin transport (Zourelidou *et al.*, 2009), it is tempting to speculate that GAs regulate phosphorylation of PINs through this family of protein kinases.

10.3.2 Gibberellin signalling from the endodermis

There is strong evidence to suggest that GAs are unequally distributed among the different root cell layers within the meristem and elongation zones. In studies in which GA signalling was specifically blocked in selected root tissues by targeted expression of the non-degradable gai-1 protein, it could be established that the endodermis is the main site of GA action to regulate root growth (Úbeda-Tomás *et al.*, 2008). Whereas expression of the gai-1 protein in epidermal, cortical or stele tissues had no effect, plants expressing gai-1 in tissues including the endodermis displayed smaller root meristems and identical root growth defects as the *gai-1* mutant (Úbeda-Tomás *et al.*, 2008). These results were confirmed in transgenic *SCR*::gai-GR plants that specifically expressed the gai-1 protein fused to the glucocorticoid receptor (GR) in the endodermis. In the absence of the dexamethasone (DEX) inducer, roots of these plants displayed normal growth, but their growth was severely reduced when germinated on DEX. Premature cessation of endodermal cell elongation was also observed to have a marked effect on expansion of the adjacent cortical and epidermal cells, consistent with a role of the endodermis as the primary GA-responsive tissue regulating growth of the different cell layers in the root.

GA signalling from the endodermis could be explained by tissue-specific GA biosynthesis, bioactive GA accumulation or an increased GA sensitivity (i.e. increased levels of GID1 GA receptors) or response (endodermal expression of DELLA-downstream effectors). Radiolabelled GAs and grafting studies with biosynthetic mutants showed that GAs move throughout the plant (Ragni *et al.*, 2011; Dayan *et al.*, 2012), although the mechanisms involved in transport of these hormones are only beginning to be unveiled (see also Chapter 4).

The exact tissues in the root where GAs accumulate was recently addressed by the development of bioactive fluorescent GA derivatives (GA-Fl), carrying a fluorescent tag attached to the tetracyclic diterpenoid C6 position of GA_3 and GA_4 (Shani *et al.*, 2013). GAs-Fl were observed to specifically accumulate in the endodermal cells of the root elongation zone, in good agreement with the reported role of the endodermis in GA signalling (Ubeda-Tomás *et al.*, 2008). GFP-RGA distribution was complementary to that of GA-FI, with strongly reduced levels of this DELLA detected in endodermal cells. Differential accumulation of GAs-Fl is actively regulated, as it is competed by non-labelled GAs, and reduced at low temperatures or after incubation with ATP synthesis inhibitors. Endodermal cells are required to drive this accumulation, since in the SHORT ROOT *shr-2* mutant, failing to form an endodermal cell layer, GA-Fl were evenly distributed in the stele. Another important finding was that ACC dramatically inhibited endodermal accumulation of GA-Fl, with accumulation being also impaired in the *ctr1-1* constitutive ethylene response mutant, suggesting that GA transport to the endodermis is regulated by ethylene (Shani *et al.*, 2013). This result may explain the delay in DELLA degradation observed in *Arabidopsis* roots following ACC treatment, as likely caused by an inhibition of GA transport (Achard *et al.*, 2003).

Noteworthy, the glucosinolate GTR1/NPF2.10 transporter was recently reported to be a multi-functional hormone transporter that transports JA-Ile and GAs, in addition to glucosinolates (Saito *et al.*, 2015). This putative GA transporter belongs to the NITRATE TRANSPORTER 1/PEPTIDE TRANSPORTER FAMILY (NPF) of transmembrane transporters, members of this family being also identified as nitrate-inhibited auxin uptake facilitators, and transporters of ABA (reviewed by Léran *et al.*, 2013). *GTR1/NPF2.10* expression responds to JA, its transcripts found to accumulate to very high levels in the stamen filaments. Loss-of-function *gtr1* mutants show reduced fertility due to having shorter stamen filaments and failure to undergo anther dehiscence, with these defects being recovered by GA application (Saito *et al.*, 2015). In *Xenopus* oocytes, GTR1 was able to transport GA_3 and JA-Ile, thus supporting a crucial role of this transporter in supplying GA for stamen development, a process that is tightly regulated by both JA and GA. These findings point to an important function of NPF proteins in GA transport, with current knowledge on this subject covered in Chapter 4. The identification of other GA transporters and characterisation of the possible mechanisms by which ethylene modulates their function will be important issues for future research. Roots proved to respond to much lower GA concentration than the shoot (Tanimoto, 2012) and as such, manipulation of GA transport may have a relevant biotechnological application in modifying the shoot:root ratio, a long-sought objective to improve nutrient uptake and drought tolerance in crops and increase yields.

10.3.3 DELLAs downstream signalling in the root

Since the elongation zone of the endodermis is the tissue controlling GA-dependent root cell elongation, genes expressed within this region are likely to be responsible for regulating GA-responsive growth. An obvious candidate is *SCARECROW-LIKE 3* (*SCL3*), a GRAS protein that lacks the GA-responsive DELLA domain and was identified as a GA-regulated gene predominantly expressed in the endodermis (Zentella *et al.*, 2007). SCL3 accumulates in the quiescent centre (QC), cortex/endodermal initials (CEI) and endodermis, with expression of this gene shown to be under control of SCARECROW (SCR) and SHORT-ROOT (SHR), two GRAS proteins that act as key regulators of stem cell renewal and radial root patterning, and whose function is required for asymmetric division of the cortex/endodermal initials, giving rise to the cortex and endodermis lineages (Cui *et al.*, 2007). Notably, while SCR is expressed in the QC and endodermis, SHR is expressed in the stele, with the protein shown to move to the adjacent cell layer, including the endodermis, CEI and QC, where it physically interacts with SCR to regulate *SCL3* transcription. The SHR-SCR complex has in fact been shown to associate to the *SCL3* promoter, with *SCL3* expression being reduced in *scr-5 shr-6* mutants, in particular in the endodermis (Heo *et al.*, 2011). *SCL3* expression is, on the other hand, substantially reduced in the presence of GA, with RGA shown to occupy the *SCL3* promoter and activate transcription of this gene (Zentella *et al.*, 2007). This suggests a function of SCL3 downstream of both the GA/DELLA and SHR/SCR pathways (Heo *et al.*, 2011), with additional studies showing that SCL3 regulates its own expression, by directly interacting with the DELLAs (Zhang *et al.*, 2011).

SCL3 acts as a positive regulator of GA signalling, likely by acting as an attenuator of DELLA proteins, since *scl3* null mutants were shown to display a reduced GA response and up-regulated levels of expression of GA biosynthetic genes, and this mutation enhance the growth defects of *ga1-3* mutants (Zhang *et al.*, 2011). In contrast, *SCL3* over-expression confers resistance to PAC and an enhanced response to GA (Zhang *et al.*, 2011). Remarkably, the root meristem size of *scl3* mutants is identical to the wild-type, indicating that GA effects on root meristem cell proliferation are independent of SCL3 function. However, PAC was observed to reduce cell elongation in the EDZ more strongly in *scl3* seedlings than in the wild-type. A role of SCL3 in the control of cell elongation was in fact demonstrated by introgressing the *pSCR:gai-GR-YFP* construct, which specifically expresses gai-1 in the endodermis, into *scl3* and *SCL3*ox lines. Upon DEX induction, root elongation was more significantly inhibited in the *scl3* than wild-type background, whereas *SCL3*ox lines were partially resistant to root growth inhibition, hence indicating that SCL3 attenuates the effects of non-degradable gai-1 in the root endodermis (Heo *et al.*, 2011).

SCL3 also plays a critical role in modulating the timing of formative cell division leading to ground tissue maturation of *Arabidopsis* roots. This involves a periclinal division of individual endodermal cells to generate an additional cell layer, termed the middle cortex (MC), which rapidly acquires a cortex character (Paquette and Benfey, 2005). *scl3* mutants show precocious periclinal cell divisions, whereas a delay in middle cortex formation is observed in *SCL3*ox lines (Heo *et al.*, 2011). Similar to *scl3* mutants, *ga1-3* and PAC-treated roots show a premature middle cortex phenotype that is suppressed by application of GA (Cui *et al.*, 2009). This phenotype is exacerbated in the double *scl3 ga1-3* mutant, while SCL3 over-expression suppresses precocious MC formation of PAC-treated or *ga1-3* roots (Heo *et al.*, 2011), indicating that SCL3 acts as an endodermis-specific integrator of GA-signalling that modulates ground tissue maturation in conjunction with the SHORT ROOT/SCARECROW (SHR/SCR) pathway (Figure 10.4b).

The mechanism of SCL3 activation by the DELLAs is beginning to be understood, with the finding that the INDETERMINATE DOMAIN (IDD) sub-family of C_2H_2 transcription factors act as transcriptional repressors and that DELLAs de-repress their gene targets by competing for TOPLESS binding to these factors. IDDs 3, 4, 5, 9 and 10 interact with DELLAs and SCL3, and directly bind the *SCL3* promoter. IDDs induce *SCL3* expression when coexpressed with DELLAs in transient reporter assays, while SCL3 competes with DELLAs for interaction with IDDs, which results in suppression of its own expression (Yoshida *et al.*, 2014). Interestingly, IDD3 and 10 had been previously identified as JACKDAW and MAGPIE, two SCR and SHR interactors also expressed in the endodermis, QC and CEI (Welch *et al.*, 2007). IDD10 expression in combination with SHR and SCR has been reported to strongly activate the *SCR* and *IDD3* promoters (Ogasawara *et al.*, 2011), with *idd10* mutants showing a precocious middle cortex formation, reminiscent of the *scl3* and *ga1-3* mutants phenotype (Welch *et al.*, 2007; Yoshida *et al*, 2014).

Yeast three hybrid and competitive pull-down experiments suggested a competition of SCL3 and DELLA for IDDs binding, which results in opposite regulation of cell growth and ground tissue maturation. In this scenario, IDD binding to DELLA proteins activates *SCL3* gene expression, with increased SCL3 levels competing with DELLA for IDD binding, and switching IDDs activity from activation to repression (Yoshida *et al.*, 2014). An interaction with GRAS proteins has been reported for several IDD family members (Welch *et al.*, 2007; Ogasawara *et al.*, 2011), including DELLAs interaction with IDD2/GAF1 (Fukazawa *et al.*, 2014) and IDD1 (Feurtado *et al.*, 2011), and likewise observed to lead to DELLA-dependent activation of the GAF1 target genes (Fukazawa *et al.*, 2014).

In summary, several GRAS proteins (DELLAs, SCL3, SHR, SCR) were shown to bind IDD transcription factors (IDDs 1, 2, 3, 4, 5, 9 and 10), interaction with these factors conferring DNA binding and therefore transcriptional regulatory capabilities on the GRAS proteins. Different combinations of GRAS and IDDs seem to exert distinct roles in developmental control, and to fully understand their exact regulatory function, it will be essential to establish the spatiotemporal pattern of accumulation of each of these protein partners, and their specific gene targets. For instance, SCL3 is expressed in the endodermis, where DELLA levels are low, as well as in the QC and CEI, where DELLAs accumulate. PAC treatment, in turn, strongly induces *SCL3* expression in the meristem, with changes in relative SCL3 and DELLA levels in these different cell types explaining the main observed function of SCL3 in elongation of the EDZ and in ground tissue maturation.

10.3.4 DELLAs promote mycorrhizal symbiosis

Legume–rhizobia (reviewed in Suzaki and Kawaguchi, 2014) and plant–mycorrhizal interactions (reviewed in Gutjahr and Parniske, 2013) are the best-studied symbiotic associations in plants. Association with these micro-organisms improves plant nutrient uptake from the soil in exchange for carbohydrates, and requires a mutual recognition, via diffusible molecules secreted by both symbionts. Mycorrhizal fungi produce mycorrhizal factors (Myc factors), whereas rhizobia produce nodulation factors (Nod factors). Recognition of both factors by the host plant activates a common symbiosis signalling pathway that promotes root colonisation and the respective formation of arbuscules and nodules, where nutrient exchange takes place (Oldroyd, 2013). To sustain both processes, roots undergo important remodelling programs, which enable microbe growth and accommodation, plants being shown to actively regulate the establishment of these symbiotic associations according to their own nutritional and developmental status.

Whereas the GA biosynthetic pathway is up-regulated during both symbiotic processes, GA signalling seems to have opposite roles in nodulation and arbuscular mycorrhization (AM). GA biosynthetic genes are up-regulated during both early (Libault *et al.*, 2010; Hayashi *et al.*, 2012) and later stages of nodulation (Kouchi *et al.*, 2004; Hayashi *et al.*, 2012), as well as upon inoculation with mycorrhizal fungi (García-Garrido *et al.*, 2010; Ortu *et al.*, 2012), GA levels increasing in fungal colonised roots (Takeda *et al.*, 2015). Increased GA biosynthesis is triggered downstream from Nod factor recognition, as mutant rhizobia defective in Nod factor production fail to induce *GA20ox* gene expression (Hayashi *et al.*, 2012).

GAs play a positive role in nodulation, with optimal GA levels shown to be required for nodule establishment. In fact, pea mutants defective in GA biosynthesis develop fewer nodules than wild-type plants (Maekawa *et al..*, 2009; Ferguson *et al.*, 2011). Grafting and GA application studies confirmed that this nodulation phenotype was indeed due to the reduced GA levels in the roots (Ferguson *et al.*, 2005, 2011). In contrast, studies in pea indicated a negative role of GA signalling in AM, with a dose-dependent inhibition of mycorrhization observed in response to GA treatment. Lower GA concentrations inhibited the formation of arbuscules, while higher concentrations fully suppressed colonisation (El Ghachtouli *et al.*, 1996). More recently, the use of *della* null mutants of *Medicago*, rice and pea elucidated a role for

DELLAs as positive regulators of AM. *Medicago della1della2* double mutants display a strong reduction in arbuscule number, without strongly affecting root colonisation. The few arbuscules formed in *della1della2* mutants developed normally, suggesting that DELLAs are specifically required to initiate arbuscule formation (Floss *et al.*, 2013). Treatment of *Medicago* roots with exogenous GAs phenocopied the *della1della2* mutant, whereas expression of a non-degradable DELLA version rescued arbuscule formation upon GA treatment (Floss *et al.*, 2013). Since *Medicago* contains three *DELLA* genes, the perturbation of AM in a triple *della* mutant could be even more severe. Rice has a single *DELLA*, *SLR1*, and in fact, rice *slr1* mutants show a severe and general reduction in root fungal colonisation (Yu *et al.*, 2014). In pea, where two *DELLA* genes have been reported, double *della* mutants show a general reduction in AM and a severe reduction in arbuscule formation (Foo *et al.*, 2013). The range of phenotypes observed in pea, rice and *Medicago* support a dual role of DELLAs in promoting colonisation of the outer cell layers and arbuscule formation in the cortex.

Although arbuscules form in cortex cells, *pDELLA1*:GUS expression was detected in the vasculature and endodermis of *Medicago* roots. Interestingly, expression of a non-degradable DELLA version under the control of the vasculature-specific PHOSPHATE TRANSPORTER 9 promoter restored arbuscular formation in the *della1della2* mutant (Floss *et al.*, 2013). This finding suggests the existence of a DELLA-dependent factor that moves from the vasculature to the cortex to promote arbuscule formation. Given that cell-to-cell movement has been reported for other GRAS family members, such as SCL23 and SHR (Rim *et al.*, 2011), the translocated factor could be DELLA itself or one of the identified DELLA-interacting GRAS proteins (see below).

AM depends on a signalling network that is also required for root nodule symbiosis (Oldroyd *et al.*, 2013). This signal transduction cascade translates the perception of microbial signalling molecules at the plasma membrane into nuclear calcium-spiking (Oldroyd *et al.*, 2013). The calcium signal activates the calmodulin-dependent kinase CCAMK/DMI3, which phosphorylates the transcription factor CYCLOPS/IPD3, responsible for transcriptional activation of genes required for symbiosis (Singh *et al.*, 2014). The phenotype of *Cyclops* mutants resembles *della1della2*, with strong impairment of arbuscule development (Messinese *et al.*, 2007), pointing to a role of DELLA and CYCLOPS in the same pathway. Indeed, a non-degradable DELLA version can restore AM development in a *cyclops* mutant (Floss *et al.*, 2013) suggesting that CYCLOPS signals upstream from DELLA, perhaps participating in DELLA stabilisation. CYCLOPS could contribute to DELLA persistence by activating a stabilising interaction partner. A yeast two-hybrid screen identified a novel GRAS protein called DELLA INTERACTING PROTEIN1 (DIP1) that interacts with the rice DELLA SLR1 (Yu *et al.*, 2014). *DIP1* expression is induced by AM colonisation, and DIP1 is required for AM development, as *DIP1*-RNAi plants reduced AM colonisation to the level of the *slr1* mutant (Yu *et al.*, 2014). Interestingly, DIP1 also interacts with REQUIRED FOR

ARBUSCULAR MYCORRHIZA (RAM1), another GRAS protein involved in mycorrhizal signalling, required both for fungal attachment to the root and for arbuscule branching (Gobbato *et al.*, 2012, 2013).

It remains to be seen whether SLR1, DIP1 and RAM1 unite in a ternary complex or whether DIP1 forms alternative complexes with SLR1 and RAM1. Although the signalling pathway is still not fully understood, a prevalent role for GRAS proteins in AM signalling is highlighted by the identification of 18 GRAS genes among the 45 transcription factors identified as AM-induced transcripts in *Lotus japonicum*. Further characterisation of one of them, RAD1, showed that it is required for arbuscule development and that it is also a RAM1 interactor (Xue *et al.*, 2015).

As previously stated, colonisation by AM must be tightly regulated by the plant in response to both developmental and external cues. Phosphate availability is one of the most important external factors, and the plants increase or impair AM according to phosphate levels in the soil (Gu *et al.*, 2011). In fact, DELLAs have been implicated in phosphate starvation response, and under high phosphate, GA levels increase and DELLAs are destabilised (Jiang *et al.*, 2007) pointing at DELLAs as integrators of hormonal and environmental cues to boost AM under appropriate conditions (Floss *et al.*, 2013).

10.4 Growth under unfavourable conditions

10.4.1 DELLAs promote resistance to abiotic stress

One of the key adaptive responses of plants to environmental stress is the reduction of growth, which requires a modification of the hormonal status and signalling (reviewed in Skirycz and Inzé, 2010; Rymen and Sugimoto, 2012). GA metabolism and signalling are strongly affected by abiotic stresses, and it is well documented that impaired growth upon abiotic stress is mediated, at least in part, by DELLA proteins. Bioactive GA levels decrease in *Arabidopsis* plants subjected to salt and cold stress (Achard *et al.*, 2006, 2008a; Magome *et al.*, 2008), and DELLA proteins accumulate upon salt, cold and osmotic stress (Achard *et al.*, 2006, 2008a, 2008b; Conti *et al.*, 2014). The physiological adjustment of GA levels under stress is presumed to have a protective role, as seen by enhanced stress tolerance of GA-deficient mutants (Achard *et al.*, 2006, 2008a; Magome *et al.*, 2004). Furthermore, like growth inhibition, stress resistance is conferred by accumulation of DELLA proteins. Quadruple *della* mutants were more severely affected by stress (Achard *et al.*, 2006, 2008b) and less sensitive to growth inhibition and flowering acceleration upon salt stress (Achard *et al.*, 2006), and the quadruple *della* mutant in a *ga1-3* background fully suppressed *ga1-3* enhanced salt tolerance (Achard *et al.*, 2008b).

A group of APETALA2/ERF (AP2/ERF) transcription factors play a central role in GA metabolism down-regulation under abiotic stress. Two transcription factors of the DREB1/CBF family, a sub-family of the AP2/ERF group, induce GA deactivation in response to salt and cold stress in *Arabidopsis* (Achard *et al.*, 2008a; Magome *et al.*, 2004, 2008). The first member of this family was identified by phenotypic analysis of the activation-tagged *ddf1* line, which displays both dwarfism and salt tolerance. The reduced levels of bioactive GAs observed in *ddf1* were assumed to confer both phenotypes, since they were lost upon GA treatment (Magome *et al.*, 2004). It was later shown that DDF1 and its homologue DDF2 effectively reduce GA levels through direct activation of *GA2ox7*, and additively increase DELLA levels by activation of *RGL3* (Magome *et al.*, 2008). Similar results were found for the *CBF1 (DREB1B)* gene (Achard *et al.*, 2008a). *Arabidopsis* plants over-expressing *CBF1* in wild-type background, but not in the *DELLA* mutant *gai-t6 rga-24*, display growth reduction and accumulate the RGA-GFP fusion protein. *CBF1* over-expression lines also showed increased expression of *GA2ox3*, *GA2ox6* and *RGL3* and decreased levels of bioactive GAs. Regulation of GA metabolism does not appear to be limited to the *DREB1/CBF* sub-family, as the *ERF* sub-family transcription factor *ERF6* has been implicated in *GA2ox6* upregulation in response to osmotic stress (Dubois *et al.*, 2013). Integration of AP2/ERF-mediated reduction in GA levels in the hormonal pathway is still unclear. The finding that EIN3 directly binds the promoter of *DREB1/CBF* transcription factors and inhibits their expression could integrate ethylene and GA signalling (Shi *et al.*, 2012).

DELLA protein accumulation under stress conditions relies therefore on specific transcriptional activation of *RGL3*, and stabilisation of all DELLAs by the low GA environment (Figure 10.5). An additional mechanism for DELLA accumulation under stress is the recent finding of *in vivo* DELLA sumoylation, which is enhanced under stress. Sumoylated DELLAs in turn bind to GID1, which contains a SUMO binding motif and binds SUMO in a GA-independent manner, resulting in GID1 sequestration and GA-independent DELLA stabilisation (Conti *et al.*, 2014; see also Chapter 6).

Stress-stabilised DELLAs inhibit cell growth through inactivation of PIF and BES1/BZR1 transcription factors (de Lucas *et al.*, 2008; Feng *et al.*, 2008; Bai *et al.*, 2012; Li *et al.*, 2012; Gallego-Bartolomé *et al.*, 2012), and may also inhibit cell division, as it has been observed that accumulation of DELLAs under mild osmotic stress triggers mitotic exit (Claeys *et al.*, 2012). However, other stress-protective roles of DELLA proteins apart from growth reduction remain still poorly understood, since AP2/ERF transcriptional up-regulation of stress-response genes and of the CBF regulon is DELLA independent (Achard *et al.*, 2008a; Dubois *et al.*, 2013). The only mechanism described so far is a direct role of DELLAs in ROS production. Fast induction of genes involved in antioxidant responses under salt stress is DELLA dependent. Moreover, ROS levels in *Arabidopsis* roots in basal conditions and in response to salt stress were increased in *della* mutants and decreased in *ga1-3* plants, and conversely, H_2O_2 induced cell death was lower in *ga1-3* and higher in *della* mutants (Achard *et al.*, 2008b). However, the identification of direct DELLA targets controlling stress responses remains elusive.

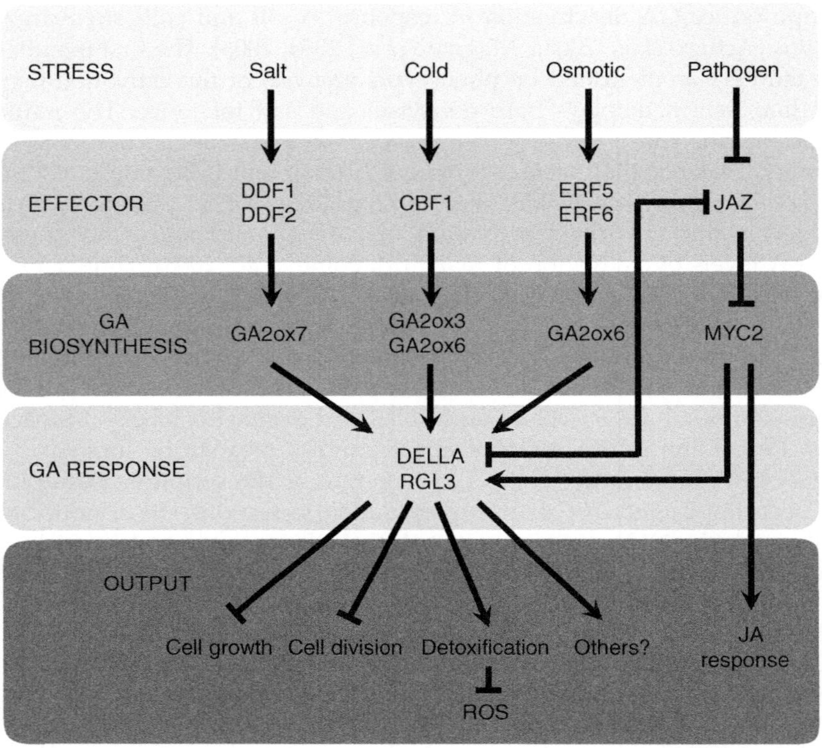

Figure 10.5 A simplified model for the role of GAs in stress response. Different abiotic stresses up-regulate AP2/ERF transcription factors, which induce the expression of GA2oxs and, in some cases, of DELLA protein RGL3. Subsequent decrease in bioactive GAs promotes DELLA accumulation. Under pathogen attack or abiotic stresses such as cold, JA-mediated JAZ protein degradation releases repression of DELLAs and MYC transcription factors. MYC2 in turn also induces expression of RGL3. Accumulation of DELLAs leads to growth cessation via inhibition of cell elongation and cell division. Other protective mechanisms include transcriptional induction of genes encoding detoxification enzymes and promotion of JA response via direct repression of JAZ activity. Arrows and T-bars indicate positive or negative regulation, respectively.

10.4.2 DELLAs and biotic stress

DELLAs also increase resistance to pathogens through extensive cross-talk with jasmonate (JA) signalling. An initial study suggested that *Arabidopsis DELLA* genes play a role in defence, because the defence elicitor flagellin appears to block plant growth by stabilising the RGA-GFP protein, and because DELLAs are needed to induce defence-related gene expression in response to methyl-JA (Navarro *et al.*, 2008). The point of GA and JA signalling pathway interaction was shown to be direct interaction between DELLA and the JAZ repressors (Hou *et al.*, 2010). DELLA interaction with the JAZ repressors competes with JAZ binding and subsequent inactivation of their downstream MYC transcription factors, which in turn drive activation of the JA-response transcriptome. Additionally, the DELLA-JAZ complex competes with DELLA-PIF interaction thus keeping the balance between growth and defence (Yang *et al.*, 2012). DELLA accumulation is in fact a direct effect of JA signalling, since JA treatment or necrotrophic pathogens strongly activate MYC2-dependent *RGL3* expression, and result in a fast and strong accumulation of the RGL3 protein (Wild *et al.*, 2012).

Yet another layer of control is the recent finding that JA response increases cold tolerance. JA response, through degradation of the JAZ proteins, promotes expression of the CBF/DREB transcriptional response. The JAZ repressors directly bind and inhibit ICE transcription factors, which are master regulators of cold responses by acting upstream of CBF/DREB1. JA-mediated activation of the CBF/DREB regulon in turn induces *GA2ox*, decreasing bioactive GA content, and further inducing *RGL3* expression (Hu *et al.*, 2013). Altogether, these data support a model according to which GA and JA pathways interact to inhibit each other and fine tune responses to environmental challenge in terms of defence and growth (Figure 10.5). Upon pathogen attack or cold stress, JA production and perception triggers degradation of some members of the JAZ family of repressors. Release of MYC2 from JAZ sequestration induces *RGL3* expression, and the RGL3 protein, in turn, binds the JAZ repressors (both JA-sensitive and stable isoforms) to inactivate them and allow full activation of the JA pathway. Conversely, under non-stress conditions, GA signalling ends in DELLA degradation, releasing JAZ restraint and therefore facilitating MYC repression by JAZ proteins.

In summary, GAs integrate hormonal signals in response to both biotic and abiotic stresses. JA-induced and ET-repressed CBF/DREB1 transcription factors down-regulate GA biosynthesis, decreasing GA content and stabilising DELLA proteins. DELLAs have a prominent role in growth suppression, which appears to be beneficial for stress tolerance. Additionally, DELLAs lead to a decrease in ROS production and bind JAZ proteins to allow full activation of the JA pathway. The possibility to uncouple DELLAs growth restraint activity from a positive role in stress and pathogen response will be a major challenge to improve crop production under stress, as in many crops growth inhibition has a negative impact on yield.

10.5 Concluding remarks

GAs mediate many aspects of plant development and distinct tissues (i.e. the shoot and root meristems) were shown to respond differently to the same GA stimulus. However, little is known about how GA signals are interpreted by individual cell types. In the root, endodermal cell layers were found to accumulate higher GA levels and coordinate root growth. Tissue-specific transcriptomic analysis will help define the competence response of different cell types. Recent work has highlighted the differential accumulation of bioactive GAs, in both different tissues and cell compartments. Live imaging methods for *in vivo* tracking of hormone concentrations, and the identification and characterisation of GA transporters will help to understand and modulate GA response in a spatial or developmental manner. The number of DELLA interactors acting as key regulators of hormonal and developmental pathways keeps on growing. Computational modelling of developmental pathways, combined with the biochemical identification of DELLA interacting partners will substantially empower future research efforts. Finally, novel mechanisms, possibly shared with other GRAS proteins, were recently proposed to explain how DELLA proteins activate gene expression. Understanding these mechanisms and integrating them in the different physiological GA responses is a major challenge for the near future.

References

Abbas, M., Alabadi, D. and Blazquez, M.A. (2013). Differential growth at the apical hook: all roads lead to auxin. *Frontiers in Plant Science* **4**:441.

Achard, P., Vriezen, W.H., Van Der Straeten, D. and Harberd, N.P. (2003). Ethylene regulates *Arabidopsis* development via the modulation of DELLA protein growth repressor function. *The Plant Cell* **15**, 2816–2825.

Achard, P., Cheng, H., De Grauwe, L., *et al.* (2006). Integration of plant responses to environmentally activated phytohormonal signals. *Science* **311**, 91–94.

Achard, P., Liao, L., Jiang, C., *et al.* (2007). DELLAs contribute to plant photomorphogenesis. *Plant Physiology* **143**, 1163–1172.

Achard, P., Gong, F., Cheminant, S., *et al.* (2008a). The cold-inducible CBF1 factor-dependent signaling pathway modulates the accumulation of the growth-repressing DELLA proteins via its effect on gibberellin metabolism. *The Plant Cell* **20**, 2117–2129.

Achard, P., Renou, J.P., Berthome, R., Harberd, N.P. and Genschik, P. (2008b). Plant DELLAs restrain growth and promote survival of adversity by reducing the levels of reactive oxygen species. *Current Biology* **18**, 656–660.

Achard, P., Gusti, A., Cheminant, S., *et al.* (2009). Gibberellin signaling controls cell proliferation rate in *Arabidopsis*. *Current Biology* **19**, 1188–1193.

Alabadí, D., Gil, J., Blazquez, M.A. and Garcia-Martinez, J.L. (2004). Gibberellins repress photomorphogenesis in darkness. *Plant Physiology* **134**, 1050–1057.

Alabadí, D., Gallego-Bartolome, J., Orlando, L., *et al.* (2008). Gibberellins modulate light signaling pathways to prevent *Arabidopsis* seedling de-etiolation in darkness. *The Plant Journal* **53**, 324–335.

An, F., Zhang, X., Zhu, Z., *et al.* (2012). Coordinated regulation of apical hook development by gibberellins and ethylene in etiolated *Arabidopsis* seedlings. *Cell Research* **22**, 915–927.

Arana, M.V., Marin-de la Rosa, N., Maloof, J.N., Blazquez, M.A. and Alabadi, D. (2011). Circadian oscillation of gibberellin signaling in *Arabidopsis*. *Proceedings of the National Academy of Sciences USA* **108**, 9292–9297.

Bai, M.Y., Shang, J.X., Oh, E., *et al.* (2012). Brassinosteroid, gibberellin and phytochrome impinge on a common transcription module in *Arabidopsis*. *Nature Cell Biology* **14**, 810–817.

Band, L.R., Wells, D.M., Larrieu, A., *et al.* (2012). Root gravitropism is regulated by a transient lateral auxin gradient controlled by a tipping-point mechanism. *Proceedings of the National Academy of Sciences USA* **109**, 4668–4673.

Barkoulas, M., Galinha, C., Grigg, S.P. and Tsiantis, M. (2007). From genes to shape: regulatory interactions in leaf development. *Current Opinion in Plant Biology* **10**, 660–666.

Blilou, I., Xu, J., Wildwater, M., *et al.* (2005). The PIN auxin efflux facilitator network controls growth and patterning in *Arabidopsis* roots. *Nature* **433**, 39–44.

Bolduc, N. and Hake, S. (2009). The maize transcription factor KNOTTED1 directly regulates the gibberellin catabolism gene ga2ox1. *The Plant Cell* **21**, 1647–1658.

Brady, S.M., Orlando, D.A., Lee, J.Y., *et al.* (2007). A high-resolution root spatiotemporal map reveals dominant expression patterns. *Science* **318**, 801–806.

Casal, J.J. (2013). Photoreceptor signaling networks in plant responses to shade. *Annual Review of Plant Biology* **64**, 403–427.

Claeys, H., Skirycz, A., Maleux, K. and Inze, D. (2012). DELLA signaling mediates stress-induced cell differentiation in *Arabidopsis* leaves through modulation of anaphase-promoting complex/cyclosome activity. *Plant Physiology* **159**, 739–747.

Conti, L., Nelis, S., Zhang, C., *et al.* (2014). Small Ubiquitin-like Modifier protein SUMO enables plants to control growth independently of the phytohormone gibberellin. *Developmental Cell* **28**, 102–110.

Cui, H. and Benfey, P.N. (2009). Interplay between SCARECROW, GA and LIKE HETEROCHROMATIN PROTEIN 1 in ground tissue patterning in the *Arabidopsis* root. *The Plant Journal* **58**, 1016–1027.

Cui, H., Levesque, M.P., Vernoux, T., *et al.* (2007). An evolutionarily conserved mechanism delimiting SHR movement defines a single layer of endodermis in plants. *Science* **316**, 421–425.

Chapman, E.J., Greenham, K., Castillejo, C., *et al.* (2012). Hypocotyl transcriptome reveals auxin regulation of growth-promoting genes through GA-dependent and -independent pathways. *PLoS One* **7**, e36210.

Cheminant, S., Wild, M., Bouvier, F., *et al.* (2011). DELLAs regulate chlorophyll and carotenoid biosynthesis to prevent photooxidative damage during seedling deetiolation in *Arabidopsis*. *The Plant Cell* **23**, 1849–1860.

Daviére, J.M., Wild, M., Regnault, T., *et al.* (2014). Class I TCP-DELLA interactions in inflorescence shoot apex determine plant height. *Current Biology* **24**, 1923–1928.

Dayan, J., Voronin, N., Gong, F., *et al.* (2012). Leaf-induced gibberellin signaling is essential for internode elongation, cambial activity, and fiber differentiation in tobacco stems. *The Plant Cell* **24**, 66–79.

de Lucas, M. and Prat, S. (2014). PIFs get BRright: PHYTOCHROME INTERACTING FACTORs as integrators of light and hormonal signals. *New Phytologist* **202**, 1126–1141.

de Lucas, M., Daviere, J.M., Rodriguez-Falcon, M., *et al.* (2008). A molecular framework for light and gibberellin control of cell elongation. *Nature* **451**, 480–484.

Dello Ioio, R., Linhares, F.S., Scacchi, E., *et al.* (2007). Cytokinins determine *Arabidopsis* root-meristem size by controlling cell differentiation. *Current Biology* **17**, 678–682.

Dello Ioio, R., Nakamura, K., Moubayidin, L., *et al.* (2008). A genetic framework for the control of cell division and differentiation in the root meristem. *Science* **322**, 1380–1384.

Depuydt, S. and Hardtke, C.S. (2011). Hormone signalling crosstalk in plant growth regulation. *Current Biology* **21**, R365–373.

Dill, A., Thomas, S.G., Hu, J., Steber, C.M. and Sun, T.P. (2004). The *Arabidopsis* F-box protein SLEEPY1 targets gibberellin signaling repressors for gibberellin-induced degradation. *The Plant Cell* **16**, 1392–1405.

Djakovic-Petrovic, T., de Wit, M., Voesenek, L.A. and Pierik, R. (2007). DELLA protein function in growth responses to canopy signals. *The Plant Journal* **51**, 117–126.

Dubois, M., Skirycz, A., Claeys, H., *et al.* (2013). Ethylene Response Factor6 acts as a central regulator of leaf growth under water-limiting conditions in *Arabidopsis. The Plant Physiology* **162**, 319–332.

El Ghachtouli, N., Martin-Tanguy, J., Paynot, M. and Gianinazzi, S. (1996). First report of the inhibition of arbuscular mycorrhizal infection of *Pisum sativum* by specific and irreversible inhibition of polyamine biosynthesis or by gibberellic acid treatment. *FEBS Letters* **385**, 189–192.

Esmon, C.A., Tinsley, A.G., Ljung, K., *et al.* (2006). A gradient of auxin and auxin-dependent transcription precedes tropic growth responses. *Proceedings of the National Academy of Sciences USA* **103**, 236–241.

Fankhauser, C. and Chen, M. (2008). Transposing phytochrome into the nucleus. *Trends in Plant Science* **13**, 596–601.

Feng, S., Martinez, C., Gusmaroli, G., *et al.* (2008). Coordinated regulation of *Arabidopsis thaliana* development by light and gibberellins. *Nature* **451**, 475–479.

Ferguson, B.J., Ross, J.J. and Reid, J.B. (2005). Nodulation phenotypes of gibberellin and brassinosteroid mutants of pea. *Plant Physiology* **138**, 2396–2405.

Ferguson, B.J., Foo, E., Ross, J.J. and Reid, J.B. (2011). Relationship between gibberellin, ethylene and nodulation in Pisum sativum. *New Phytologist* **189**, 829–842.

Feurtado, J.A., Huang, D., Wicki-Stordeur, L., *et al.* (2011). The *Arabidopsis* C2H2 zinc finger INDETERMINATE DOMAIN1/ENHYDROUS promotes the transition to germination by regulating light and hormonal signaling during seed maturation. *The Plant Cell* **23**, 1772–1794.

Floss, D.S., Levy, J.G., Levesque-Tremblay, V., Pumplin, N. and Harrison, M.J. (2013). DELLA proteins regulate arbuscule formation in arbuscular mycorrhizal symbiosis. *Proceedings of the National Academy of Sciences USA* **110**, 5025–5034.

Foo, E., Ross, J.J., Jones, W.T. and Reid, J.B. (2013). Plant hormones in arbuscular mycorrhizal symbioses: an emerging role for gibberellins. *Annals of Botany* **111**, 769–779.

Franklin, K.A., Lee, S.H., Patel, D., *et al.* (2011). PHYTOCHROME-INTERACTING FACTOR 4 (PIF4) regulates auxin biosynthesis at high temperature. *Proceedings of the National Academy of Sciences USA* **108**, 20231–20235.

Frigerio, M., Alabadi, D., Perez-Gomez, J., *et al.* (2006). Transcriptional regulation of gibberellin metabolism genes by auxin signaling in *Arabidopsis. Plant Physiology* **142**, 553–563.

Fu, X. and Harberd, N.P. (2003). Auxin promotes *Arabidopsis* root growth by modulating gibberellin response. *Nature* **421**, 740–743.

Fu, X., Richards, D.E., Fleck, B., *et al.* (2004). The *Arabidopsis* mutant sleepy1gar2-1 protein promotes plant growth by increasing the affinity of the SCF^{SLY1} E3 ubiquitin ligase for DELLA protein substrates. *The Plant Cell* **16**, 1406–1418.

Fukazawa, J., Teramura, H., Murakoshi, S., *et al.* (2014). DELLAs function as coactivators of GAI-ASSOCIATED FACTOR1 in regulation of gibberellin homeostasis and signaling in *Arabidopsis. The Plant Cell* **26**, 2920–2938.

Galinha, C., Bilsborough, G. and Tsiantis, M. (2009). Hormonal input in plant meristems: A balancing act. *Seminars in Cell and Developmental Biology* **20**, 1149–1156.

Gallego-Bartolomé, J., Alabadi, D. and Blazquez, M.A. (2011a). DELLA-induced early transcriptional changes during etiolated development in *Arabidopsis* thaliana. *PLoS One* **6**, e23918.

Gallego-Bartolomé, J., Arana, M.V., Vandenbussche, F., *et al.* (2011b). Hierarchy of hormone action controlling apical hook development in *Arabidopsis*. *The Plant Journal* **67**, 622–634.

Gallego-Bartolomé, J., Minguet, E.G., Grau-Enguix, F., *et al.* (2012). Molecular mechanism for the interaction between gibberellin and brassinosteroid signaling pathways in *Arabidopsis*. *Proceedings of the National Academy of Sciences USA* **109**, 13446–13451.

García-Garrido, J.M., Morcillo, R.J., Rodriguez, J.A. and Bote, J.A. (2010). Variations in the mycorrhization characteristics in roots of wild-type and ABA-deficient tomato are accompanied by specific transcriptomic alterations. *Molecular Plant Microbe Interactions Journal* **23**, 651–664.

Gobbato, E., Marsh, J.F., Vernie, T., *et al.* (2012). A GRAS-type transcription factor with a specific function in mycorrhizal signaling. *Current Biology* **22**, 2236–2241.

Gobbato, E., Wang, E., Higgins, G., *et al.* (2013). RAM1 and RAM2 function and expression during arbuscular mycorrhizal symbiosis and *Aphanomyces euteiches* colonization. *Plant Signaling and Behavior* **8**, e26049.

González, N., Vanhaeren, H. and Inze, D. (2012). Leaf size control: complex coordination of cell division and expansion. *Trends in Plant Science* **17**, 332–340.

Grieneisen, V.A., Xu, J., Maree, A.F., Hogeweg, P. and Scheres, B. (2007). Auxin transport is sufficient to generate a maximum and gradient guiding root growth. *Nature* **449**, 1008–1013.

Griffiths, J., Murase, K., Rieu, I., *et al.* (2006). Genetic characterization and functional analysis of the GID1 gibberellin receptors in *Arabidopsis*. *The Plant Cell* **18**, 3399–3414.

Gu, M., Chen, A., Dai, X., Liu, W. and Xu, G. (2011). How does phosphate status influence the development of the arbuscular mycorrhizal symbiosis? *Plant Signaling and Behavior* **6**, 1300–1304.

Gutjahr, C. and Parniske, M. (2013). Cell and developmental biology of arbuscular mycorrhiza symbiosis. *Annual Review of Cell and Developmental Biology* **29**, 593–617.

Hamann, T., Benkova, E., Baurle, I., Kientz, M. and Jurgens, G. (2002). The *Arabidopsis BODENLOS* gene encodes an auxin response protein inhibiting MONOPTEROS-mediated embryo patterning. *Genes and Development* **16**, 1610–1615.

Hay, A., Kaur, H., Phillips, A., *et al.* (2002). The gibberellin pathway mediates KNOTTED1-type homeobox function in plants with different body plans. *Current Biology* **12**, 1557–1565.

Hayashi, S., Reid, D.E., Lorenc, M.T., *et al.* (2012). Transient Nod factor-dependent gene expression in the nodulation-competent zone of soybean (*Glycine max* [L.] Merr.) roots. *Plant Biotechnology Journal* **10**, 995–1010.

Hedden, P. (2003). The genes of the Green Revolution. *Trends in Genetics* **19**, 5–9.

Hedden, P. and Thomas, S.G. (2012). Gibberellin biosynthesis and its regulation. *Biochemical Journal* **444**, 11–25.

Heo, J.O., Chang, K.S., Kim, I.A., *et al.* (2011). Funneling of gibberellin signaling by the GRAS transcription regulator SCARECROW-LIKE 3 in the *Arabidopsis* root. *Proceedings of the National Academy of Sciences USA* **108**, 2166–2171.

Hirano, K., Asano, K., Tsuji, H., *et al.* (2010). Characterization of the molecular mechanism underlying gibberellin perception complex formation in rice. *The Plant Cell* **22**, 2680–2696.

Hornitschek, P., Kohnen, M.V., Lorrain, S., *et al.* (2012). Phytochrome interacting factors 4 and 5 control seedling growth in changing light conditions by directly controlling auxin signaling. *The Plant Journal* **71**, 699–711.

Hou, X., Lee, L.Y., Xia, K., Yan, Y. and Yu, H. (2010). DELLAs modulate jasmonate signaling via competitive binding to JAZs. *Developmental Cell* **19**, 884–894.

Hu, J., Mitchum, M.G., Barnaby, N., *et al.* (2008). Potential sites of bioactive gibberellin production during reproductive growth in *Arabidopsis*. *The Plant Cell* **20**, 320–336.

Hu, Y., Jiang, L., Wang, F. and Yu, D. (2013). Jasmonate regulates the INDUCER OF CBF EXPRESSION-C-REPEAT BINDING FACTOR/DRE BINDING FACTOR1 cascade and freezing tolerance in *Arabidopsis*. *The Plant Cell* **25**, 2907–2924.

Huq, E., Al-Sady, B. and Quail, P.H. (2003). Nuclear translocation of the photoreceptor phytochrome B is necessary for its biological function in seedling photomorphogenesis. *The Plant Journal* **35**, 660–664.

Jaillais, Y. and Chory, J. (2010). Unraveling the paradoxes of plant hormone signaling integration. *Nature Structural and Molecular Biology* **17**, 642–645.

Jang, I.C., Henriques, R., Seo, H.S., Nagatani, A. and Chua, N.H. (2010). *Arabidopsis* PHYTOCHROME INTERACTING FACTOR proteins promote phytochrome B polyubiquitination by COP1 E3 ligase in the nucleus. *The Plant Cell* **22**, 2370–2383.

Jasinski, S., Piazza, P., Craft, J., *et al.* (2005). KNOX action in *Arabidopsis* is mediated by coordinate regulation of cytokinin and gibberellin activities. *Current Biology* **15**, 1560–1565.

Jeong, J. and Choi, G. (2013). Phytochrome-interacting factors have both shared and distinct biological roles. *Molecules and Cells* **35**, 371–380.

Jiang, C., Gao, X., Liao, L., Harberd, N.P. and Fu, X. (2007). Phosphate starvation root architecture and anthocyanin accumulation responses are modulated by the gibberellin-DELLA signaling pathway in Arabidopsis. *Plant Physiology* **145**, 1460–1470.

Josse, E.M., Gan, Y., Bou-Torrent, J., *et al.* (2011). A DELLA in disguise: SPATULA restrains the growth of the developing *Arabidopsis* seedling. *The Plant Cell* **23**, 1337–1351.

Kami, C., Lorrain, S., Hornitschek, P. and Fankhauser, C. (2010). Light-regulated plant growth and development. *Current Topics in Developmental Biology* **91**, 29–66.

Keuskamp, D.H., Sasidharan, R. and Pierik, R. (2010). Physiological regulation and functional significance of shade avoidance responses to neighbors. *Plant Signaling and Behavior* **5**, 655–662.

Kim, K., Shin, J., Lee, S.H., *et al.* (2011). Phytochromes inhibit hypocotyl negative gravitropism by regulating the development of endodermal amyloplasts through phytochrome-interacting factors. *Proceedings of the National Academy of Sciences USA* **108**, 1729–1734.

Koornneef, M. and van der Veen, J.H. (1980). Induction and analysis of gibberellin sensitive mutants in *Arabidopsis thaliana* (L.) heynh. *Theoretical and Applied Genetics* **58**, 257–263.

Kouchi, H., Shimomura, K., Hata, S., *et al.* (2004). Large-scale analysis of gene expression profiles during early stages of root nodule formation in a model legume, *Lotus japonicus*. *DNA Research* **11**, 263–274.

Lau, O.S. and Deng, X.W. (2010). Plant hormone signaling lightens up: integrators of light and hormones. *Current Opinion in Plant Biology* **13**, 571–577.

Lawit, S.J., Wych, H.M., Xu, D., Kundu, S. and Tomes, D.T. (2010). Maize DELLA proteins dwarf plant8 and dwarf plant9 as modulators of plant development. *Plant and Cell Physiology* **51**, 1854–1868.

Lee, L.Y., Hou, X., Fang, L., *et al.* (2012). *STUNTED* mediates the control of cell proliferation by GA in *Arabidopsis*. *Development* **139**, 1568–1576.

Lehman, A., Black, R. and Ecker, J.R. (1996). *HOOKLESS1*, an ethylene response gene, is required for differential cell elongation in the *Arabidopsis* hypocotyl. *Cell* **85**, 183–194.

Leivar, P. and Quail, P.H. (2011). PIFs: pivotal components in a cellular signaling hub. *Trends in Plant Science* **16**, 19–28.

Leivar, P., Monte, E., Oka, Y., *et al.* (2008). Multiple phytochrome-interacting bHLH transcription factors repress premature seedling photomorphogenesis in darkness. *Current Biology* **18**, 1815–1823.

Leivar, P., Tepperman, J.M., Monte, E., *et al.* (2009). Definition of early transcriptional circuitry involved in light-induced reversal of PIF-imposed repression of photomorphogenesis in young *Arabidopsis* seedlings. *The Plant Cell* **21**, 3535–3553.

Léran, S., Varala, K., Boyer, J.C., *et al.* (2014). A unified nomenclature of NITRATE TRANSPORTER 1/PEPTIDE TRANSPORTER family members in plants. *Trends in Plant Science* **19**, 5–9.

Li, G., Zhu, C., Gan, L., Ng, D. and Xia, K. (2015). GA_3 enhances root responsiveness to exogenous IAA by modulating auxin transport and signalling in *Arabidopsis*. *Plant Cell Reports* **34**, 483–494.

Li, H., Johnson, P., Stepanova, A., Alonso, J.M. and Ecker, J.R. (2004). Convergence of signaling pathways in the control of differential cell growth in *Arabidopsis*. *Developmental Cell* **7**, 193–204.

Li, Q.F., Wang, C., Jiang, L., *et al.* (2012). An interaction between BZR1 and DELLAs mediates direct signaling crosstalk between brassinosteroids and gibberellins in *Arabidopsis*. *Science Signaling* **5**, ra72.

Libault, M., Farmer, A., Brechenmacher, L., *et al.* (2010). Complete transcriptome of the soybean root hair cell, a single-cell model, and its alteration in response to *Bradyrhizobium japonicum* infection. *Plant Physiology* **152**, 541–552.

Lim, S., Park, J., Lee, N., *et al.* (2013). ABA-INSENSITIVE3, ABA-INSENSITIVE5, and DELLAs interact to activate the expression of *SOMNUS* and other high-temperature-inducible genes in imbibed seeds in *Arabidopsis*. *The Plant Cell* **25**, 4863–4878.

Liu, X., Chen, C.Y., Wang, K.C., *et al.* (2013). PHYTOCHROME INTERACTING FACTOR3 associates with the histone deacetylase HDA15 in repression of chlorophyll biosynthesis and photosynthesis in etiolated *Arabidopsis* seedlings. *The Plant Cell* **25**, 1258–1273.

Löfke, C., Zwiewka, M., Heilmann, I., *et al.* (2013). Asymmetric gibberellin signaling regulates vacuolar trafficking of PIN auxin transporters during root gravitropism. *Proceedings of the National Academy of Sciences USA* **110**, 3627–3632.

Maekawa, T., Maekawa-Yoshikawa, M., Takeda, N., *et al.* (2009). Gibberellin controls the nodulation signaling pathway in Lotus japonicus. *The Plant Journal* **58**, 183–194.

Magome, H., Yamaguchi, S., Hanada, A., Kamiya, Y. and Oda, K. (2004). *dwarf and delayed-flowering 1*, a novel *Arabidopsis* mutant deficient in gibberellin biosynthesis because of overexpression of a putative AP2 transcription factor. *The Plant Journal* **37**, 720–729.

Magome, H., Yamaguchi, S., Hanada, A., Kamiya, Y. and Oda, K. (2008). The DDF1 transcriptional activator upregulates expression of a gibberellin-deactivating gene, *GA2ox7*, under high-salinity stress in *Arabidopsis*. *The Plant Journal* **56**, 613–626.

Mazzella, M.A., Casal, J.J., Muschietti, J.P. and Fox, A.R. (2014). Hormonal networks involved in apical hook development in darkness and their response to light. *Frontiers in Plant Science* **5**, 52.

McGinnis, K.M., Thomas, S.G., Soule, J.D., *et al.* (2003). The *Arabidopsis SLEEPY1* gene encodes a putative F-box subunit of an SCF E3 ubiquitin ligase. *The Plant Cell* **15**, 1120–1130.

Messinese, E., Mun, J.H., Yeun, L.H., *et al.* (2007). A novel nuclear protein interacts with the symbiotic DMI3 calcium- and calmodulin-dependent protein kinase of Medicago truncatula. *Molecular Plant Microbe Interactions* **20**, 912–921.

Mitchum, M.G., Yamaguchi S., Hanada, A. *et al.* (2006). Distinct and overlapping roles of two gibberellin 3-oxidases in *Arabidopsis* development. *The Plant Journal* **45**, 804–818.

Moubayidin, L., Perilli, S., Dello Ioio, R., *et al.* (2010). The rate of cell differentiation controls the *Arabidopsis* root meristem growth phase. *Current Biology* **20**, 1138–1143.

Nakamura, H., Xue, Y.L., Miyakawa, T., *et al.* (2013). Molecular mechanism of strigolactone perception by DWARF14. *Nature Communications* **4**, 2613.

Navarro, L., Bari, R., Achard, P., *et al.* (2008). DELLAs control plant immune responses by modulating the balance of jasmonic acid and salicylic acid signaling. *Current Biology* **18**, 650–655.

Nelissen, H., Rymen, B., Jikumaru, Y., *et al.* (2012). A local maximum in gibberellin levels regulates maize leaf growth by spatial control of cell division. *Current Biology* **22**, 1183–1187.

Niwa, Y., Yamashino, T. and Mizuno, T. (2009). The circadian clock regulates the photoperiodic response of hypocotyl elongation through a coincidence mechanism in *Arabidopsis* thaliana. *Plant and Cell Physiology* **50**, 838–854.

Nozue, K., Covington, M.F., Duek, P.D., *et al.* (2007). Rhythmic growth explained by coincidence between internal and external cues. *Nature* **448**, 358–361.

Nozue, K., Harmer, S.L. and Maloof, J.N. (2011). Genomic analysis of circadian clock-, light-, and growth-correlated genes reveals PHYTOCHROME-INTERACTING FACTOR5 as a modulator of auxin signaling in *Arabidopsis*. *Plant Physiology* **156**, 357–372.

Nusinow, D.A., Helfer, A., Hamilton, E.E., *et al.* (2011). The ELF4-ELF3-LUX complex links the circadian clock to diurnal control of hypocotyl growth. *Nature* **475**, 398–402.

O'Neill, D.P. and Ross, J.J. (2002). Auxin regulation of the gibberellin pathway in pea. *Plant Physiology* **130**, 1974–1982.

O'Neill, D.P., Davidson, S.E., Clarke, V.C., *et al.* (2010). Regulation of the gibberellin pathway by auxin and DELLA proteins. *Planta* **232**, 1141–1149.

Ogasawara, H., Kaimi, R., Colasanti, J. and Kozaki, A. (2011). Activity of transcription factor JACKDAW is essential for SHR/SCR-dependent activation of SCARECROW and MAGPIE and is modulated by reciprocal interactions with MAGPIE, SCARECROW and SHORT ROOT. *Plant Molecular Biology* **77**, 489–499.

Ogawa, M., Hanada, A., Yamauchi, Y., *et al.* (2003). Gibberellin biosynthesis and response during *Arabidopsis* seed germination. *The Plant Cell* **15**, 1591–1604.

Oh, E., Yamaguchi, S., Hu, J., *et al.* (2007). PIL5, a phytochrome-interacting bHLH protein, regulates gibberellin responsiveness by binding directly to the *GAI* and *RGA* promoters in *Arabidopsis* seeds. *The Plant Cell* **19**, 1192–1208.

Oh, E., Zhu, J.Y. and Wang, Z.Y. (2012). Interaction between BZR1 and PIF4 integrates brassinosteroid and environmental responses. *Nature Cell Biology* **14**, 802–809.

Oldroyd, G.E. (2013). Speak, friend, and enter: signalling systems that promote beneficial symbiotic associations in plants. *Nature Reviews: Microbiology* **11**, 252–263.

Ortu, G., Balestrini, R., Pereira, P.A., *et al.* (2012). Plant genes related to gibberellin biosynthesis and signaling are differentially regulated during the early stages of AM fungal interactions. *Molecular Plant* **5**, 951–954.

Ottenschläger, I., Wolff, P., Wolverton, C., *et al.* (2003). Gravity-regulated differential auxin transport from columella to lateral root cap cells. *Proceedings of the National Academy of Sciences USA* **100**, 2987–2991.

Pacifici, E., Polverari, L. and Sabatini, S. (2015). Plant hormone cross-talk: the pivot of root growth. *Journal of Experimental Botany* **66**, 1113–1121.

Paquette, A.J. and Benfey, P.N. (2005). Maturation of the ground tissue of the root is regulated by gibberellin and *SCARECROW* and requires *SHORT-ROOT*. *Plant Physiology* **138**, 636–640.

Pfeiffer, A., Nagel, M.K., Popp, C., *et al.* (2012). Interaction with plant transcription factors can mediate nuclear import of phytochrome B. *Proceedings of the National Academy of Sciences USA* **109**, 5892–5897.

Proveniers, M.C. and van Zanten, M. (2013). High temperature acclimation through PIF4 signaling. *Trends in Plant Science* **18**, 59–64.

Ragni, L., Nieminen, K., Pacheco-Villalobos, D., *et al.* (2011). Mobile gibberellin directly stimulates *Arabidopsis* hypocotyl xylem expansion. *The Plant Cell* **23**, 1322–1336.

Raz, V. and Ecker, J.R. (1999). Regulation of differential growth in the apical hook of *Arabidopsis*. *Development* **126**, 3661–3668.

Reid, J.B., Davidson, S.E. and Ross, J.J. (2011). Auxin acts independently of DELLA proteins in regulating gibberellin levels. *Plant Signaling and Behavior* **6**, 406–408.

Riefler, M., Novak, O., Strnad, M. and Schmulling, T. (2006). *Arabidopsis* cytokinin receptor mutants reveal functions in shoot growth, leaf senescence, seed size, germination, root development, and cytokinin metabolism. *The Plant Cell* **18**, 40–54.

Rieu, I., Ruiz-Rivero, O., Fernandez-Garcia, N., *et al.* (2008). The gibberellin biosynthetic genes *AtGA20ox1* and *AtGA20ox2* act, partially redundantly, to promote growth and development throughout the *Arabidopsis* life cycle. *The Plant Journal* **53**, 488–504.

Rim, Y., Huang, L., Chu, H., *et al.* (2011). Analysis of *Arabidopsis* transcription factor families revealed extensive capacity for cell-to-cell movement as well as discrete trafficking patterns. *Molecules and Cells* **32**, 519–526.

Ross, J.J., O'Neill, D.P., Smith, J.J., Kerckhoffs, L.H. and Elliott, R.C. (2000). Evidence that auxin promotes gibberellin A_1 biosynthesis in pea. *The Plant Journal* **21**, 547–552.

Ruzicka, K., Simaskova, M., Duclercq, J., *et al.* (2009). Cytokinin regulates root meristem activity via modulation of the polar auxin transport. *Proceedings of the National Academy of Sciences USA* **106**, 4284–4289.

Rymen, B. and Sugimoto, K. (2012). Tuning growth to the environmental demands. *Current Opinion in Plant Biology* **15**, 683–690.

Saito, H., Oikawa, T., Hamamoto, S., *et al.* (2015). The jasmonate-responsive GTR1 transporter is required for gibberellin-mediated stamen development in *Arabidopsis*. *Nature Communications* **6**, 6095.

Sakamoto, T., Kamiya, N., Ueguchi-Tanaka, M., Iwahori, S. and Matsuoka, M. (2001). KNOX homeodomain protein directly suppresses the expression of a gibberellin biosynthetic gene in the tobacco shoot apical meristem. *Genes and Development* **15**, 581–590.

Sambade, A., Pratap, A., Buschmann, H., Morris, R.J. and Lloyd, C. (2012). The influence of light on microtubule dynamics and alignment in the *Arabidopsis* hypocotyl. *The Plant Cell* **24**, 192–201.

Sarnowska, E.A., Rolicka, A.T., Bucior, E., *et al.* (2013). DELLA-interacting SWI3C core subunit of switch/sucrose nonfermenting chromatin remodeling complex modulates gibberellin responses and hormonal cross talk in Arabidopsis. *Plant Physiology* **163**, 305–317.

Sasaki, A., Itoh, H., Gomi, K., *et al.* (2003). Accumulation of phosphorylated repressor for gibberellin signaling in an F-box mutant. *Science* **299**, 1896–1898.

Shani, E., Weinstain, R., Zhang, Y., *et al.* (2013). Gibberellins accumulate in the elongating endodermal cells of *Arabidopsis* root. *Proceedings of the National Academy of Sciences USA* **110**, 4834–4839.

Shi, Y., Tian, S., Hou, L., *et al.* (2012). Ethylene signaling negatively regulates freezing tolerance by repressing expression of CBF and type-A ARR genes in *Arabidopsis*. *The Plant Cell* **24**, 2578–2595.

Shin, J., Kim, K., Kang, H., *et al.* (2009). Phytochromes promote seedling light responses by inhibiting four negatively-acting phytochrome-interacting factors. *Proceedings of the National Academy of Sciences USA* **106**, 7660–7665.

Shin, J., Anwer, M.U. and Davis, S.J. (2013). Phytochrome-interacting factors (PIFs) as bridges between environmental signals and the circadian clock: diurnal regulation of growth and development. *Molecular Plant* **6**, 592–595.

Singh, S., Katzer, K., Lambert, J., Cerri, M. and Parniske, M. (2014). CYCLOPS, a DNA-binding transcriptional activator, orchestrates symbiotic root nodule development. *Cell Host and Microbe* **15**, 139–152.

Skirycz, A. and Inzé, D. (2010). More from less: plant growth under limited water. *Current Opinion in Biotechnology* **21**, 197–203.

Soy, J., Leivar, P., Gonzalez-Schain, N., *et al.* (2012). Phytochrome-imposed oscillations in PIF3 protein abundance regulate hypocotyl growth under diurnal light/dark conditions in *Arabidopsis*. *The Plant Journal* **71**, 390–401.

Sozzani, R. and Iyer-Pascuzzi, A. (2014). Postembryonic control of root meristem growth and development. *Current Opinion in Plant Biology* **17**, 7–12.

Spalding, E.P. (2013). Diverting the downhill flow of auxin to steer growth during tropisms. *American Journal of Botany* **100**, 203–214.

Stepanova, A.N., Robertson-Hoyt, J., Yun, J., *et al.* (2008). TAA1-mediated auxin biosynthesis is essential for hormone crosstalk and plant development. *Cell* **133**, 177–191.

Sun, J., Qi, L., Li, Y., Chu, J. and Li, C. (2012). PIF4-mediated activation of *YUCCA8* expression integrates temperature into the auxin pathway in regulating *Arabidopsis* hypocotyl growth. *PLoS Genetics* **8**, e1002594.

Sun, T.P. and Kamiya, Y. (1994). The *Arabidopsis GA1* locus encodes the cyclase ent-kaurene synthetase A of gibberellin biosynthesis. *The Plant Cell* **6**, 1509–1518.

Suzaki, T. and Kawaguchi, M. (2014). Root nodulation: a developmental program involving cell fate conversion triggered by symbiotic bacterial infection. *Current Opinion in Plant Biology* **21**, 16–22.

Takeda, N., Handa, Y., Tsuzuki, S., *et al.* (2015). Gibberellins interfere with symbiosis signaling and gene expression and alter colonization by arbuscular mycorrhizal fungi in Lotus japonicus. *Plant Physiology* **167**, 545–557.

Tanimoto, E. (2005). Regulation of root growth by plant hormones: roles for auxin and gibberellin. *Critical Reviews in Plant Sciences* **24**, 249–265.

Tanimoto, E. (2012). Tall or short? Slender or thick? A plant strategy for regulating elongation growth of roots by low concentrations of gibberellin. *Annals of Botany* **110**, 373–381.

Tepperman, J.M., Hudson, M.E., Khanna, R., *et al.* (2004). Expression profiling of phyB mutant demonstrates substantial contribution of other phytochromes to red-light-regulated gene expression during seedling de-etiolation. *The Plant Journal* **38**, 725–739.

Tepperman, J.M., Hwang, Y.S. and Quail, P.H. (2006). phyA dominates in transduction of red-light signals to rapidly responding genes at the initiation of Arabidopsis seedling de-etiolation. *The Plant Journal* **48**, 728–742.

Tian, Q., Uhlir, N.J. and Reed, J.W. (2002). *Arabidopsis* SHY2/IAA3 inhibits auxin-regulated gene expression. *The Plant Cell* **14**, 301–319.

Toledo-Ortiz, G., Huq, E. and Rodriguez-Concepcion, M. (2010). Direct regulation of phytoene synthase gene expression and carotenoid biosynthesis by phytochrome-interacting factors. *Proceedings of the National Academy of Sciences USA* **107**, 11626–11631.

Úbeda-Tomás, S., Federici, F., Casimiro, I., *et al.* (2009). Gibberellin signaling in the endodermis controls *Arabidopsis* root meristem size. *Current Biology* **19**, 1194–1199.

Úbeda-Tomás, S., Swarup, R., Coates, J., *et al.* (2008). Root growth in *Arabidopsis* requires gibberellin/DELLA signalling in the endodermis. *Nature Cell Biology* **10**, 625–628.

Vandenbussche, F., Petrasek, J., Zadnikova, P., *et al.* (2010). The auxin influx carriers AUX1 and LAX3 are involved in auxin-ethylene interactions during apical hook development in *Arabidopsis thaliana* seedlings. *Development* **137**, 597–606.

Vieten, A., Sauer, M., Brewer, P.B. and Friml, J. (2007). Molecular and cellular aspects of auxin-transport-mediated development. *Trends in Plant Science* **12**, 160–168.

Vriezen, W.H., Achard, P., Harberd, N.P. and Van Der Straeten, D. (2004). Ethylene-mediated enhancement of apical hook formation in etiolated *Arabidopsis thaliana* seedlings is gibberellin dependent. *The Plant Journal* **37**, 505–516.

Welch, D., Hassan, H., Blilou, I., *et al.* (2007). *Arabidopsis* JACKDAW and MAGPIE zinc finger proteins delimit asymmetric cell division and stabilize tissue boundaries by restricting SHORT-ROOT action. *Genes and Development* **21**, 2196–2204.

Werner, T., Motyka, V., Laucou, V., *et al.* (2003). Cytokinin-deficient transgenic *Arabidopsis* plants show multiple developmental alterations indicating opposite functions of cytokinins in the regulation of shoot and root meristem activity. *The Plant Cell* **15**, 2532–2550.

Wild, M., Daviere, J.M., Cheminant, S., *et al.* (2012). The *Arabidopsis* DELLA *RGA-LIKE3* is a direct target of MYC2 and modulates jasmonate signaling responses. *The Plant Cell* **24**, 3307–3319.

Willige, B.C., Isono, E., Richter, R., Zourelidou, M. and Schwechheimer, C. (2011). Gibberellin regulates PIN-FORMED abundance and is required for auxin transport-dependent growth and development in *Arabidopsis thaliana*. *The Plant Cell* **23**, 2184–2195.

Willige, B.C., Ogiso-Tanaka, E., Zourelidou, M. and Schwechheimer, C. (2012). WAG2 represses apical hook opening downstream from gibberellin and PHYTOCHROME INTERACTING FACTOR 5. *Development* **139**, 4020–4028.

Willige, B.C., Ahlers, S., Zourelidou, M., *et al.* (2013). D6PK AGCVIII kinases are required for auxin transport and phototropic hypocotyl bending in *Arabidopsis. The Plant Cell* **25**, 1674–1688.

Wolbang, C.M., Chandler, P.M., Smith, J.J. and Ross, J.J. (2004). Auxin from the developing inflorescence is required for the biosynthesis of active gibberellins in barley stems. *Plant Physiology* **134**, 769–776.

Xue, L., Cui, H., Buer, B., *et al.* (2015). Network of GRAS transcription factors involved in the control of arbuscule development in *Lotus japonicus. Plant Physiology* **167**, 854–871.

Yanai, O., Shani, E., Dolezal, K., *et al.* (2005). *Arabidopsis* KNOXI proteins activate cytokinin biosynthesis. *Current Biology* **15**, 1566–1571.

Yang, D.L., Yao, J., Mei, C.S., *et al.* (2012). Plant hormone jasmonate prioritizes defense over growth by interfering with gibberellin signaling cascade. *Proceedings of the National Academy of Sciences USA* **109**, E1192–1200.

Yoshida, H., Hirano, K., Sato, T., *et al.* (2014). DELLA protein functions as a transcriptional activator through the DNA binding of the indeterminate domain family proteins. *Proceedings of the National Academy of Sciences USA* **111**, 7861–7866.

Yu, N., Luo, D., Zhang, X., *et al.* (2014). A DELLA protein complex controls the arbuscular mycorrhizal symbiosis in plants. *Cell Research* **24**, 130–133.

Zádníková , P., Petrasek, J., Marhavy, P., *et al.* (2010). Role of PIN-mediated auxin efflux in apical hook development of *Arabidopsis* thaliana. *Development* **137**, 607–617.

Zentella, R., Zhang, Z.L., Park, M., *et al.* (2007). Global analysis of della direct targets in early gibberellin signaling in *Arabidopsis. The Plant Cell* **19**, 3037–3057.

Zhang, D., Jing, Y., Jiang, Z. and Lin, R. (2014). The chromatin-remodeling factor PICKLE integrates brassinosteroid and gibberellin signaling during skotomorphogenic growth in *Arabidopsis. The Plant Cell* **26**, 2472–2485.

Zhang, Y., Mayba, O., Pfeiffer, A., *et al.* (2013). A quartet of PIF bHLH factors provides a transcriptionally centered signaling hub that regulates seedling morphogenesis through differential expression-patterning of shared target genes in *Arabidopsis. PLoS Genetics* **9**, e1003244.

Zhang, Z.L., Ogawa, M., Fleet, C.M., *et al.* (2011). SCARECROW-LIKE 3 promotes gibberellin signaling by antagonizing master growth repressor DELLA in *Arabidopsis. Proceedings of the National Academy of Sciences USA* **108**, 2160–2165.

Zhong, S., Shi, H., Xue, C., *et al.* (2012). A molecular framework of light-controlled phytohormone action in *Arabidopsis. Current Biology* **22**, 1530–1535.

Zourelidou, M., Muller, I., Willige, B.C., *et al.* (2009). The polarly localized D6 PROTEIN KINASE is required for efficient auxin transport in *Arabidopsis thaliana. Development* **136**, 627–636.

Zourelidou, M., Absmanner, B., Weller, B., *et al.* (2014). Auxin efflux by PIN-FORMED proteins is activated by two different protein kinases, D6 PROTEIN KINASE and PINOID. *Elife* **3**:e02860

Annual Plant Reviews (2016) **49**, 323–358
doi: 10.1002/9781119210436.ch11 http://onlinelibrary.wiley.com

Annual Plant Reviews

Chapter 11

GIBBERELLINS AND PLANT REPRODUCTION

Andrew R.G. Plackett[1] and Zoe A. Wilson[2]
[1] *Department of Plant Sciences, University of Oxford, UK*
[2] *Department of Plant and Crop Science, University of Nottingham, UK*

Abstract: The involvement of the phytohormone gibberellin (GA) in land plant reproductive processes is ancient, but is best understood in flowering plants (angiosperms). GA acts in angiosperms to promote the transition from vegetative to reproductive development, and subsequent development of both male (stamen) and female (pistil) floral organs. Coordinating growth between these organs to promote successful fertilisation is a key regulatory function of GA. Its functions in female development are not well understood, but in stamens a number of downstream regulatory pathways have been determined. GA signalling directly regulates processes both in pollen development and in the surrounding anther tissues, particularly the tapetum. Female fertility also requires GA, which, while incompletely understood, regulates both pistil and ovule development. GA is a crucial trigger of fruit development upon fertilisation, and subsequently contributes to both the development of viable seeds and of fruiting structures.

Keywords: gibberellin, GA, reproduction, floral transition, floral development, pollen, ovule, fertility, fruit development, seed development

11.1 Introduction

The involvement of gibberellin (GA) with land plant reproduction has ancient origins: the first evidence of functional GA signalling (as defined by GA-dependent interaction between the receptor GIBBERELLIN INSENSITIVE DWARF 1 (GID1) and the DELLA class of transcription factor) observed to date is during sporogenesis of the lycophyte *Selaginella*

Annual Plant Reviews, Volume 49: The Gibberellins, First Edition.
Edited by Peter Hedden and Stephen G. Thomas.

moellendorffii (Aya *et al.*, 2011). However, this has not been seen in the bryophyte *Physcomitrella patens* (Hirano *et al.*, 2007; Yasamura *et al.*, 2007; Aya *et al.*, 2011). The functions of GA in plant reproduction have been most extensively studied in angiosperms (flowering plants), where reproductive development has become more complex and the gamete-producing haploid generation (microgametophytes and megagametophytes) has been incorporated within the organs of a specialised reproductive structure (the flower). Similarly, from the apparent initial regulation of a single process in lycophytes (Yasamura *et al.*, 2007), the role of GA has elaborated to regulate multiple aspects of angiosperm reproductive development, including the transition to flowering, establishment of the floral meristem, growth of floral organs and the development of both male and female reproductive cells (microspores and megagametophytes). The functions of GA in the reproduction of the monilophyte family, the sister group to seed plants (Pryer *et al.*, 2001) that includes ferns, have not yet been studied in detail, but chemical evidence has suggested that GA derivatives act as sex-determining antheridiogens in fern gametophytes (Yamane *et al.*, 1979; Warne and Hickok, 1989). A recent study in *Lygodium japonicum* confirmed the antheridiogen as a modified GA precursor (GA_9Me), which is taken up by immature gametophytes and processed into bioactive GA to induce male organ formation (Tanaka *et al.*, 2014). Furthermore, it is reported that application of bioactive GA to *Ceratopteris richardii* sporophytes accelerates the transition to the reproductive stage (Y. Yasamura and N. Harberd, personal communication).

The pathways through which GA regulates angiosperm reproductive development have been studied extensively in the two genetic models *Arabidopsis thaliana* (*Arabidopsis*) and rice (*Oryza sativa*), representing the diverging dicot and monocot branches of the family. In the case of pollen development, a high degree of functional conservation has been detected between these two species (reviewed in Plackett *et al.*, 2011), suggesting conservation within angiosperms. In contrast, GA regulation of fruit development has been most extensively investigated in *Arabidopsis* and tomato (*Solanum lycopersicum*), representing two divergent angiosperm fruit types (dry and fleshy, respectively). As such, this chapter will focus on evidence from *Arabidopsis*, rice and tomato to illustrate the principal functions of GA during reproductive development, although information for other species is included where appropriate.

11.2 The floral transition

11.2.1 Gibberellin promotes flowering through multiple interacting pathways

The switch to reproductive development requires the reprogramming of the vegetative shoot apical meristem (SAM) to an inflorescence meristem

(IM), which produces floral meristems (FMs) at its periphery instead of leaf primordia. This transition is regulated by a number of interacting environmental and endogenous inputs, including photoperiod, vernalisation, age and GA (reviewed in Amasino, 2010; Srikanth and Schmid, 2011). In many species, including *Arabidopsis*, the floral transition is accelerated by the action of GA, but in some species, for example pea (*Pisum sativum*), increasing GA content by chemical application or genetic manipulation delays flowering (Pharis and King, 1985; Reinecke *et al.*, 2013).

Flowering in *Arabidopsis* is promoted under permissive long day (LD) photoperiods (Searle and Coupland, 2004) through the action of CONSTANS (CO) up-regulating expression of *FLOWERING LOCUS T* (*FT*), which alone is sufficient to induce flowering (Suárez-López *et al.*, 2001). A number of *Arabidopsis* accessions and other species require a prolonged period of cold (vernalisation) to become competent to flower, antagonising the activity of the repressor FLOWERING LOCUS C (FLC) through epigenetic regulation; these signals are then integrated through *SUPPRESSOR OF OVEREXPRESSION OF CONSTANS 1* (*SOC1*) (Lee and Lee, 2010). The floral transition is also repressed through the action of microRNA156 (miR156), which negatively regulates a family of SQUAMOSA PROMOTER BINDING-LIKE (SPL) transcription factors; levels of miR156 reduce steadily with increasing plant age, with a consequent increase in *SPL* expression (Huijser and Schmid, 2011). FT and SOC1 (amongst others) orchestrate activation of important downstream transcription factors, including *APETALA1* (*AP1*), *FRUITFUL* (*FUL*) and *LEAFY* (*LFY*) to reprogram SAM identity to become an IM. *LFY* is also key to specifying subsequent FM identity; in the absence of *LFY* the transition to reproductive development still occurs, but FMs are replaced by secondary IMs (Weigel *et al.*, 1992). GA signalling promotes flowering in conjunction with these different pathways through regulating the expression of *FT*, *SOC1*, the *SPL* family and *LFY* (Figure 11.1).

Photoperiod has an important interaction with GA signalling. Under LD photoperiods, flowering is delayed in the GA-deficient mutant *ga1-3*, but not entirely blocked (Reeves and Coupland, 2001), whereas under short day (SD) photoperiods *ga1-3* does not flower unless treated with exogenous GA (Wilson *et al.*, 1992). This reflects the action of GA on two separate mechanisms, focussed on the regulation of *LFY* and *FT*, respectively. *LFY* expression is present in *ga1-3* under LD photoperiods, albeit at a reduced level, but is absent from *ga1-3* when grown under SD (Blázquez *et al.*, 1998). *LFY* is up-regulated under SD by GA at least in part through a SOC1-dependent pathway: *SOC1* is up-regulated by GA (Borner *et al.*, 2000; Moon *et al.*, 2003), can rescue *ga1-3* flowering under SD (Moon *et al.*, 2003) and directly up-regulates *LFY* in conjunction with AGAMOUS-LIKE 24 (AGL24) (Lee *et al.*, 2008). However, genetic evidence suggests that SOC1 also promotes flowering through additional, AGL24-independent pathways (Yu *et al.*, 2002), and GA can activate *LFY* expression independently of SOC1 through the GAMYB orthologue AtMYB33 (Achard *et al.*, 2004).

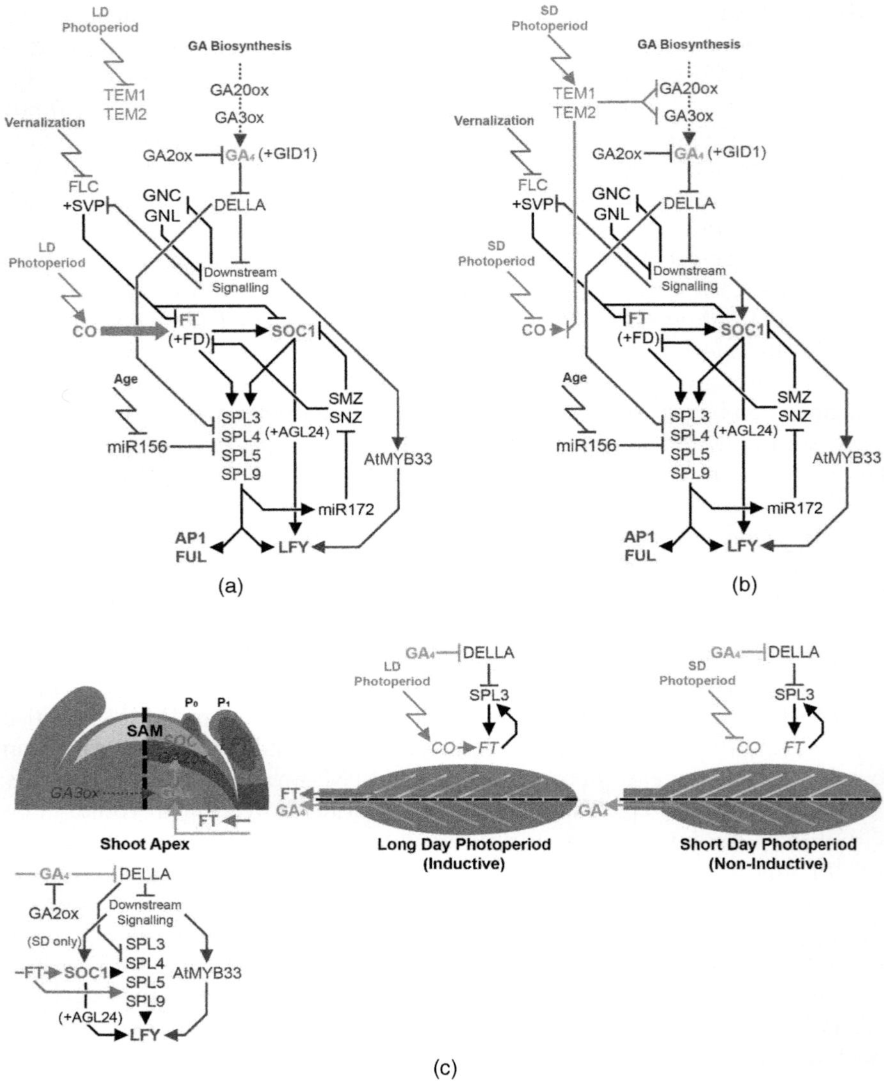
LD Photoperiod
GA Biosynthesis
TEM1
TEM2
GA20ox
GA3ox
Vernalization
GA2ox
GA4 (+GID1)
FLC
+SVP
GNC
GNL
DELLA
LD Photoperiod
Downstream Signalling
CO
FT
(+FD)
SOC1
Age
SMZ
SNZ
SPL3
SPL4
SPL5
SPL9
(+AGL24)
miR156
AtMYB33
miR172
AP1
FUL
LFY
(a)
SD Photoperiod
GA Biosynthesis
TEM1
TEM2
GA20ox
GA3ox
Vernalization
GA2ox
GA4 (+GID1)
FLC
+SVP
GNC
GNL
DELLA
SD Photoperiod
Downstream Signalling
CO
FT
(+FD)
SOC1
Age
SMZ
SNZ
SPL3
SPL4
SPL5
SPL9
(+AGL24)
miR156
AtMYB33
miR172
AP1
FUL
LFY
(b)
P0
P1
SAM
GA3ox
FT
GA4
Shoot Apex
GA4
DELLA
LD Photoperiod
SPL3
CO
FT
Long Day Photoperiod (Inductive)
GA4
DELLA
SD Photoperiod
SPL3
CO
FT
Short Day Photoperiod (Non-Inductive)
GA4
DELLA
GA2ox
Downstream Signalling
(SD only)
SPL3
SPL4
SPL5
SPL9
FT
SOC1
AtMYB33
(+AGL24)
LFY
(c)

In wheat (*Triticum* spp.), where there is a strong vernalisation requirement, it has been shown that GA signalling requires the additional presence of the vernalisation-activated AP1 homologue VERNALIZATION1 (VRN1) to up-regulate orthologues of *SOC1* and *LFY* (Pearce *et al.*, 2013). In *Arabidopsis*, the floral repressor SHORT VEGETATIVE PHASE (SVP), which acts in conjunction with the vernalisation-down-regulated repressor FLC to inhibit expression of *SOC1* and *FT* (Li *et al.*, 2008; Jang *et al.*, 2009), is itself negatively regulated by GA signalling (Li *et al.*, 2008).

LFY, *AP1* and *FUL* are each up-regulated by SPL protein activity, and have been shown to be direct transcriptional targets of *SPL3* (Yamaguchi *et al.*, 2009). *SPL3*, *-4* and *-5* have in turn been identified as downstream transcriptional targets of both SOC1 and FT (Schmid *et al.*, 2003, Jung

---

Figure 11.1 GA regulation of the *Arabidopsis* floral transition. (a) and (b) Schematic representation summarising current understanding of the interactions between the GA signalling (green), photoperiod (yellow), age (red) and vernalisation pathways (blue) in *Arabidopsis* under long day (LD; a) and short day (SD; b) photoperiods, respectively. Arrowheads and bars indicate positive and negative regulation of downstream targets, respectively. The signalling pathways shown are integrated at multiple levels. Signals under both photoperiods are integrated by expression levels of the transcription factors *FLORAL TRANSITION* (*FT*) and *SUPPRESSOR OF CONSTANS1* (*SOC1*), which promote expression of the floral identity genes *LEAFY* (*LFY*), *APETALA1* (*AP1*) and *FRUITFUL* (*FUL*) through up-regulation of *SQUAMOSA PROMOTER BINDING-LIKE* (*SPL*) transcription factors, which themselves appear to positively reinforce *FT* and *SOC1* expression through feed-forward regulation. Under permissive LD photoperiods the gene network's equilibrium is tipped towards flowering by CO-dependent up-regulation of *FT*. Under non-permissive SD photoperiods *FT* is not up-regulated and flowering instead requires GA-dependent up-regulation of *SOC1* and *LFY*. GA signalling regulates flowering through transcriptional activation of *LFY* (both via a *SOC1*-dependent pathway and independently of *SOC1* through *AtMYB33*), inhibition of SPL protein activity by direct binding of DELLA proteins, and negative regulation of environmentally-sensitive flowering inhibitors such as *SHORT VEGETATIVE PHASE* (*SVP*). (c) Tissue and organ separation of the genetic network outlined in (a) and (b). *SOC1* and *LFY* expression is restricted to the shoot apical meristem (SAM) and developing organ primordia, but are only expressed during/after the floral transition. The photoperiod signal is detected in rosette leaves, which are the site of *FT* expression and GA biosynthesis. *FT* is expressed in leaf vasculature under LD photoperiods, driven both by *CO* expression in the phloem and GA signalling. GA regulation of *SPL3* has been detected in leaf tissues and the SAM (Galvão *et al.*, 2012), suggesting that GA signalling acts through the SPL proteins at both of these sites. Under SD conditions, neither *CO* expression nor GA signalling promote *FT* expression, and levels remain low. FT and bioactive GA (LD) or bioactive GA alone (SD) synthesised in the leaves is transported to the shoot apex to promote the transition from a vegetative SAM to an inflorescence meristem (IM). Access of locally and distally-synthesised bioactive GA to the SAM is restricted by GA2ox catalytic activity. *SOC1*, *SPL* and *LFY* expression are up-regulated in the SAM to promote IM identity, leading to production of lateral floral meristems (FMs) in place of leaf primordia. SOC1 is apparently only up-regulated by GA under SD conditions. (See insert for colour representation of this figure.)

et al., 2012), which promote their expression independently of miR156 repression (Jung *et al.*, 2012). However, experimental data regarding the hierarchical position of the *SPL* genes is complex, reporting them as both downstream targets (Jung *et al.*, 2012) and upstream regulators of *FT* and *SOC1* (Yu *et al.*, 2012). The most parsimonious model consistent with this is a feed-forward loop where SPL activity inhibits the AP2-like transcription factors *SCHLAFMUTZE* (*SMZ*) and *SCHNARCGZAPFEN* (*SNZ*), which are negative regulators of both *FT* and *SOC1* (Figure 11.1a and b; Mathieu *et al.*, 2009). Negative regulation of *SMZ* and *SNZ* expression by SPL acts via the microRNA miR172 (Yant *et al.*, 2010). Importantly, DELLA proteins inhibit SPL activity through direct protein binding, as demonstrated through their inhibition of the up-regulation of *SOC1* expression by SPL9 (Yu *et al.*, 2012). Similar alteration of miRNA172 expression suggests that this regulation utilises the feed-forward circuit described above. GA induction of *SPL3*, *-4* and *-5* has been shown to be dependent on SOC1 (Jung *et al.*, 2012), although whether induction of *SOC1* by GA is likewise dependent on SPL activity is currently unclear. Thus, in the absence of *FT* or *SOC1* expression this gene network is maintained in equilibrium unless destabilised by *FT* expression under LD (Figure 11.1a) or GA-SOC1 signalling under SD (Figure 11.1b), whereupon it switches states in an apparently self-reinforcing manner.

The absolute reliance of flowering on a GA signal is over-ridden under LD photoperiods by the up-regulation of *FT*, which can itself activate *SOC1* expression (Yoo *et al.*, 2005). GA promotes *FT* expression under LD: the delayed flowering of *ga1-3* is caused by a reduction in *FT* expression, which is restored by GA treatment (Hisamatsu and King, 2008), and over-expression of *GA2ox7* reduces FT levels under LD (Porri *et al.*, 2012). Up-regulation of *FT* by GA signalling is photoperiod-dependent, with induction of *FT* by GA treatment strongly reduced under SD (Hisamatsu and King, 2008). GA-dependent promotion of *FT* expression under LD again potentially acts through the SPL feed-forward module described above, as demonstrated with *SPL3* (Figure 11.1c; Galvão *et al.*, 2012). However, GA treatment of a non-flowering *ft* mutant under LD still induced flowering (Hisamatsu and King, 2008), indicating an FT-independent effect of GA under LD photoperiods, very possibly via the parallel SOC1-LFY pathway described above.

Additional regulators of flowering also influence, or are influenced by, GA signalling, although to date their roles remain less clearly defined. In addition to a repressive effect on *FT* expression via CO (Castillejo and Pelaz, 2008), TEMPRANILLO 1 (TEM1) and TEM2 down-regulate the GA biosynthetic genes *GA 20-OXIDASE 2* (*GA20ox2*), *GA 3-OXIDASE 1* (*GA3ox1*) and *-2* (Osnato *et al.*, 2012). The *tem1 tem2* loss-of-function mutant displays an associated early flowering phenotype under LD (which is delayed in the absence of *GA3ox1*) and increased *SOC1* and *LFY* expression under both LD and SD (Osnato *et al.*, 2012). The same study found that expression of *TEM1* and *TEM2* are also photoperiod-dependent, with their expression promoted under SD (Figure 11.1b). Two other known repressors of flowering, the

GATA-type transcription factors *GATA, NITRATE-INDUCIBLE, CARBON METABOLISM INVOLVED* (*GNC*) and *GNC-LIKE* (*GNL*), are negatively regulated by GA and are themselves repressors of downstream GA signalling (Richter *et al.*, 2010).

11.2.2 Sites of gibberellin biosynthesis and action during the floral transition

The transition to flowering necessarily occurs at the SAM, but the initiation of this process occurs in other tissues. It has been demonstrated that *FT* is expressed in the phloem tissue of *Arabidopsis* leaves (Figure 11.1c; Takada and Goto, 2003) in response to vascular *CO* expression (An *et al.*, 2004) and that FT protein is subsequently transported to the SAM to initiate floral transition in both *Arabidopsis* (Corbesier *et al.*, 2007; Jaeger and Wigge, 2007; Mathieu *et al.*, 2007) and rice (Tamaki *et al.*, 2007). *LFY* expression is restricted to early and emerging *Arabidopsis* leaf primordia prior to floral transition under both LD and SD conditions (Figure 11.1c; Blázquez *et al.*, 1997), while *SOC1* is expressed in both the SAM and developing leaf primordia (Samach *et al.*, 2000). In accordance with these two distinct expression patterns, targeted expression studies in *Arabidopsis* using GA2ox and GA-resistant DELLA proteins found that GA regulation of flowering occurs in both leaves and the SAM (Galvão *et al.*, 2012; Porri *et al.*, 2012, Yu *et al.*, 2012). The relative importance of these two sites on flowering time is affected by photoperiod, corresponding to the contributions of the parallel FT and SOC1-LFY pathways under LD and SD photoperiods. Under LD conditions, blocking GA signalling in either leaves or the SAM delays flowering (Yu *et al.*, 2012); in the leaves, GA signalling increases expression of *FT* and *SPL* genes (Figure 11.1c; Galvão *et al.*, 2012; Porri *et al.*, 2012), while in the SAM, *SPL* genes, but not *SOC1*, are up-regulated by GA signalling (Porri *et al.*, 2012). Under SD conditions blocking GA signalling in leaf phloem did not affect flowering time, while a block imposed in the SAM further delayed or prevented flowering. As mentioned previously, under SD conditions up-regulation of *SOC1* is dependent on GA (Moon *et al.*, 2003), but *SOC1* expression can also be up-regulated by FT (Yoo *et al.*, 2005). Therefore GA signalling in the leaves indirectly promotes floral transition under LD through a mobile FT signal, while GA signalling in the SAM promotes flowering under both photoperiods, functioning either through *SOC1* (SD) or through the downstream *SPL* genes (LD).

Importantly, the site of GA action does not necessarily reflect that of GA biosynthesis. Under SD photoperiods, where GA signalling is an essential requirement for flowering, levels of bioactive GA_4 increase dramatically at the shoot apex (including early leaf primordia) prior to flowering, coinciding with a sharp increase in *LFY* expression in the same tissues (Eriksson *et al.*, 2006). However, no increase in GA biosynthesis (as determined by gene expression) was detected, suggesting that the necessary GA signal is synthesised elsewhere. Radiolabelling studies have demonstrated movement of bioactive GA_4 from rosette leaves to the SAM (Eriksson *et al.*, 2006). Furthermore, GA-inactivating enzymes are expressed beneath the shoot apex of *Arabidopsis*, rice and its relative, *Lolium temulentum*, during vegetative development (Figure 11.1c; Sakamoto *et al.*, 2001; Jasinski *et al.*, 2005; Zhu *et al.*, 2006; King *et al.*, 2008). In the case of *Arabidopsis*, loss of these GA2ox enzymes progressively accelerates the transition to flowering under SD conditions (Rieu *et al.*, 2008a). Therefore GA inactivation appears to have a role in regulating the timing of floral transition, by preventing bioactive GA synthesised in remote tissues from accessing the SAM.

The effects of GA on flowering and shoot elongation are separable, despite both occurring in shoot tissues. In many plant species stem internode elongation ('bolting') is triggered by the floral transition; however, this can begin either before (e.g. sugar beet (*Beta vulgaris*)) or after (e.g. *Arabidopsis*) the floral transition has occurred (Mutasa-Göttgens and Hedden, 2009). In *Arabidopsis*, *GA3ox* expression is present beneath the shoot meristem both before and after the floral transition (Mitchum *et al.*, 2006; Hu *et al.*, 2008), suggesting that additional factors inhibit cell elongation in response to GA in a stage-specific manner. The separation of stem elongation and floral induction is currently poorly understood, with possible explanations including differential DELLA interacting partners between the SAM and internode tissues, or differential sensitivity to different forms of bioactive GA. In *Arabidopsis*, both stem elongation and flowering time are more sensitive to GA_4 than to GA_1 when applied under SD (Xu *et al.*, 1997), consistent with *LFY* induction being more responsive to GA_4 (Eriksson *et al.*, 2006). In the case of *Lolium*, certain species of GA including GA_1 and GA_4 are inactivated by GA2ox enzymes surrounding the SAM, whilst GA_5 is resistant to GA2ox activity due to unsaturation on C-2 (King *et al.*, 2008). Correspondingly, GA_5 is present within the vegetative SAM and accumulates in response to a LD stimulus, whereas GA_1 and GA_4 are excluded (King *et al.*, 2001). Treatment with GA_5 was shown to have a disproportionate effect on floral induction relative to stem elongation, compared to treatment with GA_1 or GA_4 (King *et al.*, 2008). Thus selective inactivation of GA may discriminate between stem growth and floral induction in some species. In rice, the transition to reproductive development apparently alters the composition of bioactive GA produced, with GA_1 or GA_4 predominating in vegetative and reproductive tissues, respectively (Kurogochi *et al.*, 1979; Kobayashi *et al.*, 1984, 1988). The rice cytochrome P450 mono-oxygenase ELONGATED UPPERMOST INTERNODE (EUI) has been found to selectively deactivate GA_4 by 16α-17 epoxidation in elongating internodes during the reproductive phase (Zhu *et al.*, 2006), further supporting the existence of distinct functions and regulation of 13-hydroxylated and non-13-hydroxylated GAs during reproductive development in at least some plant species.

11.2.3 Gibberellin and flowering in perennial species

Although most of our understanding of the floral transition is derived from annual species, some information about the regulation of flowering in perennial species is now available. The perennial growth habit is primarily achieved through the maintenance of a sub-population of vegetative meristems despite the presence of floral inductive signals. Molecular studies of perennialism suggest that this is achieved via an elaboration of the core floral transition module identified in annual species, with particular emphasis on enhanced roles of known repressors of floral transition such as FLC (in *Arabis Alpina*; Wang *et al.*, 2009) and TERMINAL FLOWER1 (TFL1) (in Rosaceae species; reviewed in Kurokura *et al.*, 2013), an antagonist of FT action (Hanano and Goto, 2011).

In an interesting parallel, in many angiosperm perennial species GA acts as an inhibitor of flowering (Pharis and King, 1985; Wilkie *et al.*, 2008; Bangerth, 2009). GA is thought to negatively regulate seasonal flowering in Rose through promoting expression of the *TFL1* homologue *RoKSN* (Randoux *et al.*, 2012). In *Citrus* spp., where GA treatment also inhibits flowering, it has been found that GA treatment leads to reduced *FT* expression in leaves (Muñoz-Fambuena *et al.*, 2012) and in apical buds (Goldberg-Moeller *et al.*, 2013) under otherwise flowering-inductive conditions, in an apparent reversal of its effect in *Arabidopsis*. Treatment of *Citrus* trees with the GA biosynthesis inhibitor paclobutrazol (PAC) (Rademacher, 2000) under inductive conditions increases the conversion of apical meristems to flowers (Delgado *et al.*, 1986), an effect that is reversible by simultaneous GA treatment (Martínez-Fuentes *et al.*, 2013). Interestingly, increasing numbers of developing fruit negatively affect the flower-promoting effects of PAC treatment (Martínez-Fuentes *et al.*, 2013), suggesting that GA functions as part of a feedback mechanism at the whole-plant level, either as a mobile signal or as part of a localised downstream signalling response to another mobile signal.

11.3 Floral development

After the transition to flowering, the IM sequentially produces FMs in place of leaf primordia. In contrast to the indeterminate nature of the IM, development of individual flowers is both highly determinate and uniform, with a fixed pattern and chronology. In the case of *Arabidopsis*, floral development has been staged in detail (Smyth *et al.*, 1990). Extensive mutant analysis of *Arabidopsis* has demonstrated that GA has regulatory functions at multiple points throughout floral development, and is a very important determinant of successful reproduction. GA-deficient and GA-insensitive mutants in numerous monocot and dicot species are infertile, with severely underdeveloped floral organs (Koornneef and Van der Veen, 1980; Pharis and King, 1985; Cecconi *et al.*, 2002; Griffiths *et al.*, 2006; Aya *et al.*, 2009).

11.3.1 Floral patterning and early development

The earliest stages of *Arabidopsis* floral development are driven by the action of LFY, which is strongly expressed in the developing FMs from their initiation, although not in the IM itself (Weigel *et al.*, 1992). The first lateral meristems produced by the IM are not in fact flowers, but develop as inflorescence branches (Ratcliffe *et al.*, 1999). Expression of floral meristem identity genes is promoted by LFY, and recent research shows that, beyond the floral transition, GA signalling in fact delays the start of FM specification, increasing the number of inflorescence branches produced, while conversely, accumulation of DELLA protein promotes FM identity (Yamaguchi *et al.*, 2014). In *Arabidopsis*, LFY has been shown to promote depletion of bioactive GA from inflorescence tissues through up-regulation of the *EUI* homologue *EUI-LIKE P450 A1* (*ELA1*) (Zhang *et al.*, 2011; Nomura *et al.*, 2013), leading to an accumulation of DELLA protein (Yamaguchi *et al.*, 2014). It has now been shown that DELLA proteins actively promote expression of the floral identity gene *AP1* (see Section 11.2.1), binding to its promoter in association with SPL9 independently of LFY (Yamaguchi *et al.*, 2014).

Beyond FM specification, floral organs are established and emerge sequentially in concentric whorls (Figure 11.2a), with identity within each whorl determined through a complex of interacting MADS box transcription factors following the now well-established ABCE model (reviewed in Airoldi, 2010; Irish, 2010). LFY drives early expression of a number of these genes, including the B-class (petal and stamen) and C-class (stamen and carpal) factors *APETALA3* (*AP3*) and *AGAMOUS* (*AG*) (Lohmann *et al.*, 2001; Lamb *et al.*, 2002). *LFY* expression in the early FM is not regulated by GA signalling (Yu *et al.*, 2004), in contrast to its vegetative-phase expression, and continues until floral stage 5 (emergence of stamen and petal primordia).

Bioactive GA is present in early floral tissues, as evidenced by the expression of *GA 3-OXIDASE1* (*GA3ox1*), first reported in central FM tissues at stage 3 (sepal initiation), which later becomes confined to the receptacle in early floral stages (Hu *et al.*, 2008). As with *LFY*, *GA3ox* is not expressed in IM tissues. Early *GA3ox* expression is reported to continue through to floral stage 7 (stamen filament differentiation) and beyond (Mitchum *et al.*, 2006; Hu *et al.*, 2008). After *LFY* expression declines continued expression of *AP3* and *AG* becomes dependent on GA (Yu *et al.*, 2004), AG apparently establishing a positive feedback loop with *GA3ox1* (amongst others) to maintain its own expression (Gómez-Mena *et al.*, 2005). *AG* expression is required throughout floral development to maintain floral organ development, most evidently that of stamens (Ito, 2011; Ito *et al.*, 2011).

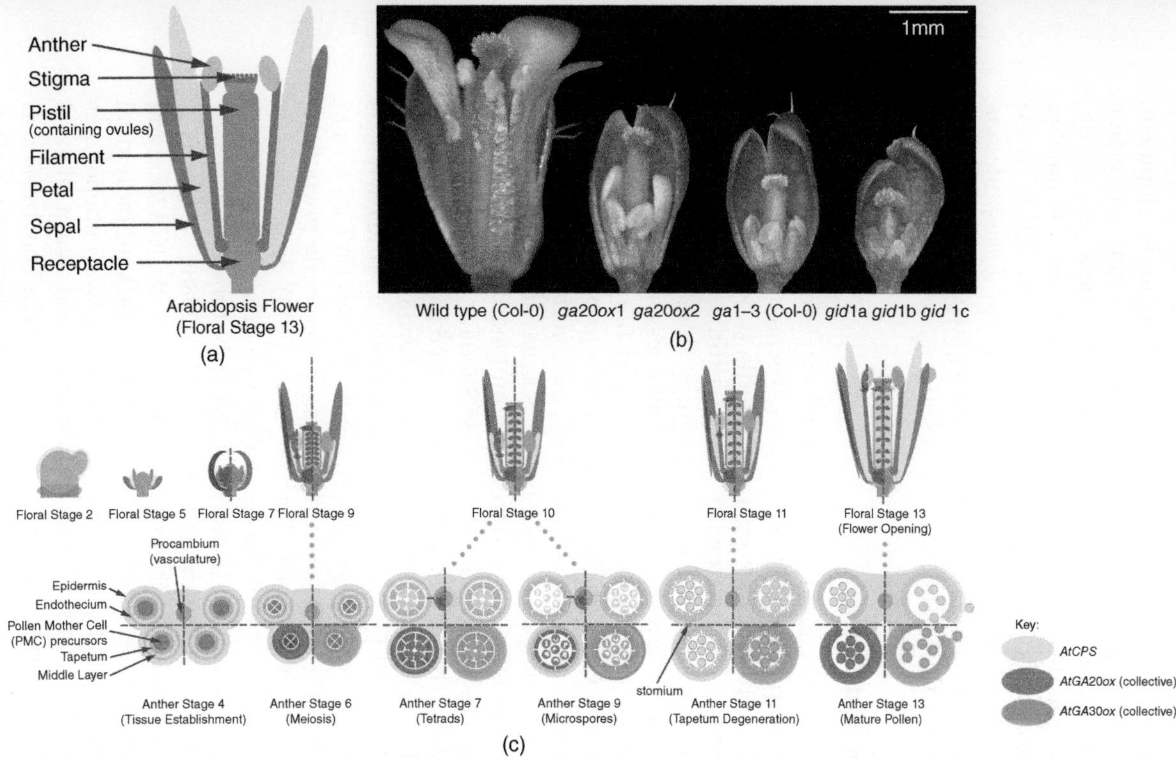

Figure 11.2 Gibberellin and *Arabidopsis* floral development. (a) Diagrammatic cross-section of an *Arabidopsis* flower, identifying major organs. (b) Comparison of *Arabidopsis* floral phenotypes from selected GA biosynthesis and signalling mutants (displaying a range of phenotypic severity) against wild type. Growth of all floral organs is reduced, with stamens and petals showing greatest sensitivity. Anthers of the *ga1-3* and *gidl1a gid1b gid1c* mutants display premature developmental arrest (Goto and Pharis, 1999; Griffiths *et al.*, 2006). (c) Sites of floral GA biosynthesis in *Arabidopsis* floral tissues during development, showing *AtCPS* (yellow) and cumulative *AtGA20ox* (blue) and *AtGA3ox* (green) gene expression. The patterns of expression shown are based on analysis of GUS reporter lines (Silverstone *et al.*, 1997; Mitchum *et al.*, 2006; Hu *et al.*, 2008; Plackett, 2012; Plackett *et al.*, 2012). *GA20ox* expression during early floral development has not been characterised. Numbers marked on the inflorescence meristem denote floral stages 2 and 3 (Smyth *et al.* 1990), respectively. The anther developmental stages shown follow the definitions of Sanders *et al.* (1999). (Adapted from Plackett *et al.* (2011). Reproduced with permission of Elsevier.) (See insert for colour representation of this figure.)

Despite these early functions, the floral plan remains essentially unchanged in *Arabidopsis* GA mutants, although some GA-dependent instability in floral organ number has been detected during the early phase of flowering, potentially due to the initial size of the developing FM (Plackett, 2012). In other species GA has a greater influence on early floral organ development. The GA-deficient tomato mutant *stamenless-2* exhibits partial conversion from stamen to carpel identity (Sawhney, 1992), while GA treatment or constitutive GA signalling in developing tomato flowers increases the number of floral organs that develop (Sawhney, 1983; Carrera *et al.*, 2012). In the monoecious species maize (*Zea mays*) loss of the *ent-copalyl diphosphate synthase (CPS)* orthologue *ANTHER EAR1* (*AN1*) induces stamen development in otherwise female-only florets (Bensen *et al.*, 1995).

11.3.2 Gibberellin and fertility

The importance of GA action on flowering plant fertility is clearly illustrated by mutants that are completely deficient in or are insensitive to GA, where both male and female reproductive organs are sterile (Koornneef and Van der Veen, 1980; Jacobsen and Olszewski, 1991; Griffiths *et al.*, 2006; Aya *et al.*, 2009). In the case of these severe phenotypes, floral organ development undergoes premature arrest (see below), while less severely deficient mutants display fertility defects caused by impaired floral organ growth (Figure 11.2b). However, GA signalling also appears to be restricted during floral development: exogenous GA treatment in *Arabidopsis* measurably reduces fertility (Jacobsen and Olszewski, 1993; Plackett *et al.*, 2014), and constitutive GA signalling in *DELLA* loss-of-function mutants causes sterility in monocot and dicot species (Lanahan and Ho, 1988; Ikeda *et al.*, 2001; Plackett *et al.*, 2014). In *Arabidopsis* this observation is complicated by ecotype-specific factors. In the Landsberg *erecta* (L*er*) ecotype, some fertility is maintained despite the loss of all five DELLA paralogues (Fuentes *et al.*, 2012), whereas in the Columbia (Col-0) ecotype loss of just two, *GA INSENSITIVE* (*GAI*) and *REPRESSOR OF ga1-3* (*RGA*), causes a pollenless, male-sterile phenotype, although female fertility is retained (Plackett *et al.*, 2014). In m-onocots, the reported sterility phenotype of the equivalent rice *slender* (*slr1*) mutant remains undescribed in any detail (Ikeda *et al.*, 2001), but the barley *slender* (*sln1*) mutant is reported as pollenless (Lanahan and Ho, 1988).

The later stages of GA biosynthesis are governed by the *GA20ox* and *GA3ox* multi-gene families (see Chapter 2), and mutants lacking individual (or multiple) paralogues from these families display semi-dwarf phenotypes associated with reduced GA biosynthesis, including reduced size of all floral organs (Hu *et al.*, 2008; Rieu *et al.*, 2008b; Plackett *et al.*, 2012). The *Arabidopsis* semi-dwarf mutant *ga20ox1 ga20ox2* displays a detectable reduction in fertility, both in the number of seeds set per silique and a failure of early flowers to set siliques (Rieu *et al.*, 2008b). This latter phenotype is associated with reduced growth of stamens relative to the pistil at flower opening

(Figure 11.2b), creating a mechanical block to pollination at anthesis. The timing of pollination is an important factor, as ovules remain receptive to fertilisation only for a limited time once mature (Vivian-Smith and Koltunow, 1999). Spontaneous recovery of fertility in later flowers without GA treatment occurs in both these mutants and in *ga1-3* (Plackett *et al.*, 2012), which has a correspondingly severe initial floral phenotype. Phenotypic recovery includes increased floral organ growth and, in the case of *ga1-3*, overcoming early arrest of pollen development. The common nature of this recovery suggests that it is downstream, or independent of GA regulation. This hypothesis is supported by detection of similar GA levels between early (infertile) and late (fertile) flowers (Hu *et al.*, 2008).

Sites of gibberellin biosynthesis and signal transduction during floral development

The sites of GA biosynthesis in *Arabidopsis* floral tissues have mostly been deduced indirectly from *GA3ox* expression patterns, using GUS reporter constructs (Figure 11.2c). Prolonged *GA3ox* expression is observed in the receptacle, but is also expressed in the stamen filament and anther from approximately floral stage 7 (Mitchum *et al.*, 2006; Hu *et al.*, 2008). Some limited expression has also been observed in the vasculature of sepals. This pattern suggests that the stamens and receptacle act as a source of bioactive GA for other floral organs. In the case of petals, this is supported by a strong correlation in size with stamens in the flowers of mutants such as *ga20ox1 ga20ox2* (Figure 11.2b; Rieu *et al.*, 2008b). No *GA3ox* expression was detected in pistils prior to pollination and the start of silique development. Expression of *CPS*, the first step in GA biosynthesis, is reported in the receptacle, stamen vasculature and anther tissues of developing flowers (Figure 11.2b; Silverstone *at al.*, 1997), but is first detected in the developing pistil only at floral stage 12, immediately prior to flower opening, where it occurs in the pistil vasculature and ovule funiculi. *GA20ox* expression was detected in the receptacle, stamen filament, anther tissues, pistil apex and sepal vasculature during floral development (Figure 11.2b; Plackett *et al.*, 2012).

DELLA expression patterns during floral development have not been comprehensively mapped in either *Arabidopsis* or rice, but expression data during the development of some critical anther tissues is now available for these species (Honys and Twell, 2004; Hirano *et al.*, 2008; Tang *et al.* 2010). Mutant analysis in *Arabidopsis* indicates that *RGA*, *GAI*, *RGA-LIKE1* (*RGL1*) and *-2* are all involved in regulating floral development to some degree (Lee *et al.*, 2002; Cheng *et al.*, 2004; Tyler *et al.*, 2004; Plackett *et al.*, 2014). An *RGL2::GUS* line indicates expression in the receptacle, sepals, petal vasculature, stamen filaments and pistil tissues, with strongest expression at the top of the style (Lee *et al.*, 2002). Interestingly, *RGL2* is apparently excluded from anther tissues, whereas *in situ* hybridisation identified *RGL1* expression specifically in the anthers of stamen primordia (Wen and Chang, 2002), suggesting functional specialisation. *RGL1* was also detected in developing ovules. A comparison between mutant phenotypes in Col-0 and *Ler* suggests that the functions (and potentially the expression patterns) of specific DELLA paralogues during floral development can vary between ecotypes (Plackett *et al.*, 2014).

Downstream of GA signalling, GAMYB has been identified as an important transcription factor during floral development of multiple species (Murray *et al.*, 2003; Kaneko *et al.*, 2004; Millar and Gubler, 2005). The interpretation of *GAMYB* expression patterns in floral tissues is complicated by negative post-transcriptional regulation by miRNA159 (Achard *et al.*, 2004; Tsuji *et al.*, 2006; Alonso-Peral *et al.*, 2010), which apparently restricts expression exclusively to anther tissues in both *Arabidopsis* and rice (Millar and Gubler, 2005; Aya *et al.*, 2009).

Male reproductive development

The two principal functions of GA during stamen development are to promote filament elongation at flower opening through increased filament cell elongation (Cheng *et al.*, 2004) and, in the anther, to promote successful pollen development and locule opening (anthesis). GA-deficient or insensitive plants demonstrate arrested anther/pollen development in numerous species (Nester and Zeevaart, 1988; Goto and Pharis, 1999; Izhaki *et al.*, 2002; Aya *et al.*, 2009). However, the precise timing of this arrest varies between the species and mutants studied. Anthers of the GA-deficient *Arabidopsis* mutant *ga1-3* undergo development arrest at the unicellular microspore stage, prior to pollen mitosis (Cheng *et al.*, 2004), whereas GA-deficient tomato mutants demonstrate an earlier arrest prior to the onset of meiosis (Nester and Zeevaart, 1988; Jacobsen and Olszewski, 1991). In the GA-deficient rice mutant *oscps1-1* normal anther development fails after microspore release from the tetrads, but defects manifest earlier in GA-insensitive *gid1-4*, where meiosis begins, but does not complete (Aya *et al.*, 2009). Inhibition of GA biosynthesis by PAC treatment of *Petunia hybrida* flowers caused a post-meiotic arrest of pollen development (Izhaki *et al.*, 2002). The mechanisms through which anther developmental arrest is enacted are not yet clearly understood, but presumably indicate the existence of 'checkpoints' during anther development that require specific conditions to be met before development proceeds. The failure of *Arabidopsis* pollen development in the absence of DELLA (Plackett *et al.*, 2014) highlights the importance of these checkpoints and the role of GA in the transition through them.

The tissues most closely associated with the failure in male fertility are pollen and the surrounding tapetum cell layer. Pollen development is highly dependent on the tapetum, both for nutrition and the deposition of pollen wall components (reviewed in Scott *et al.*, 2004). This dependency is clearly illustrated in experiments in which the tapetum is ablated in developing tobacco anthers, resulting in pollen abortion (Koltunow *et al.*, 1990). An important aspect of tapetum development is programmed cell death (PCD) to allow deposition of wall materials onto the pollen; impairing tapetum PCD progression in many mutants results in male sterility through pollen abortion (Kawanabe *et al.*, 2006; Vizcay-Barrena and Wilson, 2006). Cell-type-specific transcriptome analysis in developing rice anthers indicates active GA signalling in both the pollen mother cells (PMCs)/microspores and the tapetum from meiosis until pollen maturity, with expression of GA biosynthetic genes strongest in the tapetum immediately prior to PCD and in post-meiotic pollen at the bicellular and tricellular stages (Hirano *et al.*, 2008; Tang *et al.*, 2010). Similarly, direct and indirect experimental approaches in *Arabidopsis* collectively predict active GA biosynthesis in developing pollen from meiosis until maturity (Honys and Twell, 2004; Hu *et al.*, 2008) and in anther tissues, most strongly in the tapetum, from meiosis until tapetum degeneration (Figure 11.2c). Transmission frequencies of GA signalling mutant alleles between generations in rice and the *Arabidopsis gid1* triple mutant suggest that signalling proteins synthesised in premeiotic PMCs persist during post-meiotic development and are capable of maintaining GA signalling in haploid microspores despite the absence of a functional allele (Griffiths *et al.*, 2006; Chhun *et al.*, 2007). Thus GA signalling directly regulates both tapetum and pollen development at meiosis and during subsequent post-meiotic development. Recently the possibility has been raised that GA signalling acts to coordinate development between these two cell types: reintroduction of functional DELLA protein into a male-sterile DELLA loss-of-function background successfully rescued pollen development if expressed after the completion of meiosis in either developing pollen or the tapetum (Plackett *et al.*, 2014).

GA regulation of tapetum development has to date been most closely studied in rice including identification of a number of downstream targets of GA signalling (Figure 11.3), which acts almost entirely through the GAMYB transcription factor (Aya *et al.*, 2009). In rice GA biosynthesis and signalling mutants activation of the tapetum PCD pathway is specifically blocked (Aya *et al.*, 2009). Analysis of these mutants identified transcription factors involved in the regulation of tapetum PCD (Li *et al.*, 2006); at least one of these downstream factors, *Osc6*, contains a GAMYB-binding domain within its promoter. Other direct targets were identified in pollen wall exine biosynthesis and Ubisch body formation (Aya *et al.*, 2009), critical for successful formation of the pollen wall (Piffanelli *et al.*, 1998; Wang *et al.*, 2003). A very high degree of conservation has been detected in anther genetic pathways between rice and *Arabidopsis* (Figure 11.3; reviewed in Plackett *et al.*, 2011), suggesting that GA regulates the same pathways in *Arabidopsis* as in rice. However, these species represent only one of two types of tapetum (secretory rather than amoeboid) present in angiosperms (Huysmans, 1998) and additional functions might be present in the tapeta of other species. The tomato tapetum comprises two separate single-cell layers either side of a crescent-shaped locule, distinguished as the inner and outer tapetum, which undergo endomitosis to become binucleate at maturity (Brown, 1949). In GA-deficient *gib-1* anthers, development of the outer tapetum arrests with fewer cells than wild type and these cells also remain uninucleate,

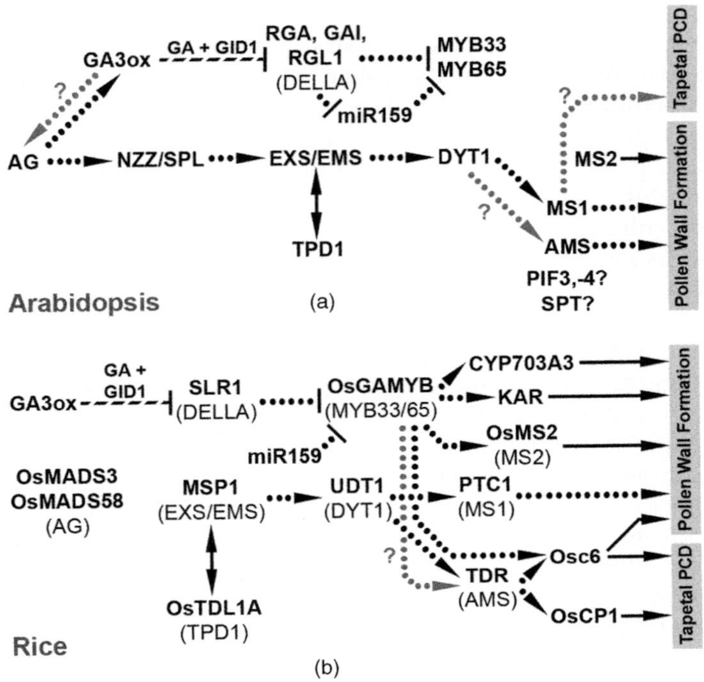

Figure 11.3 Gibberellin regulation of *Arabidopsis* tapetum development. (a) and (b) Comparison of the transcriptional networks regulating tapetum function in *Arabidopsis* (a) and rice (b) including known interactions with GA signalling. Where identified, homologous *Arabidopsis* genes are specified beneath the relevant rice gene. Dotted lines indicate transcriptional regulation, solid lines indicate protein interaction and hatched lines GA signal perception via GID1. Arrowheads and bars indicate positive and negative regulation of downstream targets, respectively. Arrows in grey indicate interactions which have not been demonstrated conclusively. (Figure is adapted from Plackett *et al.* (2011), with permission of the journal.)

but subsequently undergo endoreduplication in response to GA treatment (Jacobsen and Olszewski, 1991; van den Heuvel *et al.*, 2001). DELLA protein activity has also been shown to regulate cell cycle activity and promote endoreduplication in leaves (Claeys *et al.*, 2012). The genetic pathways regulated by GA during microspore development are currently uncharacterised. It has recently been shown that the basic helix-loop-helix (bHLH) transcription factor INDEHISCENT (IND) modulates *Arabidopsis* male fertility responses to GA: loss of IND function partially rescues phenotypic defects in mild (but not severe) GA mutants (Kay *et al.*, 2013). The precise mechanism underlying this remains undetermined, but does not appear to involve direct regulation of GA biosynthesis via *AtGA3ox1* expression (Kay *et al.*, 2013), as seen during silique development (see Section 11.4.1). *IND* expression during anther development appears to be restricted to the locule/ developing pollen, based on expression data from a GUS reporter line (Kay *et al.*, 2013).

The final phase of anther development centres on locule opening and release of mature pollen (anthesis). Anthers from the first flowers of some *Arabidopsis ga20ox* and *ga3ox* mutants are indehiscent, suggestive of late-stage developmental arrest (Hu *et al.*, 2008; Rieu *et al.*, 2008b). After a period of down-regulated expression beyond tapetum degeneration, *GA3ox* reaccumulates in anther wall tissues immediately prior to anthesis (Hu *et al.*, 2008). Jasmonate (JA) signalling apparently acts downstream of both GA and auxin during stamen maturation to promote pollen maturation, filament elongation and anther dehiscence (reviewed in Song *et al.*, 2013). These processes are dependent on the MYB transcription factors MYB21, -24 and -57, which are downstream targets of JA signalling (Cheng *et al.*, 2009; Song *et al.*, 2011). During late stamen development GA signalling promotes accumulation of JA through up-regulation of the JA biosynthesis gene *DEFECTIVE IN ANTHER DEHISCENCE1* (*DAD1*) (Cheng *et al.*, 2009), itself a direct downstream target of AGAMOUS (Ito *et al.*, 2011; see Section 11.3.1), resulting in up-regulation of the *MYB* genes in a JA-dependent manner (Cheng *et al.*, 2009). However, JA treatment of GA-deficient mutants fails to restore stamen development, despite *MYB* up-regulation (Cheng *et al.*, 2009), demonstrating that GA regulates additional, JA-independent pathways during stamen maturation. These additional pathways are currently uncharacterised.

After anthesis and the transfer of pollen to the stigma (pollination), GA also functions to promote pollen germination and pollen tube growth, as demonstrated in the rice GA biosynthesis mutants (Chhun *et al.*, 2007) and by over-expression of a *GA2ox* gene in transgenic *Arabidopsis* pollen (Singh *et al.*, 2002). Interestingly, GA treatment has also been shown to inhibit *Arabidopsis* pollen tube growth *in vitro* (Singh *et al.*, 2002), suggesting that unrestricted GA signalling is detrimental. The mechanism through which GA affects pollen tube growth, which unlike most plant cell types extends through polarised tip growth (Hepler *et al.*, 2001), appears to act through DELLA-mediated signal transduction (Swain *et al.*, 2004). Seed counts from hand-pollination experiments suggest that loss of IND and/or the MADS-box transcription factors SHATTERPROOF1 (SHP1) and SHP2 partially restores pollen tube growth in GA-deficient *35S::GA2ox* pollen (Kay *et al.*, 2013).

Female reproductive development

In angiosperms the female gametophyte is typically reduced to an embryo sac of a few highly specialised cells that develops within the tissues of the sporophytically-derived ovule (reviewed in Shi and Yang, 2011). Although not explicitly investigated to date there is mounting evidence that female reproductive development and fertility are also directly influenced by GA. The *Arabidopsis ga20ox1 ga20ox2* mutant produces fewer seeds than wild type even when manually fertilised with wild-type pollen without any obvious signs of seed abortion (Rieu *et al.*, 2008b), suggesting that ovule number is reduced in this background. This is consistent with qPCR evidence for *GA20ox2* expression in pistil tissues (Plackett *et al.*, 2012) and ovule-specific microarray analysis (Yu *et al.*, 2005; Schmidt *et al.*, 2011). Microarray analysis of developing *Arabidopsis* ovules suggests that both the GA biosynthetic and signalling pathways are expressed in these tissues (Yu *et al.*, 2005; Schmidt *et al.*, 2011), including the megaspore mother cell (MMC), which undergoes meiosis to generate four megaspores, one of which will develop into an embryo sac (Schmidt *et al.*, 2011). Interestingly, beyond the MMC neither direct laser-capture microdissection of key embryo-sac cells (Wuest *et al.*, 2010) nor indirect comparative expression profiling against embryo-sac-less mutants (Yu *et al.*, 2005; Johnston *et al.*, 2007; Jones-Rhoades *et al.*, 2007; Steffan *et al.*, 2007) have yet provided convincing evidence of GA biosynthesis or signalling occurring within the female gametophyte. This is in stark contrast to the prominent role of GA in male gametophyte development (see Section 11.3.2) and possibly reflects the evidence that female gametogenesis predates the evolution of GA signalling, which apparently first arose to regulate sporogenesis (see Section 11.1). In contrast to the apparent absence of GA-dependent female gametophyte development, it has been determined that auxin is a critical regulator of embryo-sac patterning (reviewed in Sundaresan and Alandete-Saez, 2010). *In situ* hybridisation has identified *RGL1* expression in the integuments of developing ovules, surrounding the female gametophyte (Wen and Chang, 2002), suggesting that GA signalling does occur in sporophyte-derived ovule tissues.

After ovule and female gametophyte development, female fertility in angiosperms is dependent on the pistil/gynoecium to successfully mediate fertilisation by the male gametophyte (pollen). Of particular importance are the stigma, which regulates pollen recognition and germination, and the style/transmission tract, through which pollen tubes grow to reach the ovaries. The growth of both of these tissues is responsive to GA and is enhanced in pistils of the DELLA global loss-of-function mutant (Fuentes *et al.*, 2012), which displays a constitutive GA signalling phenotype. Female fertility is reduced in this mutant, which sets fewer seeds than wild-type controls even when manually pollinated with wild-type pollen (Fuentes *et al.*, 2012). The subsequent siliques displayed ovule abortion, indicative of inefficient fertilisation. IND has been implicated in GA-mediated female fertility through analysis of reciprocal crosses (Kay *et al.*, 2013) in addition to its roles in male fertility (see Section 11.3.2) and fruit development (see Section 11.4.1). In tomato, loss of DELLA repression or GA treatment also increases style growth (Carrera *et al.*, 2012).

11.4 Seed and fruit development

Upon fertilisation, ovules develop into seeds containing a diploid embryo and triploid endosperm, while the surrounding pistil develops into a fruit (reviewed in Gillaspy *et al.*, 1993; Seymour *et al.*, 2013). In the absence of fertilisation, progression of the ovary into fruit development is repressed (Rodrigo and Garcia-Martínez, 1998; Vivian-Smith *et al.*, 2001; Vriezen *et al.*,

2008) and an unfertilised pistil eventually senesces. There is a wide range of fruit forms within the angiosperms, two of the best-studied being the *Arabidopsis* silique and the tomato berry. Amongst other signals, both GA and auxin are important regulators of fruit-set and development (reviewed by Ruan *et al.*, 2012; Sotelo-Silveira *et al.*, 2014). In *Arabidopsis*, tomato and pea continued GA signalling through either exogenous GA treatment, or loss of DELLA repression is sufficient to trigger and substantially maintain fruit development in the absence of fertilisation (parthenocarpic growth) (Ozga and Reinecke, 1999; 2003; Serrani *et al.*, 2007a, 2007b; Dorcey *et al.*, 2009; Carrera *et al.*, 2012; Fuentes *et al.*, 2012). However, GA-induced parthenocarpic *Arabidopsis* fruits stall prematurely even under continued GA treatment (Vivian-Smith and Koltunow, 1999). This has been shown to correlate with ovule senescence (Carbonell-Bejerano *et al.*, 2010), and that this onset of GA-insensitivity is influenced by ethylene signalling (Carbonell-Bejerano *et al.*, 2011). Similarly, GA-induced parthenocarpic seedless tomato fruits remain under-developed and smaller than those arising from pollination (reviewed in de Jong *et al.*, 2009a). GA signalling therefore cannot entirely complement fruit development in the absence of developing seeds in these two species.

11.4.1 Fruit development

Fruit development and maturation involves tissue growth through cell division and expansion and also the differentiation of new tissue types. As discussed previously, these processes are dependent on ovule fertilisation and continuing seed development. In both *Arabidopsis* and tomato, fertilisation up-regulates GA biosynthesis within ovule tissues, downstream of auxin biosynthesis and signalling (Serrani *et al.*, 2008; Vriezen *et al.*, 2008; Dorcey *et al.*, 2009). The up-regulation of GA biosynthesis at fertilisation was shown in *Arabidopsis* to be ovule-specific, but active GA signalling was present in both ovules and developing siliques (Figure 11.4a; Dorcey *et al.*, 2009), suggesting that GA might act as a mobile signal to promote silique-set in response to seed-set. In pea, RT-PCR and transcriptome analyses suggest that pollination/fertilisation triggers up-regulation of GA biosynthesis independently in both ovules/seeds and surrounding pericarp tissues (Ozga *et al.*, 2003; 2009). During subsequent fruit development, however, GA biosynthesis in the pericarp quickly becomes dependent on the presence of developing seeds (Ozga *et al.*, 2003). Growth of de-seeded pea pericarps can be rescued by application of either GA_1 (Ozga and Reinecke, 1999) or the auxin species 4-chloroindole-3-acetic acid (4-Cl-IAA) (Reinecke *et al.*, 1995). Regulation of pea pericarp GA biosynthesis by seeds appears to be mediated through 4-Cl-IAA, treatment with which stimulates GA biosynthesis in the pericarp through up-regulation of both *PsGA20ox1* and *PsGA3ox1* and down-regulation of the C_{19}-GA catabolic gene *PsGA2ox2* (Ozga *et al.*, 2003; 2009). Interestingly, treatment with another auxin species found in pea,

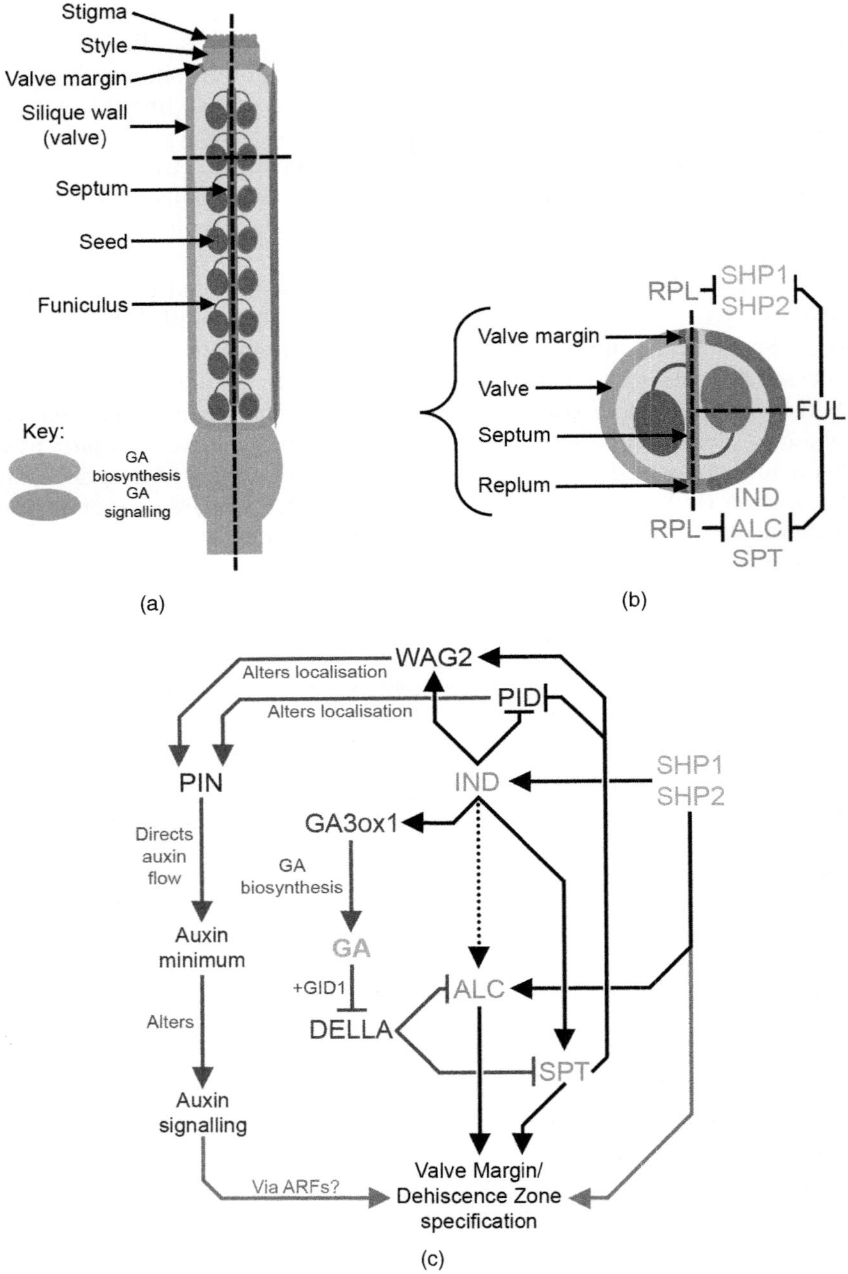
Stigma
Style
Valve margin
Silique wall
(valve)
Septum
Seed
Funiculus
Key:
GA
biosynthesis
GA
signalling
(a)
RPL
SHP1
SHP2
Valve margin
Valve
Septum
Replum
FUL
IND
RPL
ALC
SPT
(b)
WAG2
Alters localisation
PID
Alters localisation
PIN
IND
SHP1
SHP2
GA3ox1
Directs
auxin
flow
GA
biosynthesis
Auxin
minimum
GA
+GID1
ALC
DELLA
Alters
SPT
Auxin
signalling
Via ARFs?
Valve Margin/
Dehiscence Zone
specification
(c)

indole-3-acetic acid (IAA), does not significantly regulate expression of GA biosynthesis genes in the pea pericarp (Reinecke *et al.*, 1995; Ozga *et al.*, 2003; 2009) and in fact inhibits GA-dependent growth through stimulation of ethylene biosynthesis (Johnstone *et al.*, 2005). In contrast, 4-Cl-IAA treatment apparently reduces the sensitivity of pericarp tissues to ethylene (Johnstone *et al.*, 2005). These observations suggest separable signalling pathways in pea pericarps for these two naturally occurring auxin species.

Similarly, developing *Arabidopsis* siliques also directly synthesise GA, with *GA3ox* gene expression reported in the septum, valve margins and ovule funiculi (Figure 11.4a and b; Hu *et al.*, 2008; Arnaud *et al.*, 2010). *DELLA* expression patterns have not been resolved to the level of individual silique or seed tissues, but all five paralogues are expressed during silique development, with indications of differential expression patterns (Fuentes *et al.*, 2012). GA signalling regulates both silique growth and tissue differentiation processes. Mutant analysis in *Arabidopsis* showed a direct effect of *GA20ox* genes in promoting silique elongation, independent of the number of developing seeds (Rieu *et al.*, 2008b; Plackett *et al.*, 2012), and a similar seed number-independent effect on pericarp growth was recently reported in transgenic pea over-expressing *PsGA3ox1* (Reinecke *et al.*, 2013). Further genetic dissection of the *Arabidopsis GA3ox* gene family found

Figure 11.4 Gibberellin regulation of *Arabidopsis* silique development. (a) Diagrammatic longitudinal cross-section of a developing *Arabidopsis* silique, showing major organs/tissue types. Predicted sites of GA biosynthesis (green) and GA signalling (purple) are shown. Sites of GA biosynthesis are based on expression patterns of *AtGA3ox* GUS reporter lines (Mitchum *et al.*, 2006; Hu *et al.*, 2008; Arnaud *et al.*, 2010). Sites of GA signalling are predicted from phenotypes exhibited by the DELLA global mutant (Fuentes *et al.*, 2012). (b) Diagrammatic cross-section through an *Arabidopsis* silique, detailing sites of GA biosynthesis in the silique wall and expression patterns of the transcription factors regulating specification of the valve margin, *INDEHISCENT* (*IND*), *ALCATRAZ* (*ALC*), *SPATULA* (*SPT*) and *SHATTERPROOF1* (*SHP1*) and *-2*. Expression of these genes is restricted to the nascent valve margin by the action of REPLUMLESS (RPL) and FRUITFULL (FUL), which are expressed in the neighbouring replum and valve wall, respectively (Ferrándiz *et al.*, 2000; Roeder *et al.*, 2003). (c) Interaction between the GA, auxin and INDEHISCENT regulatory pathways specifying valve margin and dehiscence zone identity. Black lines indicate transcriptional regulation, red lines indicate protein action or more complex effects, where specified. Grey lines indicate indirect action, and dotted lines indicate where interactions have not been directly demonstrated. Arrowheads and bars indicate positive and negative regulation of downstream targets, respectively. *AtGA3ox1* expression is directly up-regulated by IND (Arnaud *et al.*, 2010), promoting local GA biosynthesis in the valve margin. GA signalling relieves DELLA protein repression of ALC and SPT activity, which promote valve margin specification through both auxin-dependent and auxin-independent pathways. Valve margin specification is also promoted by action of *SHP1* and *-2*, through up-regulation of *IND* and *ALC* and through an independent pathway (Liljegren *et al.*, 2000; 2004). Figure element c is adapted from Kay *et al.* (2012). Reproduced with permission from John Wiley & Sons. (See insert for colour representation of this figure.)

that GA synthesised in silique tissues and GA derived from seeds both contribute to silique elongation (Hu *et al.*, 2008). Analysis of DELLA *global* (L*er*) mutant siliques indicates that GA signalling promotes cell expansion in all cell layers within the silique wall (Figure 11.4a; Fuentes *et al.*, 2012). Interestingly, GA signalling also promoted cell division in *global* mutant siliques, but only in specific cell layers (Fuentes *et al.*, 2012). Analysis of a similar DELLA loss-of-function mutant in tomato, *procera*, suggests that GA signalling also promotes endoreduplication within tomato pericarp cells (Carrera *et al.*, 2012). Surprisingly, Fuentes *et al.* (2012) detected a potentially DELLA-independent growth-promoting effect of GA signalling. The mechanism through which this putative response functions remains unclear, but requires the GID1 receptor, the 26S proteasome and SPATULA (SPT), a growth-repressing bHLH transcription factor known to interact with GA signalling in other developmental contexts (Josse *et al.*, 2011), as well as binding with DELLA proteins *in vitro* (Gallego-Bartolomé *et al.*, 2010).

GA-deficient *Arabidopsis* siliques fail to differentiate a separation layer within the valve margin dehiscence zone, and in consequence display defects in silique dehiscence (Arnaud *et al.*, 2010). The dehiscence zone is specified by the actions of the bHLH transcription factors IND and SPT, which acts downstream of IND, to generate a localised auxin minimum (Liljegren *et al.*, 2004; Sorefan *et al.*, 2009; Girin *et al.*, 2011). ALCATRAZ (ALC) also acts downstream of IND, partially redundantly with SPT (Groszmann *et al.*, 2011), and is required to specify the separation layer within the dehiscence zone (Rajani and Sundaresan, 2001). Expression of *IND* is spatially regulated by the actions of FRUITFULL (FUL) and REPLUMLESS (RPL) (Figure 11.4b; Ferrándiz *et al.*, 2000; Roeder *et al.*, 2003). GA deficiency phenocopies the valve margin defects observed in *alc* loss-of-function mutants, and DELLA proteins bind to ALC *in vitro* (Arnaud *et al.*, 2010). *GA3ox1* expression in the valve margins is directly up-regulated by IND (Arnaud *et al.*, 2010), and it is proposed that IND acts to promote SPT and ALC function through GA signalling by triggering degradation of DELLA protein (Figure 11.4c; Arnaud *et al.*, 2010). Interestingly, SHP1 and SHP2, which also specify valve-margin identity through a separate pathway (Liljegren *et al.*, 2000), have been found to influence GA responses in pollen tube growth (see Section 11.3.2), but in this context any interaction with GA signalling remains to be determined. Tomato fruits do not dehisce at maturity, and to date much greater attention has been paid to the roles of auxin and ethylene in fruit ripening (reviewed in Seymour *et al.*, 2013). It would be very interesting to determine whether the GA-IND module is conserved between *Arabidopsis* and tomato fruit development.

Comparison of subsequent fruit development between *Arabidopsis* and tomato identifies an interesting difference. In *Arabidopsis*, *della global* mutant parthenocarpic siliques do not respond to auxin treatment (Fuentes *et al.*, 2012), suggesting that all auxin responses are mediated through GA signalling, but tomato *procera* (DELLA-deficient) parthenocarpic fruits show increased growth under auxin treatment (Carrera *et al.*, 2012). GA and auxin treatments also induce differing patterns of parthenocarpic growth in tomato (Serrani *et al.*, 2007a), including the development of pseudoembryos in response to auxin, but not GA. As such, not all regulation of tomato fruit growth by auxin is GA-dependent. In pea, 4-Cl-IAA also has GA-independent effects on pericarp growth (van Huizen *et al.*, 1996). *Solanum lycopersicum AUXIN RESPONSE FACTOR 7* (*SlARF7*), a negative regulator of parthenocarpic fruit growth in tomato (de Jong *et al.*, 2009b) appears to antagonise some of the downstream responses to both auxin and GA signalling during fruit development (de Jong *et al.*, 2011). Expression of this gene is negatively regulated by pollination and auxin treatment, but not GA (de Jong *et al.*, 2009b), suggesting that cross-talk between auxin and GA signalling during fruit development is likely to be highly complex.

11.4.2 Embryo and seed development

The transition from ovule to seed development is initiated by fertilisation of the egg cell within the female gametophyte, which develops into an embryonic sporophyte (reviewed in West and Harada, 1993). The second fertilisation event, that of the central cell, results in development of the endosperm within the seed. The embryo is dependent on nutrients mobilised from the endosperm during its early development and, in many cases, also during germination (reviewed in Lopez and Larkins, 1993; see Chapter 9). The maternally-derived ovule tissues surrounding the female gametophyte develop into the seed coat (testa). Embryo and seed development have been classified into two broad stages: morphogenesis (establishing embryo polarity, the embryo axis comprising the root and shoot apical meristems (RAM and SAM) and the immature cotyledons) and maturation (accumulation of storage reserves within the embryo and/or endosperm, followed by desiccation and entry of the embryo into a dormant state) (West and Harada, 1993).

The importance of GA to seed development can be inferred from transgenic and mutant studies in *Arabidopsis*, tomato and pea, where depletion of bioactive GA from seed tissues causes abortion at early stages of development (Swain *et al.*, 1993; Singh *et al.*, 2002; Olimpieri *et al.*, 2011). In the case of pea, seed abortion is apparently linked to GA depletion specifically in the embryo and/or endosperm rather than the testa and/or surrounding fruit tissues. Fertilisation of the lh^i GA-deficient mutant with wild-type pollen complements the abortion phenotype (Swain *et al.*, 1993), while abortion of homozygous lh^i seeds increases even if developing on a phenotypically wild-type heterozygous parent (Swain *et al.*, 1995). Early abortion correlates with a peak in bioactive GA levels during this phase of pea seed development, but has not been associated with any particular stage of embryo development (Swain *et al.*, 1993).

GA biosynthesis (as predicted by promoter *GA3ox::GUS* expression) has been reported in both the embryo and endosperm of developing *Arabidopsis* and pea seeds (Hu *et al.*, 2008; Nadeau *et al.*, 2011). Embryonic *AtGA3ox* expression occurs from the heart stage onwards, with separate tissues expressing different paralogues: *AtGA3ox1* is expressed in the nascent SAM, *AtGA3ox2* in the nascent RAM, and *AtGA3ox3* in two zones subtending the developing cotyledons either side of the embryo axis (Mitchum *et al.*, 2006; Hu *et al.*, 2008). A detailed study of developing pea seeds also identified *PsGA3ox* expression and accumulation of bioactive GA in the embryo axis, but not the developing cotyledons (Nadeau *et al.*, 2011). The localisation and timing of the accumulation of bioactive GA in the developing embryo appears to be tightly regulated. In pea the quantities of bioactive GA synthesised in seed tissues are limited through 2β-hydroxylation activity (Nadeau *et al.*, 2011). Interestingly, *PsGA2ox* expression was detected most strongly in the endosperm and seed coat, but was also present in the embryo axis and cotyledons later in development, where reduced GA biosynthesis might promote the transition from embryo growth to maturation (Nadeau *et al.*, 2011). In *Arabidopsis AtGA3ox2* expression is actively repressed in certain tissues of the developing embryo by FUSCA3 (FUC3), through direct binding to the *AtGA3ox2* promoter, and LEAFY COTYLEDON2 (LEC2), through an indirect mechanism (Curaba *et al.*, 2004). These two transcription factors are important regulators of embryo development during both embryogenesis and seed maturation (reviewed in Harada, 2001): in the absence of these genes, *AtGA3ox2* expression was detected in embryos from the heart and torpedo stages, respectively, initially with increased expression in root precursor cells, but later expanding into the epidermis and vascular tissues of the embryo (Curaba *et al.*, 2004). The significance of restricting *GA3ox* expression from these tissues during embryo development is as yet unclear.

In contrast to *Arabidopsis* vegetative tissues, which almost exclusively synthesise GA_4 (the product of the non-13-hydroxylation branch of the GA biosynthesis pathway; Talon *et al.*, 1990) (see Chapter 2), in developing *Arabidopsis* seeds GA_1 (the 13-hydroxylated equivalent) was found in equal abundance to GA_4 (Curaba *et al.*, 2004). Interestingly, loss of either *LEC2* or *FUS3* causes differential qualitative changes to GA biosynthesis in the embryo, resulting in an increase in GA_1 or GA_4, respectively (Curaba *et al.*, 2004). A study of seed development in grain-producing legumes found similarly mixed GA profiles in three out of the four species investigated (Slater *et al.*, 2013). In contrast, there is evidence in pea for preferential biosynthesis of GA_1, with the GA_4 precursor GA_9 being specifically targeted for inactivation by GA2ox enzymes, resulting in the absence of GA_4 (Nadeau *et al.*, 2011). The functional significance of synthesising different bioactive GA species in seed tissues is as yet unknown.

During seed development GA is also synthesised within the endosperm (Hu *et al.*, 2008; Nadeau *et al.*, 2011). *AtGA3ox4* is expressed throughout the early *Arabidopsis* endosperm shortly after fertilisation, becoming localised to the chalazal endosperm by the heart stage of embryo development (Hu *et al.*, 2008). In barley seeds it has been shown that the GA-responsive transcription factor HvGAMYB is expressed in the developing endosperm, where it is capable of regulating the expression of endosperm-specific genes (Diaz *et al.*, 2002). *Arabidopsis* and pea represent two different endosperm types: cellularised and liquid, respectively. GA-deficient *lh*^{i} pea seeds that do not abort grow more slowly and are smaller at maturity than wild-type (Swain *et al.*, 1993; 1995), which might reflect regulation of nutrient mobilisation or uptake by GA signalling (Swain *et al.*, 1997). Conversely, loss of DELLA activity in *Arabidopsis* seeds causes increased seed weight at maturity and is also associated with altered seed oil content (Li *et al.*, 2013). In pea, the testa has also been shown to synthesise bioactive GA, where it is thought to promote mobilisation/digestion of accumulated starch deposits for absorption by the developing embryo in conjunction with GA-induced alterations in seed coat morphology to facilitate nutrient transfer (Nadeau *et al.*, 2011). An implication of the *lh*^{i} reciprocal cross evidence outlined in the previous discussion of seed abortion is that GA synthesised in the testa cannot easily migrate to the embryo, suggesting that GA biosynthesis and signalling is regulated independently in these two tissues.

The final maturation stages of seed development involve a cessation of embryo growth, seed desiccation and entry of the embryo into dormancy, a process promoted by the phytohormone abscisic acid (ABA) to prevent precocious germination (see Chapter 9). In GA-deficient *gib1* tomato mutants, morphological indicators of the transition from embryogenesis to maturation are delayed, though probably as an indirect consequence of slower embryo development (de Castro and Hilhorst, 2006). A reduction in bioactive GA levels in seed tissues during this late phase of development has been observed in several species (White *et al.*, 2000; Nadeau *et al.*, 2011), whilst conversely a peak in ABA concentration late in seed development is associated with the onset of maturation/dormancy (White *et al.*, 2000; Weber *et al.*, 2005; de Castro and Hilhorst, 2006; Slater *et al.*, 2013).

The maintenance of dormancy by ABA and the antagonistic action of GA during germination is well-documented (see Chapter 9), and this also appears to be the case during seed maturation. ABA-deficient maize (*Zea mays*) kernels fail to enter dormancy and instead germinate during kernel development (Neill *et al.*, 1986). Inhibiting GA biosynthesis in this background suppresses precocious germination (White *et al.*, 2000), suggesting that the balance of ABA and GA levels in developing seeds regulates entry into maturation/dormancy. In maturing tomato seeds, GA deficiency has no significant effect on the timing or magnitude of peak seed ABA content (de Castro and Hilhorst, 2006), while there is evidence from mature *Arabidopsis* seeds that ABA synthesis is actively promoted by the presence of DELLA protein (Lee *et al.*, 2010). In *Arabidopsis*, the endosperm is a major contributor of ABA to maintaining seed dormancy under favourable germination conditions (Bethke *et al.*, 2007; Lee *et al.*, 2010), but at present it is unknown from what tissue the signal(s) to induce seed dormancy originate.

Acknowledgements

ARGP was supported by a Rothamsted quota studentship and is currently funded by the European Research Council; additional support was from the Biotechnology and Biological Sciences Research Council of the United Kingdom, for funding to ZAW's lab and strategic support to Rothamsted Research.

References

Achard, P., Herr, A., Baulcombe, D.C. and Harberd, N.P. (2004). Modulation of floral development by a gibberellin-regulated microRNA. *Development* **131**, 3357–3365.

Airoldi, C.A. (2010). Determination of sexual organ development. *Sexual Plant Reproduction* **23**, 53–62.

Alonso-Peral, M.M., Li, J., Li, Y. *et al.* (2010). The microRNA159-regulated *GAMYB-like* genes inhibit growth and promote programmed cell death in in Arabidopsis. *Plant Physiology* **154**, 757–771.

Amasino, R. (2010). Seasonal and developmental timing of flowering. *Plant Journal* **61**, 1001–1013.

An, H., Roussot, C., Suárez-López, P. *et al.* (2004). CONSTANS acts in the phloem to regulate a systemic signal that induces photoperiodic flowering of *Arabidopsis*. *Development* **131**, 3615–2626.

Arnaud, N., Girin, T., Sorefan, K. *et al.* (2010). Gibberellins control fruit patterning in *Arabidopsis thaliana*. *Genes and Development* **24**, 2127–2132.

Aya, K., Ueguchi-Tanaka, M., Kondo, M. *et al.* (2009). Gibberellin modulates anther development in rice via the transcriptional regulation of GAMYB. *The Plant Cell* **21**, 1453–1472.

Aya, K., Hiwatashi, Y., Kojima, M. *et al.* (2011). The gibberellin perception system evolved to regulate a pre-existing GAMYB-mediated system during land plant evolution. *Nature Communications* **2**, 544.

Bangerth, K.F. (2009). Floral induction in mature, perennial angiosperm fruit trees Similarities and discrepancies with annual/biennial plants and the involvement of plant hormones. *Scientia Horticulturae* **122**, 153–163.

Bensen, R.J., Johal, G.S., Crane, V.C. *et al.* (1995). Cloning and characterization of the maize *An1* gene. *The Plant Cell* **7**, 75–84.

Bethke, P.C., Libourel, I.G.L., Aoyama, N. *et al.* (2007). The Arabidopsis aleurone layer responds to nitric oxide, gibberellin, and abscisic acid and is sufficient and necessary for seed dormancy. *Plant Physiology* **143**, 1173–1188.

Blázquez, M.A., Soowal, L.N., Lee, I. and Weigel, D. (1997). *LEAFY* expression and flower initiation in *Arabidopsis*. *Development* **124**, 3835–3844.

Blázquez, M.A., Green, R., Nilsson, O. *et al.* (1998). Gibberellins promote flowering of *Arabidopsis* by activating the *LEAFY* promoter. *The Plant Cell* **10**, 791–800.

Borner, R., Kampmann, G., Chandler, J. *et al.* (2000). A MADS domain gene involved in the transition to flowering in *Arabidopsis*. *The Plant Journal* **24**, 591–599.

Brown, S.W. (1949). Endomitosis in the tapetum of tomato. *American Journal of Botany* **36**, 703–716.

Carbonell-Bejerano, P., Urbez, C., Carbonell, J. *et al.* (2010). A fertilization-independent developmental program triggers partial fruit development and senescence processes in pistils of Arabidopsis. *Plant Physiology* **154**, 163–172.

Carbonell-Bejerano, P., Urbez, C., Granell, A. *et al.* (2011). Ethylene is involved in pistil fate by modulating the onset of ovule senescence and the GA-mediated fruit set in *Arabidopsis*. *BMC Plant Biology* **11**, 84.

Carrera, E., Ruiz-Rivero, O., Peres, L.E.P. *et al.* (2012). Characterization of the *procera* tomato mutant shows novel functions of the SlDELLA protein in the control of flower morphology, cell division and expansion, and the auxin-signaling pathway during fruit-set and development. *Plant Physiology* **160**, 1581–1596.

Castillejo, C. and Pelaz, S. (2008). The balance between CONSTANS and TEMPRANILLO activities determines FT expression to trigger flowering. *Current Biology* **18**, 1338–1343.

Cecconi, F., Gaetani, M., Lenzi, C. and Durante, M. (2002). The sunflower dwarf mutant *dw1*: effects of gibberellic acid treatment. *Helia* **25**, 161–166.

Cheng, H., Qin, L., Lee, S. *et al.* (2004). Gibberellin regulates *Arabidopsis* floral development via repression of DELLA protein function. *Development* **131**, 1055–1064.

Cheng, H., Song, S., Xiao, L. *et al.* (2009). Gibberellin acts through jasmonate to control the expression of *MYB21*, *MYB24* and *MYB57* to promote stamen filament growth in *Arabidopsis*. *PLoS Genetics* **5**, e1000440.

Chhun, T., Aya, K., Asano, K. *et al.* (2007). Gibberellin regulates pollen viability and pollen tube growth in rice. *The Plant Cell* **19**, 3876–3888.

Claeys, H., Skirycz, A., Maleux, K. and Inzé, D. (2012). DELLA signalling mediates stress-induced cell differentiation in Arabidopsis leaves through modulation of Anaphase-Promoting Complex/cyclosome activity. *Plant Physiology* **159**, 739–747.

Corbesier, L., Vincent, C., Jang, S. *et al.* (2007). FT protein movement contributes to long-distance signalling in floral induction of *Arabidopsis*. *Science* **316**, 1030–1033.

Curaba, J., Moritz, T., Blervaque, R. *et al.* (2004). *AtGA3ox2*, a key gene responsible for bioactive gibberellin biosynthesis, is regulated during embryogenesis by *LEAFY COTYLEDON2* and *FUSCA3* in Arabidopsis. *Plant Physiology* **136**, 3660–3669.

de Castro, R.D. and Hilhorst, H.W.M. (2006). Hormonal control of seed development in GA- and ABA-deficient tomato (*Lycopersicon esculentum* Mill. cv. Moneymaker) mutants. *Plant Science* **170**, 462–470.

de Jong, M., Mariani, C. and Vriezen, W.H. (2009a). The role of auxin and gibberellin in tomato fruit set. *Journal of Experimental Botany* **60**, 1523–1532.

de Jong, M., Wolters-Arts, M., Feron, R. *et al.* (2009b). The *Solanum lycopersicum* auxin response factor 7 (SlARF7) regulates auxin signaling during tomato fruit set and development. *The Plant Journal* **57**, 160–170.

de Jong, M., Wolters-Arts, M., García-Martínez, J.L. *et al.* (2011). The *Solanum lycopersicum* AUXIN RESPONSE FACTOR 7 (SlARF7) mediates cross-talk between auxin and gibberellin signalling during tomato fruit set and development. *Journal of Experimental Botany* **62**, 617–626.

Delgado, R., Casamayor, R., Rodriguez, J.L. *et al.* (1986). Paclobutrazol effects on oranges under tropical conditions. *Acta Horticulturae* **179**, 537–544.

Diaz, I., Vicente-Carbajosa, J., Abraham, Z. *et al.* (2002). The GAMYB protein from barley interacts with the DOF transcription factor BPBF and activates endosperm-specific genes during seed development. *The Plant Journal* **29**, 453–464.

Dorcey, E., Urbez, C., Blázquez, M.A. *et al.* (2009). Fertilization-dependent auxin response in ovules triggers fruit development through the modulation of gibberellin metabolism in *Arabidopsis*. *The Plant Journal* **58**, 318–332.

Eriksson, S., Böhlenius, H., Moritz, T. and Nilsson, O. (2006). GA_4 is the active gibberellin in the regulation of *LFY* transcription and *Arabidopsis* floral transition. *The Plant Cell* **18**, 2172–2181.

Ferrándiz, C., Liljegren, S.J. and Yanofsky, M.F. (2000). Negative regulation of the *SHATTERPROOF* genes by FRUITFULL during *Arabidopsis* fruit development. *Science* **289**, 436–438.

Fuentes, S., Ljung, K., Sorefan, K. *et al.* (2012). Fruit growth in *Arabidopsis* occurs via DELLA-dependent and DELLA-independent gibberellin responses. *The Plant Cell* **24**, 3982–3996.

Gallego-Bartolomé, J., Minguet, E.G., Marín, J.A. *et al.* (2010). Transcriptional diversification and functional conservation between DELLA proteins in *Arabidopsis*. *Molecular Biology and Evolution* **27**, 1247–1256.

Galvão, V.C., Horrer, D., Küttner, F. and Schmid, M. (2012). Spatial control of flowering by DELLA proteins in *Arabidopsis thaliana*. *Development* **139**, 4072–4082.

Gillaspy, G., Ben-David, H. and Gruissem, W. (1993). Fruits: A developmental perspective. *The Plant Cell* **5**, 1439–1451.

Girin, T., Paicu, T., Stephenson, P. *et al.* (2011). INDEHISCENT and SPATULA interact to specify carpel and valve margin tissue and thus promote seed dispersal in *Arabidopsis*. *The Plant Cell* **23**, 3641–3653.

Goldberg-Moeller, R., Shalom, L., Shlizerman, L. *et al.* (2013). Effects of gibberellin treatment during flowering induction period on global gene expression and the transcription of flowering control genes in *Citrus* buds. *Plant Science* **198**, 46–57.

Gómez-Mena, C., de Folter, S., Costa, M.M.R. *et al.* (2005). Transcriptional program controlled by the floral homeotic gene *AGAMOUS* during early organogenesis. *Development* **132**, 429–438.

Goto, N. and Pharis, R.P. (1999). Role of gibberellins in the development of floral organs of the gibberellin-deficient mutant, *ga1-1*, of *Arabidopsis thaliana*. *Canadian Journal of Botany* **77**, 944–954.

Griffiths, J., Murase, K., Rieu, I. *et al.* (2006). Genetic characterization and functional analysis of the GID1 gibberellin receptors in *Arabidopsis*. *The Plant Cell* **18**, 3399–3414.

Groszmann, M., Paicu, T., Alvarez, J.P. *et al.* (2011). *SPATULA* and *ALCATRAZ* are partially redundant, functionally diverging bHLH genes required for *Arabidopsis* gynoecium and fruit development. *The Plant Journal* **68**, 816–829.

Hanano, S. and Goto, K. (2011). *Arabidopsis* TERMINAL FLOWER1 is involved in the regulation of flowering time and inflorescence development through transcription repression. *The Plant Cell* **23**, 3172–3184.

Harada, J.J. (2001). Role of *Arabidopsis LEAFY COTYLEDON* genes in seed development. *Journal of Plant Physiology* **158**, 405–409.

Hepler, P.K., Vidali, L. and Cheung, A.Y. (2001). Polarized cell growth in higher plants. *Annual Review of Cell and Developmental Biology* **17**, 159–187.

Hirano, K., Nakajima, M., Asano, K. *et al.* (2007). The GID1-mediated gibberellin perception mechanism is conserved in the lycophyte *Selaginella moellendorffii* but not in the bryophyte *Physcomitrella patens*. *The Plant Cell* **19**, 3058–3079.

Hirano, K., Aya, K., Hobo, T. *et al.* (2008). Comprehensive transcriptome analysis of phytohormone biosynthesis and signaling genes in microspore/pollen and tapetum of rice. *Plant and Cell Physiology* **49**, 1429–1450.

Hisamatsu, T. and King, R. (2008). The nature of floral signals in *Arabidopsis*. II. Roles for *FLOWERING LOCUS T* (*FT*) and gibberellin. *Journal of Experimental Botany* **59**, 3821–3829.

Honys, D. and Twell, D. (2004). Transcriptome analysis of haploid male gametophyte development in *Arabidopsis*. *Genome Biology* **5**, R85.

Hu, J., Mitchum, M.G., Barnaby, N. *et al.* (2008). Potential sites of bioactive gibberellin production during reproductive growth in *Arabidopsis*. *The Plant Cell* **20**, 320–336.

Huijser, P. and Schmid, M. (2011). The control of developmental phase transitions in plants. *Development* **138**, 4117–4129.

Huysmans, S. (1998). Orbicules in angiosperms: Morphology, function, distribution and relation with tapetum types. *The Botanical Review* **64**, 240–272.

Ikeda, A., Ueguchi-Tanaka, M., Sonoda, Y. *et al.* (2001). *slender* rice, a constitutive gibberellin response mutant, is caused by a null mutation of the *SLR1* gene, an ortholog of the height-regulating gene *GAI/RGA/RHT/D8*. *The Plant Cell* **13**, 999–1010.

Irish, V.F. (2010). The flowering of *Arabidopsis* flower development. *The Plant Journal* **61**, 1014–1028.

Ito, T. (2011). Coordination of flower development by homeotic master regulators. *Current Opinion in Plant Biology* **14**, 53–59.

Ito, T., Ng, K.-H., Lim, T.-S. *et al.* (2011). The homeotic protein AGAMOUS controls late stamen development by regulating a jasmonate biosynthetic gene in *Arabidopsis*. *The Plant Cell* **19**, 3516–3529.

Izhaki, A., Borochov, A., Zamski, E. and Weiss, D. (2002). Gibberellin regulates post-microsporogenesis processes in petunia anthers. *Physiologia Plantarum* **115**, 442–447.

Jacobsen, S.E. and Olszewski, N.E. (1991). Characterization of the arrest in anther development associated with gibberellin deficiency of the *gib-1* mutant of tomato. *Plant Physiology* **97**, 409–414.

Jacobsen, S.E. and Olszewski, N.E. (1993). Mutations at the *SPINDLY* locus of *Arabidopsis* alter gibberellin signal transduction. *The Plant Cell* **5**, 887–896.

Jaeger, K.E. and Wigge, P.A. (2007). FT protein acts as a long-range signal in *Arabidopsis*. *Current Biology* **17**, 1050–1054.

Jang, S., Torti, S. and Coupland, G. (2009). Genetic and spatial interactions between *FT*, *TSF* and *SVP* during the early stages of floral induction in *Arabidopsis*. *The Plant Journal* **60**, 614–625.

Jasinski, S., Piazza, P., Craft, J. *et al.* (2005). KNOX action in *Arabidopsis* is mediated by coordinate regulation of cytokinin and gibberellin activities. *Current Biology* **15**, 1560–1565.

Johnston, A.J., Meier, P., Gheyselinck, J. *et al.* (2007). Genetic subtraction profiling identifies genes essential for *Arabidopsis* reproduction and reveals interaction between the female gametophyte and maternal sporophyte. *Genome Biology* **8**, R204.

Johnstone, M.M.G., Reinecke, D.M. and Ozga, J.A. (2005). The auxins IAA and 4-Cl-IAA differentially modify gibberellin action via ethylene response in developing pea fruit. *Journal of Plant Growth Regulation* **24**, 214–225.

Jones-Rhoades, M.W., Borevitz, J.O. and Preuss, D. (2007). Genome-wide expression profiling of the *Arabidopsis* female gametophyte identifies families of small, secreted proteins. *PLoS Genetics* **3**, e171.

Josse, E.-M., Gan, Y., Bou-Torrent, J. *et al.* (2011). A DELLA in disguise: SPATULA restrains the growth of the developing *Arabidopsis* seedling. *The Plant Cell* **23**, 1337–1351.

Jung, J.-H., Ju, Y., Seo, P.J. *et al.* (2012). The SOC1-SPL module integrates photoperiod and gibberellic acid signals to control flowering time in *Arabidopsis*. *The Plant Journal* **69**, 577–588.

Kaneko, M., Inukai, Y., Ueguchi-Tanaka, M. *et al.* (2004). Loss-of-function mutants of the rice *GAMYB* gene impair α-amylase expression in aleurone and flower development. *The Plant Cell* **16**, 33–44.

Kawanabe, T., Ariizumi, T., Yamada-Kawai, M. *et al.* (2006). Abolition of the tapetum suicide program ruins microsporogenesis. *Plant and Cell Physiology* **47**, 784–787.

Kay, P., Groszmann, M., Ross, J.J. *et al.* (2013). Modifications of a conserved regulatory network involving INDEHISCENT controls multiple aspects of reproductive tissue development in *Arabidopsis*. *New Phytologist* **197**, 73–87.

King, R.W., Moritz, T., Evans, L.T. *et al.* (2001). Long-day induction of flowering in *Lolium temulentum* involves sequential increases in specific gibberellins at the shoot apex. *Plant Physiology* **127**, 624–632.

King, R.W., Mander, L.N., Asp, T. *et al.* (2008). Selective deactivation of gibberellins below the shoot apex is critical to flowering but not to stem elongation of *Lolium*. *Molecular Plant* **1**, 295–307.

Kobayashi, M., Yamaguchi, I., Murofushi, N. *et al.* (1984). Endogenous gibberellins in immature seeds and flowering ears of rice. *Agricultural and Biological Chemistry* **48**, 2725–2729.

Kobayashi, M., Yamaguchi, I., Murofushi, N. *et al.* (1988). Fluctuation and localisation of endogenous gibberellins in rice. *Agricultural and Biological Chemistry* **52**, 1189–1194.

Koltunow, A.M., Truettner, J., Cox, K.H. *et al.* (1990). Different temporal and spatial gene expression patterns occur during anther development. *The Plant Cell* **2**, 1201–1224.

Koornneef, M. and Van der Veen, J.H. (1980). Induction and analysis of gibberellin sensitive mutants in *Arabidopsis thaliana* (L) heyhn. *Theoretical and Applied Genetics* **58**, 257–263.

Kurogochi, S., Murofushi, N., Ota, Y. and Takahashi, N. (1979). Identification of gibberellins in the rice plant and quantitative changes of gibberellin A_{19} throughout its life-cycle. *Planta* **146**, 185–191.

Kurokura, T., Mimida, N., Battet, N.H. and Hytönen, T. (2013). The regulation of seasonal flowering in the Rosaceae. *Journal of Experimental Botany* **64**, 4131–4141.

Lamb, R.S., Hill, T.A., Tan, Q.K.-G. and Irish, V.F. (2002). Regulation of *APETALA3* floral homeotic gene expression by meristem identity genes. *Development* **129**, 2079–2086.

Lanahan, M.B. and Ho, T.-H.D. (1988). Slender barley: a constitutive gibberellin-response mutant. *Planta* **175**, 107–114.

Lee, J. and Lee, I. (2010). Regulation and function of SOC1, a flowering pathway integrator. *Journal of Experimental Botany* **61**, 2247–2254.

Lee, J., Oh, M., Park, H. and Lee, I. (2008). SOC1 translated to the nucleus by interaction with AGL24 directly regulates *LEAFY*. *The Plant Journal* **55**, 832–843.

Lee, P.K., Piskurewicz, U., Turečková, V. *et al.* (2010). A seed coat bedding assay shows that RGL2-dependent release of abscisic acid by the endosperm controls embryo growth in *Arabidopsis* dormant seeds. *Proceedings of the National Academy of the Sciences USA* **107**, 19108–19113.

Lee, S., Cheng, H., King, K.E. *et al.* (2002). Gibberellin regulates *Arabidopsis* seed germination via *RGL2*, a *GAI/RGA*-like gene whose expression is up-regulated following imbibition. *Genes and Development* **16**, 646–658.

Li, D., Liu, C., Shen, L. *et al.* (2008). A repressor complex governs the integration of flowering signals in *Arabidopsis*. *Developmental Cell* **15**, 110–120.

Li, N., Zhang, D.-S., Liu, H.-S. *et al.* (2006). The rice *TAPETUM DEGENERATION RETARDATION* gene is required for tapetum degradation and anther development. *The Plant Cell* **18**, 2999–3014.

Li, Z., Jiang, Y., Hua, S. *et al.* (2013). Characterization of seed fatty acid accumulation in DELLA mutant lines of *Arabidopsis*. *Plant Growth Regulation* **70**, 27–37.

Liljegren, S.J., Ditta, G.S., Eshed, Y. *et al.* (2000). SHATTERPROOF MADS-box genes control seed dispersal in *Arabidopsis*. *Nature* **404**, 766–770.

Liljegren, S.J., Roeder, A.H.K., Kempin, S.A. *et al.* (2004). Control of fruit patterning in *Arabidopsis* by INDEHISCENT. *Cell* **116**, 843–853.

Lohmann, J.U., Hong, R.L., Hobe, M. *et al.* (2001). A molecular link between stem cell regulation and floral patterning in *Arabidopsis*. *Cell* **105**, 793–803.

Lopez, M.A. and Larkins. B.A. (1993). Endosperm origin, development, and function. *The Plant Cell* **5**, 1383–1399.

Martínez-Fuentes, A., Mesejo, C., Muñoz-Fambuena, N. *et al.* (2013). Fruit load restricts the flowering promotion effect of paclobutrazol in alternate bearing *Citrus* spp. *Scientia Horticulturae* **151**, 122–127.

Mathieu, J., Warthmann, N., Küttner, F. and Schmid, M. (2007). Export of FT protein from phloem companion cells is sufficient for floral induction in *Arabidopsis*. *Current Biology* **17**, 1055–1060.

Mathieu, J., Yant, L.J., Mürdter, F. *et al.* (2009). Repression of flowering by the miR172 target SMZ. *PLoS Biology* **7**, e1000148.

Millar, A.A. and Gubler, F. (2005). The *Arabidopsis GAMYB-like* genes, *MYB33* and *MYB65*, are microRNA-regulated genes that redundantly facilitate anther development. *The Plant Cell* **17**, 705–712.

Mitchum, M.G., Yamaguchi, S., Hanada, A. *et al.* (2006). Distinct and overlapping roles of two gibberellin 3-oxidases in *Arabidopsis* development. *The Plant Journal* **45**, 804–818.

Moon, J., Suh, S.-S., Lee, H. *et al.* (2003). The *SOC1* MADS-box gene integrates vernalization and gibberellin signals for flowering in *Arabidopsis*. *The Plant Journal* **35**, 613–623.

Muñoz-Fambuena, N., Mesejo, C., Gonzáles-Mas, M.C. *et al.* (2012). Gibberellic acid reduces flowering intensity in sweet orange (*Citrus sinensis* (L) Osbeck) by repressing *CiFT* gene expression. *Journal of Plant Growth Regulation* **31**, 529–536.

Murray, F., Kalla, R., Jacobsen, J. and Gubler, F. (2003). A role for HvGAMYB in anther development. *The Plant Journal* **33**, 481–491.

Mutasa-Göttgens, E. and Hedden, P. (2009).Gibberellin as a factor in floral regulatory networks. *Journal of Experimental Botany* **60**, 1979–1989.

Nadeau, C.D., Ozga, J.A., Kurepin, L.V. *et al.* (2011). Tissue-specific regulation of gibberellin biosynthesis in developing pea seeds. *Plant Physiology* **156**, 897–912.

Neill, S.J., Horgan, R. and Parry, A.D. (1986). The carotenoid and abscisic acid content of viviparous kernels and seedlings of *Zea mays* L. *Planta* **169**, 87–96.

Nester, J.E. and Zeevaart, J.A.D. (1988). Flower development in normal tomato and a gibberellin-deficient (*ga-2*) mutant. *American Journal of Botany* **75**, 45–55.

Nomura, T., Hiroshi, M., Hanada, A. *et al.* (2013). Functional analysis of *Arabidopsis* CYP714A1 and CYP714A2 reveals that they are distinct gibberellin modification enzymes. *Plant and Cell Physiology* **54**, 1837–1851.

Olimpieri, I., Caccia, R., Picarella, M.E. *et al.* (2011). Constitutive co-suppression of the *GA 20-oxidase1* gene in tomato leads to severe defects in vegetative and reproductive development. *Plant Science* **180**, 496–503.

Osnato, M., Castillejo, C., Matías-Hernández, L. and Pelaz, S. (2012). *TEMPRANILLO* genes link photoperiod and gibberellin pathways to control flowering in *Arabidopsis*. *Nature Communications* **3**, 808.
Ozga, J.A. and Reinecke, D.M. (1999). Interaction of 4-chloroindole-3-acetic acid and gibberellins in early pea fruit development. *Plant Growth Regulation* **27**, 33–38.
Ozga, J.A. and Reinecke, D.M. (2003). Hormonal interactions in fruit development. *Journal of Plant Growth Regulation* **22**, 73–81.
Ozga, J.A., Yu, J. and Reinecke, D.M. (2003). Pollination-, development-, and auxin-specific regulation of gibberellin 3β-hydroxylase gene expression in pea fruit and seeds. *Plant Physiology* **131**, 1137–1146.
Ozga, J.A., Reinecke, D.M., Ayele, B.T. *et al.* (2009). Developmental and hormonal regulation of gibberellin biosynthesis and catabolism in pea fruit. *Plant Physiology* **150**, 448–462.
Pearce, S., Vanzetti, L.S. and Dubcovsky, J. (2013). Exogenous gibberellins induce wheat spike development under short days only in the presence of *VERNALIZATION1*. *Plant Physiology* **163**, 1433–1445.
Pharis, R.P. and King, R.W. (1985). Gibberellins and reproductive development in seed plants. *Annual Review of Plant Physiology* **36**, 517–568.
Piffanelli, P., Ross, J.H.E. and Murphy, D.J. (1998). Biogenesis and function of the lipidic structures of pollen grains. *Sexual Plant Reproduction* **11**, 65–80.
Plackett, A.R.G. (2012). *The Role of Gibberellin in the Reproductive Development of* Arabidopsis thaliana. PhD thesis, University of Nottingham.
Plackett, A.R.G., Thomas, S.G., Wilson, Z.A. and Hedden, P. (2011). Gibberellin control of stamen development: a fertile field. *Trends in Plant Science* **16**, 568–578.
Plackett, A.R.G., Powers, S.J., Fernandez-Garcia, N. *et al.* (2012). Analysis of the developmental roles of the *Arabidopsis* Gibberellin 20-Oxidases demonstrates that *GA20ox1*, *-2* and *-3* are the dominant paralogues. *The Plant Cell* **24**, 941–960.
Plackett, A.R.G., Ferguson, A.C., Powers, S.J. *et al.* (2014). DELLA activity is required for successful pollen development in the Columbia ecotype of *Arabidopsis*. *New Phytologist* **201**, 825–836.
Porri, A., Torti, S., Romera-Branchat, M. and Coupland, G. (2012). Spatially distinct regulatory roles for gibberellins in the promotion of flowering of *Arabidopsis* under long photoperiods. *Development* **139**, 2198–2209.
Pryer, K.M., Schneider, H., Smith, A.R. *et al.* (2001). Horsetails and ferns are a monophyletic group and the closest living relatives to seed plants. *Nature* **409**, 618–622.
Rademacher, W. (2000). Growth retardants: effects on gibberellin biosynthesis and other metabolic pathways. *Annual Review of Plant Physiology and Plant Molecular Biology* **51**, 501–531.
Rajani, S. and Sundaresan, V. (2001). The *Arabidopsis* myc/bHLH gene ALCATRAZ enables cell separation in fruit dehiscence. *Current Biology* **11**, 1914–1922.
Randoux, M., Jeauffre, J., Thouroude, T. *et al.* (2012). Gibberellins regulate the transcription of the continuous flowering regulator *RoKSN*, a rose *TFL1* homologue. *Journal of Experimental Botany* **63**, 6543– 6554.
Ratcliffe, O.J., Bradley, D.J. and Coen, E.S. (1999). Separation of shoot and floral identity in *Arabidopsis*. *Development* **126**, 1109–1120.
Reeves, P.H. and Coupland, G. (2001). Analysis of flowering time control in Arabidopsis by comparison of double and triple mutants. *Plant Physiology* **126**, 1085–1091.
Reinecke, D.M., Ozga, J.A. and Magnus, V. (1995). Effect of halogen substitution of indole- 3-acetic acid on biological activity in pea fruit. *Phytochemistry* **40**, 1361–1366.

Reinecke, D.M., Wickramarathna, A.D., Ozga, J.A. *et al.* (2013). *Gibberellin 3-oxidase* gene expression patterns influence gibberellin biosynthesis, growth, and development in pea. *Plant Physiology* **163**, 929–945.

Richter, R., Behringer, C., Müller, I.K. and Schwechheimer, C. (2010). The GATA-type transcription factors GNC and GNL/CGA1 repress gibberellin signaling downstream from DELLA proteins and PHYTOCHROME INTERACTING FACTORS. *Genes and Development* **24**, 2093–2104.

Rieu, I., Eriksson, S., Powers, S.J. *et al.* (2008a). Genetic analysis reveals that C_{19}-GA 2-oxidation is a major gibberellin inactivation pathway in *Arabidopsis*. *The Plant Cell* **20**, 2420–2436.

Rieu, I., Ruiz-Rivero, O., Fernandez-Garcia, N. *et al.* (2008b). The gibberellin biosynthetic genes *AtGA20ox1* and *AtGA20ox2* act, partially redundantly, to promote growth and development throughout the *Arabidopsis* life cycle. *The Plant Journal* **53**, 488–504.

Rodrigo, M.J. and García-Martinez, J.L. (1998). Hormonal control of parthenocarpic ovary growth by the apical shoot in pea. *Plant Physiology* **116**, 511–518.

Roeder, A.H.K., Ferrándiz, C. and Yanofsky, M.F. (2003). The role of the REPLUMLESS homeodomain proteins in patterning the *Arabidopsis* fruit. *Current Biology* **13**, 1630–1635.

Ruan, Y.L., Patrick, J.W., Bouzayen, M. *et al.* (2012). Molecular regulation of seed and fruit set. *Trends in Plant Science* **17**, 656–665.

Sakamoto, T., Kobayashi, M., Itoh, H. *et al.* (2001). Expression of a gibberellin 2-oxidase gene around the shoot apex is related to phase transition in rice. *Plant Physiology* **125**, 1508–1516.

Samach, A., Onouchi, H., Gold S.E. *et al.* (2000). Distinct roles of *CONSTANS* target genes in reproductive development of *Arabidopsis*. *Science* **288**, 1613–1616.

Sanders, P.M., Bui, A.Q., Weterings, K. *et al.* (1999). Anther developmental defects in *Arabidopsis thaliana* male-sterile mutants. *Sexual Plant Reproduction* **11**, 297–322.

Sawhney, V.K. (1983). The role of temperature and its relationship with gibberellic acid in the development of floral organs in tomato (*Lycopersicon esculentum*). *Canadian Journal of Botany* **61**, 1258–1265.

Sawhney, V.K. (1992). Floral mutants in tomato: development, physiology and evolutionary implications. *Canadian Journal of Botany* **70**, 701–707.

Schmid, M., Uhlenhaut, N.H., Godard, F. *et al.* (2003). Dissection of floral induction pathways using global expression analysis. *Development* **130**, 6001–6012.

Schmidt, A., Wuest, S.E., Vijverberg, K. *et al.* (2011). Transcriptome analysis of the *Arabidopsis* megaspore mother cell uncovers the importance of RNA helicases for plant germline development. *PLoS Biology* **9**, e1001155.

Scott, R.J., Spielman, M. and Dickinson, H.G. (2004). Stamen structure and function. *The Plant Cell* **16**, S46–S60.

Searle, I. and Coupland, G. (2004). Induction of flowering by seasonal changes in photoperiod. *The EMBO Journal* **23**, 1217–1222.

Serrani, J.C., Fos, M., Atarés, A. and García-Martínez, J.L. (2007a). Effect of gibberellin and auxin on parthenocarpic fruit growth induction in the cv. Micro-Tom of tomato. *Journal of Plant Growth Regulation* **26**, 211–221.

Serrani, J.C., Sanjuán, R., Ruiz-Rivero O. *et al.* (2007b). Gibberellin regulation of fruit set and growth in tomato. *Plant Physiology* **145**, 246–257.

Serrani, J.C., Ruiz-Rivero, O., Fos, M. and García-Martínez, J.L. (2008). Auxin-induced fruit-set in tomato is mediated in part by gibberellins. *The Plant Journal* **56**, 922–934.

Seymour, G.B., Østergaard, L., Chapman, N.H. *et al.* (2013). Fruit development and ripening. *Annual Review of Plant Biology* **64**, 219–241.

Shi, D.-Q. and Yang, W.-C. (2011). Ovule development in *Arabidopsis*: progress and challenge. *Current Opinion in Plant Biology* **14**, 74–80.

Silverstone, A.L., Chang, C.-W., Krol, E. and Sun, T.-P. (1997). Developmental regulation of the gibberellin biosynthetic gene *GA1* in *Arabidopsis thaliana*. *The Plant Journal* **12**, 9–19.

Singh, D.P., Jermakow, A.M. and Swain, S.M. (2002). Gibberellins are required for seed development and pollen tube growth in *Arabidopsis*. *The Plant Cell* **14**, 3133–3147.

Slater, S.M.H., Yuan, H.Y., Lulsdorf, M.M. *et al.* (2013). Comprehensive hormone profiling of the developing seeds of four grain legumes. *Plant Cell Reports* **32**, 1939–1952.

Smyth, D.R., Bowma, J.L. and Meyerowitz, E.M. (1990). Early flower development in *Arabidopsis*. *The Plant Cell* **2**, 755–767.

Song, S., Qi, T., Huang, H. *et al.* (2011). The jasmonate-ZIM domain proteins interact with the R2R3-MYB transcription factors MYB21 and MYB24 to affect jasmonate-regulated stamen development in *Arabidopsis*. *The Plant Cell* **23**, 1000–1013.

Song, S., Qi, T., Huang, H. and Xie, D. (2013). Regulation of stamen development by coordinated actions of jasmonate, auxin and gibberellin in *Arabidopsis*. *Molecular Plant* **6**, 1065–1073.

Sorefan, K., Girin, T., Liljegren, S.J. *et al.* (2009). A regulated auxin minimum is required for seed dispersal in *Arabidopsis*. *Nature* **459**, 583–586.

Sotelo-Silveira, M., Marsch-Martínez, N. and de Folter, S. (2014). Unraveling the signal scenario of fruit set. *Planta* **239**, 1147–1158.

Srikanth, A. and Schmid, M. (2011). Regulation of flowering time: All roads lead to Rome. *Cellular and Molecular Life Sciences* **68**, 2013–2037.

Steffan, J.G., Kang, I.-H., Macfarlane, J. and Drews, G.N. (2007). Identification of genes expressed in the *Arabidopsis* female gametophyte. *The Plant Journal* **51**, 281–292.

Suárez-López, P., Wheatley, K., Robson, F. *et al.* (2001). *CONSTANS* mediates between the circadian clock and the control of flowering in *Arabidopsis*. *Nature* **410**, 1116–1120.

Sundaresan, V. and Alandete-Saez, M. (2010). Pattern formation in miniature: the female gametophyte of flowering plants. *Development* **137**, 179–189.

Swain, S.M., Reid, J.B. and Ross, J.J. (1993). Seed development in *Pisum*: The lh^i allele reduces gibberellin levels in developing seeds, and increases seed abortion. *Planta* **191**, 482–488.

Swain, S.M., Ross, J.J., Reid, J.B. and Kamiya, Y. (1995). Gibberellins and pea seed development, expression of the lh^i, *ls* and le^{5839} mutations. *Planta* **195**, 426–433.

Swain, S.M., Reid, J.B. and Kamiya, Y. (1997). Gibberellins are required for embryo growth and seed development in pea. *The Plant Journal* **12**, 1329–1338.

Swain, S.M., Muller, A.J. and Singh, D.P. (2004). The *gar2* and *rga* alleles increase the growth of gibberellin-deficient pollen tubes in Arabidopsis. *Plant Physiology* **134**, 694–705.

Takada, S and Goto, K. (2003). TERMINAL FLOWER2, an *Arabidopsis* homolog of HETEROCHROMATIN PROTEIN1, counteracts the activation of *FLOWERING LOCUS T* by CONSTANS in the vascular tissues of leaves to regulate flowering time. *The Plant Cell* **15**, 2856–2865.

Talon, M., Koornneef, M. and Zeevaart, J.A.D. (1990). Endogenous gibberellins in *Arabidopsis thaliana* and possible steps blocked in the biosynthetic pathways of the semidwarf *ga4* and *ga5* mutants. *Proceedings of the National Academy of the Sciences USA* **87**, 7983–7987.

Tamaki, S., Matsuo, S., Wong, H.L. *et al.* (2007). Hd3a protein is a mobile flowering signal in rice. *Science* **316**, 1033–1036.

Tanaka, J, Yano, K., Aya, K. *et al.* (2014). Antheridiogen determines sex in ferns via a spatiotemporally split gibberellin synthesis pathway. *Science* **346**, 469–473.

Tang, X., Zhang, Z.-Y., Zhang, W.-J. *et al.* (2010). Global gene profiling of laser-captured pollen mother cells indicates molecular pathways and gene subfamilies involved in rice meiosis. *Plant Physiology* **154**, 1855–1870.

Tsuji, H., Aya, K., Ueguchi-Tanaka, M. *et al.* (2006). GAMYB controls different sets of genes and is differentially regulated by microRNA in aleurone cells and anthers. *The Plant Journal* **47**, 427–444.

Tyler, L., Thomas, S.G., Hu, J. *et al.* (2004). DELLA proteins and gibberellin-regulated seed development and floral development in Arabidopsis. *Plant Physiology* **135**, 1008–1019.

van den Heuvel, K.J.P.T., Barendse, G.W.M. and Wullems, G.J. (2001). Effect of gibberellic acid on cell division and cell elongation in anthers of the gibberellin deficient *gib-1* mutant of tomato. *Plant Biology* **3**, 124–131.

van Huizen, R., Ozga, J.A. and Reinecke, D.M. (1996). Influence of auxin and gibberellin on in vivo protein synthesis during early pea fruit growth. *Plant Physiology* **112**, 53–59.

Vivian-Smith, A. and Koltunow, A.M. (1999). Genetic analysis of growth-regulator-induced parthenocarpy in Arabidopsis. *Plant Physiology* **121**, 437–451.

Vivian-Smith, A., Luo, M., Chaudhury, A. and Koltunow, A. (2001). Fruit development is actively restricted in the absence of fertilization in *Arabidopsis*. *Development* **128**, 2321–2331.

Vizcay-Barrena, G. and Wilson, Z.A. (2006). Altered tapetal PCD and pollen wall development in the *Arabidopsis ms1* mutant. *Journal of Experimental Botany* **57**, 2709–2717.

Vriezen, W.H., Feron, R., Maretto, F. *et al.* (2008). Changes in tomato ovary transcriptome demonstrate complex hormonal regulation of fruit set. *New Phytologist* **177**, 60–76.

Wang, A., Xia, Q., Datla, R. and Selvaraj, G. (2003). The classical Ubisch bodies carry a sporophytically produced structural protein (RAFTIN) that is essential for pollen development. *Proceedings of the National Academy of the Sciences USA* **100**, 14487–14492.

Wang, R., Farrona, S., Vincent, C. *et al.* (2009). *PEP1* regulates perennial flowering in *Arabis alpina*. *Nature* **459**, 423–427.

Warne, T.R. and Hickok, L.G. (1989). Evidence for a gibberellin biosynthetic origin of *Ceratopteris* antheridiogen. *Plant Physiology* **89**, 535–538.

Weber, H., Borisjuk, L. and Wobus, U. (2005). Molecular physiology of legume seed development. *Annual Review of Plant Biology* **56**, 253–279.

Weigel, D., Alvarez, J., Smyth, D.R. *et al.* (1992). *LEAFY* controls floral meristem identity in *Arabidopsis*. *Cell* **69**, 843–859.

Wen, C.-K. and Chang, C. (2002). *Arabidopsis RGL1* encodes a negative regulator of gibberellin responses. *The Plant Cell* **14**, 87–100.

West, M.A.L. and Harada, J.J. (1993). Embryogenesis in higher plants: an overview. *The Plant Cell* **5**, 1361–1369.

White, C.N., Proebsting, W.M., Hedden, P. and Rivin, C.J. (2000). Gibberellins and seed development in maize. I. Evidence that gibberellin/abscisic acid balance governs germination versus maturation pathways. *Plant Physiology* **122**, 1081–1088.

Wilkie, J.D., Sedgely, M. and Olesen, T. (2008). Regulation of floral initiation in horticultural trees. *Journal of Experimental Botany* **59**, 3215–3228.
Wilson, R.N., Heckman, J.W. and Somerville, C.R. (1992). Gibberellin is required for flowering in *Arabidopsis thaliana* under short days. *Plant Physiology* **100**, 403–408.
Wuest, S.E., Vijverberg, K., Schmidt, A. *et al.* (2010). *Arabidopsis* female gametophyte gene expression map reveals similarities between plant and animal gametes. *Current Biology* **20**, 506–512.
Xu, Y.L., Gage, D.A. and Zeevaart, J.A.D. (1997). Gibberellins and stem growth in *Arabidopsis thaliana*. *Plant Physiology* **114**, 1471–1476.
Yamaguchi, A., Wu, M.-F., Yang, L. *et al.* (2009). The microRNA-regulated SBP-Box transcription factor SPL3 is a direct upstream activator of LEAFY, FRUITFULL, and APETALA1. *Developmental Cell* **17**, 268–278.
Yamaguchi, N., Winter, C.M., Wu, M.-F. *et al.* (2014). Gibberellin acts positively then negatively to control onset of flower formation in *Arabidopsis*. *Science* **344**, 638–641.
Yamane, H., Takahashi, N., Takeno, K. and Furaya, M. (1979). Identification of gibberellin A_9 methyl ester as a natural substance regulating formation of reproductive organs in *Lygodium japonicum*. *Planta* **147**, 251–256.
Yant, L., Mathieu, J., Dinh, T.T. *et al.* (2010). Orchestration of the floral transition and floral development in *Arabidopsis* by the bifunctional transcription factor APETALA2. *The Plant Cell* **22**, 2156–2170.
Yasamura, Y., Crumpton-Taylor, M., Fuentes, S. and Harberd, N.P. (2007). Step-by-step acquisition of the gibberellin-DELLA growth-regulatory mechanism during land-plant evolution. *Current Biology* **17**, 1225–1230.
Yoo, S.K., Chung, K.S., Kim, J. *et al.* (2005). *CONSTANS* activates *SUPPRESSOR OF OVEREXPRESSION OF CONSTANS 1* through *FLOWERING LOCUS T* to promote flowering in Arabidopsis. *Plant Physiology* **139**, 770–778.
Yu, H., Xu, Y., Tan, E.L. and Kumar, P.P. (2002). AGAMOUS-LIKE 24, a dosage-dependent mediator of the flowering signals. *Proceedings of the National Academy of the Sciences USA* **99**, 16336–16341.
Yu, H., Ito, T., Zhao, Y. *et al.* (2004). Floral homeotic genes are targets of gibberellin signaling in flower development. *Proceedings of the National Academy of the Sciences USA* **101**, 7827–7832.
Yu, H.-J., Hogan, P. and Sundaresan, V. (2005). Analysis of the female gametophyte transcriptome of Arabidopsis by comparative expression profiling. *Plant Physiology* **139**, 1853–1869.
Yu, S., Galvão, V.C., Zhang, Y.-C. *et al.* (2012). Gibberellin regulates the *Arabidopsis* floral transition through miR156-targeted SQUAMOSA PROMOTER BINDING-LIKE transcription factors. *The Plant Cell* **24**, 3320–3332.
Zhang, Y., Zhang, B., Yan, D. *et al.* (2011). Two *Arabidopsis* cytochrome P450 monooxygenases, CYP714A1 and CYP714A2, function redundantly in plant development through gibberellin deactivation. *The Plant Journal* **67**, 342–353.
Zhu, Y., Nomura, T., Xu, Y. *et al.* (2006). *ELONGATED UPPERMOST INTERNODE* encodes a cytochrome P450 monooxygenase that epoxidizes gibberellins in a novel deactivation reaction in rice. *The Plant Cell* **18**, 442–456.

Annual Plant Reviews (2016) **49**, 359–404 http://onlinelibrary.wiley.com
doi: 10.1002/9781119210436.ch12

Chapter 12

CHEMICAL REGULATORS OF GIBBERELLIN STATUS AND THEIR APPLICATION IN PLANT PRODUCTION

Wilhelm Rademacher

BASF SE, Global Research Crop Protection, Germany
(Contact after retirement: radifax@t-online.de)

Abstract: Gibberellins and inhibitors of their biosynthesis are widely used in modern agriculture, horticulture and viticulture. Their global market is in the range of US$ 500 million. The gibberellins GA_3, GA_4 and GA_7 are primarily used to increase fruit yield and/or quality in fruit trees and table and wine grapes. Distinct steps of the gibberellin biosynthetic pathway can be inhibited by growth retardants: chlormequat and mepiquat chloride, ancymidol, flurprimidol, paclobutrazol, uniconazole, and the fungicides tebuconazole and metconazole and daminozide, trinexapac-ethyl and prohexadione-calcium are currently used in crop production. Growth retardants reduce shoot elongation, thereby lowering the risk of lodging in cereals, rice and oilseed rape, and making ornamentals more compact. A better canopy structure with improved formation of reproductive structures is induced in cotton and peanuts. In fruit and nut trees, less pruning is required and increases in crop yield and quality are obtained.

Keywords: Gibberellins, biosynthesis inhibitors, growth retardants, plant growth regulators, agriculture, horticulture, viticulture, crop production, canopy structure, lodging

12.1 Introduction

Plant growth regulators (PGRs) (often also referred to as plant bioregulators) can be defined as natural or synthetic compounds that affect developmental

Annual Plant Reviews, Volume 49: The Gibberellins, First Edition.
Edited by Peter Hedden and Stephen G. Thomas.

or metabolic processes in higher plants in a desired way. They are not phytotoxic, play no role as a plant nutrient and are, typically, applied at low dosages. In most cases, PGRs affect the plant's hormonal status. PGRs can significantly improve the performance of crop plants and their effects are often targets for breeding, be it by conventional methods or by employing genetic engineering. However, in spite of the possibilities that breeding offers, PGRs often enable faster and better solutions to many problems. Most importantly, they allow for an active regulation of plant processes, thereby adjusting a given genotype to its particular growing conditions. In other words: plant growth regulators can be used highly flexibly for the fine-tuning of crop plants that grow at a given site under largely uncontrollable and unpredictable environmental conditions. Overviews and detailed information on uses of plant growth regulators in different crops are given in Weaver (1972), Martin (1983), Nickell (1983), Schott and Walter (1991), Hedden and Hoad (1994), Gianfagna (1995), Thomson (1999), Fletcher *et al.* (2000), Basra (2000), Petracek *et al.* (2003), Rademacher (2004a) and Rademacher and Brahm (2010).

In the global crop protection market of approximately US$ 56.7 billion (€ 42.7 billion) of annual sales worldwide in 2014 (Anonymous, 2015a), PGRs play a relatively small role and account for only some 2.5% of the sales (Rademacher, 2010a). According to rough estimations, inhibitors of gibberellin (GA) biosynthesis and GAs *per se* account for 40% and 15% of the PGR market, respectively. Very few companies are still engaged in active research on GAs or inhibitors of GA biosynthesis. Due to mounting public concerns about the safety of agrochemicals, guidelines for developing and using such products have become increasingly restrictive. PGRs do not only have a relatively small market, they are also significantly more complex to work with and above-average investments are required for research, development and market introduction. In most cases, herbicides, fungicides or insecticides are less complicated and, thus, appear better suited for investments (Rademacher, 2010a). In order to be economically viable, a new PGR must have a sufficiently large market and be sold at a price offering sufficiently attractive profitability. As a consequence, the industry has to concentrate its efforts on uses in 'big crops', which are primarily represented by wheat, rice, maize, sugar cane, soybeans, oilseed rape, cotton, apples, citrus fruits, grapevines and bananas. Developing a new plant growth regulator for a 'minor crop' is, typically, not attractive for a company and will not be pursued. To overcome this obstacle, minor uses can be made possible in most countries by employing certain programs, such as IR-4 in the USA. However, a prerequisite for approaches of this type is a valid registration of the candidate product in at least one 'major crop'.

Two main approaches to changing the levels of biologically active GAs are of current practical relevance: (1) increasing endogenous GA levels by applying additional amounts of GAs and (2) reducing endogenous GA concentrations by treatment with inhibitors of GA biosynthesis. In the following parts, such types of PGRs will be dealt with. Emphasis will be placed on compounds and the derived products that are of current use in plant cultivation. Common names will primarily be used for addressing distinct compounds. Such names are assigned by the International Organization for Standardization (ISO) and are short, distinctive, non-proprietary and widely accepted. Systematic chemical names will also be given for most compounds mentioned following the nomenclature rules of the International Union of Pure and Applied Chemistry (IUPAC). Of the product names, only the original and most widely used will be mentioned. Likewise, only key producers or main providers will be named. Much more information on such and related subjects is given in pertinent reference books, e.g. Anonymous (2012) and Anonymous (2014). Useful information is also provided on different sites in the internet, e.g. the Crop Protection Database of Farm Chemicals International: http://www.farmchemicalsinternational.com/crop-protection-database/#/ ; the database of pesticide physicochemical and ecotoxicological data provided by the University of Hertfordshire in the UK: http://sitem.herts.ac.uk/aeru/iupac/; the EU pesticide database, which, *inter alia,* indicates the registration status and maximum residue levels in plant products for EU member states: http://ec.europa.eu/sanco_pesticides/public/?event=homepage; the crop production database of the Food and Agriculture Organization of the United Nations, Statistics Division (FAOSTAT) http://faostat3.fao.org/compare/E .

12.2 Gibberellins

The main hormonal functions of GAs in higher plants are the promotion of longitudinal growth, the induction of hydrolytic enzymes in germinating seeds, the induction of bolting in long-day plants, and the promotion of fruit setting and development. Accordingly, a number of uses of GAs in crop production have become general practice. In addition, there is also considerable use to accelerate the process of malting for beer-making.

In the early 1950s, when the work of Japanese scientists on GAs from the fungus *Gibberella fujikuroi* became known in other countries, companies in the UK and in the USA started to work on both GA production by fermentation and their application on crop plants. Work at ICI's Akers Laboratories in the UK on fermentation and chemical identification yielded the first production patents in 1954. Parallel, agricultural uses for GAs were investigated at ICI's Jealott's Hill research facility. By late 1957, ICI had granted manufacturing and use licenses to Abbott Laboratories, Merck & Co., Inc., Eli Lilly & Co. and Pfizer, Inc. in the USA and to Takeda Chemical Industries in Japan (Martin, 1983; Petracek *et al.*, 2003). Syngenta (into which ICI has merged) is still a major distributor of GA preparations. Valent BioSciences Corporation emerged from Abbott Laboratories and is now part of Sumitomo Chemical

Co., Ltd. In addition to selling GA-containing products, Valent BioSciences is still actively engaged in finding new uses and improved formulations of GAs. Currently, Valent BioSciences lists several dozens of uses in plant production for its different GA preparations.

Fermentations of *G. fujikuroi* are used to produce GA_3 and a mixture of GA_4 and GA_7 on a commercial scale. A chemical synthesis of GAs is possible (Mander, 2003). However, it is highly complex and much too expensive for any commercial exploitation. Likewise, testing other GA-producing fungi did not result in competitive alternatives. Production of pure GA_4 might have been achieved, for instance, with the fungus *Sphaceloma manihoticola*, which causes the superelongation disease of cassava (Rademacher and Graebe, 1979). However, attempts made at ICI, Norsk Hydro and BASF failed because of difficulties to establish competitive large-scale fermentations (Rademacher, unpublished). The phthalimide-type compound AC 94,377 [1-(4-chloro-1,3-dihydro-1,3-dioxo-2*H*-isoindol-2-yl)cyclohexane-1-carboxamide], a relatively simple chemical structure with approximately 5 to 10% of the activity of GA_3 (Los *et al.*, 1980; Yalpani *et al.*, 1989; Rademacher, unpublished), was discovered and developed in the agricultural branch of former American Cyanamid Company (now part of BASF). It was sold under the trade name Surestem for a short time for the elongation of stems of roses (Kust, 1993), but for commercial reasons, it is no longer available. Currently,

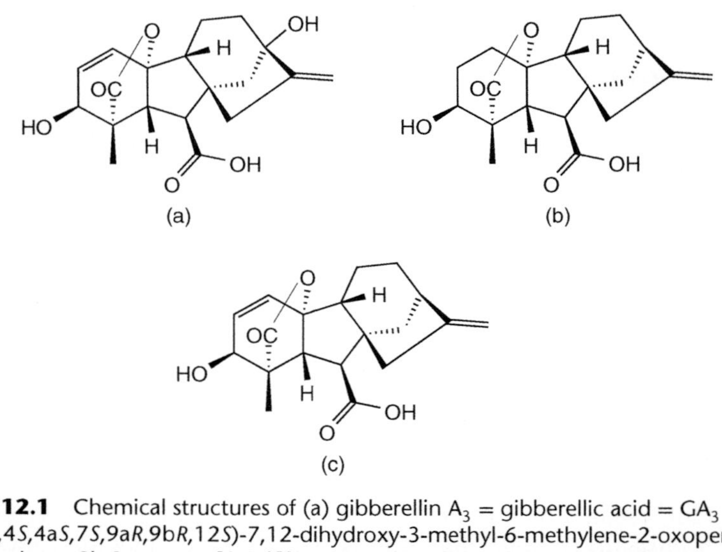

Figure 12.1 Chemical structures of (a) gibberellin A_3 = gibberellic acid = GA_3 = (3*S*,3a*R*,4*S*,4a*S*,7*S*,9a*R*,9b*R*,12*S*)-7,12-dihydroxy-3-methyl-6-methylene-2-oxoperhydro-4a,7-methano-9b,3-propeno[1,2-*b*]furan-4-carboxylic acid, (b) gibberellin A_4 = GA_4 = (3*S*,3a*R*,4*S*,4a*R*,7*R*,9a*R*,9b*R*,12*S*)-12-hydroxy-3-methyl-6-methylene-2-oxoperhydro-4a,7-methano-3,9b-propanoazuleno[1,2-*b*]furan-4-carboxylic acid and (c) gibberellin A_7 = GA_7 = (3*S*,3a*R*,4*S*,4a*R*,7*R*,9a*R*,9b*R*,12*S*)-12-hydroxy-3-methyl-6-methylene-2-oxoperhydro-4a,7-methano-9b,3-propenoazuleno[1,2-*b*]furan-4-carboxylic acid.

the majority of GAs sold globally originates from China. Several Chinese companies indicate on their websites huge supply capacities for GA_3. Prices around US$ 200.00 per kilogram for bulk quantities (>90% purity) of GA_3 are typical. Total global annual use of GA_3 is in the range of 100 tons with approximately three quarters of this used in plant production, the rest in the beer-brewing industry.

GA_3 is among the most widely used PGRs. It has found many applications in viticulture, horticulture and agriculture, e.g. in the production of seedless table grapes, in berry thinning of wine grapes, in improving citrus fruit quality, in increasing fruit size in pears and sweet cherries, and in accelerating seed germination. GA_4 is less persistent than GA_3 and GA_7 and is, therefore, better suited where too long-lasting effects are unwanted. However, due to the close chemical similarity of GA_4 with GA_7, their separation in fermentation extracts from *G. fujikuroi* is very difficult. As a result, the content of (mostly unwanted) GA_7 in different commercially available preparations varies between approximately 40% and insignificant amounts. The main use of $GA_{4/7}$ is to reduce fruit russetting in apple. The structures of the commercially available GAs are shown in Figure 12.1.

12.3 Inhibitors of gibberellin biosynthesis

The biosynthesis of GAs in higher plants is relatively well understood. A rough outline of the steps involved and the points of inhibition by the different growth retardants dealt with in this chapter is shown in Figure 12.2. GA formation can be separated into three stages, according to the nature of the enzymes involved and the corresponding localisation in the cell: (1) terpene cyclases catalyse the formation of *ent*-kaurene via *trans*-geranylgeranyl diphosphate and *ent*-copalyl diphosphate (CDP) in proplastids, (2) cytochrome P_{450}-dependent mono-oxygenases associated with membranes of the endoplasmic reticulum are involved in the steps leading from *ent*-kaurene to GA_{12}, (3) dioxygenases, which are located in the cytosol and which require 2-oxoglutarate as a cosubstrate are required for the subsequent hydroxylations into different GAs. More details on GA biosynthesis can be found in Hedden and Kamiya (1997), Yamaguchi (2008), Sponsel and Hedden (2010), Hedden (2012), Hedden and Thomas (2012), and in Chapter 2 of this volume.

Inhibitors of GA biosynthesis lead to less cell elongation and cell division, thereby making plants more compact, which may have a range of benefits in crop production. Such compounds are often referred to as 'growth retardants'. Several inhibitors of GA biosynthesis are known, some of which are used in crop production (Rademacher, 2000; Rademacher and Brahm, 2010). These compounds represent the most important group of PGRs, both in terms of commercial value and of treated area. The site of interaction of

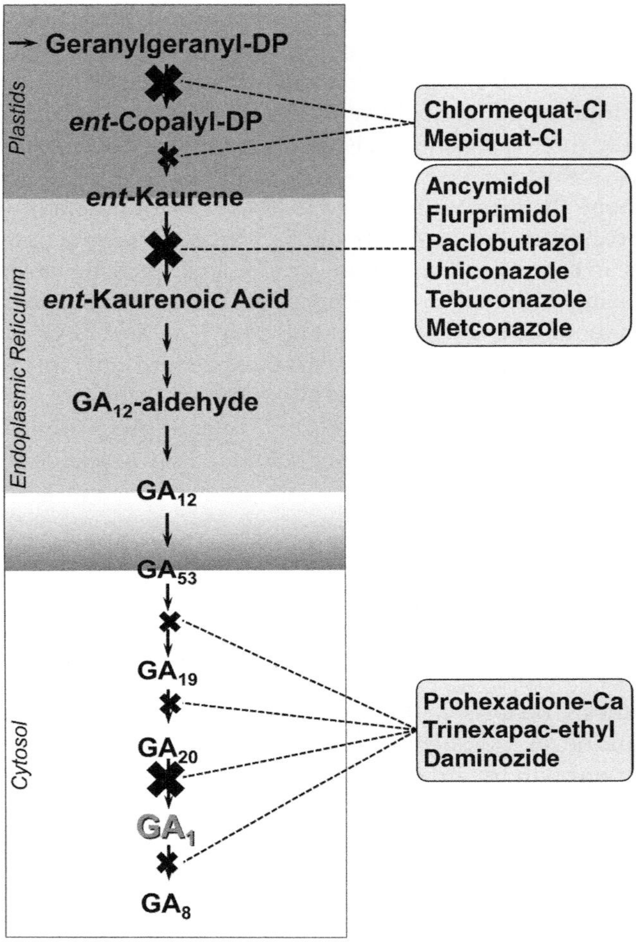

Figure 12.2 Main steps of gibberellin biosynthesis leading to biologically active GA_1 and points of inhibition by plant growth retardants. The cellular locations of the reactions is indicated by different greyscales. (The conversion of GA_{12} into GA_{53} can be located in both the endoplasmic reticulum or the cytosol.)

growth retardants with distinct steps in the biosynthetic sequence of GA formation has primarily been elucidated by using cell-free enzyme systems prepared, for instance, from *G. fujikuroi* or from immature pumpkin or pea seeds. Analysing the spectrum of GAs and their precursors from fungal cultures or intact plants treated with inhibitors has also been helpful for this purpose. Overviews on the biochemical mode of action of growth retardants have been given by Hedden (1990) and Rademacher (2000). Since that time, little additional information has become available on this subject.

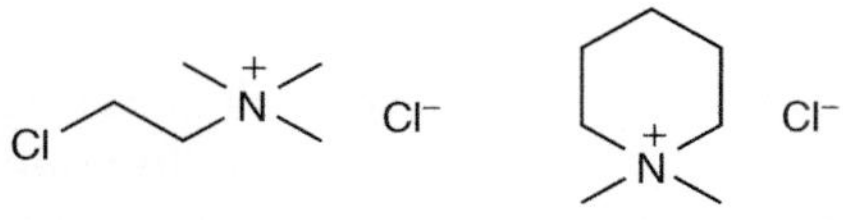

(a) (b)

Figure 12.3 Chemical structures of (a) chlormequat chloride = chlorocholine chloride (CCC) = 2-chloroethyltrimethylammonium chloride and (b) mepiquat chloride = 1,1-dimethylpiperidinium chloride.

12.3.1 Quaternary ammonium compounds

Several compounds that possess a positively charged ammonium, phosphonium or sulfonium group inhibit cyclases involved in early stages of GA biosynthesis, thereby blocking the formation of *ent*-kaurene. Out of these, the quaternary ammonium compounds chlormequat chloride and mepiquat chloride (Figure 12.3) are of practical relevance. For more 'onium-type' representatives see Rademacher (2000).

Chlormequat chloride and related compounds inhibit CDP-synthase, both in the GA-producing fungus *G. fujikuroi* and in cell-free preparations of this fungus and of higher plants. *ent*-Kaurene synthase is also inhibited, but mostly at a lower degree of activity (Shechter and West, 1969). To obtain any significant effects in cell-free preparations, relatively high concentrations of chlormequat chloride have to be used and, in some cases, the compound is even inactive (Anderson and Moore, 1967; West, 1973; Frost and West, 1977; Hedden, 1990). The same is true of mepiquat chloride: in an enzyme system derived from pumpkin (*Cucurbita maxima*) endosperm, concentrations as high as 10^{-3} M of this compound, as well as of chlormequat chloride, did not affect the spectrum of GAs and GA precursors (Hildebrandt, 1982). A possible explanation for this lack of activity could be the fact that these compounds are almost inactive in intact pumpkin plants and this may also be expected for corresponding cell-free preparations. Consequently, chlormequat chloride has been tested with enzymes derived from germinating wheat seedlings, where it gave pronounced effects (Graebe *et al.*, 1992). Chlormequat chloride lowered the levels of GA_1 in both the shoots and grains of *Triticum aestivum* (Lenton *et al.*, 1987). Likewise, it led to a dose-dependent reduction of all GAs (GA_{12}, GA_{53}, GA_{44}, GA_{19}, GA_{20}, GA_1, GA_8) present in two cultivars of *Sorghum bicolor* (Lee *et al.*, 1998). In *Eucalyptus nitens*, it caused a reduction of GA_{20} and GA_1 (Williams *et al.*, 1999).

Chlormequat chloride

Chlormequat chloride was first described in 1960 by N.E. Tolbert from Michigan State University at East Lansing in the USA to reduce shoot length in several plant species (Tolbert, 1960a,1960b; Wittwer and Tolbert, 1960). The commercial rights were held at that time by American Cyanamid Company. The significance of using chlormequat chloride as an anti-lodging agent in intense European wheat production was soon recognised (Linser *et al.*, 1961; Jung and Sturm, 1964). It was introduced under license as Cycocel® in Germany by BASF in 1965. After 50 years, this growth retardant is still the most widely used PGR in cereal production, particularly in wheat, rye, triticale and oats. Together with uses in other cultivated plants, it is, in terms of treated area, the number one PGR on a global scale.

Mepiquat chloride

After having success with chlormequat chloride, BASF developed another quaternary ammonium compound: mepiquat chloride (Zeeh *et al.*, 1974). After its introduction in 1979, mepiquat chloride became a very successful PGR, particularly in cotton. Alone or in combination with other PGRs, it is also used in other crops.

Mepiquat has been detected in samples of processed plant material, such as roasted coffee beans, roasted barley seeds, crust of bread and alfalfa pellets, even if any previous use of mepiquat chloride or other sources of contamination could be ruled out. Recent work involving coffee beans and barley seeds indicates that such findings result from a Maillard-driven degradation of lysine under dry thermal conditions and in the presence of naturally occurring trigonellin (Wermann *et al.*, 2014). Pipecolatebetain, which is structurally close to mepiquat and which occurs at relatively high concentrations in most vascular plants (Rhodes and Hanson, 1993) may also serve as a source of mepiquat under such processing conditions.

12.3.2 Compounds with a nitrogen-containing heterocycle

Distinct pyrimidines, 4-pyridines, norbornanodiazetins, imidazoles and triazoles inhibit GA formation (Rademacher, 2000). Out of these, ancymidol, flurprimidol, paclobutrazol, uniconazole and its (*E*,3*S*) isomer uniconazole-P (Figure 12.4) are of practical relevance. The triazole-type fungicides tebuconazole and metconazole induce a clear growth-retarding effect particularly in oilseed rape and are used both as fungicides and PGRs in this crop. These growth retardants act as inhibitors of cytochrome P450-dependent mono-oxygenases, which catalyse the oxidative steps from *ent*-kaurene to *ent*-kaurenoic acid and which are primarily located in the endoplasmic reticulum (Graebe, 1987 and references cited therein; Miki *et al.*, 1990). Steps lying after *ent*-kaurenoic acid, which may still involve mono-oxygenases, do not seem to be affected (Graebe, 1987). The structural feature common to all these inhibitors of *ent*-kaurene oxidation is a lone electron pair on the sp^2-hybridised nitrogen of their heterocyclic ring. In each case, this electron pair is located at the periphery of the molecule (Rademacher *et al.*, 1987) and it appears likely that it displaces oxygen from its binding site at the protoheme iron (Coolbaugh *et al.*, 1978). Evidence for such a type of interaction has been presented for ancymidol in microsome preparations of *Marah macrocarpus* (Coolbaugh and Hamilton, 1976; Coolbaugh *et al.*, 1978) and for BAS 111..W

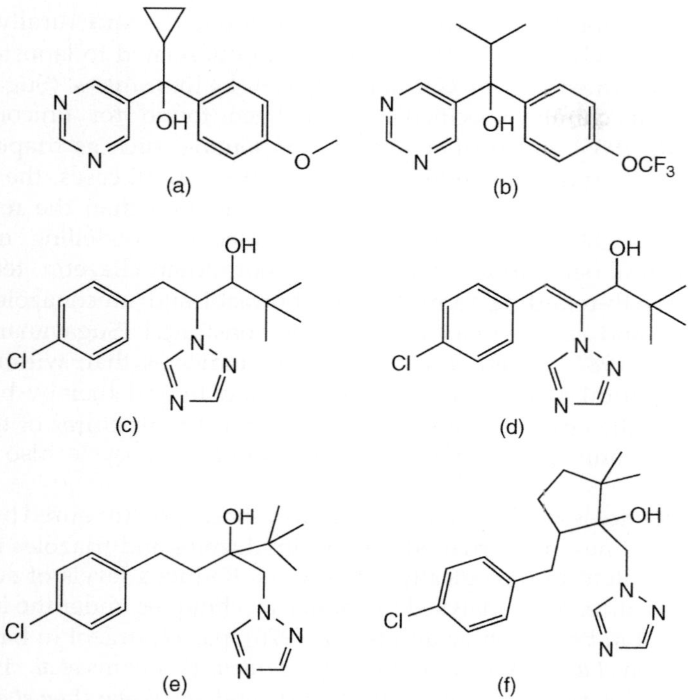

Figure 12.4 Chemical structures of (a) ancymidol = (*RS*)-α-cyclopropyl-4-methoxy-α-(pyrimidin-5-yl)benzyl alcohol, (b) flurprimidol = (*RS*)-2-methyl-1-pyrimidin-5-yl-1-(4-trifluoromethoxyphenyl)propan-1-ol, (c) paclobutrazol = (2*RS*,3*RS*)-1-(4-chlorophenyl)-4,4-dimethyl-2-(1*H*-1,2,4-triazol-1-yl)pentan-3-ol, (d) uniconazole = (*E*)-(*S*)-1-(4-chlorophenyl)-4,4-dimethyl-2-(1*H*-1,2,4-triazol-1-yl)pent-1-en-3-ol, (e) tebuconazole = (*RS*)-1-p-chlorophenyl-4,4-dimethyl-3-(1*H*-1,2,4-triazol-1-ylmethyl)pentan-3-ol and (f) metconazole = (1*RS*,5*RS*;1*RS*,5*SR*)-5-(4-chlorobenzyl)-2,2-dimethyl-1-(1*H*-1,2,4-triazol-1-ylmethyl)cyclopentanol.

(an experimental triazole-type PGR), using microsomal membranes isolated from immature pumpkin endosperm (Luster and Miller, 1993).

Depending on the presence or absence of a double bond, uniconazole-P and paclobutrazol possess one or two asymmetric carbon atoms, respectively. Since commercial paclobutrazol consists mainly of the *2RS,3RS* diastereoisomer (Sugavanam, 1984), this structure allows virtually only two enantiomers, as does uniconazole-P. Detailed experiments carried out with the optical enantiomers of paclobutrazol have shown that the *2S,3S* form exhibits more pronounced plant growth-regulatory activity and blocks GA biosynthesis more specifically, while the *2R,3R* enantiomer is more active in inhibiting sterol biosynthesis (Sugavanam, 1984; Hedden and Graebe, 1985; Burden *et al.*, 1987). Fungicidal side activities of paclobutrazol are attributed to its effect on fungal ergosterol formation (Sugavanam, 1984).

It has been demonstrated that the *2S,3S* enantiomer is structurally similar to *ent*-kaurene, whereas the *2R,3R* form is closely related to lanosterol, the respective intermediates of GA and ergosterol biosynthesis (Sugavanam, 1984). Similar chiralic specificities have been found for uniconazole-P (Izumi *et al.*, 1985) and further related compounds, such as triapenthenol (Lürssen, 1987) and inabenfide (Miki *et al.*, 1990): in all cases, the *S* enantiomer was more inhibitory to *ent*-kaurene oxidation than the respective *R* counterpart. Using computer-assisted molecular modelling methods, clear structural similarities between the norbornanodiazetin tetcyclacis and the growth-retarding forms of paclobutrazol and uniconazole-P with *ent*-kaurene and *ent*-kaurenol could be demonstrated (Sugavanam, 1984; Müller *et al.*, 1987; Katagi *et al.*, 1987). This indicates that, within limits, distinct structural features are required to bind to and thereby block the active site of the enzyme. One may assume that the structures of the other growth retardants possessing an N-containing heterocycle also fit into this scheme.

Clear evidence is available that reduction of shoot growth caused by pyrimidines, 4-pyridines, norbornanodiazetins, imidazoles and triazoles is due to a lowered content of biologically active GAs. Reduced levels of such GAs have, for instance, been analysed by modern techniques under the influence of ancymidol in beans (Shive and Sisler, 1976), paclobutrazol in barley and wheat (Lenton *et al.*, 1987) and in *Eucalyptus nitens* (Williams *et al.*, 1999) and uniconazole-P in rice (Izumi *et al.*, 1984) and *Sorghum bicolor* (Lee *et al.*, 1998).

Ancymidol and flurprimidol

Ancymidol (Tschabold *et al.*, 1970) and flurprimidol (Barrett, 1982) are closely related in structure. These pyrimidines were introduced by Elanco Products (now part of Dow AgroSciences) in 1971 and 1989, respectively. SePRO Corporation has, meanwhile, become a major distributor. The compounds are used to decrease the rate of growth in a wide range of mono- and dicotyledonous species, including perennial turf grasses, ornamental cover species, herbaceous and woody ornamentals, and deciduous and coniferous trees grown in gardens and parks. There are no registrations in plants used for human or animal nutrition.

Paclobutrazol, uniconazole and uniconazole-P

Paclobutrazol was first reported as a new and very potent PGR by Lever *et al.* (1982). Market introduction by ICI Agrochemicals (now part of Syngenta) was in 1986. The first international publication on uniconazole was by Izumi *et al.* (1984). The closely related paclobutrazol, uniconazole and uniconazole-P are very persistent, with an average half-life of approximately six months, both in plants and in the soil. They are used particularly in countries with warmer climates to control vegetative growth of fruit trees such as avocados, mangos or litchis. Lodging control in rice and the production of more compact ornamentals are further uses of these compounds.

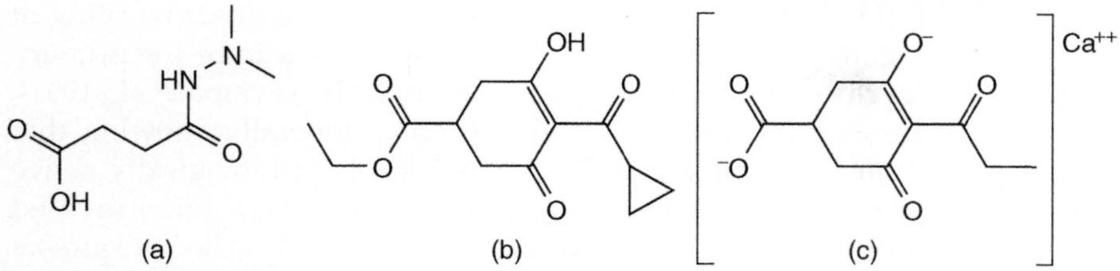

Figure 12.5 Chemical structures of (a) daminozide = *N*-(dimethylamino)succinamic acid, (b) trinexapac-ethyl = ethyl (*RS*)-4-cyclopropyl(hydroxy)methylene-3,5-dioxocyclohexanecarboxylate and (c) prohexadione-calcium = calcium 3-oxido-5-oxo-4-propionylcyclohex-3-enecarboxylate.

Tebuconazole and metconazole

In addition to blocking ergosterol biosynthesis in fungi, some triazole-type fungicides may also block GA biosynthesis in distinct plant species. This is of practical interest for using tebuconazole and metconazole to control shoot growth in oilseed rape. In Germany, metconazole is additionally in use as a PGR for ornamentals. Kuck and Berg (1986) from Bayer AG reported first on tebuconazole, which was commercialised as a fungicide starting in 1988. Metconazole was discovered by Kureha Chemical Industry Co., Ltd. in 1986 with the first international report by Sampson *et al.* (1992). The compound was jointly developed with Shell International and, later, American Cyanamid Co. and BASF AG. Market introduction (first as a fungicide, then, additionally, as a PGR) was by Cyanamid Agro in France in 1994.

12.3.3 Structural mimics of 2-oxoglutaric acid

This group is represented by the acylcyclohexanediones prohexadione-calcium and trinexapac-ethyl. Also daminozide, a succinic acid derivative, falls into this category (Figure 12.5).

The free acids prohexadione and trinexapac represent the active forms of the respective calcium salt and ethyl ester. Prohexadione is formed immediately from its calcium salt upon dissolving in water. In contrast, trinexapac-ethyl has to be saponified via biochemical processes, which may lead to delays in the onset of action, particularly when weather conditions are unfavourable (Rademacher, 2014). Prohexadione and trinexapac block soluble 2-oxoglutarate-dependent dioxygenases involved in late steps of GA biosynthesis. Studies with cell-free preparations have revealed that most steps after GA_{12} are inhibited by prohexadione and other acylcyclohexanediones (Nakayama *et al.*, 1990b, 1991; Graebe *et al.*, 1991; Griggs *et al.*, 1991; Hedden, 1991; Kamiya *et al.*, 1992; Rademacher *et al.*, 1992). Enzyme kinetic data indicate that the retardants act competitively with respect to 2-oxoglutarate, a cosubstrate for these enzymes (Griggs *et al.*, 1991; Hedden, 1991). GA 3-oxidase, which catalyses hydroxylations at position 3β (e.g.

the formation of GA_1 from GA_{20}) and also GA 2-oxidase (hydroxylating at position 2β – e.g. the conversion of GA_1 into GA_8) appear to be the primary targets of acylcyclohexanediones (Griggs *et al.*, 1991; Nakayama *et al.*, 1991). These findings are supported by analytical data, generally showing that growth reduction is accompanied by lowered levels of biologically active GAs (e.g. GA_1) and their inactive metabolites (e.g. GA_8), but increased concentrations of the inactive immediate (e.g. GA_{20}) and earlier precursors (Nakayama *et al.*, 1991; Adams *et al.*, 1992; Kamiya *et al.*, 1992; Rademacher *et al.*, 1992; Santes and García-Martínez, 1995; Brown *et al.*, 1997; Junttila *et al.*, 1997; Lee *et al.*, 1998; Na *et al.*, 2011). In selected cases, compounds like prohexadione-calcium and trinexapac-ethyl may, paradoxically, lead to increases in shoot growth, most likely by protecting endogenous active GAs from being metabolically inactivated by GA 2-oxidase (Hisamatsu *et al.*, 1998). Likewise, the inactivation of exogenously applied GA_1 by 2β-hydroxylation can be inhibited by simultaneous treatment with an acylcyclohexanedione, resulting in increased GA activity (Nakayama *et al.*, 1990a; Sponsel and Reid, 1992; see also Figure 12.2).

2-Oxoglutarate-dependent dioxygenases catalyse many different reactions in plant metabolism (Farrow and Facchini, 2014). Accordingly, some important side activities of prohexadione-calcium and trinexapac-ethyl have been detected: high dosages of these and other acylcyclohexanediones inhibit the formation of anthocyanins in flowers and other plant organs. It has been suggested that 2-oxoglutarate-dependent dioxygenases, in particular flavanone 3-hydroxylase, which is involved in the biosynthesis of anthocyanidins are targets for these growth retardants (Rademacher *et al.*, 1992). This hypothesis has been confirmed by the finding that young shoots of apple are unable to convert eriodictyol by 3-hydroxylation into flavonoids such as catechin after treatment with prohexadione-calcium. Instead, eriodictyol accumulates and large amounts of luteoliflavan, which does not normally occur in apple tissue, can be found. This shift in flavonoid metabolism is seen as the major underlying reason for reduced susceptibility of treated pome fruit trees to bacterial and fungal diseases: luteoforol, the highly reactive and unstable precursor of luteoliflavan, shows clear *in vitro* biocidal activity against a number of bacterial and fungal pathogens, including *Erwinia amylovora* and *Venturia inaequalis*, the causal agents of fire blight and apple scab, respectively (Spinelli *et al.*, 2005). Apigeninidin, luteolinidin and their derivatives, which are also 3-deoxy flavonoids, act as phytoalexins in *Sorghum bicolor* (Lo *et al.*, 1999). For more information see Römmelt *et al.* (2003), Halbwirth *et al.* (2003), Rademacher (2004b) and Halbwirth *et al.* (2006). Prohexadione-calcium and trinexapac-ethyl reduce ethylene formation in sunflower cell suspensions and in leaf disks of wheat (Grossmann, 1992). Ethylene is generated from aminocyclopropanecarboxylic acid (ACC) in a reaction catalysed by ACC oxidase. This is a dioxygenase that requires ascorbic acid as a cosubstrate, whereas 2-oxoglutaric acid and similar compounds inhibit its activity (Iturriagagoitia-Bueno *et al.*, 1996). Employing an enzyme system prepared from ripe pear fruits, it could be shown that prohexadione-calcium also inhibits ACC oxidase, presumably by displacing ascorbate from the enzyme's active site (Rademacher *et al.*, 1998).

Long after its introduction, it could also be demonstrated that daminozide interferes with GA biosynthesis. Considering structural similarities between daminozide and 2-oxoglutaric acid and newly interpreting older results from the literature, it was proposed that daminozide, like acylcyclohexanediones, would block GA formation as an inhibitor of 2-oxoglutarate-dependent dioxygenases (Rademacher, 1993). This hypothesis was later proved by working with an enzyme preparation derived from cotyledons of *Phaseolus coccineus* and by analyzing the GAs of treated peanut plants (Brown *et al.*, 1997).

Daminozide

The plant growth-regulatory activity of daminozide was first reported by Riddell *et al.* (1962) and its market introduction by Uniroyal Chemical Co., Inc. followed in 1963. Daminozide was used in several fruit tree species, particularly in apple, to reduce vegetative growth and improve fruit colouration and firmness (trade name: Alar®). Treatment of peanut plants led to more compact plants, allowing for a more efficient harvesting (trade name: Kylar®). However, due to toxicological concerns, its use on food crops was banned in 1989. The compound is still used as B-Nine® in ornamentals, with Chemtura Corp. a major supplier.

Trinexapac-ethyl and prohexadione-calcium

Trinexapac-ethyl was introduced to the scientific community by Kerber *et al.* (1990) and Adams *et al.* (1992). Ciba-Geigy AG (now Syngenta) started with sales in 1992. Prohexadione-calcium originates from Kumiai Chemical Industry Co., Ltd. and Ihara Chemical Industry Co., both of Japan. The first international publication on its properties as a growth retardant was by Nakayama *et al.* (1990a). In 1991, BASF obtained a license to develop this compound in Europe, North America and several other countries. Kumiai launched prohexadione-calcium in Japan as an anti-lodging agent in rice, while BASF started commercialisation of the compound in combination with mepiquat chloride as a stem stabiliser in cereals in 1998 and solo for use in pome fruit trees in 2000.

12.3.4 16,17-Dihydro-gibberellins

16,17-Dihydro-GAs represent the most recent group of growth retardants. A number of different structures of this type, mostly GA_5 derivatives, have been shown to reduce shoot elongation in *Lolium temulentum* (Evans *et al.*, 1994; Mander *et al.*, 1995, 1998) and other grasses (King *et al.*, 1997). Evidence is available that their growth-retarding activity is due to an inhibition of dioxygenases, which catalyse the late stages of GA metabolism, particularly GA

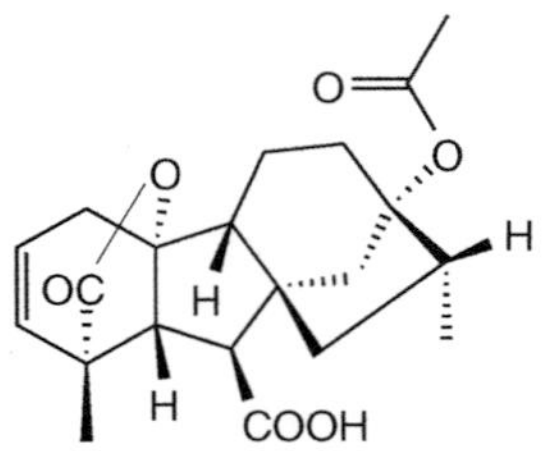

Figure 12.6 Chemical structure of *exo*-16,17-dihydro-GA_5-13-acetate.

3-hydroxylation (Takagi *et al.*, 1994; Foster *et al.*, 1997; Junttila *et al.*, 1997). Treating plants with 16,17-dihydro-GA_5 results, indeed, in changes of GA levels similar to the ones caused by acylcyclohexanediones: In *Lolium temulentum* (Junttila *et al.*, 1997) and in *Sorghum bicolor* (Foster *et al.*, 1997) the levels of GA_1 declined, whereas GA_{20} accumulated significantly.

With a view to find new anti-lodging compounds for small grains, several 16,17-dihydro-GA_5 derivatives have been systematically tested in suitable formulations. As a result of these investigations, *exo*-16,17-dihydro-GA_5-13-acetate (Figure 12.6) emerged as the most active growth retardant ever known for graminaceous plants. Under greenhouse conditions effects with as little as 500 mg per hectare can be monitored in wheat and barley. However, in order to reduce the risk of lodging under practical conditions, rates in the range of 20 g per hectare have to be used (Rademacher *et al.*, 1999). In order to explain the high biological activity it can be assumed that *exo*-16,17-dihydro-GA_5-13-acetate and related structures compete very effectively in grasses with the natural GA substrates, e.g. GA_{20}, for the respective enzymatic sites (Takagi *et al.*, 1994; King *et al.*, 2004). In contrast to graminaceous plants, *exo*-16,17-dihydro-GA_5-13-acetate and related structures are virtually inactive in reducing shoot growth in any other plant species tested (Rademacher *et al.*, 1999). In spite of the promising results obtained with *exo*-16,17-dihydro-GA_5-13-acetate, its synthesis from GA_3 in bulk quantities proved to be too expensive for commercialisation (Rademacher, unpublished).

12.4 Uses for gibberellins and inhibitors of gibberellin biosynthesis in crop production

Space limitation does not permit coverage of all practical uses that have been established for the different GAs and inhibitors of GA formation. Therefore, only the major uses and some recent additions are referred to. Detailed information on how to use a given PGR in a distinct crop can be found in the respective labels, which are provided by the distributing companies via the internet.

12.4.1 Wheat, barley, rye, oats and other small-grain cereals

The production of wheat and other small grains has undergone drastic changes since the introduction of science-based agricultural methods. This development is particularly obvious in West Europe with its maritime climate, long days at the time of grain filling and other growing conditions favourable for winter wheat. Productivity data are almost continuously available for Germany since 1878. Starting at yield levels of some 1.3 tonnes per hectare, just above 2.0 t/ha was reached prior to World War I and, after a post-war dip, again in the 1930s. However, enormous increases in yield levels could be achieved since the beginning of the 1950s: within six decades, productivity was almost quadrupled from approximately 2.0 to 7.5 t/ha. Similar degrees of intensification were reached in countries with comparable production conditions such as France and the United Kingdom (UK) (FAOSTAT; Rademacher, 2010b). Likewise, seed yield per unit of land could also be raised significantly in other small-grain species such as barley, rye, triticale, oats and spelt. It is estimated by several authors that the increases in productivity have mainly resulted from increased and better-targeted fertilisation (40–45%), followed by breeding (25–30%) and crop protection plus soil management (25–30%) (Sturm *et al.*, 1994). These factors for success are closely interconnected: dispensing, for instance, with fungicides treatments could certainly lead to yield reductions of much more than 30% under adverse production conditions. It must also be noted that the mentioned achievements have been a major prerequisite for creating modern and wealthy societies in industrialised countries with limited area available for agriculture.

With rising production intensity in Germany, UK, France and other countries, lodging became increasingly a problem in cereal cultivation in the 1950s and 1960s: heavy ears could no longer be kept upright by long stems, particularly when their leverage was increased by wind and rain. Lodging occurs mainly during the two months preceding harvest and may drastically reduce profitability through reduced yield and quality and increased costs for harvesting and grain drying (Table 12.1). If lodging occurs early (e.g. shortly after anthesis), its impact on seed yield and quality will be more intense as compared with lodging close to harvesting (Hoffmann, 1992). Under UK growing conditions, severe lodging in cereals may be expected in one out of three to four years (Baker *et al.*, 2014). It is likely that the situation is similar in other countries with high intensities of production. Assuming an average yield of 7.5 t/ha and a producer price of 180.00 €/t for wheat, a reduction in yield of 20% due to lodging is equivalent to 270.00 €/ha. Additional financial losses are likely to result from inferior grain quality and increased costs for harvesting and grain drying. The use of anti-lodging products in wheat has been banned in Sweden in 1987 in order to reduce production intensity and, thereby, lower the negative impacts agriculture may have on the environment. Not least due to this, a reduction of wheat productivity by approximately 25% has resulted in comparison to countries with similar production

Table 12.1 Negative impacts of lodging on wheat yield and quality (typical values after Pinthus, 1973; Anderson, 1979; Jung and Rademacher, 1983; Hoffmann, 1992; Easson *et al.*, 1993; Berry *et al.*, 2004; Baker *et al.*, 2014 and after results of BASF field trials).

Parameter	Effect
Total grain yield	Decreased by 10–40% (up to 80% in extreme cases)
1000-grain-weight	Decreased by 8–15%
Crude protein content of seeds	Relative increase by 3–20%
Carbohydrate content of seeds	Relative decrease by 10–17%
Milling quality	Decreased
Baking quality	Decreased
Presence of mycotoxins	Significantly increased risk
Costs for harvesting	Increased by up to 50%
Costs for grain drying	Increased by 20–30%

conditions such as Denmark, Germany or the UK (data from FAOSTAT). A few years ago, the ban on PGR use in Swedish wheat production was lifted.

Two forms of lodging can be differentiated: (1) stem lodging occurs when heavy wind and rainfall exert a force that breaks the stem base. Often, stem lodging is found after a severe thunderstorm. Eyespot, caused by *Pseudocercosporella herpotrichoides*, and other foot rot diseases may intensify the risk of stem lodging. (2) Root lodging is typically observed when, after several days of rainfall, the plant's root system is unable to keep the stem, with its heavy, water-soaked ear, upright. The risk of both forms of lodging is strongly influenced by cultivar and husbandry factors, including sowing date, seed rate, drilling depth and rate of nitrogen application. In spite of this knowledge, the use of anti-lodging products has become an integral part of the production system in order to secure seed yield and quality. These products reduce stem length, thereby lowering the leverage of the ear and other upper plant parts. Increased stability results also from histological changes caused in the stems (Petry *et al.*, 1989). Evidence is also available that anti-lodging agents lead to increases in root growth, thereby providing better anchorage against falling over and enabling plants to absorb water and nutrients more effectively (Rademacher, 2015).

Breeding for short-strawed varieties has only partly contributed to stem stabilisation under production conditions targeted for high yield and quality. It is suggested that the optimum mature height of winter wheat in the UK is close to 80 cm. Shorter stems would have a negative impact on light interception, encourage leaf diseases and make harvesting more difficult (Flintham *et al.*, 1997; Austin, 1999; Berry *et al.*, 2004). Consequently, breeders are relying to a considerable extent on stem shortening 'when needed' by means of PGRs. This is also reflected by the fact that on approximately 34% of the land in Germany devoted to seed propagation in 2015 for wheat, barley, rye, triticale, oats or spelt varieties with a lodging susceptibility rated from 'medium' to 'very high' (grades 5 to 9) were grown. Even grade 4 varieties ('low to medium'), which are propagated on approximately 42% of the area (Anonymous, 2015b), are candidates for anti-lodging treatments, particularly when grown at high intensity. This situation is similar in the other major cereal-producing countries in Europe.

After its market introduction in 1965, chlormequat chloride was the first PGR to be used on a large scale as an anti-lodging agent in cereal production. Meanwhile, additional PGRs have been introduced for stem stabilisation and lodging control in small grains. Combinations of chlormequat chloride or mepiquat chloride with trinexapac-ethyl or prohexadione-calcium, either by tank-mixing or by using ready-mix formulations, currently represent the best technical solutions for lodging control: chlormequat chloride and mepiquat chloride act at relatively low temperatures and may be used early in the season. Their onset of activity is relatively slow, but long-lasting. Complementing these compounds, trinexapac-ethyl and prohexadione-calcium act comparatively quickly, but are relatively short-lived. Furthermore, they require somewhat higher temperatures for activity. When chlormequat chloride is applied at early stages of tillering, it increases the number of fertile tillers in addition to reducing stem length. This can be of special interest after winter losses of plants. Barley is less responsive to chlormequat chloride and mepiquat chloride than wheat, rye, triticale or oats. Therefore, products containing trinexapac-ethyl and prohexadione-calcium or the ethylene-releasing ethephon are preferentially used to reduce the risk of lodging in this crop. Additional information on lodging in small grains and on the use of anti-lodging agents in cereal production can be found in Easson *et al.* (1993), Berry *et al.* (2004), Rademacher (2010b), Berry and Spink (2012) and Baker *et al.* (2014). Different anti-lodging products are listed in Table 12.2.

Area- and value-wise, stem shortening in small grains (and in intense rice production in Japan and South Korea) to reduce the risk of lodging is the main application of PGRs worldwide. Indeed some 25% of the global PGR sales (equalling approximately € 270 million) are represented by stem stabilisers. The usage of such products is general practice in countries with intense production of wheat, barley, rye, triticale and oats, such as France, Germany and the UK. For instance, 89% of the winter wheat, 76% of the winter barley, 73% of the oats and 100% of the winter rye acreage were treated with anti-lodging products in the UK in 2012. Chlormequat chloride and trinexapac-ethyl rank third and seventh, respectively, in terms of treated area of all crop protectants in all field crops in this country (Garthwaite *et al.*, 2013).

Anti-lodging products for small-grain cereals play only a minor role in countries where these crops are generally grown at relatively low levels of intensity. In the USA, Canada and Australia, climatic factors, in particular unfavourable temperatures and shortage of moisture, represent the main restrictions. Since sufficient arable land is available, significant surpluses for export can still be produced on a national level. However, the use of

Table 12.2 GA3 and inhibitors of GA biosynthesis used in cereal and rice production.

Crop species	Use	Active ingredient (s)	Main trade name (s)	Main producer(s) and/or supplier(s)
Rice	Acceleration of germination and seedling development	GA_3	Release®	Valent BioSciences
Wheat, Rye, Oats, Triticale	Reduced risk of lodging	Chlormequat chloride	Cycocel®	BASF
Wheat		Chlormequat chloride + Imazaquin	Meteor®	BASF
Barley, Wheat, Rye, Triticale		Mepiquat chloride + Ethephon	Terpal®	BASF
Wheat, Barley, Rye, Oats, Triticale, Spelt		Trinexapac-ethyl Prohexadione-Ca + Mepiquat chloride	Moddus® Medax® / Canopy®	Syngenta BASF
Rice		Uniconazole-P Uniconazole-P in coated fertilizer granules Paclobutrazol Prohexadione-Ca	Lomica® Rakuichi® Smarect® Viviful®	Sumitomo Sumitomo Syngenta Kumiai

PGRs is indispensable when wheat or other small grains are grown more intensively in areas where this is possible (Rademacher, 2010b). To date, chlormequat chloride and ethephon represent only niche markets in Canada. Ethephon has recently been withdrawn, whereas trinexapac-ethyl has just been registered as an anti-lodging agent in the USA. A survey with agronomists working for major rural supply companies across Australian grain-growing regions has recently been conducted: only 20% of the participants recommended the use of PGRs for crop management in wheat (Acuna, 2014). Chlormequat chloride and ethephon are available for cereal lodging control in Australia. The registration of trinexapac-ethyl is expected for 2015.

12.4.2 Rice

In direct-seeded rice, application of GA_3 as a seed dressing is relatively common. This treatment significantly improves germination percentage, seedling emergence and seedling height and is especially important at sub-optimal temperatures (Dunand, 1992).

As in small grains, lodging can also be a severe problem in intense rice production (Jennings and Sornchai, 1964; Moody, 1986; Nishiyama, 1986; Yoshinaga, 2005; Shibata and Takebayashi, 2007; Salassi *et al.*, 2013). However, modern semi-dwarf cultivars produce relatively high yields, while being largely lodging-resistant. In spite of this option, many farmers still prefer to grow tall but lodging-susceptible varieties, which are tastier and achieve a higher price. For instance, Japanese consumers prefer rice from the traditional long-strawed cultivar 'Koshihikari', which is grown on approximately 35% of the Japanese rice-producing area. Without treatment, 'Koshihikari' typically reaches a final shoot length around 110 cm and is very prone to lodging (Shibata and Takebayashi, 2007). As with cereal grains, several husbandry methods can be employed to minimise lodging in rice. Reducing overgrowth of stem and leaves by applying PGRs at mid-tillering stage reduces lodging incidence and gives a generally improved plant stature (Yim *et al.*, 1997). In 1988 some 12% of the Japanese rice-growing area was treated with stem stabilisers (Schott and Walter, 1991). One may assume that this percentage has at least remained constant.

Whereas stem stabilisers are applied to cereal plants by spraying the leaves, granules for throwing into the paddy field are preferred in rice production, particularly in Japan and South Korea. In order to be absorbed via the roots, such stem stabilisers have to be relatively persistent. Accordingly, preparations based on long-lived uniconazole-P and paclobutrazol are the main active ingredients used as anti-lodging agents in this crop. Prohexadione-calcium, which has to be spray-applied, is only of minor importance (Table 12.2).

12.4.3 Sugarcane

Ethephon, glyphosate or other herbicides are often used as chemical ripeners in sugarcane production. They have to be applied via aircraft or ground-operated booms. By rapidly reducing the sink demand of young and growing plant parts, sucrose storage within the stalk is accelerated leading to high harvest yields. Even better effects can be achieved with trinexapac-ethyl (Resende *et al.*, 2000; Rixon, 2007). Meanwhile, trinexapac-ethyl is registered in Brazil, Australia, the USA and other countries for use in this crop.

12.4.4 Pasture and turf grasses

GA_3 has found some use in the USA and other countries to stimulate shoot growth in pasture grasses (Matthew *et al.*, 2009). However, much more interest is directed towards reducing shoot elongation. Here, inhibitors of GA biosynthesis are important in high-intensity fine turf, particularly on golf courses. A main reason is to reduce vertical leaf growth, which

Table 12.3 PGRs used in pasture and turf grasses.

Crop species	Use	Active ingredient(s)	Main trade name(s)	Main producer(s) and/or supplier(s)
Forage grasses	Stimulation of shoot growth	GA_3	ProGibb®	Valent BioSciences
Grasses grown for seed	Reduced risk of lodging	Trinexapac-ethyl	Palisade®	Syngenta
	Reduced risk of lodging Suppression of annual bluegrass	Prohexadione-Ca	Apogee®	BASF
Turf grasses	Improved turf quality Less need for mowing Suppression of annual bluegrass	Flurprimidol Paclobutrazol Prohexadione-Ca	Cutless® Trimmit® Regalis®	SePRO Syngenta BASF
	Improved turf quality Less need for mowing	Trinexapac-ethyl	Primo® Maxx	Syngenta

leads to smoother and more uniform playing surfaces. Darker leaf colour, intensified root growth, reduced water consumption, seed head suppression of unwanted annual bluegrass *(Poa annua)*, and, not least, less need for mowing are additional benefits. Trinexapac-ethyl, paclobutrazol, flurprimidol and different combinations of these retardants are the main PGRs used for this purpose in the USA. Similar products for growth regulation in fine turf grasses are available in several other countries. Prohexadione-calcium is available as a PGR for use on turf grasses in Germany. A recent survey of growth regulators in turfgrasses is available from March *et al.* (2013).

As in cereal grain production, lodging may also be a problem when grasses are grown for seed production. Trinexapac-ethyl and prohexadione-calcium are the main active ingredients used to reduce this risk, particularly in the US state of Oregon, where grass grown for seed is a major business. Prohexadione-calcium has a small advantage in performance: most likely due to its more immediate action after application (Rademacher, 2014), it inhibits seedhead formation in unwanted *Poa annua* (Beam and Askew, 2005). PGRs used in pasture and turfgrasses are listed in Table 12.3.

Table 12.4 PGRs used in oilseed rape.

Crop species	Use	Active ingredient(s)	Main trade name(s)	Main producer(s) and/or supplier(s)
Oilseed rape	Improved winter survival	Tebuconazole	Folicur®	Bayer CropScience
	Improved shoot structure	Metconazole	Caramba®	BASF
	Reduced risk of lodging	Metconazole + Mepiquat chloride	Caryx®	BASF
	Control of pathogens	Paclobutrazol + Difenoconazole	Toprex®	Syngenta
	Improved shoot structure Reduced risk of lodging	Trinexapac-ethyl	Moddus®	Syngenta

12.4.5 Oilseed rape

Winter oilseed rape (*Brassica napus, ssp. napus*) has become an important oilseed crop in many European countries and elsewhere. It can be kept from too intensive growth in late autumn, thereby making it less vulnerable to freezing and desiccation in winter. Later in its development, yield losses due to lodging may occur, which can also be reduced by stem-shortening agents (Kightley, 2001; Berry and Spink, 2009; Baker *et al.*, 2014). The leading compounds used are the triazoles tebuconazole and metconazole, which are marketed for this purpose in France, the UK, Germany and several other European countries. Recent introductions are the combination of metconazole with mepiquat chloride and paclobutrazol with the fungicide difenoconazole (Table 12.4). Tebuconazole and metconazole are primarily used as fungicides in a number of crop plants including oilseed rape. Their shoot growth-reducing activity is restricted to oilseed rape and a few other species. Figure 12.7 demonstrates that they produce a pronounced reduction in GA levels in oilseed rape.

12.4.6 Cotton

Cotton is a perennial plant, which, however, is cultivated in most countries in an annual cycle (Edmisten *et al.*, 2010). In its native habitat, cotton plants do not die in the autumn, but continue to grow until environmental conditions become too restrictive. Another growth characteristic associated with its perennial nature is its indeterminate fruiting habit. Rather than

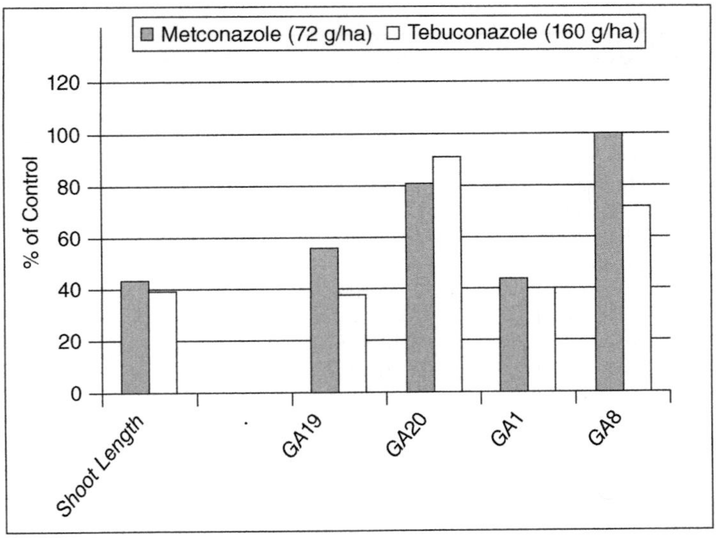

Figure 12.7 Effects of metconazole (applied as Caramba®) and tebuconazole (applied as Folicur®) on shoot growth and GA levels in glasshouse-grown seedlings of oilseed rape, cv. 'Akela'. Absolute values in control plants: Length of shoot axis 6.3 cm; GA_{19} 3.9 µg/kg, GA_{20} 22.7 µg/kg, GA_1 9.8 µg/kg, GA_8 5.3 µg/kg of shoot dry weight. Other GAs detected: GA_{29}, GA_{34}, GA_{44}. (S.J. Croker, P. Hedden and W. Rademacher, unpublished)

flowering during a distinct period following vegetative growth, cotton plants simultaneously produce vegetative and fruiting structures. In order to enable high yield and quality formation and to allow efficient mechanical harvesting, intense usage of PGRs has become a standard practice in many cotton-producing countries. The regime typically comprises control of vegetative growth by mepiquat-containing products, defoliation by thidiazuron and boll-opening by ethephon.

The growth-retardant mepiquat chloride was commercially introduced for vegetative growth control in cotton in the USA in 1980 and has since become a cornerstone of modern cotton production (Kerby *et al.*, 1996; Edmisten *et al.*, 2010). Instead of or in addition to mepiquat chloride, chlormequat chloride is used for the same purpose in some countries, for instance in Australia. Another variant is mepiquat pentaborate, which was brought to the US market in 2003 (Stapleton and Via, 2003). Due to its more rapid uptake (Rademacher, unpublished) and the nutritive value of the contained boron, the pentaborate form of mepiquat leads to better plant performance under distinct growing conditions as compared to its chloride salt (e.g. Norton and Borrego, 2006). Mepiquat chloride is also available in combination with kinetin or cyclanilide (Table 12.5). However, in the majority of cases the

Table 12.5 Inhibitors of GA biosynthesis used in cotton production.

Crop species	Use	Active ingredient(s)	Main trade name(s)	Main producer(s) and/or supplier(s)
Cotton	Management of shoot growth Earlier harvesting Improved fibre quality	Mepiquat chloride	Pix®	BASF/Arysta LifeScience
		Mepiquat chloride + Kinetin	Mepex® Gin Out	Nufarm
		Mepiquat chloride + Cyclanilide	Stance®	Bayer CropScience
		Mepiquat pentaborate	Pentia®	BASF

different mepiquat-containing products on the market give comparable results (e.g. Jost *et al.*, 2006).

Treatment of cotton plants with mepiquat starting at the beginning of flowering reduces the intensity of new growth and, thereby, improves the sink strength of the first six to ten fruiting branches. This is of great importance because the bolls retained in this part of the plant will give the highest and earliest yields. The shifting of assimilates into the older fruiting structures is at the expense of younger fruits, which the plant is continuously forming, even late in the season and which are unlikely to contribute to yield at mechanical harvesting. Plants treated with mepiquat produce higher yields and can, typically, be harvested three to ten days earlier than untreated plants. Earliness is of great importance because harvesting can often be performed prior to periods of rainfall. This would also reduce the incidence of fungal diseases. Further benefits from short-season production may result from savings in late-season irrigation and insecticide costs. Finally, decreases in quantity and quality of the lint due to weathering are reduced in the oldest, first-opened cotton bolls. Valuable contributions on different aspects of using mepiquat chloride in cotton have been presented by Cathey and Meredith (1988), Kerby *et al.* (1986, 1996), Cook and Kennedy (2000) and Gwathmey and Clement (2010).

12.4.7 Peanuts

The foliage of peanut plants is still green at harvesting such that excessive vine growth may reduce digging efficiency. Prohexadione-calcium, which is registered for this use in the USA, retards vegetative growth and improves the visibility of rows, resulting in improved harvesting efficiency. Pod yield and kernel quality may also be improved (Jordan *et al.*, 2001, 2008).

12.4.8 Opium poppy

Lodging is also a problem in poppy cultivation. Trials in Tasmania, Australia, where a major portion of the global legal opium poppy production is located, were therefore conducted with trinexapac-ethyl and prohexadione-calcium to overcome this problem. Surprisingly, it was found that treatments with these compounds did not only improve lodging resistance, but also changed the alkaloid spectrum in the harvested plant material in a desirable way: more thebain, which is of higher value, is formed at the expense of lower-valued oripavine (Cotterill, 2005). It has been postulated that a 2-oxoglutarate-dependent dioxygenase, which catalyses the conversion of thebain into oripavine is blocked (Hagel and Facchini, 2010; Dean, 2011). Trinexapac-ethyl is now registered in Australia for use in opium poppy.

12.4.9 Fruit trees growing in temperate climate

Gibberellins and growth retardants have been used for many years in the cultivation of pome and stone fruit trees such as apples, pears, peaches, plums and cherries, which are typically grown in temperate climates. It is much more difficult to apply PGRs in such perennial fruit crops than in annual arable crops. Mistakes made in one year may often lead to problems in the years following. On the other hand and in contrast to field crops, fruits, typically, represent a higher-value crop and, hence, allow the use of more elaborate and expensive products. Recent overviews on different PGR uses in fruit production have been presented by Petracek *et al.* (2003), Greene (2010), Rademacher and Brahm (2010) and Looney and Jackson (2011). Table 12.6 gives an overview of uses for GAs and inhibitors of GA biosynthesis in fruit trees.

Gibberellins

In pears, parthenocarpic fruit formation can be achieved with GA_3. This may lead to an increased fruit set and is particularly important when the generative part of the flower has been damaged by frost or when there has been poor pollination. Some varieties, e.g. 'Williams' and 'Abate Fetel', are more responsive than others. GA_3 may also be used in sweet cherries to produce brighter-coloured, firmer fruits with increased size. The mixture of GA_4 and GA_7 is often used by apple growers to reduce fruit russetting, a superficial disorder in which the fruit surface is interrupted by raised corky outgrowths. Evidence is available that the GA_4 component is the primary active ingredient for the control of russetting, whereas GA_7 rather inhibits flowering and

Table 12.6 GAs and inhibitors of GA biosynthesis used in temperate fruit trees.

Crop species	Use	Active ingredient(s)	Main trade name(s)	Main producer(s) and/or supplier(s)
Apples	Reduction of russetting	$GA_{4/7}$	Regulex®	Syngenta
			ProVide®	Valent BioSciences
		GA_4 (only traces of GA_7)	Novagib®	Fine Agrochemicals
	Improved fruit shape and size	$GA_{4/7}$ + 6-benzyladenine	Promalin®	Valent BioSciences
	Control of shoot growth	Paclobutrazol	Cultar®	Syngenta
	Control of shoot growth Improved fruit set and yield Control of fire blight	Prohexadione-Ca	Regalis®/ Apogee®	BASF
Pears	Parthenocarpic fruit set	GA_3 and $GA_{4/7}$	Berelex® and Regulex®	Syngenta
			Gibb 3 and Gibb Plus	Globachem
	Control of shoot growth Improved fruit set Control of fire blight	Prohexadione-Ca	Regalis®	BASF
	Control of shoot growth	Paclobutrazol	Cultar®	Syngenta
Peaches, Nectarines, Plums	Control of shoot growth	Paclobutrazol	Cultar®	Syngenta
Sweet cherries	Improved fruit quality	GA_3	ProGibb®	Valent BioSciences
	Control of shoot growth	Prohexadione-Ca	Apogee®	BASF

reduces return bloom (Carlson and Crovetti, 1990). Hence, preparations low in GA_7 may have an advantage for this use. Combined with the cytokinin benzyladenine, $GA_{4/7}$ is also used to improve size and shape of apples fruits. Fruit elongation and development of more prominent calyx lobes in Red Delicious apples are of special interest in North America.

Inhibitors of gibberellin biosynthesis

Proper tree growth management is of major concern in commercial pome fruit production: avoiding excessive shoot growth will induce earlier flowering and fruiting in young trees. Older trees have to be restricted to their allocated space, thereby reducing crowding and shading. The crowns of fruit trees should be sufficiently open to allow good light penetration to the inner parts of the canopy, thereby improving photoproductivity and fruit colouration. Additionally, efficient crop protection is significantly facilitated in such trees. Since the beginning of professional apple and pear production, several techniques have been employed to avoid excessive shoot growth: different types of dwarfing rootstocks and scions have become available, particularly for apple. Different cultivars may also show significant differences in shoot vigour. Dormant and summer pruning are the main cultural practices for shoot control in addition to regulation of fruit set. Other methods include: root pruning, root restriction, stem girdling or stem sawing, limb bending, breaking or wounding, and restrictive fertilisation and irrigation. However, each of these methods is cost-intensive and/or bears a high risk of failure. Furthermore, part of the trees' assimilates or potential assimilates are lost.

Chemical regulation of shoot growth has been practiced over many years by using distinct inhibitors of GA biosynthesis. However, health concerns about daminozide have led to the ban of this compound in edible crops. Additional negative attitudes towards using PGRs for shoot growth regulation resulted from unacceptably high residues of chlormequat in pear fruits, due to excessive use in this crop. In essence, only paclobutrazol remained registered in some countries at the end of the 1990s as a regulator of shoot growth in pome and stone fruits. However, this compound is extremely persistent. Its half-life in the plant and in the soil is in the range of six months. Application is via spraying or as a soil drench. In order to avoid effects on following crops, the label for the UK recommends to withhold using paclobutrazol for up to seven years if an orchard is due for grubbing. Consequently, paclobutrazol is currently registered for use in fruit trees in the UK and Spain as the only European countries. Likewise, there is no legal use of this compound in fruit trees in the US or in Canada.

A new option for controlling growth of fruit trees became available with the introduction of prohexadione-calcium, which has a half-life in plants of approximately 10 to 14 days and of less than one day in microbially active soil. Due to its simultaneous effects on the formation of GAs, ethylene and flavonoids, treated pome fruit trees benefit in several ways (Rademacher *et al.*, 1998; Byers and Yoder, 1999; Greene, 1999; Owens and Stover, 1999; Unrath, 1999; Yoder *et al.*, 1999; Basak and Rademacher, 2000; Rademacher and Kober, 2003; Rademacher *et al.*, 2004; Costa *et al.*, 2006; Rademacher *et al.*, 2006):

- Figure 12.8 demonstrates that GA_{19} and GA_{20}, inactive precursors of GA_1, accumulate in apple shoots after application of prohexadione-calcium. In contrast, levels of GA_1, the main active GA in apple shoots, decline. The result is reduced shoot elongation. The labour needed for dormant and summer pruning in treated trees is typically reduced by approximately 30%. Furthermore, a more open canopy allows better light penetration into the central parts of the crown leading to improved fruit quality. Spray application of crop protectants is also made more efficient.

- The inhibition of ethylene formation may be employed to increase fruit set. Since assimilates no longer needed for shoot growth are available, fruit size and internal quality, as well as return bloom do not suffer, provided that over-cropping is avoided by appropriate dosaging and timing.
- Blocking flavanone 3-hydroxylase activity in pome fruit shoots leads to a *de novo* formation of 3-deoxyflavonoids, such as luteoforol and luteoliflavan (Halbwirth *et al.*, 2003; Römmelt *et al.*, 2003; Rademacher, 2004b). Luteoforol shows biocidal activity against several bacterial and fungal pathogens, including *Erwinia amylovora* and *Venturia inaequalis*, the causal agents of fire blight and apple scab, respectively (Spinelli *et al.*, 2005). The triggering of such phytoalexin-like compounds explains why treated pome fruit trees are significantly less affected by a number of diseases. This induction of defence is of particular value to control shoot fire blight, a disease caused by the bacterium *Erwinia amylovora*, which is primarily controlled by using antibiotics, a treatment that is highly controversial (Stockwell, 2014). The action of fungicides is promoted by this enhanced resistance, but also by the more open canopy resulting from reduced levels of active GAs.
- Apple and pear trees treated with prohexadione-calcium are also less affected by insect pests, such as aphids, psyllids and leafhoppers (Brisset *et al.*, 2005; Paulson *et al.*, 2005; Leahy *et al.*, 2006), which synergises the action of insecticides. The underlying biochemical mode of action could also be changes in the spectrum of flavonoids, which may, for instance, repel sucking and chewing insects. However, paclobutrazol-treated trees are also less affected by insect pests (Campbell *et al.*, 1989). This indicates that a thicker epidermis or thicker cell walls resulting from reduced elongation growth may also be the reason for less insect attack.

12.4.10 Fruit and nut trees growing in subtropical and tropical climates

Uses of GA_3 and growth retardants in fruit and nut trees growing in warm climates are listed in Table 12.7. GA_3 is of major interest in most citrus-growing countries for use with a variety of citrus species and varieties (El-Otmani *et al.*, 2000; Coggins and Lovatt, 2014). It is mainly applied to increase fruit set, delay harvesting and improve fruit quality. In Navel oranges, rind aging may be delayed. Growers can spray part of their groves to allow sequential harvesting after picking fruits from non-treated blocks early. A delayed harvest may also be used to 'store' citrus fruits (which are non-climacteric) on the tree

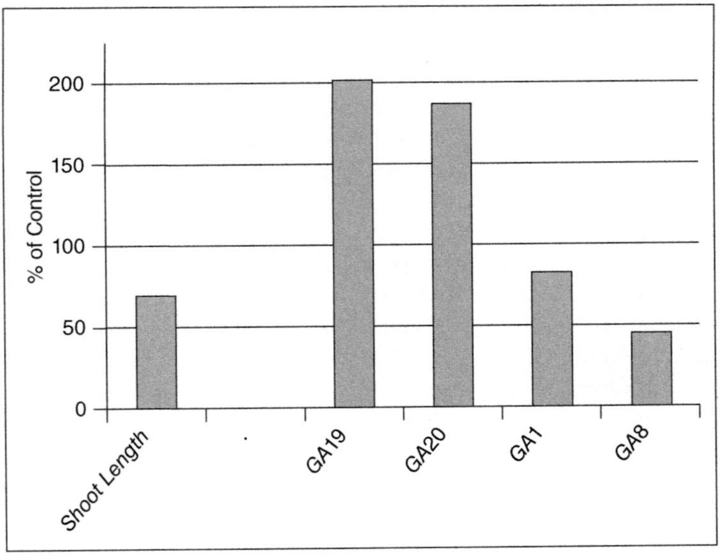

Figure 12.8 Effect of 25 g/ha of prohexadione-calcium (applied as Regalis®) on shoot growth and GA levels in glasshouse-grown plantlets of apple, cv. 'Jonagold' on rootstock M9. Absolute values in control plants: Growth increment of shoot after treatment 36 cm; GA_{19} 30.6 µg/kg, GA_{20} 10.6 µg/kg, GA_1 4.0 µg/kg, GA_8 7.8 µg/kg of shoot dry weight. GA_{29} was also detected. (After Croker *et al.*, 2001)

until a better market window opens. Fruits from treated trees will also display less rind disorders (e.g. rind staining, water spotting, puffy rind, sticky rind). 'Creasing' or 'puff and crease' is an important rind disorder, which is of particular concern in Navel and Valencia oranges and in Satsuma mandarins. The disorder occurs when the rind tissue (orange-coloured flavedo layer plus epidermis) continues to stretch when the albedo layer (white tissue under the rind) has stopped growing. As a result, some parts of the fruit surface appear inflated ('puffy'), whereas other areas are indented (creased). Puffiness occurs also in grapefruits. Application of GA_3, together with certain cultural practices, is employed to reduce the incidence of this disorder. GA_3 may also be used to increase fruit set and yield in Navel and Valencia oranges, as well as in clementines, tangelos and tangerines.

In several countries (e.g. Australia, South Africa, India and Mexico), paclobutrazol and uniconazole or uniconazole-P are used to control excessive shoot growth in fruit and nut trees, such as avocado, mango, litchi, pecan and macadamia. This facilitates tree management (e.g. pruning, application of crop protectants, fruit picking) and more trees may be grown per unit area. Increases in fruit yield and quality are often observed. The longevity of the compounds can be seen as an advantage since less persistent compounds may be degraded too rapidly under the given climatic conditions.

Table 12.7 Uses of GA_3, uniconazole-P and paclobutrazol on sub-tropical and tropical fruit and nut trees.

Crop species	Use	Active ingredient(s)	Main trade name(s)	Main producer(s) and/or supplier(s)
Citrus	Increased fruit set and yield Delay of fruit maturation Improved fruit quality	GA_3	ProGibb®	Valent BioSciences
Avocados	Control of shoot growth Improved fruit shape and size	Uniconazole-P Paclobutrazol	Sunny® Cultar®	Sumitomo Syngenta
Mangos, Litchis Pecan and Macadamia nuts	Control of shoot growth			

Application of paclobutrazol and uniconazole-P is either via spraying the foliage or as a soil drench. As with paclobutrazol in pome and stone fruits, accurate timing and dosage are important. It is recommended not to carry out soil drenches for the last three years before grubbing an orchard in order to avoid growth retardation in subsequently planted crops. Background information on this type of PGR use is given by Yeshitela *et al.* (2004), Menzel and Le Lagadec (2014) and Pires and Yamanishi (2014).

12.4.11 Grapevines

Testing of GA_3 on wine and table grapes started in the late 1950s (Weaver, 1958; Weaver *et al.*, 1962) and led to the first major practical uses of this new plant hormone. As far as can be judged, the main use of GA_3 on a global scale is still in grapevines, particularly in seedless table grapes, where its use has become a standard practice. Table grapes without seeds are attractive to consumers. However, their small size represents a problem for commercialisation. GA_3, applied at the right time and dosage for a given variety, is used to overcome this problem. In general, treatment at approximately 20 mm of cluster length may be used to 'stretch' the rachis, application at 30% to 80% cap (calyptra) fall reduces berry set and later treatments increase berry size. Under advanced production conditions (e.g. in California, Italy or Chile), seedless varieties may be grown on more than 80% of the area where they are virtually all treated with GA_3. For additional reading see Dokoozlian and Peacock (2001) and Casanova *et al.* (2009).

Several varieties of (seeded) wine grapes tend to form very dense clusters. At veraison, the pressure exerted by the ripening and expanding adjacent berries causes wounding and leakage of juice, which may lead to bunch rot (caused by *Botrytis cinerea*) and also to sour rot (caused by different bacteria and yeasts). This is of particular concern to vintners, when rainfall keeps the clusters moist, thereby facilitating the spreading of diseases. GA_3 is used to elongate the rachis and reduce berry set. The window for application lies between a cluster length of approximately 7.5 cm and the end of flower opening. Earlier treatments lead primarily to a stretching of the inflorescence, whereas later timings cause berry thinning or the formation of non-seeded shot berries. Berry abortion may result from too high GA levels, a combination of endogenous hormone produced by the developing seeds and that applied externally. Many varieties, for instance of the Pinot family and Chardonnay, respond relatively well to treatment with GA_3. However, Riesling, Sauvignon Blanc and other varieties suffer from poor induction of shoot and flower buds for the next season and, therefore, are not suitable for GA_3 treatment (Petgen, 2005; Molitor *et al.*, 2012). Recently, an alternative strategy was developed through the application of prohexadione-calcium. Trials at BASF in Germany with a view to controlling excessive shoot growth were without success since too high dosages were required. However, in the course of these investigations, an interesting effect on berry-thinning was observed after applying moderate dosages early in the season. This phenomenon was systematically pursued and prohexadione-calcium is now available for this use in Germany and Austria. Independently, Lo Giudice *et al.* (2003, 2004) at Virginia State University also observed that prohexadione-calcium reduced berry set and suggested that this could be of practical interest. However, there was no follow-up in the USA. Mid flower opening, when the caps on 20% to 80% of the flowers have abscised, is recommended as an ideal timing for prohexadione-calcium application, which overlaps with the time window for GA_3. At first glance, it may appear paradoxical that an inhibitor of GA biosynthesis is giving effects equivalent to those of an active GA. First attempts to find out the underlying mechanisms remained inconclusive (Böll *et al.*, 2009). However, it appears likely that prohexadione-calcium blocks the inactivation of active GAs present at the time of treatment. Its continued activity is likely to be the cause of berry thinning. Different from applying the relatively persistent GA_3, this effect will be relatively short-lived and will not cause negative effects for the following season. This explanation corresponds well with the detailed analytical data on the presence of different GAs in developing grape berries (Giacomelli *et al.*, 2013): GA_1 and GA_4, which are both biologically active, give a clear peak at anthesis and decline sharply

Table 12.8 Uses of GA_3, prohexadione-calcium and mepiquat chloride on grapevines.

Crop Species	Use	Active ingredient(s)	Main trade name(s)	Main producer(s) and/or supplier(s)
Seedless table grapes	Cluster stretching Berry enlargement	GA_3	Activol®	Valent BioSciences
Seeded wine grapes	Cluster stretching Berry thinning	GA_3	ProGibb®	
	Berry thinning	Prohexadione-Ca	Regalis®	BASF
Table and wine grapes	Control of shoot growth	Mepiquat chloride	Pix® Frastar®	BASF Nisso

thereafter. Prohexadione-calcium is likely to inhibit the hydroxylation of these GAs via GA 2-oxidase into inactive GA_8 and GA_{34}, respectively. GA-like effects resulting from treatment with prohexadione-calcium has been reported for *Matthiola incana*. The authors suggest that inactivation of existing GAs by GA 2-oxidase is blocked by prohexadione-calcium as the underlying mechanism (Hisamatsu *et al.*, 1998).

Control of vegetative growth is another objective in the production of table and wine grapes. Excessive cane growth is particularly observed in warmer climates, when there is ample supply of water. As a result, shading and insufficient air circulation often lead to reductions in yield and quality. Additionally, berry production will severely suffer from more intense competition for assimilates from shoot growth. Mepiquat chloride is in use for vegetative growth control, for instance, in Spain, Japan and South Korea (Lim *et al.*, 2004); chlormequat chloride is used in India. An overview on the mentioned compounds and their uses is given in Table 12.8.

12.4.12 Ornamentals

GA_3 has found some use in the production of ornamentals, when longer stems or peduncles are desirable. Increased flowering is also induced by GA_3 in certain species. However, reducing shoot elongation and promoting lateral branching and flowering in ornamental and bedding plants is of much greater relevance: compact and dense plants require less space in a greenhouse, they need less water for irrigation, they have an increased shelf life, but, above all, they sell better because of their dark green leaves, which is generally associated with better quality. A wide assortment of growth retardants is currently available to ornamental growers in the USA. These products are based on chlormequat chloride, daminozide, ancymidol, flurprimidol, paclobutrazol and uniconazole-P (Whipker and Latimer, 2013). Detailed use recommendations for a large number of ornamental species and bedding plants raised under greenhouse conditions are given by Whipker (2013). The spectrum of active ingredients available in EU member countries consists primarily of chlormequat chloride, paclobutrazol and daminozide. In Germany metconazole and prohexadione-calcium are additionally allowed for use in ornamentals. The latter compound should not be used in plants with red or blue flowers because of its interference with anthocyanin formation.

Inhibitors of GA biosynthesis, in particular flurprimidol and paclobutrazol, are also being used in the USA and several other countries to reduce the growth of woody and non-woody ornamentals in gardens or parks. Paclobutrazol and flurprimidol often serve as tools to arborists to limit the size and growth of trees and shrubs in power line and utility rights-of-way corridors. Tree growth regulation is regularly applied in high visibility locations such as parks, historic downtowns, residential areas and other places, where trees have a cultural value and where pruning and trimming is difficult to conduct or unwanted (Chaney, 2005). Whereas paclobutrazol is applied via soil injection or soil drenching, flurprimidol is typically administered via stem injection. Table 12.9 gives an overview on uses of GA_3 and different inhibitors of GA biosynthesis in ornamentals.

Table 12.9 GA_3 and inhibitors of GA biosynthesis used in ornamentals.

Crop species	Use	Active ingredient(s)	Main trade name(s)	Main producer(s) and/or supplier(s)
Ornamentals	Promotion of shoot growth Longer stems or peduncles Intensified flowering	GA_3	ProGibb®	Valent BioSciences
Bedding plants Flowering and foliage plants Bulb plants Woody and non-woody landscape ornamentals	More compact shoots Leaves with darker green Less space required for production	Chlormequat chloride Daminozide Ancymidol Flurprimidol Paclobutrazol Uniconazole-P Metconazole Prohexadione-Ca	Cycocel® B-Nine® A-Rest® Topflor® Bonzi® Sumagic® Caramba® Regalis®	BASF/OHP OHP SePRO SePRO Syngenta Sumitomo BASF BASF

12.4.13 Hybrid seed production

Hybrid cultivars have become common in maize and several other species. However, no cost-efficient hybrid seed production systems exist so far for a number of other crop plants, including wheat. A prerequisite for hybrid breeding is tight pollination control, which avoids self-fertilisation and provides viable pollen from the 'male' plant at the right time and at the right place to fertilise the 'female' plant. GA_3 is used by breeders in several plant species to coordinate the development of the 'male' (fertile) and the 'female' (male sterile) plant for crossing. For instance in rice, GA_3 increases the emergence of the 'female' panicles from the leaf sheath, thereby improving the ability to accept pollen from the 'male' plant (Jagadeeswari *et al.*, 2004; Gavino *et al.*, 2008).

12.5 Outlook

GAs and inhibitors of GA biosynthesis are well established in agriculture, horticulture and viticulture. For the time being, the existing products are suitable for the applications for which they were developed. Therefore and because of the enormous costs involved, it will be difficult to introduce new active ingredients. However, it is likely that additional markets or additional uses for the known PGRs will be found. For instance, climate changes in several European countries have significantly raised the risk of over-growth in winter cereals in late autumn. Therefore, it appears overdue that autumn applications of anti-lodging products are registered in order to avoid winter damage. Prohexadione-calcium is well suited to assist in the production of high-quality strawberry transplants (Reekie *et al.*, 2007). In the same crop, it may also be used to reduce runner formation under long-day conditions, thereby enhancing flower induction and berry yield in the following year (Hytönen *et al.*, 2008). Introducing new combinations of registered PGRs may also offer new and improved solutions. For instance, prohexadione-calcium plus trinexapac-ethyl combine immediate and longer-lasting actions, respectively, in graminaceous species (Rademacher and Strobel, 2010; Rademacher, 2014). Also, active ingredients may lose registration due to toxicological concerns, in which case it is likely that substitutes will be needed. Furthermore, it is possible that a competitive synthesis can be found for *exo*-16,17-dihydro-GA_5-13-acetate. Because of its high specificity, this compound might represent a compound with 'ultra-safe' toxicological features.

References

Acuna, T.B., Merry, A., Dean, G. *et al.* (2014). Plant growth regulators in broad acre crops. In *Proceedings of the 2014 South Australian GRDC Grains Research Update for*

Advisors, Adelaide, 25–26 February, 2014. (eds. J. Crane, and C. James), ORM Communications, Bedigo, pp. 83–87.

Adams, R., Kerber, E., Pfister, K. and Weiler, E.W. (1992). Studies on the action of the new growth retardant CGA 163'935 (cimectacarb). In *Progress in Plant Growth Regulation, Proceedings of the 14th International Conference on Plant Growth Substances, Amsterdam* (eds. C.M. Karssen, L.C. van Loon and D. Vreugdenhil), Kluwer Academic Publishers, Dordrecht, pp. 818–827.

Anderson, A.S. (1979). Plant growth retardants: Present and future use in food production. In *Plant Regulation and World Agriculture* (ed. T.K. Scott), Plenum Press, New York, pp. 251–277.

Anderson, J.D. and Moore, T.C. (1967). Biosynthesis of (–)-kaurene in cell-free extracts of immature pea seeds. *Plant Physiology* **42**, 1527–1534.

Anonymous (2012). *The Pesticide Manual*, 16th edn., BCPC Publications, British Crop Protection Council, Alton, Hampshire.

Anonymous (2014). *2014 Crop Protection Handbook*, Meister Media Worldwide, Willoughby, OH, USA.

Anonymous (2015a). *Global Crop Protection Market 2014*, Phillips McDougall, Crop Protection and Biotechnology Consultants, Edinburgh (quoted in: Annual Report 2014/2015, German Crop Protection Association, Frankfurt).

Anonymous (2015b). *Descriptive Variety List: Cereals, Maize, Oil and Fibre Plants, Pulse Crops, Beets, Catch Crops* (2015), Bundessortenamt, Hannover (in German), [Online], Available: http://www.bundessortenamt.de/internet30/index.php?id=164 [15 Mar 2015]

Austin, R.B. (1999). Yield of wheat in the United Kingdom. Recent advances and prospects. *Crop Science* **39**, 1604–1610.

Baker, C.J., Sterling, M. and Berry, P. (2014). A generalised model of crop lodging. *Journal of Theoretical Biology* **363**, 1–12.

Barrett, J.E. (1982). Chrysanthemum height control by ancymidol, PP333, and EL-500 dependent on medium composition [growth regulators, pine bark, daminozide]. *HortScience* **17**, 896–897.

Basak, A. and Rademacher, W. (2000). Growth regulation of pome and stone fruit trees by use of prohexadione-Ca. *Acta Horticulturae* **514**, 41–50.

Basra, A.S. (ed.) (2000). *Plant Growth Regulators in Agriculture and Horticulture*, Food Products Press, New York.

Beam, J.B. and Askew, S.D. (2005). Prohexadione calcium effects on bermudagrass, Kentucky bluegrass, perennial ryegrass, and zoysiagrass. *International Turfgrass Society Research Journal* **10**, 286–295.

Berry, P.M. and Spink, J.H. (2009). Understanding the effect of a triazole with anti-gibberellin activity on the growth and yield of oilseed rape (*Brassica napus*). *Journal of Agricultural Science* **147**, 273–285.

Berry, P.M. and Spink, J. (2012). Predicting yield losses caused by lodging in wheat. *Field Crops Research* **137**, 19–26.

Berry, P.M., Sterling, M., Spink, J.H. *et al.* (2004). Understanding and reducing lodging in cereals. *Advances in Agronomy* **84**, 217–271.

Böll, S., Hofmann, H. and Schwappach, P. (2009). Rebsortenspezifische Unterschiede in Fertilitätsparametern und ihre Bedeutung für die Anwendung der Wachstumsregulatoren GA_3 und Prohexadion-Calcium. *Mitteilungen Klosterneuburg* **59**, 209–216.

Brisset, M.N., Chartier, R., Didelot, F. *et al.* (2005). Inducteurs de défenses naturelles des plantes: Potentalités contre les bioagresseurs du pommier et du poirier. *Phytoma* **581**, 20–24.

Brown, R.G.S., Kawaide, H., Yang, Y.Y. *et al.* (1997). Daminozide and prohexadione have similar modes of action as inhibitors of late stages of gibberellin metabolism. *Physiologia Plantarum* **101**, 309–313.

Burden, R.S., Carter, G.A., Clark, T. *et al.* (1987). Comparative activity of the enantiomers of triadimenol and paclobutrazol as inhibitors of fungal growth and plant sterol and gibberellin biosynthesis. *Pesticide Science* **21**, 253–267.

Byers, R.E. and Yoder, K.S. (1999). Prohexadione-calcium inhibits apple, but not peach, tree growth, but has little influence on apple fruit thinning or quality. *HortScience* **34**, 1205–1209.

Campbell, C.A.M., Easterbrook, M.A. and Fisher, A.J. (1989). Effect of the plant growth regulators paclobutrazol and chlormequat chloride on pear psyllid (*Cacopsylla pyricola* (Foerster)) and pear rust mite (*Epitrimerus piri* (Nal.)). *Journal of Agricultural Science* **64**, 561–564.

Carlson, R.D. and Crovetti, A.J. (1990). Commercial uses of gibberellins and cytokinins and new areas of applied research. In: *Plant Growth Substances 1988, Proceedings of the 13th International Conference on Plant Growth Substances* (eds. R.P. Pharis and S.B. Rood), Springer-Verlag, Berlin, pp. 604–610.

Casanova, L., Casanova, R., Moret, A. and Agusti, M. (2009). The application of gibberellic acid increases berry size of 'Emperatriz' seedless grape. *Spanish Journal of Agricultural Research* **7**, 919–927.

Cathey, G.W. and Meredith, W.R. (1988). Cotton response to planting date and mepiquat chloride. *Agronomy Journal* **80**, 463–466.

Chaney, W.R. (2005). *Growth retardants: A promising tool for managing urban trees.* Publication FNR-252-W, Purdue Extension, Purdue University, Purdue, 5 p.

Coggins, C.W. and Lovatt, C.J. (2014). Plant growth regulators. In: *Citrus Production Manual* (eds. L. Ferguson and E.E. Grafton-Cardwell), University of California, Agricultural and Natural Resources, Publication 3539, Richmont, pp. 215–226.

Cook, D.R. and Kennedy, C.W. (2000). Early flower bud loss and mepiquat chloride effects on cotton yield distribution. *Crop Science* **40**, 1678–1684.

Coolbaugh, R.C. and Hamilton, R. (1976). Inhibition of *ent*-kaurene oxidation and growth by α-cyclopropyl-α-(p-methoxyphenyl)-5-pyrimidine methyl alcohol. *Plant Physiology* **57**, 245–248.

Coolbaugh, R.C., Hirano, S.S. and West, C.A. (1978). Studies on the specificity and site of action of α-cyclopropyl-α-[p-methoxyphenyl]-5-pyrimidine methyl alcohol (ancymidol), a plant growth regulator. *Plant Physiology* **62**, 571–576.

Costa, G., Andreotti, C., Spinelli, F. and Rademacher, W. (2006). Prohexadione-Ca: More than a growth regulator for pome fruit trees. *Acta Horticulturae* **727**, 107–116.

Cotterill, P. (2005). Method of altering the alkaloid composition in poppy plants. International Patent Application WO/2005/107436.

Croker, S.J., Hedden, P. and Rademacher, W. (2001). Effects of prohexadione-Ca on GA levels in apple shoots. In *Proceedings of the 27th Annual Meeting of the Plant Growth Regulation Society of America*, (ed. R. Dunand), The Plant Growth Regulation Society of America, LaGrange, p. 259.

Dean, G.J. (2011). *Modifying Poppy Growth and Alkaloid Yield with Plant Growth Regulators.* Master Thesis, University of Tasmania, Hobart.

Dokoozlian, N.K. and Peacock, W.L. (2001). Gibberellic acid applied at bloom reduces fruit set and improves size of 'Crimson Seedless' table grapes. *HortScience* **34**, 706–709.

Dunand, R.T. (1992). Enhancement of seedling vigor in rice (*Oryza sativa* L.) by seed treatment with gibberellic acid. In *Progress in Plant Growth Regulation. Proceedings of the 14th International Conference on Plant Growth Substances,* (eds. C.M. Karssen, L.C. van Loon and D. Vreugdenhil), Kluwer Academic Publishers, Dordrecht, pp. 835–841.

Easson, D.L., White, E.M. and Pickles, S.J. (1993). The effects of weather, seed rate and cultivar on lodging and yield in winter wheat. *Journal of Agricultural Science* **121**, 145–156.

Edmisten, K.L., Yelverton, F.H., Spears, J.F. *et al.* (eds) (2010). *2010 Cotton Information,* North Carolina State University, Raleigh.

El-Otmani, M., Coggins, C.W., Agusti, M. and Lovatt, C.J. (2000). Plant growth regulators in citriculture: World current uses. *Critical Reviews in Plant Science* **19**, 395–447.

Evans, L.T., King, R.W., Mander, L.N. *et al.* (1994). The differential effects of C-16,17-dihydro gibberellins and related compounds on stem elongation and flowering in *Lolium temulentum*. *Planta* **193**, 107–114.

Farrow, S.C. and Facchini, P.J. (2014). Functional diversity of 2-oxoglutarate/Fe(II)-dependent dioxygenases in plant metabolism. *Frontiers in Plant Science* **5**:524.

Fletcher, R.A., Gilley, A., Sankhla, N. and Davis, T.D. (2000). Triazoles as plant growth regulators and stress protectants. *Horticultural Reviews* **24**, 55–138.

Flintham, J.E., Boerner, A., Worland, A.J. and Gale, M.D. (1997). Optimizing wheat grain yield: Effects of Rht (gibberellin-insensitive) dwarfing genes. *Journal of Agricultural Science* **128**, 11–25.

Foster, K.R., Lee, I.J,, Pharis, R.P. and Morgan, P.W. (1997). Effects of ring D-modified gibberellins on gibberellin levels and development in selected *Sorghum bicolor* maturity genotypes. *Journal of Plant Growth Regulation* **16**, 79–87.

Frost, R.G. and West, C.A. (1977). Properties of kaurene synthetase from *Marah macrocarpus*. *Plant Physiology* **59**, 22–29.

Garthwaite, D.G., Hudson, S., Barker, I. *et al.* (2013). Arable crops in the United Kingdom 2012. *Pesticide usage survey report No. 250*. Food and Environment Research Agency, York. [Online], Available: https://secure.fera.defra.gov.uk/pusstats/surveys/documents/arable2012.pdf [15 Mar 2015]

Gavino, R.B., Pi, Y. and Abon, C.C. (2008). Application of gibberellic acid (GA_3) in dosage for three hybrid rice seed production in the Philippines. *Journal of Agricultural Technology* **4**, 183–192.

Giacomelli, L., Rota-Stabelli, O., Masuero, D. *et al.* (2013). Gibberellin metabolism in *Vitis vinifera* L. during bloom and fruit set: functional characterization and evolution of grapevine gibberellin oxidases. *Journal of Experimental Botany* **64**, 4403–4419.

Gianfagna, T. (1995). Natural and synthetic growth regulators and their use in horticultural and agronomic crops. In *Plant Hormones* (ed. P.J. Davies), Kluwer Academic Publishers, Dordrecht, pp. 751–773.

Graebe J.E. (1987). Gibberellin biosynthesis and control. *Annual Review of Plant Physiology* **38**, 419–465.

Graebe, J.E., Lange, T., Pertsch, S. and Stöckl, D. (1991). The relationship of different gibberellin biosynthetic pathways in *Cucurbita maxima* endosperm and embryos and the purification of a C-20 oxidase from the endosperm. In *Gibberellins* (eds. N. Takahashi, B.O. Phinney and J. MacMillan), Springer-Verlag, New York, pp. 51–61.

Graebe, J.E., Böse, G., Grosselindemann, E. *et al.* (1992). The biosynthesis of *ent*-kaurene in germinating seeds and the function of 2-oxoglutarate in gibberellin biosynthesis. In *Progress in Plant Growth Regulation, Proceedings of the 14th International Conference on Plant Growth Substances* (eds. C.M. Karssen, L.C. van Loon and D. Vreugdenhil), Kluwer Academic Publishers, Dordrecht, pp. 545–554.

Greene, D.W. (1999). Tree growth management and fruit quality of apple trees treated with prohexadione-calcium (BAS 125). *HortScience* **34**, 1209–1212.

Greene, D.W. (2010). The development and use of plant bioregulators in tree fruit production. *Acta Horticulturae* **884**, 31-40.

Griggs, D.L., Hedden, P., Temple-Smith, K.E. and Rademacher, W. (1991). Inhibition of gibberellin 2ß-hydroxylases by acylcyclohexanedione derivatives. *Phytochemistry* **30**, 2513–2517.

Grossmann, K. (1992). Plant growth retardants: Their mode of action and benefit for physiological research. In: *Progress in Plant Growth Regulation, Proceedings of the 14th International Conference on Plant Growth Substances* (eds. C.M. Karssen, L.C. van Loon and D. Vreugdenhil), Kluwer Academic Publishers, Dordrecht, pp. 788–797.

Gwathmey, C.O. and Clement, J.D. (2010). Alteration of cotton source-sink relations with plant population density and mepiquat chloride. *Field Crops Research* **116**, 101–107.

Hagel, J.M. and Facchini, P.J. (2010). Dioxygenases catalyze the O-demethylation steps of morphine biosynthesis in opium poppy. *Nature Chemical Biology* **6**, 273–275.

Halbwirth, H., Fischer, T.C., Römmelt, S. *et al.* (2003). Induction of antimicrobial 3-deoxyflavonoids in pome fruit trees controls fire blight. *Zeitschrift für Naturforschung* **58c**, 765–770.

Halbwirth, H., Fischer, T.C., Schlangen, K. et al. (2006). Screening for inhibitors of 2-oxoglutarate-dependent dioxygenases: Flavanone 3ß-hydroxylase and flavonol synthase. *Plant Science* **171**, 194–205.

Hedden, P. (1990). The action of plant growth retardants at the biochemical level. In *Plant Growth Substances 1988.* (eds. R.P. Pharis and S.B. Rood), Springer-Verlag, Berlin, pp. 322–332.

Hedden, P. (1991). Gibberellin biosynthetic enzymes and the regulation of gibberellin concentrations. In *Gibberellins.* (eds. N Takahashi, B.O. Phinney and J. MacMillan), Springer-Verlag, New York, pp. 94–105.

Hedden, P. (2012). Gibberellin biosynthesis. In *eLS.* John Wiley and Sons, Ltd., Chichester. http://www.els.net [doi: 10.1002/9780470015902.a0023720]

Hedden, P. and Graebe, J.E. (1985). Inhibition of gibberellin biosynthesis by paclobutrazol in cell-free homogenates of *Cucurbita maxima* endosperm and *Malus pumila* embryos. *Journal of Plant Growth Regulation* **4**, 111–122.

Hedden, P. and Hoad, G.V. (1994). Growth regulators and crop productivity. In: *Mechanisms of Plant Growth and Improved Productivity.* (ed. *A.S. Basra)*, pp. 173-98. Marcel Dekker Inc., New York.

Hedden, P. and Kamiya, Y. (1997). Gibberellin biosynthesis: Enzymes, genes and their regulation. *Annual Review of Plant Physiology and Plant Molecular Biology* **48**, 431–460.

Hedden, P. and Thomas, S.G. (2012). Gibberellin biosynthesis and its regulation. *Biochemical Journal* **444**, 11–25.

Hildebrandt, E. (1982). Der Einfluß von Ausgewählten Wachstumshemmern auf die Gibberellinbiosynthese in Zellfreien Systemen. Diploma Thesis, University of Göttingen, Göttingen.

Hisamatsu T., Koshioka, M. Kubota, S. and King, R.W. (1998). Effect of gibberellin A_4 and GA biosynthesis inhibitors on growth and flowering of stock [*Matthiola incana* (L.) R. Br.]. *Journal of the Japanese Society for Horticultural Science* **67**, 537–543.

Hoffmann, G. (1992). Use of plant growth regulators in arable crops: Survey and outlook. In: *Progress in Plant Growth Regulation, Proceedings of the 14th International Conference on Plant Growth Substances* (eds. C.M. Karssen, L.C. van Loon and D. Vreugdenhil), Kluwer Academic Publishers, Dordrecht, pp. 571–577.

Hytönen, T., Mouhu, K., Koivu, I. and Junttila, O. (2008). Prohexadione-calcium enhances the cropping potential and yield of strawberry. *European Journal of Horticultural Science* **73**, 210–215.

Iturriagagoitia-Bueno, T., Gibson, E.J., Schofield, C.J. and John, P. (1996). Inhibition of 1-aminocyclo-propane-1-carboxylate oxidase by 2-oxoacids. *Phytochemistry* **43**, 343–349.

Izumi, K., Yamaguchi, I., Wada, A. *et al.* (1984). Effects of a new plant growth retardant *(E)*-1-(4-chlorophenyl)-4,4-dimethyl-2-(1,2,4-triaxol-1-yl)-penten-3-ol (S-3307) on the growth and gibberellin content of rice plants. *Plant and Cell Physiology* **25**, 611–617.

Izumi, K., Kamiya, Y., Sakurai, A. *et al.* (1985). Studies of sites of action of a new plant growth retardant *(E)*-1-(4-chlorophenyl)-4,4-dimethyl-2-(1,2,4-triaxol-l-yl)-1-penten -3-ol (S-3307) and comparative effects of its stereoisomers in a cell-free system from *Cucurbita maxima*. *Plant and Cell Physiology* **26**, 821–827.

Jagadeeswari, P., Sharma, S.P. and Dadlani, M. (2004). Effect of different chemicals on traits favouring outcrossing and optimization of GA_3 for seed production of cytoplasmic male sterile line in hybrid rice. *Seed Science and Technology* **32**, 473–483.

Jennings, P.R. and Sornchai, S. (1964). The influence of lodging on rice yields. *International Rice Commission Newsletter* **13**, 10–21.

Jordan, D.L., Beam, J.B., Johnson, P.D. and Spears, J.F. (2001). Peanut response to prohexadione calcium in three seeding rate-row pattern planting systems. *Agronomy Journal* **93**, 232–236.

Jordan, D.L., Nuti, R.C., Beam, J.B. *et al.* (2008). Peanut (*Arachis hypogaea* L.) cultivar response to prohexadione calcium. *Peanut Science* **35**, 101–107.

Jost, P., Whitaker, J., Brown, S.M. and Bednarz, C. (2006). Use of plant growth regulators as a management tool in cotton. *Bulletin 1305*, The University of Georgia Cooperative Extension, Athens, 12 p.

Jung, J. and Sturm, H. (1964). Wachstumsregulierende Wirkung von Chlorcholinchlorid (CCC). *Landwirtschaftliche Forschung* **17**, 1–9.

Jung, J. and Rademacher, W. (1983). Plant growth regulating chemicals – cereal grains. In *Plant Growth Regulating Chemicals: Volume I* (ed. L.G. Nickell), CRC Press, Boca Raton. pp. 253–271.

Junttila, O., King, R.W., Poole, A. *et al.* (1997). Regulation in *Lolium temulentum* of the metabolism of gibberellin A_{20} and gibberellin A_1 by 16,17-dihydro GA_5 and by the growth retardant, LAB 198 999. *Australian Journal of Plant Physiology* **24**, 359–369.

Kamiya, Y., Nakayama, I. and Kobayashi, M. (1992). Useful probes to study the biosynthesis of gibberellins. In: *Progress in Plant Growth Regulation, Proceedings of the 14th International Conference on Plant Growth Substances* (eds. C.M. Karssen, L.C. van Loon and D. Vreugdenhil), Kluwer Academic Publishers, Dordrecht, pp. 555–565.

Katagi, T., Mikami, N., Matsuda, T. and Miyamoto, J. (1987). Structural studies of the plant growth regulator uniconazole (ES pure) and computer-aided analysis of its interaction with cytochrome P-450. *Journal of Pesticide Science* **12**, 627–633.

Kerber, E., Leypoldt, G. and Seiler, A. (1990). CGA 163'935, a new plant growth regulator for small grain cereals, rape and turf. In *Proceedings Brighton Crop Protection Conference: Weeds, 1989*, The British Crop Protection Council, Farnhem, pp. 83–88.

Kerby, T.A., Hake, K. and Keeley, M. (1986). Cotton fruiting modification with mepiquat chloride. *Agronomy Journal.* **78**, 907–912.

Kerby, T.A., Weir, B.L. and Keeley, M.P. (1996). In *Cotton Production Manual* (eds. S.J. Hake, T.A. Kerby and K.D. Hake), Publication No. 3352, University of California, Oakland, pp. 294–304.

Kightley, S. (2001). Lodging control in winter rape. *Plant Varieties and Seeds* **14**, 171–180.

King, R.W., Blundell, C., Evans, L.T. *et al.* (1997). Modified gibberellins retard growth of cool-season turfgrasses. *Crop Science* **37**, 1878–1883.

King, R.W., Junttila, O., Mander, L.N. and Beck, E.J. (2004). Gibberellin structure and function: biological activity and competitive inhibition of gibberellin 2- and 3-oxidases. *Physiologia Plantarum* **120**, 287–297.

Kuck, K.H. and Berg, D. (1986). Systemische Eigenschaften und biochemischer Wirkungsmechanismus eines neuen Azolfungizids. *Mitteilungen aus der Biologischen Bundesanstalt für Land- und Forstwirtschaft Berlin-Dahlem* **232**, 196.

Kust, C.A. (1993). Effects of Surestem plant growth modifier on stem number, length, and quality of hybrid tea roses grown in greenhouses. *HortScience* **28**, 548.

Leahy, K., Greene, D.W., Autio, W.R. and Norelli, J.L. (2006). Potato leafhoppers, fire blight and prohexadione-Ca: A look at their interaction under field conditions. *Acta Horticulturae* **704**, 249–251.

Lee, I.J., Foster, K.R. and Morgan, P.W. (1998). Effect of gibberellin biosynthesis inhibitors on native gibberellin content and floral initiation in *Sorghum bicolor*. *Journal of Plant Growth Regulation* **17**, 185–195.

Lenton, J.R., Hedden, P. and Gale, M.D. (1987). Gibberellin insensitivity and depletion in wheat - consequences for development. In *Hormone Action in Plant Development: A Critical Appraisal*, (eds. G.V. Hoad, J.R. Lenton, M.B. Jackson and R.K. Atkin), Butterworths, London, pp. 145–160.

Lever, B.G., Shearing, S.J. and Batch, J.J. (1982). PP333-a new broad spectrum growth retardant, In *Brighton Crop Protection Conference - Weeds 1982, Volume 1*, British Crop Protection Council, Croydon, pp. 3–10.

Lim, S.C., Kim, S.K., Kim, Y.H. *et al.* (2004). Vine growth and fruit quality of 'Kyoho' grapes as affected by mepiquat chloride and GA. *Acta Horticulturae* **653**, 145–149.

Linser, H., Mayer, H.H. and Bodo, G. (1961). Über die Wirkung von Chlorcholinchlorid auf Sommerweizen. *Bodenkultur* **12**, 279–280.

Lo, S.C.C., de Verdier, K. and Nicholson, R.L. (1999). Accumulation of 3-deoxyanthocyanidin phytoalexins and resistance to *Collelotrichum sublineolum* in sorghum. *Physiology and Molecular Plant Pathology* **55**, 263–273.

Lo Giudice, D., Wolf, T.K. and Marini, R.P. (2003). Vegetative response of *Vitis vinifera* to prohexadione-calcium. *HortScience* **38**, 1435–1438.

Lo Giudice, D., Wolf, T.K. and Zoecklein, B.W. (2004). Effects of prohexadione-calcium on grape yield components and fruit and wine composition. *American Journal of Enology and Viticulture* **55**, 73–83.

Looney, N. and Jackson, D. (2011). Use of bioregulators in fruit production. In *Temperate and Subtropical Fruit Production* (eds. D. Jackson, N.E. Looney, M. Morley-Bunker and G. Thiele), CABI, Wallingford, Oxfordshire, pp. 92–100.

Los, M., Kust, C.A., Lamb, G. and Diehl, R.E. (1980). Phthalimides as plant growth regulators. *HortScience* **15**, 22.

Lürssen, K. (1987). The use of inhibitors of gibberellin and sterol biosynthesis to probe hormone action. In *Hormone Action in Plant Development* (eds. G.V. Hoad, J.R. Lenton, M.B. Jackson and R.K. Atkin), Butterworths, London, pp. 133–144.

Luster, D.G. and Miller, P.A. (1993). Triazole plant growth regulator binding to native and detergent-solubilized plant microsomal cytochrome P450. *Pesticide Biochemistry and Physiology* **46**, 27–39.

Mander, L. (2003). Twenty years of gibberellin research. *Natural Products Reports* **20**, 49–69.

Mander, L.N., Camp, D., Evans, L.T. *et al.* (1995). Designer gibberellins: the quest for specific activity. *Acta Horticulturae* **394**, 45–55.

Mander, L.N., Sherburn, M., Camp, D. *et al.* (1998). Effects of D-ring modified gibberellins on flowering and growth of *Lolium temulentum*. *Phytochemistry* **49**, 2195–2206.

March, S.R., Martins, D. and McElroy, J.S. (2013). Growth inhibitors in turfgrass. *Planta Daninha* **31**, 733–747.

Martin, G.C. (1983). Commercial uses of gibberellins. In *The Biochemistry and Physiology of gibberellins* (ed. A. Crozier), Praeger Publishers, New York, pp. 395–444.

Matthew, C., Hofmann, W.A. and Osborne, M.A. (2009). Pasture response to gibberellins: a review and recommendations. *New Zealand Journal of Agricultural Research* **52**, 213–225.

Menzel, C.M. and Le Lagadec, M.D. (2014). Increasing the productivity of avocado orchards using high-density plantings: A review. *Scientia Horticulturae* **177**, 21–36.

Miki, T., Kamiya, Y., Fukazawa, M. *et al.* (1990). Sites of inhibition by a plant-growth regulator, 4′-chloro-2′-(α-hydroxybenzyl)-isonicotinanilide (inabenfide), and its related compounds in the biosynthesis of gibberellins. *Plant and Cell Physiology* **31**, 201–206.

Molitor, D., Behr, M., Hoffmann, L and Evers, D. (2012). Research note: Benefits and drawbacks for pre-bloom applications of gibberellic acid (GA_3) for stem elongation in Sauvignon Blanc. *South African Journal of Enology and Viticulture* **33**, 198–202.

Moody, K. (1986). Role of plant growth regulators in tropical rice cultivation. In *Plant Growth Regulators in Agriculture* (eds. P. Macgregor, Y. Saito and R. Wang), Food and Fertilizer Technology Center for the Asian and Pacific Region, Taipei, pp. 140–147.

Müller, U., Huxley, P. and Ebert, E. (1987). Computer assisted molecular modelling (CAMM) – A tool for structure-activity considerations of inhibitors of the enzymatic oxidation of *ent*-kaurene to *ent*-kaurenoic acid. In *Pesticide Science and Biotechnology, Proceedings of the Sixth International Congress on Pesticide Chemistry 1986.* (eds. R. Greenhalgh and T.R. Roberts), Blackwell, Oxford, pp. 69–72.

Na, C.I., Hamayun, M., Khan, A.L. *et al.* (2011). Influence of prohexadione-calcium, trinexapac-ethyl and hexaconazole on lodging characteristic and gibberellin biosynthesis of rice (*Oryza sativa* L.). *African Journal of Biotechnology* **10**, 13097–13106.

Nakayama, I., Miyazawa, T., Kobayashi, M. *et al.* (1990a). Effects of a new plant growth regulator prohexadione calcium (BX-112) on shoot elongation caused by exogenously applied gibberellins in rice (*Oryza sativa* L.) seedlings. *Plant and Cell Physiology* **31**, 195–200.

Nakayama, I., Kamiya, Y., Kobayashi, M. *et al.* (1990b). Effect of a plant growth regulator, prohexadione, on the biosynthesis of gibberellins in cell-free systems derived from immature seeds. *Plant and Cell Physiology* **31**, 1183–1190.

Nakayama, I., Miyazawa, T., Kobayashi, M. *et al.* (1991). Studies on the action of the plant growth regulators BX-112, DOCHC, and DOCHC-Et. In *Gibberellins* (eds.

N. Takahashi, B.O. Phinney and J. MacMillan), Springer-Verlag, New York, pp. 311–319.

Nickell, L.G. (ed.) (1983). *Plant Growth Regulating Chemicals, Volumes 1 and 2*, CRC Press, Boca Raton.

Nishiyama, I. (1986). Lodging of rice plants and countermeasures. In *Plant Growth Regulators in Agriculture* (eds. P. Macgregor, Y. Saito and R. Wang), Food and Fertilizer Technology Center for the Asian and Pacific Region, Taipei, pp. 152–163.

Norton, E.R. and Borrego, H.J. (2006). Evaluation of plant growth regulator formulations in Arizona cotton production systems. In *2006 Cotton Report (Series P-145)*, The University of Arizona, Tucson, pp. 143–152.

Owens, C.L. and Stover, E. (1999). Vegetative growth and flowering of young apple trees in response to prohexadione-calcium. *HortScience* **34**, 1194–1196.

Paulson, G.S., Hull, L.A. and Biddinger, D.J. (2005). Effect of a plant growth regulator prohexadione-calcium on insect pests of apple and pear. *Journal of Economic Entomology* **98**, 423–431.

Petgen, M. (2005). Gibberellin-Einsatz zur Qualitätsregulierung. *Schweizer Zeitschrift für Obst- und Weinbau* **2005 (7)**, 6–9.

Petracek, P.D., Silverman, F.P. and Greene, D.W. (2003). A history of commercial plant growth regulators in apple production. *HortScience* **38**, 937–942.

Petry, W., Rademacher, I. and Kühbauch, W. (1989). Messung und Dokumentation mikromorphologischer Veränderungen in Pflanzen am Beispiel CCC-behandelten Sommerweizens mit Hilfe der quantitativen Bildanalyse. *Journal of Agronomy and Crop Science* **162**, 73–80.

Pinthus, M.J. (1973). Lodging in wheat, barley and oats: the phenomenon, its causes and preventative measures. *Advances in Agronomy* **25**, 209–263.

Pires, M.C. and Yamanishi, O.K. (2014). Girdling combined with paclobutrazol boosted yield of 'Bengal' lychee in Brazil. *Acta Horticulturae* **1042**, 189–195.

Rademacher, W. (1993). On the mode of action of acylcyclohexanediones - a new type of plant growth retardant with possible relationships to daminozide. *Acta Horticulturae* **329**, 31–34.

Rademacher, W. (2000). Growth retardants: effects on gibberellin biosynthesis and other metabolic pathways. *Annual Review of Plant Physiology and Plant Molecular Biology* **51**, 501–531.

Rademacher, W. (2004a). Recent situation and trends in global plant bioregulator utilization. *Regulation of Plant Growth and Development* **39**, 142–151.

Rademacher, W. (2004b). Prohexadione-Ca induces resistance to fireblight and other diseases. *EPPO Bulletin* **34**, 383–388.

Rademacher, W. (2010a). Dealing with plant bioregulators: An industrial view. *Acta Horticulturae* **884**, 717–724.

Rademacher, W. (2010b). Control of lodging in intense European cereal production. In *Proceedings of the 36th Annual Meeting of the Plant Growth Regulation Society of America* (ed. B. Whipker), The Plant Growth Regulation Society of America, LaGrange, pp. 61–69.

Rademacher, W. (2014). Prohexadione-Ca and trinexapac-ethyl: Similarities in structure but differences in biological action. *Acta Horticulturae* **1042**, 33–41.

Rademacher, W. (2015). Determination of root growth in field-grown and PGR-treated winter wheat. In *2014 Meeting Summary of the 41st Annual Meeting of the Plant Growth Regulation Society of America* (ed. C. Sams), The Plant Growth Regulation Society of America, Alexandria, 2 p.

Rademacher, W. and Brahm, L. (2010). *Plant growth regulators*. In *Ullmann's Encyclopedia of Industrial Chemistry*, Wiley-VCH, Weinheim, 14 p.

Rademacher, W. and Graebe, J.E. (1979). Gibberellin A_4 produced by *Sphaceloma manihoticola*, the cause of the superelongation disease of cassava (*Manihot esculenta*). *Biochemical and Biophysical Research Communications* **91**, 35–40.

Rademacher, W. and Kober, R. (2003). Efficient use of prohexadione-Ca in pome fruits. *European Journal of Horticultural Science* **68**, 101–107.

Rademacher, W. and Strobel, D. (2010). Use of acylcyclohexanedione carboxylic acid or salts thereof in combination with acylcyclohexanedione carboxylic acid esters for improving the development of graminaceous plants. International Patent WO/2010/034813.

Rademacher, W., Fritsch, H., Graebe, J.E. *et al.* (1987). Tetcyclacis and triazole-type plant growth retardants: their influence on the biosynthesis of gibberellins and other metabolic processes. *Pesticide Science* **21**, 241–252.

Rademacher, W., Temple-Smith, K.E., Griggs, D.L. and Hedden, P. (1992). The mode of action of acylcyclohexanediones - a new type of plant growth retardant. In *Progress in Plant Growth Regulation, Proceedings of the 14th International Conference on Plant Growth Substances* (eds. C.M. Karssen, L.C. van Loon and D. Vreugdenhil), Kluwer Academic Publishers, Dordrecht, pp. 571–577.

Rademacher, W., Speakman, J.B., Evans R.R. *et al.* (1998). Prohexadione-Ca - a new plant growth regulator for apple with interesting biochemical features. In *Proceedings of the 25th Annual Meeting of The Plant Growth Regulation Society of America* (ed. W.E. Shafer), The Plant Growth Regulation Society of America, LaGrange, pp. 113–118.

Rademacher, W., Pharis, R.P. and Mander, L.N. (1999). Agricultural use of new GA derivatives. *Japanese Journal of Crop Science* **68 (S)**, 362–367.

Rademacher, W., van Saarloos, K., Garuz Porte, J.A. *et al.* (2004). Impact of prohexadione-Ca on the vegetative and reproductive performance of apple and pear trees. *European Journal of Horticultural Science* **69**, 221–228.

Rademacher, W., Spinelli, F. and Costa, G. (2006). Prohexadione-Ca: Modes of action of a multifunctional plant bioregulator for fruit trees. *Acta Horticulturae* **727**, 97–106.

Reekie, J.Y., Struik, P.C., Hicklenton, P.R. and Duval, J.R. (2007). Dry matter partitioning in a nursery and a plasticulture fruit field of strawberry cultivars 'Sweet Charlie' and 'Camarosa' as affected by prohexadione-calcium and partial leaf removal. *European Journal of Horticultural Science* **72**, 122–129.

Resende, P.A.P., Soares, J.E. and Hudetz, M. (2000). MODDUS®, a plant growth regulator and management tool for sugarcane production in Brazil. *International Sugar Journal* **102**, 5–9.

Rhodes, D. and Hanson, A.D. (1993). Quaternary ammonium and tertiary sulfonium compounds in higher plants. *Annual Review of Plant Physiology and Plant Molecular Biology* **44**, 357–384.

Riddell, J.A., Hageman, H.A., J'Anthony, C.M. and Hubbard, W.L. (1962). Retardation of plant growth by a new group of chemicals. *Science* **136**, 391.

Rixon, C.M., Di Bella, L.P., Kingston, G. *et al.* (2007). Moddus - a sugar enhancer. *Proceedings of the Australian Society of Sugar Cane Technology* **29**, 318–327.

Römmelt, S., Zimmermann, N., Rademacher, W. and Treutter, D. (2003). Formation of novel flavonoids in apple (*Malus×domestica*) treated with the 2-oxoglutarate-dependent dioxygenase inhibitor prohexadione-Ca. *Phytochemistry* **64**, 709–716.

Salassi, M.E., Deliberto, M.A., Linscombe, S.D. *et al.* (2013). Impact of harvest lodging on rough rice milling yield and market price. *Agronomy Journal* **105**, 1860–1867.

Sampson, A.J., Cazenave, A., Laffranque, J.P. *et al.* (1992). Metconazole, an advance in disease control in cereals and other crops. In *Brighton Crop Protection Conference: Pests and Diseases*, The British Crop Protection Council, Farnhem, pp. 419–426.

Santes, C.M. and García-Martínez, J.L. (1995). Effect of the growth retardant 3,5-dioxo-4-butyryl-cyclohexane carboxylic acid ethyl ester, an acylcyclohexanedione compound, on fruit growth and gibberellin content of pollinated and unpollinated ovaries in pea. *Plant Physiology* **108**, 517–523.

Schott, P.E. and Walter, H. (1991). Bioregulators: Present and future fields of application. In *Plant Biochemical Regulators* (ed. H. Gausman), Marcel Dekker, New York, pp. 247–321.

Shechter, I. and West, C.A. (1969). Biosynthesis of gibberellins. IV. Biosynthesis of cyclic diterpenes from trans-geranylgeranyl pyrophosphate. *Journal of Biological Chemistry* **244**, 3200–3209.

Shibata, H. and Takebayashi, Y. (2007). Development of a forefront one-shot basal application fertilizer 'Rakuichi®' for paddy rice. *Sumitomo Kagaku* **2007-I**, 1-13. (English translation)

Shive, J.B. and Sisler, H.D. (1976). Effects of ancymidol (a growth retardant) and triarimol (a fungicide) on the growth, sterols, and gibberellins of *Phaseolus vulgaris* (L.). *Plant Physiology* **57**, 640–644.

Spinelli, F., Speakman, J.B., Rademacher, W. *et al.* (2005). Luteoforol, a flavan 4-ol, is induced in pome fruits by prohexadione-calcium and shows phytoalexin-like properties against *Erwinia amylovora* and other pathogens. *European Journal of Plant Pathology* **112**, 133–142.

Sponsel, V.M. and Reid, J.B. (1992). The effect of the growth retardant LAB 198 999 and its interaction with gibberellins A_1, A_3, and A_{20} in fruit growth of tall and dwarf peas. In: *Progress in Plant Growth Regulation, Proceedings of the 14th International Conference on Plant Growth Substances,* (eds. C.M. Karssen, L.C. van Loon and D. Vreugdenhil), Kluwer Academic Publishers, Dordrecht, pp. 578–584.

Sponsel, V.M. and Hedden, P. (2010). Gibberellin biosynthesis and inactivation. In *Plant Hormones, Biosynthesis, Signal Transduction, Action!* (ed. P.J. Davies), 3rd edn, Springer, Dordrecht, pp. 578–584.

Stapleton, G.S. and Via, J.D. (2003). Pentia™ (BAS 130): next generation plant growth regulator for enhanced growth control, boll retention and cotton lint yield. In *Proceedings of the Beltwide Cotton Conference* (eds. P. Dugger and D. Richter), National Cotton Council, Memphis, p. 1794.

Stockwell, V.O. (2014). Overview of concerns surrounding antibiotic use for control of fire blight. *Acta Horticulturae* **1056**, 39–42.

Sturm, H., Buchner, A. and Zerulla, W. (1994). *Gezielter Düngen*. DLG-Verlags-GmbH, Frankfurt.

Sugavanam, B. (1984). Diastereoisomers and enantiomers of paclobutrazol: their preparation and biological activity. *Pesticide Science* **15**, 296–302.

Takagi, M., Pearce, D.W., Janzen, L.M. and Pharis R.P. (1994). Effect of *exo*-16,17 -dihydro-gibberellin A_5 on gibberellin A_{20} metabolism in seedlings of dwarf rice (*Oryza sativa* L. cv. Tan-ginbozu). *Plant Growth Regulation* **15**, 207–213.

Thomson, W.T. (1999). *Agricultural Chemicals, Book III, Miscellaneous Agricultural Chemicals,* Thomson Publications, Fresno.

Tschabold, E.E., Taylor, H.M., Davenport J.D. *et al.* (1970). A new plant growth regulator. *Plant Physiology* **46**, S-19.

Tolbert, N.E. (1960a). (2-Chloroethyl)-trimethylammoniumchloride and related compounds as plant growth substances. I. Chemical structure and bioassay. *Journal of Biological Chemistry* **235**, 475–479.

Tolbert, N.E. (1960b). (2-Chloroethyl)-trimethylammoniumchloride and related compounds as plant growth substances. II. Effect on growth of wheat. *Plant Physiology* **35**, 380–385.

Unrath, C.R. (1999). Prohexadione-Ca: A promising chemical for controlling vegetative growth of apples. *HortScience* **34**, 1197–1200.

Weaver, R.J. (1958). Effect of gibberellic acid on fruit set and berry enlargement in seedless grapes of *Vitis vinifera*. *Nature* **181**, 851–852.

Weaver, R.J. (1972). *Plant Growth Substances in Agriculture*. W. H. Freeman and Company, San Francisco.

Weaver, R.J., Kasimatis, A.N. and McCune, S.B. (1962). Studies with gibberellin on wine grapes to decrease bunch rot. *American Journal of Enology and Viticulture* **13**, 78–82.

Wermann, S., Theurillat, V., Verzegnassi, L. *et al.* (2014). N,N-dimethylpiperidinium (mepiquat) Part 2. Formation in roasted coffee and barley during thermal processing. *Food Additives and Contaminants: Part A* **31**, 234–241.

West, C.A. (1973). Biosynthesis of gibberellins, In *Biosynthesis and Its Control in Plants* (ed B.V. Millborrow), Academic Press, New York, pp. 143–169.

Whipker, B.E. (2013). Growth regulators for floriculture crops in greenhouses. In *Grower Talks, 2013 Plant Growth Regulator Guide,* Ball Publishing, West Chicago, pp. 8–37.

Whipker, B.E. and Latimer, J. (2013). Wide assortment of available PGRs. In *Grower Talks, 2013 Plant Growth Regulator Guide,* Ball Publishing, West Chicago, pp. 2–4.

Williams, D.R., Ross, J.J., Reid, J.B. and Potts, B.M. (1999). Response of *Eucalyptus nitens* seedlings to gibberellin biosynthesis inhibitors. *Plant Growth Regulation* **27**, 125–129.

Wittwer, S.H. and Tolbert, N.E. (1960). (2-Chloroethyl)-trimethylammoniumchloride and related compounds as plant growth substances. III. Effect on growth and flowering of the tomato. *American Journal of Botany* **47**, 560–565.

Yalpani, N., Suttle, J.C., Hultstrand, J.F. and Rodaway, S.J. (1989). Competition for in vitro [^{3}H]gibberellin A_4 binding in cucumber by substituted phthalimides. *Plant Physiology* **91**, 823–828.

Yamaguchi, S. (2008). Gibberellin metabolism and its regulation. *Annual Review of Plant Biology* **59**, 225–251.

Yeshitela, T., Robbertse, P.J. and Stassen, P.J.C. (2004). Effects of various inductive periods and chemicals on flowering and vegetative growth of 'Tommy Atkins' and 'Keitt' mango (*Mangifera indica*). *New Zealand Journal of Crop and Horticultural Science* **32**, 209–215.

Yim, K.O., Kwon, Y.W. and Bayer, D.E. (1997). Growth responses and allocation of assimilates of rice seedlings by paclobutrazol and gibberellin treatment. *Journal of Plant Growth Regulation* **16**, 35–41.

Yoder, K.S., Miller, S.S. and Byers, R.E. (1999). Suppression of fireblight in apple shoots by prohexadione-calcium following experimental and natural inoculation. *HortScience* **34**, 1202–1204.

Yoshinaga, S. (2005). Improved lodging resistance in rice (*Oryza sativa* L.) cultivated by submerged direct seeding using a newly developed hill seeder. *Japanese Agricultural Research Quarterly* **39**, 147–152.

Zeeh, B., König, K.H. and Jung, J. (1974). Development of a new plant growth regulator with biological activity related to CCC. *Kemia (Helsinki)* **9**, 621–623.

Annual Plant Reviews (2016) **49**, 405–430 http://onlinelibrary.wiley.com
doi: 10.1002/9781119210436.ch13

Chapter 13

GENETIC CONTROL OF GIBBERELLIN METABOLISM AND SIGNALLING IN CROP IMPROVEMENT

Andrew L. Phillips
Plant Biology and Crop Science Department, Rothamsted Research, UK

Abstract: Many developmental processes regulated by gibberellin (GA) impact on crop yield and quality, and genetic variants affected in GA biosynthesis or signalling have been widely exploited in agriculture. Notably, these include alleles of *Rht-1* and *Sd-1* that confer a semi-dwarf habit to wheat and rice, respectively, underpinning the Green Revolution that increased yields in the latter half of the twentieth century. The chapter discusses the utility of various alleles at these and other GA loci in controlling stature, mainly in cereal crops. The scope for the development of additional variants in the GA pathway through mutagenesis or transgenic approaches is also reviewed.

Keywords: Gibberellin, mutant, semi-dwarfs, stature, yield, Green Revolution, wheat, rice

13.1 Introduction

The earlier chapters of this book illustrate the wide range of developmental processes in plants in which gibberellins (GAs) are involved; many of these have impacts on agricultural and horticultural traits including architecture, fertility, disease resistance and end-use properties. Consequently, significant modification of GA status or responses is practised through either chemical intervention, as described in the previous chapter, or through the exploitation or creation of genetic variation in GA signalling components. Finally,

Annual Plant Reviews, Volume 49: The Gibberellins, First Edition.
Edited by Peter Hedden and Stephen G. Thomas.

it should also be recognised that control of plant growth and development in horticultural species through manipulation of environmental factors such as temperature, light quantity and quality may also be mediated by the GA pathway, and the molecular basis for some of these interactions is explored elsewhere in this book.

The selection and utilisation of genetic variation in GA-related traits pre-dates knowledge of the GA hormones themselves and of their biosynthetic and signalling pathway. Hence, most agriculturally important loci that encode GA components were exploited in their respective crop species long before the underlying mechanism for their action was known. Furthermore, as the most conspicuous role of GA signalling is in elongation growth of leaves and stems, which in turn are significant determinants of both yield and crop husbandry, it is the roles of GA in these architecture traits that have been elucidated more comprehensively and will be discussed in most detail.

13.2 The *REDUCED HEIGHT-1* (*Rht-1*) alleles of wheat

Even before the advent of scientific plant breeding at the start of the last century, crop height is likely to have been an important character in wheat. Peng and colleagues have suggested that the tall stature of most wheat landraces indicates that early farmers are more likely to have selected, directly or indirectly, for increased height and vigour in order to increase biomass and yield (Peng *et al.*, 2011). Indeed, there is convincing evidence that wheat stature and yield are, in general, positively correlated (Law *et al.*, 1978; Griffiths *et al.*, 2012). However, the widespread use of nitrogen fertilisers from the 1920s onward resulted in a need to reduce crop height in order to control lodging, itself a consequence of the increase in grain yield and stem growth that were promoted by enhanced nutrition. Initial efforts to reduce height through selection while maintaining yield were only partially successful (Law *et al.*, 1978). However, the generally positive correlation between stature and yield was broken by the introduction of semi-dwarfing alleles of the *REDUCED HEIGHT-1* (*Rht-1*) homoeologous series of genes into plant breeding programmes in the 1960s. *Rht-B1b* (originally named *Rht1*) and *Rht-D1b* (originally *Rht2*) (Figure 13.1) originated in Japan in the variety Norin-10; their history is discussed in detail by Wilhelm *et al.* (2013a). These semi-dwarfing alleles confer a reduction in overall crop height of 12–15% and 16–20%, respectively, when in the homozygous state, with a smaller semi-dominant effect in heterozygotes (Flintham *et al.*, 1997). Although introduced to combat lodging as described above, these alleles also were found significantly to increase yields, and were rapidly taken up, not only by plant breeders in the developed world, but also by Norman Borlaug at CIMMYT in Mexico where they were introduced into varieties targeted at developing countries in South America and Asia. In combination

Figure 13.1 Near-isogenic lines of wheat containing homozygous alleles of *Rht-B1* and *Rht-D1* in the cv. Mercia background. Original nomenclature: Tall (*RhtA1a*/*Rht-B1a*/ *Rht-D1a*) = *rht*; *Rht-B1d* = *Rht1*: *Rht-D1b* = *Rht2*; *Rht-B1c* = *Rht3*; *Rht-D1c* = *Rht10*; *Rht-B1d* = from cv. *Saitama 27*; *Rht-B1e* = from cv. *Bezostaya*.

with higher fertiliser rates, more widespread use of pesticides and increased mechanisation, the *Rht-1* alleles made a major contribution to the worldwide rise in wheat yields over the following decades, and became synonymous with the Green Revolution. In 2010, it was estimated that 90% of recent wheat varieties cultivated in the USA contained either *Rht-B1b* or *Rht-D1b* (Guedira *et al.*, 2010).

The *Rht-B1b* and *Rht-D1b* alleles exert their positive effect on yield through increases in grain number per ear despite reducing individual grain weights (Flintham *et al.*, 1997; Rebetzke and Richards, 2000). Although leaf area is reduced, photosynthetic capacity per unit leaf area is elevated and therefore there is little effect on total biomass, whereas the reduction in straw mass results in an increased harvest index. How this is achieved, in comparison to most height-reducing alleles that compromise yield, is still unclear. It seems likely that increased photosynthate partitioning to the developing ear during stem elongation is responsible for the survival of a greater number of florets, resulting in the increase in grain number. It is clear that the scale of the positive effect of these alleles depends upon the genetic background into which they are placed, with a more pronounced effect in taller cultivars; indeed, in very tall backgrounds a combination of *Rht-B1b* and *Rht-D1b*, which are located on homoeologous chromosomes of hexaploid wheat, was found to be optimal (Flintham *et al.*, 1997) while in short backgrounds such as

```
                                       DELLA                                          VHYNP
AtGAI     19: NEEDDGNGMDELLAVLGYKVRSSEMADVAQKLEQLEVMMSNV---------QEDDLSQLATETVHYNPAELYTWLDSMLTDLNPP :94
OsSLR1    30: GAAGEEEDVDELLAALGYKVRSSDMADVAQKLEQLEMAMGMGGV-SAPGAADDGFVSHLATDTVHYNPSDLSSWVESMLSELNAP :113
ZmDWARF8  29: AGEQEEEDVDELLAALGYKVRSSDMADVAQKLEQLEMAMGMGGVGGAGATADDGFVSHLATDTVHYNPSDLSSWVESMLSELNAP :113
HvSLN1    30: SEAGEGEEVDELLAALGYKVRASDMADVAQKLEQLEMAMGMGG-----PAPDDGFATHLATDTVHYNPTDLSSWVESMLSELNAP :109
TaRht-B1  31: AAAGEGEEVDELLAALGYKVRASDMADVAQKLEQLEMAMGMGGV-GAGAAPDDSFATHLATDTVHYNPTDLSSWVESMLSELNAP :114
TaRht-D1  29: AAAGEGEEVDELLAALGYKVRASDMADVAQKLEQLEMAMGMGGV-GAGAAPDDSFATHLATDTVHYNPTDLSSWVESMLSELNAP :112
                                   ▲
```

Figure 13.2 Locations of nonsense, missense and deletion mutations in the DELLA/VHYNP domains of DELLA genes from model and crop species that yield dwarf or semi-dwarf phenotypes. Altered sites are highlighted in black: *AtGAI*: a 17 amino acid deletion in the DELLA domain of the *Arabidopsis GAI* gene in the *gai-1* mutant; *OsSLR1*: independent missense mutations in *RICE SLENDER1* in the lines *slr1-d2* (Val to Met), *slr1-d3* (Leu to Phe) and *slr1-d1* (Met to Lys); *ZmDWARF8*: a 12 amino acid deletion in the maize *DWARF8* gene in the line *D8-2023*; *HvSLN1*: a missense mutation, *SLN1d* (Gly to Glu), in the *SLENDER1* gene of barley; *TaRht-B1*: sites of introduction of stop codons in the wheat *REDUCED HEIGHT* alleles *Rht-B1b, Rht-B1d* and *Rht-B1e; TaRht-D1*: site of the introduced stop codon in *Rht-D1b* and *RhtD1c*. A triangle shows the position of the 30-amino acid insertion in *Rht-B1c*. Residue numbers are displayed at the start and end of each sequence fragment.

Mercia there may be no change or even a reduction in yield (Gooding *et al.*, 2012). Similarly, the use of more extreme dwarf alleles at these loci such as *Rht-B1c* (originally *Rht3*) and *Rht-D1c* (*Rht10*) generally has a negative effect on biomass and, hence, on grain yield.

The mechanism for the reduction in stem elongation in *Rht-1* semi-dwarf lines was investigated soon after their widespread adoption. It was shown that the *Rht-B1c* severe dwarf had highly elevated levels of bioactive GAs (Radley, 1970), and both dwarf and semi-dwarf lines had reduced responses to applied GA (Gale and Marshall, 1973). The similarity between the wheat lines and semi-dominant GA-insensitive mutants of other species allowed Peng and colleagues (Peng *et al.*, 1999) to use synteny and sequence homology to clone *Rht-1* and to demonstrate that the homoeologous series of genes were orthologous to the *GAI* gene of *Arabidopsis*. *GAI* and other DELLA genes encode repressors of GA-regulated processes such as elongation growth, and missense or in-frame insertions in the 'DELLA' domain (Figure 13.2) result in reduced sensitivity to GA, as discussed in detail in Chapter 6. By analogy with *Arabidopsis*, rice and barley it seems highly probable that GA signalling in wheat proceeds through GA-mediated interaction of the wild-type Rht-1 proteins with the GID1 receptor, ubiquitination by the $SCF^{GID2/SLY1}$ complex and degradation by the 26S proteasome, resulting in the release of Rht-1-mediated growth repression. In wheat, although interaction between Rht-1 and GID1 has been observed in yeast two-hybrid experiments (Pearce *et al.*, 2011), Rht-1 protein has yet to be detected in wheat tissues, due to lack of the appropriate molecular tools.

The exact basis for the reduction in GA sensitivity in the semi-dwarf *Rht-B1b* and *Rht-D1b* lines remains somewhat unresolved. Peng and colleagues showed that both alleles are the result of mutations in the *Rht-1* coding region that result in the introduction of stop codons within the N-terminal 'DELLA' domain (Figure 13.2). In both transcripts these stop codons are followed by multiple AUG codons which could allow reinitiation of translation, resulting in the production of an N-terminally truncated protein lacking part of the 'DELLA' domain as well as the short, prematurely aborted N-terminal peptide. Yeast two hybrid assays have indicated that neither the predicted N-terminal fragment of Rht-1 nor the longer reinitiation product could interact with GID1, irrespective of the presence of GA (Pearce *et al.*, 2011). As the latter N-terminally truncated polypeptide contains the downstream domains of the Rht-1 protein that are thought to be active in protein interactions and control of gene expression (see Chapter 7), it is likely that this product accumulates in elongating tissues and suppresses growth, even in the presence of GA; however, this has not yet been demonstrated *in planta*.

In addition to *Rht-B1b* and *Rht-D1b*, several other alleles of the *Rht-1* genes that affect height to a greater or lesser degree have been identified, as illustrated in Figure 13.1. The alternative semi-dwarfing alleles *Rht-B1d* and *Rht-B1e* have seen some limited commercial use, while the more severe dwarfing alleles, such as *Rht-B1c* and *Rht-D1c*, have not so far been exploited as they reduce biomass and yield (Flintham *et al.*, 1997), and also fertility may be compromised, particularly under abiotic stress (Alghabari *et al.*, 2014). These additional *Rht*-1 alleles have been characterised by sequence analysis, revealing that several appear to be derivatives of *Rht-B1b* and *Rht-D1b* in that they contain the same mutations in the 'DELLA' domain. Thus, *Rht-B1d*, which confers a slightly weaker dwarfing effect than *Rht-B1b*, contains the same introduced stop codon; no other sequence polymorphisms have yet been found to explain the difference in height (Pearce *et al.*, 2011). *Rht-D1c* similarly carries the *Rht-D1b* stop codon, but in this case the severe dwarf phenotype appears to be due to higher expression of the mutant allele as the result of an increase in copy number (Pearce *et al.*, 2011), possibly a tandem duplication (Li *et al.*, 2012). In contrast, *Rht-B1c* and *Rht-B1e* contain novel mutations. The mutation in *Rht-B1e* introduces a translation stop three codons before the *Rht-B1b* mutation (Figure 13.2) but confers a significantly more severe phenotype, possibly as a result of more efficient reinitiation of translation at the downstream AUG codons. *Rht-B1c*, which gives a severe dwarfing effect, contains a 2-kbp retrotransposon insertion that is partially removed during splicing, resulting in an in-frame, 90-bp insertion in the region encoding the 'DELLA' domain that prevents the translated protein from interacting with the GID1 receptor (Pearce *et al.*, 2011; Wen *et al.*, 2013). Although *Rht-B1c* also contains a 197-bp insertion in its promoter this does not affect expression (Wen *et al.*, 2013) and the severe dwarfing conferred by this allele is most likely due to efficient translation of the transcript with its in-frame insertion compared to inefficient reinitiation of translation after termination in *Rht-B1b*, *Rht-D1b*, *Rht-B1d* and *Rht-B1e*. Finally, although an *Rht-1* homoeologue has been identified on the A genome of wheat and has been shown to be expressed at similar levels to *Rht-B1* and *Rht-D1* (Pearce *et al.*, 2011), no dwarfing or semi-dwarfing alleles of this gene have so far been reported; indeed, in a broad survey of *Rht-1* haplotypes, the *Rht-1A* gene was shown to have unusually low diversity with over three-quarters of all accessions studied having a single, wild-type variant (Wilhelm *et al.*, 2013b), which might indicate positive selection for the wild-type allele.

A novel strategy to identify alternative dwarfing alleles of *Rht-1* using an internal suppressor screen was developed by Chandler and Harding (2013). These authors carried out mutagenesis with sodium azide on a line containing the severe dwarf *Rht-B1c* allele and screened M_2 lines in the field for increased height of seedlings and of mature plants. A sample of 60 'over-growth' mutants was sequenced in the *Rht-B1* gene which identified 19 derivative alleles of *Rht-B1c*. These novel alleles fell into two classes: four mutants, of nearly wild-type (ie *Rht-B1a*) height, had acquired premature stop codons and the truncated protein had presumably lost the ability to suppress growth. Although in a diploid species such as rice, such loss-of-function alleles result in an elongated, slender phenotype (e.g. rice *slr1*; Ikeda *et al.* (2001)), in wheat the loss of the semi-dominant *Rht-B1c* allele restores the normal, wild-type stature, due to the action of the two remaining functional *Rht-1* homoeologues. The other 15 novel suppressor alleles of *Rht-B1c* showed varying degrees of restoration of the wild-type tall phenotype and were shown to contain missense mutations that result in amino acid substitutions at positions within the C-terminal two-thirds of the Rht-1 protein that is thought to be important for downstream interactions and growth repression. These novel alleles confer a range of final heights and will provide useful variation in crop stature that can be exploited in breeding.

13.2.1 Pleiotropic effects of *Rht-1* alleles

The commercially useful *Rht-1* alleles achieve their beneficial effects on lodging and harvest index through partial suppression of stem elongation, as described above. However, wheat, in common with other grasses, possesses only a single paralogue of this gene, in contrast to dicot species such as *Arabidopsis*, which has five 'DELLA' paralogues with distinct, but overlapping, functions in different GA-regulated processes (reviewed in Daviere and Achard, 2013). Hence, although wheat has three homoeologues of *Rht-1*, these are expressed at similar levels in all tissues examined (Pearce *et al.*, 2011) and therefore dwarfing or semi-dwarfing mutations in these genes will affect GA signalling in all developmental processes in which these hormones are involved. In some cases such pleiotropic effects are beneficial: for example, *Rht-B1b* and *Rht-D1b* protect against high α-amylase levels in mature grain that would otherwise reduce bread-making quality (Mrva and Mares, 1996; Tan *et al.*, 2010), presumably by interfering with GA signalling in the aleurone layer or scutellum. However, some pleiotropic effects of *Rht-1* dwarfing mutations have negative consequences for agronomic performance. In addition to the reduction in leaf elongation (Appleford and

Lenton, 1991), the *Rht-1* semi-dwarfs are also affected in emergence due to shorter coleoptiles. This is more serious under water-limited agriculture where deep sowing is commonplace, as the crop can suffer from poor establishment (Rebetzke *et al.*, 2005). Under such conditions, alternative dwarfing alleles that have greater early vigour would have improved emergence and also more rapid leaf growth to reduce water evaporation from the soil (Spielmeyer *et al.*, 2007); it seems likely that alleles of genes other than *Rht-1* will be required to achieve such targeted effects.

GAs are important in root development, and there is some evidence on the effects of *Rht-1* alleles on root architecture. Wojciechowski *et al.* (2009) studied root growth in gel chambers, soil columns and in the field, but found no significant effect of the semi-dwarfing alleles, and the dwarfing alleles *Rht-B1c* and *Rht-D1c* yielded both increased and decreased root lengths depending on the cultivation conditions. Bai and colleagues (Bai *et al.*, 2013) assessed seedling roots using a paper roll system and found that both *Rht-B1c* and *Rht-D1c* had lower total root lengths and reduction in other seminal root parameters, but the only significant effect of the semi-dwarf allele *Rht-D1b* was on root dry weight. Such limited effects of *Rht-1* alleles may reflect a high endogenous sensitivity of roots to bioactive GAs.

DELLA proteins have also been shown to play a role in the response to pathogens, possibly via interaction with the ROS pathway or with the stress hormones methyl jasmonate, salicylic acid and ethylene. Saville and colleagues (Saville *et al.*, 2012) tested the resistance of *Rht-1* dwarf and semi-dwarf lines of wheat to a range of pathogens and showed that a severe dwarf containing both the *Rht-B1c* and *Rht-D1b* alleles was more susceptible to biotrophic fungal pathogens, such as *Botrytis graminis*, which was attributed to a delay in ROS-induced cell death. Conversely, semi-dwarf lines were more resistant to the necrotrophic phase of disease spread by *Fusarium graminearum*, which might similarly be due to effects on the ROS-cell death pathway or through interactions with jasmonate signalling (Saville *et al.*, 2012), a mechanism discussed in Chapter 8.

GA signalling is important for fertility, being involved in programmed cell death of the tapetum during microspore production in the anther (reviewed in Plackett *et al.*, 2011) . As a result, severely GA-deficient or insensitive mutants in many plant species are partially or totally male-sterile. While *Rht-1* alleles do not appear to severely affect wheat pollen production under normal conditions, Law and Worland (1984) reported that fertility was reduced under heat stress at booting. However, a more recent, and more thorough, assessment suggests that such effects may be relatively minor and confined to extreme dwarfs such as *Rht-D1c* (Alghabari *et al.*, 2014).

As noted above, semi-dwarfing alleles achieve increases in yield despite a reduction in individual grain weights (Flintham *et al.*, 1997). Although these authors suggest that the effect on grain size is due to increased competition between florets, an equally likely explanation is that GA is directly involved in regulating grain expansion, which is thereby limited in the partially GA-insensitive dwarf and semi-dwarf lines. In support of this latter hypothesis, the developing wheat grain is a rich source of GAs (Gaskin *et al.*, 1980) and all components of the GA perception and signal transduction pathway are expressed in the maternal grain envelope tissues (A. Huttly, S. P. Vaughan and A.L. Phillips, unpublished). This has provoked interest in alternative dwarfing alleles or transgenes that could control lodging and maintain the increase in grain numbers without affecting grain size, but so far these remain elusive.

13.2.2 *Rht-1* orthologues in other crop species

Rht-1, as a central component of the GA signalling pathway, has direct orthologues in all vascular plants; mutants in these DELLA genes have been identified in various crops that confer either GA-insensitivity (due to loss of interaction with GID1) or GA-independent growth (due to loss of downstream growth repression). The rice orthologue of *Rht-1*, *SLENDER1* (*SLR1*), was originally identified as a loss-of-function overgrowth mutant in a γ-irradiated population (Ikeda *et al.*, 2001). Although the *slender* mutant itself is unlikely to be of any agronomic value, screening a rice population mutagenised with *N*-methyl-*N*-nitrosourea identified three semi-dwarf lines whose lesion mapped to *SLR1*. These three lines were shown to contain missense mutations in the N-terminal DELLA/VHYNP domain (Figure 13.2; see also Chapter 6) resulting in reduced degradation of the SLR-1 protein and a dominant dwarf phenotype (Asano *et al.*, 2009) analogous to that of the wheat *Rht-1* dwarfs described above. Although the dwarfing effect was considered too strong to be commercially useful, the authors indicate that it should be possible to identify weaker alleles of *SLR1* conferring a more subtle phenotype by TILLING in mutagenised populations of rice.

Maize contains two *DELLA* paralogues, *DWARF8* (*D8*) and *DWARF9* (*D9*). There are a number of dwarf alleles of *D8* that result from small deletions or insertions in the DELLA/VHYNP domain, such as *d8-2023* (Figure 13.2) (Peng *et al.*, 1999), all having a severe or very severe effect on stature (Winkler and Freeling, 1994). Only a single GA-insensitive allele of *D9* is known, *d9-1*, that contains an indel and a variety of missense substitutions throughout the protein; over-expression in *Arabidopsis* of a series of chimaeric genes between the wild-type and variant indicated that a C-terminal amino acid substitution, E600K, was responsible for the loss of GA sensitivity (Lawit *et al.*, 2010), suggesting that regions of the protein other than the DELLA/VHYNP domain could confer such a phenotype.

As in rice, loss-of-function mutants in the *SLENDER1* (*SLN1*) gene of barley have an extreme elongated phenotype (Chandler and Robertson, 1999) and negatively affect both performance and grain quality. This gene was subsequently shown to be the barley orthologue of wheat *Rht-1* and a wider search for dwarf mutants identified line M640 that proved to be due to a missense mutation in the DELLA domain of *SLN1* (Chandler *et al.*, 2002). This

DELLA mutation (Figure 13.2) confers reduced GA sensitivity to both elongation growth and α-amylase production by the seed aleurone layer, but no exploitation of this allele in crop breeding has been reported.

Oilseed rape (*Brassica napus*) is a relatively undeveloped crop in breeding terms and has many weedy traits, including a tendency to lodge, and therefore there is interest in identifying semi-dwarfing genes. A dwarf, GA-insensitive mutant identified in *B. rapa* was shown to be due to a missense mutation in *BrRGA1*, a DELLA gene that is the orthologue of *RGA* in *Arabidopsis* (Muangprom *et al.*, 2005). The affected residue was located in the C-terminal half of BrRGA1 and resulted in accumulation of the DELLA protein indicating, as with maize *d9-1*, that residues outside the N-terminal DELLA/VHYNP domains could confer a GA-insensitive phenotype. Introgression of this allele, *BrRGA1-d*, into *B. napus* resulted in a significant increase in lodging resistance coupled with no loss of seed yield (Muangprom *et al.*, 2006) indicating its potential for breeding. A dwarfing mutation, *ds-1*, in one of the native homoeologues of *RGA* in *B. napus* has also been identified (Liu *et al.*, 2010). The semi-dwarf phenotype of *ds-1* is due to a missense mutation affecting the VHYNP domain of BnRGA that prevents interaction with the GID1 GA receptor, and this allele may similarly be useful in modifying height and controlling lodging in oilseed rape.

13.3 The *SEMI-DWARF-1* (*SD-1*) alleles of rice

Semi-dwarf lines of rice have had an even more dramatic effect on worldwide cereal yields than those in wheat. Efforts to increase rice yields through the use of nitrogen fertilisers led, as in wheat, to severe lodging due to increased seed weight and culm length. The introduction by the International Rice Research Institute of the short-statured variety IR8 in the 1960s and its widespread adoption throughout Asia led to doubling of rice yields in a few short years (the origin of the term 'Green Revolution') and led to both India and the Philippines becoming net exporters of rice. The reduced height of IR8 was due to the introduction of a major, recessive dwarfing gene from the Korean *indica* rice variety Deo-Geo-Woo-Gen (DGWG). Similar breeding efforts in California in *japonica* rice led to the selection of a semi-dwarf line, Calrose 76, derived by gamma irradiation, whose semi-dwarfing gene was designated as *semidwarf-1* (*sd-1*) and was shown to be allelic to the gene from DGWG, as were several other independent mutants from the Calrose population (Rutger, 1983).

The identity of the *SD-1* gene was investigated in parallel by groups in Australia and Japan, both of whom showed that lines containing the recessive semi-dwarfing allele from DGWG accumulated GA_{53}, the initial substrate of the GA biosynthetic enzyme, GA 20-oxidase (Monna *et al.*, 2002; Sasaki *et al.*, 2002; Spielmeyer *et al.*, 2002), suggesting that GA20ox activity was reduced in these lines. These groups also showed that the gene *OsGA20ox2*, encoding one of the isozymes of GA20ox, was very closely linked to the *SD-1* locus and that all *sd-1* semi-dwarf lines, including DGWG and Calrose 76, had lesions in *OsGA20ox2*. The *sd-1* allele from DGWG was found to have a 383-bp deletion spanning the first intron and parts of the first two exons, while other independent *sd-1* alleles consisted of missense mutations in the coding region (Sasaki *et al.*, 2002; Spielmeyer *et al.*, 2002). These results confirmed that the rice *SD-1* locus encodes OsGA20ox2.

It has been shown that selection for *sd-1* alleles was involved in the domestication of *japonica* rice from *Oryza rufipogon* approximately 10,000 years ago. QTL analysis in a population developed from an *indica* x *japonica* cross identified a locus on chromosome 1S with a major effect on culm length (Asano *et al.*, 2011). The QTL was mapped to a region containing *OsGA20ox2* and the *japonica* variety, Nipponbare, was shown to contain two non-synonymous SNPs in the coding region. Heterologous expression of *OsGA20ox2* in *E. coli* to determine the effect of the missense substitutions on enzyme activity showed that introduction of either or both *japonica* SNPs into the *indica* cDNA resulted in significant loss of GA20ox activity of the recombinant protein. This indicates that the *SD-1* gene from *japonica* rice contains functional polymorphisms in *OsGA20ox2* that result in a measurable reduction in stature. A lower level of genetic diversity in the *SD-1* region of the *japonica* genome compared with the corresponding regions of *indica* rice and *O. rufipogon* landraces confirmed that that acquisition of this *SD-1* semi-dwarfing allele by *japonica* rice was due to artificial selection, presumably for reduced height, during domestication (Asano *et al.*, 2011). However, this allele has a relatively minor effect on height, as illustrated by the size of the QTL effect detected in the *indica* × *japonica* population, and by the fact that the Calrose 76 *sd-1* semi-dwarf mutant is in a *japonica* background that already contains these prehistoric *SD-1* SNPs (Spielmeyer *et al.*, 2002).

Rice researchers and breeders continue to mine the *SD-1* locus for additional dwarfing alleles. Casella *et al.* (2013) generated an EMS-population of a European *japonica* rice and screened this by TILLING in the first two exons of the *OsGA20ox2* gene, and identified several novel mutations. A nonsense mutation in exon 1 and a missense mutation in exon 2 each gave significant reductions in final height, with a lesser effect in the heterozygous state. RNA interference has also been used to generate variation in stature by suppression of *OsGA20ox2*: three RNAi lines of *japonica* rice showed reduced expression of the native *OsGA20ox2* gene, resulting in lower levels of bioactive GA and a strong effect on final height, with reductions of 53–58% of wild-type height (Qiao *et al.*, 2007).

Orthologues of *SD-1*/*OsGA20ox2* are found in all sequenced grass genomes (www.phytozome.net), with maize having two closely related copies due to the relatively recent genome duplication in that species, and wheat having three homoeologues derived from its progenitor genomes (Pearce *et al.*, 2015). In wheat, it is unlikely that loss-of-function alleles in one homoeologue would have a dramatic effect of stature as the *GA20ox2* gene function would be maintained by the other two functional homoeologues. However, as alternative semi-dwarfing alleles may be of interest in wheat (see Section 10.2.1 above), investigating the effects of RNAi suppression of the *TaGA20ox2* genes should be worth pursuing as a first step towards identifying and combining alleles in all three homoeologues. In barley, there is strong evidence that the *sdw/denso* locus that controls plant height, and also influences yield and quality, may be *GA20ox2*. The *sdw/denso* gene maps to the long arm of chromosome 3H in the syntenic position of *SD-1/OsGA20ox2*, and sequencing this gene in tall and semi-dwarf lines revealed a SNP in the second intron (Jia *et al.*, 2009). It seems unlikely that this SNP significantly affects GA20ox activity *in planta*, but measurement of transcript levels in a doubled haploid population segregating for *denso* identified an eQTL for *HvGA20ox2* expression that was coincident with the semi-dwarfing locus (Jia *et al.*, 2011). It seems likely that sequencing outside the transcribed region of *HvGA20ox2* in *sdw* and *denso* lines will identify further polymorphisms that are responsible for the reduced expression of this gene and its effect on stature.

13.4 The *ELONGATED UPPERMOST INTERNODE* (*EUI*) gene of rice

The reproductive tissues of many species are highly active in GA biosynthesis: the anthers of rice, for example, are an extremely rich source of GA_4 (Kobayashi *et al.*, 1988), the main bioactive GA in the non-13-hydroxylation pathway (see Chapter 2). This anther-derived GA is important in rice stem elongation, as revealed by cytoplasmic male-sterile (CMS) lines that have shortened upper internodes. This has severe consequences for the production of hybrid rice as the reduction in final internode elongation in the CMS female parent inhibits full emergence of the panicle from the leaf sheath, preventing efficient pollination by the male parent. This can be ameliorated by application of GA_3 to promote elongation and emergence and thereby increase seed set, but the GA treatment has both economic costs and impacts on grain quality. As a genetic solution to this problem, Rutger and Carnahan (1981) introduced the *elongated uppermost internode* gene (*eui*), which promoted internode elongation and thereby increased panicle emergence and seed set.

As introgression of the recessive *eui* allele into CMS lines could phenocopy treatment with bioactive GA, it was suspected that the wild-type *EUI* gene might encode a component involved in GA turnover or signalling. Zhu *et al.* (2006) measured GA levels in the uppermost internode of wild-type and *eui* plants and showed that the latter had dramatically increased levels of the bioactive GAs, GA_1 and GA_4; the non-13-hydroxy-GA was especially abundant, despite being undetectable in the wild-type. These authors cloned the *EUI* gene by fine mapping of independent alleles and showed that it encoded a cytochrome P450 enzyme of unknown function. To identify the catalytic function of EUI, the gene was expressed in yeast and microsomal fractions assayed for activity against GA substrates. EUI was shown to catalyse 16α,17-epoxidation of the double bond of the non-13-hydroxylated GAs, GA_{12}, GA_9 and GA_4 (Zhu *et al.*, 2006), but 13-hydroxylated GAs were poor substrates. It was also demonstrated that the 16α,17-epoxidation product of GA_4 was much reduced in biological activity and that therefore EUI was involved in inactivation of 13-H GAs. The highest level of expression of *EUI* in wild-type plants is in the panicle and upper internode, and the relative abundance of 16α,17-epoxy-GA products of GA_{12} and GA_9 (identified in the hydrated, diol form) in wild-type tissues, but low levels of products derived from GA_4 might suggest that a major role is in controlling the levels of bioactive GAs being produced *in situ* as well as inactivating GA_4 that is exported from the anther. *EUI* is also expressed in other tissues, and may have additional effects on rice development: for example, a role in gravitropic responses in roots has been described (Zhang *et al.*, 2008).

13.5 Commercially useful alleles of other genes from the gibberellin pathway

At least 20 *REDUCED HEIGHT* (*Rht*) genes with major effects on plant height have been identified in wheat, but of these only four are known to involve the GA pathway: as described above (Section 13.2.1), *Rht1* (renamed *Rht-B1b*), *Rht2* (*Rht-B1d*), *Rht3* (*Rht-B1c*) and *Rht10* (*Rht-D1c*) involve lesions in homoeologues of the *Rht-1* DELLA genes. To characterise the remaining *Rht* loci, Ellis *et al.* (2004) assessed the GA responsiveness of leaf elongation in tall and dwarf near-isogenic lines for each locus. In addition to the four *Rht-1* alleles, *Rht11* and *Rht17* were shown to have a reduced response to applied GA, possibly suggesting a role in GA perception or signal transduction; *Rht16*, *Rht18* and *Rht-Leeds* conferred an increased response to GA compared to tall lines, suggesting a reduction in GA levels in the dwarf lines; while the remaining lines, including *Rht8* and *Rht12*, did not affect leaf elongation and showed no difference in GA response between tall and short lines, possibly indicating that these lines may affect later developmental processes.

Rht8 has been widely used in continental European breeding programmes as it confers a semi-dwarf habit without negatively affecting yield, but has good seedling vigour, important for deep sowing in water-restricted environments (Rebetzke *et al.*, 1999). To investigate a possible link with GA signalling, Gasperini *et al.* (2012) measured the levels of several GAs in *Rht8*, but found no difference between tall and short isolines. *Rht8* lines showed a slightly reduced response to *epi*-brassinolide, but fine mapping of *Rht8* did not identify any promising candidates from the BR pathway (Gasperini *et al.*, 2012), thus the mechanism underlying its dwarfing effect remains unclear.

Rht12 confers a dominant, GA-responsive dwarf phenotype (Worland *et al.*, 1994). Although this allele has a strong effect, with a nearly 50% reduction in final height, there is some interest in breeding applications, as seedling vigour is maintained, spikelet fertility is increased and despite reduced grain size there may be, depending on background, no loss of yield (Chen *et al.*, 2013). *Rht12* shows other characteristics of GA-deficient mutants, such as dark green leaves and late flowering, and as external GA application almost completely rescues the dwarf phenotype it seems highly likely that bioactive GA levels are reduced in the mutant, but this has yet to be demonstrated.

Loss-of-function mutants affecting the early part of the GA pathway have been identified in several species and generally have a severe dwarf phenotype, as these enzymes are often encoded by single-copy genes, without the genetic redundancy present in the later steps in the pathway (see Chapter 2). Semi-dwarfing alleles that might have commercial applications have rarely been identified, possibly because even relatively large changes in the levels of early intermediates such as GA_{12} may be eliminated by the homeostatic mechanism that operates through feedback regulation of the *GA20ox*, *GA3ox* and *GA2ox* genes (see Chapters 2, 3 and 7) – although the widespread utility of growth retardants that act on the early steps in the pathway would counter this argument. One exception is the *d35* mutation in the Tan-Ginbozu variety of rice, which was the most commercially successful rice variety in 1950s Japan. Analysis of early GA intermediates in Tan-Ginbozu suggested that the lesion reduced the activity of *ent*-kaurene oxidase (KO) (Ogawa *et al.*, 1996) and it was subsequently shown that *D35* is one of a family of five tandemly-arranged *KO-like* genes in rice (Itoh *et al.*, 2004). Only *OsKO1* and *OsKO2* could rescue the semi-dwarf phenotype of Tan-Ginbozu, suggesting that these were involved in GA biosynthesis while other *KO-like* genes may be involved in phytoalexin biosynthesis. Of the two genes involved in GA biosynthesis, only *OsKO2* was expressed in vegetative tissues, suggesting that the mutation in *d35* represents a weak, reduced function allele of this gene.

Although mutations in the *GA3ox1* gene of *Arabidopsis* confers a semi-dwarf phenotype (*ga4*; Chiang *et al.*, 1995), few useful *GA3ox* lesions have been described, possibly because in most species GA 3-oxidation is probably not a limiting step in production of the bioactive form. A notable exception is the dwarfing allele of pea, *le*, which was one of the characters used by Gregor Mendel to develop the laws of heredity. Wild-type pea plants have a height of over 1 m, and the *le* allele confers a reduction in internode length to provide a dwarf habit that needs less plant support. The *le* dwarf was shown to respond to GA_1, but not GA_{20}, suggesting a loss of GA3ox activity (Ross and Reid, 1987), and sequencing of the *le* allele revealed a missense mutation in the coding region of *PsGA3ox1* (Lester *et al.*, 1997; Martin *et al.*, 1997) that reduced activity of the heterologous expression product by ~95%. Other alleles of *le* are known, such as *le-3* that has an intermediate phenotype or le^d that generates a more severe dwarfing effect;

the *le-3* allele is similarly due to a missense mutation, while *le^d* is a frameshift mutation in the *le* background that leads to complete loss of activity. Only the original *le* allele is known to be exploited, where it is widely used in commercial and garden varieties.

In addition to major genes affecting stature there are a multitude of alleles in all crops with more minor effects, some of which probably involve the GA pathway. For example, analysis of QTL for seedling vigour in rice identified an 81-kbp interval containing the *OsGA20ox1* gene (Abe *et al.*, 2012). Quantification of the levels of GA intermediates in near-isogenic lines suggested a reduction in GA 20-oxidase activity, providing further evidence for the identity of the causal polymorphism. Similarly, a QTL for plant height in maize that acts through effects in internode elongation was shown to be due to polymorphism in *ZmGA3ox2* (Teng *et al.*, 2013), which encodes the enzyme catalysing the last step in GA biosynthesis and is also probably synonymous with the *DWARF1* gene, a severe allele of which yields a GA-sensitive dwarf that is blocked in the conversion of GA_{20} to GA_1 (Spray *et al.*, 1996; Chen *et al.*, 2014). In wheat, Griffiths *et al.* (2012) carried out a meta-analysis of QTL for height across a number of populations, and although some of these QTL coincide with mapped positions of genes from the GA pathway (A.L. Phillips, unpublished), none of these relationships have so far have been confirmed.

The GA receptor, GID1, was first identified in rice as an extreme dwarf, infertile, GA-insensitive mutant (Ueguchi-Tanaka *et al.*, 2005). Although several alleles of *OsGID1* have been identified, most have a similarly severe phenotype; the most severe alleles are completely unresponsive to applied GA, and even milder alleles may have fertility defects (Ueguchi-Tanaka *et al.*, 2007) and so probably have little use in breeding. In contrast, Chandler *et al.* (2008) isolated a series of GA-sensitivity (*gse*) mutants of barley, 16 of which were shown to have point mutations in the *HvGID1* gene and which showed a range of effects on height and fertility. All these *gse* mutants set seed, even in the homozygous state, and most retained some responsiveness of leaf elongation and grain α-amylase production to applied GA_3, albeit at relatively high concentrations. However, the potential for use of the barley *gse*/*gid1* mutants in breeding has yet to be explored.

GID2 encodes an F-box component that is part of the $SCF^{SLY1/GID2}$ complex involved in GA-dependent polyubiquitination of DELLA proteins, resulting in DELLA degradation and release of growth repression (see Chapter 6). The original *gid2* mutant of rice has an extreme dwarf phenotype of no commercial interest (Sasaki *et al.*, 2003), but Zhang *et al.* (2011) identified a semi-dwarf mutant with a small insertion in the coding region of *OsGID2*; this was identified in the *eui* background where it conferred a reduction in panicle length, but also in leaf sheath elongation, reducing the overall height of the *eui* line, although still allowing full panicle emergence (see Section 13.2.3 above). The allele may therefore be of some interest in hybrid rice breeding.

13.6 Transgenic approaches to manipulation of gibberellin-dependent processes in crops

The modification of GA status or sensitivity through manipulation of components of the biosynthesis, inactivation or signalling provides a way to develop genetic variation not available in natural or mutagenised populations. Examples include the suppression of multiple paralogues or homoeologues of genes by RNAi or targeted over-expression using tissue-specific or developmentally regulated promoters. While many examples exist of transgenic manipulation of the GA pathway to yield tall or dwarf plants, this section will focus on those that might lead to applications in agriculturally or horticulturally relevant traits.

13.6.1 Cereals

As described earlier in this chapter, semi-dwarfing genes with major effects provide a means of controlling height in the small-grain cereals to avoid lodging under high nitrogen applications. Dwarfing alleles of genes from the GA biosynthetic and signalling pathway, such as *SD-1* in rice and *Rht-1* in wheat, appear to be unique in providing reduced height without compromising biomass and yield, encouraging a belief that other genes from this pathway, targeted through transgenesis, could generate useful phenotypes.

The *SD-1* (*OsGA20ox2*) gene that provides the most widely used allele for height reduction in rice is one of four *GA20ox* paralogues in that species (Sakamoto *et al.*, 2004). As the complete loss-of-function allele of *sd-1* that was derived from Deo-Geo-Woo-Gen provides only a partial dwarfing effect, one or more of the remaining three *GA20ox* genes must be contributing to GA biosynthesis in rice stems, either *in situ* or by production of mobile GAs in other tissues. Oikawa *et al.* (2004) generated transgenic lines of rice expressing RNAi constructs targeted against *OsGA20ox1* and achieved a 13–18% reduction in height compared with wild-type cv. Nipponbare. The authors noted that *OsGA20ox1* is also expressed at high levels in reproductive tissues, but did not investigate fertility in the transgenic lines (Oikawa *et al.*, 2004).

In contrast to the overlapping expression pattern of *GA20ox* paralogues in rice, *OsGA3ox2* is the only *GA3ox* gene expressed in vegetative tissues and, therefore, loss-of-function mutants in this gene have a severe dwarf phenotype (eg. *d18 –dy*, also known as Waito C (Itoh *et al.*, 2001)). However, anti-sense suppression of *OsGA3ox2* yielded a semi-dwarf habit, presumably due to only partial suppression of the target gene by the transgene (Itoh *et al.*, 2002). As with the *OsGA20ox1* RNAi lines, these anti-sense lines may have some utility in the control of lodging. However, it should be noted that the search for semi-dwarfing genes in rice that maintain yield has, in many independent screens, almost invariably identified mutants that are allelic to *sd-1* (Rutger, 1983), perhaps implying that most other simple lesions affecting the GA pathway may have pleiotropic effects that compromise their effectiveness.

An alternative strategy to reducing crop height through manipulation of GA levels is via over-expression of *GA2ox* genes to increase rates of GA turnover. Over-expression of *GA2ox* using a constitutive promoter such as maize *Ubi-1* or CaMV-35S can deliver a range of phenotypes from severe to semi-dwarfs depending on the level of expression, and can be achieved using either Class I/II or Class III *GA2ox* genes (Sakamoto *et al.*, 2003; Appleford *et al.*, 2007; Huang *et al.*, 2010), indicating that reduction in GA levels can be achieved through turnover of both C_{19}-GAs and their C_{20}-GA precursors. Expression of a runner bean *PcGA2ox1* coding sequence in wheat under the *Ubi-1* promoter not only achieved a reduction in stature through expression in vegetative tissue, but also its expression in grain resulted in a reduction in grain α-amylase levels (Appleford *et al.*, 2007) that might be beneficial in reducing the deleterious effects of preharvest sprouting or pre-maturity amylase on grain quality. Targeting *GA2ox* expression to appropriate tissues of the grain should allow such beneficial effects without severe negative effects on leaf expansion and ear emergence that are otherwise a feature of low GA levels in vegetative and flowering tissues. Thus targeting of *GA2ox* expression can limit the range of effects to specific tissues; this was elegantly demonstrated by Sakamoto and colleagues, who used the *OsGA3ox2* promoter to drive ectopic expression of *OsGA2ox1* in vegetative tissues of rice, thereby confining transgene expression to such tissues (Sakamoto *et al.*, 2003). The resulting plants expressing the *OsGA3ox2:OsGA2ox1* construct all showed a semi-dwarf habit with normal flowering time and grain numbers, in contrast to the severe dwarf phenotype of plants with *GA2ox1* under the rice actin promoter, which flowered later and had significantly fewer grains per panicle. Such targeted expression of *GA2ox* genes could be used to prove the effects of stem growth on yield and could conceivably improve on existing semi-dwarfing alleles in rice, wheat and other crops.

13.6.2 Other crop species

Semi-dwarfism is a target trait in a number of non-cereal crops and horticultural plants, and transgenic intervention in GA metabolism or signalling has been reported in species as diverse as apple, chrysanthemum, *Solanum* and turf grass. In apple, both suppression of an *MsGA20ox* paralogue by sense cosuppression and over-expression of the wild-type *Arabidopsis GAI* genes were successful in reducing stem elongation (Bulley *et al.*, 2005; Zhu *et al.*, 2008), although the more dwarfed *GAI* lines also had reduced numbers of adventitious roots, illustrating the possible negative effects of using constitutive promoters. However, effects on roots are of little relevance in a grafted species such as apple, as demonstrated by Bulley *et al.* (2005) whose *MsGA20ox1* suppression lines maintained their dwarfed habit even when grafted onto invigorating rootstocks.

Sugar beet (*Beta vulgaris*) is a biennial crop that is harvested before flowering in the second year. However, exposure to low temperatures and long days in the first year of growth can lead to flowering and consequent loss of yield. As GAs are known to play a role in floral induction, Mutasa-Gottgens and colleagues assessed the effects of manipulating both GA levels and sensitivity through expression of *GA2ox* and *GAI* genes. In both cases the resulting transgenic lines had a dwarfed habit and flowered later than controls (Mutasa-Gottgens *et al.*, 2009). Both sets of lines eventually flowered, albeit with shorter stems, suggesting that GA signalling is not absolutely required for bolting in this species. As sugar beet plants can be de-vernalised by a long, non-inductive period, delayed flowering should be agronomically favourable in the field.

In the potted plant and cut flower market, reducing stem height without affecting flower number or size is important, and often practised using growth retardants (see Chapter 11) or manipulation of light and temperature. Bhattacharya and colleagues generated dwarf lines of woody nightshade (*Solanum dulcamara*) through over-expression of a *GA20ox* gene from pumpkin that encodes an enzyme producing mainly C-20-carboxylate products rather than the C_{19} precursor of bioactive GAs (Curtis *et al.*, 2000). This activity diverts GA substrates from the main bioactive pathway into inactive products such as GA_{25}. The transgene strongly suppressed stem growth with weaker effects on leaf and flower size, generating an attractively bushy phenotype. Similar effects in this species can be achieved by over-expression of *GA2ox* genes to increase GA turnover (Dijkstra *et al.*, 2008). However, other species may respond differently: Otani *et al.* (2013) expressed a *GA2ox* gene from *Torenia fournieri* under the CaMV35S promoter in the liliaceous monocot *Tricyrtis* and achieved a 70–90% reduction in height. However, in this case leaf and flower size were also severely affected, again demonstrating that targeted expression of *GA2ox* genes may be required to achieve the desired phenotype. A range of scale of effects can also be achieved by transgenic expression of GA pathway genes under their own promoters: transfer of the *Arabidopsis GAI* gene into *Chrysanthemum* yielded a range of height effects (Petty *et al.*, 2003), as in apple, presumably due to position effects on the transgene.

There is some interest in controlling growth of amenity grasses in order to improve ground cover and to reduce mowing. Two groups have reported promising results of over-expression of *GA2ox* genes in different grasses. Constitutive expression of *AtGA2ox1* in bahiagrass (*Paspalum notatum*) resulted in a reduction in elongation growth and also a substantial increase in tiller number that improved ground cover (Agharkar *et al.*, 2007), similar to the effects on tillering observed in *GA2ox* over-expressing lines of wheat (Appleford *et al.*, 2007). Studzinska and colleagues similarly over-expressed *PcGA2ox1* in creeping bentgrass (*Agrostis stolonifera*) and although ground cover was not improved under full sun, the transgenic lines showed better performance than control lines under shaded conditions (Studzinska *et al.*, 2012).

Although most emphasis in the examples above has been on reducing GA levels to achieve a dwarf or semi-dwarf phenotype, there are cases where increased production of GAs might be beneficial. Here the selection of target gene is important: although increasing the expression of genes early in the pathway might be expected to increase flux through the pathway, attempts to achieve this in *Arabidopsis* by over-expression of *CPS* and *KS* were not successful: although the level of *ent*-kaurene increased spectacularly, there was no change in bioactive GAs levels due to limited flux through later steps in the pathway (Fleet *et al.*, 2003). In most species, GA20ox appears to be rate-limiting, as judged by the balance of substrates and products and by the results of over-expression (e.g. Coles *et al.*, 1999). Alternatively, reduction in GA2ox activity would be expected to increase GA levels through reducing turnover, but this enzyme is encoded by a family of 7–10 genes with overlapping expression patterns and down-regulation of a single gene, or even multiple genes, may have little effect (e.g. Rieu *et al.*, 2008). Finally, a novel enzyme (DES) from the GA pathway in the fungus *Gibberella fujikuroi* has been identified that catalyses the desaturation of GA_4 to GA_7; as the biologically active product of this activity is unsaturated at C-2 it is not a substrate for GA2ox enzymes and therefore cannot be inactivated by this route. Over-expression of *GfDES* therefore raises bioactive GA levels and increases height, as demonstrated in transgenic lines of several plant species (Bhattacharya *et al.*, 2012). The resistance of 1,2-didehydro-GAs such as GA_7 to inactivation by GA 2-oxidation was illustrated by the rescue of severely dwarfed plants over-expressing the runner bean Pc*GA2ox* by crossing with *GfDES*-expressing lines .

One possible application of GA over-production is in promoting the growth of fruit, particularly as a route to parthenocarpic production (see Chapter 11). The degree to which GA application or biosynthesis can maintain parthenocarpic fruit expansion varies between species, but over-expression of a citrus *GA20ox* under the CaMV-35S promoter in a normally fertile tomato variety resulted in over-growth of the stigma and a lack of fertilisation in a proportion of the flowers (Garcia-Hurtado *et al.*, 2012). However, these unfertilised flowers produced normal, if somewhat small, fruit, whereas no fruit was produced from unfertilised wild-type flowers. There may be applications for GA over-production in other parthenocarpic or stenocarpic systems, although it must be remembered that over-expressing one step in the biosynthetic pathway will have no effect unless the rest of the pathway is present and active in that tissue, or possibly in neighbouring tissues, depending on the extent of movement of GAs and intermediates.

Manipulation of GA levels in forest species was first suggested as a possible route to extending the growth season of trees in far Northern latitudes. Over-expression of an *Arabidopsis GA20ox* under the CaMV-35S promoter in hybrid aspen resulted in increased growth rates (Eriksson *et al.*, 2000). In contrast, over-expression of a *GA3ox* gene had little effect on phenotype, illustrating the importance of selecting the appropriate step in the pathway for manipulation (Israelsson *et al.*, 2004). An important feature of the GA20ox-over-expressing lines was an increase in the number and length of xylem fibre cells, important traits for paper production. Exogenous GA application is also known to promote fibre initiation and elongation in cotton (Gokani and Thaker, 2002), and over-expression of a cotton *GA20ox* in that species, also under CaMV-35S promoter, resulted in an increase in the number and length of the fibres produced by the ovule and leading to an increase in quality of the cotton boll, although it also led to an increase in stem length and a decrease in flower number (Xiao *et al.*, 2006). Clearly, over-expression of *GA20ox* targeted to developing ovaries or flowers would be required for such manipulation of GA status to be commercially useful. An alternative approach might be to silence *GA2ox* genes expressed in the ovary wall, as suppression of *GA2ox* in tobacco was found to be even more effective than over-expression of *GA20ox* in increasing stem xylem fibres in tobacco (Dayan *et al.*, 2010).

13.7 Conclusions

The introduction of semi-dwarfing alleles acting through the GA biosynthetic and signalling pathways that increased harvestable yield in cereals under high nitrogen application was a major success of plant breeding in the last century and underpinned the Green Revolution. However, few other dwarfing alleles, whether or not from the GA pathway, appear to be able to achieve the effect of reducing stem height to prevent lodging while maintaining biomass and yield. Further work on understanding the changes in development and metabolism that allow *Rht-1* and *sd-1* to achieve these effects are clearly required. Similarly, manipulation of GA status or sensitivity through over-expression or suppression of components of the GA pathway is now commonplace and can generate a range of beneficial phenotypes in various crop species. However, in most cases the use of constitutive promoters also produces a wide variety of negative pleiotropic effects that make this first generation of hormonally modified plants unsuitable, in the main, for commercial release. There is a clear requirement for tissue-specific, developmentally regulated and/or inducible promoters to drive transgene expression to provide targeted modification of GA metabolism and signalling, that will both expand our understanding of the role of GA in determining crop performance and will also help to achieve even greater increases in yield and quality.

Acknowledgements

The author receives support from the 20:20 Wheat© Institute Strategic Programme at Rothamsted Research funded by the Biotechnology and Biological Sciences Research Council of the United Kingdom.

References

Abe, A., Takagi, H., Fujibe, T. *et al.* (2012). *OsGA20ox1*, a candidate gene for a major QTL controlling seedling vigor in rice. *Theoretical and Applied Genetics* **125**, 647–657.

Agharkar, M., Lomba, P., Altpeter, F. *et al.* (2007). Stable expression of *AtGA2ox1* in a low-input turfgrass (*Paspalum notatum* Flugge) reduces bioactive gibberellin levels and improves turf quality under field conditions. *Plant Biotechnology Journal* **5**, 791–801.

Alghabari, F., Lukac, M., Jones, H.E. and Gooding, M.J. (2014). Effect of Rht alleles on the tolerance of wheat grain set to high temperature and drought stress during booting and anthesis. *Journal of Agronomy and Crop Science* **200**, 36–45.

Appleford, N.E.J. and Lenton, J.R. (1991). Gibberellins and leaf expansion in near-isogenic wheat lines containing *rht1* and *rht3* dwarfing alleles. *Planta* **183**, 229–236.

Appleford, N.E.J., Wilkinson, M.D., Ma, Q. *et al.* (2007). Decreased shoot stature and grain α-amylase activity following ectopic expression of a gibberellin 2-oxidase gene in transgenic wheat. *Journal of Experimental Botany* **58**, 3213–3226.

Asano, K., Hirano, K., Ueguchi-Tanaka, M. *et al.* (2009). Isolation and characterization of dominant dwarf mutants, *Slr1-d*, in rice. *Molecular Genetics and Genomics* **281**, 223–231.

Asano, K., Yamasaki, M., Takuno, S. *et al.* (2011). Artificial selection for a green revolution gene during japonica rice domestication. *Proceedings of the National Academy of Sciences USA* **108**, 11034–11039.

Bai, C.H., Liang, Y.L. and Hawkesford, M.J. (2013). Identification of QTLs associated with seedling root traits and their correlation with plant height in wheat. *Journal of Experimental Botany* **64**, 1745–1753.

Bhattacharya, A., Kourmpetli, S., Ward, D.A. *et al.* (2012). Characterization of the fungal gibberellin desaturase as a 2-oxoglutarate-dependent dioxygenase and its utilization for enhancing plant growth. *Plant Physiology* **160**, 837–845.

Bulley, S.M., Wilson, F.M., Hedden, P. *et al.* (2005). Modification of gibberellin biosynthesis in the grafted apple scion allows control of tree height independent of the rootstock. *Plant Biotechnology Journal* **3**, 215–223.

Casella, L., Greco, R., Bruschi, G. *et al.* (2013). TILLING in European rice: hunting mutations for crop improvement. *Crop Science* **53**, 2550–2562.

Chandler, P.M. and Harding, C.A. (2013). 'Overgrowth' mutants in barley and wheat: new alleles and phenotypes of the 'Green Revolution' DELLA gene. *Journal of Experimental Botany* **64**, 1603–1613.

Chandler, P.M. and Robertson, M. (1999). Gibberellin dose-response curves and the characterization of dwarf mutants of barley. *Plant Physiology* **120**, 623–632.

Chandler, P.M., Marion-Poll, A., Ellis, M. and Gubler, F. (2002). Mutants at the Slender1 locus of barley cv Himalaya. molecular and physiological characterization. *Plant Physiology* **129**, 181–190.

Chandler, P.M., Harding, C.A., Ashton, A.R. *et al.* (2008). Characterization of gibberellin receptor mutants of barley (*Hordeum vulgare* L.). *Molecular Plant* **1**, 285–294.

Chen, L., Phillips, A.L., Condon, A.G. *et al.* (2013). GA-responsive dwarfing gene *Rht12* affects the developmental and agronomic traits in common bread wheat. *Plos One* **8**.

Chen, Y., Hou, M.M., Liu, L.J. *et al.* (2014). The maize *DWARF1* encodes a gibberellin 3-oxidase and is dual localized to the nucleus and cytosol. *Plant Physiology* **166**, 2028–U1267.

Chiang, H.H., Hwang, I. and Goodman, H.M. (1995). Isolation of the *Arabidopsis GA4* locus. *The Plant Cell* **7**, 195–201.

Coles, J.P., Phillips, A.L., Croker, S.J. *et al.* (1999). Modification of gibberellin production and plant development in *Arabidopsis* by sense and antisense expression of gibberellin 20-oxidase genes. *The Plant Journal* **17**, 547–556.

Curtis, I.S., Ward, D.A., Thomas, S.G. *et al.* (2000). Induction of dwarfism in transgenic *Solanum dulcamara* by over-expression of a gibberellin 20-oxidase cDNA from pumpkin. *The Plant Journal* **23**, 329–338.

Daviere, J.M. and Achard, P. (2013). Gibberellin signaling in plants. *Development* **140**, 1147–1151.

Dayan, J., Schwarzkopf, M., Avni, A. and Aloni, R. (2010). Enhancing plant growth and fiber production by silencing GA 2-oxidase. *Plant Biotechnology Journal* **8**, 425–435.

Dijkstra, C., Adams, E., Bhattacharya, A. *et al.* (2008). Over-expression of a gibberellin 2-oxidase gene from *Phaseolus coccineus* L. enhances gibberellin inactivation and induces dwarfism in Solanum species. *Plant Cell Reports* **27**, 463–470.

Ellis, M.H., Rebetzke, G.J., Chandler, P. *et al.* (2004). The effect of different height reducing genes on the early growth of wheat. *Functional Plant Biology* **31**, 583–589.

Eriksson, M.E., Israelsson, M., Olsson, O. and Moritz, T. (2000). Increased gibberellin biosynthesis in transgenic trees promotes growth, biomass production and xylem fiber length. *Nature Biotechnology* **18**, 784–788.

Fleet, C.M., Yamaguchi, S., Hanada, A. *et al.* (2003). Overexpression of *AtCPS* and *AtKS* in Arabidopsis confers increased *ent*-kaurene production but no increase in bioactive gibberellins. *Plant Physiology* **132**, 830–839.

Flintham, J.E., Borner, A., Worland, A.J. and Gale, M.D. (1997). Optimizing wheat grain yield: effects of (gibberellin-insensitive) dwarfing genes. *Journal of Agricultural Science* **128**, 11–25.

Gale, M.D. and Marshall, G.A. (1973). Insensitivity to gibberellin in dwarf wheats. *Annals of Botany* **37**, 729–735.

Garcia-Hurtado, N., Carrera, E., Ruiz-Rivero, O. *et al.* (2012). The characterization of transgenic tomato overexpressing gibberellin 20-oxidase reveals induction of parthenocarpic fruit growth, higher yield, and alteration of the gibberellin biosynthetic pathway. *Journal of Experimental Botany* **63**, 5803–5813.

Gaskin, P., Kirkwood, P.S., Lenton, J.R. *et al.* (1980). Identification of gibberellins in developing wheat grain. *Agricultural and Biological Chemistry* **44**, 1589–1593.

Gasperini, D., Greenland, A., Hedden, P. *et al.* (2012). Genetic and physiological analysis of *Rht8* in bread wheat: an alternative source of semi-dwarfism with a reduced sensitivity to brassinosteroids. *Journal of Experimental Botany* **63**, 4419–4436.
Gokani, S.J. and Thaker, V.S. (2002). Role of gibberellic acid in cotton fibre development. *Journal of Agricultural Science* **138**, 255–260.
Gooding, M.J., Addisu, M., Uppal, R.K. *et al.* (2012). Effect of wheat dwarfing genes on nitrogen-use efficiency. *Journal of Agricultural Science* **150**, 3–22.
Griffiths, S., Simmonds, J., Leverington, M. *et al.* (2012). Meta-QTL analysis of the genetic control of crop height in elite European winter wheat germplasm. *Molecular Breeding* **29**, 159–171.
Guedira, M., Brown-Guedira, G., Van Sanford, D. *et al.* (2010). Distribution of *Rht* genes in modern and historic winter wheat cultivars from the eastern and central USA. *Crop Science* **50**, 1811–1822.
Huang, J., Tang, D., Shen, Y. *et al.* (2010). Activation of gibberellin 2-oxidase 6 decreases active gibberellin levels and creates a dominant semi-dwarf phenotype in rice (*Oryza sativa* L.). *Journal of Genetics and Genomics* **37**, 23–36.
Ikeda, A., Ueguchi-Tanaka, M., Sonoda, Y. *et al.* (2001). *slender* rice, a constitutive gibberellin response mutant, is caused by a null mutation of the *SLR1* gene, an ortholog of the height-regulating gene *GAI/RGA/RHT/D8*. *The Plant Cell* **13**, 999–1010.
Israelsson, M., Mellerowicz, E., Chono, M. *et al.* (2004). Cloning and overproduction of gibberellin 3-oxidase in hybrid aspen trees. Effects of gibberellin homeostasis and development. *Plant Physiology* **135**, 221–230.
Itoh, H., Ueguchi-Tanaka, M., Sentoku, N. *et al.* (2001). Cloning and functional analysis of two gibberellin 3β-hydroxylase genes that are differently expressed during the growth of rice. *Proceedings of the National Academy of Sciences USA* **98**, 8909–8914.
Itoh, H., Ueguchi-Tanaka, M., Sakamoto, T. *et al.* (2002). Modification of rice plant height by suppressing the height-controlling gene, *D18*, in rice. *Breeding Science* **52**, 215–218.
Itoh, H., Tatsumi, T., Sakamoto, T. *et al.* (2004). A rice semi-dwarf gene, *Tan-Ginbozu (D35)*, encodes the gibberellin biosynthesis enzyme, *ent*-kaurene oxidase. *Plant Molecular Biology* **54**, 533–547.
Jia, Q.J., Zhang, J.J., Westcott, S. *et al.* (2009). GA 20-oxidase as a candidate for the semidwarf gene *sdw1/denso* in barley. *Functional & Integrative Genomics* **9**, 255–262.
Jia, Q.J., Zhang, X.Q., Westcott, S. *et al.* (2011). Expression level of a gibberellin 20-oxidase gene is associated with multiple agronomic and quality traits in barley. *Theoretical and Applied Genetics* **122**, 1451–1460.
Kobayashi, M., Yamaguchi, I., Murofushi, N. *et al.* (1988). Fluctuation and localization of endogenous gibberellins in rice. *Agricultural and Biological Chemistry* **52**, 1189–1194.
Law, C.N. and Worland, A.J. (1984). An effect of temperature on the fertility of wheats containing the dwarfing genes *Rht1, Rht2,* and *Rht3.* In *Plant Breeding Institute Annual Report*, pp. 69–71: Plant Breeding Institute, Cambridge.
Law, C.N., Snape, J.W. and Worland, A.J. (1978). Genetic relationship between height and yield in wheat. *Heredity* **40**, 133–151.
Lawit, S.J., Wych, H.M., Xu, D.P. *et al.* (2010). Maize DELLA proteins dwarf plant8 and dwarf plant9 as modulators of plant development. *Plant and Cell Physiology* **51**, 1854–1868.
Lester, D.R., Ross, J.J., Davies, P.J. and Reid, J.B. (1997). Mendel's stem length gene (*Le*) encodes a gibberellin 3β-hydroxylase. *The Plant Cell* **9**, 1435–1443.

Li, Y.Y., Xiao, J.H., Wu, J.J. *et al.* (2012). A tandem segmental duplication (TSD) in green revolution gene *Rht-D1b* region underlies plant height variation. *New Phytologist* **196**, 282–291.

Liu, C., Wang, J.L., Huang, T.D. *et al.* (2010). A missense mutation in the VHYNP motif of a DELLA protein causes a semi-dwarf mutant phenotype in *Brassica napus*. *Theoretical and Applied Genetics* **121**, 249–258.

Martin, D.N., Proebsting, W.M. and Hedden, P. (1997). Mendel's dwarfing gene: cDNAs from the *Le* alleles and function of the expressed proteins. *Proceedings of the National Academy of Sciences USA* **94**, 8907–8911.

Monna, L., Kitazawa, N., Yoshino, R. *et al.* (2002). Positional cloning of rice semidwarfing gene, *sd-1*: Rice 'Green revolution gene' encodes a mutant enzyme involved in gibberellin synthesis. *DNA Research* **9**, 11–17.

Mrva, K. and Mares, D.J. (1996). Expression of late maturity α-amylase in wheat containing gibberellic acid insensitivity genes. *Euphytica* **88**, 69–76.

Muangprom, A., Thomas, S.G., Sun, T.P. and Osborn, T.C. (2005). A novel dwarfing mutation in a green revolution gene from *Brassica rapa*. *Plant Physiology* **137**, 931–938.

Muangprom, A., Mauriera, I. and Osborn, T.C. (2006). Transfer of a dwarf gene from *Brassica rapa* to oilseed *B. napus*, effects on agronomic traits, and development of a 'perfect' marker for selection. *Molecular Breeding* **17**, 101–110.

Mutasa-Gottgens, E., Qi, A.M., Mathews, A. *et al.* (2009). Modification of gibberellin signalling (metabolism & signal transduction) in sugar beet: analysis of potential targets for crop improvement. *Transgenic Research* **18**, 301–308.

Ogawa, S., Toyomasu, T., Yamane, H. *et al.* (1996). A step in the biosynthesis of gibberellins that is controlled by the mutation in the semidwarf rice cultivar Tan-Ginbozu. *Plant and Cell Physiology* **37**, 363–368.

Oikawa, T., Koshioka, M., Kojima, K. *et al.* (2004). A role of *OsGA20ox1*, encoding an isoform of gibberellin 20-oxidase, for regulation of plant stature in rice. *Plant Molecular Biology* **55**, 687–700.

Otani, M., Meguro, S., Gondaira, H. *et al.* (2013). Overexpression of the gibberellin 2-oxidase gene from *Torenia fournieri* induces dwarf phenotypes in the liliaceous monocotyledon *Tricyrtis* sp. *Journal of Plant Physiology* **170**, 1416–1423.

Pearce, S., Saville, R., Vaughan, S.P. *et al.* (2011). Molecular characterization of *Rht-1* dwarfing genes in hexaploid wheat. *Plant Physiology* **157**, 1820–1831.

Pearce, S.P., Huttly, A.K., Prosser, I.M. *et al.* (2015). Heterologous expression and transcript analysis of gibberellin biosynthetic genes of grasses reveals novel functionality in the GA3ox family. *BMC Plant Biology* **15**, 130.

Peng, J.H.H., Sun, D.F. and Nevo, E. (2011). Domestication evolution, genetics and genomics in wheat. *Molecular Breeding* **28**, 281–301.

Peng, J.R., Richards, D.E., Hartley, N.M. *et al.* (1999). 'Green revolution' genes encode mutant gibberellin response modulators. *Nature* **400**, 256–261.

Petty, L.M., Harberd, N.P., Carre, I.A. *et al.* (2003). Expression of the *Arabidopsis gai* gene under its own promoter causes a reduction in plant height in chrysanthemum by attenuation of the gibberellin response. *Plant Science* **164**, 175–182.

Plackett, A.R.G., Thomas, S.G., Wilson, Z.A. and Hedden, P. (2011). Gibberellin control of stamen development: a fertile field. *Trends in Plant Science* **16**, 568–578.

Qiao, F., Yang, Q., Wang, C.L. *et al.* (2007). Modification of plant height via RNAi suppression of *OsGA20ox2* gene in rice. *Euphytica* **158**, 35–45.

Radley, M. (1970). Comparison of endogenous gibberellins and response to applied gibberellin of some dwarf and tall wheat cultivars. *Planta* **92**, 292–300.
Rebetzke, G.J. and Richards, R.A. (2000). Gibberellic acid-sensitive dwarfing genes reduce plant height to increase kernel number and grain yield of wheat. *Australian Journal of Agricultural Research* **51**, 235–245.
Rebetzke, G.J., Richards, R.A., Fischer, V.M. and Mickelson, B.J. (1999). Breeding long coleoptile, reduced height wheats. *Euphytica* **106**, 159–168.
Rebetzke, G.J., Bruce, S.E. and Kirkegaard, J.A. (2005). Longer coleoptiles improve emergence through crop residues to increase seedling number and biomass in wheat (*Triticum aestivum* L.). *Plant and Soil* **272**, 87–100.
Rieu, I., Eriksson, S., Powers, S.J. *et al.* (2008). Genetic analysis reveals that C_{19}-GA 2-oxidation is a major gibberellin inactivation pathway in *Arabidopsis*. *The Plant Cell* **20**, 2420–2436.
Ross, J.J. and Reid, J.B. (1987). Internode length in *Pisum*: a new allele at the *Le* locus. *Annals of Botany* **59**, 107–109.
Rutger, J.N. (1983). Applications of induced and spontaneous mutation in rice breeding and genetics. *Advances in Agronomy* **36**, 383–413.
Rutger, J.N. and Carnahan, H.L. (1981). A 4th genetic element to facilitate hybrid cereal production: a recessive tall in rice. *Crop Science* **21**, 373–376.
Sakamoto, T., Miura, K., Itoh, H. *et al.* (2004). An overview of gibberellin metabolism enzyme genes and their related mutants in rice. *Plant Physiology* **134**, 1642–1653.
Sakamoto, T., Morinaka, Y., Ishiyama, K. *et al.* (2003). Genetic manipulation of gibberellin metabolism in transgenic rice. *Nature Biotechnology* **21**, 909–913.
Sasaki, A., Ashikari, M., Ueguchi-Tanaka, M. *et al.* (2002). Green revolution: A mutant gibberellin-synthesis gene in rice: new insight into the rice variant that helped to avert famine over thirty years ago. *Nature* **416**, 701–702.
Sasaki, A., Itoh, H., Gomi, K. *et al.* (2003). Accumulation of phosphorylated repressor for gibberellin signaling in an F-box mutant. *Science* **299**, 1896–1898.
Saville, R.J., Gosman, N., Burt, C.J. *et al.* (2012). The 'Green Revolution' dwarfing genes play a role in disease resistance in *Triticum aestivum* and *Hordeum vulgare*. *Journal of Experimental Botany* **63**, 1271–1283.
Spielmeyer, W., Hyles, J., Joaquim, P. *et al.* (2007). A QTL on chromosome 6A in bread wheat (*Triticum aestivum*) is associated with longer coleoptiles, greater seedling vigour and final plant height. *Theoretical and Applied Genetics* **115**, 59–66.
Spielmeyer, W., Ellis, M.H. and Chandler, P.M. (2002). Semidwarf (*sd-1*), "Green Revolution" rice, contains a defective gibberellin 20-oxidase gene. *Proceedings of the National Academy of Sciences USA* **99**, 9043–9048.
Spray, C.R., Kobayashi, M., Suzuki, Y. *et al.* (1996). The *dwarf-1* (*d1*) mutant of *Zea mays* blocks 3 steps in the gibberellin-biosynthetic pathway. *Proceedings of the National Academy of Sciences USA* **93**, 10515–10518.
Studzinska, A.K., Gardner, D.S., Metzger, J.D. *et al.* (2012). The effect of *PcGA2ox* overexpression on creeping bentgrass (*Agrostis stolonifera* L.): performance under various light environments. *Hortscience* **47**, 280–284.
Tan, M.K., Verbyla, A.P., Cullis, B.R. *et al.* (2010). Genetics of late maturity α-amylase in a doubled haploid wheat population. *Crop & Pasture Science* **61**, 153–161.
Teng, F., Zhai, L.H., Liu, R.X. *et al.* (2013). *ZmGA3ox2*, a candidate gene for a major QTL, qPH3.1, for plant height in maize. *The Plant Journal* **73**, 405–416.
Ueguchi-Tanaka, M., Ashikari, M., Nakajima, M. *et al.* (2005). *GIBBERELLIN INSENSITIVE DWARF1* encodes a soluble receptor for gibberellin. *Nature* **437**, 693–698.

Ueguchi-Tanaka, M., Nakajima, M., Katoh, E. *et al.* (2007). Molecular interactions of a soluble gibberellin receptor, GID1, with a rice DELLA protein, SLR1, and gibberellin. *The Plant Cell* **19**, 2140–2155.
Wen, W., Deng, Q.Y., Jia, H.Y. *et al.* (2013). Sequence variations of the partially dominant DELLA gene *Rht-B1c* in wheat and their functional impacts. *Journal of Experimental Botany* **64**, 3299–3312.
Wilhelm, E.P., Boulton, M.I., Barber, T.E.S. *et al.* (2013a). Genotype analysis of the wheat semidwarf *Rht-B1b* and *Rht-D1b* ancestral lineage. *Plant Breeding* **132**, 539–545.
Wilhelm, E.P., Mackay, I.J., Saville, R.J. *et al.* (2013b). Haplotype dictionary for the *Rht-1* loci in wheat. *Theoretical and Applied Genetics* **126**, 1733–1747.
Winkler, R.G. and Freeling, M. (1994). Physiological genetics of the dominant gibberellin-nonresponsive maize dwarfs, *dwarf-8* and *dwarf-9*. *Planta* **193**, 341–348.
Wojciechowski, T., Gooding, M.J., Ramsay, L. and Gregory, P.J. (2009). The effects of dwarfing genes on seedling root growth of wheat. *Journal of Experimental Botany* **60**, 2565–2573.
Worland, A.J., Sayers, E.J. and Borner, A. (1994). The genetics and breeding potential of *Rht12*, a dominant dwarfing gene in wheat. *Plant Breeding* **113**, 187–196.
Xiao, J.H., Li, H.X., Zhang, J.H. *et al.* (2006). Dissection of GA 20-oxidase members affecting tomato morphology by RNAi-mediated silencing. *Plant Growth Regulation* **50**, 179–189.
Zhang, S.B., Chen, R.X., Zhao, L. *et al.* (2011). A novel semidwarf mutant *02428ha* in rice is caused by reduced gibberellin sensitivity. *Acta Physiologiae Plantarum* **33**, 113–122.
Zhang, Y.Y., Zhu, Y.Y., Peng, Y. *et al.* (2008). Gibberellin homeostasis and plant height control by *EUI* and a role for gibberellin in root gravity responses in rice. *Cell Research* **18**, 412–421.
Zhu, L.H., Li, X.Y. and Welander, M. (2008). Overexpression of the *Arabidopsis gai* gene in apple significantly reduces plant size. *Plant Cell Reports* **27**, 289–296.
Zhu, Y.Y., Nomura, T., Xu, Y.H. *et al.* (2006). *ELONGATED UPPERMOST INTERNODE* encodes a cytochrome P450 monooxygenase that epoxidizes gibberellins in a novel deactivation reaction in rice. *The Plant Cell* **18**, 442–456.

Annual Plant Reviews (2016) **49**, 431–436
doi: 10.1002/9781119210436.Appendix

http://onlinelibrary.wiley.com

Annual Plant Reviews

Appendix

THE STRUCTURES OF THE GIBBERELLINS

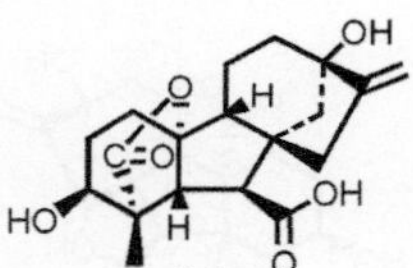

GA$_1$

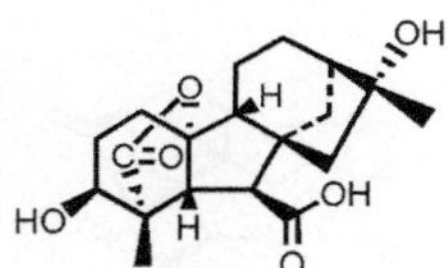

GA$_2$

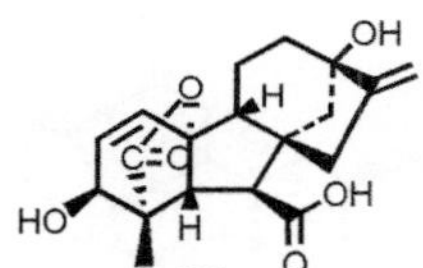

GA$_3$

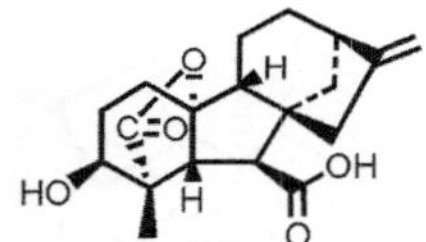

GA$_4$

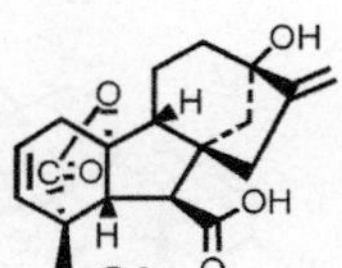

GA$_5$

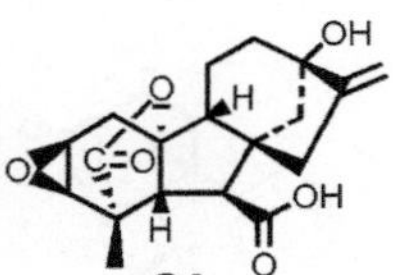

GA$_6$

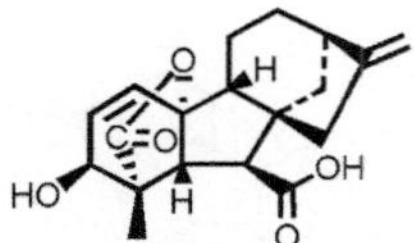

GA$_7$

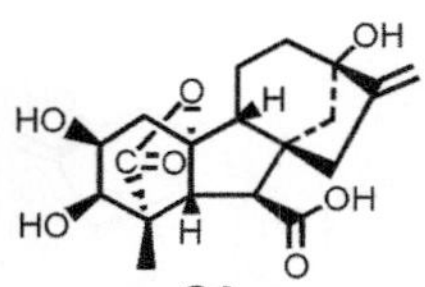

GA$_8$

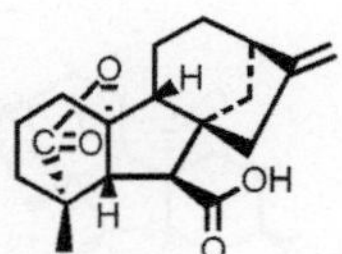

GA$_9$

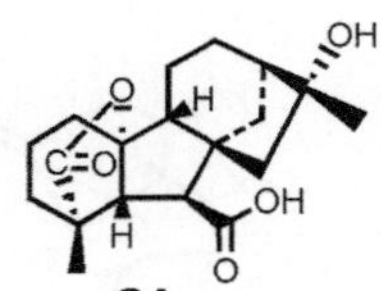

GA$_{10}$

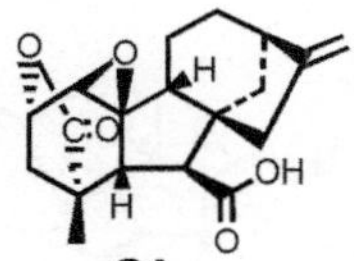

GA$_{11}$

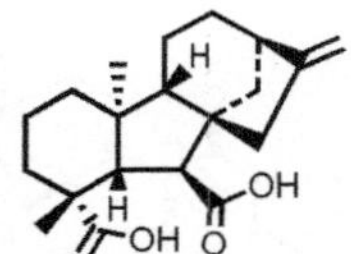

GA$_{12}$

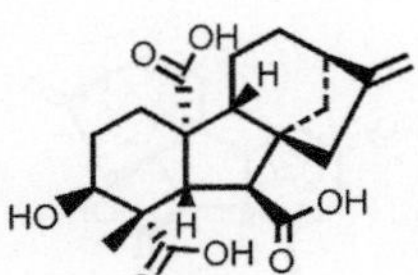

GA$_{13}$

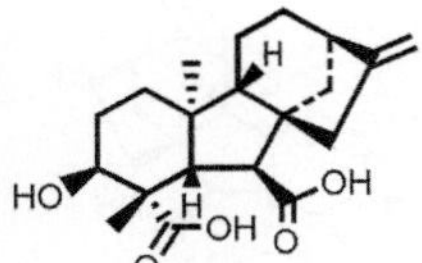

GA$_{14}$

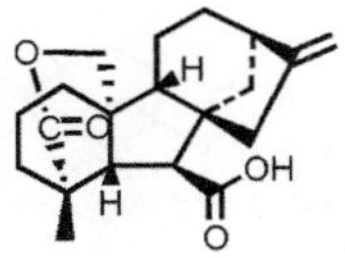

GA$_{15}$

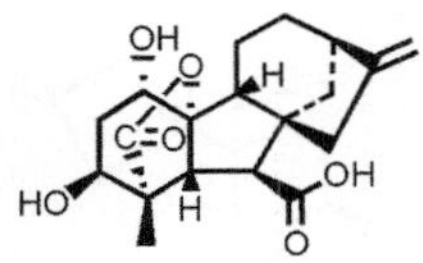

GA$_{16}$

Annual Plant Reviews, Volume 49: The Gibberellins, First Edition.
Edited by Peter Hedden and Stephen G. Thomas.

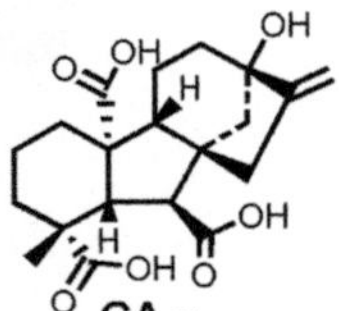
GA$_{17}$

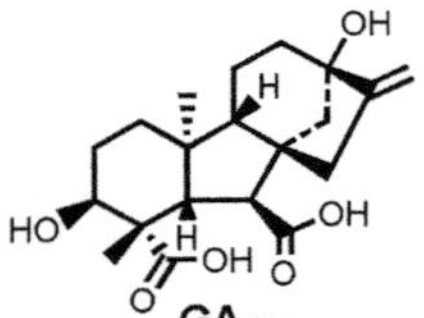
GA$_{18}$

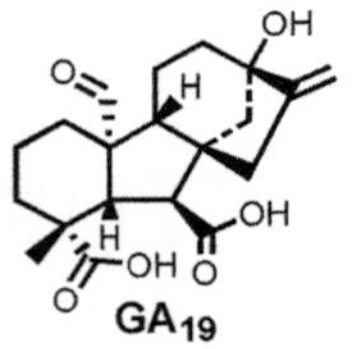
GA$_{19}$

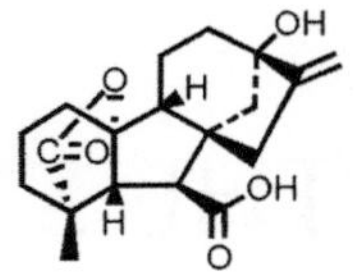
GA$_{20}$

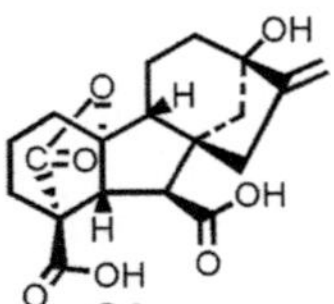
GA$_{21}$

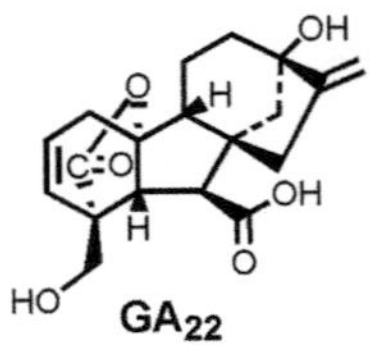
GA$_{22}$

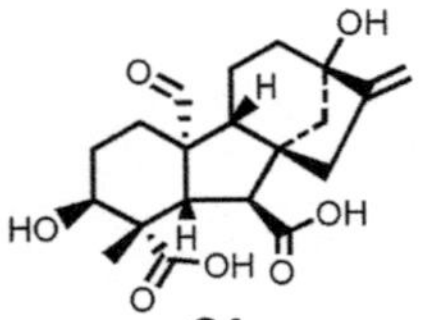
GA$_{23}$

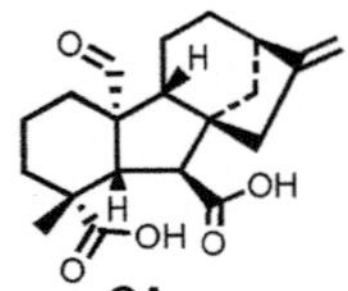
GA$_{24}$

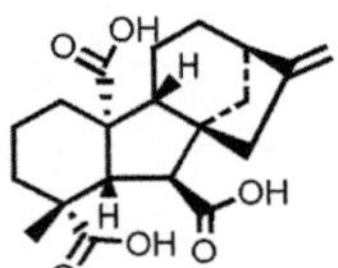
GA$_{25}$

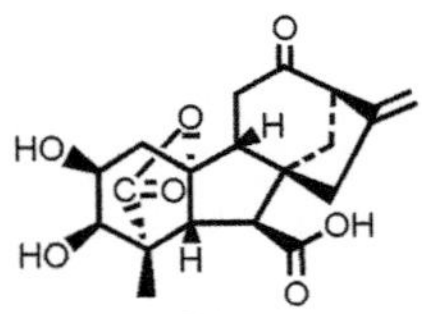
GA$_{26}$

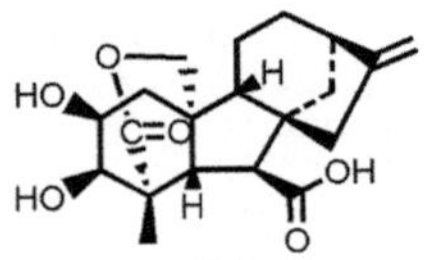
GA$_{27}$

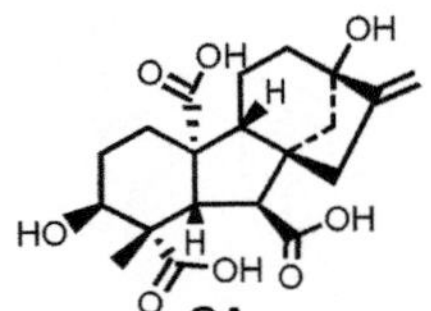
GA$_{28}$

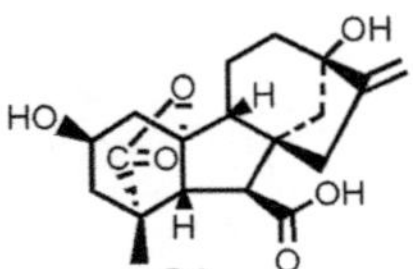
GA$_{29}$

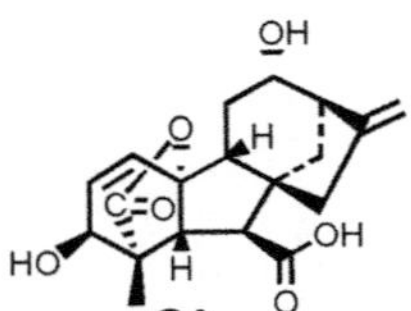
GA$_{30}$

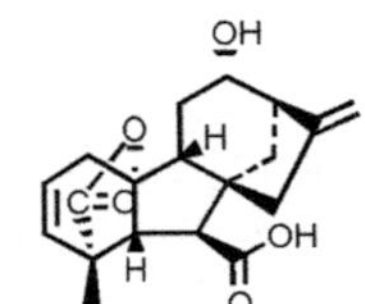
GA$_{31}$

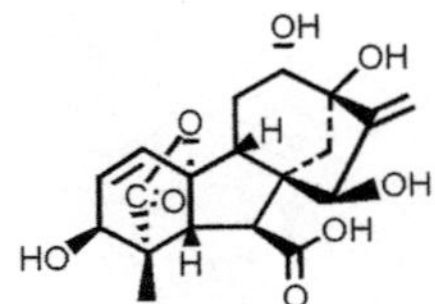
GA$_{32}$

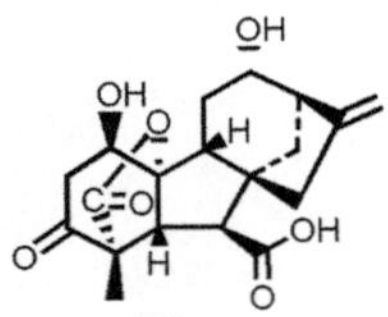
GA$_{33}$

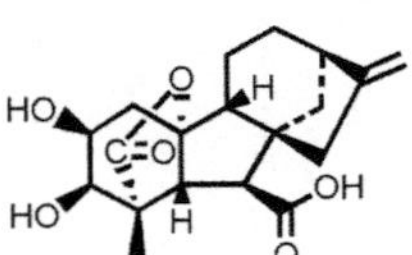
GA$_{34}$

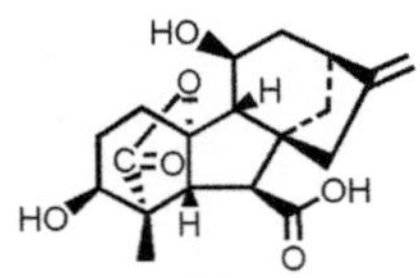
GA$_{35}$

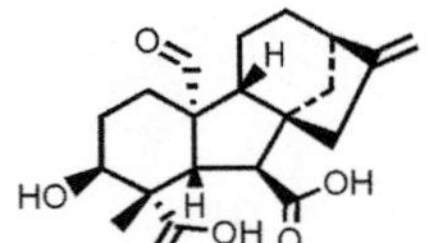
GA$_{36}$

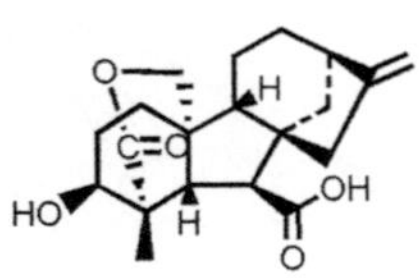
GA$_{37}$

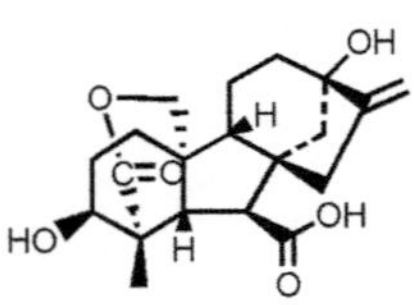
GA$_{38}$

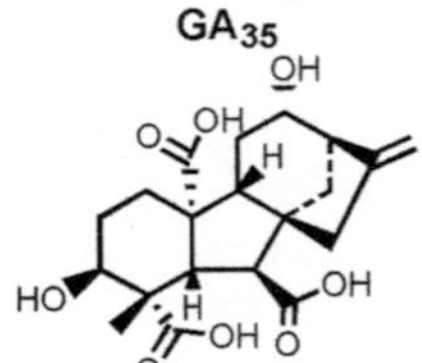
GA$_{39}$

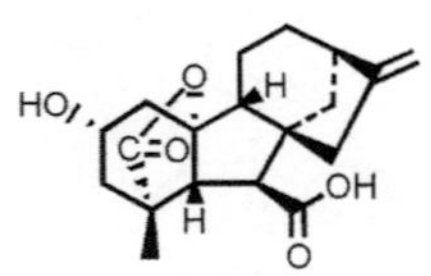
GA$_{40}$

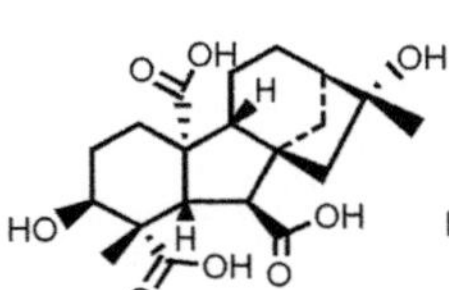
GA$_{41}$

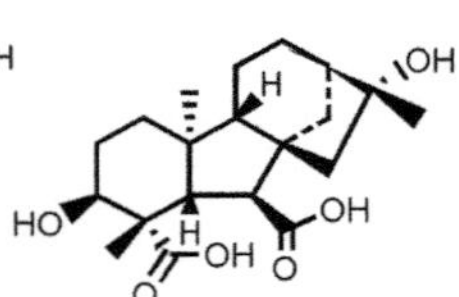
GA$_{42}$

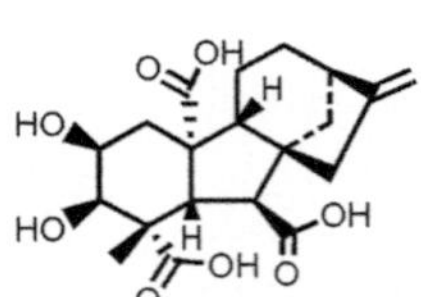
GA$_{43}$

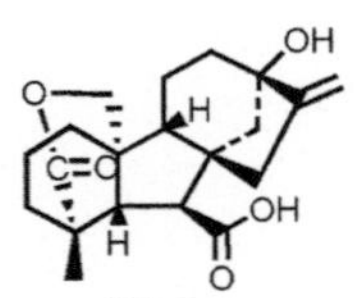
GA$_{44}$

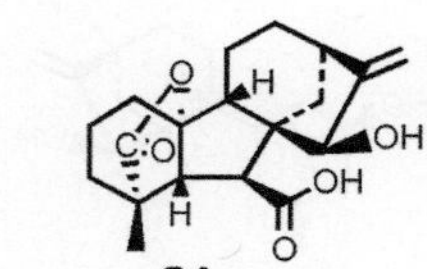
GA45

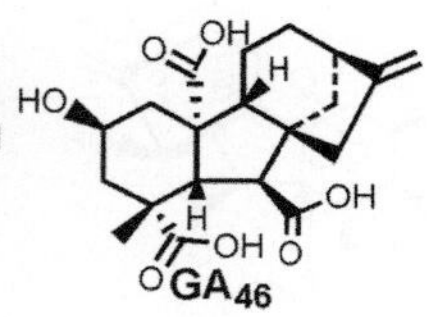
GA46

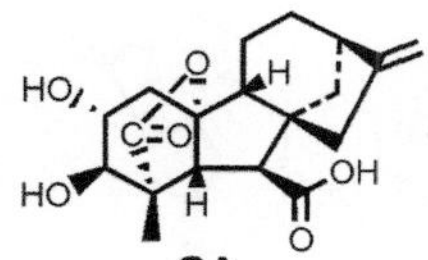
GA47

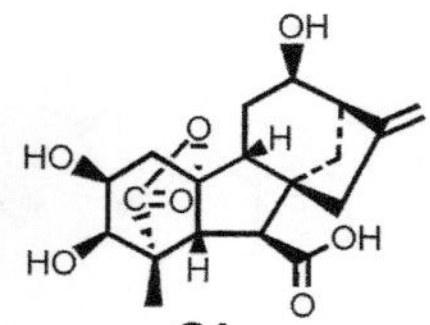
GA48

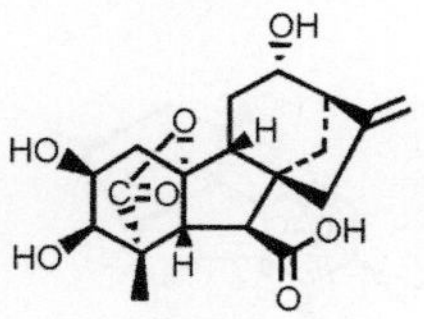
GA49

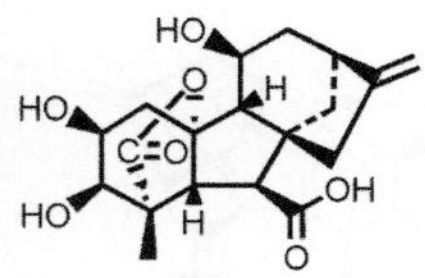
GA50

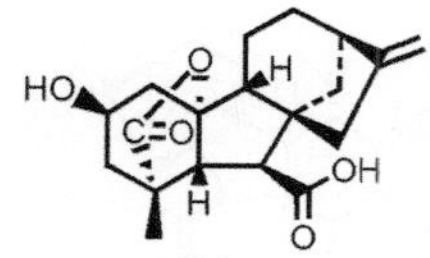
GA51

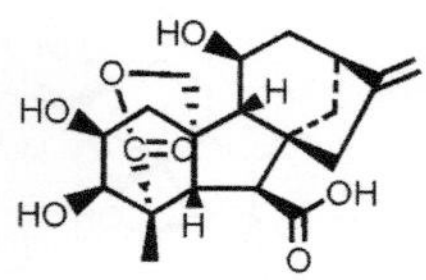
GA52

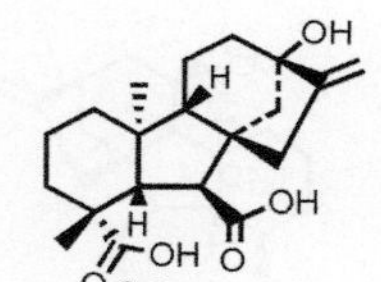
GA53

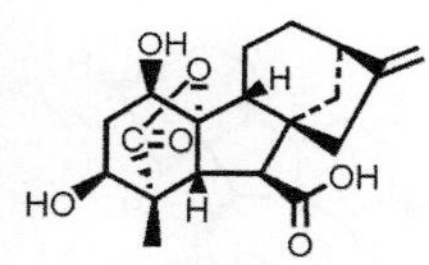
GA54

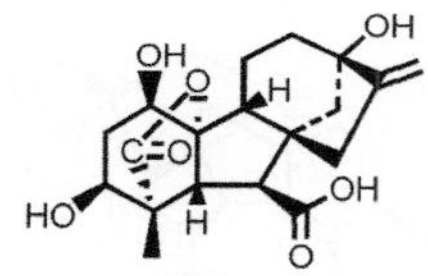
GA55

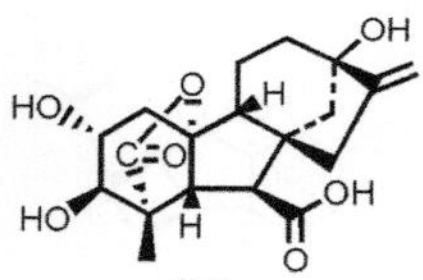
GA56

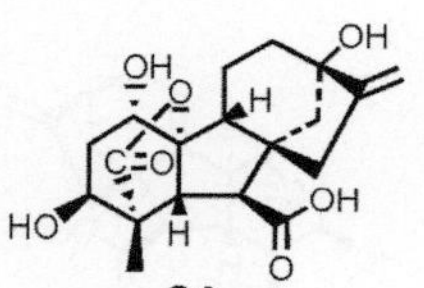
GA57

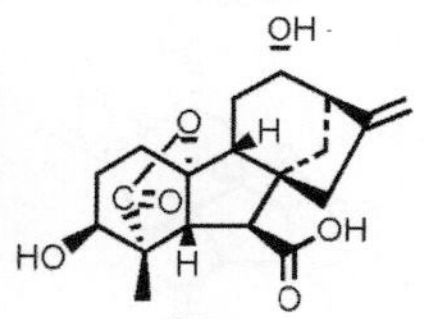
GA58

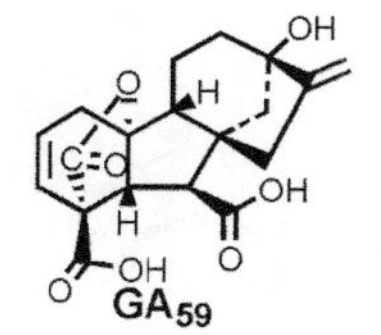
GA59

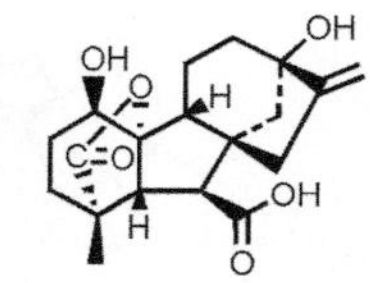
GA60

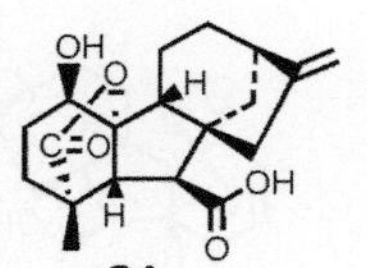
GA61

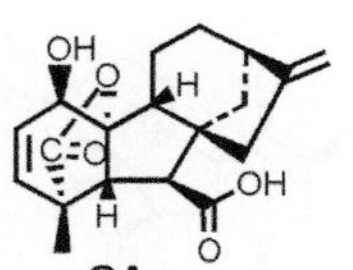
GA62

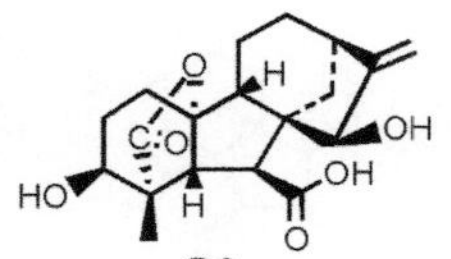
GA63

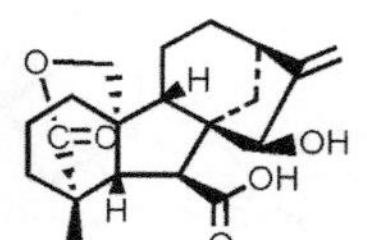
GA64

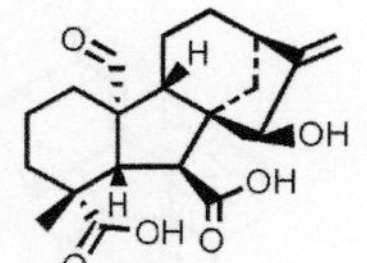
GA65

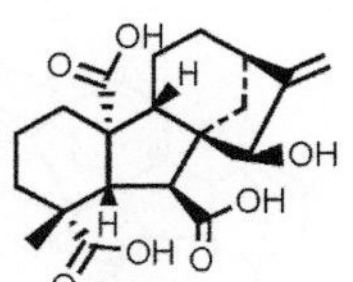
GA66

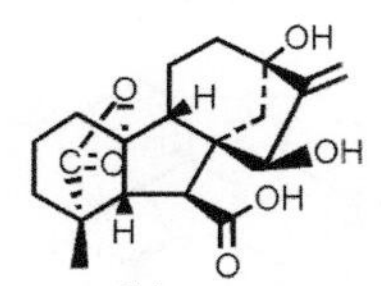
GA67

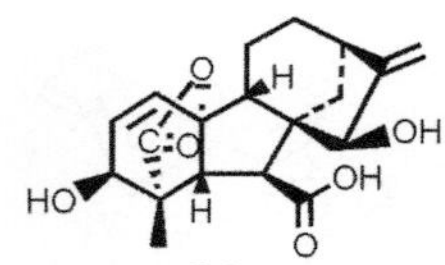
GA68

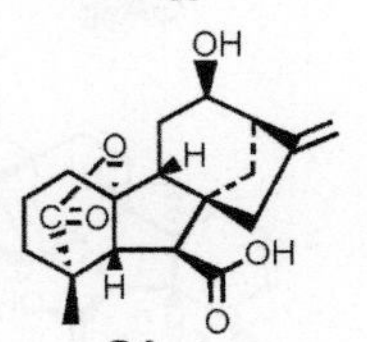
GA69

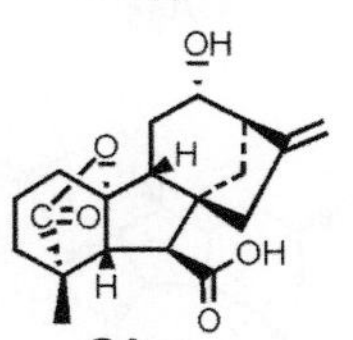
GA70

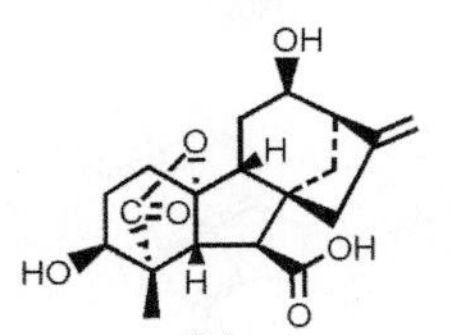
GA71

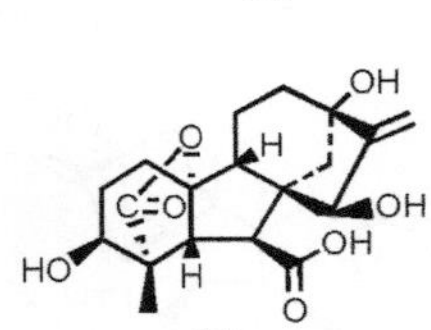
GA72

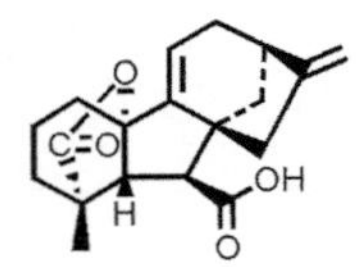
GA$_{73}$

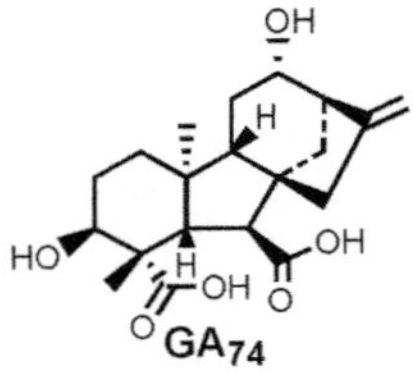
GA$_{74}$

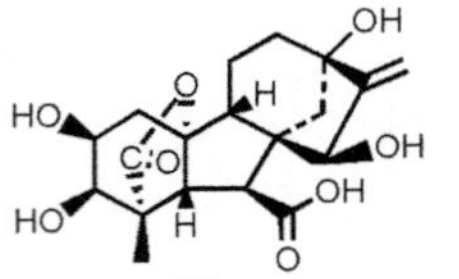
GA$_{75}$

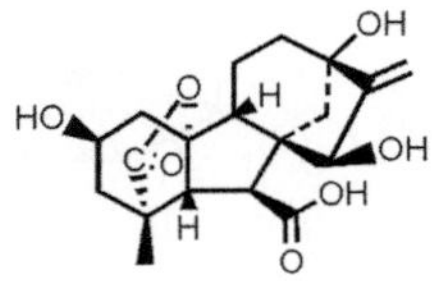
GA$_{76}$

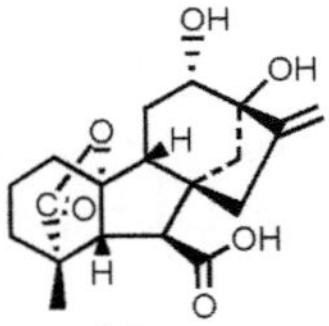
GA$_{77}$

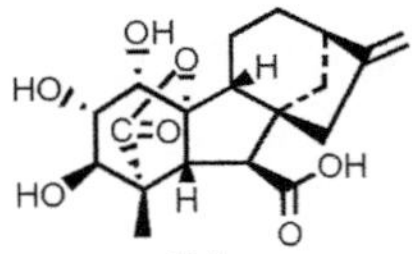
GA$_{78}$

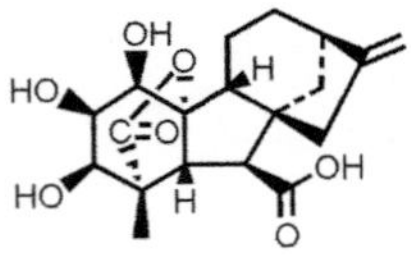
GA$_{79}$

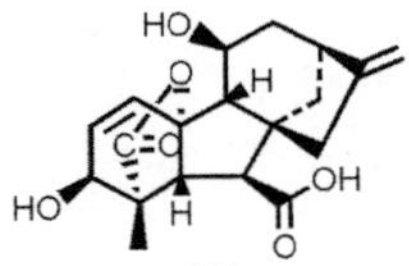
GA$_{80}$

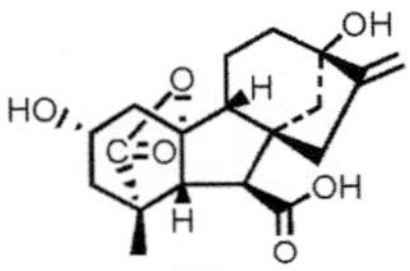
GA$_{81}$

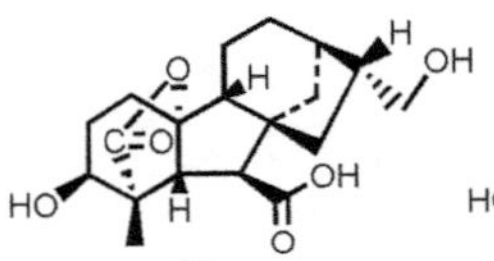
GA$_{82}$

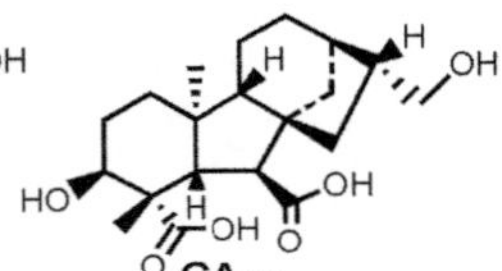
GA$_{83}$

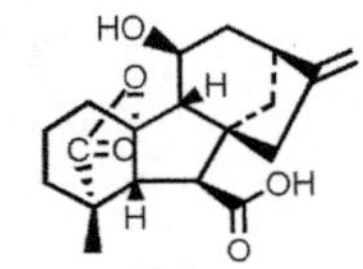
GA$_{84}$

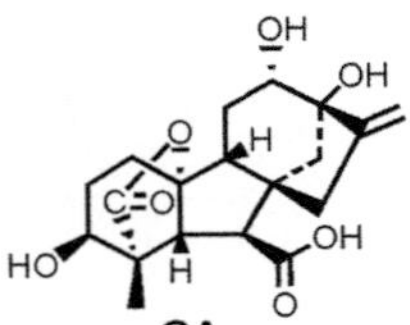
GA$_{85}$

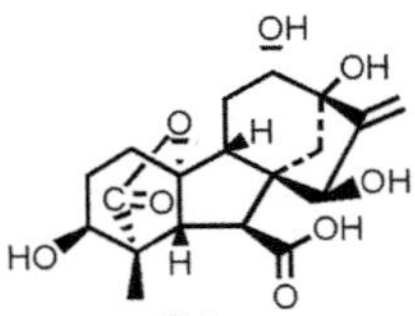
GA$_{86}$

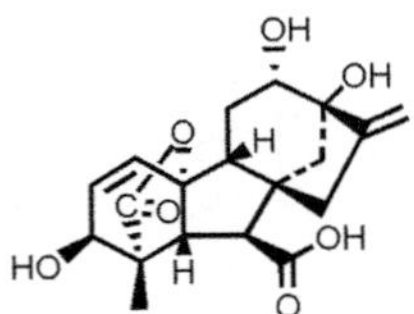
GA$_{87}$

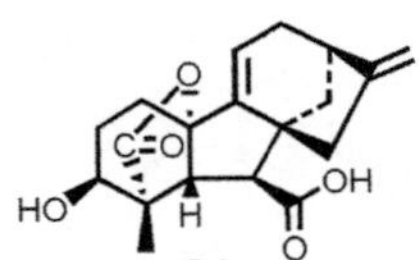
GA$_{88}$

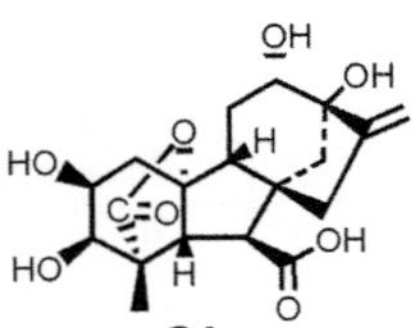
GA$_{89}$

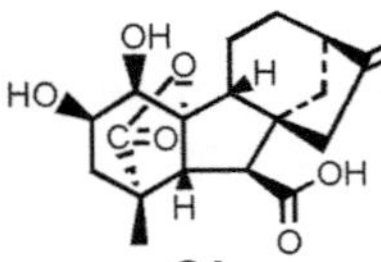
GA$_{90}$

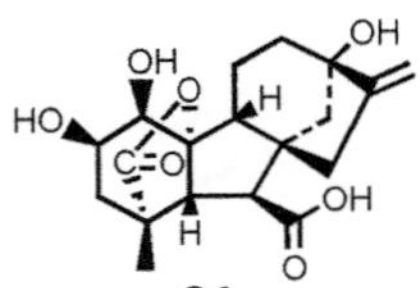
GA$_{91}$

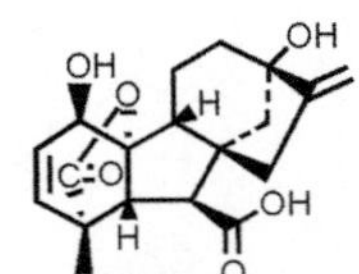
GA$_{92}$

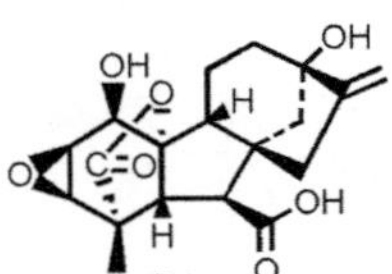
GA$_{93}$

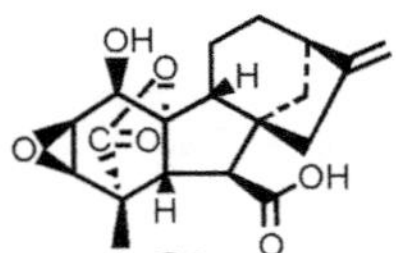
GA$_{94}$

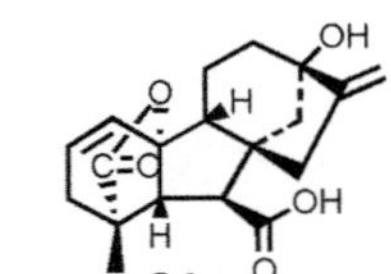
GA$_{95}$

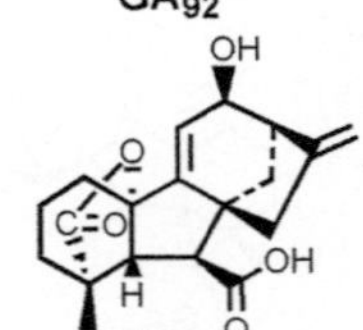
GA$_{96}$

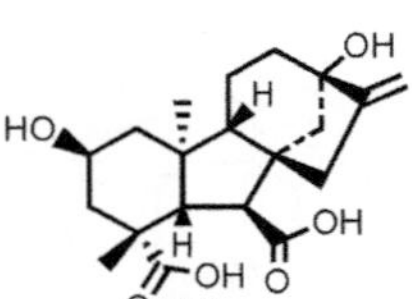
GA$_{97}$

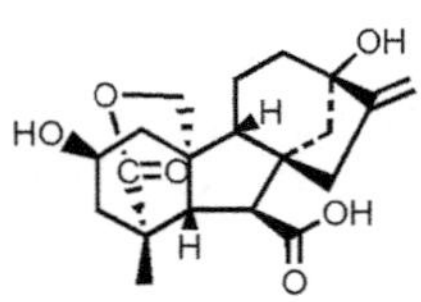
GA$_{98}$

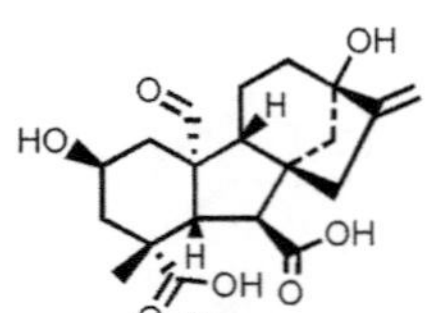
GA$_{99}$

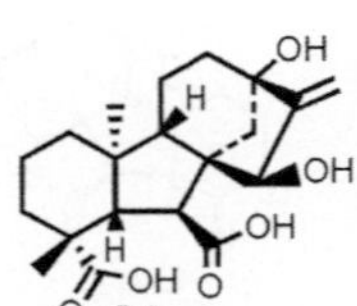
GA$_{100}$

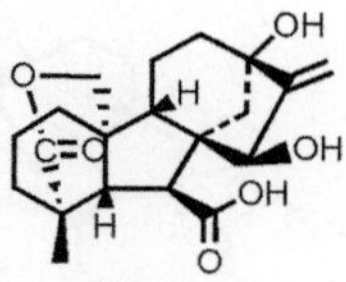

GA_{101}

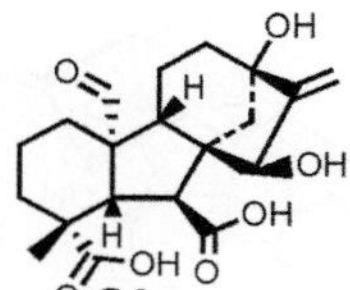

GA_{102}

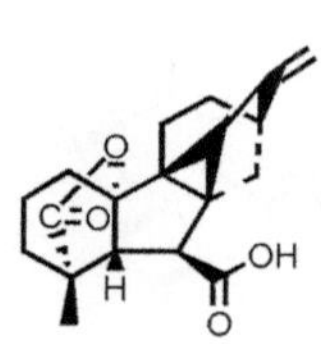

GA_{103}

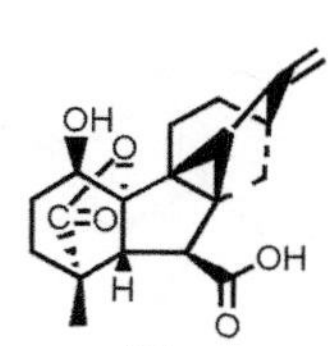

GA_{104}

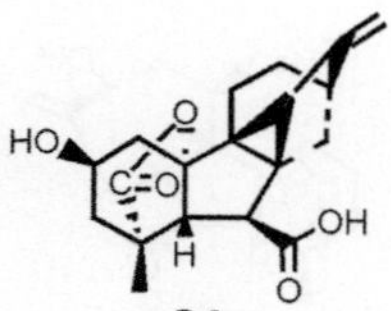

GA_{105}

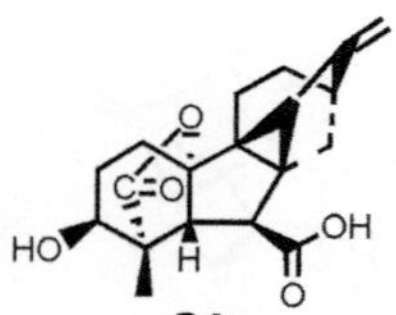

GA_{106}

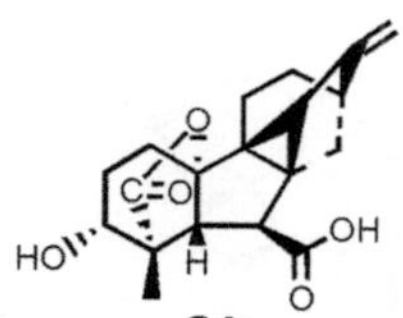

GA_{107}

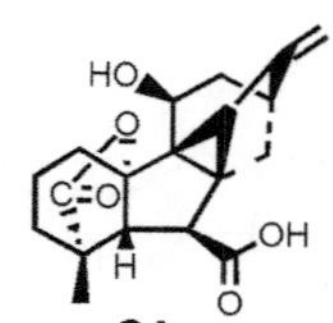

GA_{108}

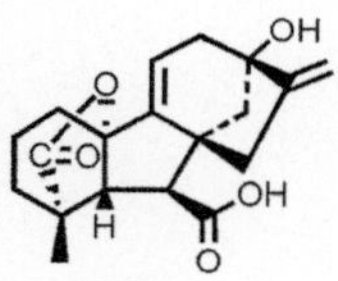

GA_{109}

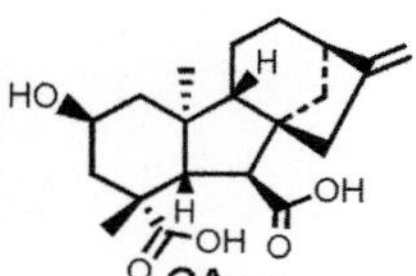

GA_{110}

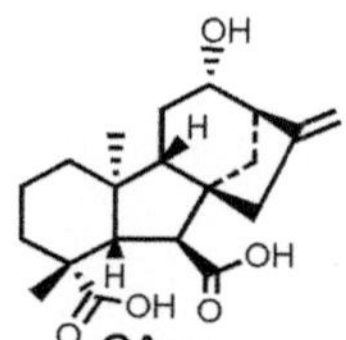

GA_{111}

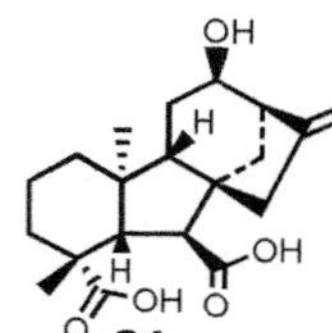

GA_{112}

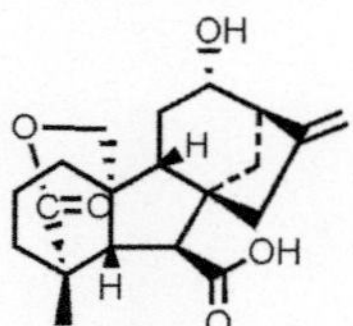

GA_{113}

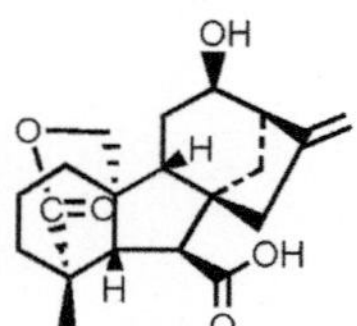

GA_{114}

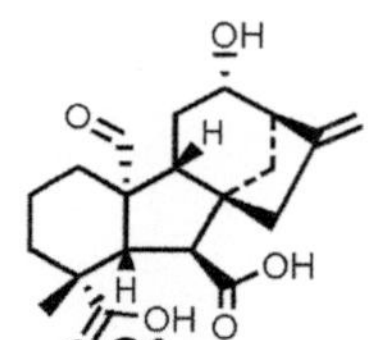

GA_{115}

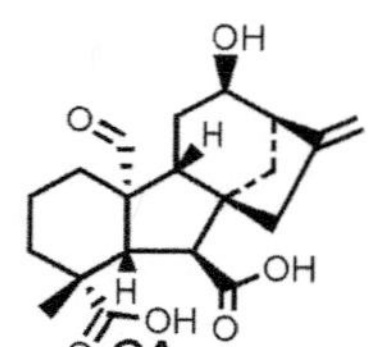

GA_{116}

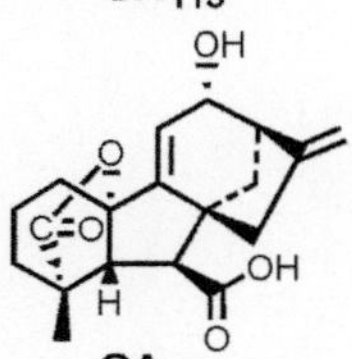

GA_{117}

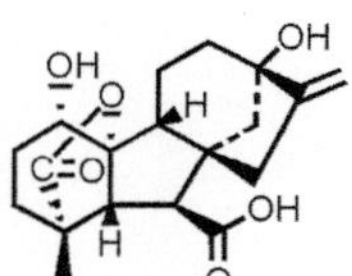

GA_{118}

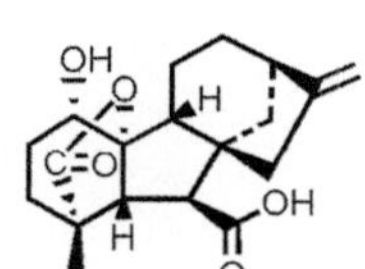

GA_{119}

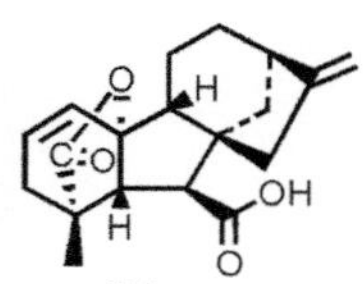

GA_{120}

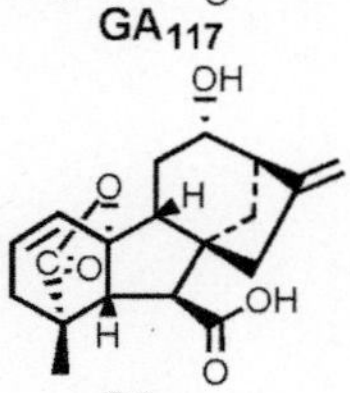

GA_{121}

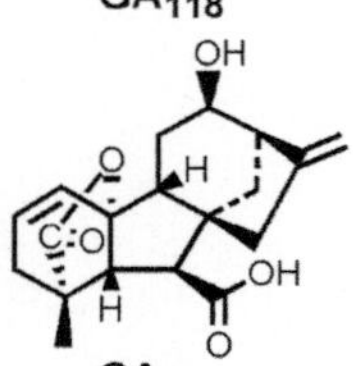

GA_{122}

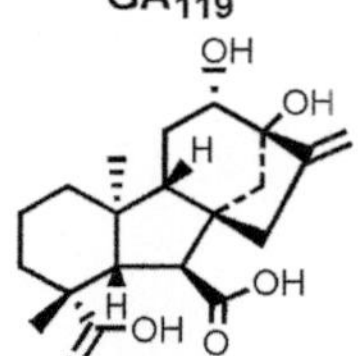

GA_{123}

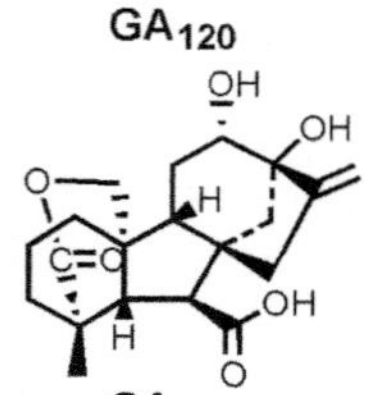

GA_{124}

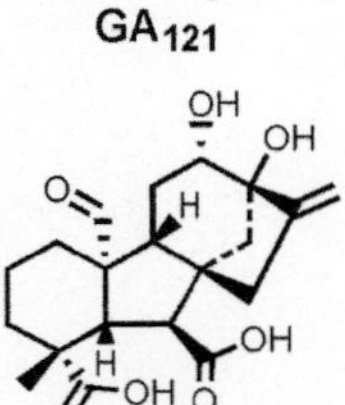

GA_{125}

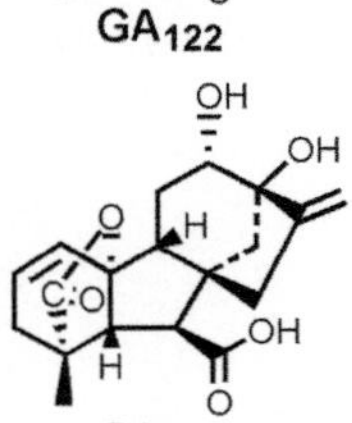

GA_{126}

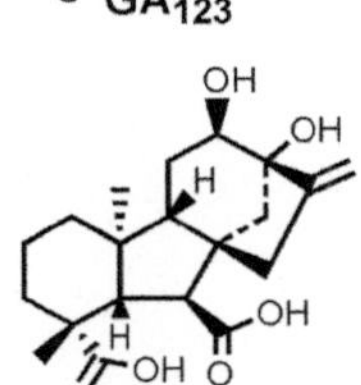

GA_{127}

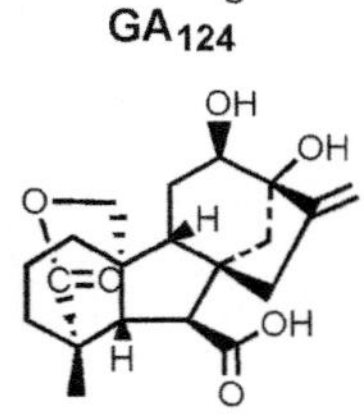

GA_{128}

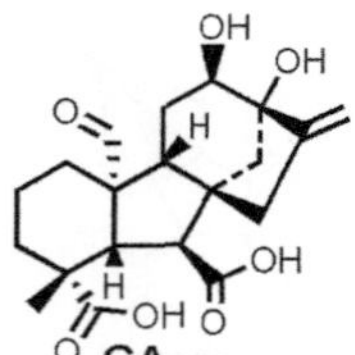

GA$_{129}$

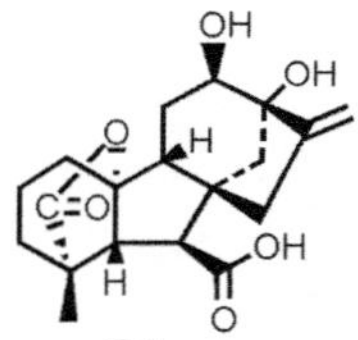

GA$_{130}$

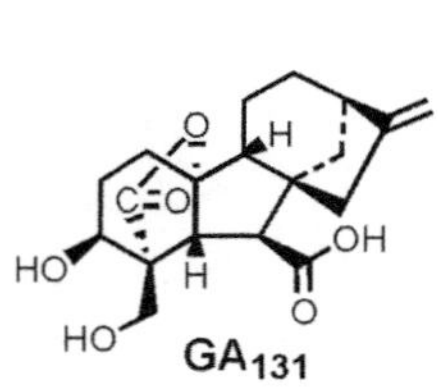

GA$_{131}$

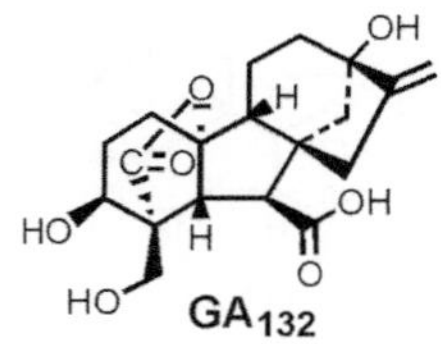

GA$_{132}$

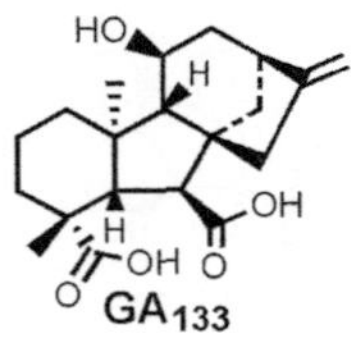

GA$_{133}$

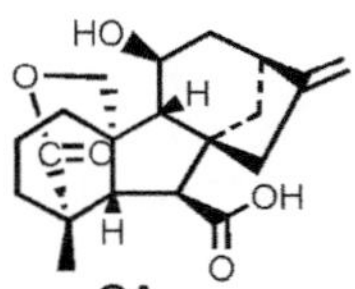

GA$_{134}$

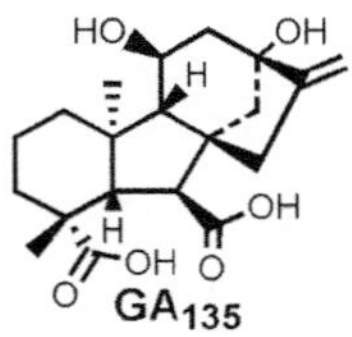

GA$_{135}$

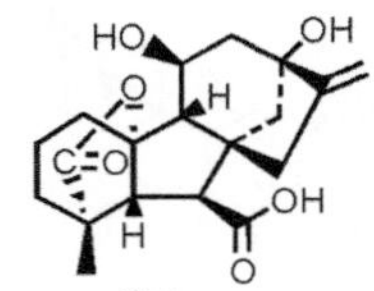

GA$_{136}$

Index

ABC transporters 111
ABI *see* ABSCISIC ACID INSENSITIVE
abiotic stress 58, 85–6, 270–3, 308–9, 409
abscisic acid (ABA) 55, 83, 97, 205, 216, 232–4, 244, 347
  seed germination and 259–61, 265, 269–71, 273
  stability of DELLA and 237
  transporters 97, 111
ABSCISIC ACID INSENSITIVE3 (ABI3) 205, 216, 244, 246, 269, 271, 288
ABSCISIC ACID INSENSITIVE4 (ABI4) 270, 271
ABSCISIC ACID INSENSITIVE5 (ABI5) 205, 216, 244, 269–70, 271, 288
AC 94,377 [1(4-chloro-1, 3-dihydro-1, 3-dioxo-2H-isoindol-2-yl) cyclohexane-1-carboxamide] 362
ACC *see* 1-aminocyclopropane-1-carboxylic acid; 1-aminocyclopropane-1-carboxylic acid oxidase; 1-aminocyclopropane-1-carboxylic acid synthase
N-acetylglucosamine, *O*-linked (*O*-GlcNAc) 169–72, 177
ACS5/ETO2 297
ACS8 (ACC SYNTHASE8) 297
AGAMOUS 51, 82, 332, 339
agriculture *see* production
AHK3 (*ARABIDOPSIS* HISTIDINE KINASE 3) 300
ALC (ALCATRAZ) 157, 202, 344
aleurone (cereal) 17–18, 21, 100, 161, 190, 213–16, 264
allelochemicals 273–6
1-aminocyclopropane-1-carboxylic acid (ACC) 236, 265, 267, 303, 370
1-aminocyclopropane-1-carboxylic acid oxidase (ACC oxidase) 297
1-aminocyclopropane-1-carboxylic acid synthase (ACC synthase; ACS8) 297
ammonium compounds, quaternary 365–6
α-amylase 17, 18, 100, 213, 214, 215, 420
ancymidol 77, 366, 367, 369, 390
anther development 216, 336, 338–9
antheridiogen 142
antheridium formation 142, 143
anthocyanins 370, 390
antilodging compounds *see* lodging
AP1 (APETALA1) 203, 205, 325, 327, 332
AP2 (APETALA2)/ERF (ETHYLENE RESPONSE FACTOR) family of transcription factors 58, 85, 86, 295, 308–9
apical hook formation 201, 203, 241, 242, 295–7
apical meristem
  root 284, 289, 298
  shoot (SAM) 51, 81–2, 289–90, 329–30
apoplast 97, 98, 131, 274, 275
apple (and apple trees) 363, 370, 371, 382–3, 384, 385, 420, 421
ARABIDOPSIS HISTIDINE KINASE 3 (AHK3) 300
ARABIDOPSIS RESPONSE REGULATORS (ARRs) 205, 300
Arabidopsis silique 51, 102, 334, 341, 343–4

Annual Plant Reviews, Volume 49: The Gibberellins, First Edition.
Edited by Peter Hedden and Stephen G. Thomas.

Arabidopsis thaliana
inactivation enzymes 76
as model system 9, 15
seed germination 254–61, 265–9, 271–3, 274
arbuscular mycorrhization (AM) 306–8
ARFs (AUXIN RESPONSE FACTORs) 54, 83, 204, 296, 345
ARRs (*ARABIDOPSIS* RESPONSE REGULATORS) 205, 300
ATP-dependent transporters 111
auxin(s) (incl. IAA) 54–5, 230–1
apical hook formation and 296–7
GA inactivation and 83
GA synthesis and 54–5, 230–1
hypocotyl elongation and 292
leaf size and 289, 290
root development and 298–302
stability of DELLAs and 235–6
transport 245–6
polar 296, 298, 302
AUXIN RESPONSE FACTORs (ARFs) 54, 83, 204, 296, 345
avocado 386, 387
Azospirillum 133–4, 138
A. lipoferum 133–4, 138

Bacillus 133
B. subtilis 274
bacteria 133–7
biosynthesis in 133–9
symbiotic 134–7, 306
bakanae (foolish seedling) disease 2, 122, 131
barley 373–6
GAs and GA synthesis inhibitors in production of 373–6
SLN1 (SLENDER1) 21, 77, 194, 211–12, 215, 412
basic leucine zipper (bZIP) 84, 198, 205
bean seeds, extraction from 4
bending response 236
see also lodging
berry sets (grapes) 387, 388, 389
BES1 (BRI1-EMS-SUPPRESSOR 1) 203, 204, 293, 309
Beta vulgaris (sugar beet) 421
bHLH 24, 200, 201, 202, 292, 344
biosynthesis 37–71
allelochemical interference of 273–6
bacteria 133–9
fungi 122–33
growth and 286, 295, 301, 302, 304, 306
higher plants 37–71
regulation 50–9
sites *see subheading below*
historical studies 6–17
inhibitors 359–403
uses 372–91
lower plants 139–43
other hormones affecting 54–5, 230–4, 246
in reproduction
embryo and seed development 345–6, 347
floral development 335–6
floral transition 329–30
fruit development 341–3
male reproductive development 336, 337, 338, 339
seed germination and 255–6, 257, 259, 261, 269, 270, 273–6
sites (in higher plants) 49–50
floral development and 335–6
floral transition and 329–30
temporal and spatial patterns 255–6
transport analysis and patterns of 101–3
biotic stress 87, 273–6, 310–11
BOIs (BOTRYTIS SUSCEPTIBLE1 INTERACTORs) 208, 270
bolting *see* internode elongation
BOTRYTIS SUSCEPTIBLE1 INTERACTORs (BOIs) 208, 270
Bradyrhizobium japonicum 122, 134–8, 140, 143, 144
Brassica napus (oilseed rape), GAs and GA synthesis inhibitors in production of 379
brassinosteroids (BRs) 55, 96, 203, 204, 234, 237–40, 288, 293
BRASSINOZALE-RESISTANT1 (BZR1) 157, 203, 204, 209, 212, 237–9, 240, 246, 288, 293, 309
breeding
cereals for stem stabilisation 374–5
hybrid seed production 391

BRI1-EMS-SUPPRESSOR 1 (BES1) 203, 204, 293, 309
bZIP 84, 198, 205
BZR1 (BRASSINOZALE-RESISTANT1) 157, 203, 204, 209, 212, 237–9, 240, 246, 288, 293, 309

cambium 105, 106, 109
carotenoids 40, 292, 293
CBF/DREB (C-REPEAT-BINDING FACTOR/DEHYDRATION RESPONSE ELEMENT BINDING FACTOR) 85, 86, 309, 311
cells, monitoring of concentration in and between 110–11
cereals (small grain) 373–6, 419–20
  aleurone 17–18, 21, 100, 161, 190, 213–16, 264
  GAs and GA synthesis inhibitors and their uses 376–9
    antilodging 373–6, 406, 410, 412
  germination of grain 17, 190, 213–16, 264
  transgenic approaches 419–20
  see also specific cereals
chalazal endosperm 256–7, 259, 263
chemical regulators of GA status and their applications in plant production 359–403
cherries, sweet 362, 382, 383
CHLH (CONDITIONAL CHLORINA) 291, 293
chlormequat chloride 365–6
  cereals 375–6
  cotton 380
  grapevines 389
  ornamentals 390
  structure 365
4-chloroindole-3-acetic acid (4-Cl IAA) 230, 341, 343, 345
chlorophyll 40, 292–3
CHOTTO1 55, 270
chromatin 197, 198, 208–9, 288
citrus fruit trees 385
CKIs (CYCLIN-DEPENDENT KINASE INHIBITORS) 300–1
CKX3 (CYTOKININ OXIDASE 3) 290
CO (CONSTANS) 28, 325
coat dormancy release 259, 265, 276
COI1 (Skp1, Cullin, F-box; Skp1) 21, 167, 168, 242, 243
commercial production *see* production
concentrations (levels) of GAs
  approaches to using regulators for increasing or decreasing 360–1
  interactions involving effects of other hormones on 230–4
  intra-/intercellular, monitoring 110–11
CONDITIONALCHLORINA (CHLH) 291, 293
CONSTANS (CO) 28, 325
CONSTITUTIVELY PHOTOMORPHOGENIC 1 (COP1) 94, 292
COP1 (CONSTITUTIVELY PHOTOMORPHOGENIC 1) 94, 292
ent-copalyl diphosphate (CPP) 40–2
ent-copalyl diphosphate synthase (CPS) 40–2, 50, 123, 126, 129–31, 134–5, 139, 157, 286, 334, 422
cortex/endodermal initials (CEI) 304, 305, 306
cotton 423
  GAs and GA synthesis inhibitors in production of 379–80
cotyledons 291, 292, 305, 368
  GA transport and 104
C-REPEAT-BINDING FACTOR (CBF)/DEHYDRATION RESPONSE ELEMENT BINDING FACTOR (DREB) 85, 86, 309, 311
crop production *see* production
CRYPTOCHROME (CRY) 56, 57, 83, 84, 191, 192
CUL1 (CULLIN1) 168
  see also SCF
cyclanilide and mepiquat chloride, cotton production 381
CYCLIN-DEPENDENT KINASE INHIBITORS (CKIs) 300–1
CYCLOPS 307
cytochrome P450 mono-oxygenases (CYPs) 42–5
  fungi 125
  inactivation of GA and 76, 78–9, 83

cytokinin (CK) 82, 205, 290, 300–1
CYTOKININ OXIDASE 3 (CKX3) 290
cytoplasmic male-sterile lines 415

D6 PROTEIN KINASE (D6PK) 302
DAD1 (DEFECTIVE IN ANTHER DEHISCENCE1) 339
daminozide 369, 371, 384, 390
DDF1 (DWARF AND DELAYED FLOWERING1) 85, 86, 309
DEFECTIVE IN ANTHER DEHISCENCE1 (DAD1) 339
DEHYDRATION RESPONSE ELEMENT BINDING FACTOR (DREB)/ C-REPEAT-BINDING FACTOR (CBF) 85, 86, 309, 311
DELAY OF GERMINATION1 (DOG1) 85, 264
DELLA(s) (proteins/repressors)
  degradation/destruction 24, 132, 153–227, 286, 292, 301, 302, 303
    GA signalling without 175–7
    in relief of growth repression 193–4
  down (negative)-regulation 153–88
  fruit development and 344
  functional diversification of DELLA genes 194–7
  GA homoeostasis and 53, 83
  GA-independent 173–5
  GID1 and 162–6, 245, 408
  in growth and development 153–227, 286–312
  historical studies 21–6
  lower plants 141–2
  phosphorylation 169–72
  protein–protein interaction *see* protein–protein interactions
  Rht-1 alleles and
    orthologues of 412–13
    wheat and 408, 409
  seed germination and 268–70
  stability 235–7, 246
DELLA INTERACTING PROTEIN1 (DIP1) 307–8
denso gene 60, 415
Deo-Geo-Woo-Gen (DGWG) 413–14, 419
desaturase (DES) 123, 125, 126, 129, 133, 134, 422
16,17-dihydro-gibberellins 371–2
dimethylallyl diphosphate (DMAPP) 40
dioxygenases 286
  2-oxoglutarate-dependent (2ODDs) 45–9, 50, 75, 76, 79, 125, 126, 138, 144, 230, 238, 363, 369, 370, 371, 382
DIP1 (DELLA INTERACTING PROTEIN1) 307–8
diterpene(s) 6, 9, 45, 138, 139, 141, 143
diterpene cyclases 123, 138
DNA transcription *see* transcription; transcription factors; transcriptome
Dofs 56, 198
DOG1 (DELAY OF GERMINATION1) 85, 264
dormancy
  maintenance 347
  release 259, 265, 276
downstream regulatory proteins *see* DELLAs
DREB/CBF 85, 86, 309, 311
DWARF AND DELAYED FLOWERING1 (DDF1) 85, 86, 309
dwarf mutants/dwarf lines/dwarfism (in general) 168, 170, 172, 416, 417, 418
  historical studies 4, 14, 15, 19, 21
  maize 4, 14, 15, 19, 20, 39, 45, 52, 156, 191, 412, 414, 418, 420
  see also semi-dwarfism
DWARF-1 and *dwarf-1* 4, 14
DWARF-8 and *dwarf-8* 19, 20, 51, 412
DWARF-9 412

E1 ubiquitin activating enzyme 167
E2 ubiquitin conjugating enzyme 167
E3 ubiquitin ligase 21, 167, 184, 208, 211, 286
EARLY FLOWERING1 (EL1) 170, 171, 172

EARLY FLOWERING3 and EARLY FLOWERING4 (ELF3 and ELF4) 295
EGL3 (ENHANCER OF GLABRA 3) 202–3, 208
EIL1 242, 293, 297
EIN3 (ETHYLENE INSENSITIVE 3) 203–4, 212, 240–1, 246, 288, 295, 296, 297, 309
EL1 (EARLY FLOWERING1) 170, 171, 172
ELF3 and ELF4 295
ELONGATED HYPOCOTYL 5 (HY5) 84
ELONGATED UPPERMOST INTERNODE (EUI) 45, 78, 79, 83, 87, 330, 332, 415–16
elongation/extension growth 202, 245–6
  auxins and 231
  stem *see* internode; stem
embryo growth and development 253, 275, 345–7
endodermis 302–3, 304, 305, 306
endoplasmic reticulum (ER) 101, 363, 366
endosperm
  extraction from 10
  micropylar *see* micropylar endosperm
  peripheral (PE) 256, 257, 259, 263
  weakening and rupture 254, 256, 259, 261, 269, 274, 275, 276
ENHANCER OF GLABRA 3 (EGL3) 202–3, 208
ENHYDROUS (ENY) 206
environmental factors 55–9, 84–7
  GA inactivation 84–7
  GA synthesis 55–9
  unfavourable (abiotic stress) 58, 85–6, 270–3, 308–9, 409
ENY (ENHYDROUS) 206
enzymes
  biosynthetic 39–49
  Fusarium fujikuroi GA biosynthesis 122–6
  historical studies 11, 16
  inactivating 75–80
9-*cis*-epoxycarotenoid dioxygenase gene (*NCED4*) 270–1
ERF/AP2 family (of transcription factors) 58, 85, 86, 295, 308–9
ethylene 231–2, 233, 240–2, 246
  DELLAs and 240–2
    stability of 236
  fruit trees and inhibition of formation of 385
  growth and 288, 295, 296–7, 303
  hypocotyl elongation and 295
  seed germination and 265, 271
ETHYLENE INSENSITIVE 3 (EIN3) 203–4, 212, 240–1, 246, 288, 295, 296, 297, 309
EUI (ELONGATED UPPERMOST INTERNODE) 45, 78, 79, 83, 87, 330, 332, 415–16
evolution
  bacterial 137–9
  fungal 128–31
  lower plants 141–3

F-box (sub-units/components/proteins) 21, 24, 166, 167–8, 169, 176, 242, 418
factors I and II 4–5
female reproductive development, gametophyte 339, 340, 345
fermentations, *Fusarium* (*Gibberella*) *fujikuroi* 361, 362, 363
ferns 142–3, 324
fertility 334–40
  Rht-1 alleles and 411
fibre and its formation 105–7, 111, 423
flavanone 3-hydroxylase 370, 385
FLC (FLOWERING LOCUS C) 325, 327, 331
flower 323–34
  development (floral development) 106, 158
    historical studies 28
  patterning 332–4
  transition to 324–31
FLOWERING LOCUS C 325, 327, 331
FLOWERING LOCUS T 28, 56, 325
flurprimidol 366, 367, 368, 378, 390
foolish seedling (bakanae) disease 2, 122, 131
forage (pasture) grasses, GAs and GA synthesis inhibitors in production of 377–9
forest trees 423

fruit
  development 340–5
  parthenocarpy 341, 344–5, 382, 383, 422
fruit trees
  temperate climate 382–5
  tropical climates 385–7
FRUITFUL (FUL) 325–7
FT (FLOWERING LOCUS T) 28, 56, 325
FUL (FRUITFUL) 325–7
function/role (and physiological responses)
  DELLAs
    diversification 194–7
    relationships to structure 210–13
  GAs 26–9
    bacterial 137–8
    fungal, in plant infection 131–2
fungi
  biosynthesis 122–33
  historical studies 2, 7, 10, 16
  as model system 6–7
  mycorrhizal 306–8
fungicides 373, 379, 389
  triazole 366
FUS3 (FUSCA3) 51, 82, 271, 346
Fusarium 128–31, 132, 138
  evolution 128–31
  F. fujikuroi (*Gibberella fujikuroi*) 2, 6–7, 9, 10, 11, 13, 16–17, 122–33, 361
    fermentations 361, 362, 363
  F. graminearum 128, 411
FUSCA3 (FUS3) 51, 82, 271, 346

GA INSENSITIVE (GAI and *gai-1*) 19–20, 25, 53, 157, 160, 170, 174, 175, 192, 196, 198, 206, 238, 268–70, 292, 295, 297, 302, 304, 408, 420, 421
GA-INSENSITIVE DWARF1 *see* GID1
GA-INSENSITIVE DWARF2 (GID2) 21, 22, 172, 176, 210, 211, 286, 418
GA methyltransferase 77
GA oxidases *see* oxidases
GA response elements (GREs; GARE) 18, 214, 215
ga1 (and *ga1-3*) 15, 20, 157, 158, 176, 190, 239, 240, 298, 300–2, 305, 308, 309, 325, 335
GA_1 and GA_2, structure 431
GA_3 (gibberellic acid) 2, 3–4, 7, 10, 15
  production/use/application 363
    fruit trees
  and nut trees in tropical climates 385–6
  in temperate climates 382
    grapevines 387–8, 389
    hybrid seed production 391
    ornamentals 389, 390
    pasture and turf grasses 377–8
  structure 362, 431
GA_4
  desaturase 125, 126, 133
  production/use/application 363
    fruit trees in temperate climates 382–3
  structure 362, 431
GA_5 and GA_6, structure 431
GA_7
  production and applications 363
    fruit trees in temperate climates 382
  structure 362, 431
GA_8 to GA_{11}, structure 431
GA_{12} 45, 104–5, structure 431
GA_{12}-aldehyde 7, 8, 9, 11, 12, 14, 44, 125, 126, 128
GA_{13} to GA_{136}, structure 431–6
GAF1 (GAI-ASSOCIATED FACTOR1) 53, 59, 206, 305
GAI (and *gai-1*) 19–20, 25, 53, 157, 160, 170, 174, 175, 192, 196, 198, 206, 238, 292, 295, 297, 302, 304, 408, 420, 421
  seed germination and 268–70
GAI-ASSOCIATED FACTOR1 (GAF1) 53, 59, 206, 305
gametophyte 324
  female 339, 340, 345
  male (pollen), development 213–16, 336–7, 340
GAMYB 18, 190, 213–16, 336, 337, 347
garden plants (ornamentals) 389–90, 421
GAREs (GA response elements; GREs) 18, 214, 215
gas chromatography-mass spectrometry (GC-MS), historical studies 9, 14, 15

GATA-type transcription factors 201, 325
genes
cereal aleurone cell gene expression 214–16
crop species 405–29
GA biosynthesis
bacteria 134–7
fungi 122–6
geranylgeranyl diphosphate (GGPP) 7, 39–43, 50, 101, 123
ent-kaurene formation from 40–2
formation 39–40
geranylgeranyl diphosphate synthase gene (*GGS2/GGPS*) 123, 126, 129, 130, 131, 132, 134, 135
germination 253–84
cereal grain 17, 190, 213–16, 264
seed 253–84
DELLAs and 268–70
historical studies 6, 26
light and 56, 261–2, 269, 271, 274, 275
spore (moss) 140
GGPP *see* geranylgeranyl diphosphate
GGS2 (geranylgeranyl diphosphate synthase) gene 123, 126, 129, 130, 131, 132
Gibberella fujikuroi see Fusarium
gibberellic acid *see* GA_3
gibberellin *see* GA
GID1 (GA-INSENSITIVE DWARF1) receptor 18, 22–4, 100, 110, 112, 154, 155, 156, 159–66, 168, 172, 173–4, 175, 176, 193–4, 210–11, 216, 245
crop species and 408, 409, 412, 418
DELLAs and 162–6, 245, 408
lower plants 141–2, 143
seed germination and 264–7, 268, 275, 276
vegetative growth and 286, 295, 309
GID1-Like (*GID1L*) genes 141
GID2 (GA-INSENSITIVE DWARF2) 21, 22, 172, 176, 210, 211, 286, 418
GLABRA 1 (GL1) 203
GLABRA 3 (GL3) 202, 203, 208
glucose conjugation 80
glucosinolate GTR1/NPF2.10 transporter 303
glycosylation 80
GNC (NITRATE-INDUCIBLE CARBON-METABOLISM INVOLVED) 201, 329
grafting experiments 103–4, 111
grain (cereal), germination 17, 190, 213–16, 264
graminaceous plants 372, 391
grape-growing 387–9
GRAS (family/domain/proteins) 20–1, 24, 53, 154–5, 166, 192, 194, 199, 210–13, 218, 238, 286
vegetative growth and 286, 305, 307–8
grasses
amenity 421–2
pasture and turf, GAs and GA synthesis inhibitors in production of 377–9
gravitropism, root 301
Green Revolution 20, 156, 286, 405, 407, 413, 423
GREs (GA response elements; GAREs) 18, 214, 215
growth and development 153–227, 285–322
DELLAs and 153–227, 286–312
embryo 253, 275, 345–7
historical studies 25–6, 27
hormones (other than GA) in 83
regulation/control 50–1, 153–87, 285–322
chemicals involved (PGRs), *see subheading below*
down-regulating DELLA repressors of 153–87
GA inactivation and 81–2
inhibitors of inhibitors 191
regulators (PGRs; plant growth regulators/bioregulators) 359–63, 363, 366, 367, 368, 369, 372, 374, 375, 376, 377, 391
cotton 379
fruit trees 382, 384, 387
grapevine 387, 389
oilseed rape 379
pasture and turf grasses 378
retardants 359, 363–6

growth and development (*continued*)
in cotton 380
6,17-dihydro-gibberellins 371–2
in fruit trees 382, 385
ornamental plants 390
2-oxoglutaric acid mimics 369, 370
in pasture and turf grasses 378
stem *see* stem
transport and its significance for secondary growth 104–7
GRS (GAI Related Sequence) 19, 20
GTR1 transporter 91, 97, 303

Hd3a 28
height, reduced *see* dwarf mutants; semi-dwarfism
heterocyclic compounds, nitrogen-containing 366–9
higher plants, biosynthesis of GAs *see* biosynthesis
histone 208, 209, 293
historical studies 1–36, 361
HLS1 (HOOKLESS 1) 203, 241–2, 296, 297
homoeostasis, GA 61–4, 82–3
HOOKLESS 1 (HLS1) 203, 241–2, 296, 297
Hordeum vulgare (barley), GAs and GA synthesis inhibitors in production of 373–6
hormone(s) (other than GA) 229–52
biosynthesis of GAs and 54–5, 230–4, 246
inactivation of GAs and 83
root development 298–302
signalling (involving GA) and 234–7
hormone-sensitive lipases (HSLs) 23, 161
horticulture *see* production
hub proteins 217, 218
HY5 (ELONGATED HYPOCOTYL 5) 84
hybrid seed production 391
hydrocinnamic acid (HCA) 274
2β-hydroxylation 12, 13, 49, 75, 83, 346, 370
3β-hydroxylation 8, 9, 12–14, 16, 48, 52, 77, 83, 125, 126, 128, 275, 370, 372
13-hydroxylation 9, 10, 11, 12, 13, 14, 15, 45, 48, 49, 59, 78–9, 97, 126, 133, 135–7, 138, 330
4-hydroxy-3-methylbutenyl diphosphate (HMBPP) 40
hypocotyl(s) 238–40, 290–5
elongation/expansion 103, 106, 201, 202, 235, 239, 290–5
hypocotyl–radicle axis (RAD) 254, 255, 256–9, 261, 263, 265, 268, 269, 275, 276

IAA *see* auxin
IAA19 209, 231
IDD (INDETERMINATE DOMAIN) transcription factors 51, 53–4, 205, 206, 288, 305
in vitro systems, historical studies 9, 10, 11, 16
in vivo studies, historical 11–12, 16
inactivation of GA 73–93
regulation 80–7
temperature and 85, 273
INDEHISCENT (IND) 202, 238, 339, 340, 344
INDETERMINATE DOMAIN (IDD) transcription factors 51, 53–4, 205, 206, 288, 305
indica rice 413, 414
indole-3-acetic acid *see* auxin
infection (of plants), GA role 131–2
inflorescence meristem (IM) 324–5, 325, 327, 331, 332
inhibitors of inhibitors (of growth and development) 191
insecticide synergy with prohexadione-calcium 385
intercellular concentration monitoring 110–11
internode elongation (bolting) 233, 330, 415, 418
historical studies 18, 26–7
intracellular concentration monitoring 110–11
IPD3 307
IPT (isopentenyltransferase) genes 82, 290
isopentenyl diphosphate (IPP) 16, 40, 123

isopentenyl transferase (IPT) genes 82, 290
isoprenyl diphosphate synthase 134, 135

japonica rice 413, 414
jasmonate/jasmonic acid (JA) 97, 196, 203, 206–8, 242–3, 303, 339
  biotic stress and 87, 310–11
jasmonate ZIM-domain (JAZ) proteins 132, 157, 196, 203, 206–8, 212, 242–3, 288, 310–11
JAZ proteins 132, 157, 196, 203, 206–8, 212, 242–3, 288, 310–11
Jungermannia 139, 140

KAR 216
ent-kaurene 7, 11, 15, 39–42, 44, 96–7, 101, 123, 124, 125, 134, 135, 137, 138, 139–41, 256, 286, 363, 366, 368
ent-kaurene oxidase 42–4, 125, 256, 286, 417
ent-kaurene synthase (KS) 7, 15, 42, 123, 126, 130, 131, 133, 134, 135, 137, 139, 140, 143, 256, 286, 365, 422
ent-kaurenoic acid 7–9, 11, 15, 44, 79, 125, 126–8, 140, 141, 366
ent-kaurenoic acid oxidase (KAO) 44, 45, 101, 104, 125, 257, 264, 286
β-ketoacyl reductase (KAR) 216
kinetin + mepiquat chloride, cotton production 381
KIP-RELATED PROTEIN (KRP) 300
KNOX 51, 81, 82, 290
Koshihikari 377
KRP (KIP-RELATED PROTEIN) 300

LA 83, 191, 192
LE and *le* 14, 235–6, 417–18
leaf/leaves
  GA synthesis 49
  GA transport 104–5, 107–9
  size 289–90
LEAFY (LFY) 28, 325–6, 328, 329, 330, 332
LEAFY COTYLEDON2 (LEC2) 51, 82, 346
LEC2 (LEAFY COTYLEDON2) 51, 82, 346
Lepidium sativum, seed germination 261–3, 264, 268, 270, 271, 273, 274, 275
LFY (LEAFY) 28, 325–6, 328, 329, 330, 332
light 56–8, 84–7, 201–2
  GA inactivation and 84–7
  GA synthesis and 56–8
  seed germination and 56, 261–2, 269, 271, 274, 275
lipases, hormone-sensitive (HSLs) 23, 161
liverworts 139, 140, 143
lodging (and its prevention)
  chemicals used 371, 372
  in oilseed rape 379, 413
  in pasture and turf grasses 378
  in rice 377, 413
  in small-grain cereals 373–6, 406, 410, 412
lower plants 139–43
luteoliflavan and luteoforol 370, 385
LUX/ARRHYTHMO 295
lycophytes 139, 141–2, 143, 323–4
Lygodium 142–3, 324

MADS box proteins 51, 82, 198, 332, 339
Magnaporthe oryzae 87, 129, 130
maize 82, 129, 289
  dwarfs 4, 14, 15, 19, 20, 39, 45, 52, 156, 191, 412, 414, 418, 420
  transgenic approaches 420
male 336–9
  reproductive development 336–9
    gametophyte (pollen) 213–16, 336–7, 340
  sterility 102, 334, 336, 391, 411
    cytoplasmic 415
Medicago 306–7
mepiquat chloride 365, 366
  cereals 375, 376
  cotton 380–1
  grapevines 389
  oilseed rape 379, 380
  prohexadione-calcium combined with 371
  structure 365

mepiquat pentaborate, cotton 380, 381
meristem
apical *see* apical meristem
floral (FM) 325, 332, 334
inflorescence (IM) 324–5, 325, 327, 331, 332
root 298, 300, 301, 304
metabolism
crop species 405–30
historical studies 6–17
seed germination and 254–64, 273, 274, 275, 276
metconazole 366, 369, 379, 380, 390
methyl erythritol phosphate (MEP) pathway 16, 38, 39–40
methyltransferase 77
mevalonate/mevalonic acid (MVA) pathway 10, 16, 39, 40, 122
micropylar endosperm 254, 255, 256, 259, 261–3, 265, 268, 275, 276
microtubules (MTs) 25, 209–10, 218, 289
molecular recognition features (MoRFs) 165
mosses 122, 139, 140, 141, 143
movement (transport) 95–119
MYB 18, 24, 203, 214, 215, 243, 339
MYC2 132, 202, 203, 206, 207, 208, 212, 243, 288, 311
mycorrhiza 306–8
Myrigalone A (MyA) 274–6, 276
N-acetylglucosamine, *O*-linked (*O*-GlcNAc) 169–72, 177

NCED4 270–1
NITRATE-INDUCIBLE CARBON-METABOLISM INVOLVED (GNC) 201, 329
NITRATE TRANSPORTER 1/PEPTIDE TRANSPORTER FAMILY (NPF) 303
nitrogen-containing heterocyclics 366–9
nodulation 306
NODULATION SIGNALING PATHWAY1 (NSP1) 154–5
non-ribosomal peptide synthase (NRPSs)-derived secondary metabolites 133
NPF (NITRATE TRANSPORTER 1/PEPTIDE TRANSPORTER FAMILY) 303
NSP1 (NODULATION SIGNALING PATHWAY) 154–5
NTH15 51, 82
nut trees 385–7

O-linked *N*-acetylglucosamine (O-GlcNAc) 169–72, 177
oats, GAs and GA synthesis inhibitors in production of 373–6
oilseed rape
GAs and GA synthesis inhibitors in production of 379
Rht-1 orthologues 413
opium poppy 382
oranges 385, 386
ornamentals 389–90, 421
Oryza see rice
OTS1 and OTS2 (OVERLY TOLERANT TO SALT1 and 2) 174–5
OVERLY TOLERANT TO SALT1 and 2 (OTS1 and OTS2) 174–5
ovule 335–6, 339, 340, 341, 345
oxidases 16, 102, 125, 137, 140, 230, 231, 257, 263, 286
GA 2-oxidase (GA2ox) 51, 52, 55, 75–7
crop genetic manipulation and 419, 421, 422, 423
grapevines and 389
homeostasis of GA and 82–3
light and 56, 84–5
other hormones and 83
seed development and 82
shoot apical meristem and 81–2
stress and 58, 85–6, 308–9
temperature and 85, 273
transport and 102
GA 3-oxidase (GA3ox) 48–9, 51, 52, 53, 55
crop breeding and 417, 418, 419, 420
light and 56, 57, 84
other hormones and 54, 55, 83
reproduction and 328, 332, 335, 338, 339, 341, 343, 344, 345–6
seed germination and 256, 257, 259, 261, 263, 269, 271–3, 275

temperature and 58, 85
transport and 102, 103
GA 12α-oxidase 75, 76
GA 13-oxidase 78–80, 83, 87
GA 16,17-oxidase (GA16, 17ox) 75, 78, 87
GA 20-oxidase (GA20ox) 46–8, 50–9, 231
crop breeding and 413–15, 418, 419, 420, 421, 422, 423
reproduction and 328, 334, 335, 339, 343
seed germination and 256, 257, 261, 273
transport and 101, 102
oxidative reactions of GA biosynthesis pathways in bacteria 135–7
2-oxoglutarate-dependent dioxygenases (2ODDs) 45–9, 50, 75, 76, 79, 125, 126, 138, 144, 230, 238, 363, 369, 370, 371, 382
2-oxoglutarate (2-oxoglutaric acid) 363
mimics 369–71

paclobutrazol 175, 192, 231, 236, 366, 367, 368
fruit and nut trees 383, 384, 385, 386, 387
oilseed rape 379
ornamentals 390
pasture and turf grasses 378
PACLOBUTRAZOL RESISTANCE 1 (PRE1) 209
parthenocarpy 341, 344–5, 382, 383, 422
pasture grasses, GAs and GA synthesis inhibitors in production of 377–9
peanuts 371, 381–2
pear (and pear trees) 363, 382, 384, 385
perception, GA 100–1, 112, 141, 153–87
perennial species, flowering 331
peripheral endosperm, (PE) 256, 257, 259, 263
petals 50
petiole 108, 109
elongation 202, 286, 291, 293
PFD (prefoldin) 25, 209–10
Phaeosphaeriae 122, 124, 128, 130, 143
PHOSPHATE TRANSPORTER-9 307
4′-phosphopantheteinyl-transferase-encoding gene (*PPT1*) 133
phosphorylation
DELLAs 169–72
PINs 302
photomorphogenesis 56, 141, 201, 290, 292
photoperiod 56–8, 84, 201
floral transition and 325, 328, 329
Physcomitrella patens 122, 139, 140, 141, 143, 324
physiological responses *see* function
phytochrome(s) (PHYs) 291–2
phytochrome-interacting factor(s) *see* PIFs
PHYTOCHROME-INTERACTING FACTOR3-LIKE2 (PIL2) 202
PHYTOCHROME-INTERACTING FACTOR3-LIKE5 (PIL5) 56, 58, 84, 269
PHYTOENE SYNTHASE (PSY) 293
PICKLE (PKL) 209, 288
PIF(s) (phytochrome-interacting factors) 25, 84, 157, 159, 200, 201, 202, 204, 207–8, 209, 238, 243, 270
vegetative growth and 287–8, 292–5, 297
PIF3-LIKE 2 (PIL2) 202
PIF3-LIKE 5 (PIL5) 56, 58, 84, 269
PIL2 (PIF3-LIKE 2) 202
PIL5 (PIF3-LIKE 5) 56, 58, 84, 269
PIN-FORMED (PIN) 245, 289, 296, 297, 300, 301–2
pipecolatebetain 366
PKL (PICKLE) 209, 288
pleiotropic effects of *RHT-1* alleles 410–12
polar auxin transport 296, 298, 302
pollen development 213–16, 336–7, 340
poly-hydroxylated GAs 13–14
polyketide synthase 133
pome fruit trees 384, 385
PORC (NADPH: PROTOCHLOROPHYLLIDE OXIDOREDUCTASE C) 201
PPT1 (4′-phosphopantheteinyl-transferase-encoding gene) 133

PRE1 (PACLOBUTRAZOL RESISTANCE 1) 209
prefoldin (PFD) 25, 209–10
production, commercial (plant - crops/agriculture/horticulture) 359–429
  chemical regulators of GA status and their use in 359–403
  fungal strain improvement 132–3
prohexadione-calcium 369, 370, 371, 391
  cereals and rice 375, 376, 377
  fruit trees 383, 384, 385
  grapevines 388–9
  opium poppy 382
  ornamentals 390
  pasture and turf grasses 378
  peanuts 381–2
promoters and DELLAs 204–6
  and growth 288
proteasome 84, 110, 154, 166–9, 171, 176, 193, 201, 264, 286, 344, 408
protein–protein interactions, DELLAs in 212, 237–45, 246, 247
  GID1 and 162–6
  SCL3 and 206
protochlorophyllide 292–3
PROTOCHLOROPHYLLIDE OXIDOREDUCTASE (POR) 201, 292
PSY (PHYTOENE SYNTHASE) 293

quantitative trait loci (QTL) analysis 270, 414, 418
quaternary ammonium compounds 365–6
quiescent centre (QC) 299, 304, 305, 306

radicle protrusion 254, 256, 259
  see also hypocotyl–radicle axis
RAM1 (REQUIRED FOR ARBUSCULAR MYCORRHIZA) 307–8
RAP2.3 (RELATED TO AP2.3) 203–4, 241
reactive oxygen species (ROS) 87, 275, 309, 311, 411
receptors 18, 23–4
  GID1 *see* GID1
REDUCED HEIGHT (*RHT*) GENES *see RHT*
RELATED TO AP2.3 (RAP2.3) 203–4, 241
REPLUMLESS (RPL) 344
reproduction 323–58
REQUIRED FOR ARBUSCULAR MYCORRHIZA (RAM1) 307–8
RGA (and *rga*; repressor of *ga1–3*) 20, 24, 110, 155, 157, 158, 159, 165, 174, 175, 190, 192, 196, 197, 197–8, 205, 235, 236, 237, 268–70, 300, 413
RGA-LIKE (RGL1/RGL2/RGL3) 21, 157–9, 160, 165–6, 170, 194, 196, 197, 209, 238, 243, 268, 335
  abiotic stress and 308
  biotic stress and 311
  floral development and 158, 335
  seed germination and 268–9
RGL *see* RGA-LIKE
rhizobacteria 122, 133–7, 274
  free-living 133–4
rhizobial/rhizobacterial, symbiotic 134–7, 306
RHT 406–13
  Rht-1 20, 51, 406–13
    pleotropic effects of *RHT-1* alleles 410–12
    Rht-B1b (originally *Rht1*) 20, 406, 407, 408, 409, 410, 416
    Rht-B1c (originally *Rht3*) 51, 58, 408, 409, 410, 411, 416
    Rht-B1d 407, 409, 416
    Rht-B1e 409
    Rht-D1b (originally *Rht2*) 20, 406, 407, 408, 409, 410, 411
    Rht-D1c (originally *Rht10*) 408, 409, 411, 416
  Rht8 416
  Rht12 416, 417
rice (*Oryza*) 376–9, 413–15
  GAs and GA synthesis inhibitors in production of 376–9
    lodging prevention 377, 413
  inactivation enzymes 76
  SEMI-DWARF (*SD-1*) alleles 413–15
  SLR1 (SLENDER1) 21, 23, 154, 162, 166, 170, 194, 198, 199, 210–11, 212, 233–4, 245, 307–8, 334, 412

roots 298–304
development 298–304
Rht-1 alleles 411
lodging (vs stem lodging) 374
meristem 298, 300, 301, 304
transport in 109–10
RPL (REPLUMLESS) 344
russetting in apples 363, 382, 383
rye, GAs and GA synthesis inhibitors in production of 373–6

SCARECROW (SCR) 304
SCARECROW-LIKE 3 (SCL3) 53, 205–6, 304–5, 306
SCARECROW-LIKE 27 (SCL27) 201
SCF (Skp1, Cullin, F-box) 21, 167, 168, 242, 243
SCHLAFMUTZE (SMZ) 328
SCHNARCGZAPFEN (SNZ) 328
SCL3 (SCARECROW-LIKE 3) 53, 205–6, 304–5, 306
SCL27 (SCARECROW-LIKE 27) 201
SCR (SCARECROW) 304
scutellum epithelium 48, 49, 100
SD-1 (SEMI-DWARF) 413–15, 419, 420, 423
sdw gene 20, 415
second messengers *see* signalling
seed(s)
development 82, 345–7
extraction from 4, 12
germination *see* germination
grasses grown from 378
hybrid, production 391
transport in 100–1
seedless grapes 387, 389
Selaginella 122, 141, 142, 143, 323–4
SEMI-DWARF (SD-1) 413–15, 419, 420, 423
semi-dwarfism 334, 423
cereals 20, 48, 155–6, 406, 408–10, 411–12, 413–15, 417, 419–20
non-cereal crops 420, 422
sexual reproduction 323–58
SHATTERPROOF1 and 2 (SHP1 and SHP2) 331
shoot 288–97
apical meristem (SAM) 51, 81–2, 289–90, 329–30
chemicals used in regulation of growth 384
development 288–97
SHOOTMERISTEMLESS (STM) 81, 82, 290
SHORT HYPOCOTYL2 (SHY2) 300, 301
SHORT-ROOT (SHR) 298, 304, 305, 307
SHORT VEGETATIVE PHASE (SVP) 56–7, 327, 327, 327
SHR (SHORT-ROOT) 298, 304, 305, 307
SHY2 (SHORT HYPOCOTYL2) 300, 301
SIAMESE (SIM) 289, 300
signalling (incl. signal transduction and second messengers) 153–87, 234–45, 246–7
crop species 406, 408, 410, 411, 412, 416, 419
downstream, regulation 191–2, 275
root 304–6
growth and 286–300
historical studies 17–25
hormones (other than GA) involved in 54, 234–45
lower plants 139–43
perception of signal 100–1, 112, 141, 153–87
seed germination and 264–70, 275, 276
source and sink tissues (and orientation of signal flow) 107–10
timeline of significant advances 178
silique 51, 102, 334, 341, 343–4
SIM (SIAMESE) 289, 289, 300
SIM (SUMO-Interaction-Motif) 174
sink tissues, GA 107–10, 112
skotomorphogenesis 25, 201, 204, 209
SKP1 168
see also SCF
SLEEPY1 (SLY1) 21, 24, 166, 167, 168–9, 172, 176, 177
SLENDER1
barley (SLN1) 21, 77, 194, 211–12, 215, 412
rice (SLR1) 21, 23, 154, 162, 166, 170, 194, 198, 199, 210–11, 212, 233–4, 245, 307–8, 334, 412
SLN1 (SLENDER1) 21, 77, 194, 211–12, 215, 412

SLR1 (SLENDER1) 21, 23, 154, 162, 166, 170, 194, 198, 199, 210–11, 212, 233–4, 245, 307–8, 334, 412
SLR1-Like1 (SLRL1) 166, 194, 233
SLRL1 (SLR1-Like1) 166, 194, 233
SLY1 (SLEEPY1) 21, 24, 166, 167, 168–9, 172, 176, 177
Small Ubiquitin-like Modifier modification (SUMOylation) 172, 174–5, 309
SMZ (SCHLAFMUTZE) 328
SNE (SNEEZY) 168–9
SNZ (SCHNARCGZAPFEN) 328
SOC1 (SUPPRESSOR OF OVEREXPRESSION OF CONSTANS 1) 325–8, 329
SOMNUS (SOM) 56, 57, 84, 205, 269, 270, 288
spatiotemporal expression of GA metabolism, seed germination and 254–64, 273, 274, 275
SPATULA (SPT) 58, 202, 269, 293, 344
Sphaceloma manihoticola 121, 122, 124, 126–8, 129, 130, 143, 144, 362
SPINDLY (SPY) 171, 172, 176–7
SPL *see* SQUAMOSA PROMOTER BINDING-LIKE
spore germination (moss) 140
SPT (SPATULA) 58, 202, 269, 293, 344
SPY (SPINDLY) 171, 172, 176–7
SQUAMOSA PROMOTER BINDING-LIKE (SPL) 325–8, 329
  SPL9 203, 205, 328, 332
stamen development 50, 102, 303, 334, 335, 336, 339
stature, reduced *see* dwarf mutants; semi-dwarfism
stem
  elongation/extension/development 28–9, 49, 104, 105, 106, 157, 158–9, 160, 170, 176, 245, 246, 330, 415
    hypocotyl 103, 106, 201, 202, 235, 239, 290–5
    internode *see* internode
    Rht-1 and reduction in 408, 410
  length mutants *see* dwarf mutants; semi-dwarfism
  lodging
    differentiation from root lodging 374
    prevention 375, 377
  petiole 202, 286, 291, 293
  transport in 108, 109
sterility, male *see* male
STM (SHOOTMERISTEMLESS) 81, 82, 290
stress 58–9, 270–6, 308–11
  abiotic 58, 85–6, 270–3, 308–9, 409
  biotic 87, 273–6, 310–11
  tolerance 85, 86, 87, 308, 309, 311
strigolactones 235, 244–5
  receptor D14 218, 245
structures
  GA 431–6
    GID1a binding and 161–2
    historical studies 3–4
STUNTED 289
SUBMERGENCE-1 (SUB-1) 233
sugar 112–13
  conjugation 80, 112–13
  transport 112–13
sugar beet 421
sugar cane, GAs and GA synthesis inhibitors in production of 377
SUMOylation 172, 174–5, 309
SUPPRESSOR OF OVEREXPRESSION OF CONSTANS 1 (SOC1) 325–8, 329
suspensor 50
SVP (SHORT VEGETATIVE PHASE) 56–7, 327, 327, 327
sweet cherries 362, 382, 383
SWI3C (SWITCH3C) 157, 209, 288
symbiosis
  mycorrhizal 306–8
  rhizobial/rhizobacterial 134–7, 306
symplast 98, 101, 110, 112, 131
synthesis *see* biosynthesis

TA (TATCCA) box 214, 215
tapetum development 336–9
TAR2 (TRYPTOPHAN AMINOTRANSFERASE RELATED 2) 296
TATCCA (TA) box 214, 215
tebuconazole 366, 379, 380

TEM1 (TEMPRANILLO 1) 328
temperate climate, fruit trees 382–5
temperature 58, 85, 270–3
  GA inactivation and 85, 273
  GA synthesis and 58
temporospatial expression of GA metabolism, seed germination and 254–64, 273, 274, 275
TEMPRANILLO 1 (TEM1) 328
TERMINAL FLOWER1 (TFL1) 331
terpene cyclases 7, 15, 42, 286
TFL1 (TERMINAL FLOWER1) 331
thermoinhibition 270–3
tomato
  fruit development 341, 344–5
  seed development 345, 347
TOPLESS RELATED (TPR) 53, 205, 206
TOPP4 (TYPE ONE PROTEIN PHOSPHATASE4) 170–2
TPR (TOPLESS RELATED) 53, 205, 206
transcription, DELLAs as regulators of 189–227
transcription factors 51, 198–209
  DELLA and 156–7, 200–4
  GA synthesis and 51, 52, 56, 58–9
  growth and development and 198–209, 216
  inactivation processes 82, 84, 85, 86
  interactions between GA and other hormones 237–8
  reproduction and 325, 327, 328, 329, 332, 336, 337, 338, 344, 346
  seed development and germination and 82, 269–70
  vegetative growth and 288, 295, 305, 308, 309, 311
transcriptome 197–8
  seed germination and 256
transgenic approaches, crop species 419
translocation along plant bodies 96–100
transport
  auxins 245–6
  GAs 95–119
triazole fungicides 366, 369
trinexapac/trinexapac-ethyl 369, 370, 371
  oilseed rape 379
  opium poppy 382
  pasture and turf grasses 378
  small-grain cereals 375, 376
  sugarcane 370
Triticum aestivum see wheat
tropical climates, fruit trees 385–7
TRYPTOPHAN AMINOTRANSFERASE RELATED 2 (TAR2) 296
turf grasses, GAs and GA synthesis inhibitors in production of 377–9
TYPE ONE PROTEIN PHOSPHATASE4 (TOPP4) 170–2

ubiquitin and ubiquitination 21, 24, 52, 154, 166–9, 174, 175, 193, 208, 211, 286, 408, 418
uniconazole (and uniconazole-P) 366, 367, 368
  fruit and nut tress 386, 387
  fruit ornamentals 390
  rice 376, 377

vernalisation 27, 325, 327
viticulture (growing grape) 387–9

wheat (*Triticum aestivum*)
  GAs and GA synthesis inhibitors in production of 373–6
  RHT-1 alleles *see RHT*
wine grapes 387, 388, 389

XERICO 24
xylem expansion, hypocotyl 103, 106

YABBY 52

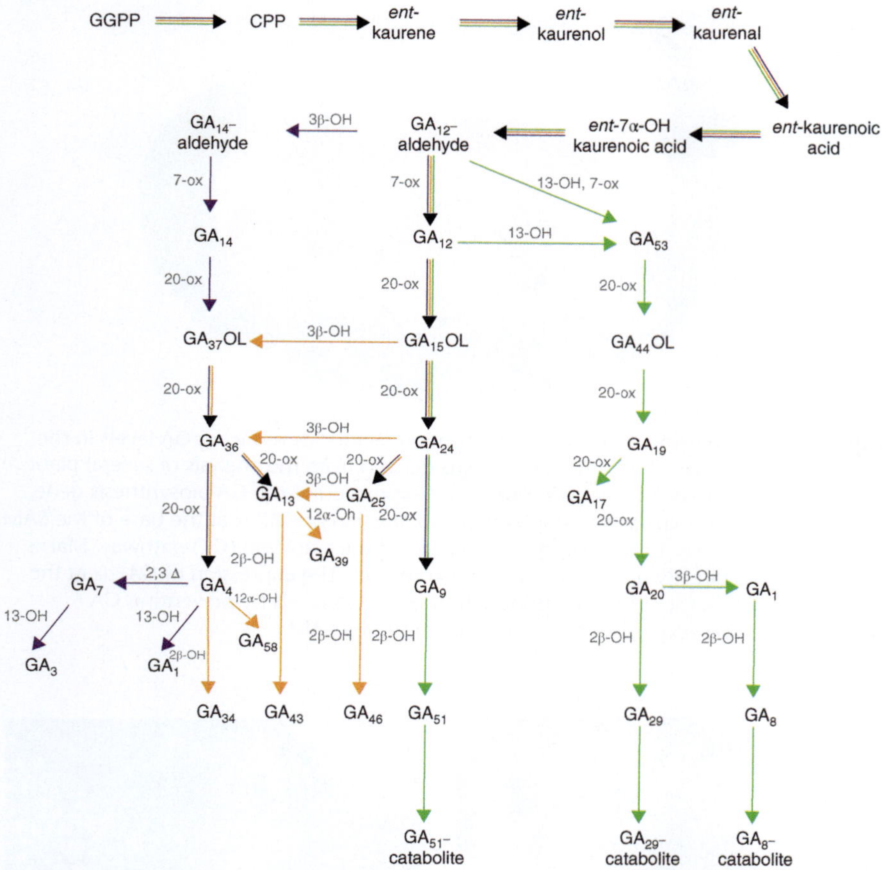

Figure 1.3 Composite scheme showing the conversion of GGPP to the first-formed GA, GA_{12}-aldehyde, and the predominant GA metabolic pathways from GA_{12}-aldehyde in three model systems: *Gibberella* (early 3-hydroxylation pathway, purple), pumpkin (late 3-hydroxylation pathway, orange), and pea (early 13-hydroxylation pathway, and non-hydroxylation pathway, green). Note that GA_1 occurs on both left and right sides of the scheme. GGPP geranylgeranyl diphosphate, CPP copalyl diphosphate, OL open lactone (CH_2OH at C-20).

Annual Plant Reviews, Volume 49: The Gibberellins, First Edition.
Edited by Peter Hedden and Stephen G. Thomas.

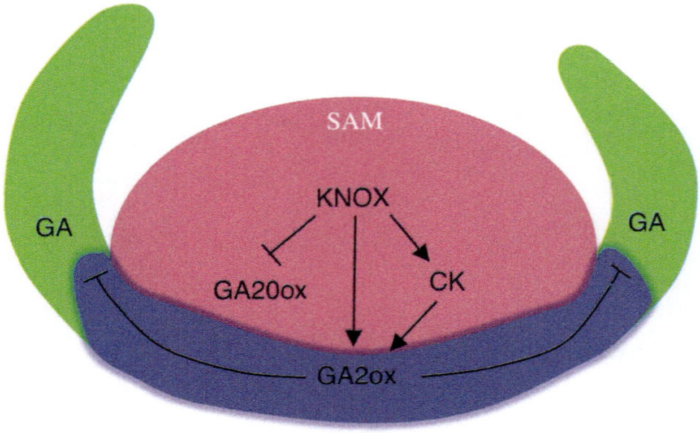

Figure 3.3 Involvement of KNOX transcription factors for reducing GA levels in the SAM. Summary of the role of KNOX proteins derived from the analysis of several plant species is shown. KNOX proteins suppress the expression of the GA biosynthesis gene, *GA20ox*. KNOX proteins also up-regulate the expression of *GA2ox* at the base of the SAM. This is in part likely to be mediated by activation of the cytokinin (CK) pathway. Maize KNOX proteins directly up-regulate *GA2ox* expression. The expression of *GA2ox* at the base of the SAM is thought to contribute to the low GA levels by preventing GA synthesised in the developing leaves from entering the SAM.

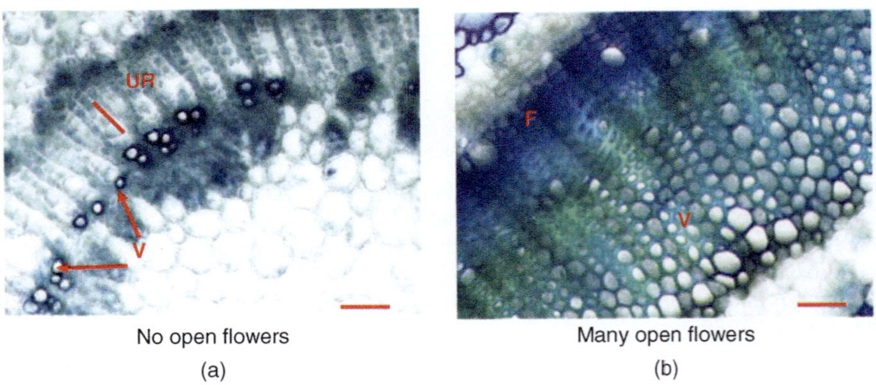

Figure 4.2 Flowers induce cambial activity and fibre formation. Cross sections in inflorescence stalks. (a) Depiction of the vasculature before flower maturation. A large undifferentiated cell region (UR) develops under young flowers. (b) Once the flowers in the inflorescence mature and their petals open, xylem cell differentiation occurs swiftly. In the inner xylem (first inside cells to differentiate) only vessels develop (V), while the later cells, closer to the cambium, form fibres (F). This depicts the gradient of hormonal effects from the developing flower. Beginning with high auxin to GA ratios (inducing vessels), and ending with high GA to auxin ratios, thereby inducing fibre formation. (Roni Aloni and Jonathan Dayan, unpublished.)

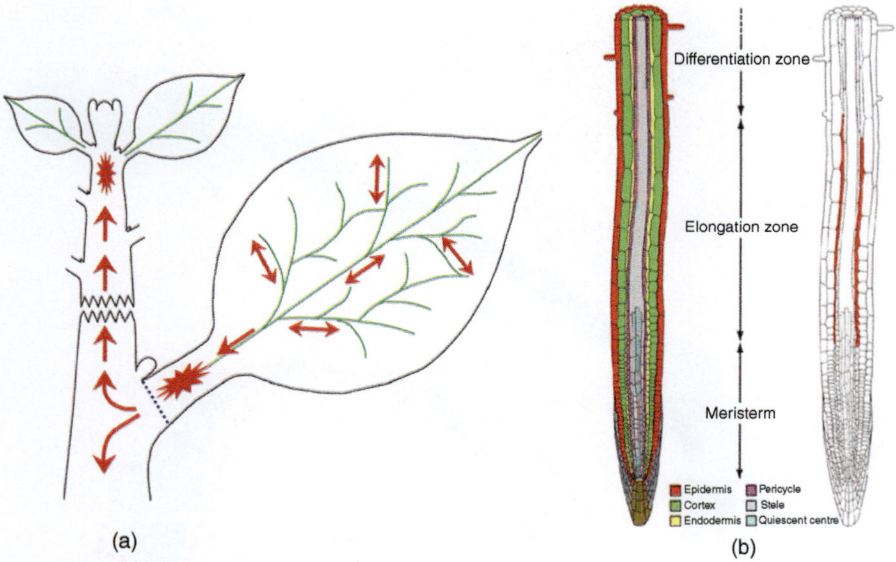

Figure 4.3 The source and translocation of the gibberellin signal. (a) An illustration modelling the translocation of a leaf-derived GA signal. The signal originates in developing leaves. Its flow is non-polar in the leaf blade and becomes polar only in the lower mid-vein towards the stem (arrows mark flow orientation). The unique anatomy at the base of the petiole (dotted line) potentially retards the flow which induces a local maximum (star), thereby acting as the leaf's elongation driving force. Although the signal flows in both directions along the stem, its upward movement from developing leaves reaches the young internodes (star) and induces stem elongation at the shoot apex. Throughout the flow along the stem, the signal results in bioactive GA accumulation that controls cambial activity and fibre differentiation. This illustration was originally published in Dayan *et al.* (2012). (b) Schematic illustrations of (left) the tissue organisation and zones within the *Arabidopsis* primary root and (right) the cellular distribution of the hormone signal GA. The concentric layers of epidermal, cortical, endodermal, pericycle and stele (vascular) tissues are colour coded (see key); the GA distribution is denoted in red. This illustration was originally published in Band and Bennett (2013).

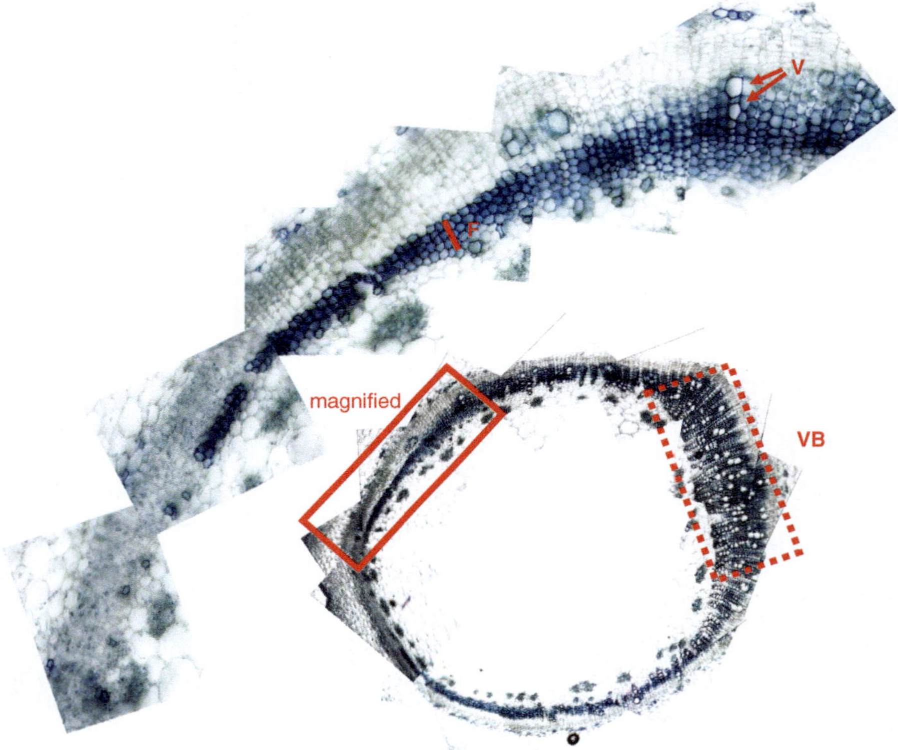

Figure 4.4 Diffusion of gibberellin signalling from a developing leaf. Effects of one leaf in otherwise de-leafed plants. Apical leaves were excised before becoming mature. A young leaf kept intact for 3.5 weeks, was cross sectioned in the internode beneath the leaf. Vessels form immediately beneath the leaf at the vascular bundle (VB). A fibre-cell layer is produced around the bundle and diminishes as the distance from the bundle grows. This layer resembles interfasicular fibers (here only one fasicular bundle has developed, indicated VB in the view of the whole cross section). (Roni Aloni and Jonathan Dayan, unpublished.)

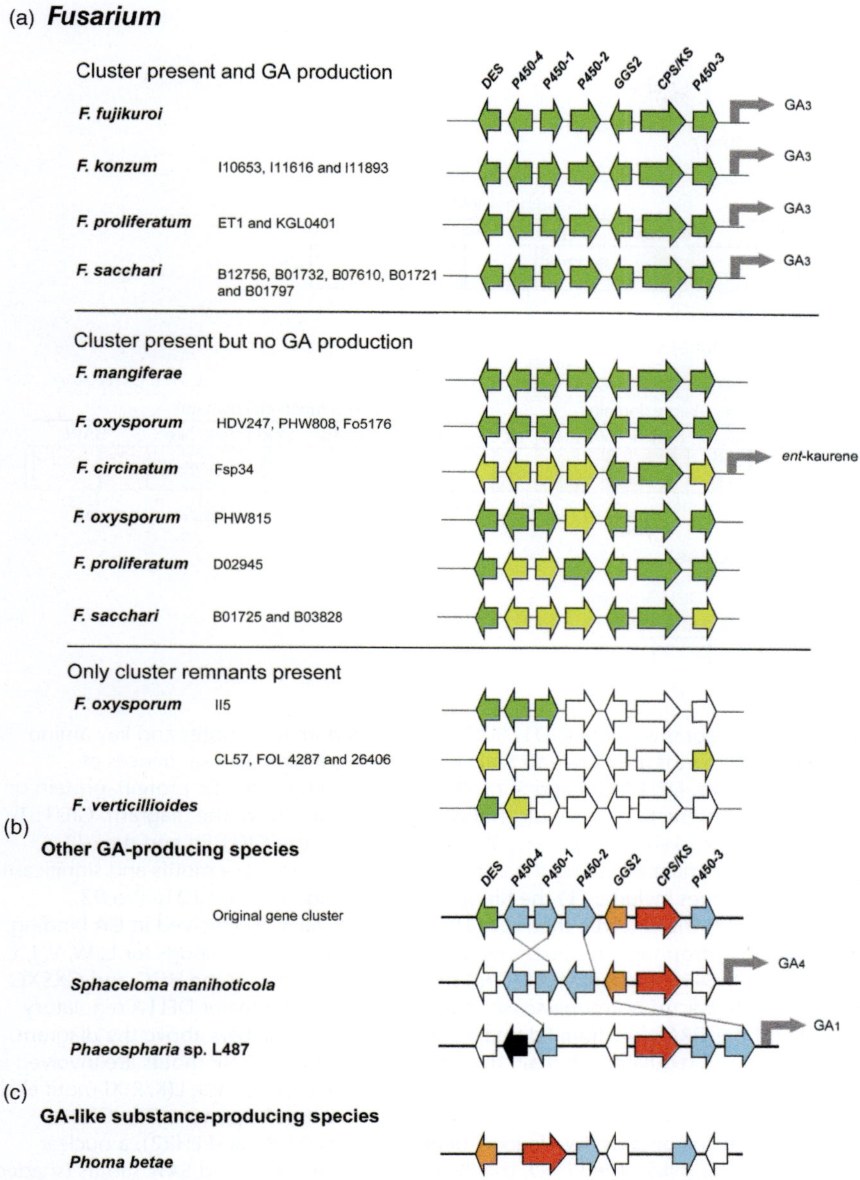

Figure 5.2 Comparison of GA biosynthetic gene clusters in fungal genomes. (a) Presence of GA gene clusters or parts of it in the genomes of *Fusarium* species. Green arrows represent functional genes or genes without obvious mutations. Yellow arrows represent genes with frame shift mutations, early stop codons or amino acid substitutions leading to non-functional enzymes. White arrows represent genes missing in the genome of this *Fusarium* isolate. (b) Functional GA gene clusters in the genomes of distantly related fungal species. (c) Aphidicolin biosynthetic gene cluster in *Phoma betae*. The cluster contains a similar gene pair consisting of a GGS2-like gene and a CPS/KS-like diterpene cyclase gene, both transcribed from the same bidirectional promoter.

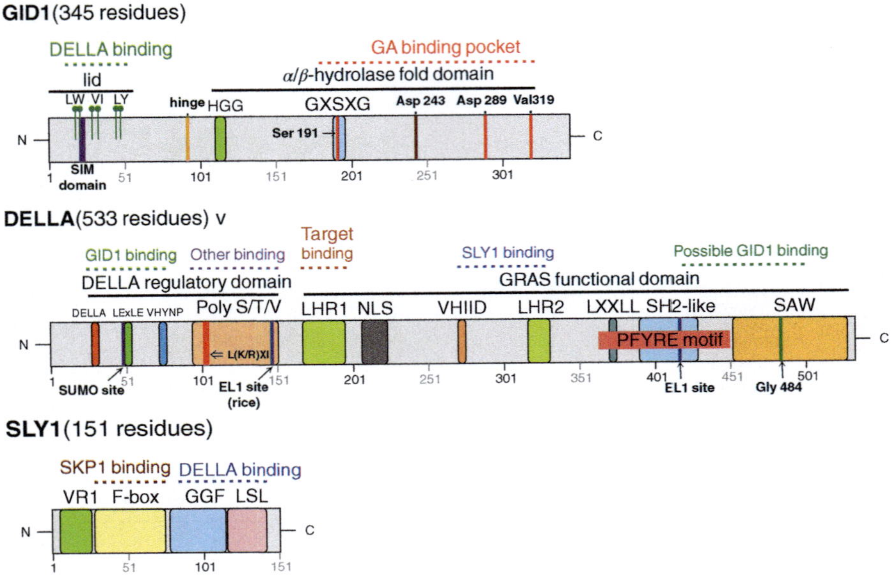

Figure 6.2 Diagrams of the GID1, SLY1 and DELLA domains, motifs and key amino acid residues. Proteins are drawn to scale based on the amino acid sequences of *Arabidopsis* GID1a, DELLA GAI and SLY1. Regions involved in specific protein–protein or protein–ligand interactions are indicated with dotted bars above the diagram. GID1: Two major domains marked by black bars above diagram are the GID1 lid and the α/β hydrolase fold domain including the core GA-binding pocket. Key motifs and significant amino acid residues include: (1) the hinge residue (orange bar for GID1a Pro 92, OsGID1 P99), (2) the catalytic triad (Ser 191, Asp 289, Val 319) involved in GA binding, (3) the six lid hydrophobic residues involved in DELLA-binding (lollipops for L, W, V, I, L, Y), (4) the SUMO-Interaction-Motif (SIM) domain (WVLI) and (5) the HGG and GXSXG motifs characteristic of hormone-sensitive lipases. DELLA: The major DELLA regulatory domain and the GRAS functional domain are marked by black bars above the diagram. Within the DELLA regulatory domain the DELLA, LExLE and VHYNP motifs are involved in GID1 binding (shaded boxes), and the poly S/T/V motif contains the L(K/R)XI motif likely involved in binding an undetermined 'other' GA signalling component. The GRAS functional domain contains: two leucine heptad repeats (LHR1 and LHR2), a nuclear localisation signal (NLS), the VHIID, the PFYRE, LXXLL, SH2-like, and SAW motifs (shaded boxes). Significant residues are marked by full-height bars. SLY1: SLY1 and GID2 contain the F-box domain that binds SKP1, the GGF and LSL motifs involved in DELLA-binding, and a variable region (VR1).

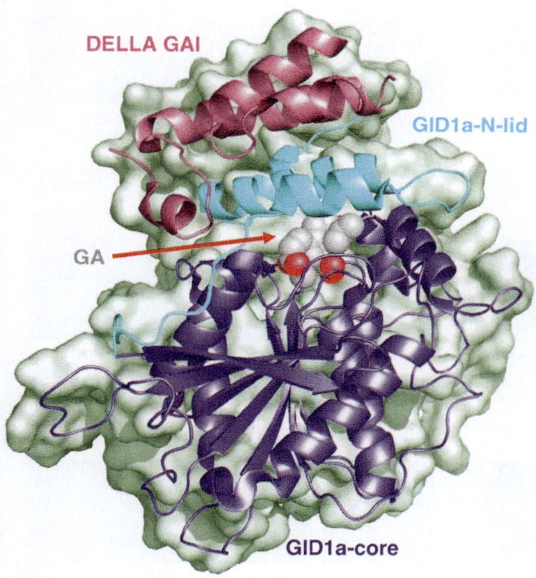

Figure 6.4 The GID1a-GA_3-DELLA complex based on the 1.8 angstrom crystal structure (Murase *et al.*, 2008). Ribbon representation of GID1a in complex with DELLA GAI and GA_3. The N-terminal GAI DELLA domain residues 11–113 (pink) is shown in complex with GID1a residues 1–344. The GID1a N-terminal extension or lid domain (GID1a-N-lid) is shown in blue and the GID1a α/β core domain in purple (GID1a-core). The GA_3 molecule (arrow) is shown in its binding pocket as a space-filling model where carbon is grey and oxygen red. (Figure was kindly provided by Toshio Hakoshima.)

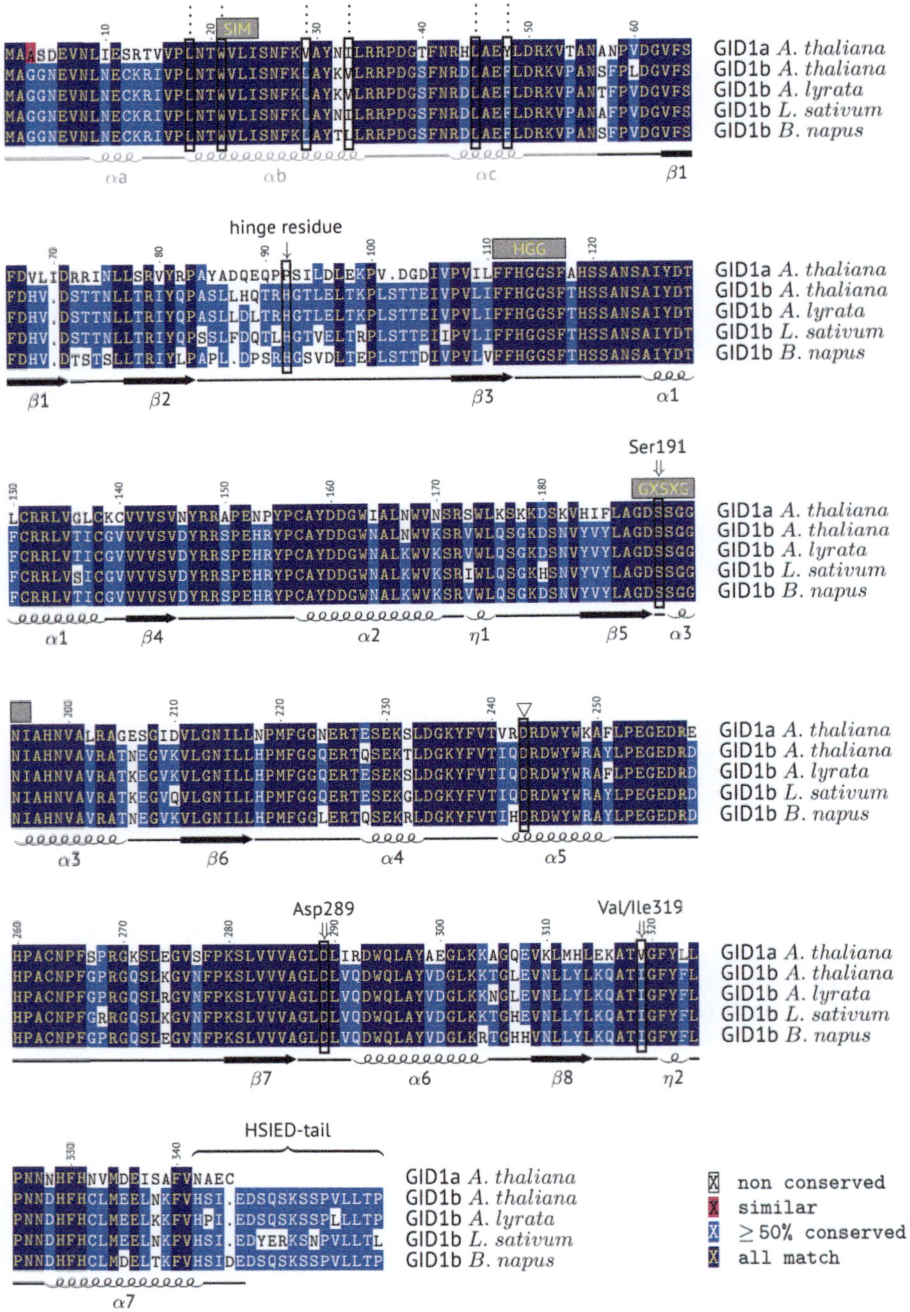

SIM
HGG
hinge residue
Ser191
GXSXG
Asp289
Val/Ile319
HSIED-tail
GID1a A. thaliana
GID1b A. thaliana
GID1b A. lyrata
GID1b L. sativum
GID1b B. napus
αa
αb
αc
α1
α2
α3
α4
α5
α6
α7
β1
β2
β3
β4
β5
β6
β7
β8
η1
η2
non conserved
similar
≥ 50% conserved
all match

Figure 6.5 Predicted amino acid sequence alignment of *Arabidopsis thaliana* GID1a and GID1b with GID1b-type homologues from *Arabidopsis lyrata, Lepidum sativum* and *Brassica napus* showing GID1b-type specific regions of homology using ClustalΩ for alignment and TeXshade package in LaTeX (Beitz 2000; Sievers *et al.*, 2011). Amino acid residue numbers are based on GID1a. Significant residues (boxed) include the DELLA interacting residues in the lid (four dots mark these six residues), the 'hinge residue' in a loop of the lid that differs between GID1ac- and GID1b-type receptors, the catalytic triad involved in binding GA (Ser 191, Asp 289, and Val/Ile 319), and the negatively charged Asp 243 that likely reduces affinity for GA_1 and GA_3 compared to GA_4 (downward facing triangle). Significant motifs are the SUMO-Interaction-Motif (SIM), the HGG motif and the GXSXG motif (grey box). Predicted secondary structures are presented below the aligned sequences as a solid line (loop), spiral (α indicates α-helix, η indicates 3_{10}-helix) and block arrow (β-strand); where the lid containing α-helices αa, αb and αc is grey and the core domain is black. Some regions are conserved only among GID1b-type receptors, such as hinge residue and the C-terminal HSIED-tail (bracket).

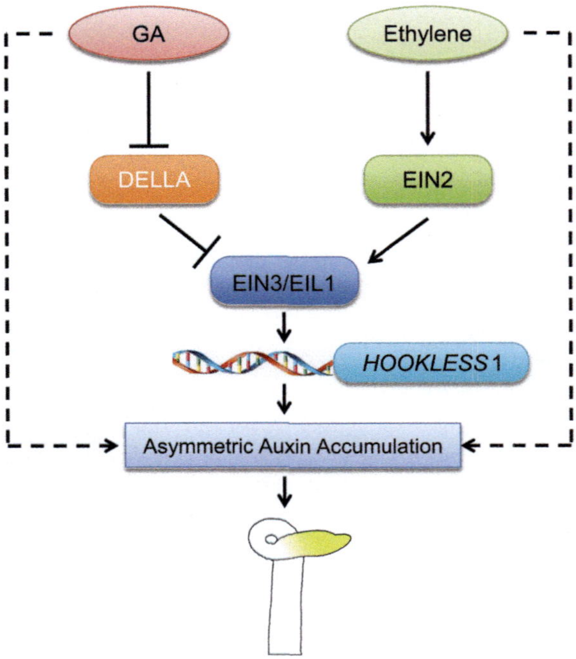

Figure 8.6 GAs and ethylene interact to regulate apical hook formation. Both ethylene and GAs induce *HOOKLESS1* (*HLS1)* transcription through the activity of ETHYLENE INSENSITIVE 3/EIN3-LIKE 1 (EIN3/EIL1). Ethylene stabilizes EIN3/EIL1, while GAs remove the inhibitory effect of DELLA proteins on that protein. It has also been proposed (as indicated by the dotted lines) that GAs and ethylene also initiate *HLS1*-independent pathways to regulate hook curvature. (An *et al.* (2012). Reproduced with permission from Nature Publishing Group.)

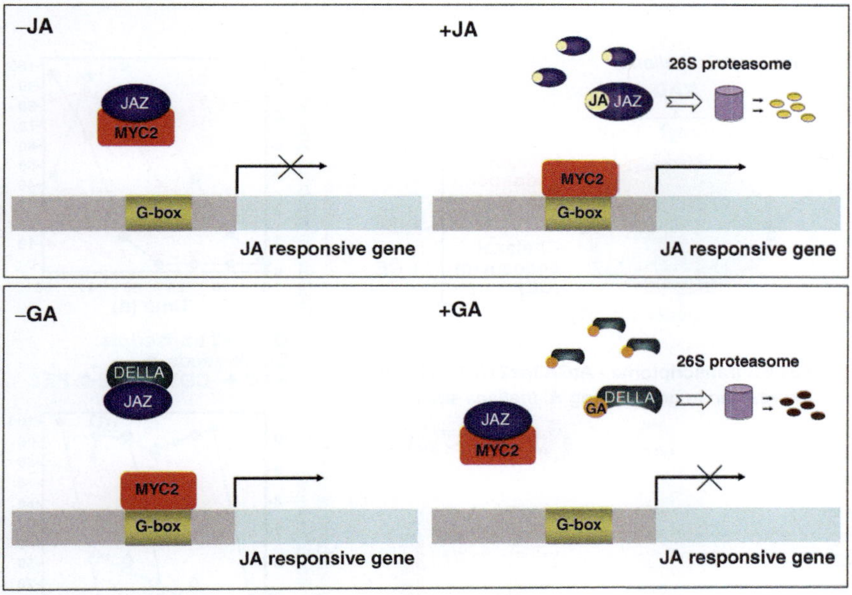

Figure 8.7 DELLAs compete with MYC2 for binding to JAZ, modulating JA signalling. When JA levels are low, JAZ proteins interact with MYC2, inhibiting its activity as a transcriptional activator. When JA levels are high, JAZ proteins are destabilised and MYC2 is released to activate the expression of JA-responsive genes. DELLA proteins compete with MYC2 for binding to JAZ; when GA levels are low, stabilised DELLAs bind to JAZ, leaving unbound MYC2 to activate JA-responsive genes. When GA is present, the DELLAs are degraded and MYC2 is able to bind to JAZs without competition, thus repressing the expression of JA-responsive genes. (Hou *et al.* (2010). Reproduced with permission from Elsevier.)

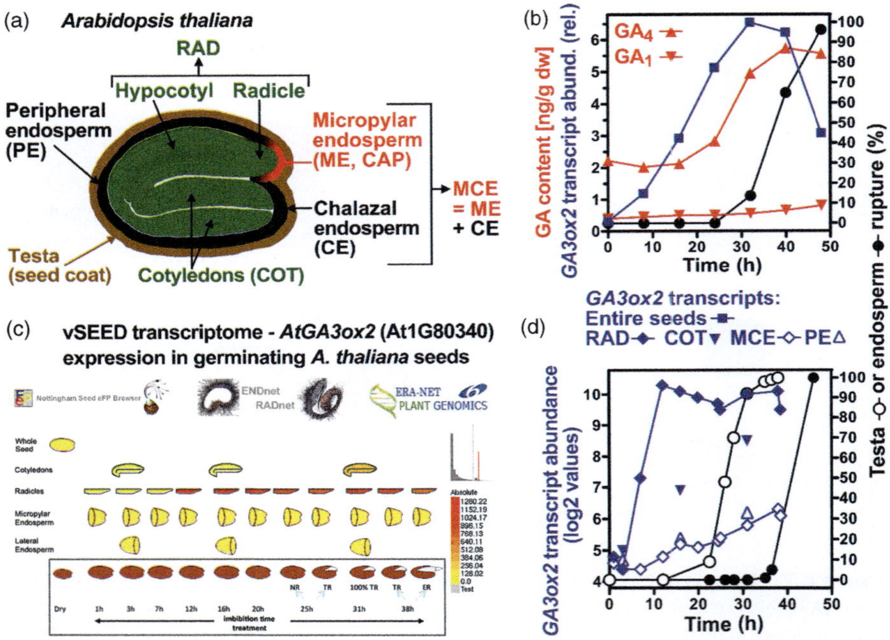

Figure 9.1 Spatio-temporal expression of the GA-biosynthetic gene *AtGA3ox2* during *Arabidopsis thaliana* seed germination. (a) Structure of a mature seed showing key seed compartments, including RAD (lower one-third of the hypocotyl/radicle axis, embryo growth zone) and CAP/ME (micropylar endosperm). (b) *AtGA3ox2* transcript abundance and bioactive GA_1 and GA_4 contents during seed germination. Note the increase in GA_4 content during late germination. (c) Seed compartment-specific transcriptome analysis during seed germination conducted by the ERA-NET Plant Genomics Consortium vSEED (eFP Browser results from vseed.nottingham.ac.uk). (d) Spatio-temporal *AtGA3ox2* expression during seed germination in relation to the kinetics of testa and endosperm rupture. (Results from (b) Ogawa *et al.*, 2003 and (d) Dekkers *et al.*, 2013.)

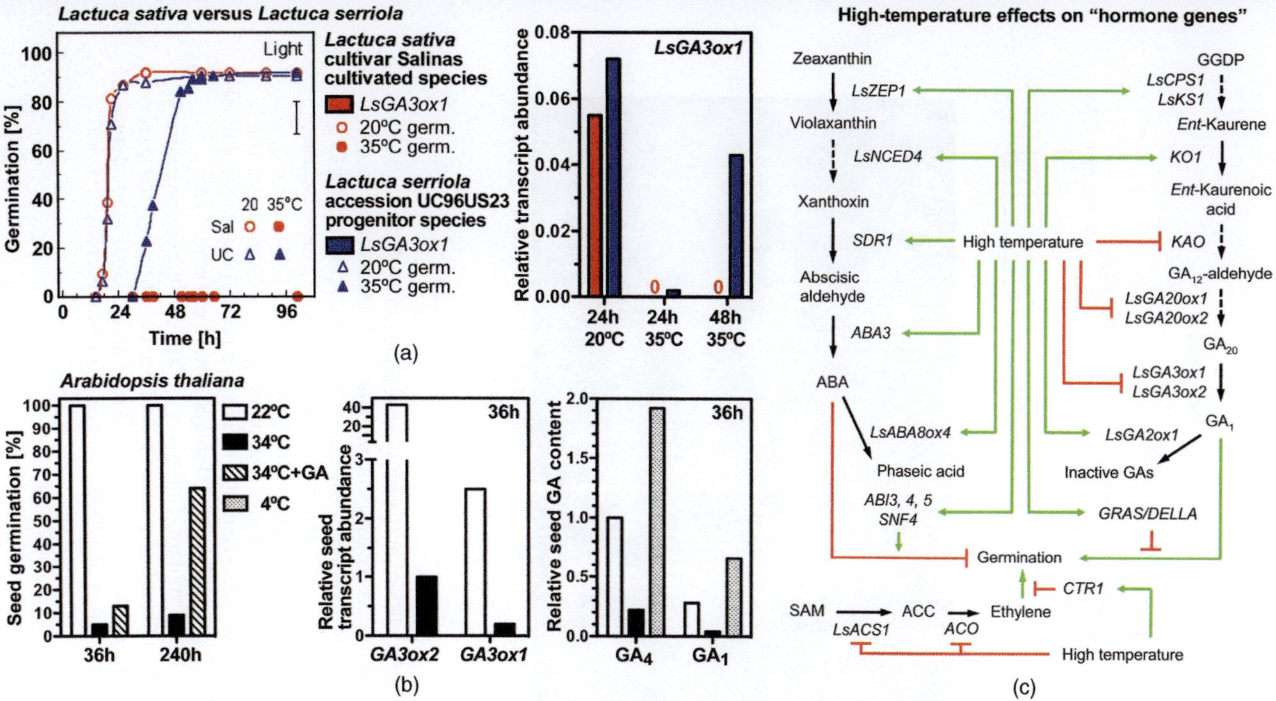

Figure 9.6 Hormonal aspects of thermoinhibition of lettuce and *Arabidopsis* seed germination. (a) The effect of supra-optimal temperature on germination (*left*) and *LsGA3ox1* gene expression (*right*) during germination of *Lactuca sativa* cv. Salinas (cultivated lettuce) and *Lactuca serriola* UC96US23 (thermotolerant lettuce). Note that compared to 20 °C, thermoinhibion of *L. sativa* occurs at 35 °C and is associated with inhibited *LsGA3ox1* expression, while the gene is expressed in thermotolerant *L. serriola* seeds (results from Argyris *et al.*, 2008).
(b) Thermoinhibition of *Arabidopsis* seeds imbibed in the light occurs at 34 °C and is alleviated by GA addition to the medium (*left*). The effect of temperature on the expression of the GA biosynthetic genes *AtGA3ox1* and *AtGA3ox2* and on the seed bioactive GA_1 and GA_4 contents is shown (*right*). Note that thermoinhibition is associated with a decrease, while cold-stratification (4 °C) is associated with an increase in the seed GA contents (results from Toh *et al.*, 2008 and from Yamauchi *et al.*, 2004). (c) Promoting (arrows) and inhibiting (bars) effects of high temperature on metabolism and signalling genes of hormones (GA, ABA and ethylene) involved in lettuce seed germination (from Argyris *et al.*, 2008).

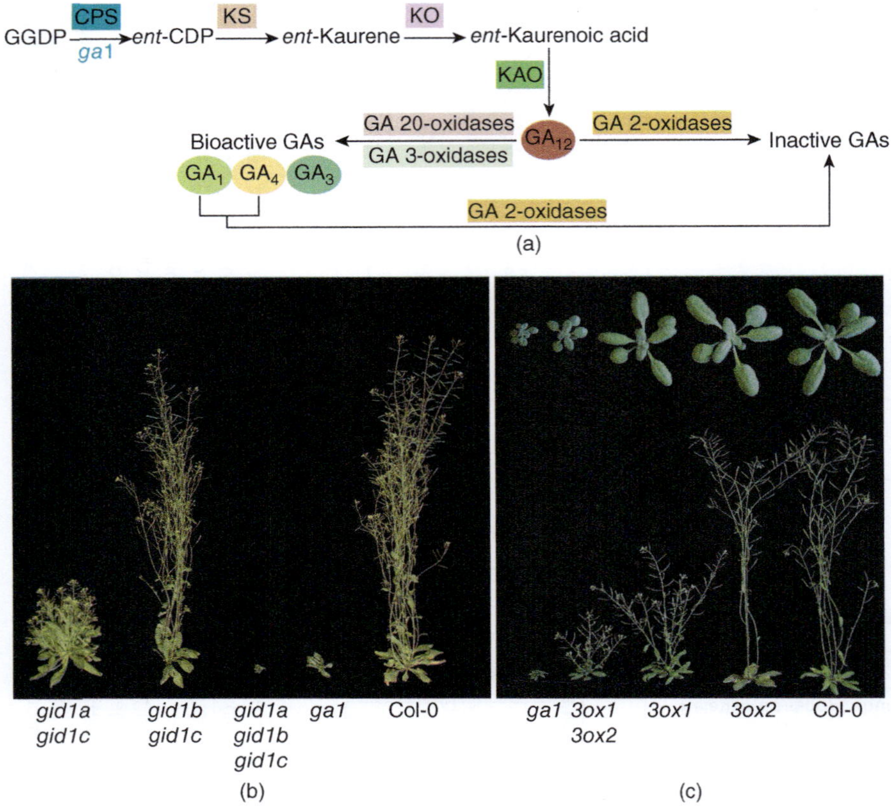

Figure 10.1 Phenotype of GA-deficient and insensitive mutants in *Arabidopsis thaliana*. (a) Schematic representation of the GA biosynthetic pathway and deactivation by GA 2-oxidase. GGDP, geranylgeranyl diphosphate; *ent*-CDP *ent*-copalyl diphosphate; CPS *ent*-copalyl diphosphate synthase; KS, *ent*-kaurene synthase; KO, *ent*-kaurene oxidase; KAO, *ent*-kaurenoic acid oxidase. (b) Phenotype of the *gid1* mutants. GID1, the gibberellin receptor, is encoded by three partially redundant genes in *Arabidopsis*. The semi-dwarf phenotype of *gid1a gid1c* double mutants reveals a predominant role of these two genes in stem elongation. Triple *gid1* mutant displays a more extreme dwarf phenotype than the biosynthetic *ga1* mutant. Griffiths *et al.* (2006). Reproduced with permission of American Society of Plant Biologists.) (c) Additive phenotype of the *ga3ox* mutants. Among the four *ga3ox* genes in *Arabidopsis*, *GA3ox1* and *GA3ox2* control stem elongation and branching. (Mitchum *et al.* (2006). Reproduced with permission from John Wiley & Sons.)

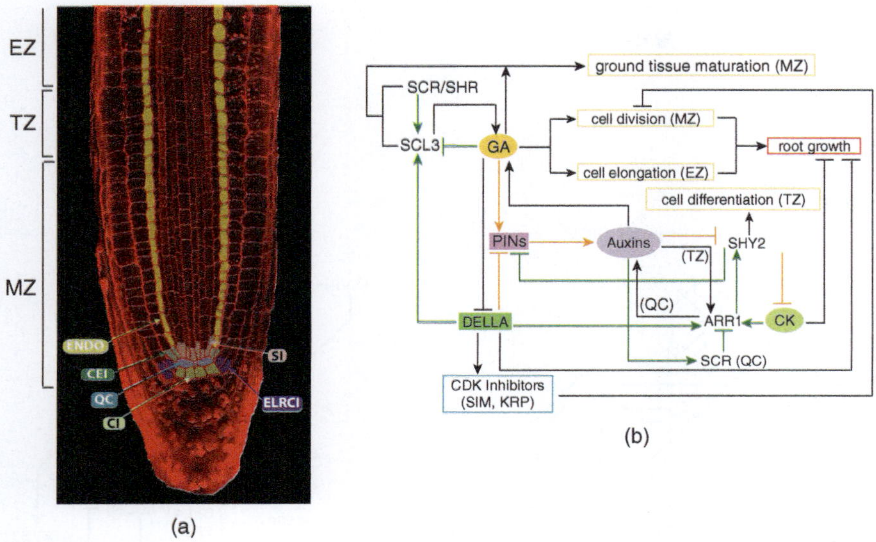

Figure 10.4 GA regulation of root development. (a) Longitudinal view of the *Arabidopsis* root tip. False-coloured confocal microscope image showing tissue layers and different developmental zones: MZ = meristematic zone, TZ = transition zone, EZ = elongation zone, QC = quiescent centre, CI = columela initials, CEI = cortex/endodermis initials, SI = stele initials, ELRCI = epidermis/lateral root cap initials, ENDO = endodermis. Gibberellins (GA) act in the endodermis of the meristematic and elongation zone (yellow cells), not affecting directly the stem cell niche, which includes the stem cells together with the quiescent centre (QC). (b) Schematic representation of the role of GAs in root growth. During the first 3 days post-germination (dpg) GA levels increase promoting cell division and elongation in detriment of cell differentiation and consequently, meristem growth. 5 dpg GA levels decrease and DELLA proteins increase in the TZ. As a consequence, DELLAs induce the expression of *ARR1* keeping SHY2 levels high, repressing auxin flow and thus balancing cell division with differentiation. Additionally, a DELLA-dependent accumulation of CDK inhibitors of the SIM and KRP families, promotes a reduction in cell division. Cytokinins (CK) in the TZ induce cell differentiation through up-regulation of SHY2, mediated by ARR1. On the other hand, auxins promote SHY2 degradation, thus sustaining PIN activity and cell division in the proximal meristem. Auxins also positively regulate GA biosynthesis. The resultant increased GA signalling induces polar auxin transport, and represses *ARR1*. In the QC, SCR represses *ARR1* expression and therefore, the synthesis of auxins. The auxins produced in the QC are transported through PIN action to the TZ where they induce *ARR1* expression promoting in this zone cell differentiation. In the endodermis of the MZ, SCL3, a positive regulator of GA signalling, and a DELLA-induced gene, is under the control of SCR and SHR, regulating together with GAs the ground tissue maturation. (The picture of the RAM was generously donated by Dr. Enrique Rojo.)

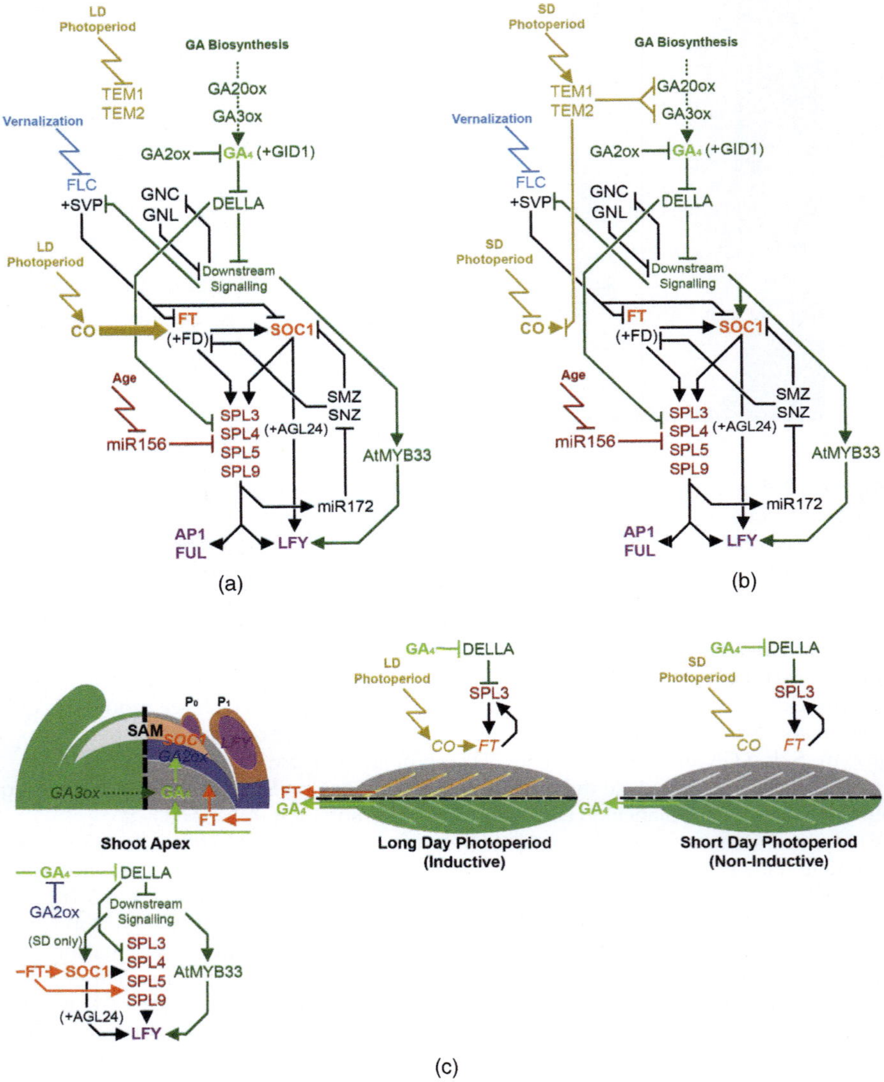
LD Photoperiod
TEM1
TEM2
GA Biosynthesis
GA20ox
GA3ox
Vernalization
GA2ox
GA4 (+GID1)
FLC
GNC
GNL
+SVP
DELLA
LD Photoperiod
Downstream Signalling
FT
CO
(+FD)
SOC1
Age
SMZ
SNZ
SPL3
SPL4
SPL5
SPL9
(+AGL24)
miR156
AtMYB33
miR172
AP1
FUL
LFY
(a)
SD Photoperiod
TEM1
TEM2
GA Biosynthesis
GA20ox
GA3ox
Vernalization
GA2ox
GA4 (+GID1)
FLC
GNC
GNL
+SVP
DELLA
SD Photoperiod
Downstream Signalling
FT
CO
(+FD)
SOC1
Age
SMZ
SNZ
SPL3
SPL4
SPL5
SPL9
(+AGL24)
miR156
AtMYB33
miR172
AP1
FUL
LFY
(b)
P0
P1
SAM
SOC1
LFY
GA2ox
GA3ox
GA4
FT
Shoot Apex
GA4
DELLA
LD Photoperiod
SPL3
CO
FT
Long Day Photoperiod (Inductive)
GA4
DELLA
SD Photoperiod
SPL3
CO
FT
Short Day Photoperiod (Non-Inductive)
GA4
DELLA
GA2ox
Downstream Signalling
(SD only)
SPL3
SPL4
SPL5
SPL9
FT
SOC1
AtMYB33
(+AGL24)
LFY
(c)

Figure 11.1 GA regulation of the *Arabidopsis* floral transition. (a) and (b) Schematic representation summarising current understanding of the interactions between the GA signalling (green), photoperiod (yellow), age (red) and vernalisation pathways (blue) in *Arabidopsis* under long day (LD; a) and short day (SD; b) photoperiods, respectively. Arrowheads and bars indicate positive and negative regulation of downstream targets, respectively. The signalling pathways shown are integrated at multiple levels. Signals under both photoperiods are integrated by expression levels of the transcription factors *FLORAL TRANSITION* (*FT*) and *SUPPRESSOR OF CONSTANS1* (*SOC1*), which promote expression of the floral identity genes *LEAFY* (*LFY*), *APETALA1* (*AP1*) and *FRUITFUL* (*FUL*) through up-regulation of *SQUAMOSA PROMOTER BINDING-LIKE* (*SPL*) transcription factors, which themselves appear to positively reinforce *FT* and *SOC1* expression through feed-forward regulation. Under permissive LD photoperiods the gene network's equilibrium is tipped towards flowering by CO-dependent up-regulation of *FT*. Under non-permissive SD photoperiods *FT* is not up-regulated and flowering instead requires GA-dependent up-regulation of *SOC1* and *LFY*. GA signalling regulates flowering through transcriptional activation of *LFY* (both via a *SOC1*-dependent pathway and independently of *SOC1* through *AtMYB33*), inhibition of SPL protein activity by direct binding of DELLA proteins, and negative regulation of environmentally-sensitive flowering inhibitors such as *SHORT VEGETATIVE PHASE* (*SVP*). (c) Tissue and organ separation of the genetic network outlined in (a) and (b). *SOC1* and *LFY* expression is restricted to the shoot apical meristem (SAM) and developing organ primordia, but are only expressed during/after the floral transition. The photoperiod signal is detected in rosette leaves, which are the site of *FT* expression and GA biosynthesis. *FT* is expressed in leaf vasculature under LD photoperiods, driven both by *CO* expression in the phloem and GA signalling. GA regulation of *SPL3* has been detected in leaf tissues and the SAM (Galvão *et al.*, 2012), suggesting that GA signalling acts through the SPL proteins at both of these sites. Under SD conditions, neither *CO* expression nor GA signalling promote *FT* expression, and levels remain low. FT and bioactive GA (LD) or bioactive GA alone (SD) synthesised in the leaves is transported to the shoot apex to promote the transition from a vegetative SAM to an inflorescence meristem (IM). Access of locally and distally-synthesised bioactive GA to the SAM is restricted by GA2ox catalytic activity. *SOC1*, *SPL* and *LFY* expression are up-regulated in the SAM to promote IM identity, leading to production of lateral floral meristems (FMs) in place of leaf primordia. SOC1 is apparently only up-regulated by GA under SD conditions.

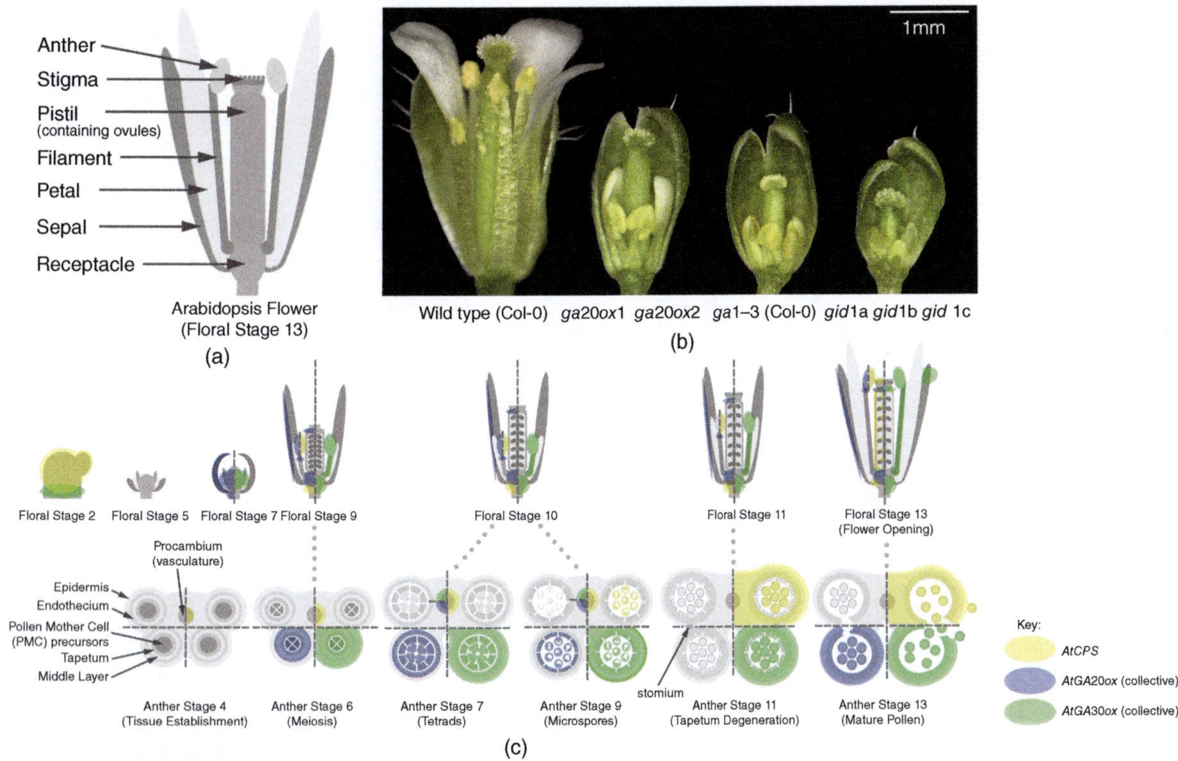

Figure 11.2 Gibberellin and *Arabidopsis* floral development. (a) Diagrammatic cross-section of an *Arabidopsis* flower, identifying major organs. (b) Comparison of *Arabidopsis* floral phenotypes from selected GA biosynthesis and signalling mutants (displaying a range of phenotypic severity) against wild type. Growth of all floral organs is reduced, with stamens and petals showing greatest sensitivity. Anthers of the *ga1-3* and *gidl1a gid1b gid1c* mutants display premature developmental arrest (Goto and Pharis, 1999; Griffiths *et al.*, 2006). (c) Sites of floral GA biosynthesis in *Arabidopsis* floral tissues during development, showing *AtCPS* (yellow) and cumulative *AtGA20ox* (blue) and *AtGA3ox* (green) gene expression. The patterns of expression shown are based on analysis of GUS reporter lines (Silverstone *et al.*, 1997; Mitchum *et al.*, 2006; Hu *et al.*, 2008; Plackett, 2012; Plackett *et al.*, 2012). *GA20ox* expression during early floral development has not been characterised. Numbers marked on the inflorescence meristem denote floral stages 2 and 3 (Smyth *et al.* 1990), respectively. The anther developmental stages shown follow the definitions of Sanders *et al.* (1999). (Adapted from Plackett *et al.* (2011). Reproduced with permission of Elsevier.)

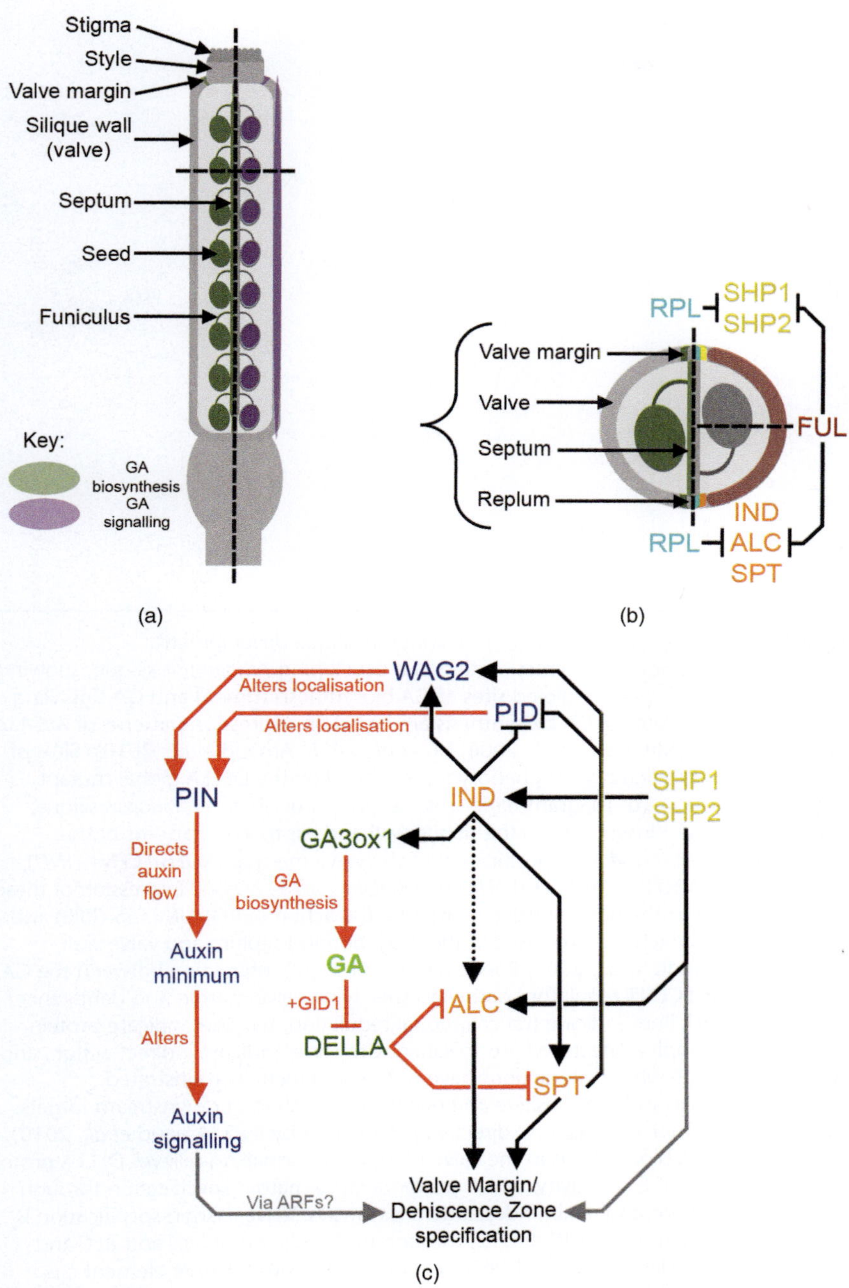
Stigma
Style
Valve margin
Silique wall
(valve)
Septum
Seed
Funiculus
Key:
GA
biosynthesis
GA
signalling
(a)
RPL
SHP1
SHP2
Valve margin
Valve
Septum
Replum
FUL
IND
RPL
ALC
SPT
(b)
WAG2
Alters localisation
PID
Alters localisation
PIN
IND
SHP1
SHP2
GA3ox1
Directs
auxin
flow
GA
biosynthesis
GA
Auxin
minimum
+GID1
ALC
DELLA
Alters
SPT
Auxin
signalling
Valve Margin/
Dehiscence Zone
specification
Via ARFs?
(c)

Figure 11.4 Gibberellin regulation of *Arabidopsis* silique development. (a) Diagrammatic longitudinal cross-section of a developing *Arabidopsis* silique, showing major organs/tissue types. Predicted sites of GA biosynthesis (green) and GA signalling (purple) are shown. Sites of GA biosynthesis are based on expression patterns of *AtGA3ox* GUS reporter lines (Mitchum *et al.*, 2006; Hu *et al.*, 2008; Arnaud *et al.*, 2010). Sites of GA signalling are predicted from phenotypes exhibited by the DELLA global mutant (Fuentes *et al.*, 2012). (b) Diagrammatic cross-section through an *Arabidopsis* silique, detailing sites of GA biosynthesis in the silique wall and expression patterns of the transcription factors regulating specification of the valve margin, *INDEHISCENT* (*IND*), *ALCATRAZ* (*ALC*), *SPATULA* (*SPT*) and *SHATTERPROOF1* (*SHP1*) and *-2*. Expression of these genes is restricted to the nascent valve margin by the action of REPLUMLESS (RPL) and FRUITFULL (FUL), which are expressed in the neighbouring replum and valve wall, respectively (Ferrándiz *et al.*, 2000; Roeder *et al.*, 2003). (c) Interaction between the GA, auxin and INDEHISCENT regulatory pathways specifying valve margin and dehiscence zone identity. Black lines indicate transcriptional regulation, red lines indicate protein action or more complex effects, where specified. Grey lines indicate indirect action, and dotted lines indicate where interactions have not been directly demonstrated. Arrowheads and bars indicate positive and negative regulation of downstream targets, respectively. *AtGA3ox1* expression is directly up-regulated by IND (Arnaud *et al.*, 2010), promoting local GA biosynthesis in the valve margin. GA signalling relieves DELLA protein repression of ALC and SPT activity, which promote valve margin specification through both auxin-dependent and auxin-independent pathways. Valve margin specification is also promoted by action of *SHP1* and *-2*, through up-regulation of *IND* and *ALC* and through an independent pathway (Liljegren *et al.*, 2000; 2004). Figure element c is adapted from Kay *et al.* (2012). Reproduced with permission from John Wiley & Sons.

Printed and bound by CPI Group (UK) Ltd, Croydon, CR0 4YY
17/09/2026
14964295-0001